Beginning AutoCAD 2015

EXERCISE WORKBOOK

by
Cheryl R. Shrock
Professor, retired
Drafting Technology
Orange Coast College, Costa Mesa, Ca.
Autodesk Authorized Author

Updated for AutoCAD 2015
by
Steve Heather
Former Lecturer of
Mechanical Engineering &
Computer Aided Design

INDUSTRIAL PRESS

ISBN 978-0-8311-3497-6

Printed in the United States of America

Limits of Liability and disclaimer of Warranty
The author and publisher make no warranty of any kind, expressed or implied, with regard to the documentation contained in this book.

Industrial Press Inc.
32 Haviland Street, Unit 2C
South Norwalk, CT. 06854

10 9 8 7 6 5 4 3 2 1

Many thanks are due to Cheryl Shrock for allowing me to continue on with her Exercise Workbook series. And special thanks to John Carleo, Editorial Director of Industrial Press, for having faith in me.

Steve Heather

AutoCAD Books by Cheryl R. Shrock:

For information about these books visit: www.industrialpress.com

Table of Contents

Introduction

Lesson 1

Lesson 2

Lesson 3

Lesson 4

Lesson 5

Lesson 6

Lesson 7

Lesson 8

Lesson 9

Lesson 10

Lesson 11

Lesson 12

Lesson 13

Lesson 14

Lesson 15

Lesson 16

Lesson 17

Lesson 18

Lesson 19

Lesson 20

Lesson 21

Lesson 22

Lesson 29

Lesson 30

Appendix

Index

INTRODUCTION

About this workbook

Beginning AutoCAD® 2015, Exercise Workbook is designed for classroom instruction or self-study. There are 30 lessons. Each lesson starts with step-by-step instructions followed by exercises designed for practicing the commands you learned within that lesson.

You may find the order of instruction in this workbook somewhat different from most textbooks. The approach I take is to familiarize you with the drawing commands first. After you are comfortable with the drawing commands, you will be taught to create your own setup drawings. This method is accomplished by supplying you with a drawing, "**2015-workbook-helper.dwg**". This drawing is preset and ready for you to open and use. For the first 8 lessons you should not worry about settings, **you just draw.**

Important
How to get the drawing listed above?
The file "2015-workbook-helper.dwg" should be downloaded from our website:

http://new.industrialpress.com/ext/downloads/acad/2015-workbook-helper.dwg

Enter the address into your web browser and the download will start automatically.

Printing

The exercises in the workbook have been designed to be printed on any **8-1/2 X 11 (Letter size) printer**. You will be able to print your exercises and view your results as you progress through the lessons. If you would like to configure a large format printer refer to **Appendix-A to "Add a Printer / Plotter**".

About the Authors

Cheryl R. Shrock is a recently retired Professor and Chairperson of Computer Aided Design at Orange Coast College in Costa Mesa, California. She is also an Autodesk® registered author. Cheryl began teaching CAD in 1990. Previous to teaching, she owned and operated a commercial product and machine design business where designs were created and documented using CAD. This workbook is a combination of her teaching skills and her industry experience.

Steve Heather is a former Lecturer of Mechanical Engineering and Computer Aided Design in England, UK. For the past 6 years he has been a Beta Tester for Autodesk®, testing the latest AutoCAD® software. Previous to teaching and for more than 30 years, he worked as a Precision Engineer in the Aerospace and Defense industries.

Steve can be contacted for questions or comments at: *steve.heather@live.com*

CONFIGURING YOUR SYSTEM

AutoCAD ® allows you to customize it's configuration. While you are using this workbook it is necessary for you to make some simple changes to your configuration so our configurations are the same. This will ensure that the commands and exercises work as expected. The following instructions will guide you through those changes.

1. Start AutoCAD®
   a. To start AutoCAD, follow the instructions on pages 1-2 through to 1-4. Then return to this page and go to step 2.

2.
   a. Type: ***options*** anywhere within the main drawing area then press the **<enter>** key. (not case sensitive)
   b. Type: ***options*** on the Command Line then press the **<enter>** key. (not case sensitive)

The text that you type will appear in the Dynamic Input box or on the Command Line, as shown below:

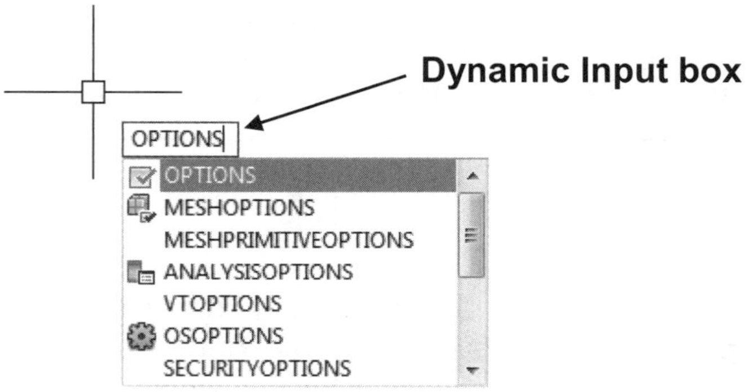

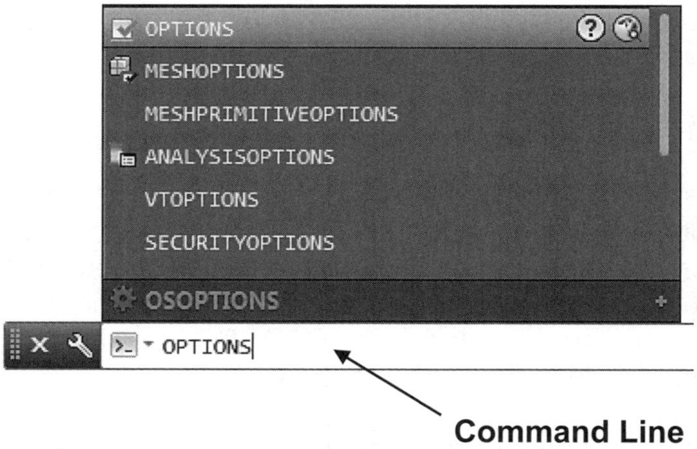

Dynamic Input box

Command Line

NOTE:

AUTOCAD LT USERS:
You may find that some of the settings appear slightly different.
But they are mostly the same.

Configuration Settings

3. Select the ***Display*** tab and change the settings on your screen to match the dialog box below.

Options

Current profile: <<Unnamed Profile>> Current drawing: Drawing1.dwg

Files | Display | Open and Save | Plot and Publish | System | User Preferences | Drafting | 3D Modeling | Selection | Profiles | Online

Window Elements

Color scheme: Dark

- [ ] Display scroll bars in drawing window
- [ ] Use large buttons for Toolbars
- [x] Resize ribbon icons to standard sizes
- [x] Show ToolTips
  - [x] Show shortcut keys in ToolTips
  - [x] Show extended ToolTips
  - 2 Number of seconds to delay
- [x] Show rollover ToolTips
- [x] Display File Tabs

Colors... Fonts...

Layout elements

- [x] Display Layout and Model tabs
- [x] Display printable area
- [x] Display paper background
  - [x] Display paper shadow
- [x] Show Page Setup Manager for new layouts
- [ ] Create viewport in new layouts

Display resolution

1000 Arc and circle smoothness

8 Segments in a polyline curve

0.5 Rendered object smoothness

4 Contour lines per surface

Display performance

- [ ] Pan and zoom with raster & OLE
- [x] Highlight raster image frame only
- [x] Apply solid fill
- [ ] Show text boundary frame only
- [ ] Draw true silhouettes for solids and surfaces

Crosshair size

5

Fade control

Xref display

50

In-place edit and annotative representations

70

OK Cancel Apply Help

You may select different colors for your display

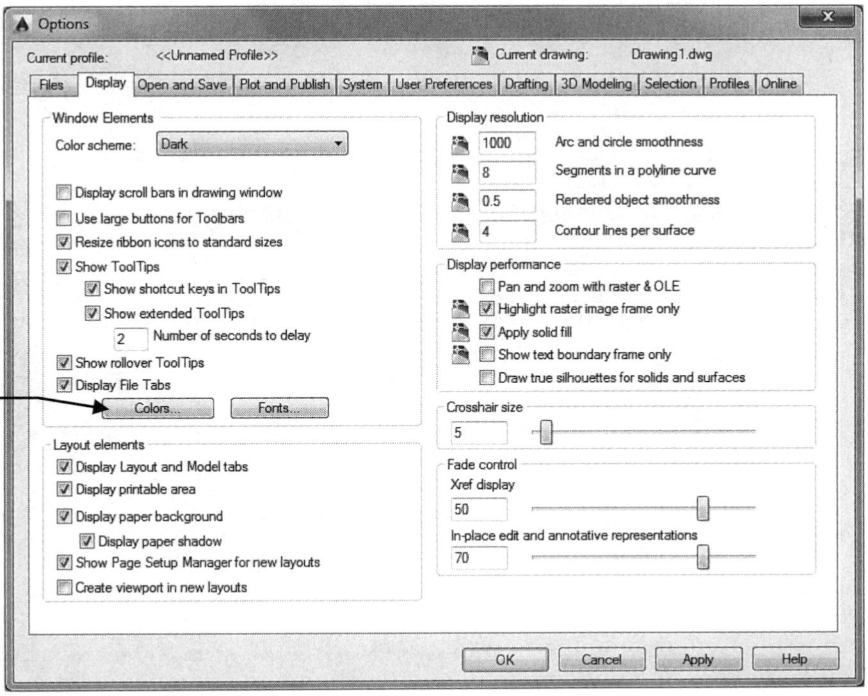

4. Select the ***Open and Save*** tab and change the settings on your screen to match the dialog box below.

Options

Current profile: <<Unnamed Profile>> Current drawing: Drawing1.dwg

Files | Display | Open and Save | Plot and Publish | System | User Preferences | Drafting | 3D Modeling | Selection | Profiles | Online

File Save

Save as:

AutoCAD 2013 Drawing (*.dwg)

- [x] Maintain visual fidelity for annotative objects
- [x] Maintain drawing size compatibility

Thumbnail Preview Settings...

50 Incremental save percentage

File Safety Precautions

- [x] Automatic save
  - 10 Minutes between saves
- [x] Create backup copy with each save
- [ ] Full-time CRC validation
- [ ] Maintain a log file

ac$ File extension for temporary files

Security Options...

- [x] Display digital signature information

File Open

9 Number of recently-used files

- [ ] Display full path in title

Application Menu

9 Number of recently-used files

External References (Xrefs)

Demand load Xrefs:

Enabled with copy

- [x] Retain changes to Xref layers
- [x] Allow other users to Refedit current drawing

ObjectARX Applications

Demand load ObjectARX apps:

Object detect and command invoke

Proxy images for custom objects:

Show proxy graphics

- [x] Show Proxy Information dialog box

OK Cancel Apply Help

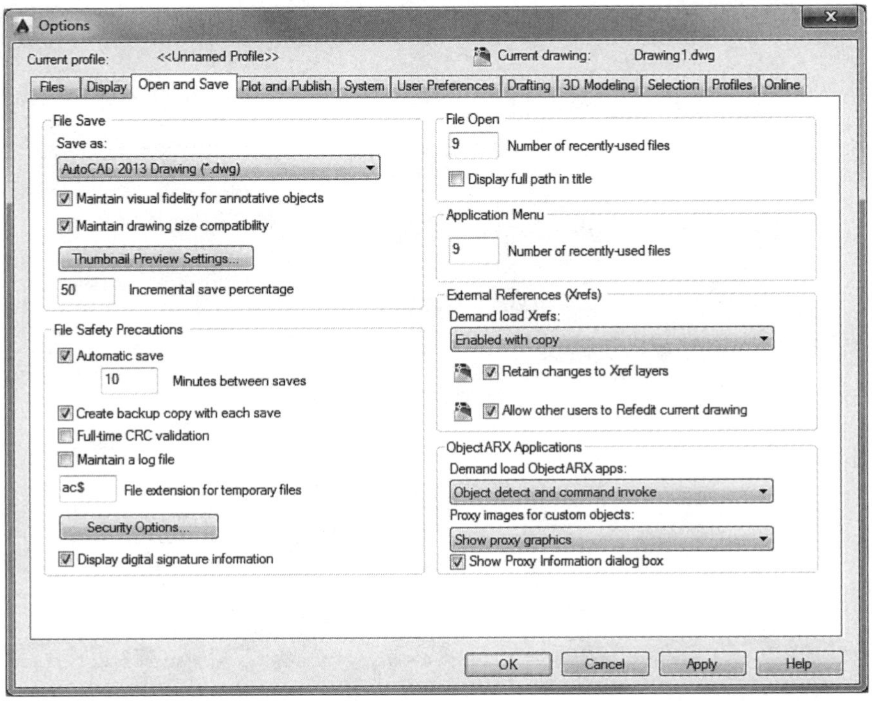

Configuration Settings

5. Select the ***Plot and Publish*** tab and change the settings on your screen to match the dialog box below.

You may use this setting or select the actual name of the printer

Options
Current profile: <<Unnamed Profile>>
Current drawing: Drawing1.dwg
Files | Display | Open and Save | Plot and Publish | System | User Preferences | Drafting | 3D Modeling | Selection | Profiles | Online
Default plot settings for new drawings
Use as default output device
Default Windows System Printer.pc3
Use last successful plot settings
Add or Configure Plotters...
Plot to file
Default location for plot to file operations:
C:\Users\Steve\documents
Background processing options
Enable background plot when:
Plotting
Publishing
Plot and publish log file
Automatically save plot and publish log
Save one continuous plot log
Save one log per plot
Auto publish
Automatic Publish
Automatic Publish Settings...
General plot options
When changing the plot device:
Keep the layout paper size if possible
Use the plot device paper size
System printer spool alert:
Always alert (and log errors)
OLE plot quality:
Automatically select
Use OLE application when plotting OLE objects
Hide system printers
Specify plot offset relative to
Printable area
Edge of paper
Plot Stamp Settings...
Plot Style Table Settings...
OK | Cancel | Apply | Help

6. Select the ***System*** tab and change the settings on your screen to match the dialog box below.

Options
Current profile: <<Unnamed Profile>>
Current drawing: Drawing1.dwg
Files | Display | Open and Save | Plot and Publish | System | User Preferences | Drafting | 3D Modeling | Selection | Profiles | Online
Hardware Acceleration
Graphics Performance
Automatically check for certification update
Current Pointing Device
Current System Pointing Device
Accept input from:
Digitizer only
Digitizer and mouse
Touch Experience
Display touch mode ribbon panel
Layout Regen Options
Regen when switching layouts
Cache model tab and last layout
Cache model tab and all layouts
General Options
Hidden Messages Settings
Display OLE Text Size Dialog
Beep on error in user input
Allow long symbol names
Help
Access online content when available
InfoCenter
Balloon Notifications
Security
Executable File Settings
dbConnect Options
Store Links index in drawing file
Open tables in read-only mode
OK | Cancel | Apply | Help

Configuration Settings

7. Select the ***User Preferences*** tab and change the settings on your screen to match the dialog box below.

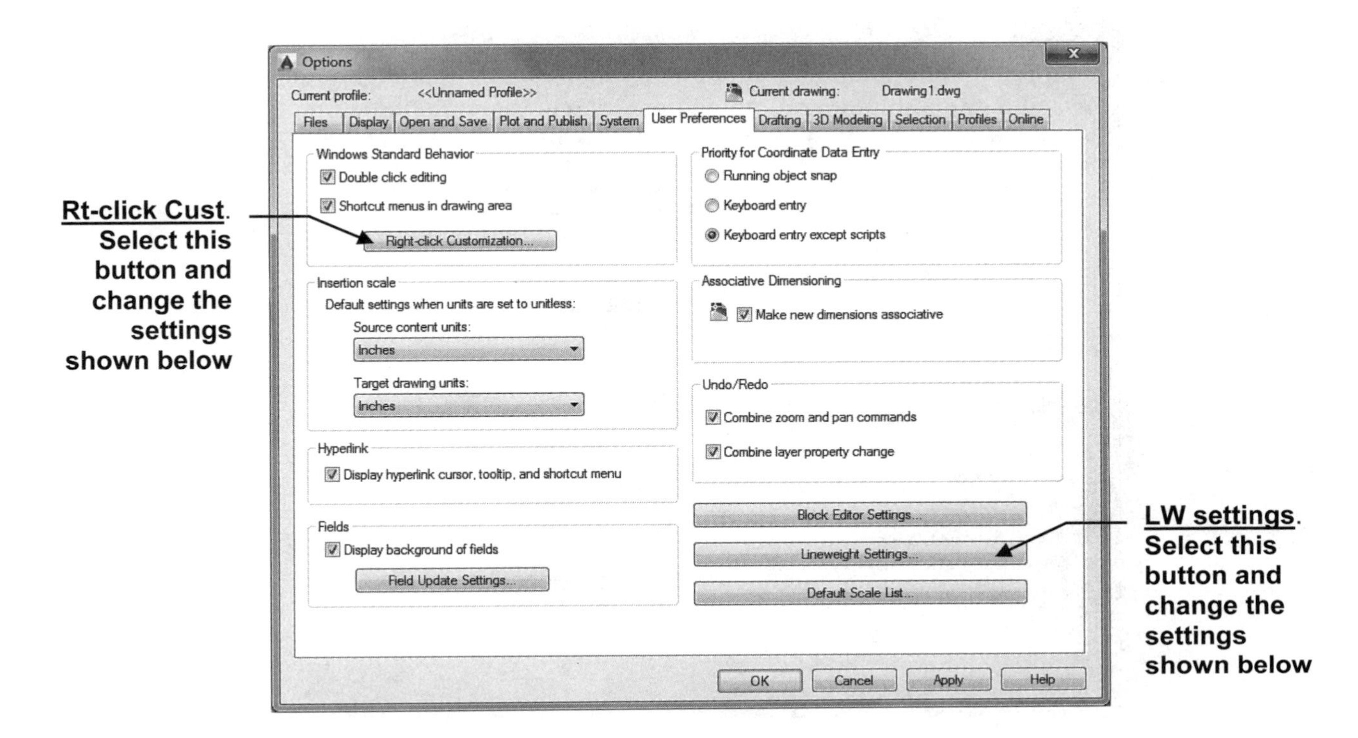

8. After making the setting changes shown below select **Apply & Close** button.

Right-click Customization **Lineweight Settings**

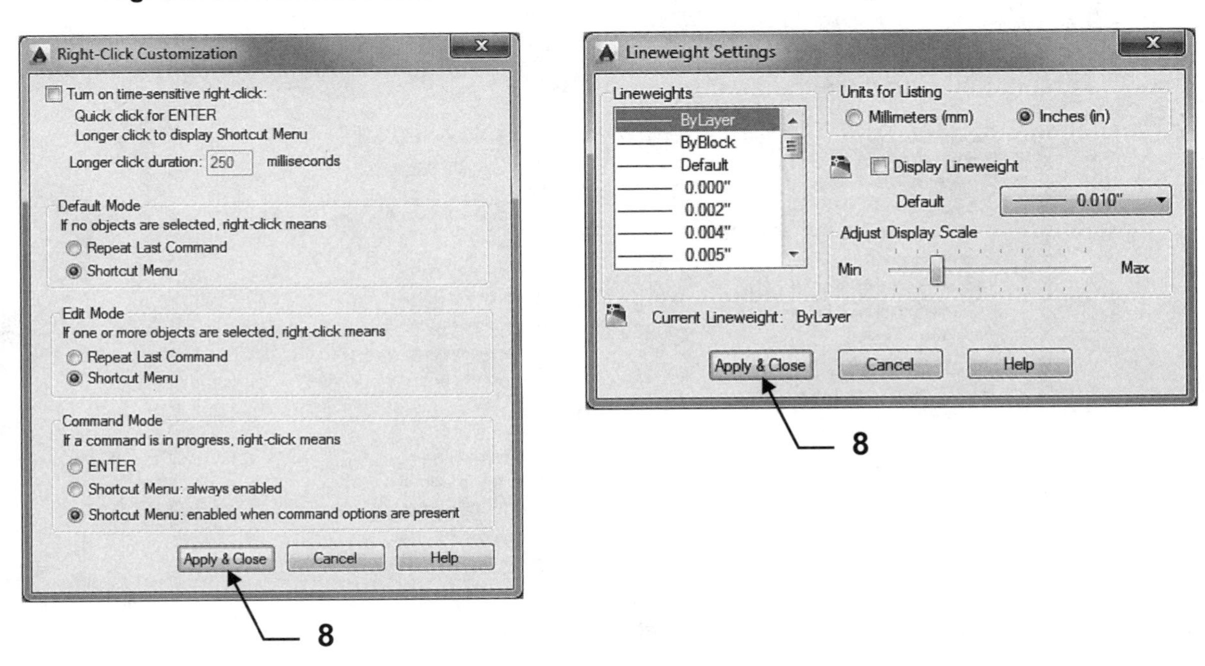

Configuration Settings

9. Select the ***Drafting*** tab and change the settings on your screen to match the dialog box below.

Options
Current profile: <<Unnamed Profile>>
Current drawing: Drawing1.dwg
Files | Display | Open and Save | Plot and Publish | System | User Preferences | Drafting | 3D Modeling | Selection | Profiles | Online

AutoSnap Settings
- [x] Marker
- [x] Magnet
- [x] Display AutoSnap tooltip
- [x] Display AutoSnap aperture box

Colors...

AutoTrack Settings
- [x] Display polar tracking vector
- [x] Display full-screen tracking vector
- [x] Display AutoTrack tooltip

Alignment Point Acquisition
- (•) Automatic
- ( ) Shift to acquire

AutoSnap Marker Size

Aperture Size

Object Snap Options
- [x] Ignore hatch objects
- [x] Ignore dimension extension lines
- [x] Ignore negative Z object snaps for Dynamic UCS
- [ ] Replace Z value with current elevation

Drafting Tooltip Settings...
Lights Glyph Settings...
Cameras Glyph Settings...

OK | Cancel | Apply | Help

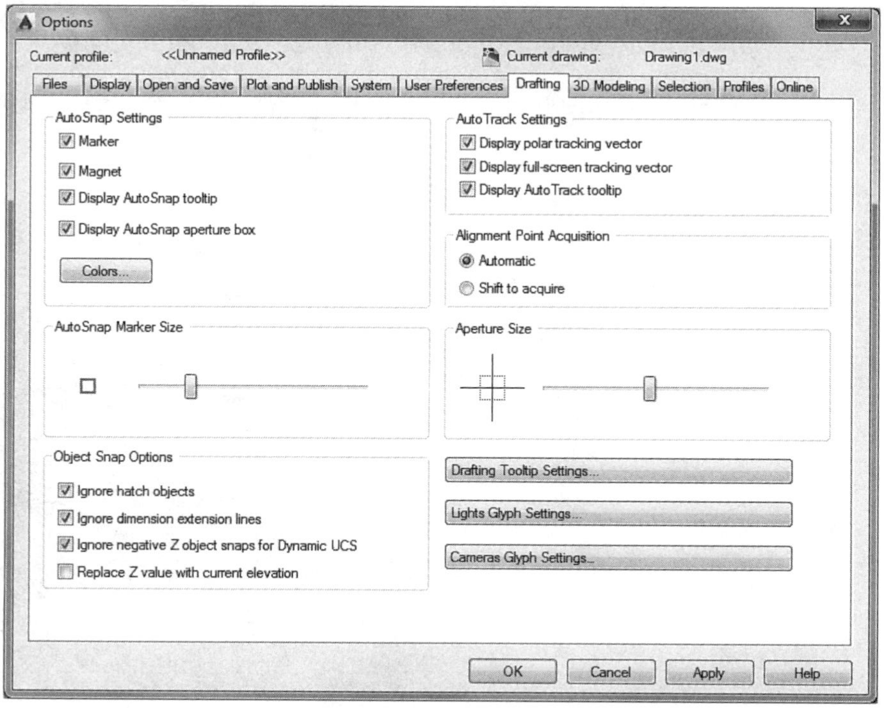

10. Select the ***Selection*** tab and change the settings on your screen to match the dialog box below. (Note: 3D Modeling tab was skipped)

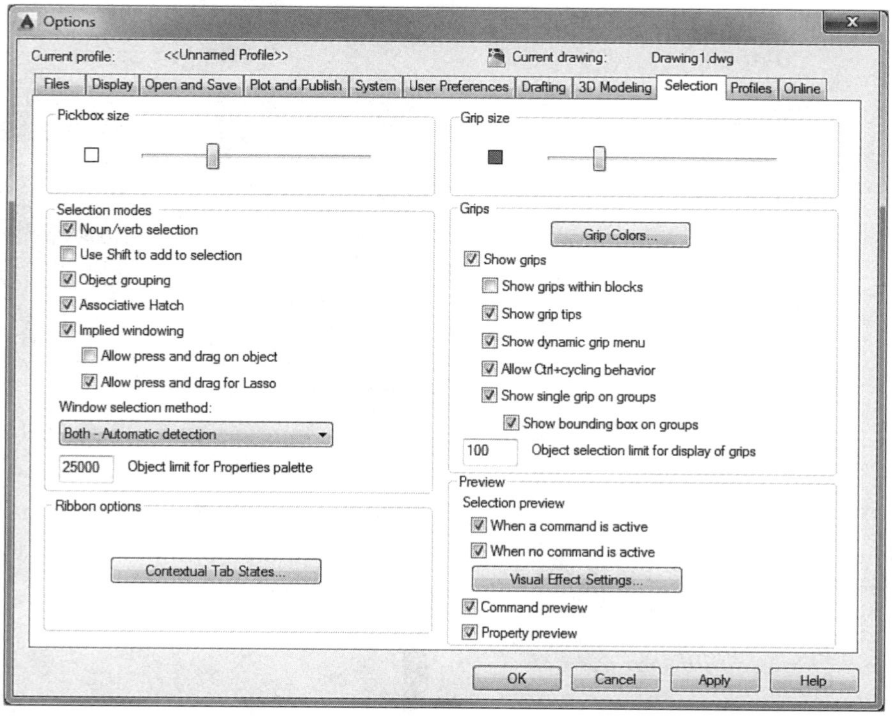

Options
Current profile: <<Unnamed Profile>>
Current drawing: Drawing1.dwg
Files | Display | Open and Save | Plot and Publish | System | User Preferences | Drafting | 3D Modeling | Selection | Profiles | Online

Pickbox size

Grip size

Selection modes
- [x] Noun/verb selection
- [ ] Use Shift to add to selection
- [x] Object grouping
- [x] Associative Hatch
- [x] Implied windowing
  - [ ] Allow press and drag on object
  - [x] Allow press and drag for Lasso

Window selection method:
Both - Automatic detection
25000 Object limit for Properties palette

Grips
Grip Colors...
- [x] Show grips
  - [ ] Show grips within blocks
  - [x] Show grip tips
  - [x] Show dynamic grip menu
  - [x] Allow Ctrl+cycling behavior
  - [x] Show single grip on groups
    - [x] Show bounding box on groups

100 Object selection limit for display of grips

Ribbon options
Contextual Tab States...

Preview
Selection preview
- [x] When a command is active
- [x] When no command is active

Visual Effect Settings...
- [x] Command preview
- [x] Property preview

OK | Cancel | Apply | Help

Configuration Settings

11. Select the ***Apply*** button.

12. Select the ***OK*** button.

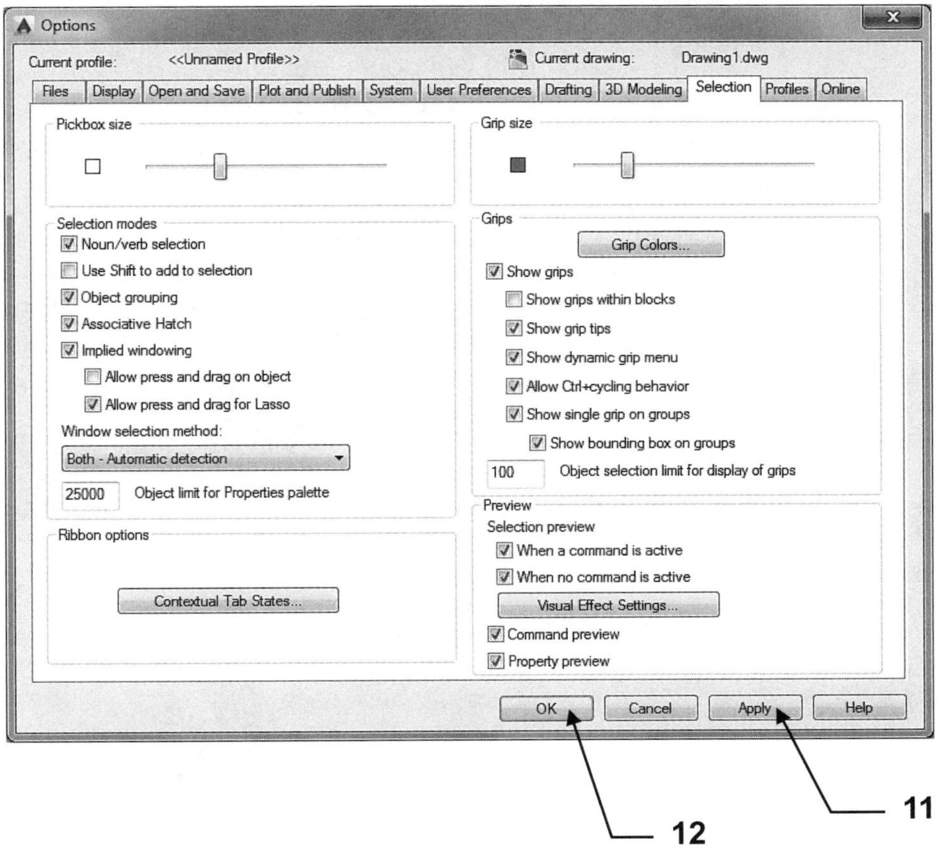

13. Now you should be back to the AutoCAD screen.

AutoCAD 2015 System Requirements for 32-bit

Description	Requirement
Operating System	• Microsoft Windows 7® Enterprise • Microsoft Windows 7® Ultimate • Microsoft Windows 7® Professional • Microsoft Windows 7® Home Premium • Microsoft Windows 8/8.1® • Microsoft Windows 8/8.1® Pro • Microsoft Windows 8/8.1® Enterprise
Browser	Internet Explorer® 9.0 or later
Processor	Intel Pentium 4 or AMD Athlon™ Dual Core, 3.0 GHz or Higher with SSE2 technology
Memory	2 GB RAM (3 GB Recommended)
Display Resolution	1024 x 768 (1600 x 1050 or higher recommended) with True Color
Disk Space	Installation 6.0 GB
Pointing Device	MS-Mouse Compliant
Media	Download and installation from DVD
.NET Framework	.NET Framework 4.5
Additional requirements for large datasets, point clouds, and 3D modeling	Intel Pentium 4 processor or AMD Athlon™, 3.0 GHz or greater or Intel or AMD Dual Core processor, 2.0 GHz or greater 3 GB RAM 6 GB free hard disk available not including installation requirements 1280 x 1024 True color video display adapter 128 or greater, Pixel Shader 3.0 or greater, Direct3D® capable workstation class graphics card Note: 64-bit operating systems are recommended if you are working with large datasets, point clouds and 3D Modeling - please refer to the AutoCAD 2015 64-bit System Requirements on the next page

Continued on the next page...

AutoCAD 2015 System Requirements for 64-bit

Description	Requirement
Operating System	• Microsoft Windows 7® Enterprise • Microsoft Windows 7® Ultimate • Microsoft Windows 7® Professional • Microsoft Windows 7® Home Premium • Microsoft Windows 8/8.1® • Microsoft Windows 8/8.1® Pro • Microsoft Windows 8/8.1® Enterprise
Browser	Internet Explorer® 9.0 or later
Processor	AMD Athlon™ 64 with SSE2 technology AMD Opteron™ with SSE2 technology Intel Xeon® with Intel EM64T support and SSE2 technology Intel Pentium® 4 with Intel EM64T support and SSE2 technology
Memory	2 GB RAM (8 GB Recommended)
Display Resolution	1024 x 768 (1600 x 1050 or higher recommended) with True Color
Disk Space	Installation 6.0 GB
Pointing Device	MS-Mouse Compliant
Media	Download and installation from DVD
.NET Framework	.NET Framework 4.5
Additional requirements for 3D modeling	8 GB RAM or more 6 GB free hard disk available not including installation requirements 1280 x 1024 True color video display adapter 128 or greater, Pixel Shader 3.0 or greater, Direct3D® capable workstation class graphics card

Customizing your Wheel Mouse

A Wheel mouse has two or more buttons and a small wheel between the two topside buttons. The default functions for the two top buttons and the Wheel are as follows:
Left Hand button is for **input** and can't be reprogrammed.
Right Hand button is for **Enter** or the **shortcut menu**.
The Wheel may be used to Zoom and Pan or Zoom and display the Object Snap menu. You will learn more about this later.

The following describes how to select the Wheel functions. After you understand the functions, you may choose to change the setting.
To change the setting you must use the **MBUTTONPAN** variable.

MBUTTONPAN setting 1: (Factory setting)

ZOOM	Rotate the wheel forward to zoom in Rotate the wheel backward to zoom out
ZOOM EXTENTS	Double click the wheel to view entire drawing
PAN	Press the wheel and drag the mouse to move the drawing on the screen.

MBUTTONPAN setting 0:

ZOOM	Rotate the wheel forward to zoom in Rotate the wheel backward to zoom out
OBJECT SNAP	Object Snap menu will appear when you press the wheel

To change the setting:

1. Type: **mbuttonpan <enter>**
2. Enter **0** or **1 <enter>**

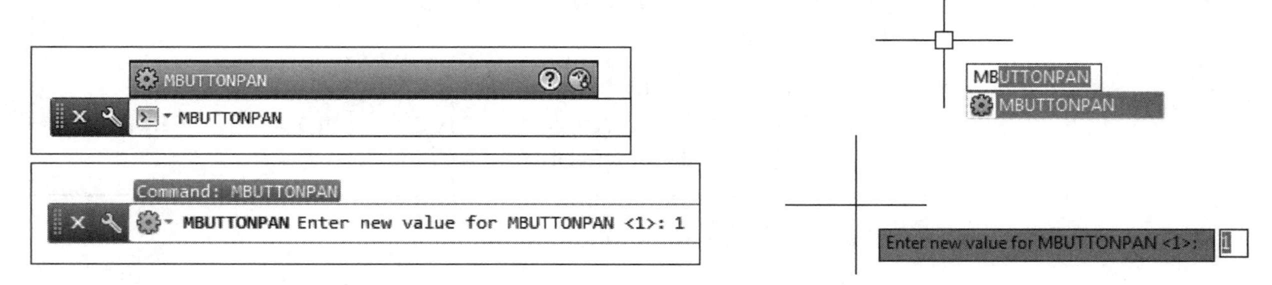

Command Line **Dynamic Input**

LEARNING OBJECTIVES

After completing this lesson, you will be able to:

1. Start AutoCAD program
2. Understand AutoCAD Workspaces
3. Recognize all of the features in the AutoCAD Window

LESSON 1

STARTING AUTOCAD

To Start AutoCAD use one of the 2 methods below.
(Be patient it may take a few minutes to load. It is a large program)

1. Select **Start / All Programs / Autodesk / AutoCAD 2015**
   or
   Double click on the **AutoCAD 2015 desktop shortcut icon**.

2. The **"New Tab"** page should appear.

Note: Your computer should be connected to the Internet.

The **New Tab** page contains two sliding content pages called **LEARN** and **CREATE**.

CREATE: The Create page (shown below) is displayed by default and allows you to start a **New** drawing, **Open** an existing drawing or template, **Explore** sample drawings, view **Recent** documents or access **Online Services**.

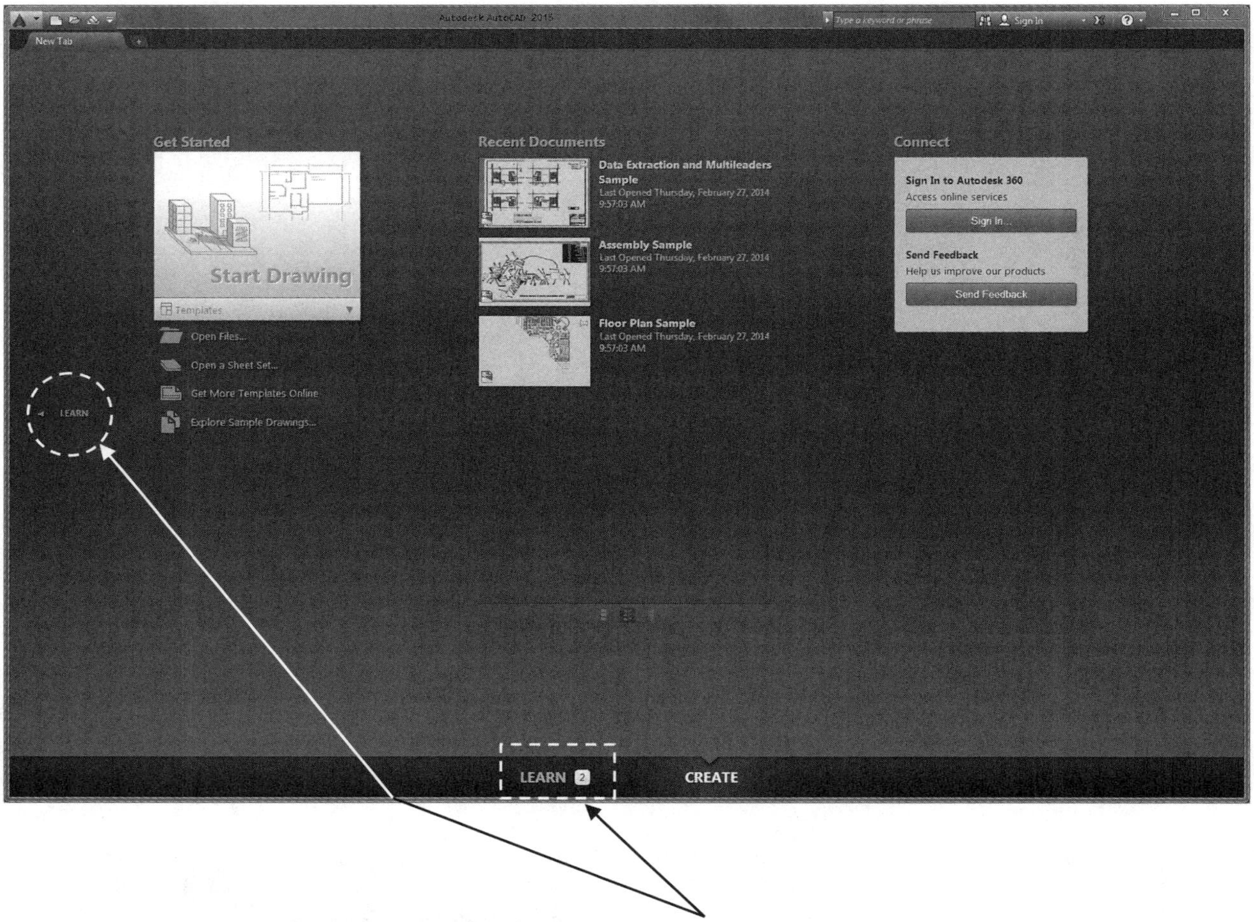

You can access the **LEARN** page by clicking on **LEARN** at the bottom of the page or on the left-hand side of the page.

Continued on the next page...

STARTING AUTOCAD....continued

You can start a new drawing using the default template by clicking on the **Start Drawing** tool, or you can access a list of available templates by clicking on the Templates down ▼ arrow.

Note: The last template you use will become the new default template and is highlighted in the template dropdown list.

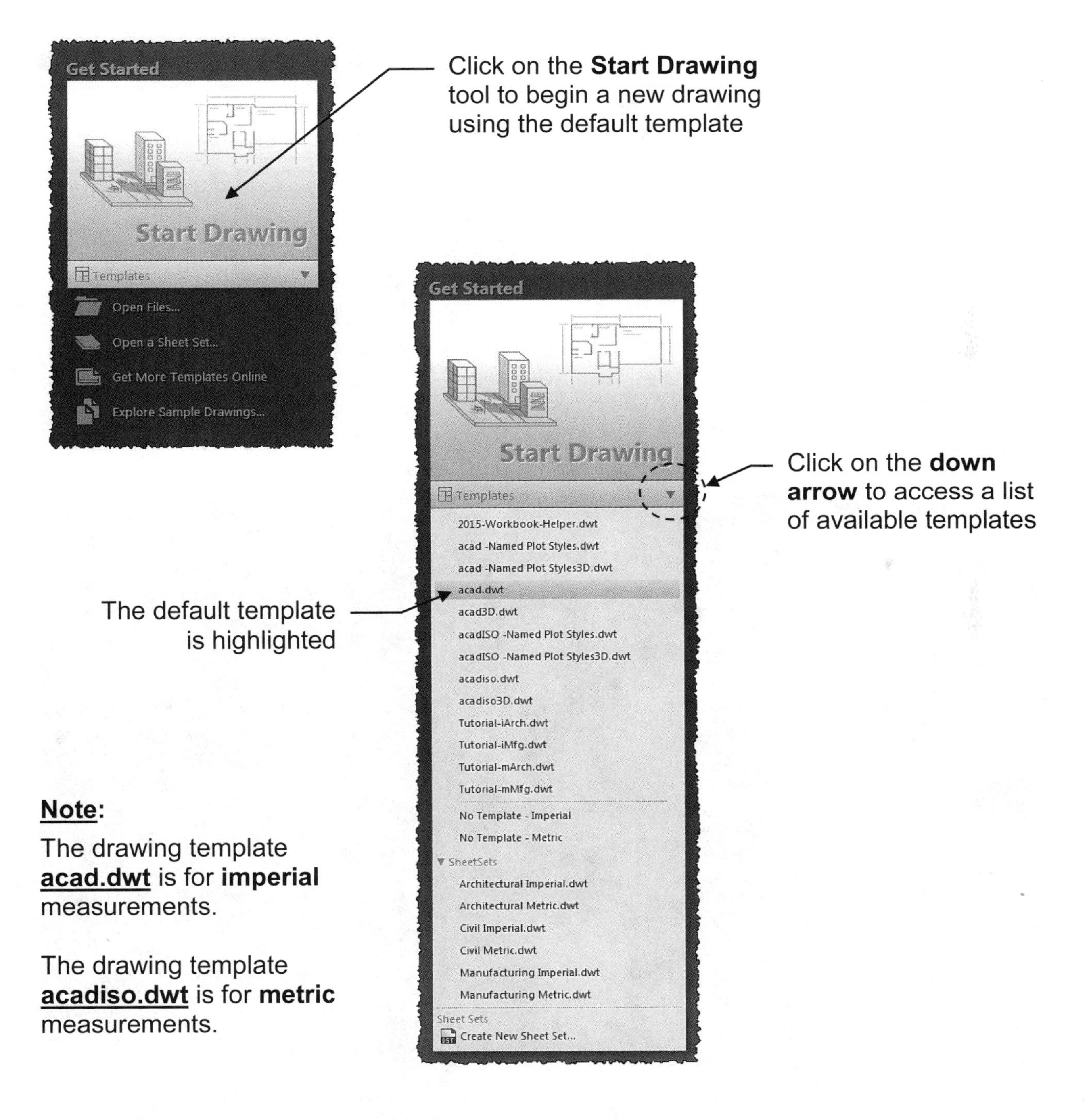

Note:

The drawing template **acad.dwt** is for **imperial** measurements.

The drawing template **acadiso.dwt** is for **metric** measurements.

Continued on the next page...

STARTING AUTOCAD....continued

Click on the **Start Drawing** tool to open the default template and to open the main AutoCAD drawing workspace.

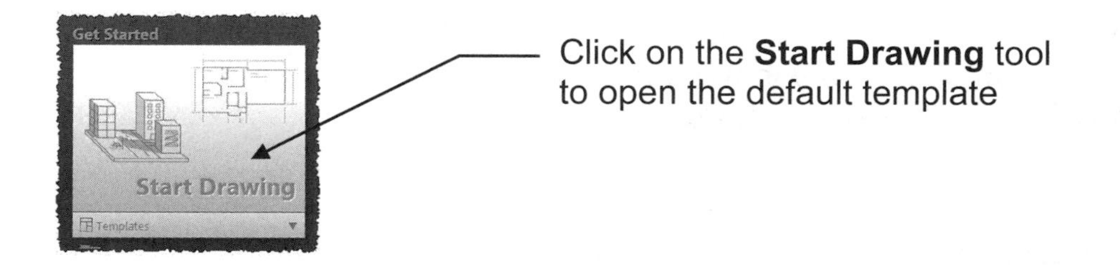

Note: If your screen does not appear as shown below go to the **Intro section** of this workbook and follow the steps for configuring AutoCAD to match the workbook configuration.

The following pages will describe each area and element.

I know you are anxious to start drawing but be patient and read the remainder of this lesson. It is very important that you understand and are familiar with AutoCAD's interface.

Note: You can close the **Design Feed** palette by clicking on the "**X**". This will not be used until later in the workbook.

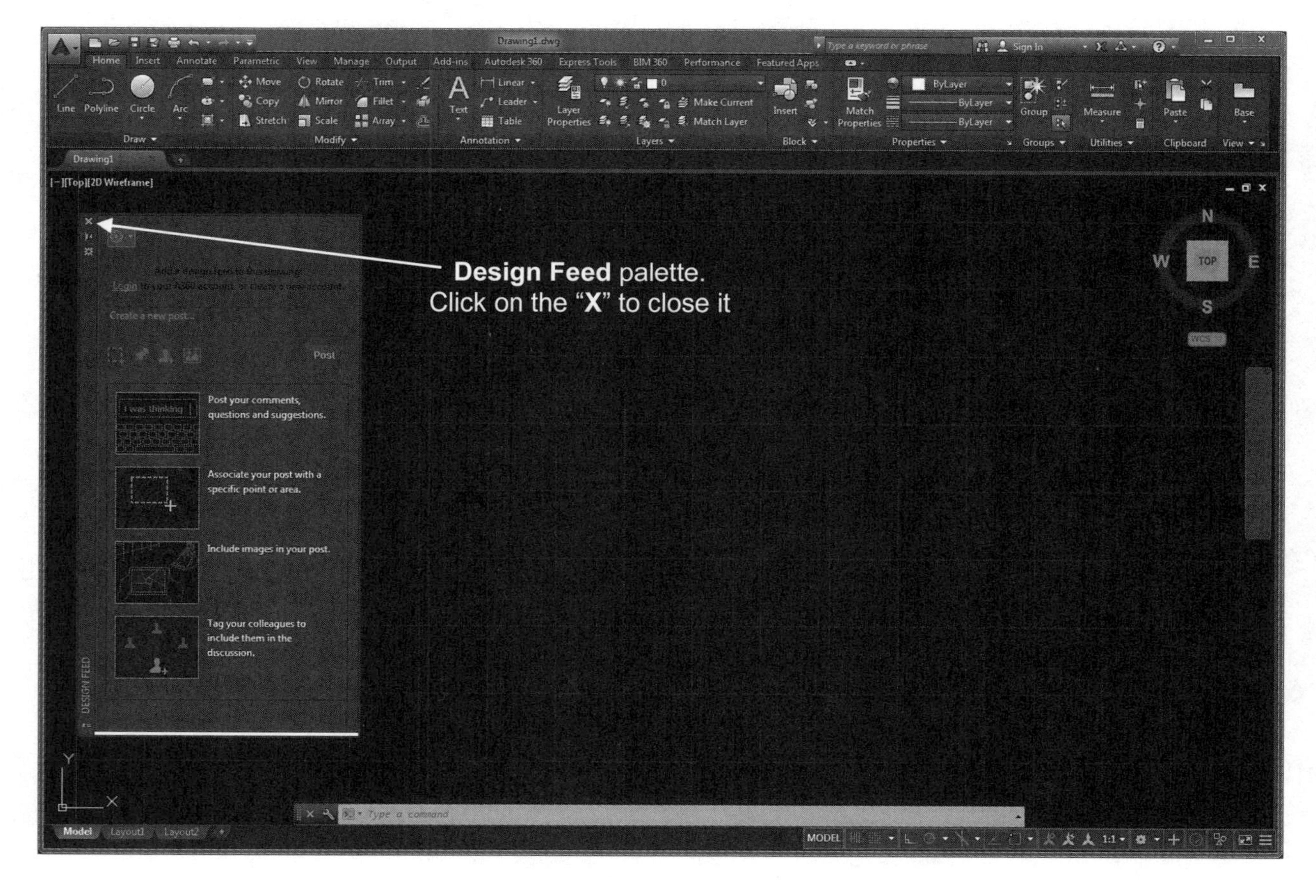

AUTOCAD APPLICATION WINDOW

The AutoCAD Application Window default workspace
Drafting & Annotation, is shown below.

Important: I have changed my 2D background color to white for this workbook. Yours may be another color.

AutoCAD 2015 comes with new Dark and Light color themes, on some of the images within this book you may see the ribbon color showing the Light theme, this is just for clarity. You may choose the Light theme if you wish.

You may change the color of many areas using: **Options / Display tab / Colors button**

If the remainder of your screen does not appear as shown below go to the **Intro section** of this workbook and follow the steps for configuring AutoCAD to match the workbook configuration.

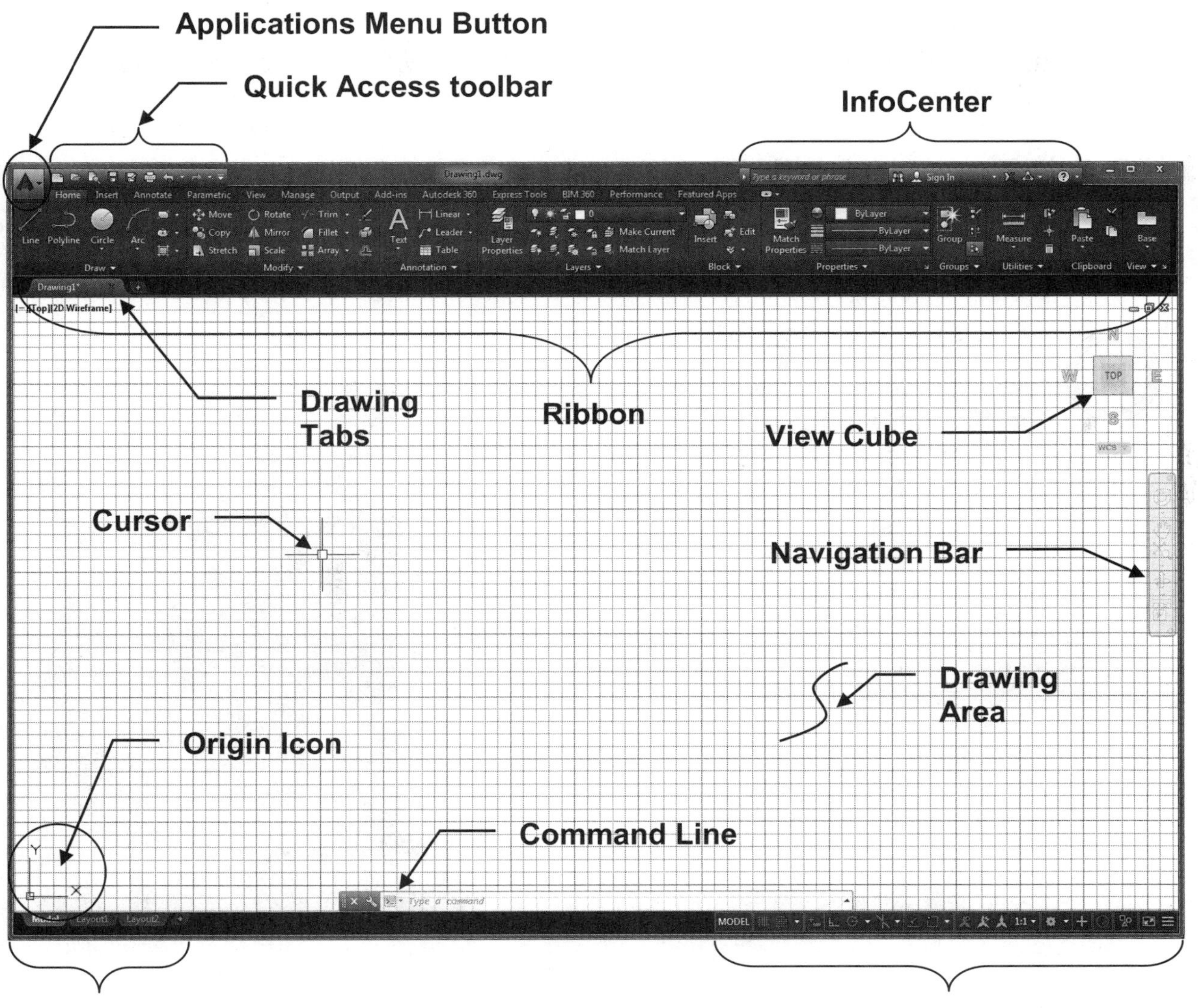

A description of each area is shown on the following pages.

APPLICATION WINDOW DESCRIPTIONS

WORKSPACE

Workspaces control the display of ribbons, tabs, menus, toolbars, and palettes. AutoCAD gives you the option of deciding how you would like them displayed.
When you use a workspace, only the menus, toolbars, and palettes that are relevant to a task are displayed. For example if you selected the 3D Modeling workspace only 3D menus, toolbars and palettes would be displayed.

There are 3 preset Workspaces.

Drafting & Annotation (shown on the previous page)
This workspace is the default display. It displays the necessary ribbons, tabs, menus, tools and palettes for 2D drafting. We will be using this workspace for all lessons within this workbook.

3D Basics
This 3D Basics workspace provides a simple workspace with the most basic tools for creating and visualizing 3D Models.

3D Modeling
The 3D Modeling workspace provides access to the vast array of 3D tools in AutoCAD.

HOW TO SELECT A WORKSPACE

1. Selecting a workspace is easy. Select the ▼ on the **Workspace Switching** icon located on the Status Bar at the bottom right corner of the screen.

2. Select one of the workspace names displayed.

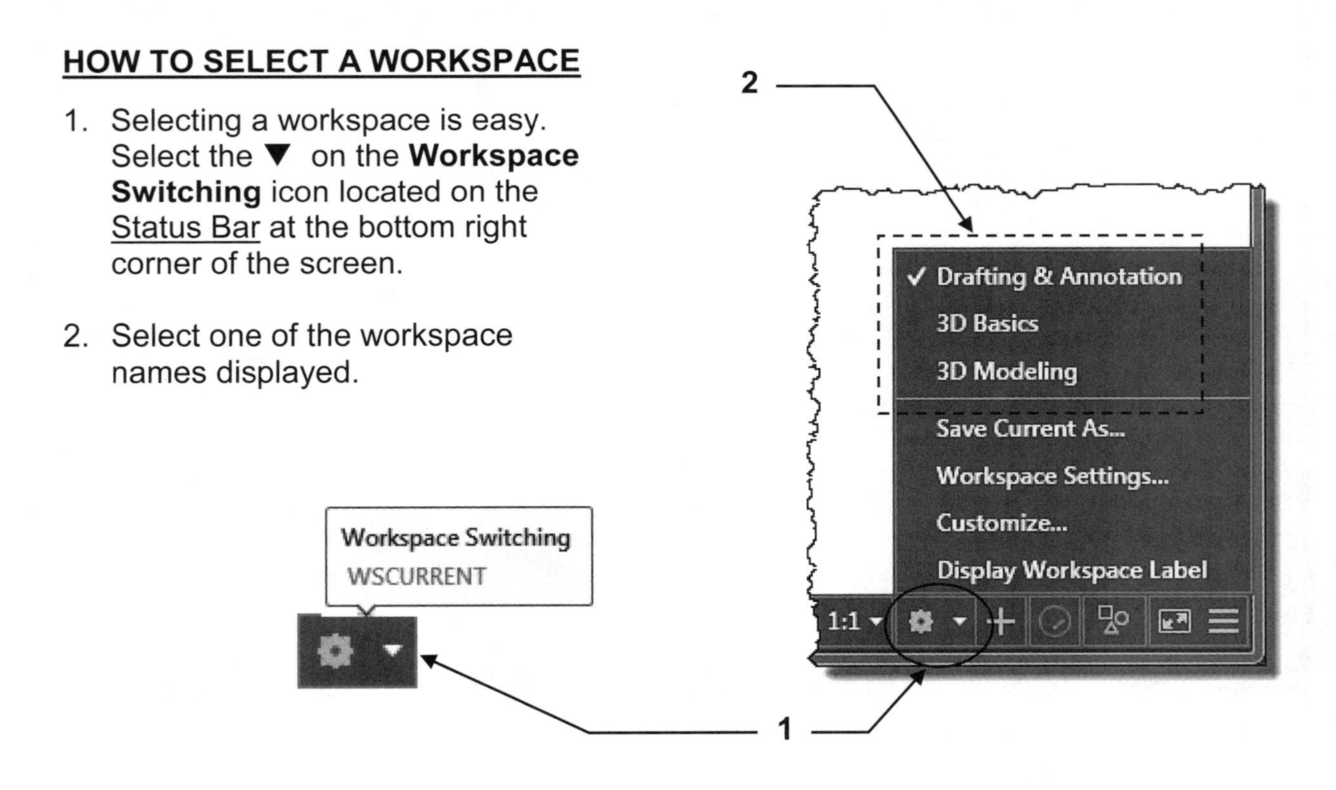

Note: We will be using the "Drafting & Annotation" Workspace.

APPLICATION WINDOW DESCRIPTIONS

APPLICATION MENU

The Application Menu provides easy access to common tools
Each of the tools will be discussed later in the workbook.

1. Click on the **Application Menu button** in the upper left corner of the AutoCAD display screen. (The big **"A"**)

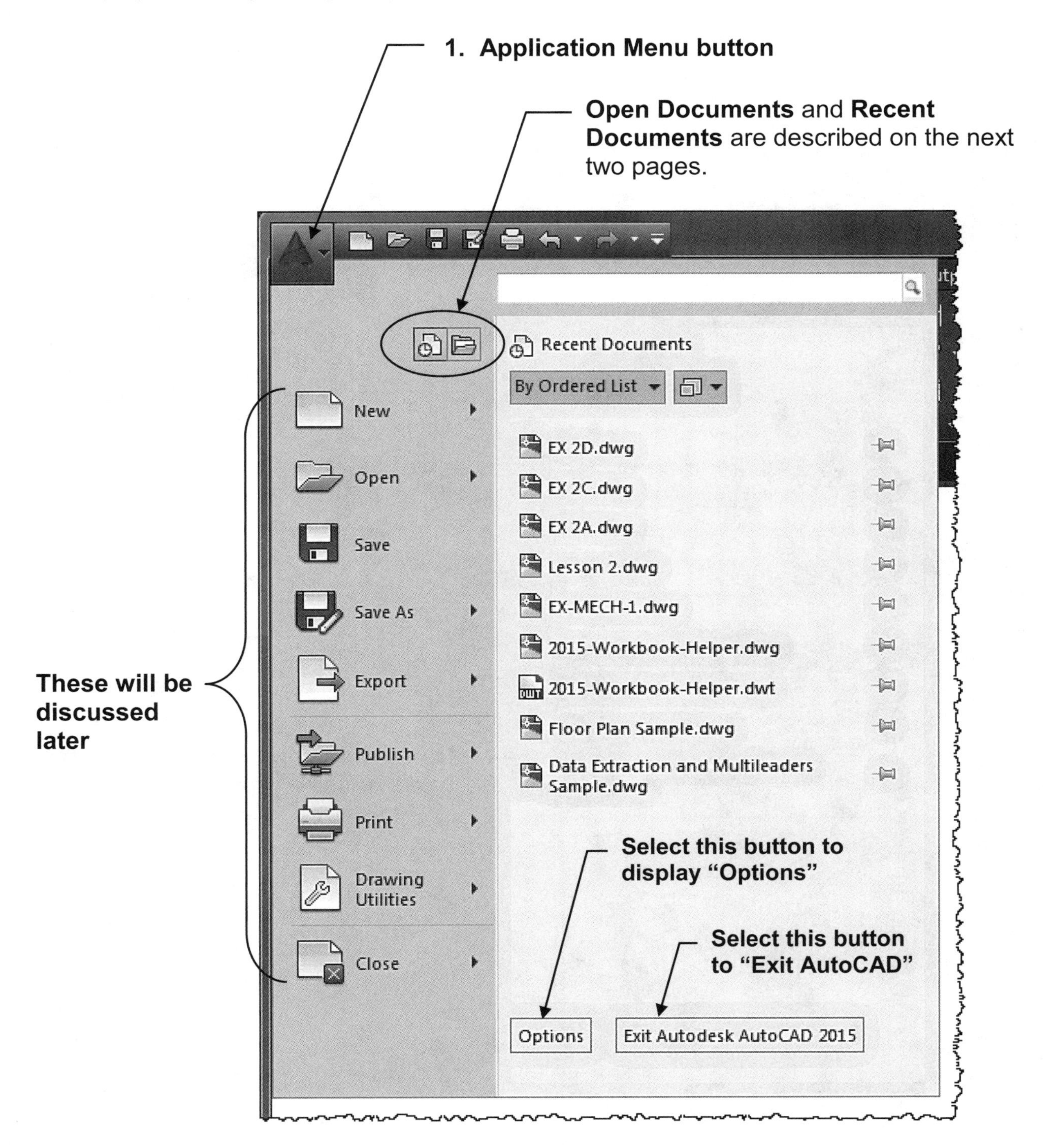

APPLICATION WINDOW DESCRIPTIONS....continued

Open Documents

First let me emphasis **this is not a method to "open" a drawing file**.

Open Documents is a list of all documents that are **already open** within AutoCAD.

Display choice: The list of documents may be displayed as icons or images.
If you hover the cursor over a document name a preview image will appear.

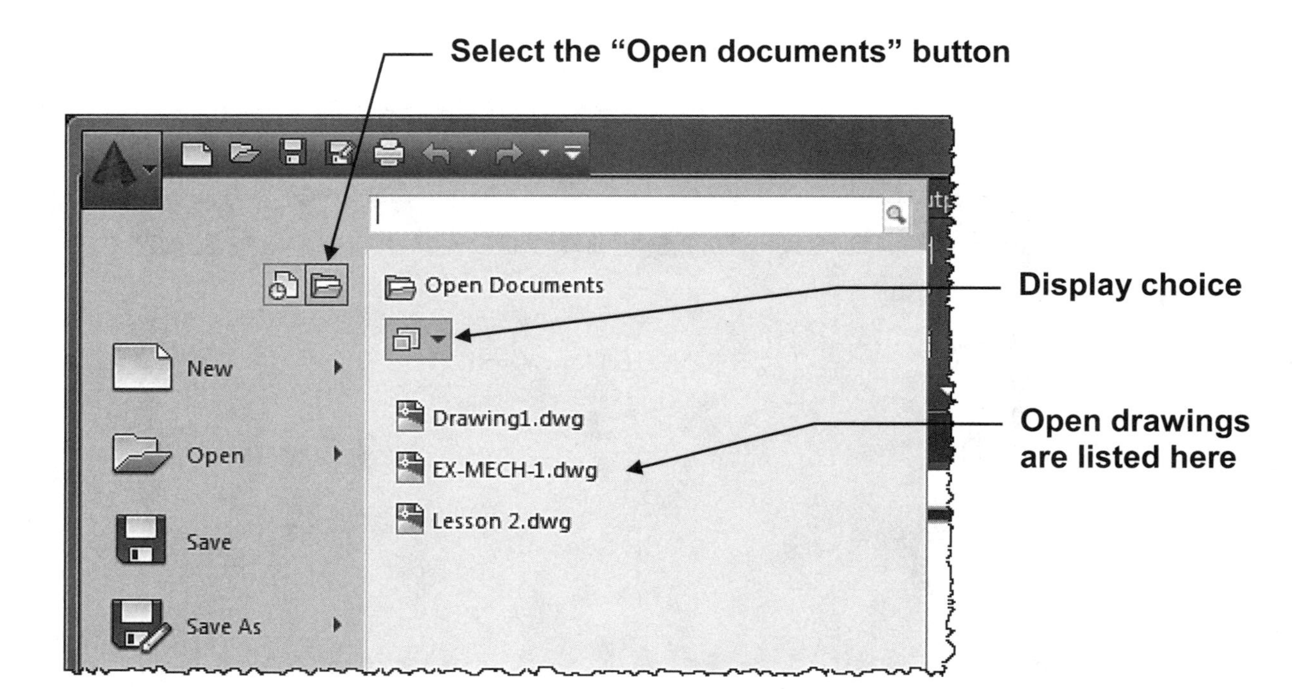

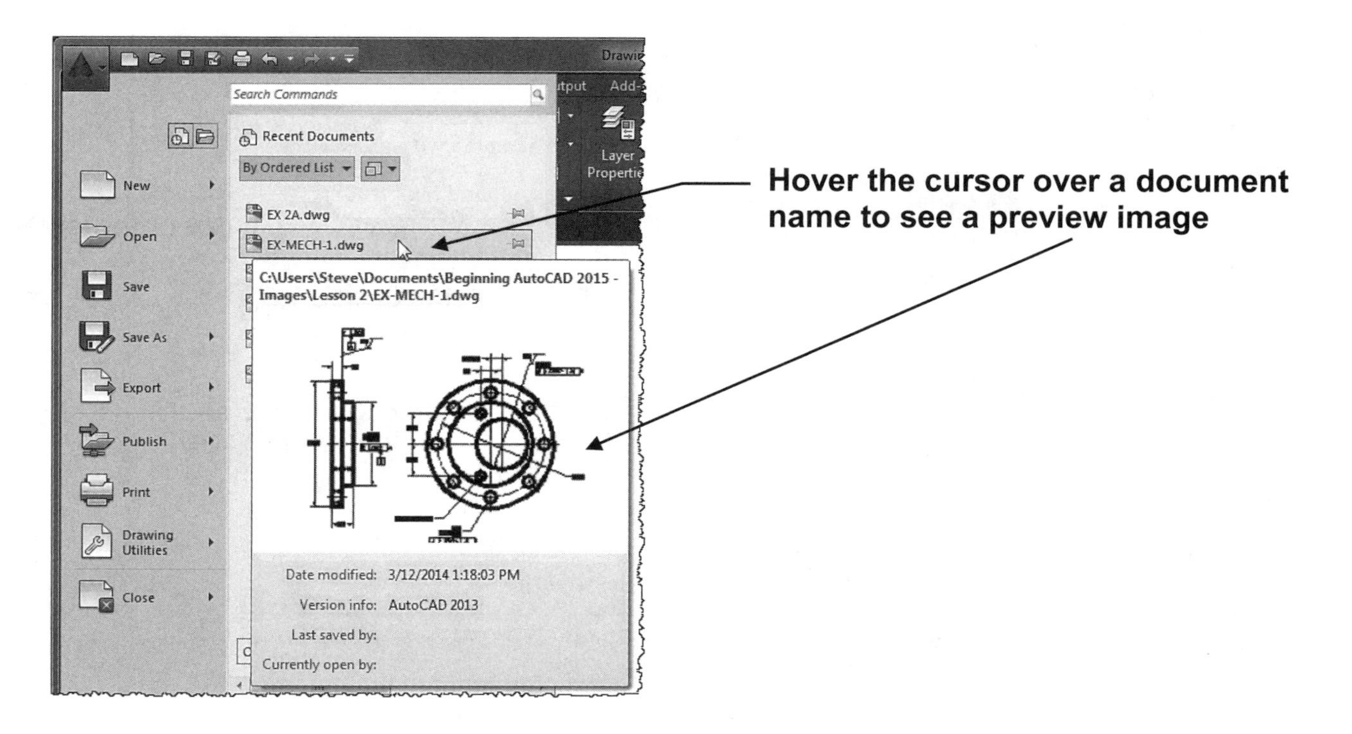

APPLICATION WINDOW DESCRIPTIONS....continued

View Recent Documents

When you select **Recent Documents** a list of the recently viewed documents will appear.

Display choices: This list may be displayed as icons or images and may be sorted in an ordered list or grouped by date or file type.

If you hover the cursor over a document name a preview image will appear.

Pinned Files

You can keep a file listed regardless of files that you save later using the push pin button to the right. The file is displayed at the bottom of the list until you turn off the push pin button.

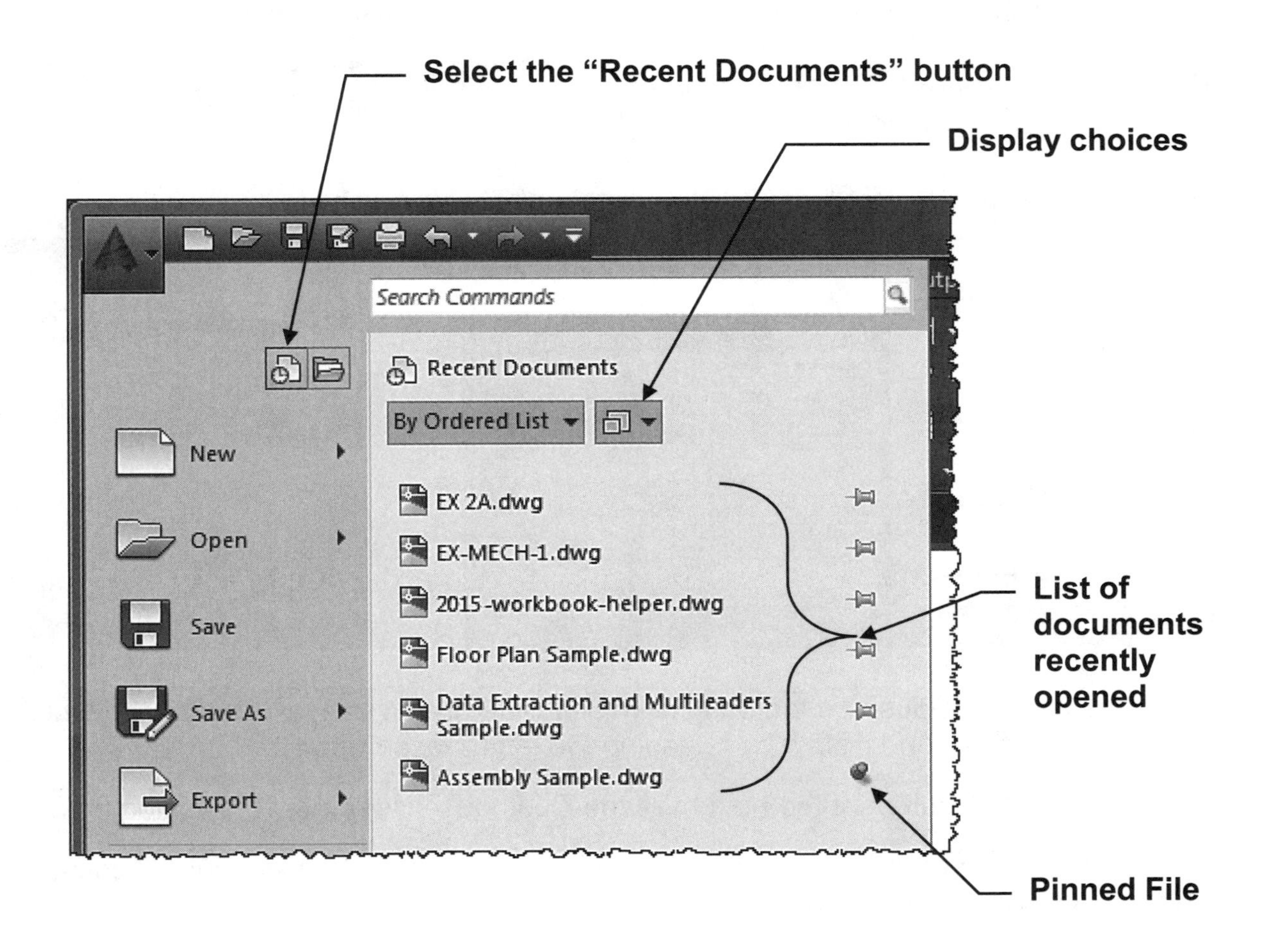

Quick Access Toolbar

The **Quick Access Toolbar** is located in the top left corner of the AutoCAD window. It includes the most commonly used tools, such as New, Open, Save, Save as, Print, Undo and Redo.

How to Customize the Quick Access Toolbar

You can add tools with the Customize User Interface dialog box.

For Example:
After you have completed Lesson 4 you will find that you will be using **"Zoom All"** often. So I add the "**Zoom All**" tool to the Quick Access Toolbar. If you would like to add the "**Zoom All**" tool, or any other tool, to your Quick Access Toolbar follow the steps below.

1. Place the Cursor on the Quick Access Toolbar and press the right mouse button.
2. Select **"Customize Quick Access Toolbar..."** from the menu.
3. Scroll through the list of Commands to **"Zoom, All"**

Remove from Quick Access Toolbar
Add Separator
Customize Quick Access Toolbar
Show Quick Access Toolbar below the Ribbon

2

Customize User Interface
Customizations in All Files
To add a command, drag a command from the Command List pane to the Quick Access toolbar, a toolbar, or tool palette.
Command List:
All Commands Only

Command	Source
Zoom, All	ACAD
Zoom, Center	ACAD
Zoom, Dynamic	ACAD

OK Cancel Apply Help

3
5

4. Press the Left mouse button on "**Zoom, All**" and drag it to a location on the Quick Access Toolbar and drop it by releasing the left mouse button.
5. Select the **OK** button at the bottom of the Customize User Interface dialog box.

The Customize User Interface dialog box will disappear and the new Quick Access Toolbar is saved to the current Workspace.

To Remove a tool:
Place the cursor on the tool to remove and press the right mouse button.
Select **Remove from Quick Access Toolbar.**

APPLICATION WINDOW DESCRIPTIONS....continued

RIBBON

The **RIBBON** provides access to the AutoCAD tools.
The **TABS** contain multiple **PANELS**. Each **PANEL** contains multiple tools.
When you select a **TAB** a new set of **PANELS** will appear.

Additional Tools: If you select the ▼ symbol, at the bottom of the panel, the panel will expand to access additional tools.

Control the display of Tabs and Panels

Right click on the Ribbon and select which Tabs or Panels you choose to display.
The check mark confirms the tab or panel is already displayed.

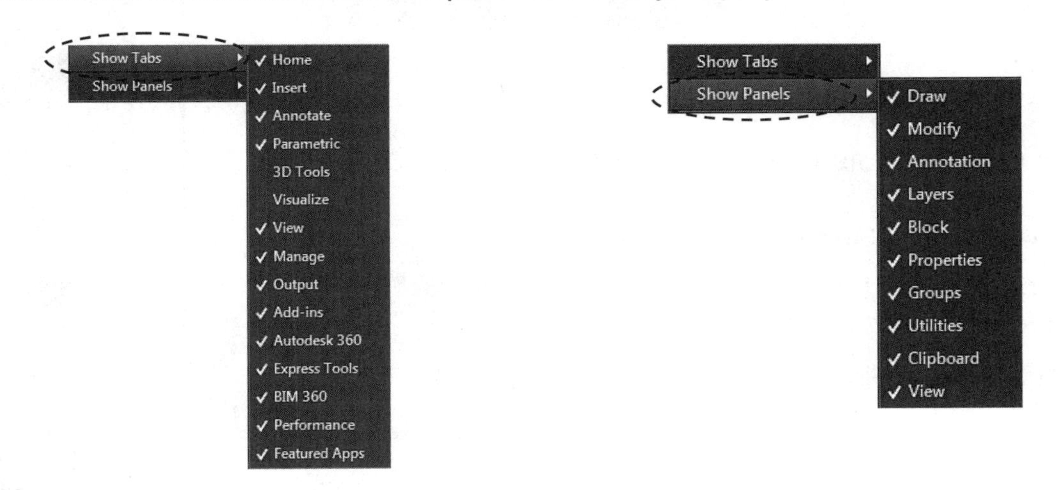

Control the TAB order

If you would like to change the order of the tabs, click and drag the tab horizontally to the new location.

Floating PANELS

If you prefer to separate a Panel from the Ribbon you may drag the panel off the Ribbon to a new location on the screen.

STATUS BAR

The Status Bar is located on the bottom of the screen. It displays the current settings. These settings can be turned ON or OFF by clicking on one of the buttons or by pressing a corresponding function key, F2, F3 etc.

When an icon is turned on it will display a neon blue in color.

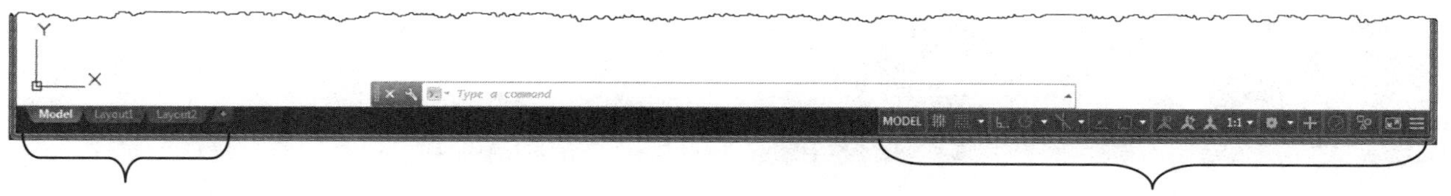

Model and Layout Tabs
Discussed fully in Lesson 26

Status Bar Icons

Status Bar Icons

The status bar provides you with a set of commonly used drawing tools like grid display, snap, object snap and isometric drafting. You can choose to remove some or all of them, or you can choose to add more tools.

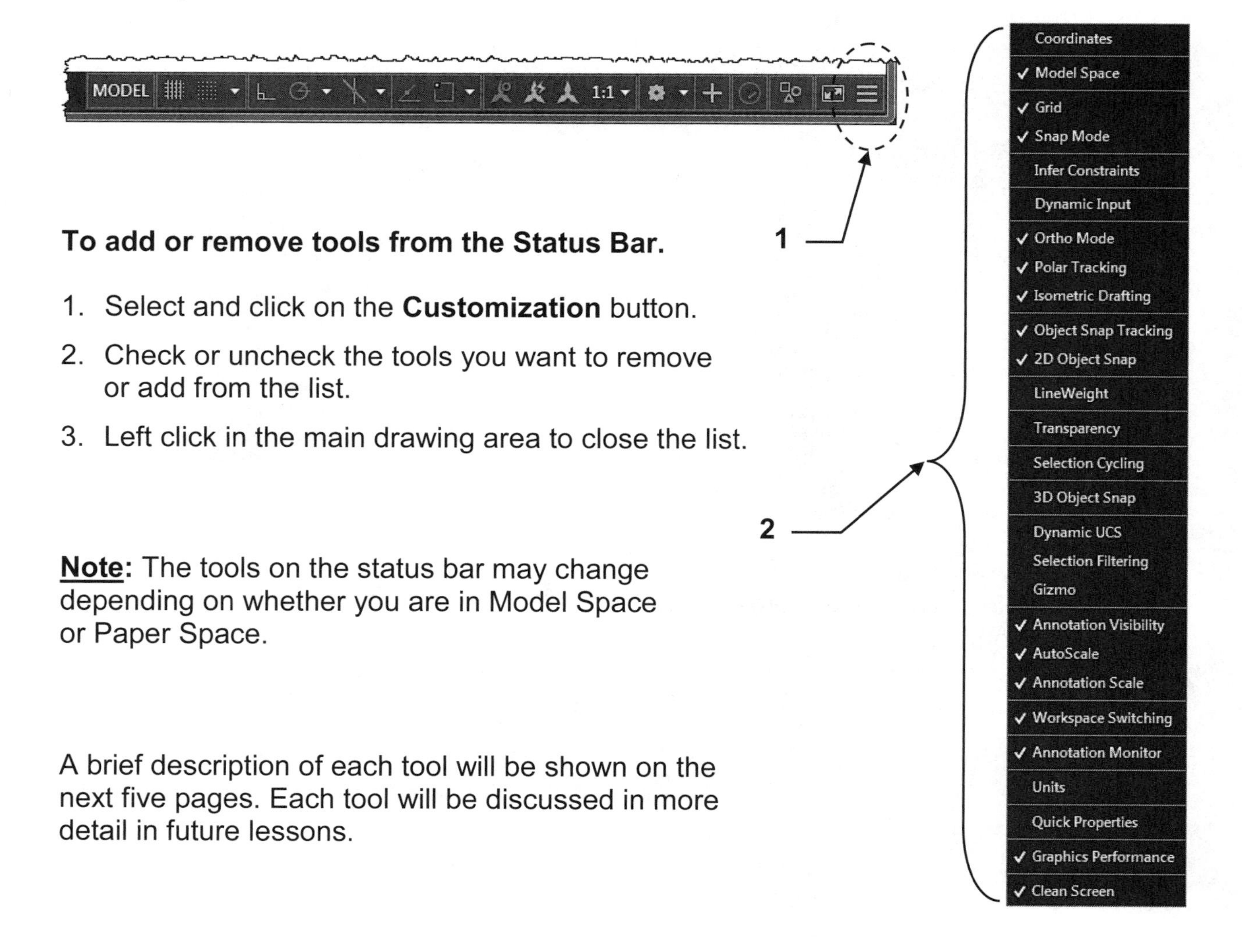

To add or remove tools from the Status Bar.

1. Select and click on the **Customization** button.
2. Check or uncheck the tools you want to remove or add from the list.
3. Left click in the main drawing area to close the list.

<u>**Note**</u>: The tools on the status bar may change depending on whether you are in Model Space or Paper Space.

A brief description of each tool will be shown on the next five pages. Each tool will be discussed in more detail in future lessons.

APPLICATION WINDOW DESCRIPTIONS....continued

STATUS BAR TOOL BUTTON DESCRIPTIONS

I have enabled all the tool buttons and broken them down into two sections starting from the left-hand side.

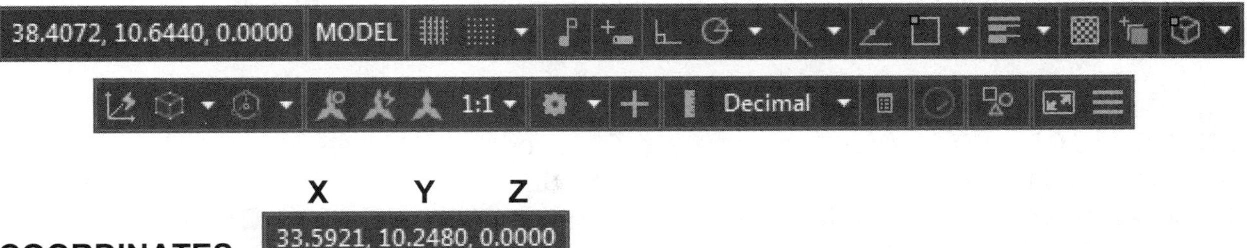

X Y Z
33.5921, 10.2480, 0.0000

COORDINATES
The coordinates display the location of the cursor in reference to the Origin. The Origin is currently in the lower left corner of the drawing area. These numbers will change as you move the cursor.

First set of numbers represents the horizontal movement of the cursor (**X-axis**).
Second set of numbers represents the vertical movement of the cursor (**Y-axis**).
Third set of numbers represents the **Z-axis** which is used for 3D and not discussed.

MODEL MODEL
The **MODEL** / **PAPER** button allows you to work in either model space or paper space without leaving the layout tab. When you switch to a layout tab this button automatically switches to **PAPER**.

GRID (You may also use **F7** to toggle ON or OFF)
The criss-cross lines in the Drawing Area are called the Grid. It is only a drawing aid and will not print. The default spacing is 1 unit of measurement. You may change the Grid spacing at any time by typing **DS** then press **<enter>**, then select the **Snap and Grid** tab from the **Drafting Settings** dialog box.

SNAP MODE (You may also use **F9** to toggle ON or OFF)
Increment Snap controls the incremental movement of the cursor. If it is **ON** the cursor will "snap" in an incremental movement. If it is **OFF** the cursor will move smoothly. You may set the increments by clicking on the down ▼ arrow and selecting **Snap Settings**.

You can also choose whether to use **Grid Snap** or **Polar Snap** on the same menu. →

Polar Snap
✓ Grid Snap
Snap Settings...

INFER CONSTRAINTS (Note: Not used in this workbook)
Inferred Geometric Constraints automatically applies coincident constraints for Endpoint, Midpoint, Center, Node, and Insertion object snaps.

APPLICATION WINDOW DESCRIPTIONS....continued

STATUS BAR TOOL BUTTON DESCRIPTIONS....CONTINUED

DYNAMIC INPUT (You may also use **F12** to toggle ON or OFF)
When **Dynamic Input** is **ON**, you can enter coordinate values in tooltips near the cursor. More on this in Lesson 11.

ORTHO MODE (You may also use **F8** to toggle ON or OFF)
Ortho restricts the movement of the cursor to Horizontal or Vertical. When Ortho is **ON** the cursor moves only in the horizontally or vertically. When Ortho is **OFF** the cursor moves freely in any direction.

POLAR TRACKING  (You may also use **F10** to toggle ON or OFF)
Polar Tracking restricts cursor movement to specified increments along a polar angle. More on this in Lesson 11.

You may set the increments by clicking on the down ▼ arrow and selecting **Tracking Settings**, or you can select one of the predefined angles from the list.

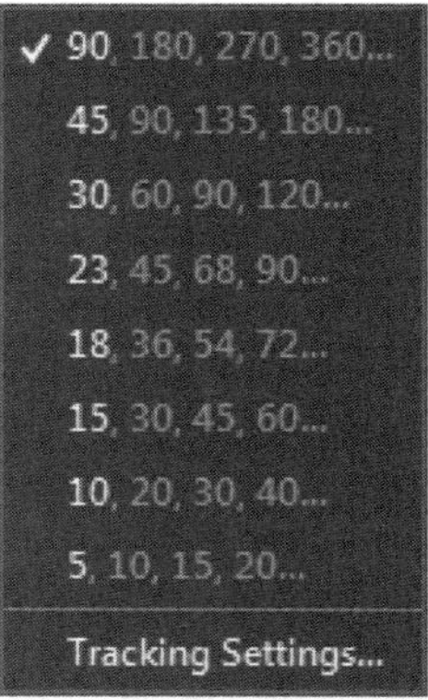

ISOMETRIC DRAFTING 
Isometric drawing allows you to simulate a 3D object by aligning along 3 axes, these are **Top**, **Right** and **Left**, called **Isoplanes**. When the button is enabled you can toggle the Isoplanes by pressing the **F5** key. Isometric Drafting is discussed in the Advanced AutoCAD Workbook.

You may also change Isoplanes by clicking the down ▼ arrow and selecting from the menu.

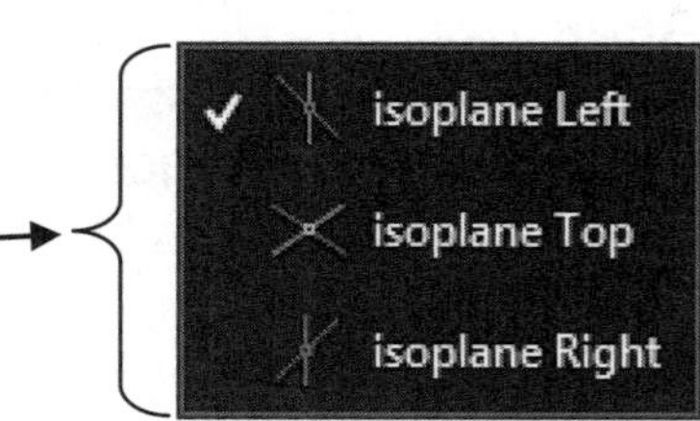

OBJECT SNAP TRACKING (You may also use **F11** to toggle ON or OFF)
Object Snap Tracking controls the display of object snap reference lines, AutoSnap marker, tooltip and magnet.

APPLICATION WINDOW DESCRIPTIONS....continued

STATUS BAR TOOL BUTTON DESCRIPTIONS....CONTINUED

2D OBJECT SNAP  (You may also use **F3** to toggle ON or OFF)
When 2D Object Snap is **ON** the cursor will "snap" to preset locations on 2D objects. More on this in Lesson 4.

You may also add or remove Object Snaps by clicking the down ▼ arrow and selecting from the menu.

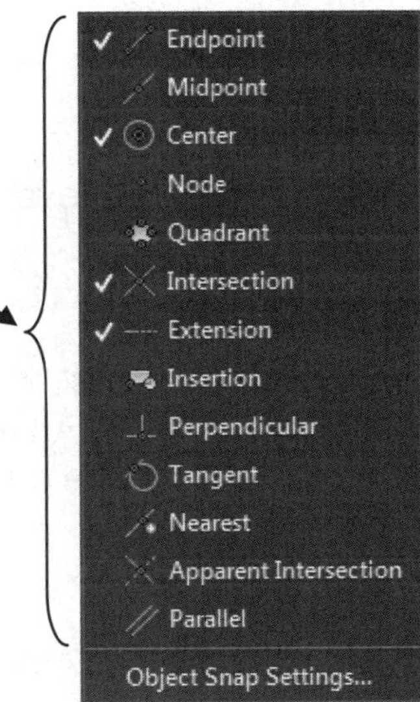

LINEWEIGHT
Lineweight displays the width assigned to each object. When it is **ON** the lineweights are visible. You can change the lineweight settings by clicking on the down ▼ arrow and selecting **Lineweight Settings**. More on this in Lesson 3.

TRANSPARENCY
When Transparency Show/Hide is **ON** all transparent layers will be displayed. If it is **OFF** no layers will display as transparent. More on this in Lesson 3.

SELECTION CYCLING 
Selection Cycling allows you to select objects that are overlapping. This is most useful when creating 3 dimensional models discussed in the Advanced AutoCAD Workbook.

3D OBJECT SNAP  (You may also use **F4** to toggle ON or OFF)
When 3DOsnap is **ON** the cursor will "snap" to preset locations on 3D objects. This option will be discussed in the "Advanced AutoCAD Workbook".

DYNAMIC UCS (You may also use **F6** to toggle ON or OFF)
Dynamic User Coordinate System changes the grid plane to follow the XY plane of the dynamic UCS. Used for 3D, refer to the Advanced AutoCAD Workbook.

STATUS BAR TOOL BUTTON DESCRIPTIONS....CONTINUED

SELECTION FILTERING

Selection filtering allows you to filter whether certain faces, edges, vertices or solid history subobjects are highlighted when you roll over them, very useful in complex 3D.

GIZMO

Gizmo tools help you move, rotate, or scale an object or set of objects along a 3D Plane, and discussed in the Advanced AutoCAD Workbook.

ANNOTATION VISIBILITY

When switched **ON** the Annotation Visibility tool displays or hides the visibility of annotation objects at the current scale.

AUTOSCALE

When switched **ON** the AutoScale tool Automatically updates annotative objects to support the annotation scale when the annotation scale is changed.

ANNOTATION SCALE

The Annotation Scale tool displays the current annotation scale. You can change the scale by clicking on the down ▼ arrow and selecting from the list of predefined scales or you may create a custom scale. You can also display the scale in percentages by selecting **Percentages** from the list.

WORKSPACE SWITCHING

Workspace Switching allows you to change the workspace environment, you can choose between "Drafting & Annotation", 3D Basics and 3D Modeling. You can change the Workspace by clicking on the down ▼ arrow and selecting from the list.

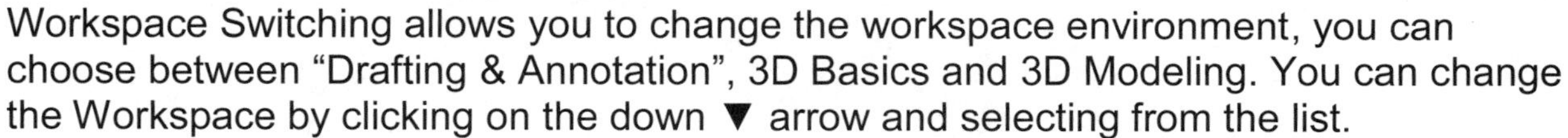

ANNOTATION MONITOR

Provides feedback regarding the state of associative annotations when using parametric dimensioning. This option is discussed in the "Advanced AutoCAD Workbook".

UNITS

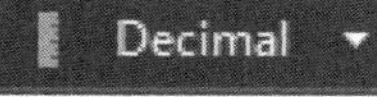

The Units tool allows you to change the display style of the Drawing Units. You can choose between Decimal, Architectural, Engineering, Fractional and Scientific. You can change the drawing unit display by clicking on the down ▼ arrow and selecting from the list.

STATUS BAR TOOL BUTTON DESCRIPTIONS....CONTINUED

QUICK PROPERTIES 

If **ON**, Quick Properties displays the properties of the object selected. If **OFF** the Quick Properties box will not appear. More on this in Lesson 12.

GRAPHICS PERFORMANCE 

The Graphics Performance tool examines your graphics card and 3D display driver and determines whether to use software acceleration or hardware acceleration. You can change the performance settings by right-clicking on the tool button and selecting **Graphics Performance**, then change any settings required in the dialog box.

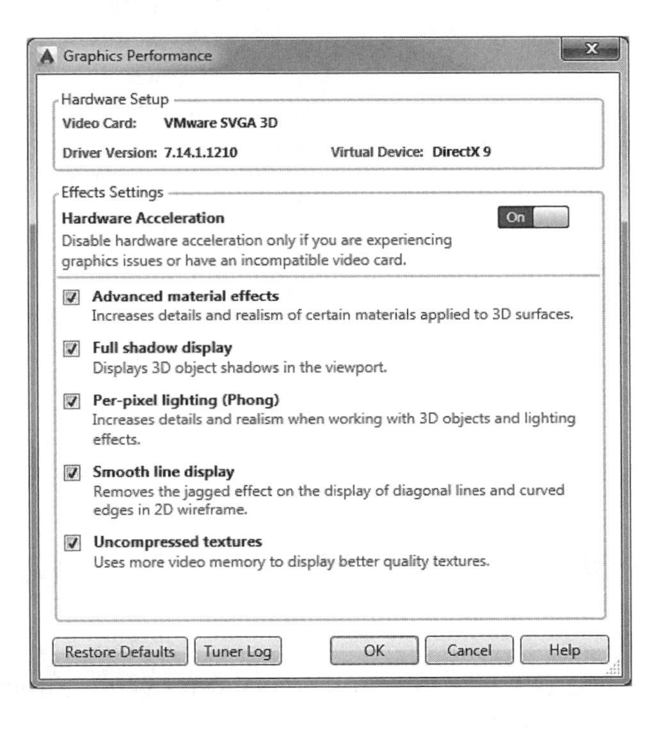

ISOLATE OBJECTS 

You can choose to isolate objects by keeping them visible on the screen, all other objects will be hidden. Or you can choose to hide objects. To isolate or hide objects, left click on the **Isolate Objects** tool button and select either **Isolate Objects** or **Hide Objects** from the list. To restore all hidden objects left click on the **Isolate Objects** tool button and select **End Object isolation** from the list.

CLEAN SCREEN (You may also use **Ctrl+0** to toggle ON or OFF)

When Clean Screen is selected it will hide all tool palettes, windows and ribbons from the screen leaving you with a larger drawing area to work with. You can restore all the palettes, windows and ribbons by selecting the Clean screen tool button again.

APPLICATION WINDOW DESCRIPTIONS....continued

FLOATING COMMAND LINE

When you first start AutoCAD, and if the software has not been modified, the **Command Line** will be displayed at the bottom of the screen, as shown below.

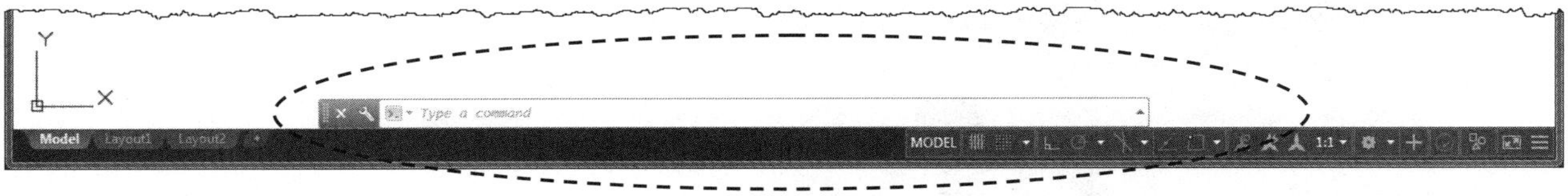

This is where AutoCAD will prompt you for information and you will enter commands, values and select options. Basically this is how you communicate with AutoCAD.

You may "dock" the command line at the top or bottom of the AutoCAD window or let it float in the drawing area.

To move the command line, place the cursor on the left end grip, press the left mouse button and drag the command line to a desired location.

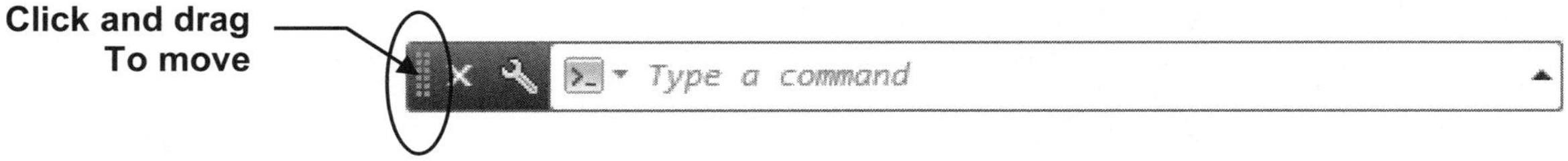

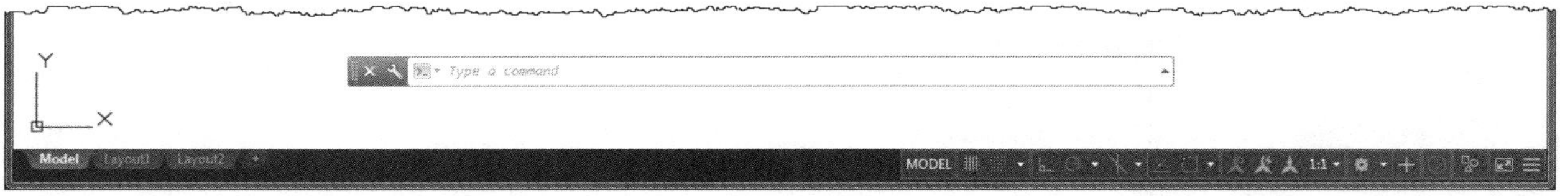

Click and drag To move

To "dock" the command line drag it to the top or bottom of the drawing area. It will snap to the edge. You can't dock the command line to the sides.

You may also drag it below the drawing area as shown below.

Note: You may toggle the Command Line **ON** and **OFF** using **Ctrl + 9**.

COMMAND LINE

How to enter a command on the Command Line.

1. Type the first letter of a command, such as **c** for **circle.**

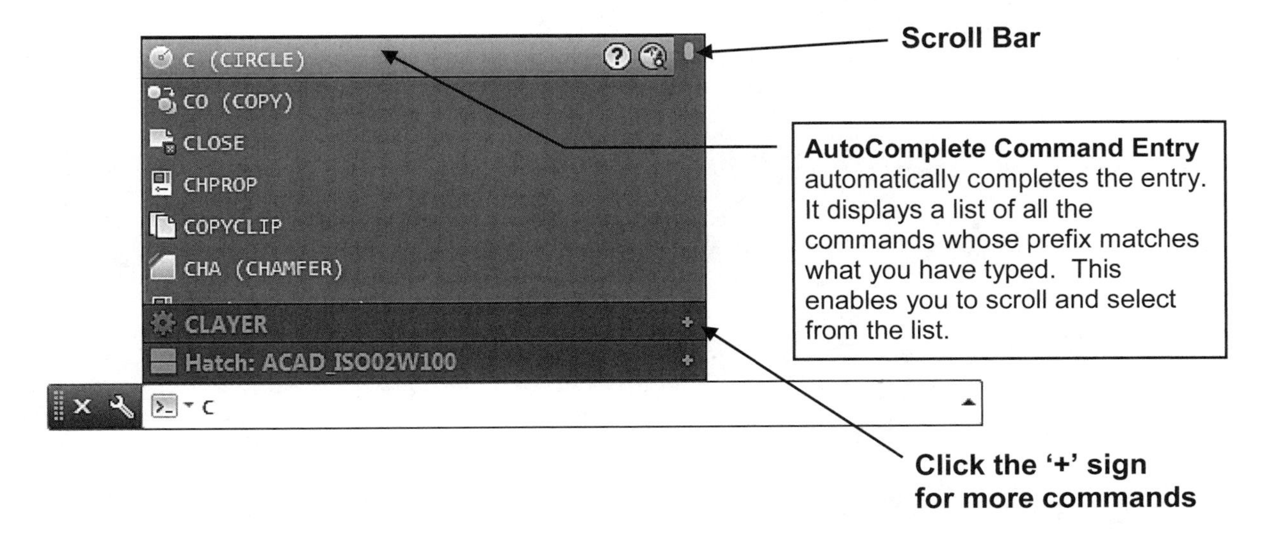

2. A list of commands that begin with the letter **c** will appear. Select the desired command from the list.

3. When you enter a **command** such as Circle, the **prompt** and **options** will be displayed on the command line.

4. The **prompt** for Circle command asks you to:

   "*Specify center point for circle*" or *[3P/2P/Ttr (tan tan radius)]*:

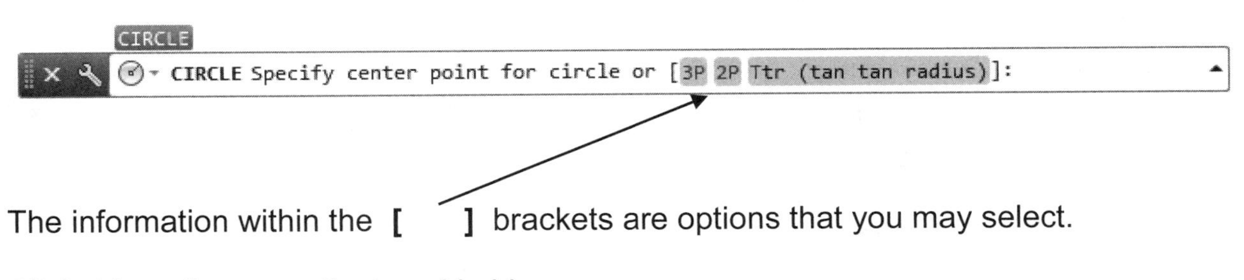

The information within the **[ ]** brackets are options that you may select.

Clickable options are displayed in blue.
Options displayed in Black must be typed or selected from the option menu.

This will be discussed more in Lesson 2. Or for more advanced Command Line options, see Appendix-D.

APPLICATION WINDOW DESCRIPTIONS....continued

FLOATING COMMAND LINE

Command and Prompt history

As you enter commands AutoCAD records them as "history". You may display this history by pressing **F2** or the up arrow at the right hand end of the command line.

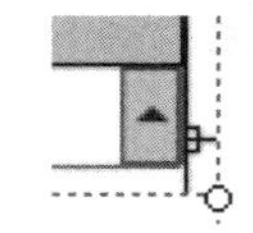

```
Command: *Cancel*
Command: *Cancel*
Command: *Cancel*
Command: CLIPROMPTLINES
Enter new value for CLIPROMPTLINES <3>:
Command: *Cancel*
Command: *Cancel*
Command: _.INPUTSEARCHDELAY
Enter new value for INPUTSEARCHDELAY <300>: 2
Requires an integer between 100 and 10000.
Enter new value for INPUTSEARCHDELAY <300>: 200
Command: C
CIRCLE
Specify center point for circle or [3P/2P/Ttr (tan tan radius)]:
Specify radius of circle or [Diameter]:
Automatic save to C:\Users\Steve\appdata\local\temp\Drawing1_1_1_2856.sv$ ...
Command:
Command:
Command: _.erase 1 found
```

Command Line tools

Recent commands tool displays recently selected commands.

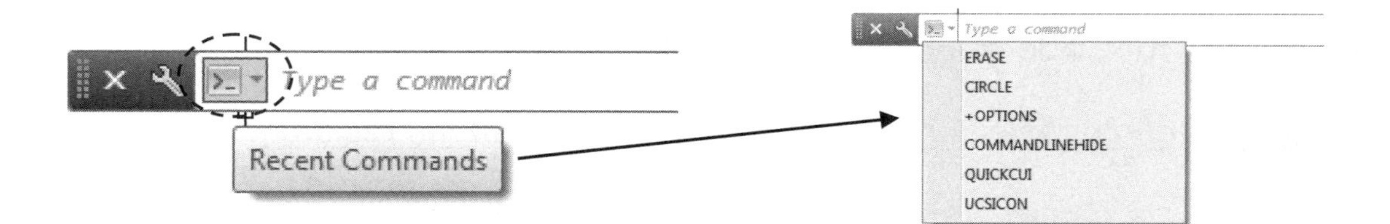

Customize tool allows you to select options for the AutoComplete by selecting '**Input Settings**'. You can also control how many lines of history are displayed and the degree of transparency for the Command Line.

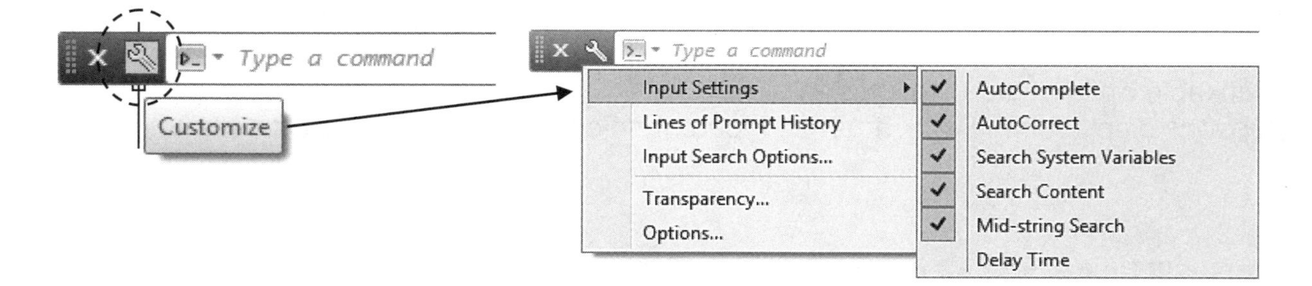

DYNAMIC INPUT

Dynamic Input is another method of inputting commands, values and selecting options.

To use Dynamic Input you must turn **ON** the **Dynamic Input** button in the Status Bar, shown on page 1-14.

If you choose to use Dynamic Input the command will be entered in the tooltip box beside the cursor.

How to enter a command using Dynamic Input.

1. Place the cursor in the Drawing Area.
2. Type the first letter of a command, such as **c** for **circle.**
3. A list of commands that begin with the letter **c** will appear. Select the command from the list.

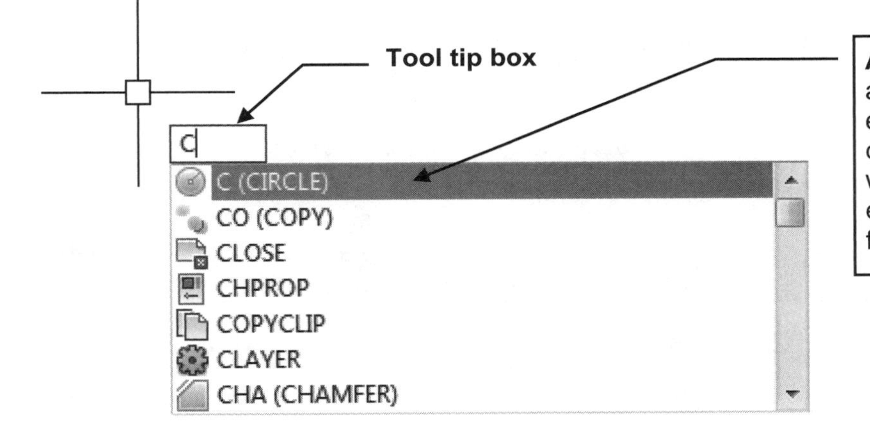

AutoComplete Command Entry automatically completes the entry. It displays a list of all the commands whose prefix matches what you have typed. This enables you to scroll and select from the list.

4. If you press the ↓ down arrow the options will appear below the prompt.

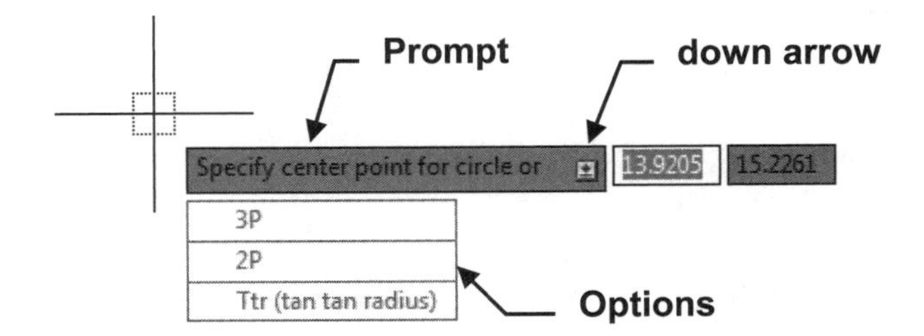

Notice the command entry is being displayed on the command line also.

Using the Command Line or Dynamic Input is **your choice**.

This will be discussed more in Lesson 2. Or for more advanced Command Line options, see Appendix-D.

APPLICATION WINDOW DESCRIPTIONS....continued

DRAWING AREA
The Drawing Area is the large open area of the screen. This is where you will draw. Consider this your paper.
The color of this area can be changed using **Options / Display tab / Color**

ORIGIN Icon
The Origin icon or UCS icon indicates the location of the Origin.
The Origin is where the coordinates X, Y and Z originate.
The X and Y coordinates for the Origin is 0, 0.
This will be discussed more in future Lessons.

CURSOR
The Cursor is located within the Drawing Area. The movement of the pointing device, such as a mouse, controls the movement of the cursor. You will use the cursor to locate points, make selections and draw objects.
The size can be changed using **Options / Display tab / Crosshair Size**.

INFOCENTER
The InfoCenter is a tool to search for information. It is located in the upper right corner of the screen.

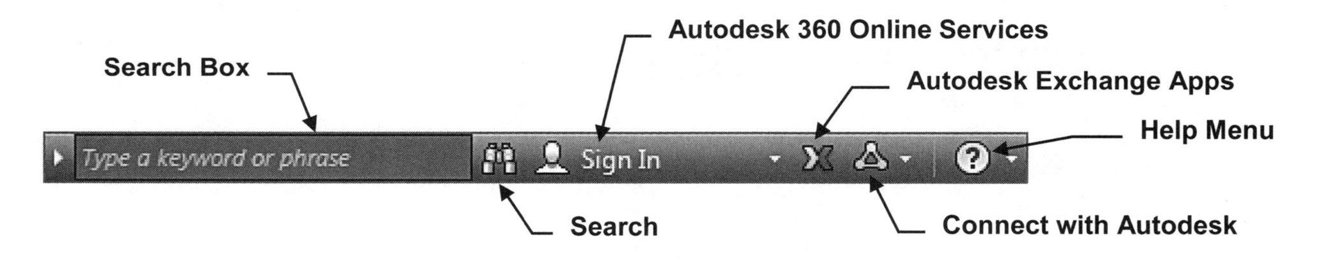

Search Box
The InfoCenter allows you to search for information by typing key words or a question in the "**Help Box**". After typing press **<enter>**

Search
Displays multiple search options.

Autodesk 360 Online Services
Sign in to Autodesk Online to access services that integrate with your desktop software.

Autodesk Exchange Apps
Displays the Autodesk Exchange window.

Help Menu
Displays the Help Window

VIEWCUBE and NAVIGATION BAR

The ViewCube and the Navigation Bar are used primarily in the 3D mode. They enable you to view and rotate the 3D Model.

We will not be using these tools in this Workbook.
Refer to the "Advanced" workbook.

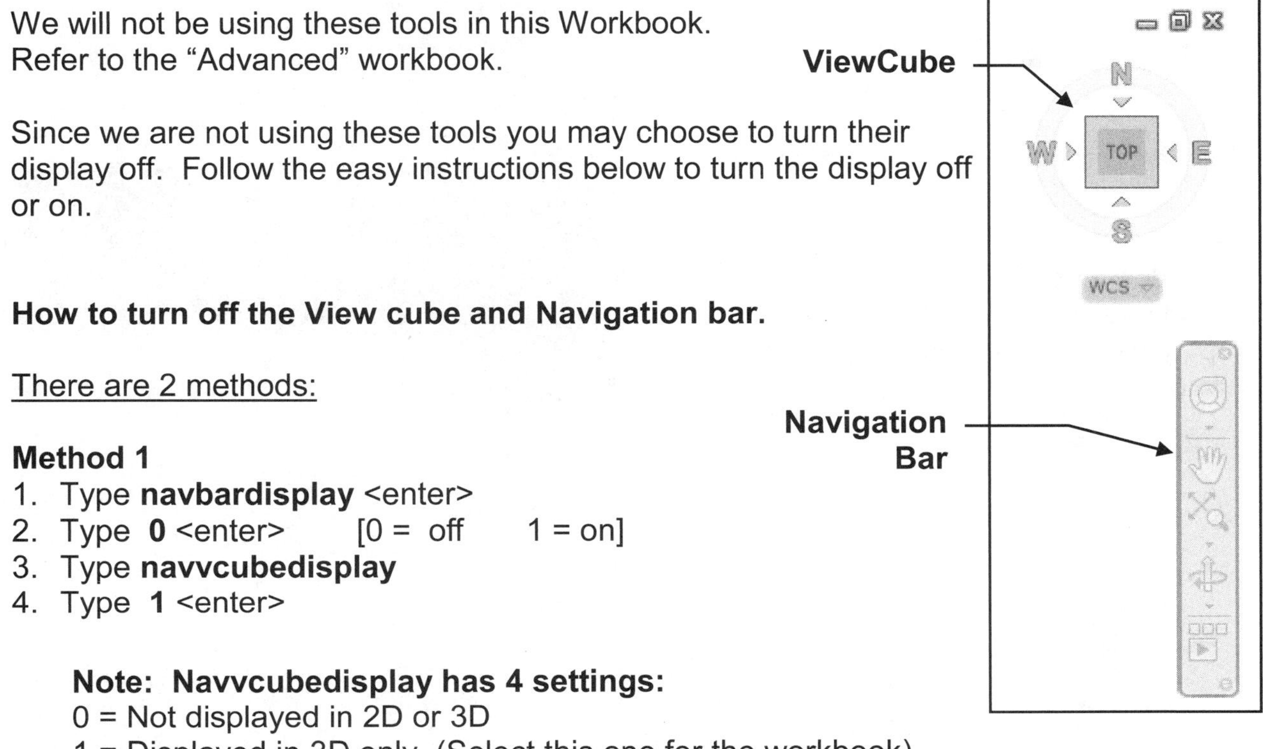

Since we are not using these tools you may choose to turn their display off. Follow the easy instructions below to turn the display off or on.

How to turn off the View cube and Navigation bar.

<u>There are 2 methods:</u>

Method 1

1. Type **navbardisplay** <enter>
2. Type **0** <enter> [0 = off 1 = on]
3. Type **navvcubedisplay**
4. Type **1** <enter>

Note: Navvcubedisplay has 4 settings:
0 = Not displayed in 2D or 3D
1 = Displayed in 3D only (Select this one for the workbook)
2 = Displayed in 2D only
3 = Displayed in both 2D and 3D

Method 2.

1. Select the **View** tab
2. Left click on the **ViewCube** and **Navigation Bar** buttons.

<u>**Note**</u>: The buttons will be a Neon Blue in color when switched on.

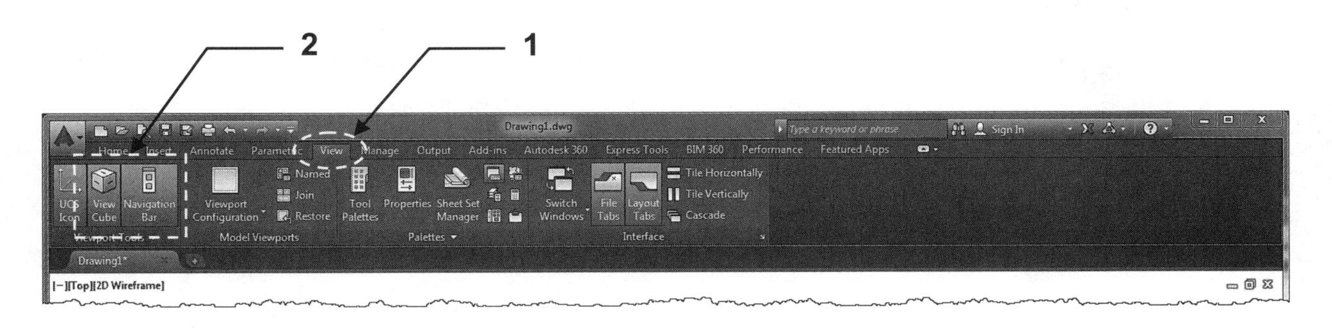

TOOLTIP HELP

Basic Tooltip
When you hover your cursor over a tool an initial Tooltip will appear telling you the name of the tool with a brief description.

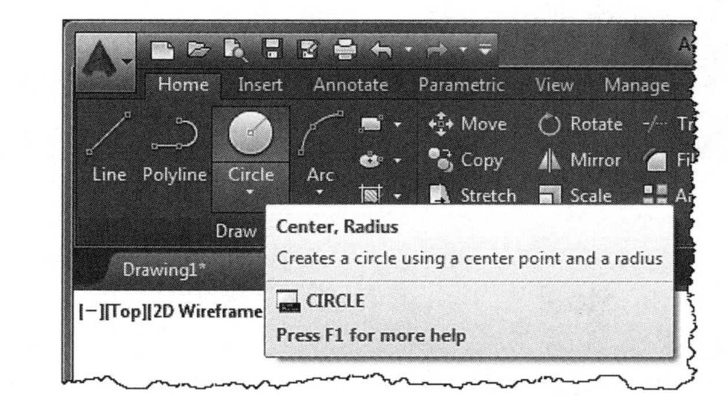

Extended Tooltip
If you hover just a little longer, a graphic display directly from the Help system, will appear.

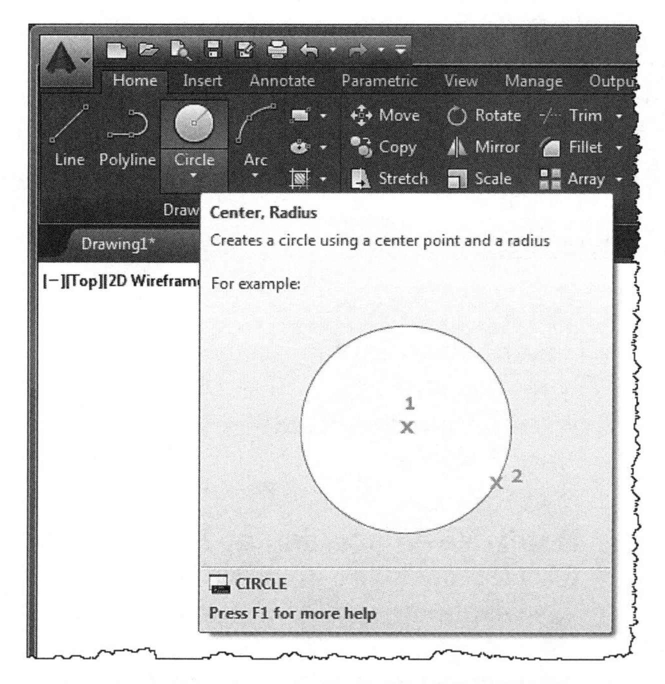

How to turn off Tooltips
After you become familiar with AutoCAD you will want to turn these off. Or you may just want to delay the extended Tooltips.

1. Type **options** and press **<enter>**.

2. Select the **Display** tab

3. Uncheck boxes

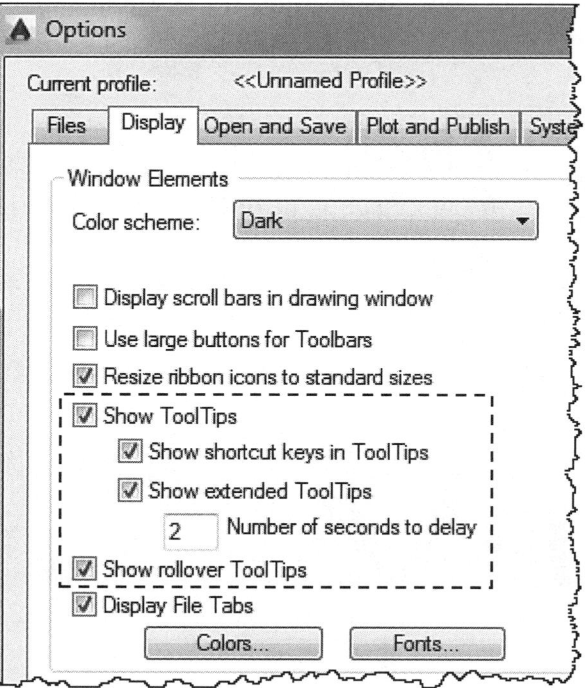

LEARNING OBJECTIVES

After completing this lesson, you will be able to:

1. Create and use a Template
2. Select a Command
3. Draw, Select and Erase objects
4. Start a New drawing
5. Open an Existing Drawing
6. Open Multiple Drawings
7. Save, Backup and Recover a drawing
8. Exit AutoCAD

LESSON 2

CREATE A TEMPLATE

The first item on the learning agenda is **how to create a template file** from a drawing file. **This is important:** You will need this template to complete Lessons 2 through 8.

First you need to download a drawing file.

A. Type the website address shown below into your web browser, then press **<enter>**

http://new.industrialpress.com/ext/downloads/acad/2015-workbook-helper.dwg

B. The 2015-workbook-helper file will download automatically.

C. Save the downloaded file to your desktop.

Now you will create a template. (This will be a very easy task.)

1. Start AutoCAD, if you haven't already. (Refer to page 1-2)

2. Select the **OPEN** tool from the **Quick Access Toolbar**. (Refer to page 1-10)

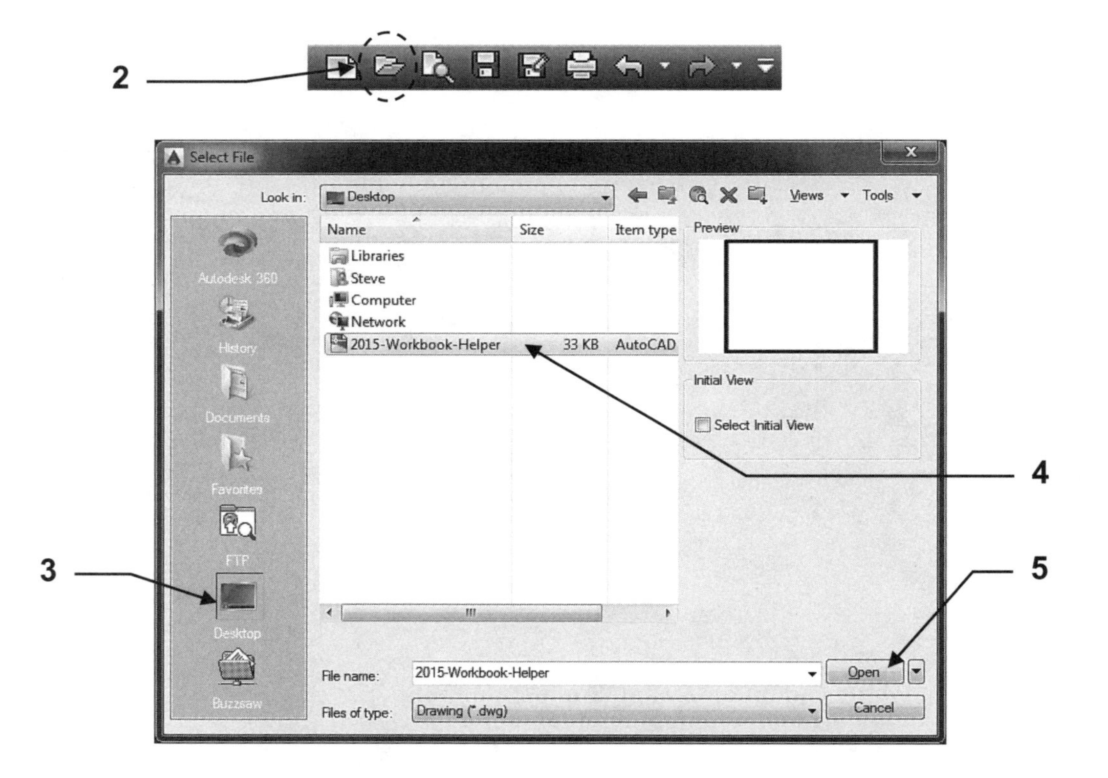

3. Select the **Desktop** directory

4. Select the **2015-workbook-helper.dwg**

5. Select the **Open** button located in the lower right corner.

CREATE A TEMPLATE....continued

Your screen should appear as shown below.

I created the **Rectangular shape** that appears in the drawing area. I have designed the exercises that follow to fit on an 11 X 8.5 sheet of paper so you can easily print them on any letter size printer. The Rectangle represents an 11 X 8.5 sheet of paper. While completing the exercises within this workbook please try to draw all objects within this rectangle.

The criss-cross lines are **Grids**. I have set them to display every 1 inch vertically and horizontally. You will learn more about Grids in Lesson 3. For now notice that the grids are 11 horizontally and 8.5 vertically. Grids are merely a visual aid and will not print. The size may be changed at any time and they may be turned **ON** or **OFF** easily by selecting the **"Grid"** button on the status line or **F7**. (Refer to page 1-13)

The next step is to create a template from this drawing.
Continue on to step 6 on the next page.

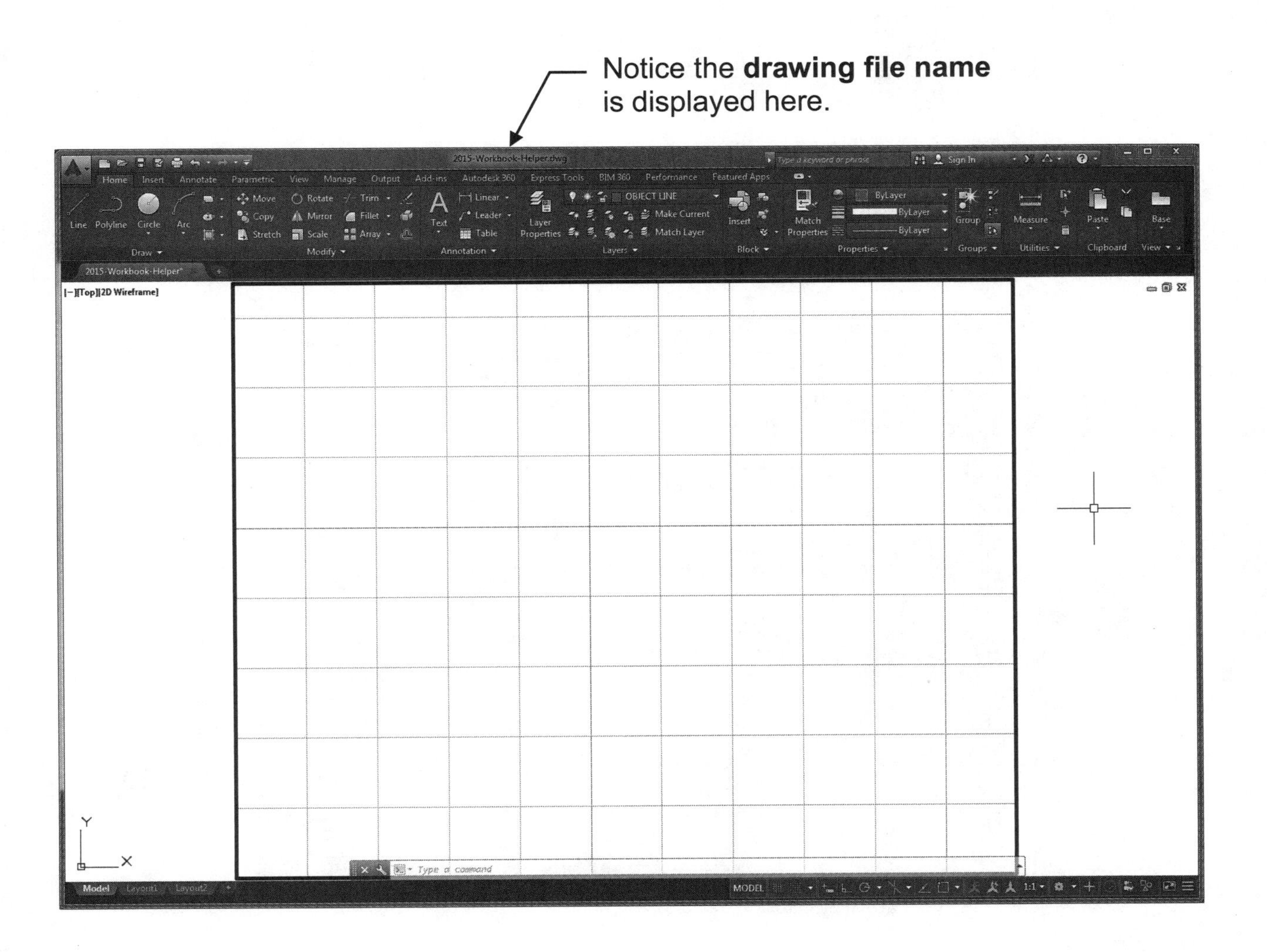

Continued on the next page...

CREATE A TEMPLATE....continued

6. Select the **"Application Menu"** ▼

7. Select **Save As** " ▸" (Click on arrow not words Save As)

8. Select **"Drawing Template".**

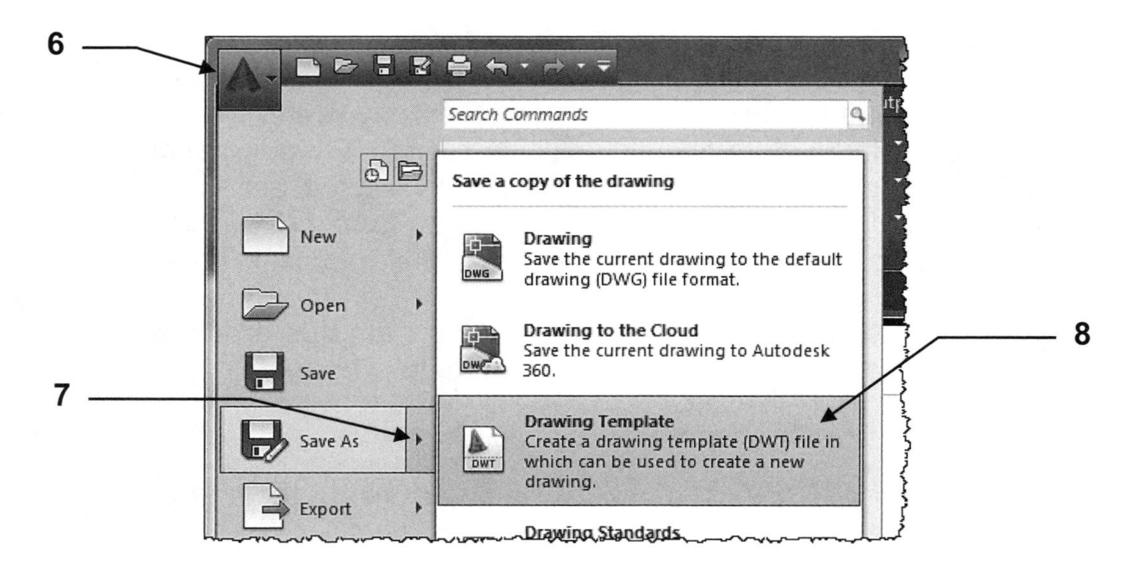

9. The name of the new file should already be highlighted in the "**File name**" box, if it's not just type in; **2015-Workbook Helper** in the **File name** box.
   Do not type the extension .dwt, AutoCAD will add it automatically.

10. Select the **Save** button.

Continued on the next page...

CREATE A TEMPLATE....continued

11. Type the description as shown below.

12. Select **OK** button.

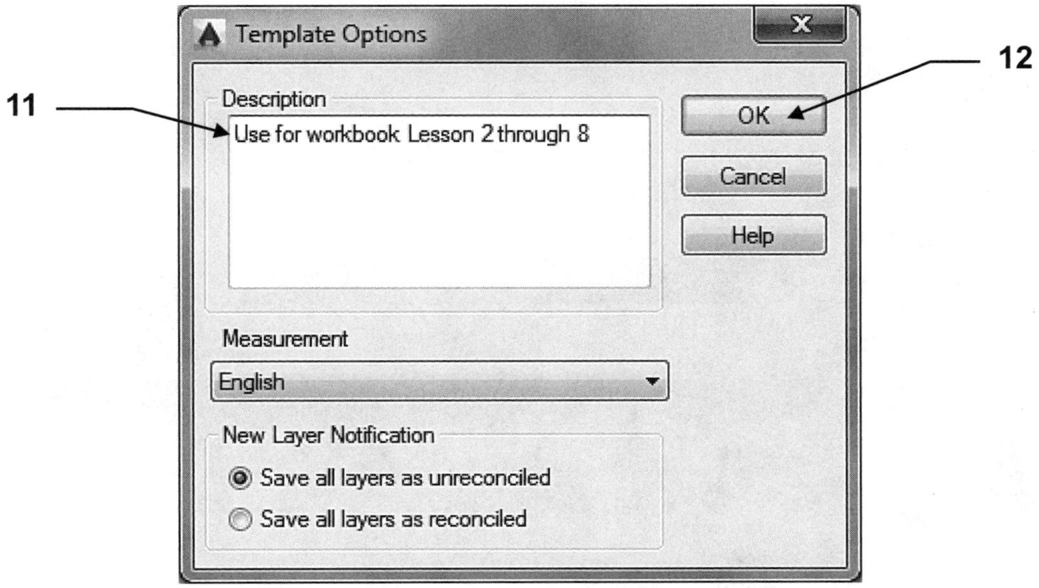

Template Options

Description

Use for workbook Lesson 2 through 8

OK

Cancel

Help

Measurement

English

New Layer Notification

Save all layers as unreconciled

Save all layers as reconciled

11

12

Now you have a template to use for Lessons 2 through 8.

At the beginning of each exercise you will be instructed to start a **NEW** drawing using the **2015-Workbook Helper.dwt.**

Using a template as a master setup drawing is very good CAD management.

More on using the template on the next page

USING A TEMPLATE

The template that you created from the previous pages will be used for lessons 2 through 8. Many variables have been preset in this template. This will allow you to start drawing immediately. You will learn how to set those variables before you complete this workbook, but for now you will concentrate on learning the AutoCAD commands and hopefully have some fun.

TO USE A TEMPLATE

1. Select the **NEW** tool from the **Quick Access Toolbar.**

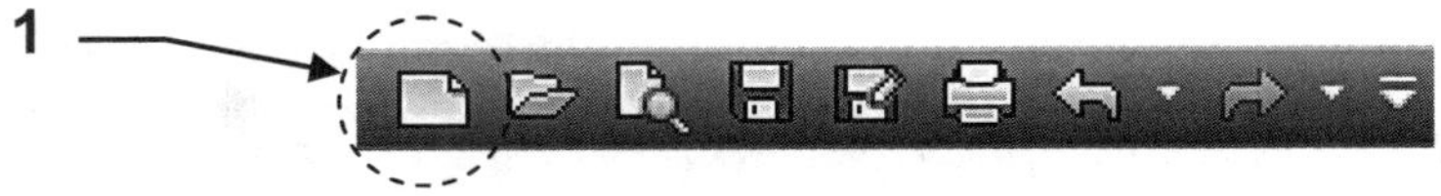
1

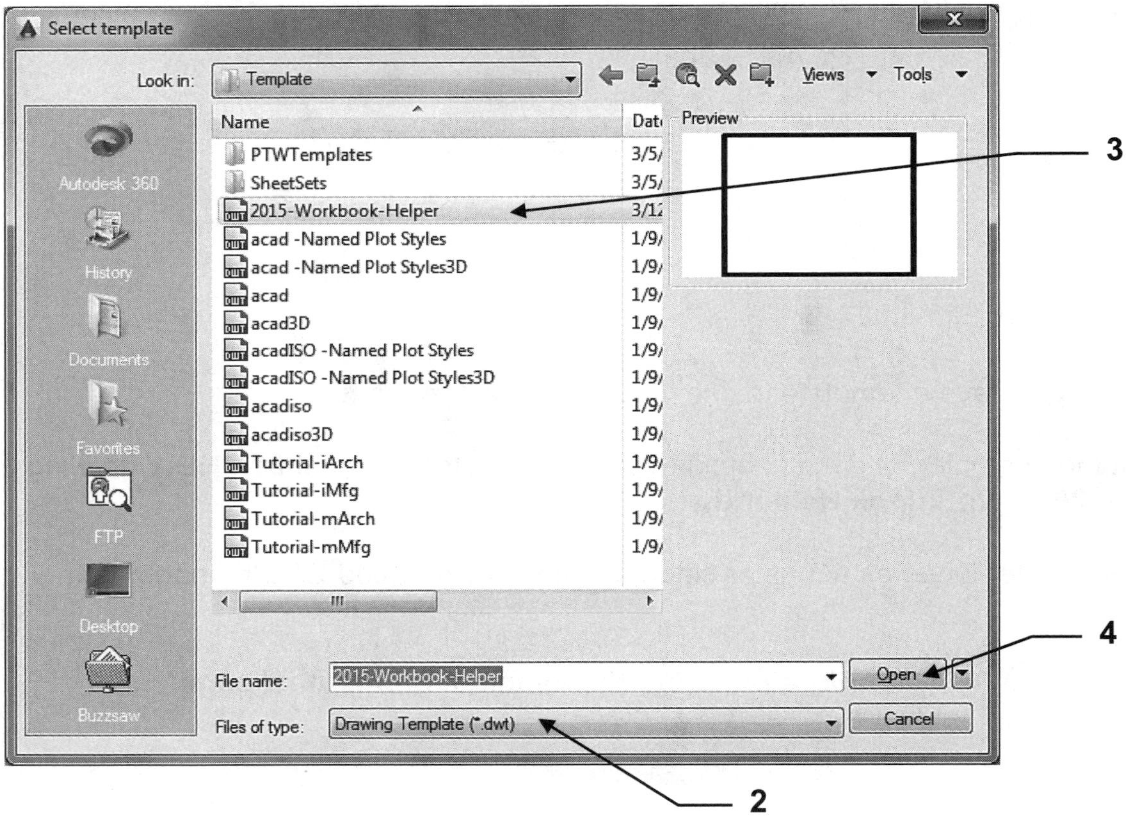
Select template
Look in:
Template
Views
Tools
Name
Dat
Preview
PTWTemplates
SheetSets
2015-Workbook-Helper
acad -Named Plot Styles
acad -Named Plot Styles3D
acad
acad3D
acadISO -Named Plot Styles
acadISO -Named Plot Styles3D
acadiso
acadiso3D
Tutorial-iArch
Tutorial-iMfg
Tutorial-mArch
Tutorial-mMfg
Autodesk 360
History
Documents
Favorites
FTP
Desktop
Buzzsaw
File name:
2015-Workbook-Helper
Files of type:
Drawing Template (*.dwt)
Open
Cancel
3
4
2

2. Select **Drawing Template [*.dwt]** from the **"Files of type"** if not already selected.

3. Select the **2015-Workbook Helper.dwt** from the list of templates.

   Note: If you do not have this template, refer to page 2-2 for instructions.

4. Select the **Open** button.

HOW TO SELECT A COMMAND

AutoCAD provides you with 2 different methods for selecting commands.
One is **selecting a tool from the Ribbon**, the other is **typing the command**. Both methods will accomplish the same end result. You decide which method you prefer. An example of method 1 is shown below. Method 2 is on the next page.

Method 1: Selecting a tool from the Ribbon

1. First select a **tab** such as **Home**.
2. Locate the correct **Panel** such as **Draw**.
3. Select a **tool** such as **Circle**

1. Tab

2. Panel

3. Tool

Note:
If the tool includes multiple types it will have a down-arrow ▼

If you select the down-arrow a sub-menu will appear.

Select the desired type such as **2-Point**.

The latest selection will then become the current displayed tool because AutoCAD assumes that you may need that tool again.

Sub-Menu

Continued on the next page...

HOW TO SELECT A COMMAND….continued

Method 2: Keyboard entry

You may type commands on the **Command line** (Shown below) or in the **Dynamic Input tooltip** (Shown on the next page)
It depends on whether you have Dynamic Input On or Off.

COMMAND LINE

How to enter a command on the Command Line.

1. Place the cursor in the Command Line area. (Important)
2. Type the first letter of a command, such as **c** for **circle.**

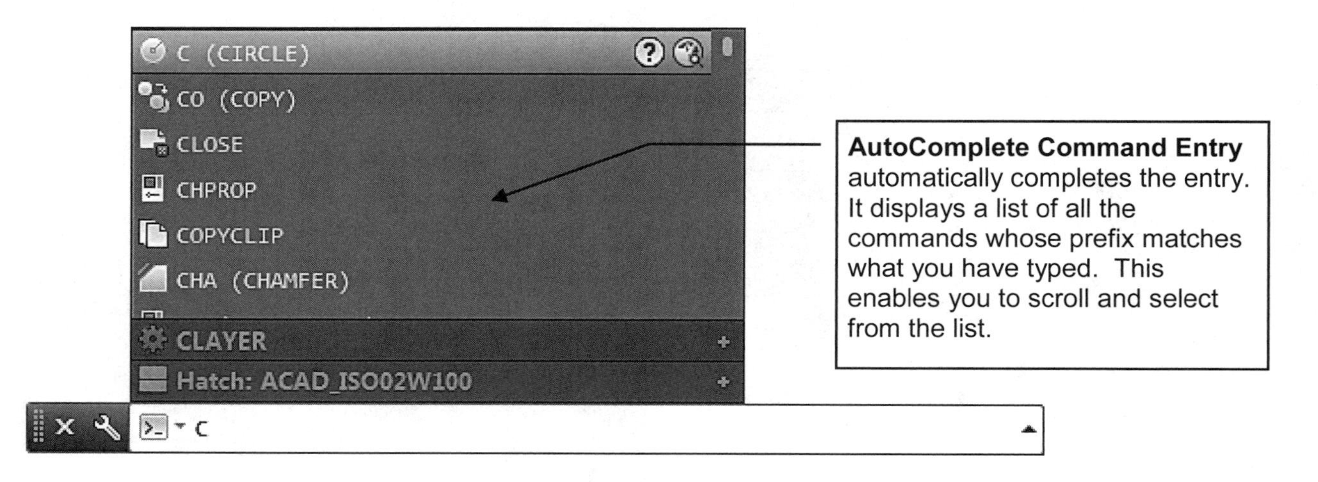

3. A list of commands that begin with the letter **c** will appear. Select the command from the list.
4. When you enter a **command** such as Circle the **prompt** and **options** will be displayed on the command line.

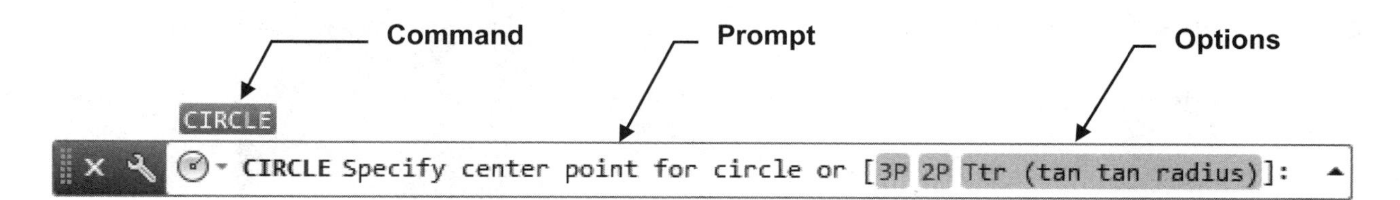

5. The **prompt** for Circle command asks you to:

“***Specify center point for circle***” or **[** ***3P/2P/Ttr (tan tan radius)*** **]:**

The information within the **[ ]** brackets are options that you may select.

HOW TO SELECT A COMMAND....continued

Method 2: Keyboard entry....continued

DYNAMIC INPUT

Dynamic Input is another method of inputting commands, values and select options.

To use Dynamic Input you must turn **ON** the **DYNAMIC INPUT** button in the Status Bar, shown on page 1-14.

If you choose to use Dynamic Input the command will be entered in the tooltip box beside the cursor.

How to enter a command using Dynamic Input.

1. Place the cursor in the Drawing Area. (Important)
2. Type the first letter of a command, such as **c** for **circle.**
3. A list of commands that begin with the letter **c** will appear.
   Select the command from the list.

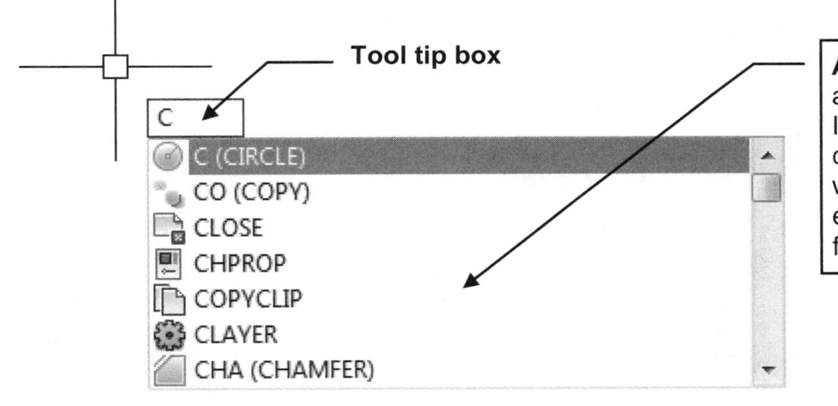

AutoComplete Command Entry automatically completes the entry. It displays a list of all the commands whose prefix matches what you have typed. This enables you to scroll and select from the list.

4. If you press the ↓ down arrow the options will appear below the prompt.

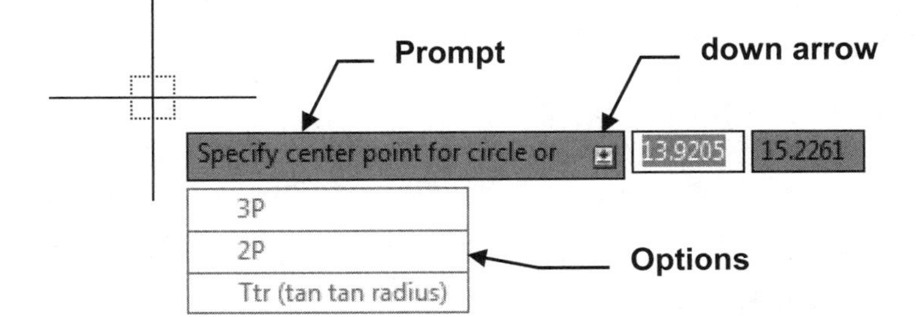

Notice the command entry and prompts are being displayed on the command line also.

Using the Command Line or Dynamic Input is **your choice**.

DRAWING LINES

A **Line** can be **one segment** or a **series of connected segments.**
But each segment is an individual object.

One Segment
One object

Series of connected Segments
5 objects

<u>Start the Line command using one of the following methods:</u>

Ribbon = Home tab / Draw Panel /
or
Keyboard = L <enter>

Line

<u>**Lines** are drawn by specifying the locations for each endpoint.</u>
Move the cursor to the location of the **"First"** endpoint **(1)** then press the left mouse button and release. (Click and release, do Click and Drag) Move the cursor again to the **"next"** endpoint **(2)** and press the left mouse button. Continue locating **"next"** endpoints until you want to stop drawing lines.

<u>There are 2 ways to **Stop drawing a line**:</u>
Press the **<enter>** key **or** press the **<Space Bar>**

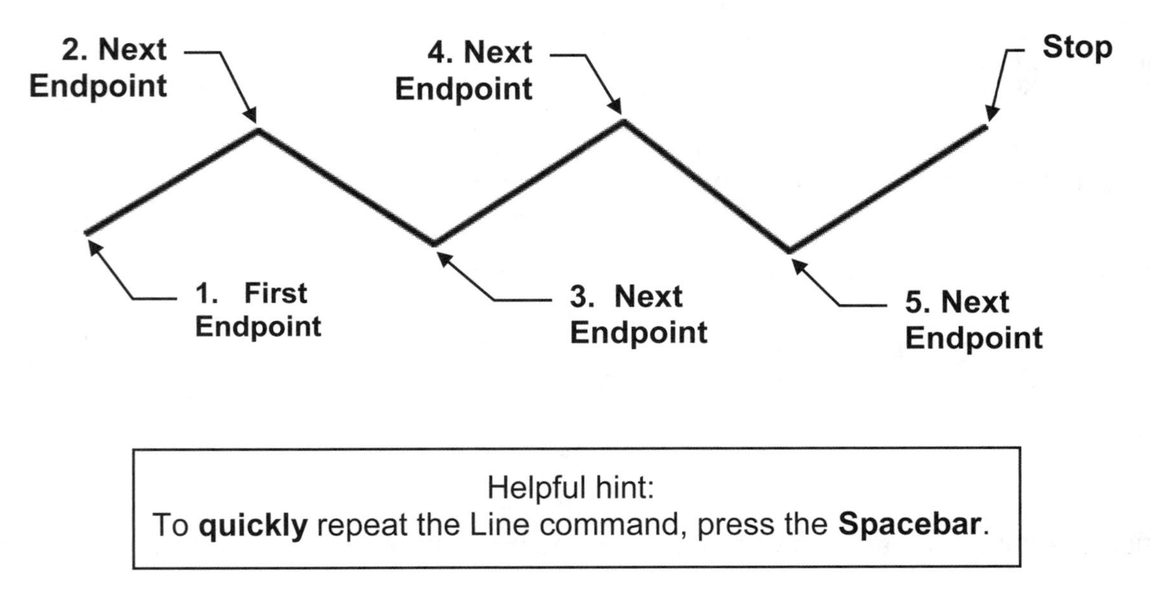

> Helpful hint:
> To **quickly** repeat the Line command, press the **Spacebar**.

Continued on the next page...

DRAWING LINES....continued

Horizontal and Vertical Lines

To draw a Line perfectly Horizontal or Vertical select the **Ortho** mode by selecting the **Ortho** button on the Status Bar or pressing the **F8** key.

Try the following example:

1. Select the **Line** command. (Refer to the previous page)
2. Place the **First endpoint** anywhere in the drawing area.
3. Turn **Ortho ON** by selecting the **Ortho** button or **F8**. (The "**Ortho**" button will change to a neon blue when ON.)
4. Move the cursor to the right and press the left mouse button to place the **next endpoint.** (The line should appear perfectly horizontal.)
5. Move the cursor down and press the left mouse button to place the next endpoint. (The line should appear perfectly vertical)
6. Now turn **Ortho OFF** by selecting the **Ortho** button. (The "**Ortho**" button will change to gray when OFF.)
7. Now move the cursor up and to the right on an angle (the line should move freely now) and press the left mouse button to place the next endpoint.

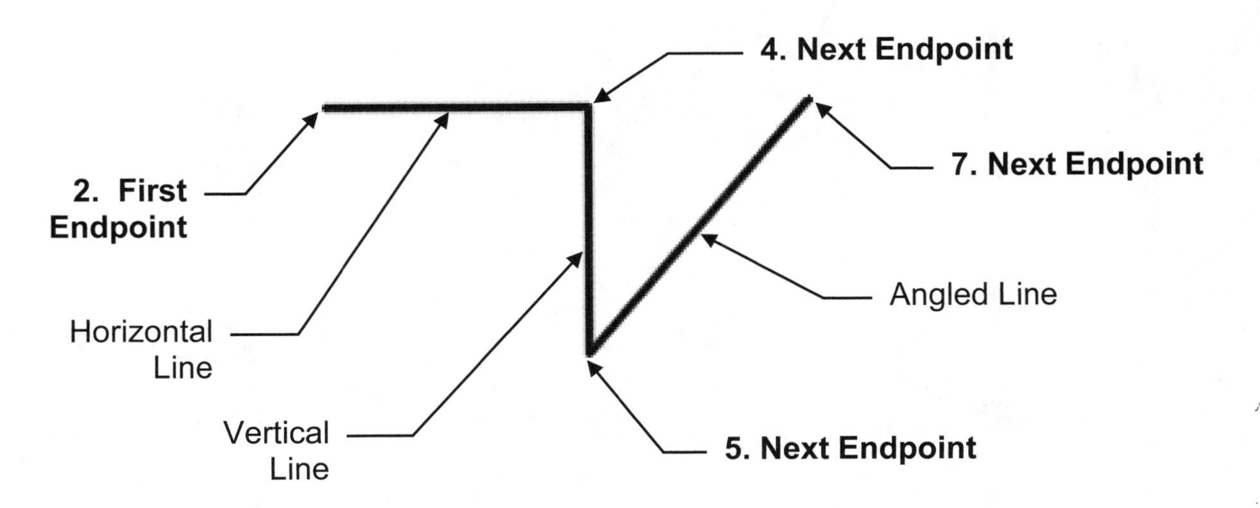

Ortho can be turned ON or OFF at any time while you are drawing. It can also be turned ON or OFF temporarily by holding down the **Shift** key. Release the Shift key to resume.

Continued on the next page...

DRAWING LINES....continued

Closing Lines

If you have drawn 2 or more line segments, the **endpoint of the last line segment** can be connected automatically to the **first endpoint** using the **Close** option.

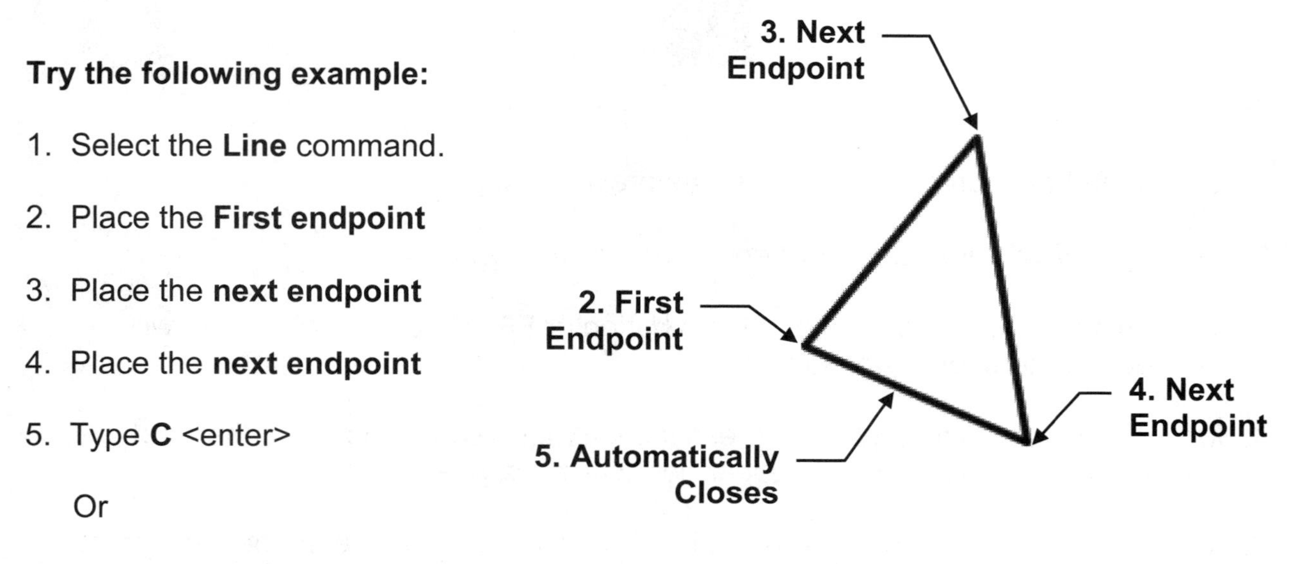

Try the following example:

1. Select the **Line** command.
2. Place the **First endpoint**
3. Place the **next endpoint**
4. Place the **next endpoint**
5. Type **C** <enter>

   Or

5. Press the **right** mouse button and select **Close** from the **Shortcut menu.**

What is the Shortcut Menu?
The **Shortcut menu** gives you quick access to command options.

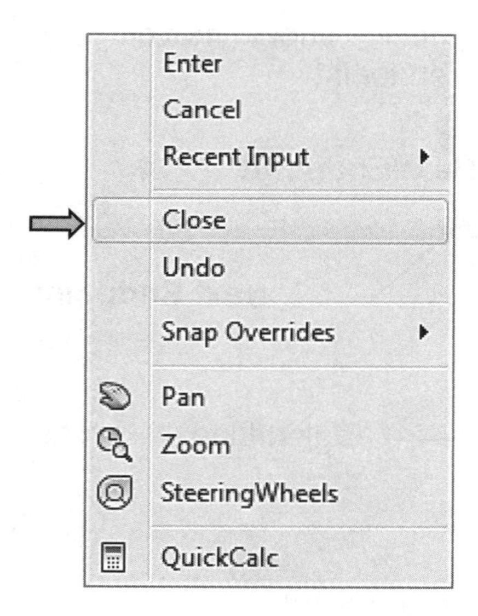

Using the Shortcut menu:
Press the right mouse button.
The shortcut menu will appear.
Select an option.

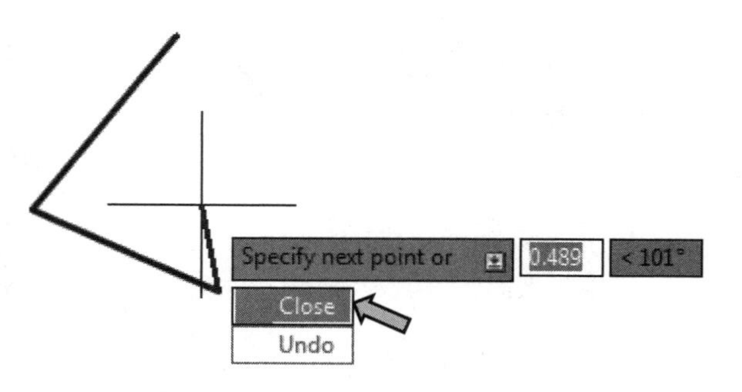

Using the Dynamic Input down arrow:
You may use the right mouse button or press the down arrow ↓ and the options will appear below the Dynamic Input prompt.

METHODS OF SELECTING OBJECTS

Many AutoCAD commands prompt you to "**select objects**". This means select the objects that you want the command to effect. There are 3 methods.

Method 1. <u>Pick</u>, is very easy and should be used if you have only 1 or 2 objects to select. **Method 2. <u>Window selection</u>**, is a little more difficult but once mastered it is extremely helpful and time saving. **Method 3. <u>Lasso Selection</u>**, is a little more difficult than Window Selection but again, once mastered it is very useful and will save you time. Practice the following examples.

<u>Method 1. PICK</u>:

First start a command such as ERASE. (Press E <enter>)
Next you will be prompted to "**Select Objects**", place the cursor (pick box) on the object but do not press the mouse button yet. The object will highlight. This appearance change is called "Rollover Highlighting". This gives you a preview of which object AutoCAD is recognizing. Press the left mouse button to actually select the highlighted object.

<u>Method 2. WINDOW selection</u>: <u>C</u>rossing and <u>W</u>indow

Crossing:
Place your cursor in the area **up** and to the **right** of the objects that you wish to select (**P1**) and press the left mouse button. (**Do not** hold the mouse button down. Just press and release) Then move the cursor **down** and to the **left** (**P2**) and press the left mouse button again. (Note: The window will be ***green*** and outer line is ***dashed***.)
Only the objects that this window **<u>crosses</u>** will be selected.

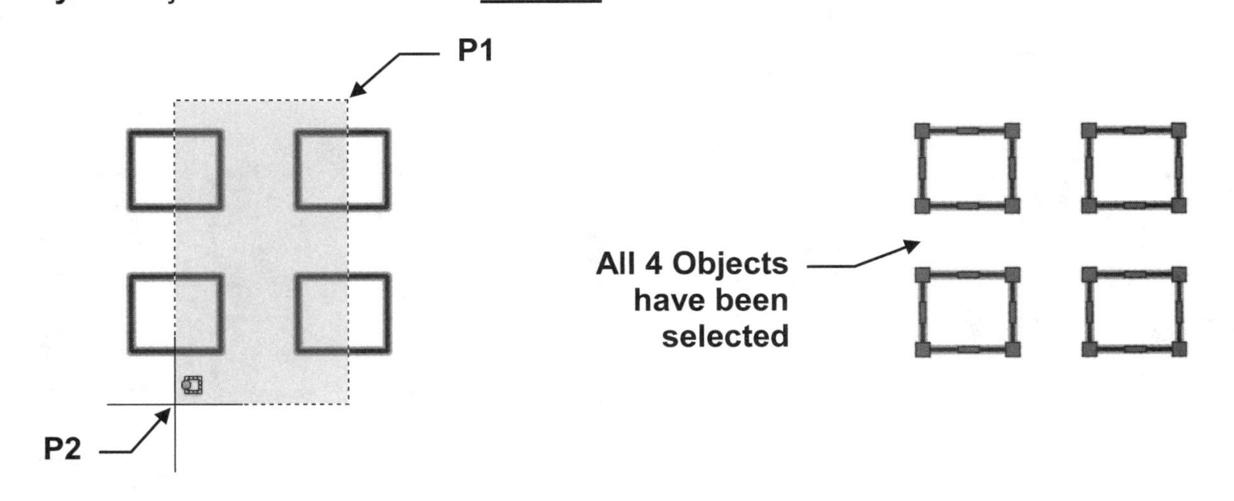

In the example above, all 4 rectangles have been selected because the Crossing Window **<u>crosses</u>** a portion of each.

Continued on the next page...

METHODS OF SELECTING OBJECTS....continued

Window:
Place your cursor in the area **up** and to the **left** of the objects that you wish to select (**P1**) and press the left mouse button (**Do not** hold the mouse button down. Just press and release.) Then move the cursor **down** and to the **right** of the objects (**P2**) and press the left mouse button. (Note: The window will be ***blue*** and outer line is ***solid***.)
Only the objects that this window **completely enclosed** will be selected.

In the example below, only 2 rectangles have been selected.
(The other 2 rectangles are **not** completely enclosed in the **Window**.)

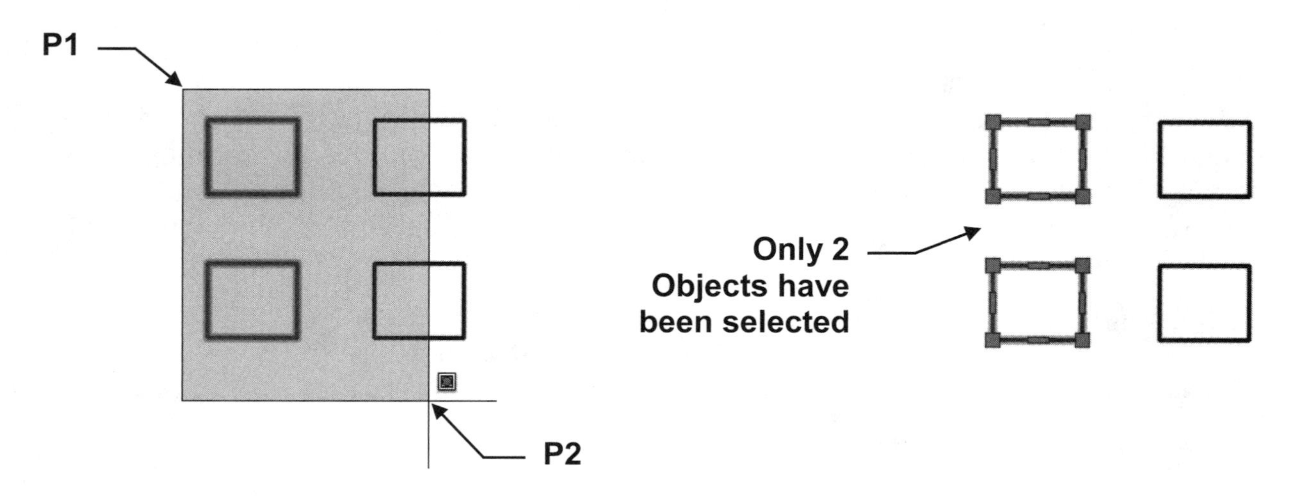

Method 3. LASSO Selection: Crossing, Window and Fence

Crossing:
Place your cursor in the area **up** and to the **right** of the objects that you wish to select (**P1**) then press and **hold** the left mouse button. (**Do not** release the mouse button.) Then move the cursor in an anti-clockwise direction until you have crossed the objects you want to select (**P2**) then release the left mouse button. (Note: The Lasso window will be ***green*** and outer line is ***dashed***.)
Only the objects that the Lasso window **crosses** will be selected.

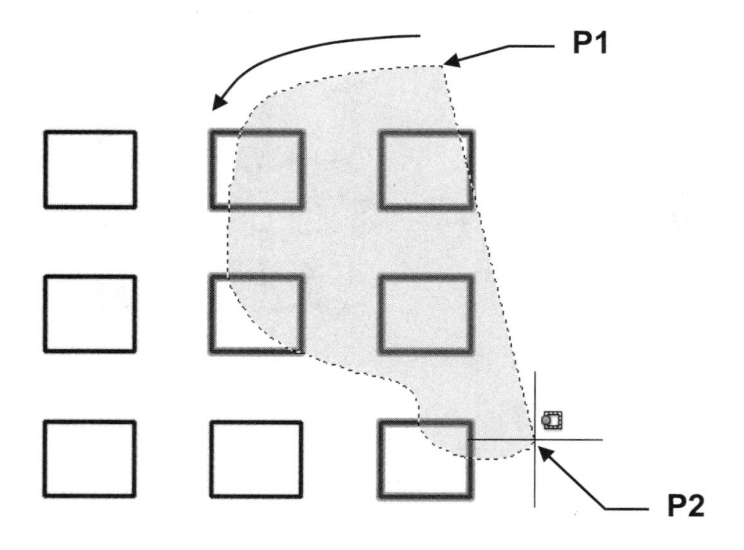

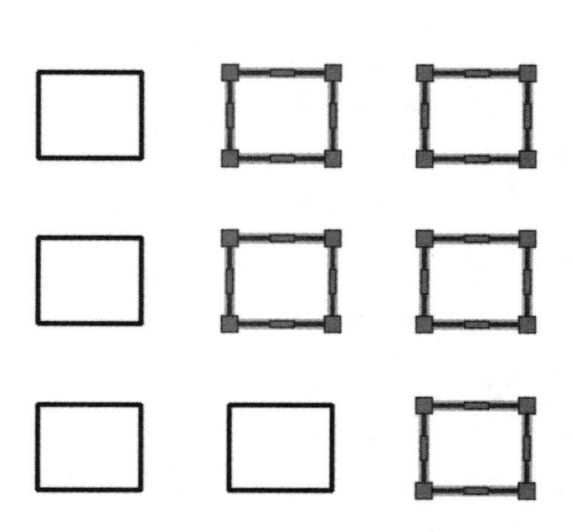

Continued on the next page...

METHODS OF SELECTING OBJECTS....continued

Window:
Place your cursor in the area **up** and to the **left** of the objects that you wish to select (**P1**) then press and **hold** the left mouse button. (**Do not** release the mouse button.) Then move the cursor in a clockwise direction until you have completely enclosed the objects you want to select (**P2**) then release the left mouse button. (Note: The window will be ***blue*** and outer line is ***solid***.) **Only** the objects that this window **completely enclosed** will be selected.

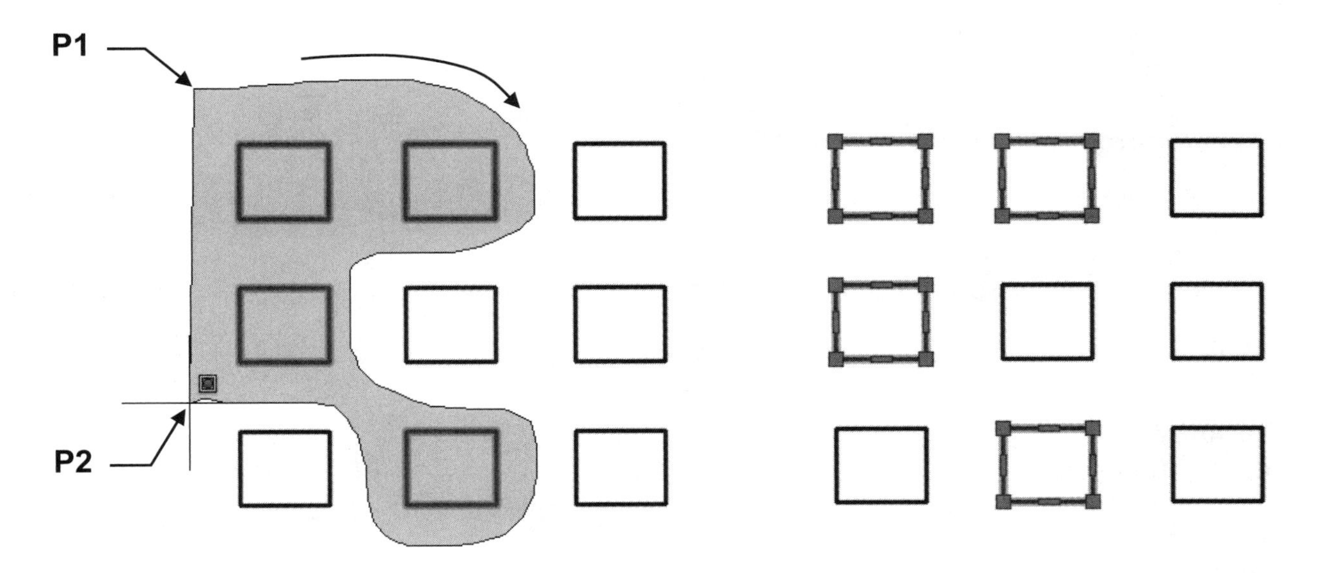

Fence:
With the **Fence** option of the **Lasso** selection you can place the mouse cursor in any position you choose. For this example place your cursor at (**P1**) then press and **hold** the left mouse button. (**Do not** release the mouse button.) Move the mouse until you see either the green or blue lasso, then press the **Spacebar** until you see just a **Dashed Fence Line**. Move the mouse over the objects you want to select (**P2**) then release the left mouse button. Only the objects that the Fence line **crosses** will be selected.

Note: You may have to press the **Spacebar** twice to activate the **Fence Line** option.

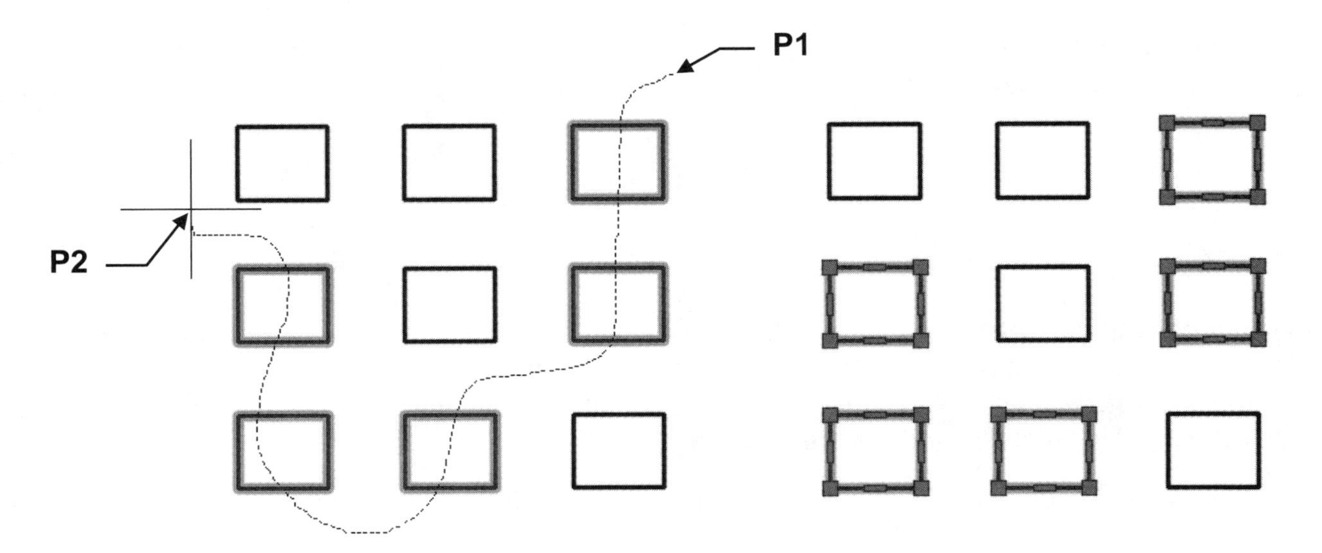

ERASE

There are 3 methods to erase (delete) objects from the drawing.
They all work equally well. You decide which one you prefer to use.

Method 1
Select the **Erase** command first and then select the objects.

Example:
1. Start the **Erase** command using one of the following:

   Ribbon = Home tab / Modify panel /
   or
   Keyboard = E <enter>

2. Select objects: ***Pick one or more objects***
   Select objects: ***Press <enter> and the objects selected will disappear.***

Method 2
Select the Objects first and then the **Delete** Key.

Example:
1. Select the object to be erased.
2. Press the **Delete** Key.

Method 3
Select the Objects first and then select Erase command from the Shortcut Menu.

Example:
1. Select the object to be erased.
2. Press the right Mouse button.
3. Select **Erase** from the Shortcut Menu using the left mouse button.

Note: Very Important
If you want the erased objects to return, select the **Undo tool** from the **Quick Access Toolbar**. This will **Undo** the last command.
More about Undo and Redo on the next page.

UNDO and REDO

The **UNDO** and **REDO** tools allow you to undo or redo previous commands.
For example, if you erase an object by mistake, you can UNDO the previous "erase" command and the object will reappear. So don't panic if you do something wrong. Just use the UNDO command to remove the previous commands.

The **Undo** and **Redo** tools are located in the **Quick Access Toolbar**.

Note:
You may UNDO commands used during a work session until you close the drawing.

How to use the Undo tool.

1. Draw a line, circle and a rectangle.

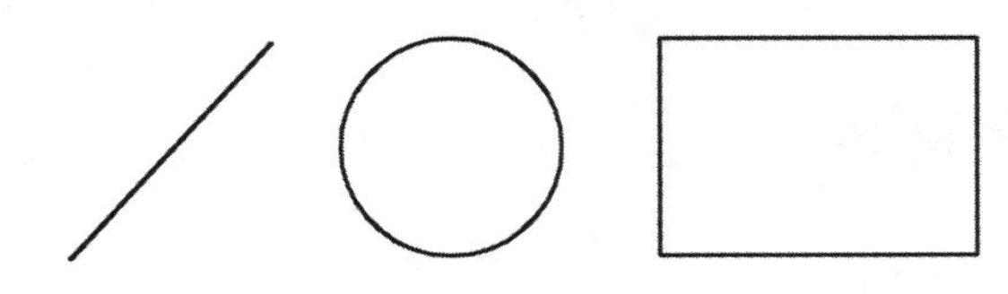

Your drawing should look approximately like this.

2. Next Erase the Circle and the Rectangle.

(The Circle and the Rectangle disappear.)

3. Select the **UNDO** arrow.

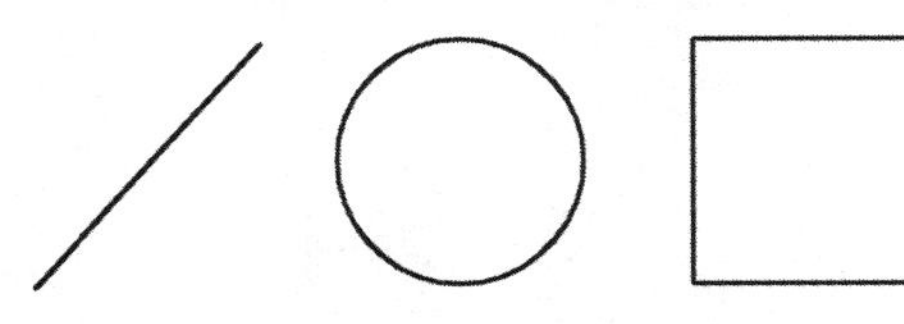

You have now deleted the ERASE command operation.
As a result the erased objects reappear.

How to use the Redo command:
Select the **REDO** arrow and the Circle and Rectangle will disappear again.

STARTING A NEW DRAWING

Starting A <u>New</u> Drawing means that you want to start with a previously created Template file. That is why I taught you "how to create a template" at the beginning of this lesson. You will use the **2015-Workbook Helper.dwt** template each time you are instructed to **Start a <u>New</u> Drawing**.

*Note: Do not use the **New** tool if you want to **open** an **existing drawing**.*
*Refer to page 2-19 to **OPEN** an existing drawing file.*

HOW TO START A NEW DRAWING

1. Select the **NEW** tool from the **Quick Access Toolbar.**

2. Select the **2015-Workbook Helper.dwt** from the list of templates.

   Note: If you do not have this template, refer to page 2-2 for instructions.

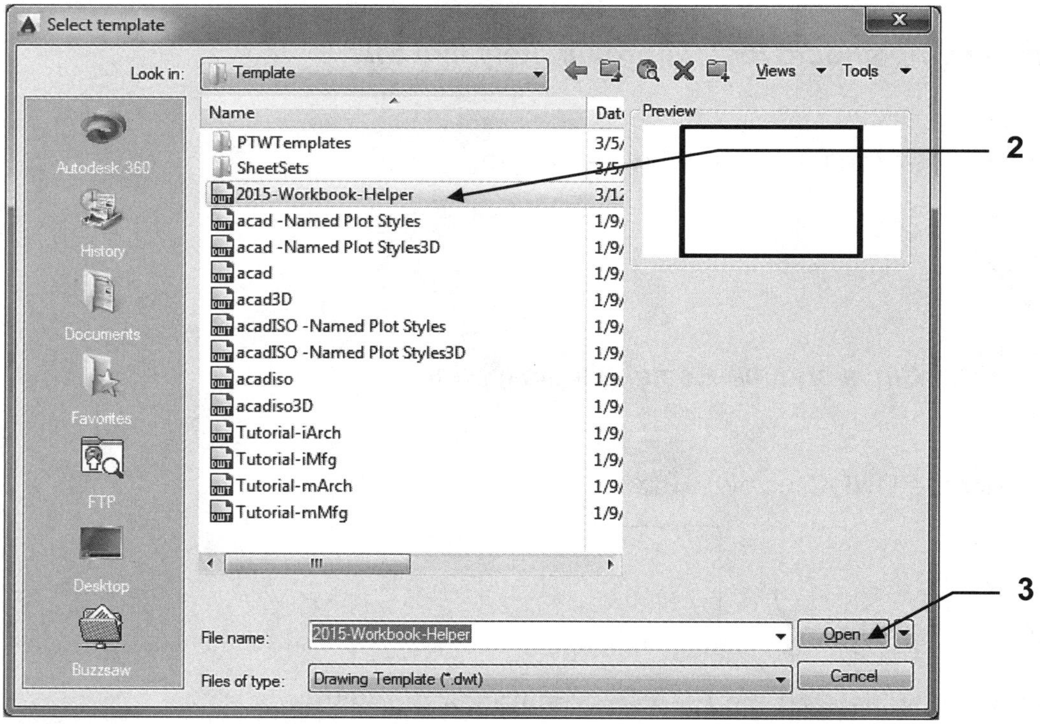

3. Select the **Open** button.

OPENING AN EXISTING DRAWING FILE

Opening an **Existing Drawing File** means that you would like to open, on to the screen, a drawing that has been previously created and saved. Usually you are opening it to continue working on it or you need to make some changes.

1. Select the **OPEN** tool on the **Quick Access Toolbar**.

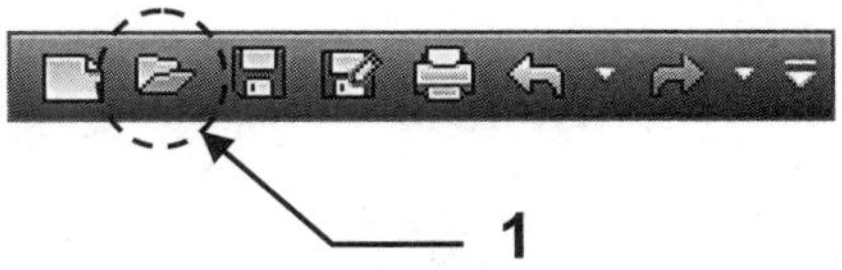

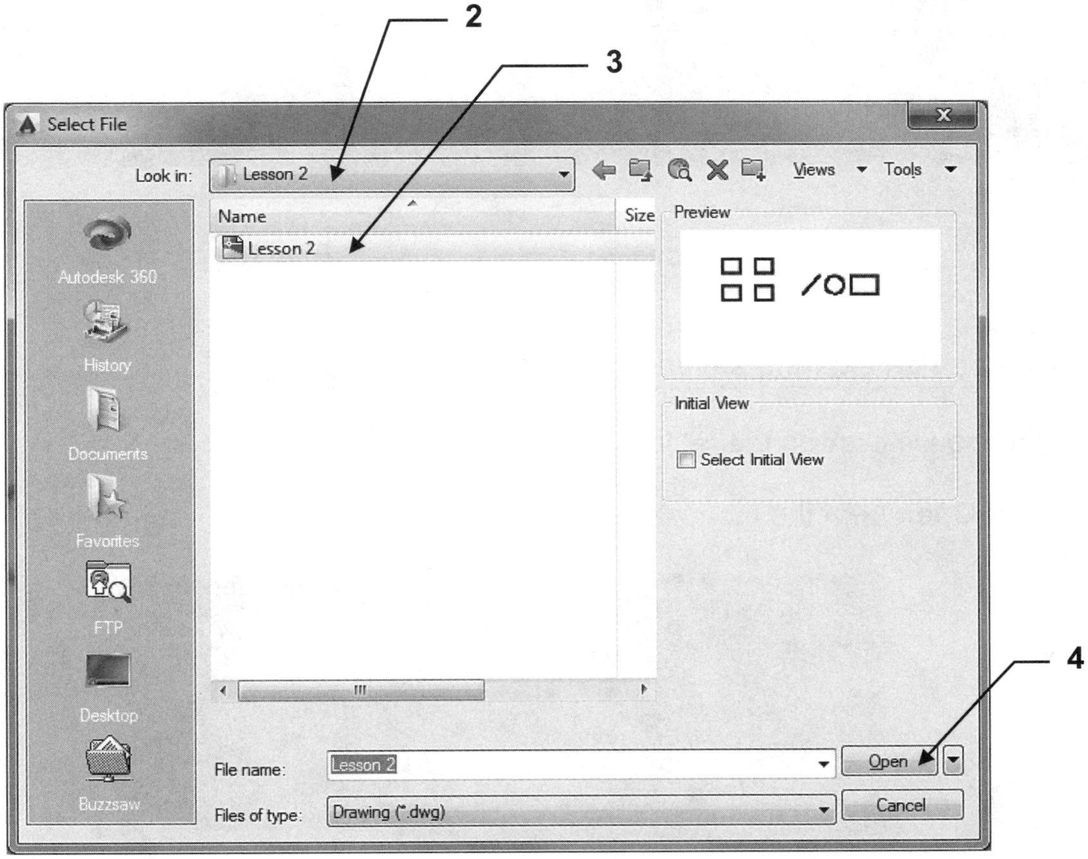

2. Locate the Directory and Folder in which the file had previously been saved.

3. Select the File that you wish to OPEN.

4. Select the **Open** button.

OPEN MULTIPLE FILES

The **File Tabs** tool allows you to have multiple drawings open at the same time. If the File Tabs tool is switched on (on by default), you can open existing saved drawings or create new ones.

The **File Tabs** tool is located on the **Interface** panel of the **View** tab, and is a Neon Blue color when switched on.

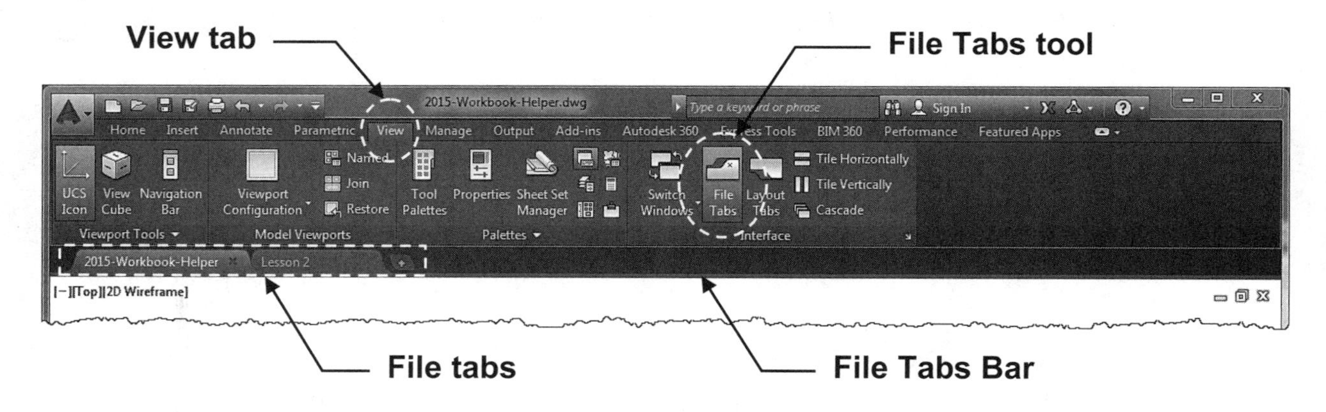

How to open an existing saved drawing from the File Tabs

1. Right mouse click on the '**+**' icon.

2. Select **Open** from the menu.

3. Locate the Directory and Folder for the previously saved file. (Refer to page 2-19)

4. Select the File you wish to open.

5. Select the **Open** button.

OPEN MULTIPLE FILES....continued

How to open a new drawing from the Files Tab.

1. Right mouse click on the '**+**' icon. (Refer to page 2-20)
2. Select **Drawing Template (*.dwt)** from the **Files of type** drop-down list.
3. Select the Template you require.
4. Select the **Open** button.

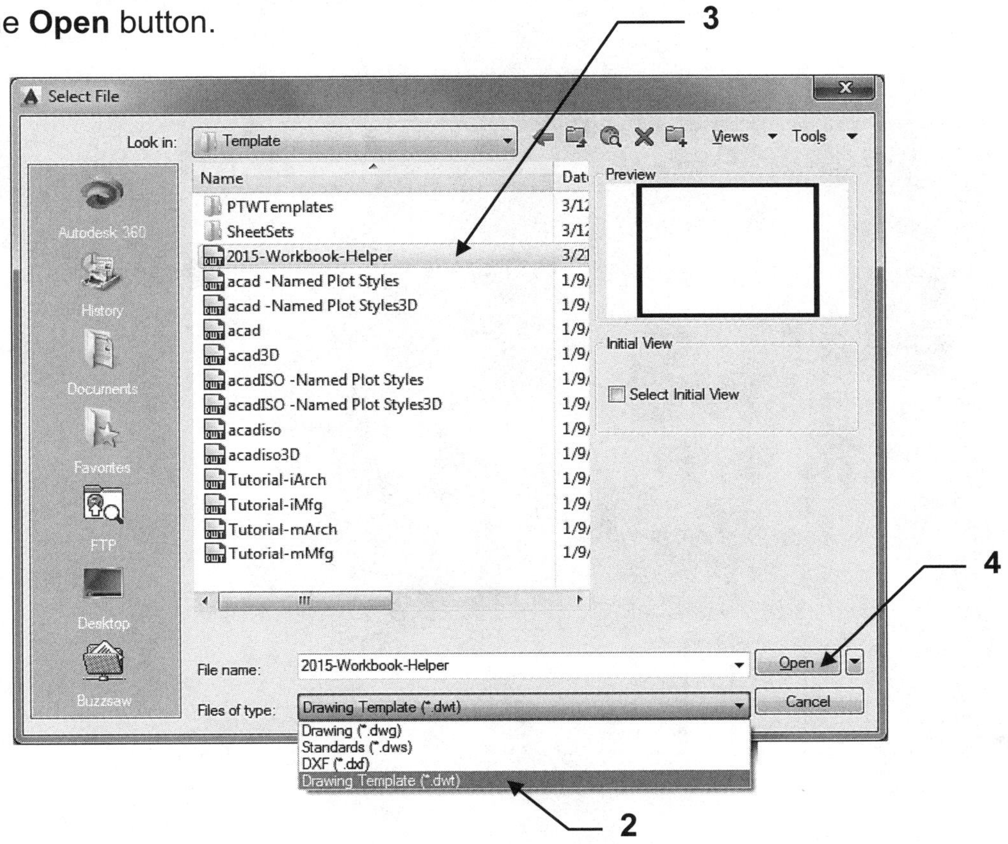

Note:
If you right mouse click on any **File Tab** a menu appears with various options, including closing all open drawing tabs except the one you just clicked on.

You can also select a **New Tab** page where you can access online resources and the Learn and Create pages. (Refer to page 1-2) You can also left mouse click on the '**+**' icon to access the **New Tab** page.

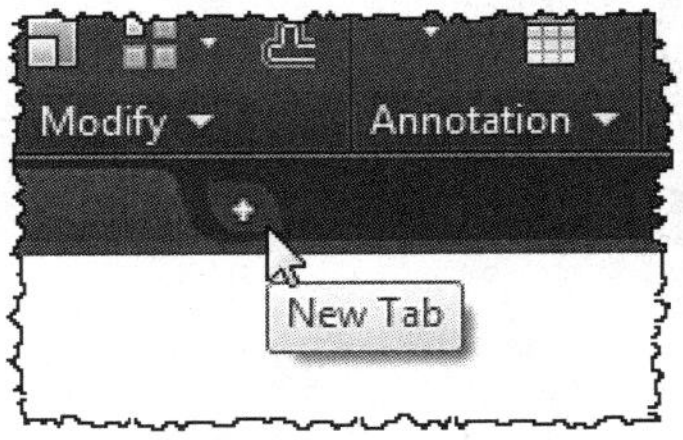

OPEN MULTIPLE FILES....continued

The File Tabs drawing previews allow you to quickly change between open drawings. If you hover your mouse over any open File Tab, a preview of the Model and the Layout tabs are displayed. You can click on any of the previews to take you to that particular open drawing or view.

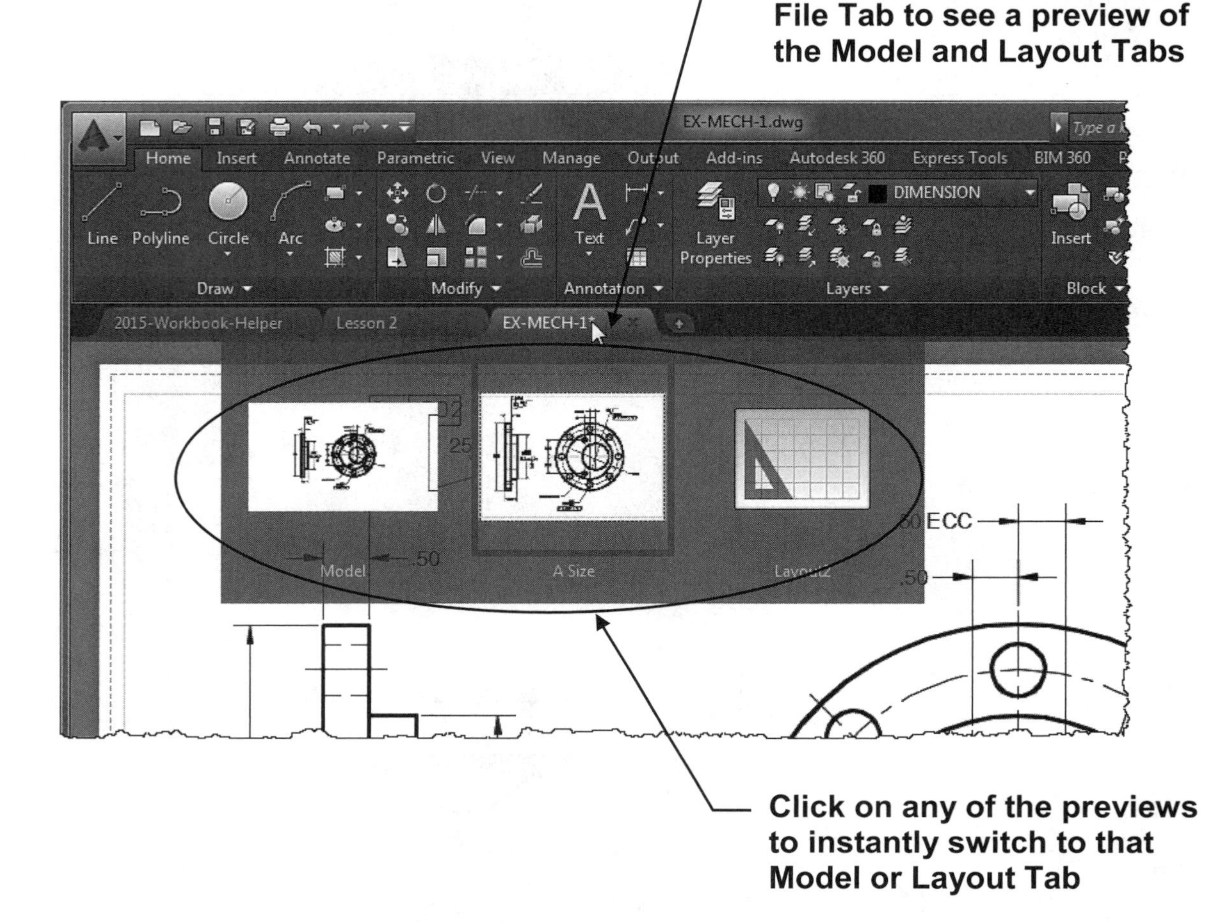

If an asterisk is displayed on a File Tab it means that particular drawing has not been saved since it was last modified. The asterisk will disappear when the drawing has been saved.

Draw | Modify | Annotation

2015-Workbook-Helper | Lesson 2* | EX-MECH-1*

–][Top][2D Wireframe]

These two drawings are showing the asterisk and have not been saved

SAVING A DRAWING FILE

After starting a new drawing, it is best practice to save it immediately. Learning how to save a drawing correctly is almost more important than making the drawing. If you can't save correctly, you will lose the drawing and hours of work.

There are 2 commands for saving a drawing: **Save** and **Save As**.
I prefer to use **Save As**.

The **Save As** command always pauses to allow you to choose where you want to store the file and what name to assign to the file. This may seem like a small thing, but it has saved me many times from saving a drawing on top of another drawing by mistake.

The **Save** command will automatically save the file either back to where you retrieved it or where you last saved a previous drawing. Neither may be the correct destination. And may replace a file with the same name. So play it safe, use **Save As** for now.

1. Select the **Saveas** command using one of the following:

Quick Access Toolbar =
or
Application Menu = Save As / Drawing
or
Keyboard = SA <enter> Saveas

2. Select the "Save In" location.
(This is where the file will be saved.)

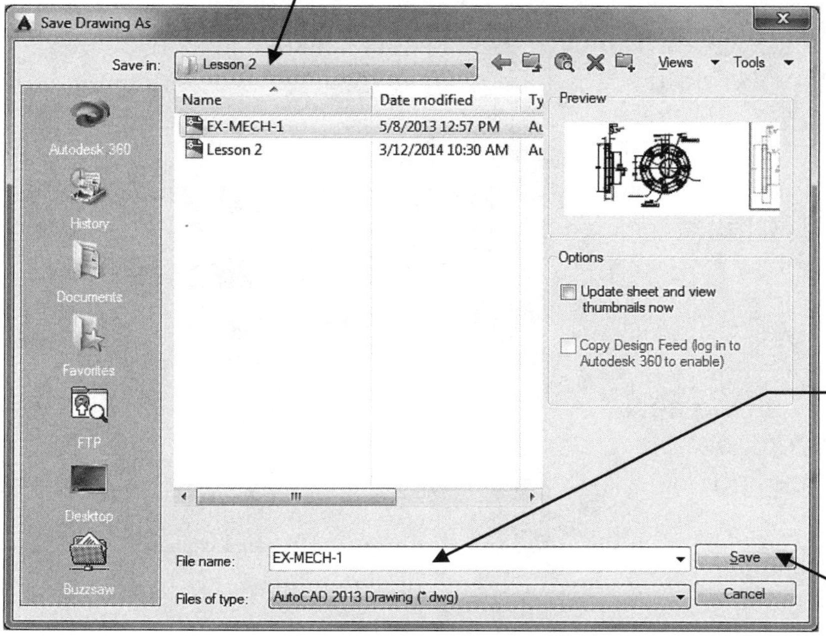

3. Type the new File name here.

4. Select the "Save" button.

AUTOMATIC SAVE

AUTOMATIC SAVE

If you turn the automatic save option ON, your drawing is saved at specified time intervals. These temporary files are automatically deleted when a drawing closes normally. The default save time is every 10 minutes. You may change the save time Intervals and where you would prefer the Automatic Save files to be saved.

How to set the Automatic Save intervals

1. Type **options** <enter>
2. Select the **Open and Save** tab.
3. Enter the desired **minutes between saves**.
4. Select the **OK** button.

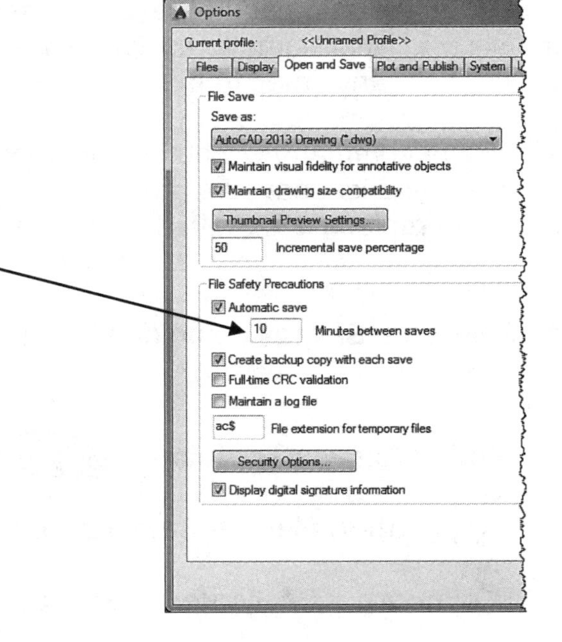

How to change the Automatic Save location

1. Type **options** <enter>
2. Select the **Files** tab.
3. Locate the **Automatic Save File Location** and click on the **"+"** to display the **path**.
4. Double click on the path.
5. Browse to locate the Automatic Save Location desired and highlight it.
6. Select **OK**.

(The browse box will disappear and the new location path should be displayed under the Automatic Save File Location heading)

7. Select **OK** to accept the change.

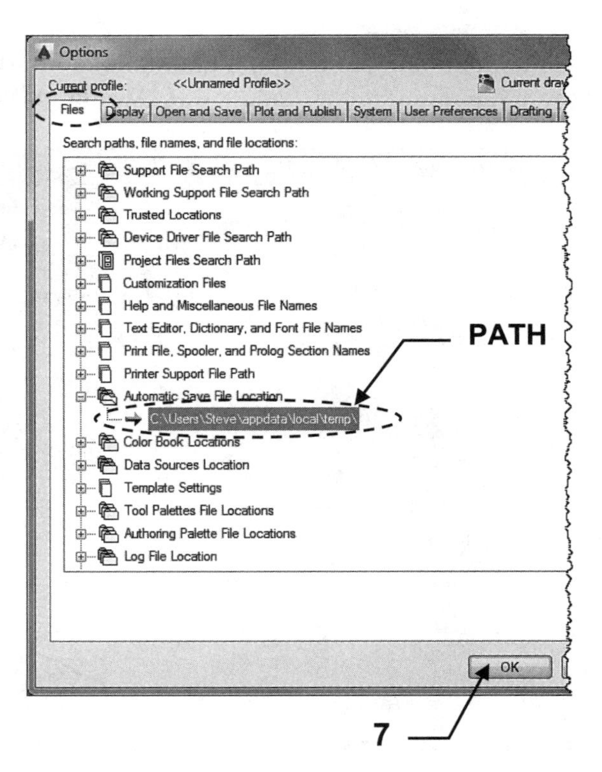

BACK UP FILES and RECOVER

BACK UP FILES

When you save a drawing file, Autocad creates a file with a **.dwg** extension.
For example, if you save a drawing as **12b**, Autocad saves it as **12b.dwg**.
The next time you save that same drawing, Autocad replaces the old with the new and renames the old version **12b.bak**. The old version is now a <u>back up</u> file.
(Only 1 backup file for each drawing file is stored.)

How to open a back up file:
You can't open a **".bak"** file.
It must first be renamed with a **".dwg"** file extension.

How to view the list of back up files:
The backup files will be saved in the same location as the drawing file.
You must use Windows Explorer to locate the .bak files.

How to rename a back up file:
1. Right click on the file name.
2. Select "Rename".
3. Change the .bak extension to .dwg and press <enter>.

RECOVERING A DRAWING

In the event of a program failure or a power failure any open files should be saved automatically. (Refer to page 2-24)

When you attempt to re-open the drawing the **Drawing Recovery Manager** will display a list of all drawing files that were open at the time of a program or system failure. You can preview and open each .dwg or .bak file to choose which one should be saved as the primary file.

EXITING AUTOCAD

To safely exit AutoCAD follow the instructions below.

1. Save all open drawings.

2. Start the **EXIT** procedure using one of the following.

    Ribbon = None
    or
    Application Menu = Exit Autodesk AutoCAD 2015
    or
    Keyboard = Exit <enter>

If any changes have been made to the drawing since the last **Save As**, the warning box shown below will appear asking if you want to ***SAVE THE CHANGES***?

Select ***YES, NO*** *or* ***CANCEL***.

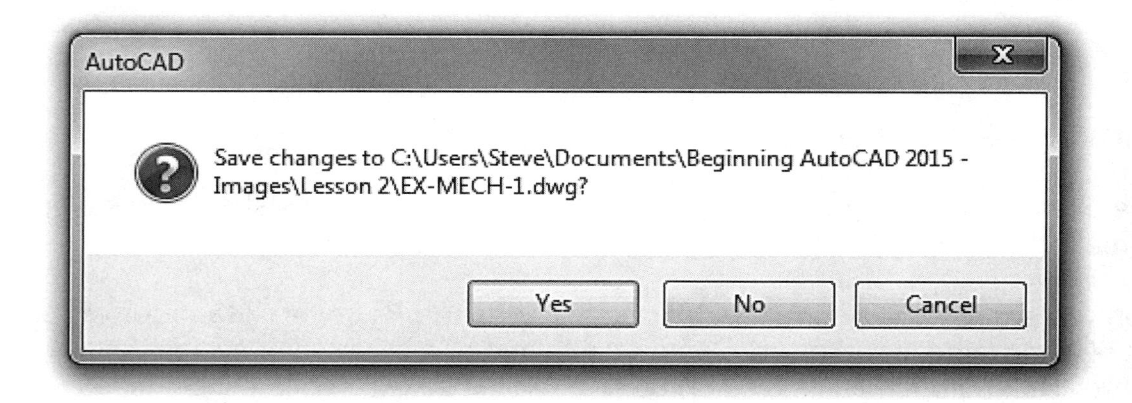

EXERCISE 2A

INSTRUCTIONS:

1. Start a **NEW** file using **2015-Workbook Helper.dwt.**
2. **Draw** the objects below using:
    A. **Line** command
    B. **Ortho** (F8) **ON** when drawing Horizontal and Vertical lines.
    C. **Ortho** (F8) **OFF** when drawing lines on an Angle.
    D. Turn **Increment Snap** (F9) **ON**
    E. Turn **Osnap** (F3) **OFF**
    F. Turn **Grid** (F7) **ON**
    G. Use the **Close** option
3. **Save** the drawing as: **EX2A**

EXERCISE 2B

INSTRUCTIONS:

1. **OPEN EX2A**, if not already open.
2. **Erase** the missing Lines as shown.
   A. Turn **Osnap** (F3) **OFF** (It will be easier to move the cursor accurately)
3. **Save** the drawing as: **EX2B**

EXERCISE 2C

INSTRUCTIONS:

1. Start a **NEW** file using **2015-Workbook Helper.dwt.**
2. **Draw** the objects below using:
    A. **Line** command
    B. **Ortho** (F8) **ON** when drawing Horizontal and Vertical lines.
    C. **Ortho** (F8) **OFF** when drawing lines on an Angle.
    D. Turn **Increment Snap** (F9) **ON**
    E. Turn **Osnap** (F3) **OFF**
    F. Turn **Grid** (F7) ON
    G. Use the **Shift Key** to toggle Ortho ON and OFF
3. **Save** the drawing as: **EX2C**

EXERCISE 2D

INSTRUCTIONS:

1. Start a **NEW** file using **2015-Workbook Helper.dwt.**
2. **Draw** the objects below using:
   A. **Line** command
   B. **Ortho** (F8) **ON** when drawing Horizontal and Vertical lines.
   C. **Ortho** (F8) **OFF** when drawing lines on an Angle.
   D. Turn **Increment Snap** (F9) **ON**
   E. Turn **Osnap** (F3) **OFF**
   F. Turn **Grid** (F7) ON
   G. Use the **Shift Key** to toggle Ortho ON and OFF
3. **Save** the drawing as: **EX2D**

LEARNING OBJECTIVES

After completing this lesson, you will be able to:

1. Create a Circle using 6 different methods
2. Create Rectangles with chamfers, fillets, width and rotation.
3. Set Grids and Increment Snap
4. Draw using Layers
5. Control Layers
6. Create Layers

LESSON 3

CIRCLE

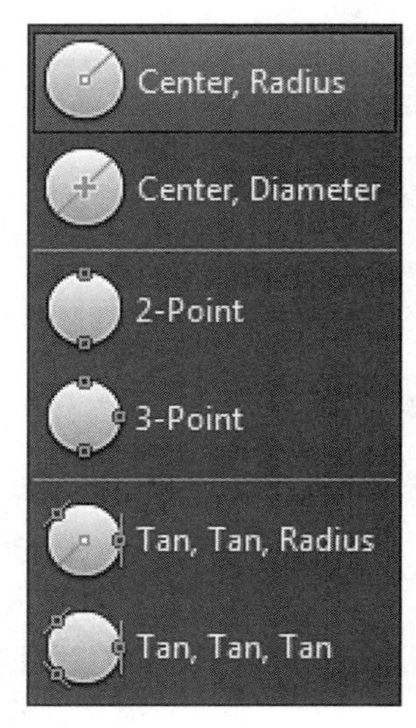

There are 6 options to create a circle.

The default option is "**Center, radius**". (Probably because that is the most common method of creating a circle.)

We will try the **"Center, radius"** option first.

1. Start the **Circle** command by using one of the following:

   Ribbon = Home tab / Draw panel /
   or
   Keyboard = C <enter>

2. The following will appear on the command line:
   Command: _circle Specify center point for circle or [3P/2P/Ttr (tan tan radius)]:

3. Locate the center point for the circle by moving the cursor to the desired location in the drawing area **(P1)** and press the left mouse button.

4. Now move the cursor away from the center point and you should see a circle forming.

5. When the circle is the size desired **(P2)**, press the left mouse button, or type the radius and then press <enter>.

Note: To use one of the other methods described below, first select the Circle command, then select one of the other Circle options.

Center, Radius: (Default option)

1. Specify the center (P1) location.
2. Specify the Radius (P2).

   (Define the Radius by moving the cursor or typing radius)

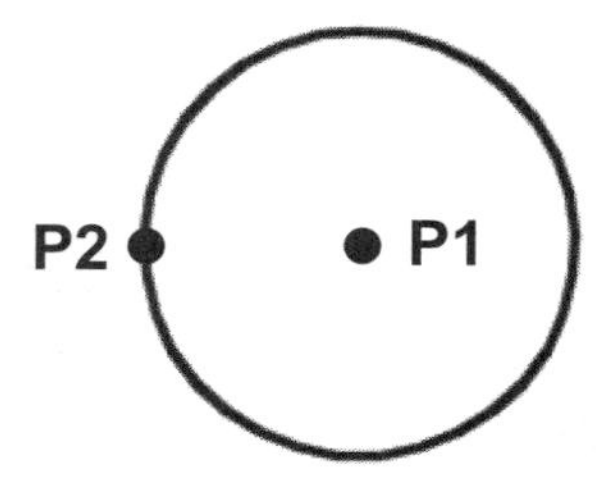

Center, Diameter:

1. Specify the center (P1) location.
2. Specify the Diameter (P2). (Define the Diameter by moving the cursor or typing Diameter)

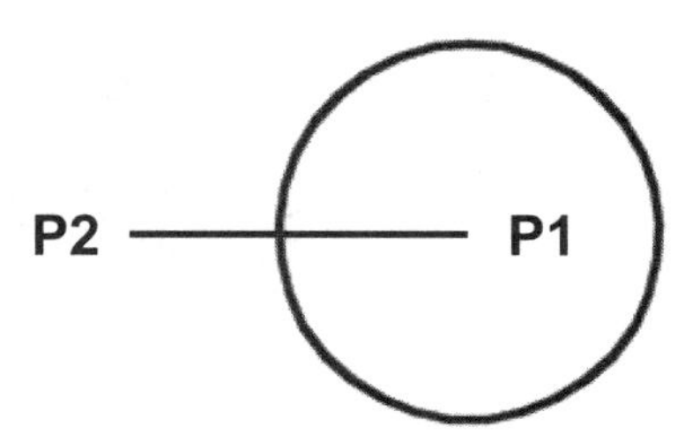

CIRCLE....continued

2 Points:

1. Select the 2 point option
2. Specify the 2 points (P1 and P2) that will determine the Diameter.

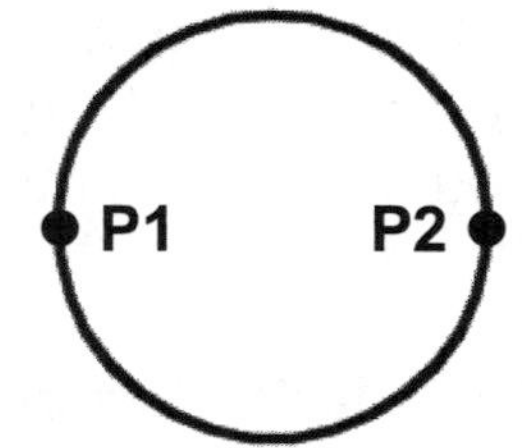

3 Points:

1. Select the 3 Point option
2. Specify the 3 points (P1, P2 and P3) on the circumference. The Circle will pass through all three point

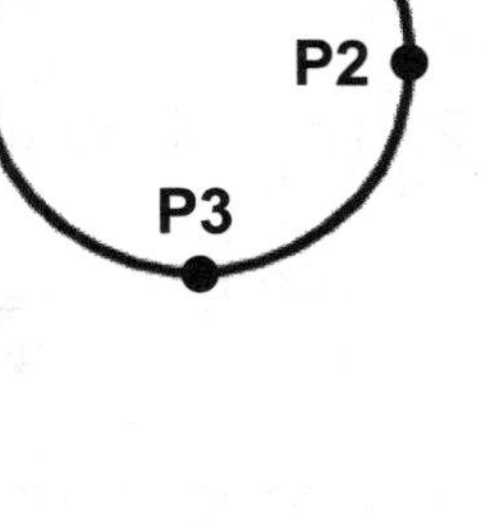

Tangent, Tangent, Radius:

1. Select the Tangent, Tangent, Radius option .
2. Select two objects (P1 and P2) for the Circle to be tangent to by placing the cursor on the object and pressing the left mouse button
3. Specify the radius.

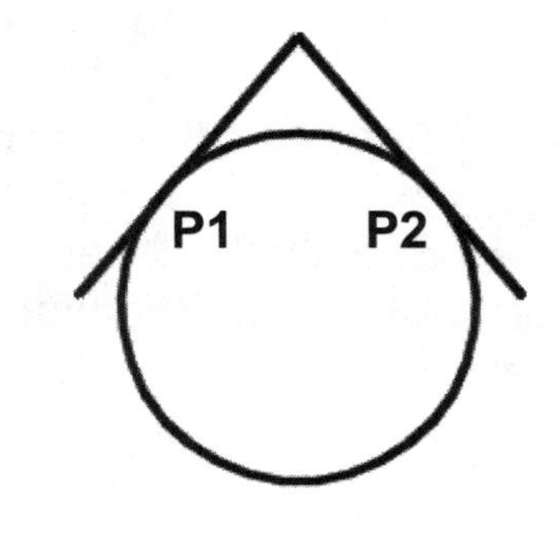

Tangent, Tangent, Tangent:

1. Select the Tangent, Tangent, Tangent option
2. Specify three objects (P1, P2 and P3) for the Circle to be tangent to by placing the cursor on each of the objects and pressing the left mouse button.
   (AutoCAD will calculate the diameter.)

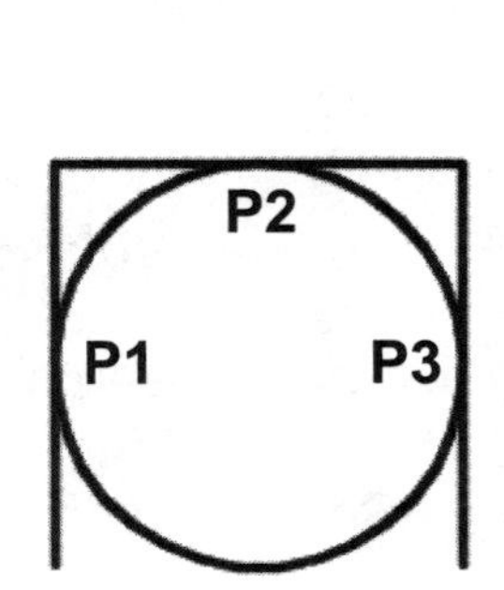

RECTANGLE

A Rectangle is a closed rectangular shape. It is one object not 4 lines.
You can specify the length, width, area, and rotation options.
You can also control the type of corners on the rectangle—fillet, chamfer, or square and the width of the Line.

First, let's start with a simple Rectangle using the cursor to select the corners.

1. Start the **RECTANGLE** command by using one of the following:

   Ribbon = Home tab / Draw panel /
   or
   Keyboard = REC <enter>

2. The following will appear on the command line:

   Command: _rectang
   Specify first corner point or [Chamfer/Elevation/Fillet/Thickness/Width]:

3. Specify the location of the first corner by moving the cursor to a location (**P1**) and then press the left mouse button.

   The following will appear on the command line:

   Specify other corner point or [Area / Dimensions / Rotation]:

4. Specify the location of the **diagonal** corner (**P2**) by moving the cursor diagonally away from the first corner (**P1**) and pressing the left mouse button.

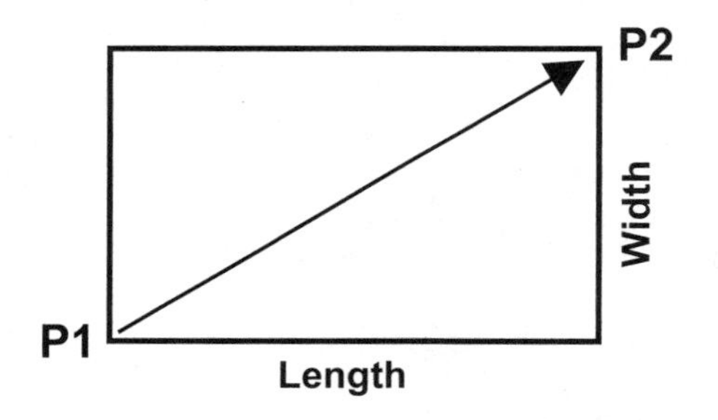

OR

4. Type **D <enter>** (or click on the blue letter "D")

   Specify length for rectangles <0.000>: ***Type the desired length <enter>.***

   Specify width for rectangles <0.000>: ***Type the desired width <enter>.***

   Specify other corner point or [Dimension]: ***move the cursor up, down, right or left to specify where you want the second corner relative to the first corner and then press <enter> or press left mouse button.***

RECTANGLE....continued

OPTIONS: Chamfer, Fillet and Width

Note: the following options are <u>only</u> available <u>before</u> you place the **first corner** of the Rectangle.

CHAMFER

A chamfer is an angled corner. The Chamfer option automatically draws all 4 corners with chamfers simultaneously and all the same size. You must specify the distance for each side of the corner as distance 1 and distance 2.

Example: A Rectangle with dist1 = .50 and dist2 = .25

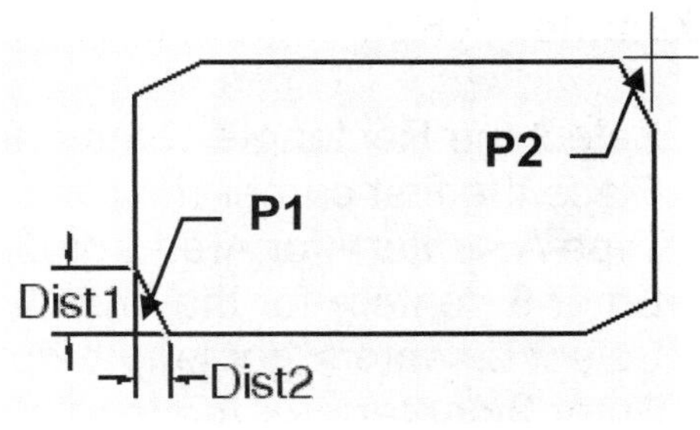

1. Select the Rectangle command
2. Type C <enter> (or click on the blue letter "C")
3. Enter .50 for the first distance
4. Enter .25 for the second distance
5. Place the first corner (P1)
6. Place the diagonal corner (P2)

FILLET

A fillet is a rounded corner. The fillet option automatically draws all 4 corners with fillets (all the same size). You must specify the radius for the rounded corners.

Example: A Rectangle with .50 radius corners.

1. Select the Rectangle command
2. Type F <enter>(or click on the blue letter "F")
3. Enter .50 for the radius.
4. Place the first corner (P1)
5. Place the diagonal corner (P2)

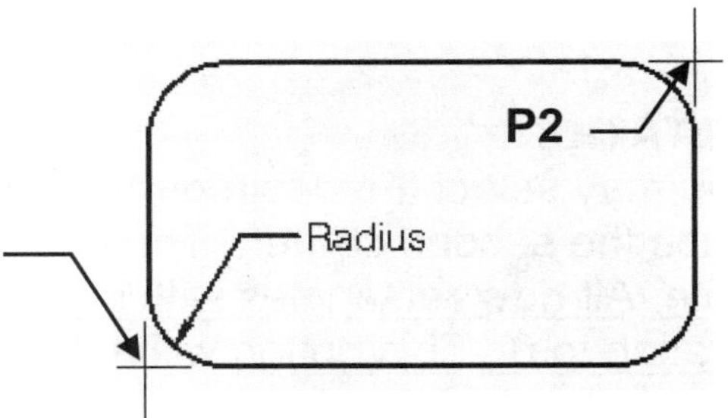

Note: You must set Chamfer and Fillet back to "0" before defining the width. Unless you want fat lines <u>and</u> Chamfered or Filleted corners.

WIDTH

Defines the width of the rectangle lines.
Note: Do not confuse this with the "Dimensions" Length and Width.
Width makes the lines appear fatter.

Example: A Rectangle with a width of .50

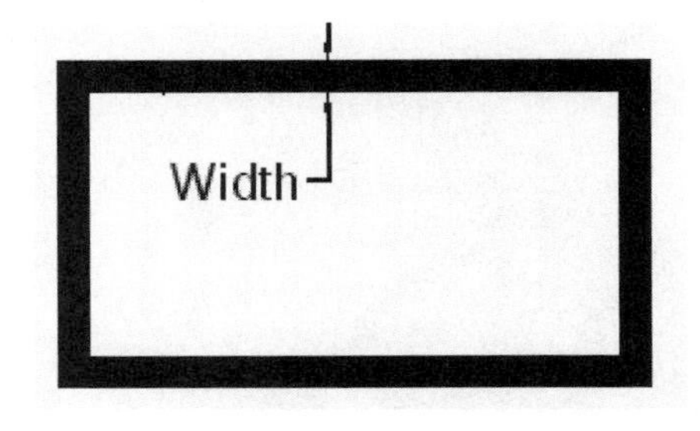

1. Select the Rectangle command
2. Type W <enter> (or click on the blue letter "W")
3. Enter .50 for the width.
4. Place the first corner (P1)
5. Place the diagonal corner (P2)

RECTANGLE....continued

OPTIONS: Area and Rotation

Note: the following options are available **AFTER** you place the **first corner** of the Rectangle.

AREA

Creates a Rectangle using the AREA and either a LENGTH or a WIDTH. If the Chamfer or Fillet option is active, the area includes the effect of the chamfers or fillets on the corners of the rectangle.

Example: A Rectangle with an Area of 6 and a Length of 2.

1. Select the Rectangle command
2. Place the first corner (P1)
3. Type A <enter> for Area. (or click on blue "A")
4. Enter 6 <enter> for the Area
5. Select L <enter> for length option (or click on blue "L")
6. Enter 2 <enter> for the length

(The width will automatically be calculated)

Width

P1

Length

ROTATION

You may select the desired rotation angle after you place the first corner and before you place the second corner. The base point (pivot point) is the first corner.
Note: All **new** rectangles within the drawing will also be rotated unless you reset the rotation to 0. This option will not effect rectangles already in the drawing.

Example: A Rectangle with a rotation angle of 45 degrees.

1. Select the Rectangle command
2. Place the first corner (P1)
3. Type R <enter> for rotation. (or click on blue "R")
4. Enter 45 <enter>
5. Place the diagonal corner (P2)

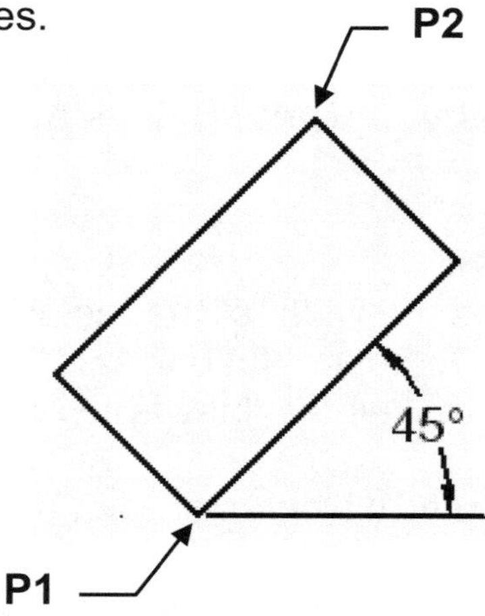

GRID and INCREMENT SNAP

<u>**GRID**</u> is the criss-cross lines in the drawing area. The grid is only a drawing aid to assist you in aligning objects and visualizing the distances between them.
The Grid <u>will not plot</u>. (Refer to page 1-13)

<u>**INCREMENT SNAP**</u> controls the movement of the cursor. If it is **OFF** the cursor will move smoothly. If it is **ON**, the cursor will jump in an ***incremental*** movement.
(Refer to page 1-13)

The **DRAFTING SETTINGS** dialog box allows you to set **INCREMENT SNAP** and **GRID** spacing. You may change the Grid Spacing and Increment Snap at anytime while creating a drawing. The settings are only drawing aids to help you visualize the size of the drawing and control the movement of the cursor.

1. Select **DRAFTING SETTINGS** by using one of the following:

   Keyboard = DS <enter>

   Status Bar = Right Click on SNAP or GRID button and select SETTINGS.

2. The dialog box shown below will appear.

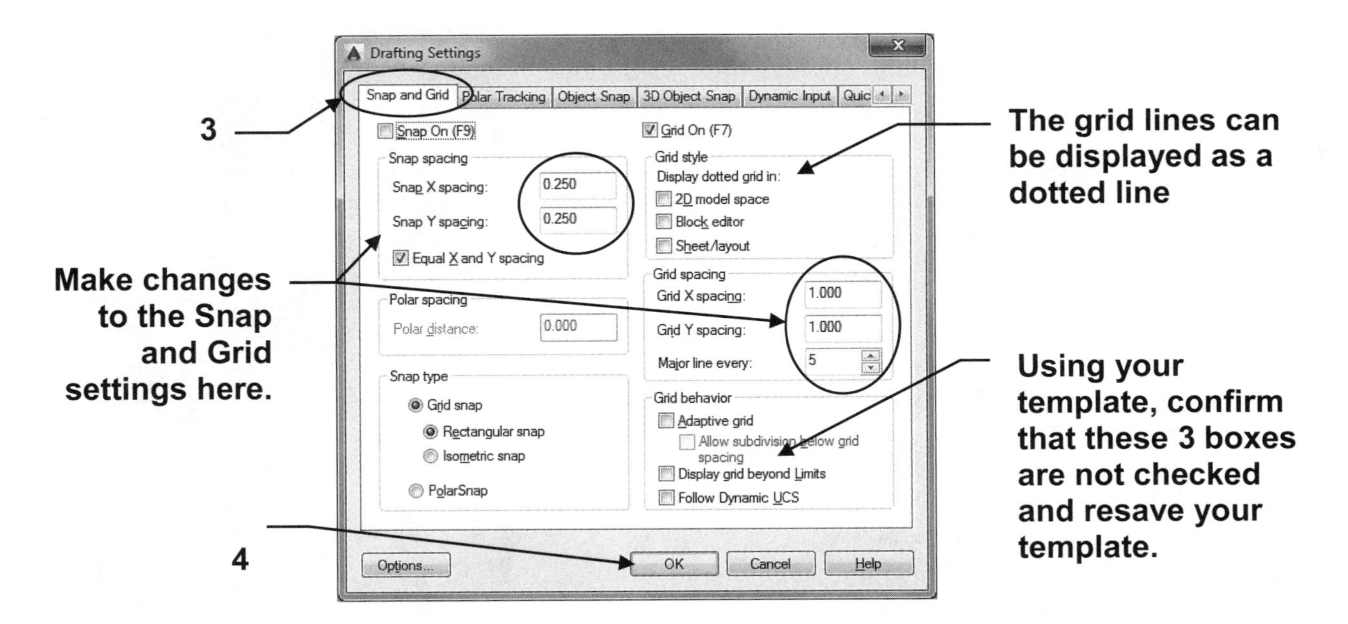

3. Select the **"Snap and Grid"** tab.

4. Make your changes and select the **OK** button to save them.
   If you select the **CANCEL** button, your changes will **not** be saved.

LAYERS

A **LAYER** is like a transparency. Have you ever used an overhead light projector? Remember those transparencies that are laid on top of the light projector? You could stack multiple sheets but the projected image would have the appearance of one document. Layers are basically the same. Multiple layers can be used within one drawing.

The example, on the right, shows 3 layers.
One for annotations (text), one for dimensions and one for objects.

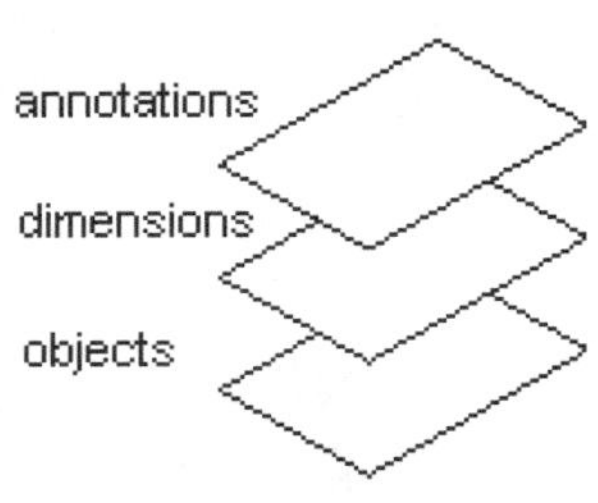

HOW TO USE LAYERS
First you select the layer and then you draw the objects.
Always select the layer first and then draw the objects.

It is good "drawing management" to draw related objects on the same layer.
For example, in an architectural drawing, you would select layer "walls" and then draw the floor plan.
Then you would select the layer "Electrical" and draw the electrical objects.
Then you would select the layer "Plumbing" and draw the plumbing objects.
Each layer can then be controlled independently.
If a layer is Frozen, it is not visible. When you Thaw the layer it becomes visible again.
(Refer to the following pages for detailed instructions for controlling layers.)

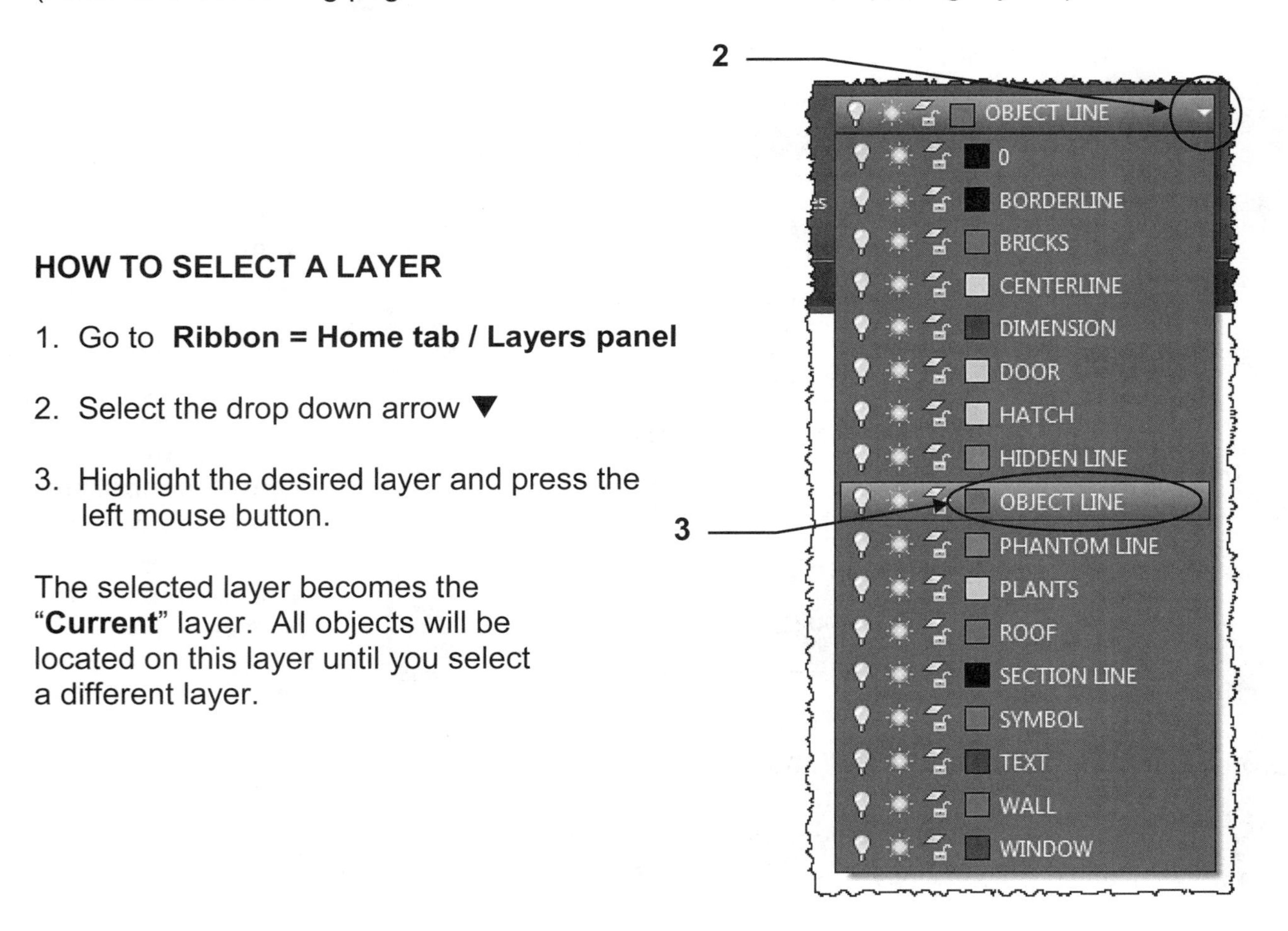

HOW TO SELECT A LAYER

1. Go to **Ribbon = Home tab / Layers panel**
2. Select the drop down arrow ▼
3. Highlight the desired layer and press the left mouse button.

The selected layer becomes the "**Current**" layer. All objects will be located on this layer until you select a different layer.

CONTROLLING LAYERS

The following controls can be accessed using the Layer drop down arrow ▼.

ON or OFF
If a layer is **ON** it is **visible**. If a layer is **OFF** it is **not visible**.
Only layers that are **ON** can be **edited** or **plotted**.

FREEZE or THAW
Freeze and **Thaw** are very similar to On and Off. A Frozen layer is not visible and a Thawed layer is visible. Only thawed layers can be edited or plotted.

Additionally:

a. Objects on a Frozen layer **cannot** be accidentally erased
b. When working with large and complex drawings, freezing saves time because frozen layers are not **regenerated** when you zoom in and out.

LOCK or UNLOCK
Locked layers are visible but cannot be edited.
They are visible so they **will** be plotted.

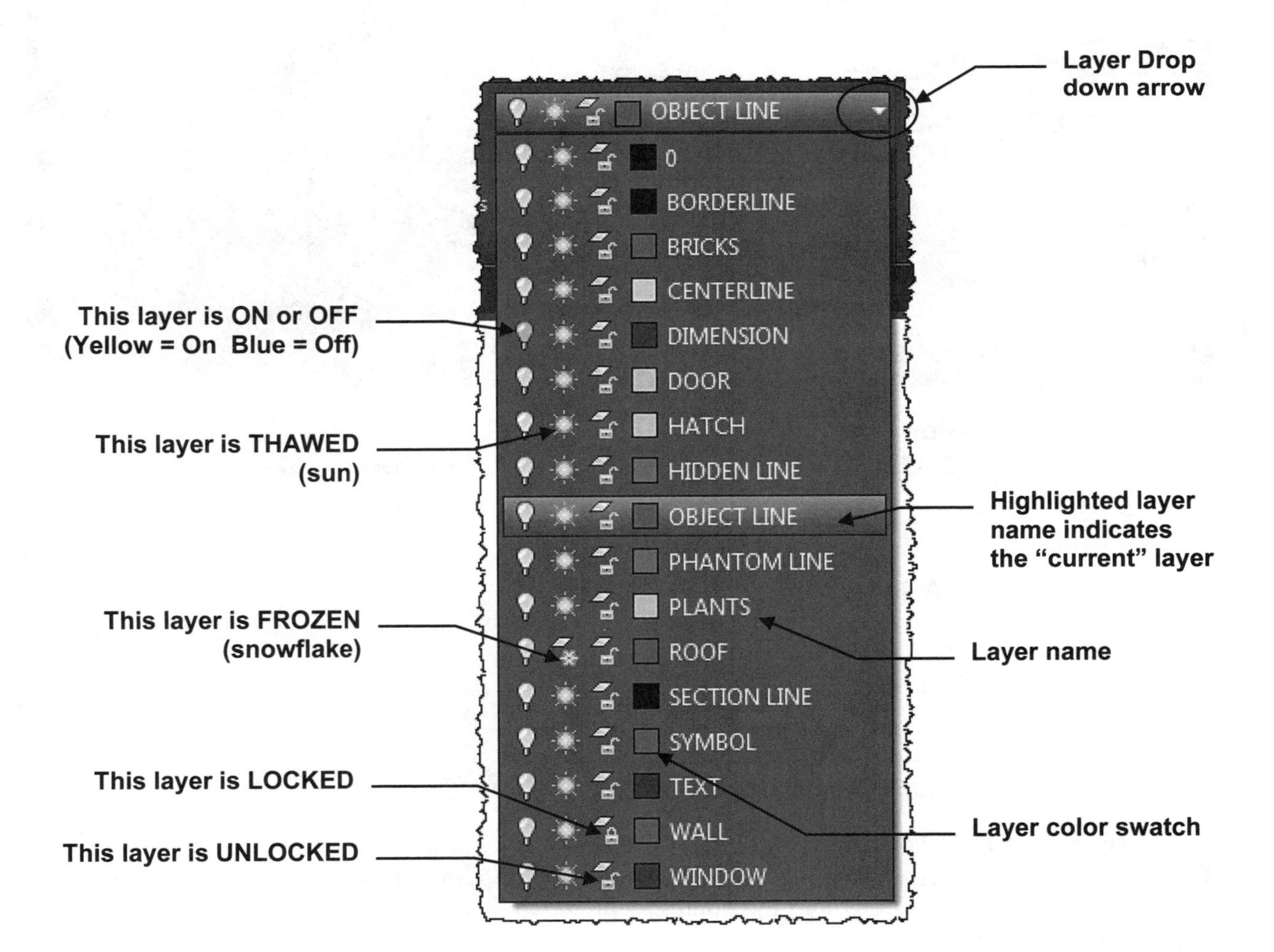

CONTROLLING LAYERS....continued

To access the following options you must use the **Layer Properties Manager**. You may also access the options listed on the previous page within this dialog box.

To open the **Layer Properties Manager** use one of the following.

Ribbon = Home tab / Layers panel /
or
Keyboard = LA <enter>

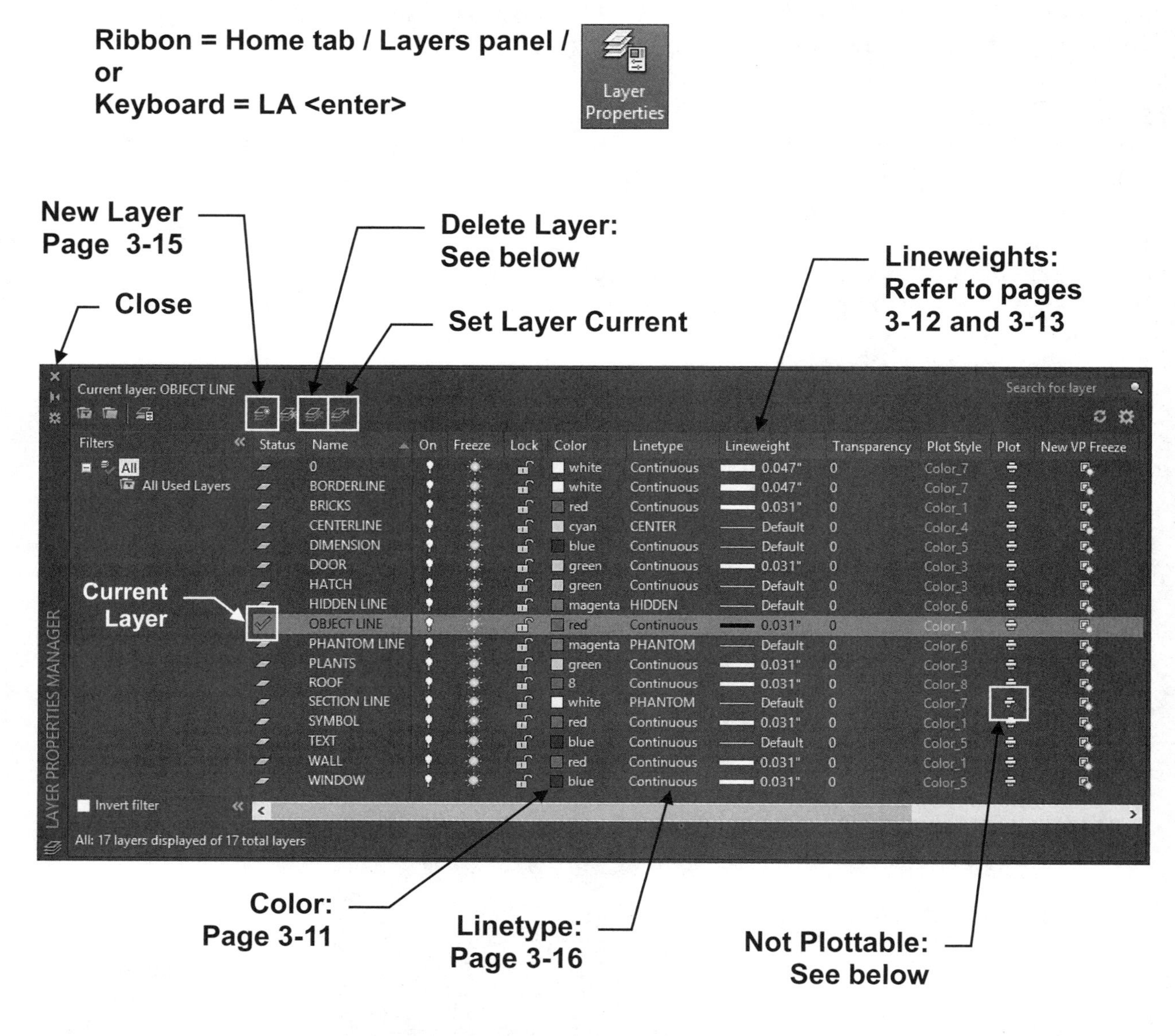

HOW TO DELETE AN EXISTING LAYER

1. Highlight the layer name to be deleted.
2. Select the **Delete Layer** tool.

Or

1. Highlight the layer name to be deleted.
2. Right click and select **Delete Layer**

PLOT or NOT PLOTTABLE

This tool prevents a layer from plotting even though it is visible within the Drawing Area. A Not Plottable layer will not be displayed when using Plot Preview. If the Plot tool has a slash the layer will not plot.

LAYER COLOR

Color is not merely to make a pretty display on the screen. Layer colors can help define objects. For example, you may assign color Green for all doors. Then, at a glance, you could identify the door and the layer by their color.

Here are some additional things to consider when selecting the colors for your layers.

Consider how the colors will appear on the paper.
(Pastels do not display well on white paper.)

Consider how the colors will appear on the screen.
(Yellow appears well on a black background but not on white.)

How to change the color of a layer.

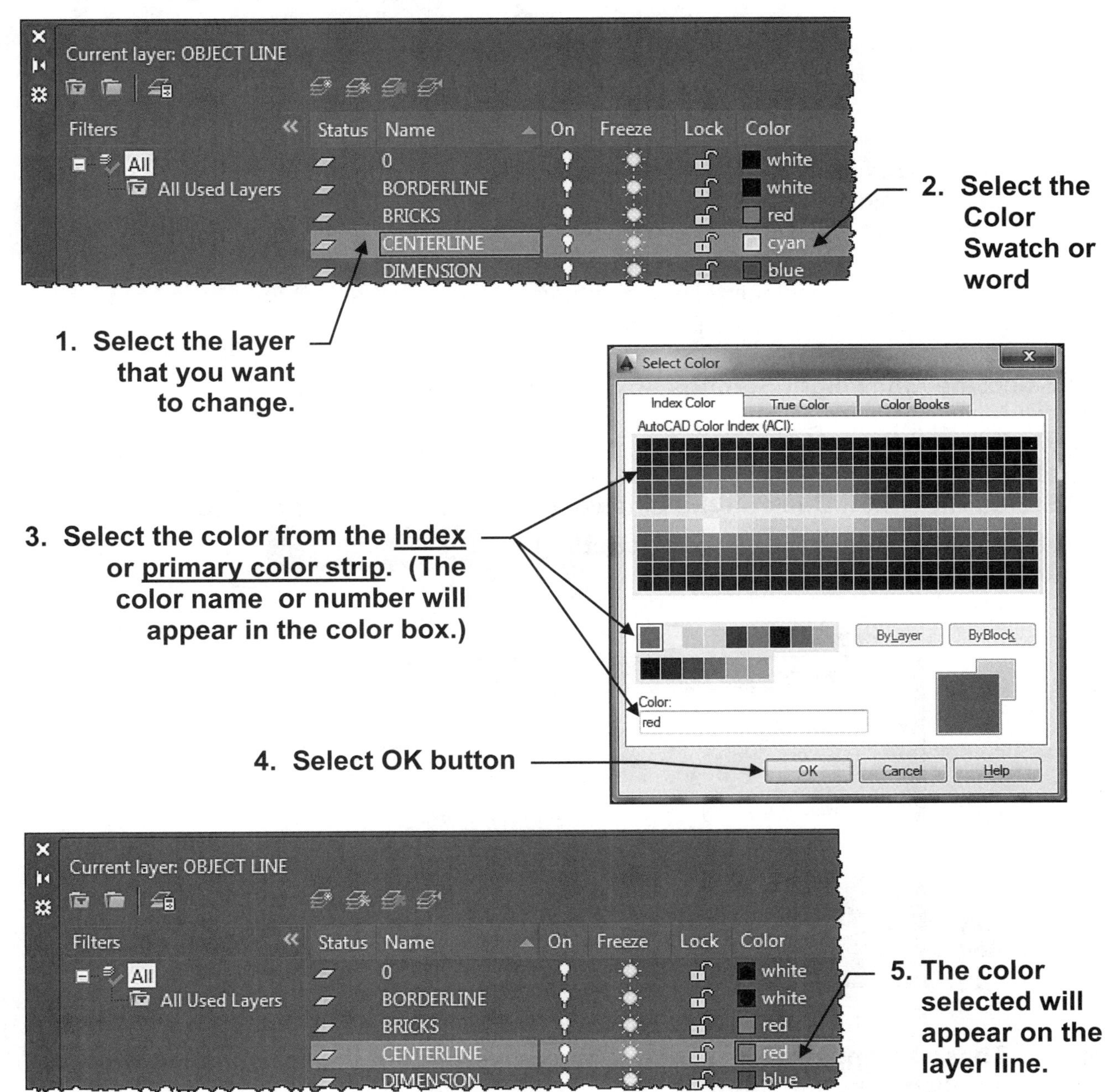

LINEWEIGHTS

A **Lineweight** means "**how heavy or thin is the object line**".

It is "good drawing management" to establish a contrast in the lineweights of entities.

In the example below the rectangle has a heavier lineweight than the dimensions. The contrast in lineweights makes it easier to distinguish between entities.

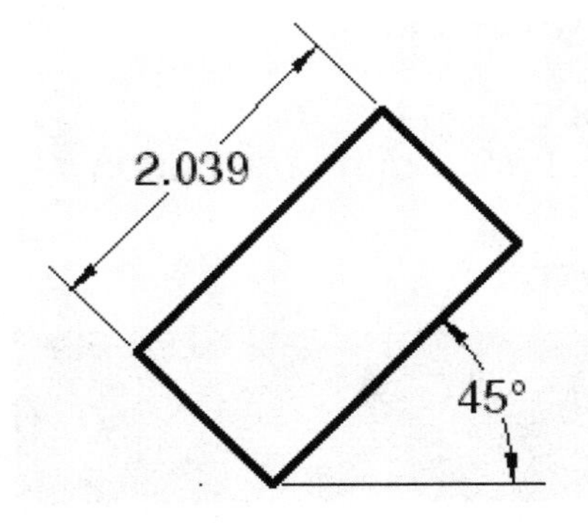

LINEWEIGHT SETTINGS

Lineweights are plotted with the exact width of the lineweight assigned.
But you may adjust how they are displayed on the screen. (Refer to #4 below)

IMPORTANT: Before assigning lineweights you should first select the **Units for Listing** and **Adjust Display Scale** as shown below.

1. Select the **Lineweight Settings** box using one of the following:

   Keyboard = lw <enter>
   or
   Status Bar = Left Click on the Lineweight button down arrow and then left click on Lineweight Settings. Lineweight Settings...

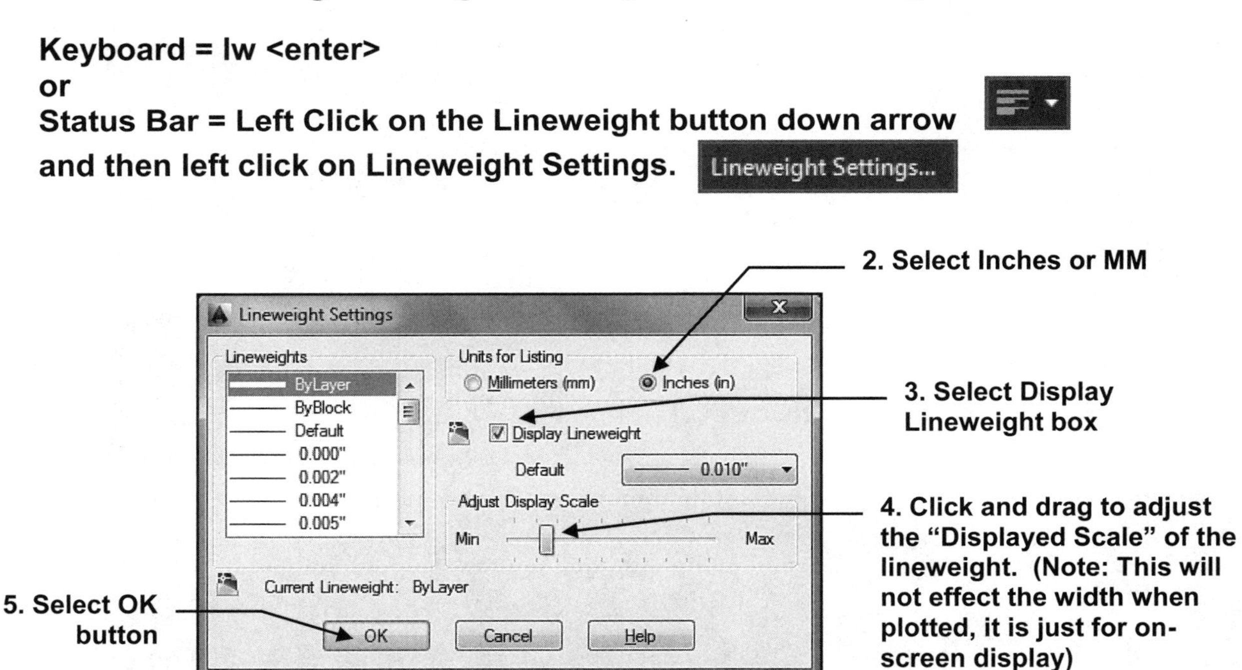

NOTE: These settings will be saved to the computer not the drawing and will remain until you change them.

ASSIGNING LINEWEIGHTS

Note: Before assigning **Lineweights** to Layers make sure your **Lineweight settings** (Units for listing and Adjust Display scale) are correct. Refer to the previous page.

ASSIGNING LINEWEIGHTS TO LAYERS

1. Select the Layer Properties Manager using one of the following:

   Ribbon = Home tab / Layers panel /
   or
   Keyboard = LA <enter>

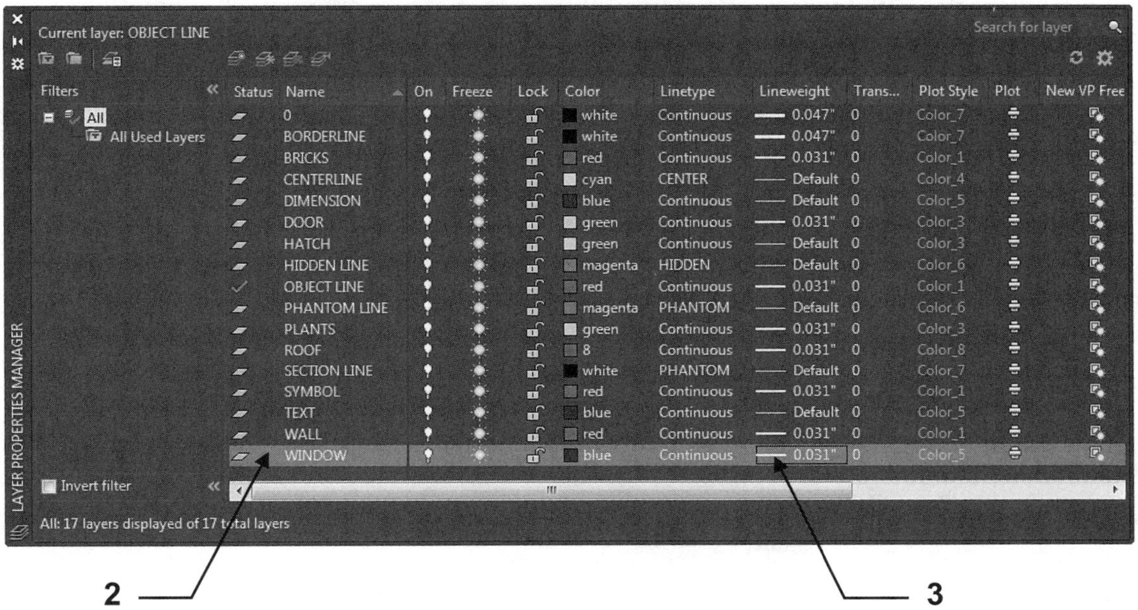

2. Highlight a Layer (Click on the name)

3. Click on the Lineweight for that layer.

4. Scroll and select a Lineweight from the list.

5. Select the **OK** button.

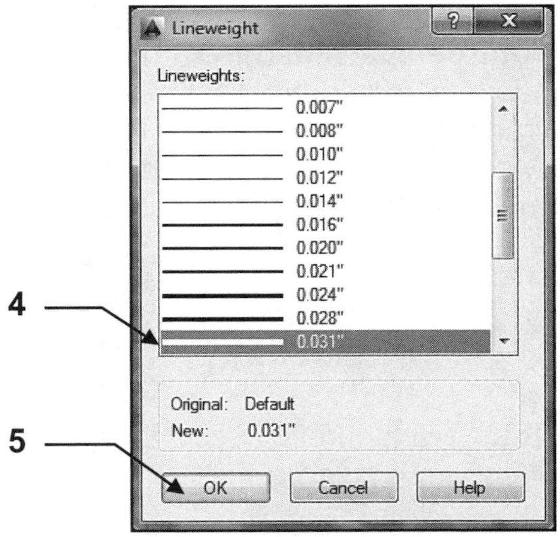

Note:
Lineweight selections will be saved within the **current** drawing and will not effect any other drawing.

TRANSPARENCY

Each layer may be assigned a transparency percentage from 0 to 90 percent. 0 would not be transparent at all and 90 would be 90% transparent.

ASSIGNING TRANSPARENCY TO LAYERS

1. Select the Layer Properties Manager using one of the following:

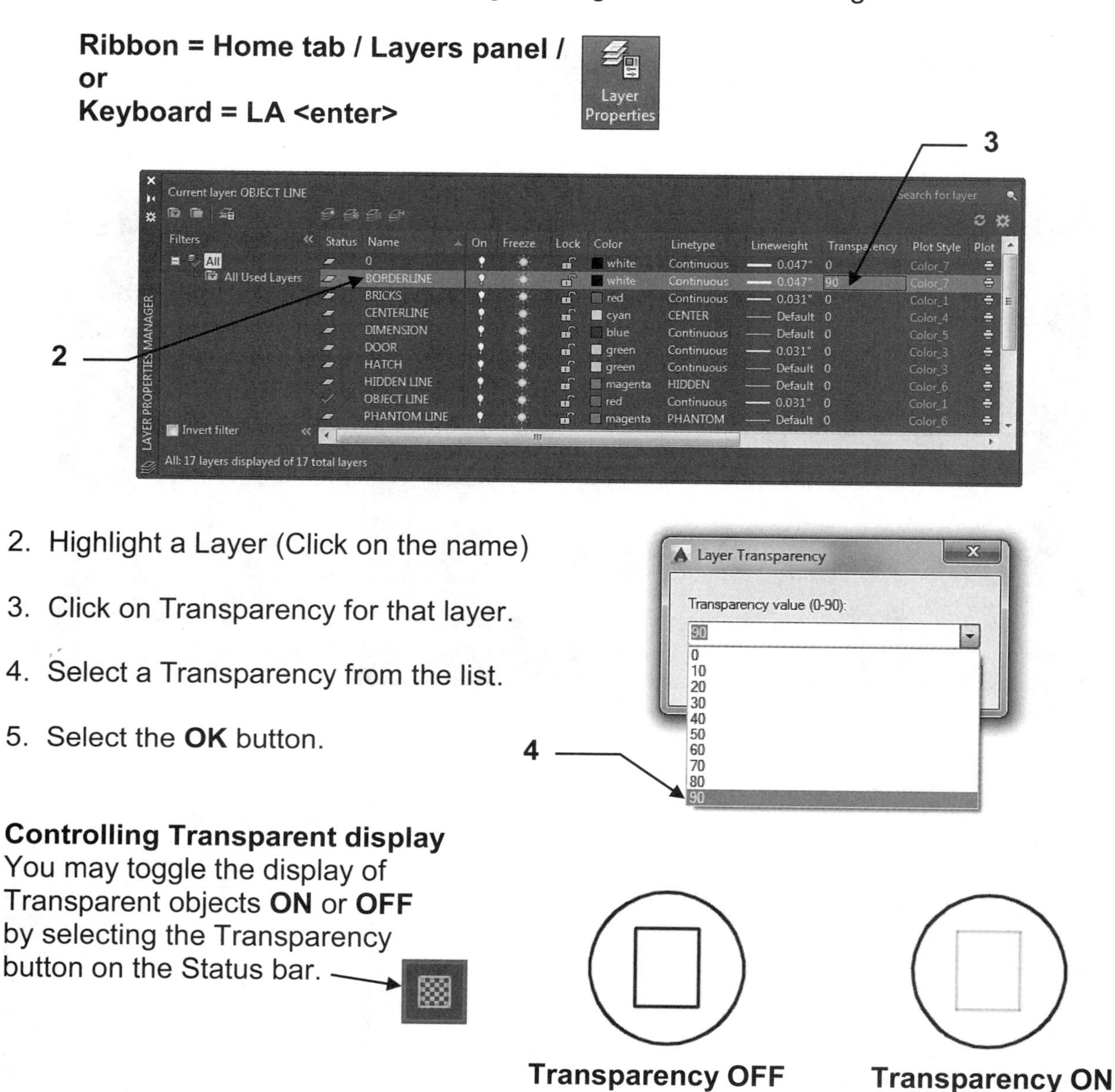

Ribbon = Home tab / Layers panel /
or
Keyboard = LA <enter>

2. Highlight a Layer (Click on the name)

3. Click on Transparency for that layer.

4. Select a Transparency from the list.

5. Select the **OK** button.

Controlling Transparent display
You may toggle the display of Transparent objects **ON** or **OFF** by selecting the Transparency button on the Status bar.

Note: Transparency selections will be saved within the **current** drawing and will not effect any other drawing.

Plotting Transparent Objects
Plotting transparency is disabled by default. To plot transparent objects, check the Plot transparency option in either the Plot dialog box or the Page Setup dialog box.
This will be discussed in Lesson 26

CREATING NEW LAYERS

Using layers is an important part of managing and controlling your drawing. It is better to have too many layers than too few. You should draw like objects on the same layer. For example, place all doors on the layer "door" or centerlines on the layer "centerline".

When you create a new layer you will assign a **name, color, linetype, lineweight, transparency** and whether or not it should **plot**.

1. Select the Layer command using one of the following:

    Ribbon = Home tab / Layers panel /
    or
    Keyboard = LA <enter>

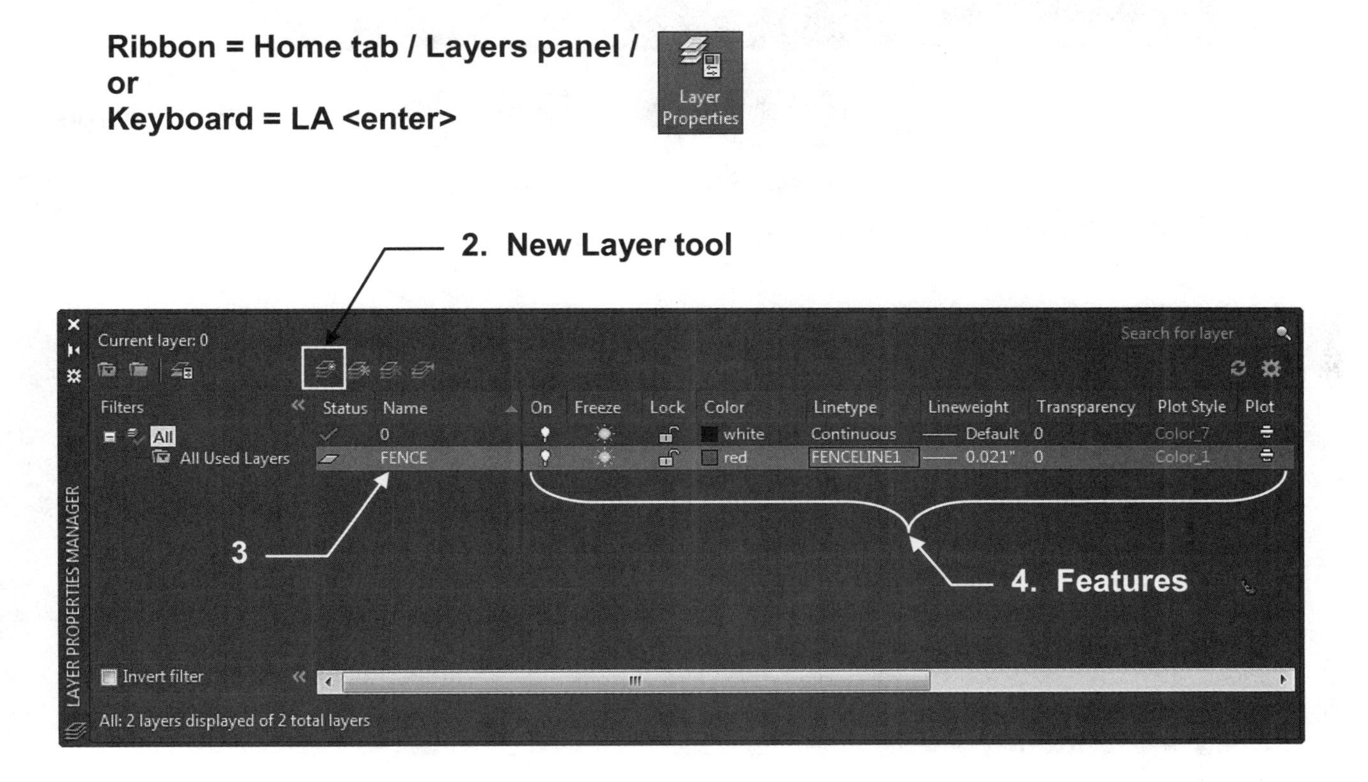

2. Select the **New Layer tool** and a new layer will appear.

3. Type the new layer **name** and press <enter>

4. Select any of the **features** and a dialog box will appear.

Features:
Refer to the previous pages for controlling and selecting color, lineweights and transparency.

Refer to the next page for Linetype.

LOADING and SELECTING LAYER LINETYPES

In an effort to conserve data within a drawing file, AutoCAD automatically loads only one linetype called “**continuous**”. If you would like to use other **linetypes**, such as “dashed” or “fenceline”, you must **Load** them into the drawing as follows:

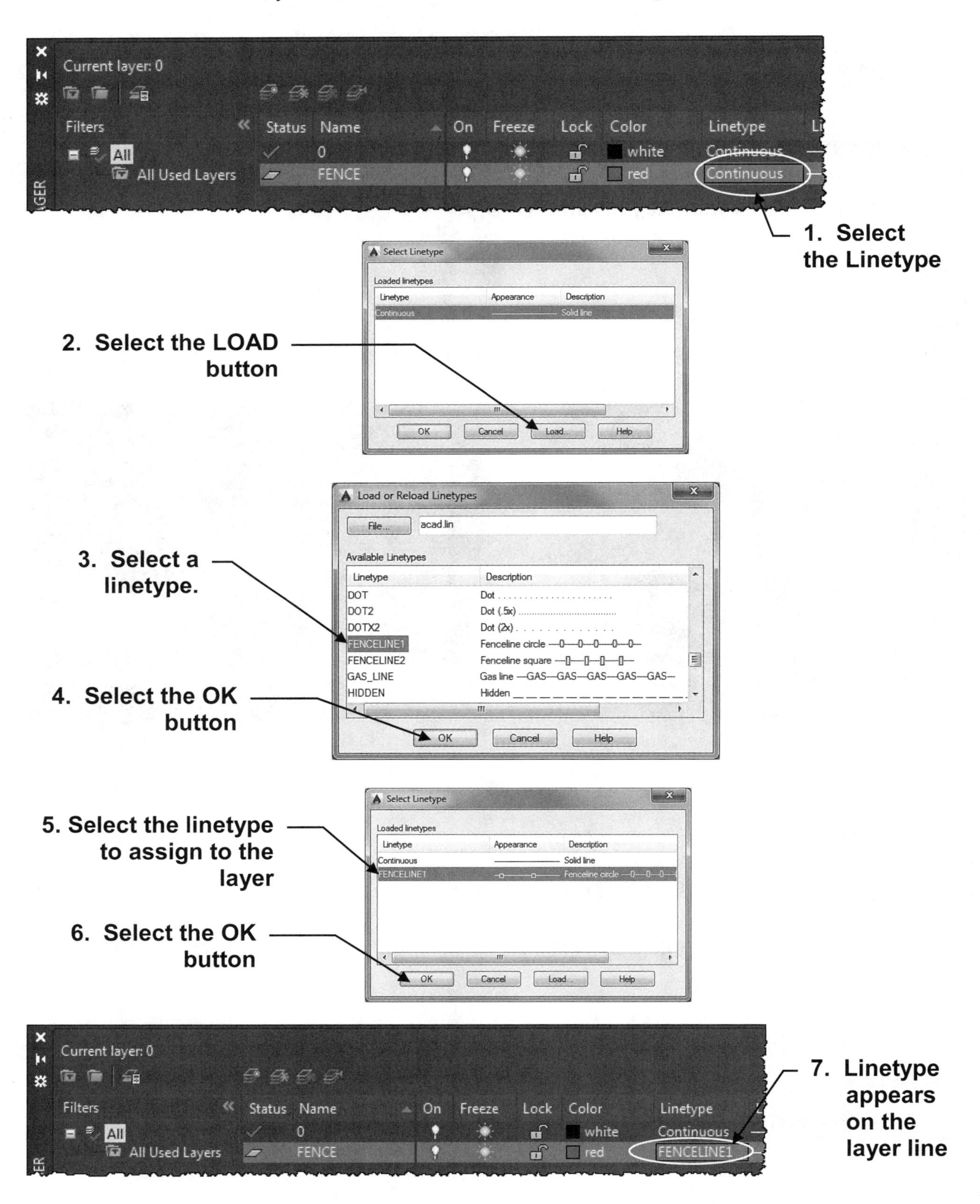

EXERCISE 3A

INSTRUCTIONS:

1. Start a **NEW** file using **2015-Workbook Helper.dwt.**
2. **Draw** the **LINES** below using:
    A. **Line** command
    B. **Ortho** (F8) **ON**.
    C. Turn **Increment Snap** (F9) **ON**
3. Select the appropriate layer, then select the Line command, then draw a line.
4. Select the appropriate layer, then select the Line command, then draw a line.
5. Select the appropriate layer, then select the Line command, then draw a line.

    Are you noticing a pattern here?
6. **Save** the drawing as: **EX3A**

Do not enter text.

You will learn Text soon.

BORDERLINE

CENTERLINE

DIMENSION

DOOR

HATCH

HIDDEN LINE

OBJECT LINE

PHANTOM LINE

PLANTS

ROOF

SECTION LINE

TEXT

WALL

WINDOW

EXERCISE 3B

INSTRUCTIONS:

1. Start a **NEW** file using **2015-Workbook Helper.dwt.**
2. Change the **Increment Snap to .20** and **Grid spacing to .40**

   If you have the Snap and Grids set correctly it will be easy.
3. Make sure **SNAP** and **GRID** status bar buttons are **ON**.
4. Draw the objects below using **layer Object Line**.

   Review pages 3-2 through 3-6 for Circle and Rectangle options.
5. **Do not dimension**, you will learn dimensioning soon.
6. **Save** the drawing as: **EX3B**

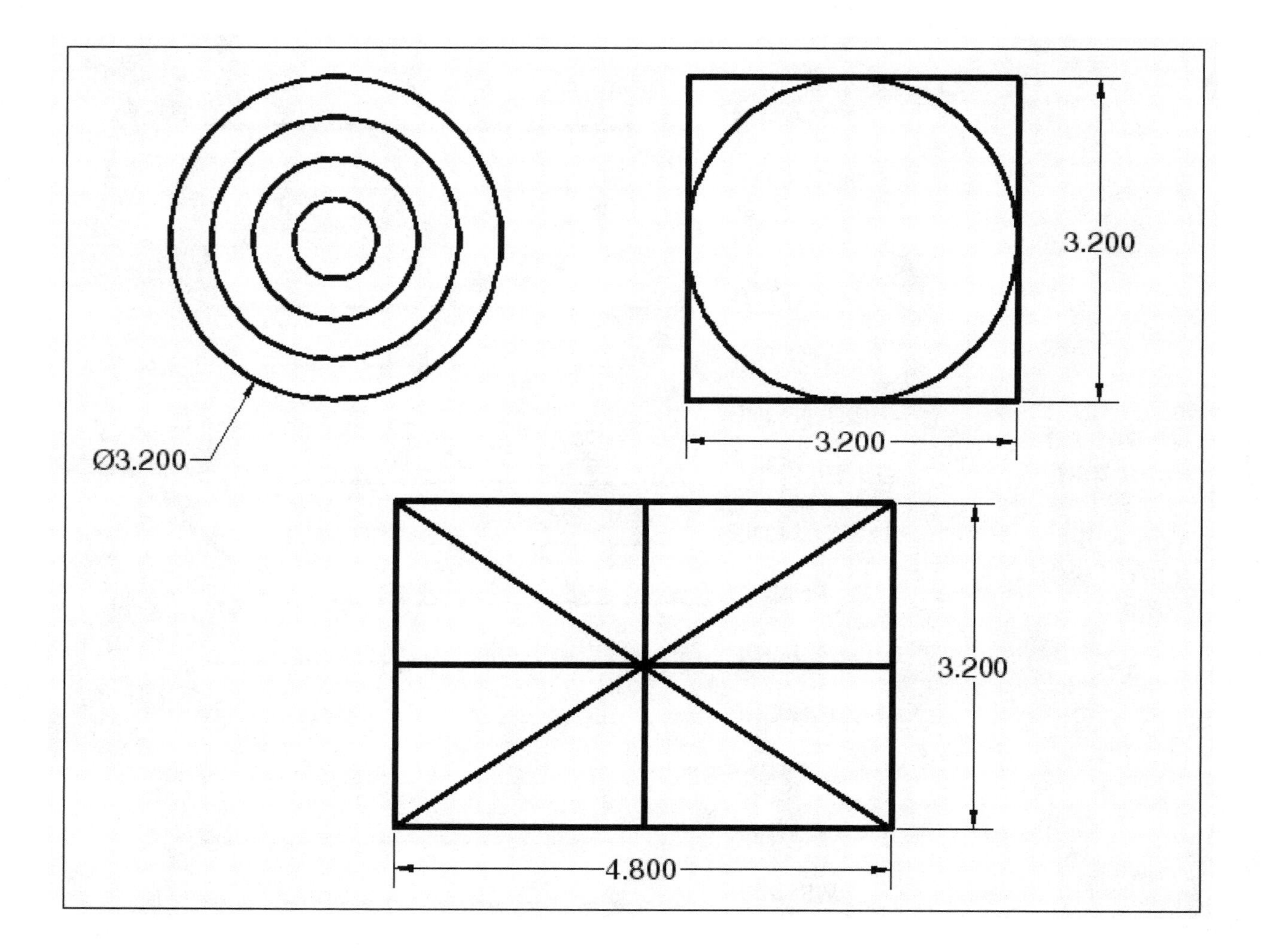

EXERCISE 3C

INSTRUCTIONS:

1. Start a **NEW** file using **2015-Workbook Helper.dwt.**
2. **Draw** the **Rectangles** below using the options:

   Dimension, Chamfer, Fillet, Width and Rotation
3. Use layer **Object Line.**
4. **Save** the drawing as: **EX3C**

Rectangle Dimensions: Length = 3 Width = 2

Chamfer = Dist1 = .50 Dist2 = .50

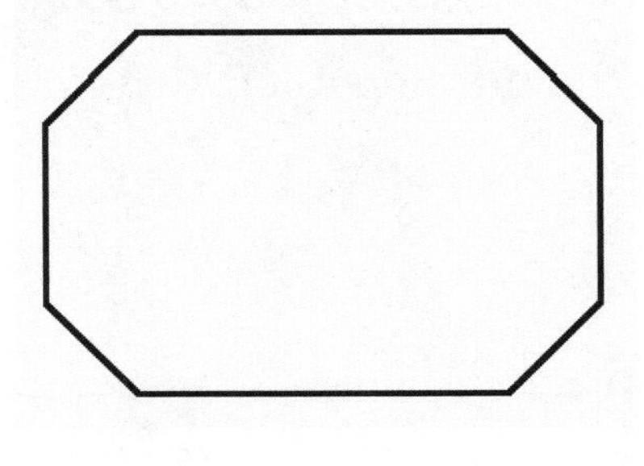

Rectangle Dimensions: Length = 3 Width = 2

Fillet = .50

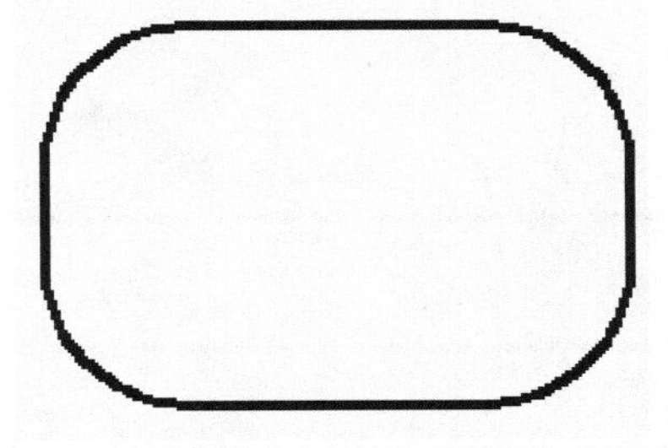

Rectangle Dimensions: Length = 3 Width = 2

Rotation = 45

Chamfer = 0

Fillet = 0

Rectangle Dimensions: Length = 3 Width = 2

Width = .200

Rotation = 0

Chamfer = 0

Fillet = 0

EXERCISE 3D

INSTRUCTIONS:

1. Start a **NEW** file using **2015-Workbook Helper.dwt.**
2. **Important, important: Draw** the objects below using the following Layers:

   Roof, Wall, Window, Door and Plants
3. You can change Snap and Grid settings to whatever you like.
4. You decide when to turn Ortho and Snap ON or OFF.
5. All objects must be placed accurately. All lines must intersect exactly.
6. **Save** the drawing as: **EX3D**

EXERCISE 3E

INSTRUCTIONS:

1. Open **EX3D.** (Use OPEN not NEW to open an existing drawing)
2. Make layer **Object** "current"

   (I am having you do that because you can't Freeze or Lock a layer that is current and I want to make sure that none of the layers that you used in the drawing is current.)
3. **Freeze** the following layers: Window and Plants (Do not use erase)
4. **Lock** layer Roof
5. Try to erase any of the roof lines.

   You can't because the Roof layer is Locked.
6. **Save** the drawing as: **EX3E**

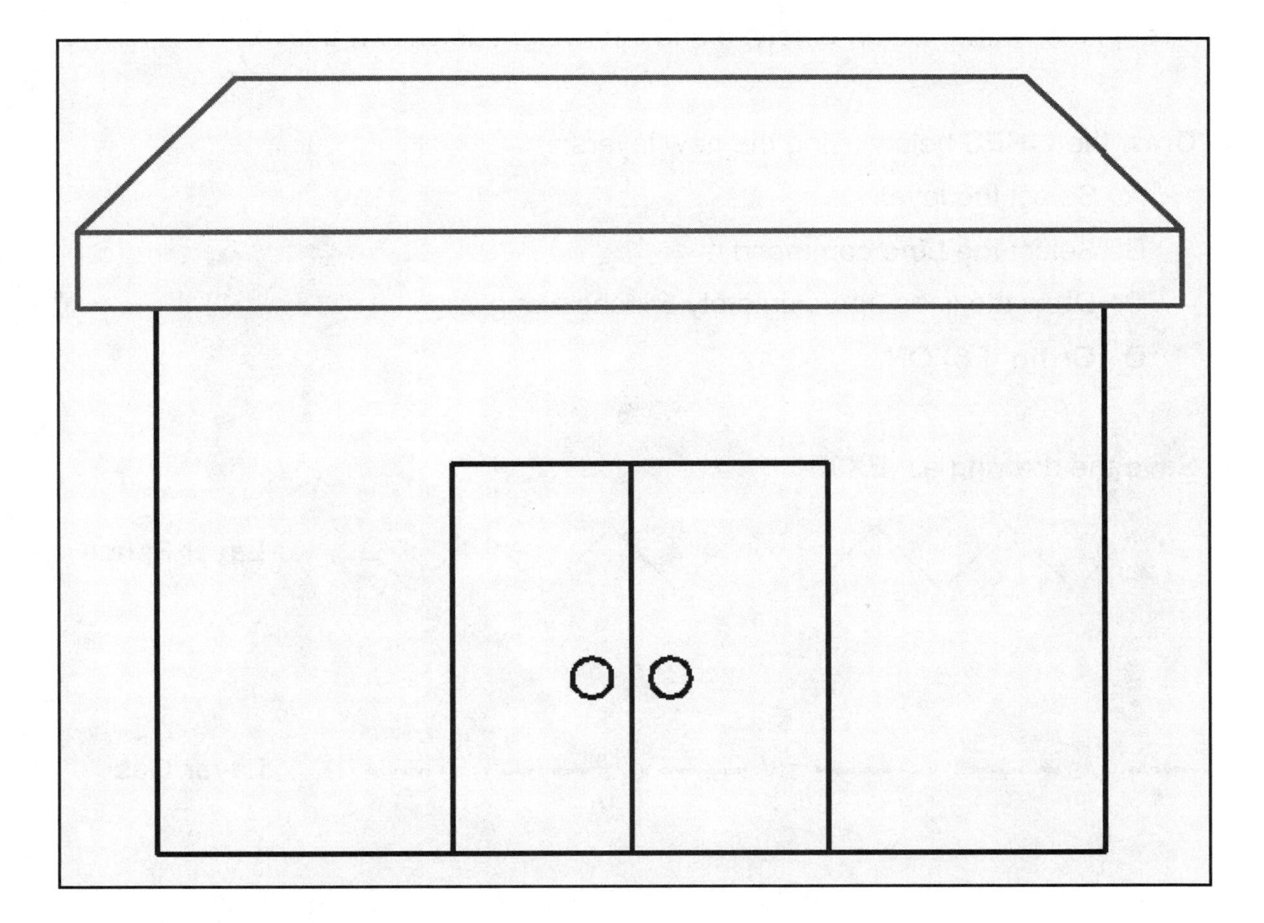

EXERCISE 3F

INSTRUCTIONS:

1. Start a new file using template **Acad** **(do not select Acad3D)**
   (Located in the Template list 3 names below your 2015-Workbook Helper template)
2. Load Linetypes **Zigzag, Gas Line and Dashed** (Refer to page 3-16)

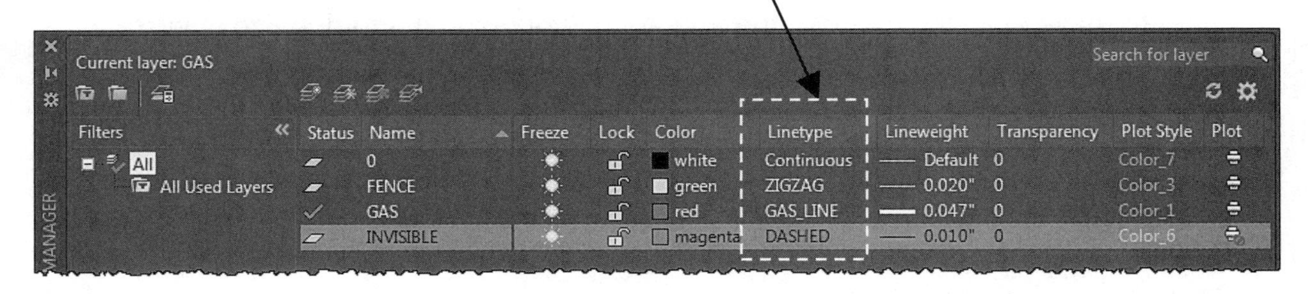

3. Create 3 layers named **Fence, Gas and Invisible.** (Refer to page 3-15)
   Assign the Name, Color, Linetype and Lineweight as shown above.

4. **Draw** the **LINES** below using the new layers:
   A. Select the layer
   B. Select the **Line** command
   C. Draw the lines approximately 5 inches long.
   D. **Ortho** (F8) **ON**.

5. **Save** the drawing as: **EX3F**

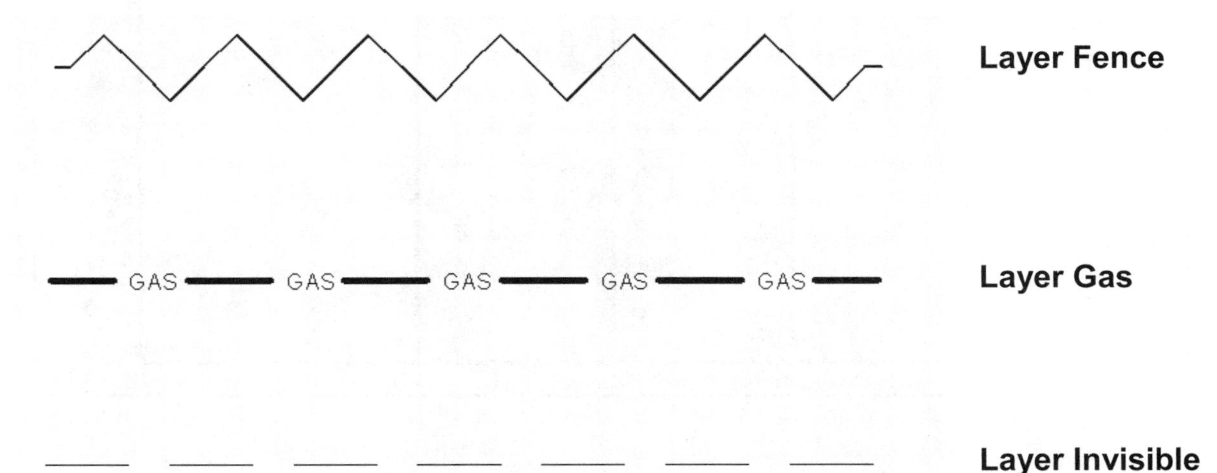

EXERCISE 3G

INSTRUCTIONS:

1. **Open EX3B**
2. Open the **Layer Properties Manager** (Refer to page 3-10)
3. Change Transparency of the Object Line layer to **70** (Refer to Page 3-14)

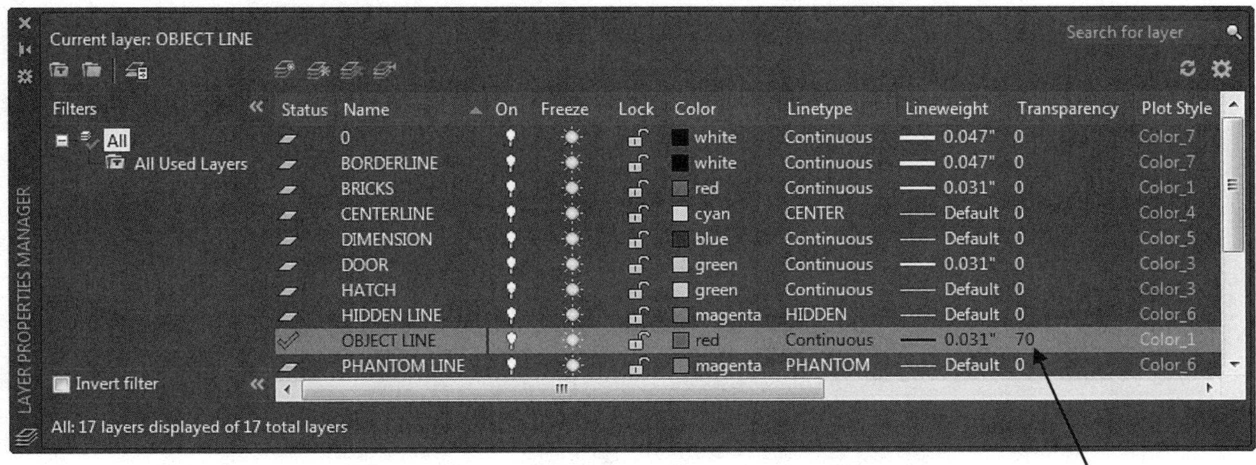

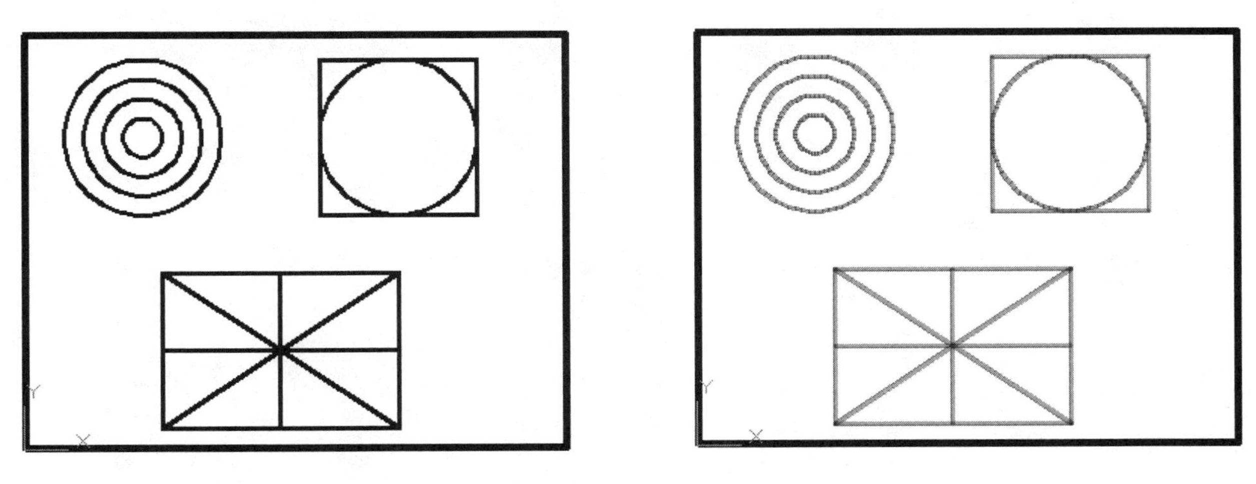

Transparency = OFF

Transparency = ON

4. Select the **Transparency** status button to show or hide Transparency.
5. **Save** the drawing as: **EX3G**

Notes:

LEARNING OBJECTIVES

After completing this lesson, you will be able to:

1. Understand Object Snap
2. Use Running Object Snap
3. Use the Zoom options to view the drawing
4. Change the Drawing Limits
5. Select the Units of Measurement and Precision

LESSON 4

OBJECT SNAP

In Lesson 3 you learned about Increment Snap. Increment Snap enables the cursor to move in an incremental movement. So you could say your cursor is "snapping to increments" preset by you.

Now you will learn about Object Snap. If Increment Snap snaps to increments, what do you think Object Snap snaps to? That's right; "objects". Object snap enables you to snap to "objects" in very specific and accurate locations on the objects.
For example, the endpoint of a line or the center of a circle.

How to select from the Object Snap Menu

1. You must select a command, such as LINE, before you can select Object Snap.

2. While holding down the shift key, press the right mouse button The menu shown below should appear.

3. Highlight and press left mouse button to select.

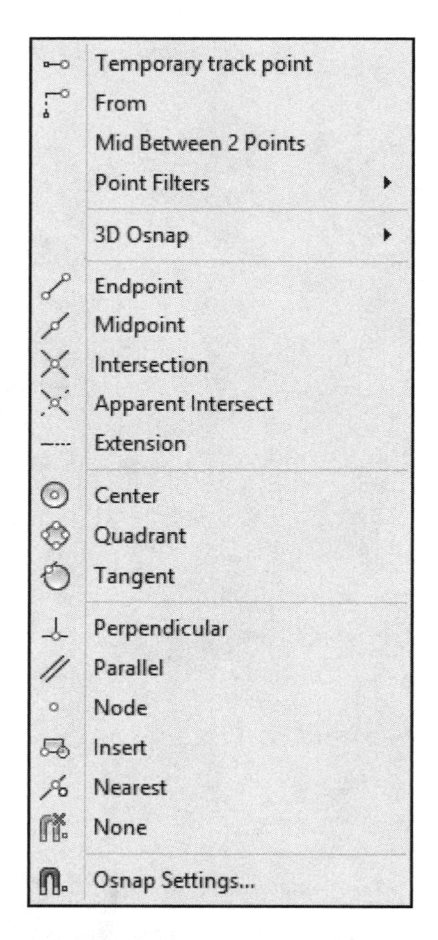

The following object snaps will be discussed in this Lesson:
Endpoint, Midpoint, Intersection, Center, Quadrant, Tangent and Perpendicular.
Refer to their descriptions on the next page.

The remaining will be discussed in future lessons.

OBJECT SNAP....continued

Object Snap Definitions
Object snap is used when AutoCAD prompts you to place an object. Object snap allows you to place objects very accurately.

<u>A step by step example of "How to use object snap" is shown on the next page</u>.

Note: You may type the **3 bold letters** shown rather than select from the menu.

ENDpoint — Snaps to the closest endpoint of a Line, Arc or polygon segment. Place the cursor on the object close to the end and the cursor will snap like a magnet to the end of the line.

MIDpoint — Snaps to the middle of a Line, Arc or Polygon segment. Place the cursor anywhere on the object and the cursor will snap like a magnet to the midpoint of the line.

INTersection — Snaps to the intersections of any two objects. Place the cursor directly on top of the intersection or select one object and then the other and Autocad will locate the intersection.

CENter — Snaps to the center of an Arc, Circle or Donut. Place the cursor on the object, or at the approximate center location and the cursor will snap like a magnet to the center.

QUAdrant — Snaps to a 12:00, 3:00, 6:00 or 9:00 o'clock location on a circle or ellipse. Place the cursor on the circle near the desired quadrant location and the cursor will snap to the closest quadrant.

TANgent — Calculates the tangent point of an Arc or Circle. Place the cursor on the object as near as possible to the expected tangent point.

PERpendicular — Snaps to a point perpendicular to the object selected. Place the cursor anywhere on the object then pull the cursor away from object and press the left mouse button.

How to use OBJECT SNAP

The following is an example of attaching a line segment to previously drawn vertical lines. The new line will start from the upper endpoint, to the midpoint, to the lower endpoint.

1. Turn Off **SNAPMODE, ORTHOMODE** and **OBJECT SNAP** on the Status Bar. (Gray is OFF)

2. Select the **Line** command.

3. Draw two vertical lines as shown below (they don't have to be perfectly straight)

4. Select the **Line** command again.

5. Hold the shift key down and press the right mouse button.

6. Select the Object snap **Endpoint** from the object snap menu.

7. Place the cursor close to the upper endpoint of the left hand line.

   The cursor should snap to the end of the line like a magnet. A little square and an "Endpoint" tooltip are displayed.

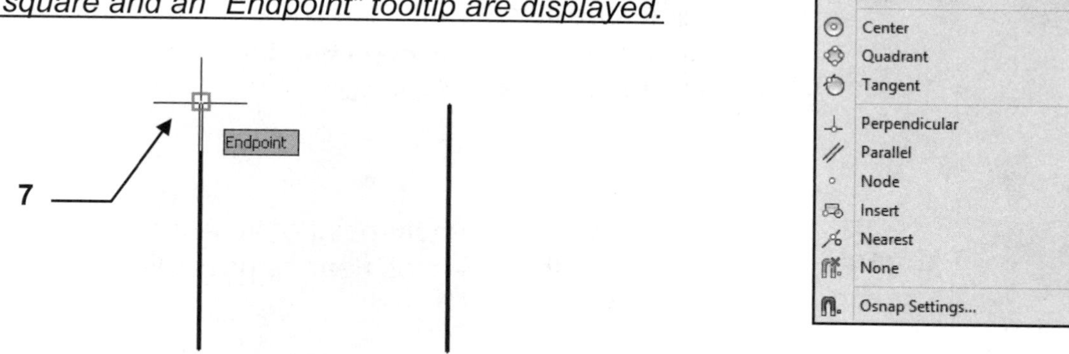

8. Press the left mouse button to attach the new line to the upper endpoint of the previously drawn vertical line. (Do not end the Line command yet.)

Continued on the next page...

How to use OBJECT SNAP....continued

9. Now hold the shift key down and press the right mouse button and select the **Midpoint** object snap option.

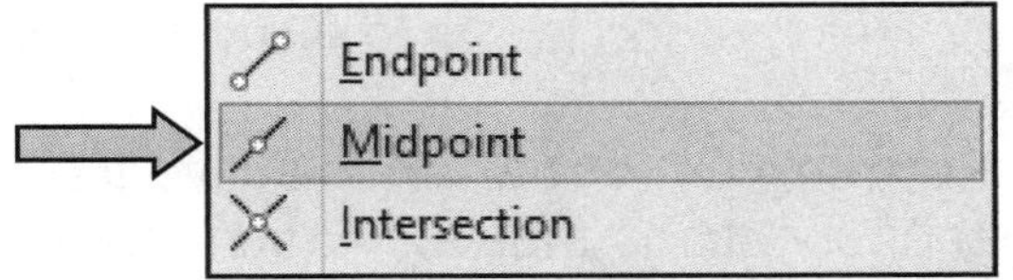

10. Move the cursor to approximately the middle of the right hand vertical line.

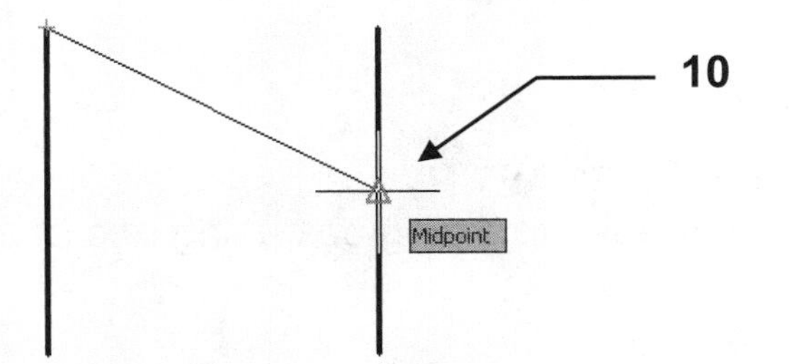

The cursor should snap to the midpoint of the line like a magnet.
A little triangle with a "Midpoint" tool tip are displayed.

11. Press the left mouse button to attach the new line to the midpoint of the previously drawn vertical line. (Do not end the Line command yet.)

12. Now hold the shift key down and press the right mouse button and select the object snap **Endpoint** again.

13. Move the cursor close to the lower endpoint of the left hand vertical line.

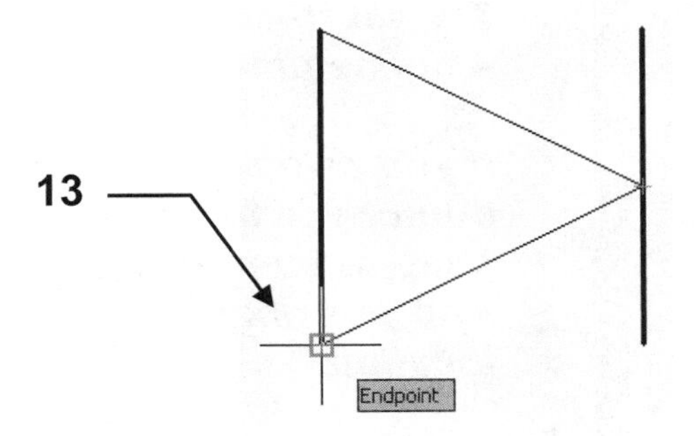

The cursor should snap to the end of the line like a magnet.
A little square and a tooltip are displayed.

14. Press the left mouse button to attach the new line to the endpoint of the previously drawn vertical line.

15. Stop the Line command and disconnect by pressing **<enter>.**

RUNNING OBJECT SNAP

Selecting Object Snap is not difficult but AutoCAD has provided you with an additional method to increase your efficiency by allowing you to preset frequently used object snap options. This method is called **RUNNING OBJECT SNAP.**

When **Running Object Snap** is active the cursor will automatically snap to any preset object snap locations thus eliminating the necessity of invoking the object snap menu for each locations.

First you must set the running object snaps, and second, you must turn ON the Running Object Snap option.

SETTING RUNNING OBJECT SNAP

1. Select the **Running Object Snap** dialog box using one of the following:

   Keyboard = DS <enter>
   or
   Status Bar = Left Click on OBJECT SNAP button down arrow ▾ and select Object Snap Settings.

2. Select the **Object Snap** tab.

3. Select the Object Snaps desired.
   (In the example below object snap ***Endpoint***, ***Midpoint*** *and* ***Intersection*** have been selected.

Note:
Try not to select more than 3 or 4 at one time.

If you select too many the cursor will flit around trying to snap to multiple snap locations. And possibly snap to the wrong location.

You will lose control and it will confuse you.

4. Select the **OK** button.

5. Turn ON the **OBJECT SNAP** button on the Status Bar. (blue is ON)

RUNNING OBJECT SNAP....continued

6. Now try drawing the line from the endpoint to the midpoint again **but this time do not select the "Object Snap Menu".** Just move the cursor close to the endpoint and the cursor will automatically snap to the end of the line.

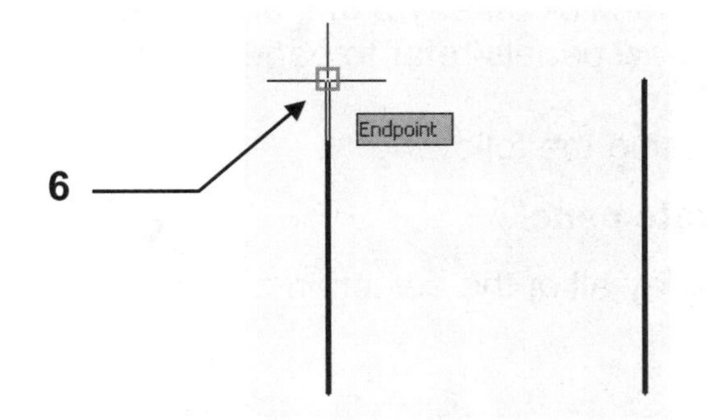

7. Move the cursor to approximately the middle of the right hand vertical line and the cursor will automatically snap to the midpoint of the line.

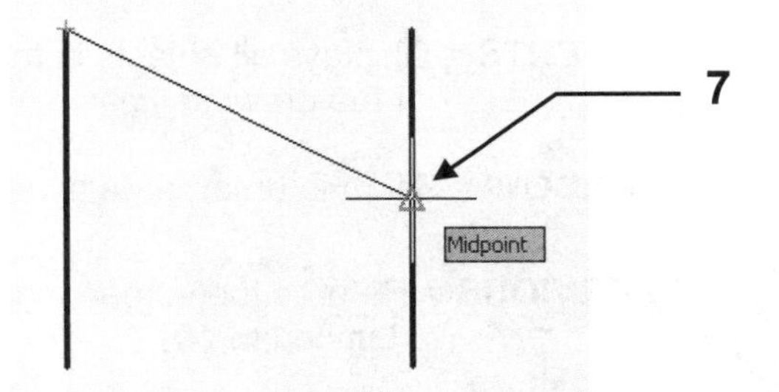

8. Move the cursor close to the lower endpoint of the left hand vertical line and the cursor will automatically snap to the lower endpoint of the left line.

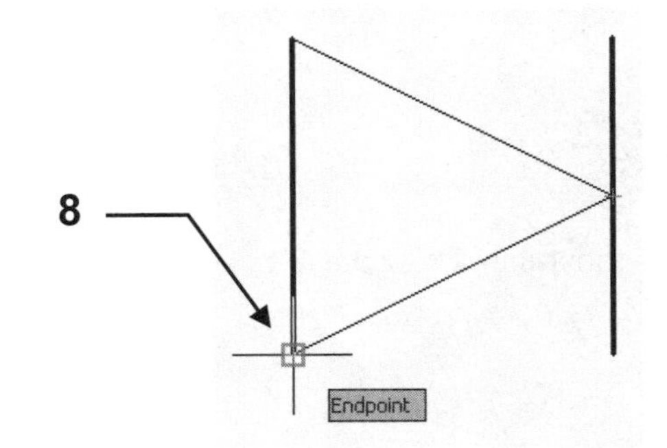

Running Object snap is very handy but remember do not select more than 3 or 4 at a time. The selections will fight each other and you may end up snapping to a location that you did not want.
If you wish to snap to a location that is not preset merely select the Object Snap Menu, as shown on page 4-2, and select the one you want. Running Object Snap and Object snap work together very well but it may take a little practice.

ZOOM

The **ZOOM** command is used to move closer to or farther away from an object. This is called Zooming In and Out.

The Zoom commands are located on the **Navigate** panel of the **View** tab and are **<u>off</u>** by default. Select the **View** tab then right-click on any panel and select **Show Panels**, activate the **Navigate** panel. (To show panels refer to page 1-11)

1. Select the Zoom command by using the following:

    Ribbon = View tab / Navigate panel

2. Select the ▾ down arrow to display all of the selections.

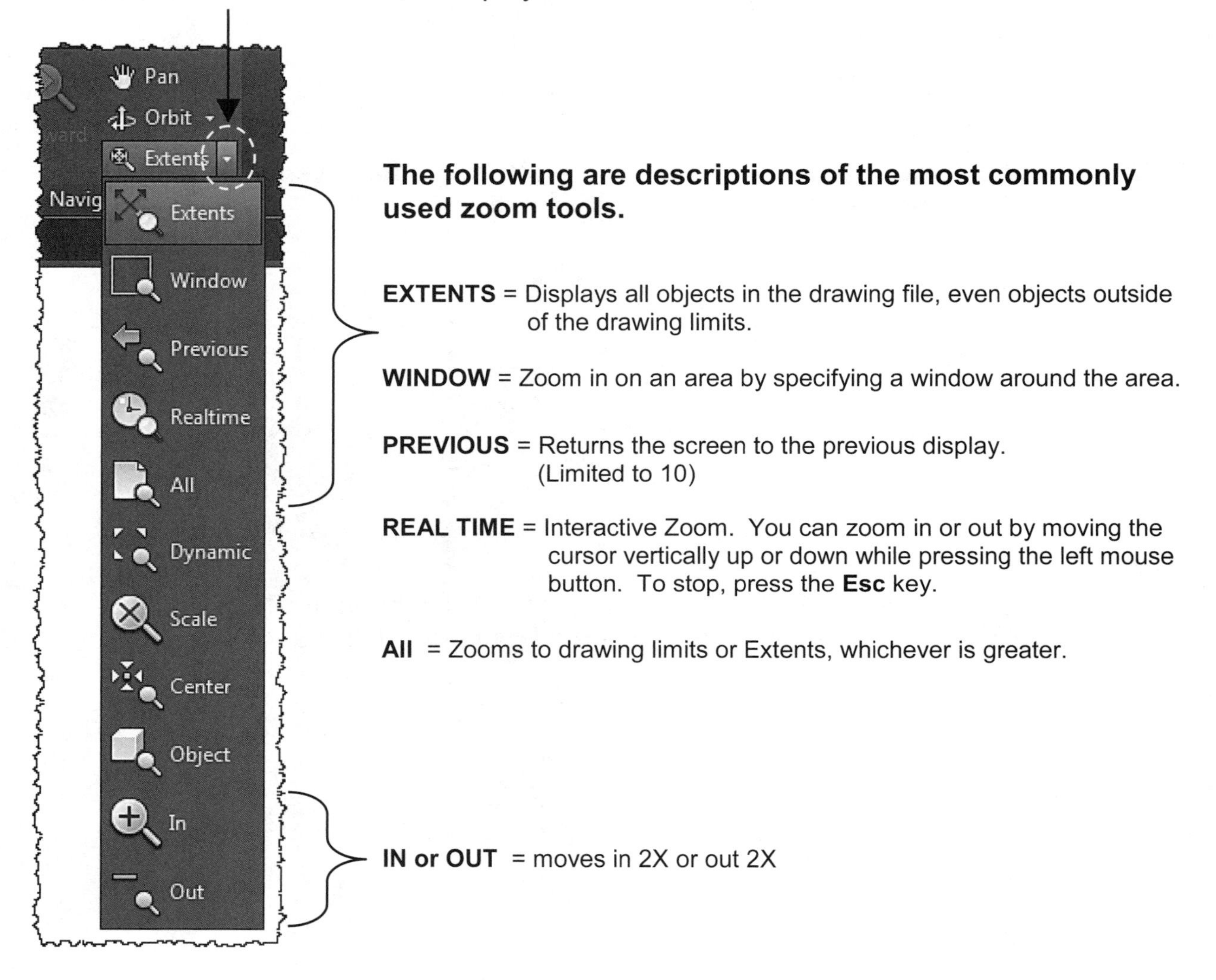

The following are descriptions of the most commonly used zoom tools.

EXTENTS = Displays all objects in the drawing file, even objects outside of the drawing limits.

WINDOW = Zoom in on an area by specifying a window around the area.

PREVIOUS = Returns the screen to the previous display. (Limited to 10)

REAL TIME = Interactive Zoom. You can zoom in or out by moving the cursor vertically up or down while pressing the left mouse button. To stop, press the **Esc** key.

All = Zooms to drawing limits or Extents, whichever is greater.

IN or OUT = moves in 2X or out 2X

You may also select the Zoom commands using one of the following:

Right Click and select Zoom from the Short cut menu.
(Refer to Intro-5 for “right-click” settings)

Keyboard = Z <enter> Select from the options listed..

ZOOM....continued

How to use ZOOM / WINDOW

1. Select Zoom / Window (Refer to previous page)

2. Create a window around the objects you want to enlarge.
   (Creating a "window" is a similar process to drawing a rectangle. It requires a first corner and then diagonal corner)

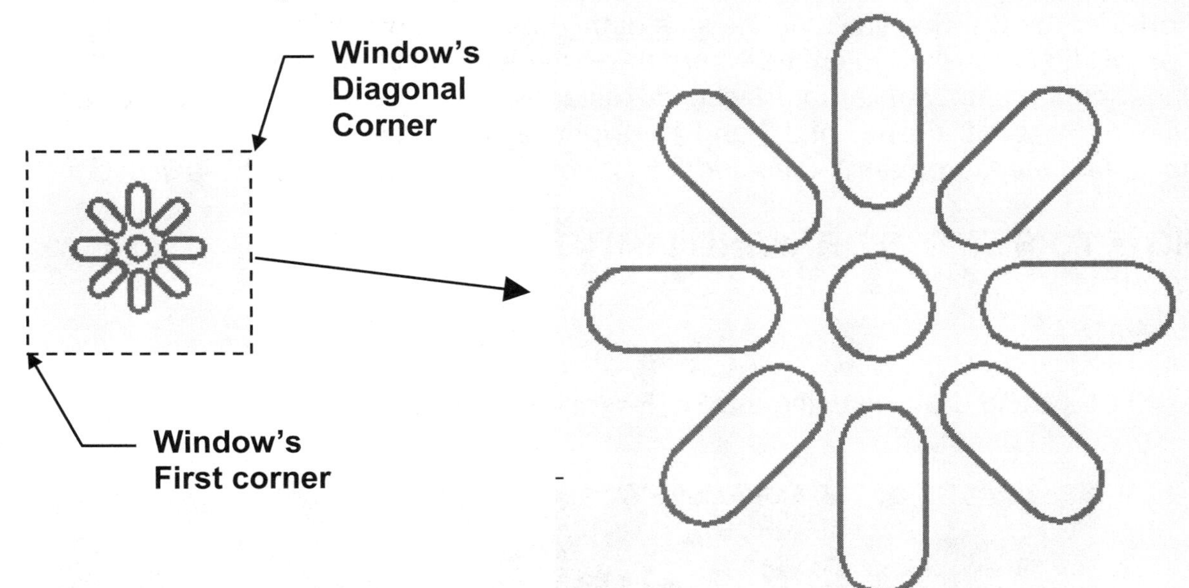

Magnified to this view
Note: the objects have been magnified. But the actual size has not changed.

How to return to Original View

1. Type: **Z** <enter> **A** <enter> (This is a shortcut for Zoom / All)

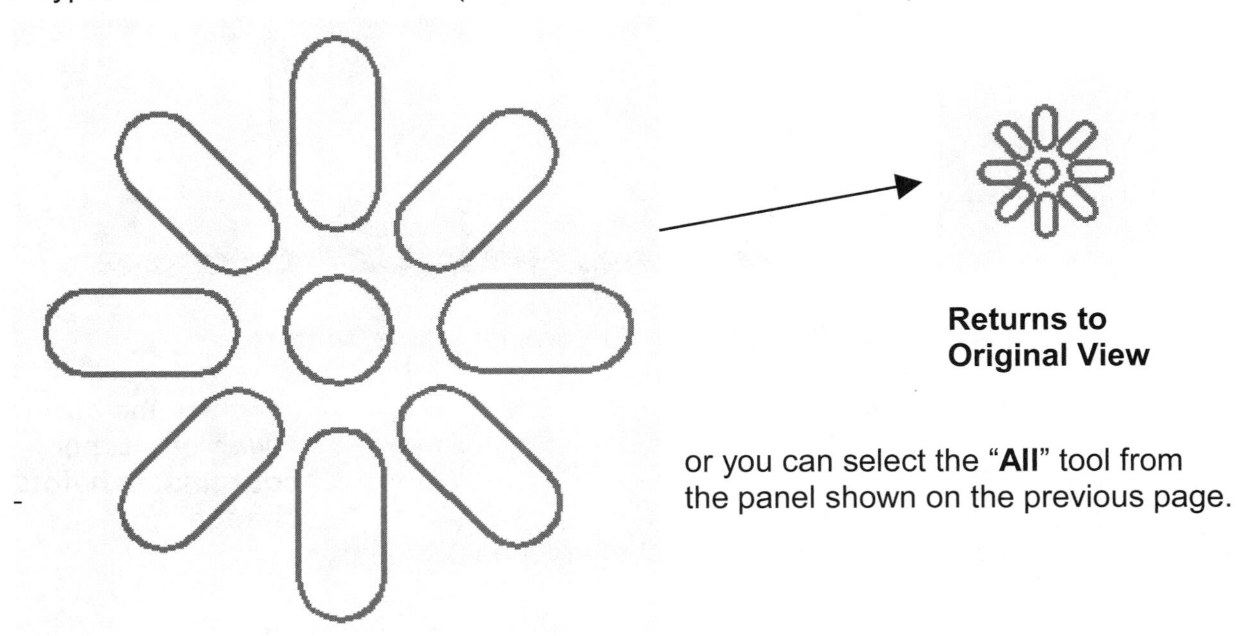

or you can select the "**All**" tool from the panel shown on the previous page.

DRAWING SETUP

When drawing with a computer, you must "set up your drawing area" just as you would on your drawing board if you were drawing with pencil and paper. You must decide what size paper you will need, what Units of measurement you will use (feet and inches or decimals, etc) and how precise you need to be. In CAD these decisions are called "Setting the **Drawing Limits, Units** and **Precision".**

DRAWING LIMITS

Consider the drawing limits as the size of the paper you will be drawing on.
You will first be asked to define where the lower left corner should be placed, then the upper right corner, similar to drawing a Rectangle. An 11 x 8.5 piece of paper would have a **lower left corner** of 0,0 and an **upper right corner** of 11, 8.5. *(11 is the horizontal measurement X-axis and 8.5 is the vertical measurement Y-axis.)*

HOW TO SET THE DRAWING LIMITS

EXAMPLE

1. Start a **NEW** drawing using the **2015-Workbook Helper** template. (Refer to page 2-6)

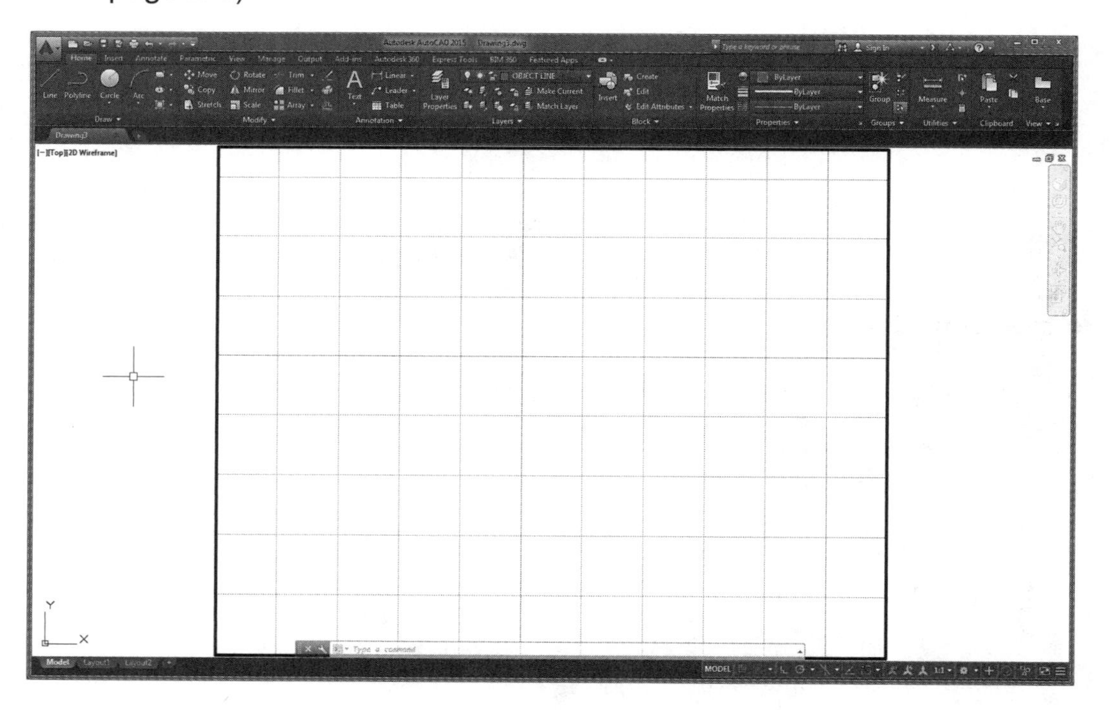

2. Select the **DRAWING LIMITS** command by typing: **Limits <enter>**

3. The following prompt will appear:

    Command: '_limits
    Reset Model space limits:
    Specify lower left corner or [ON/OFF] <0.0000,0.0000>:

    Displays the current lower left corner coordinates before change

Continued on the next page...

DRAWING SETUP....continued

4. Type the X,Y coordinates **0, 0 <enter>** for the new lower left corner location of your piece of paper .

5. The following prompt will appear:

   Specify upper right corner <12.0000, 9.0000>:

Displays the current upper right corner coordinates <u>before</u> change

6. Type the X,Y coordinates **36,24 <enter>** for the new upper right corner of your piece of paper.

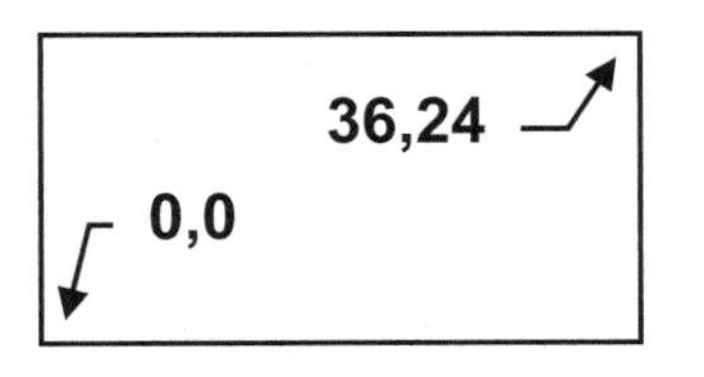

Note: visually the screen has not changed. Do the next step and it will.

7. **<u>This next step is very important</u>:**

   Type **Z <enter> A <enter>** to make the screen display the new drawing limits. (This is the shortcut for Zoom / All. Refer to page 4-9)

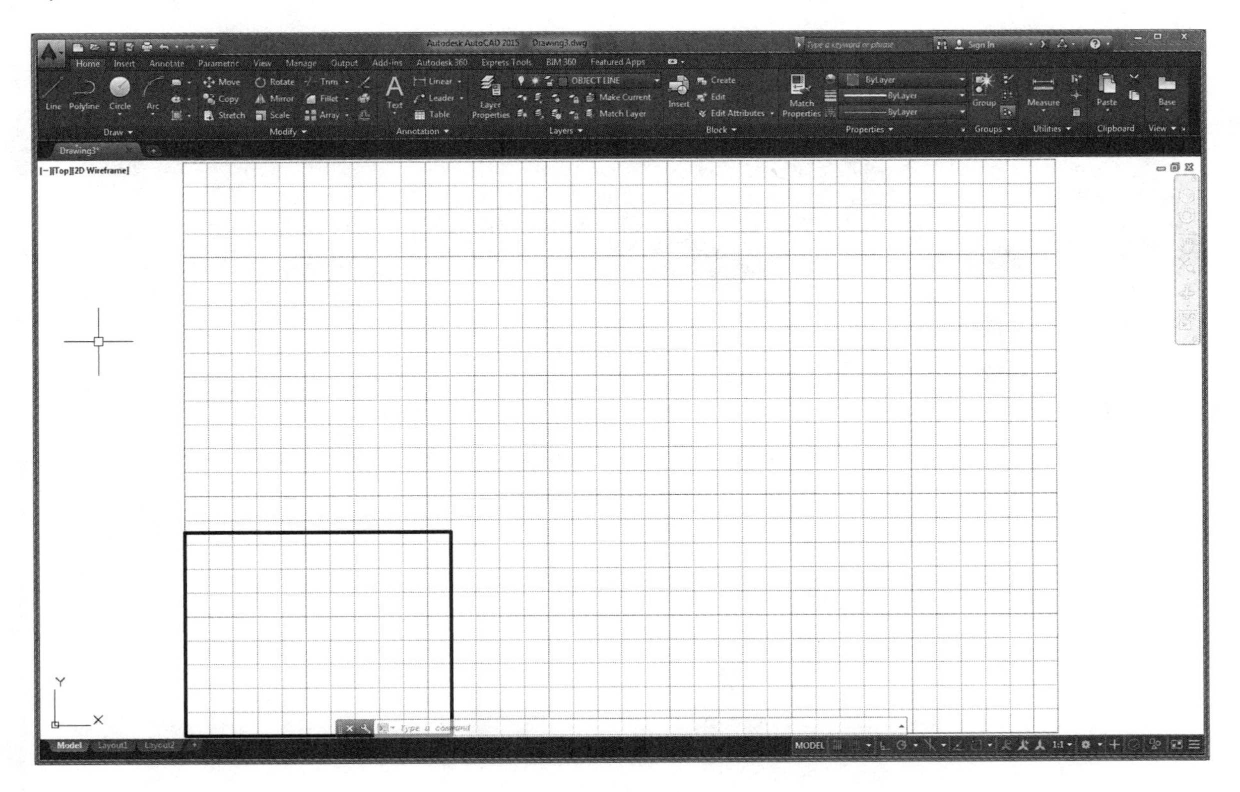

The drawing limits (area) is now 36 wide X 24 high. The rectangle did not change size or location. But the drawing area around it got larger so the rectangle appears to have moved and gotten smaller. Think about it.

Grids within Limits

If you have your **Grid behavior** setting ☐ **Display grid beyond Limits** is turned Off (no check mark) the grids will only be displayed within the Limits that you set.

Drafting Settings

Snap and Grid | Polar Tracking | Object Snap | 3D Object Snap | Dynamic Input | Quic

Snap On (F9)

Snap spacing

Snap X spacing: 0.250

Snap Y spacing: 0.250

Equal X and Y spacing

Polar spacing

Polar distance: 0.000

Snap type

Grid snap

Rectangular snap

Isometric snap

PolarSnap

Grid On (F7)

Grid style

Display dotted grid in:

2D model space

Block editor

Sheet/layout

Grid spacing

Grid X spacing: 1.000

Grid Y spacing: 1.000

Major line every: 5

Grid behavior

Adaptive grid

Allow subdivision below grid spacing

Display grid beyond Limits

Follow Dynamic UCS

Options... | OK | Cancel | Help

Off ☐

Grids displayed within Limits only.

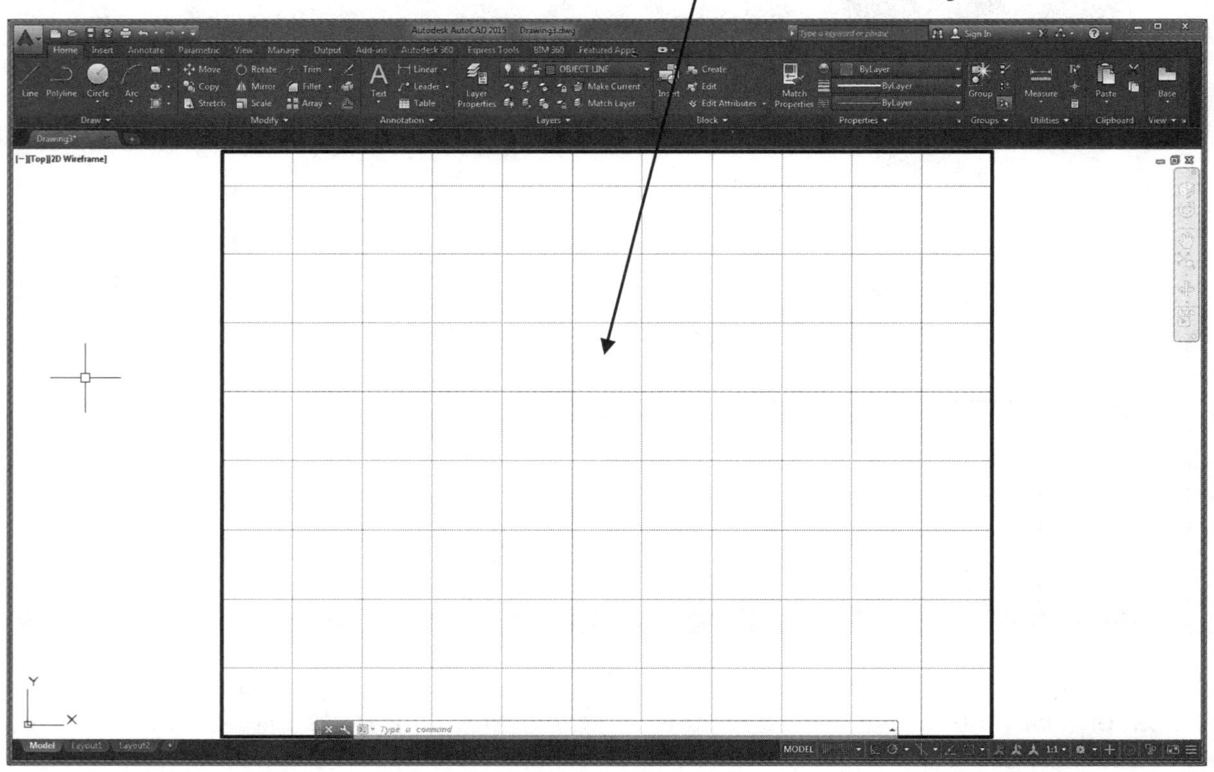

DRAWING SETUP....continued

Grids beyond Limits

If you have your **Grid behavior** setting ☑ **Display grid beyond Limits** is turned **ON** (check mark) the grids will be displayed beyond the Limits that you set.

ON ☑

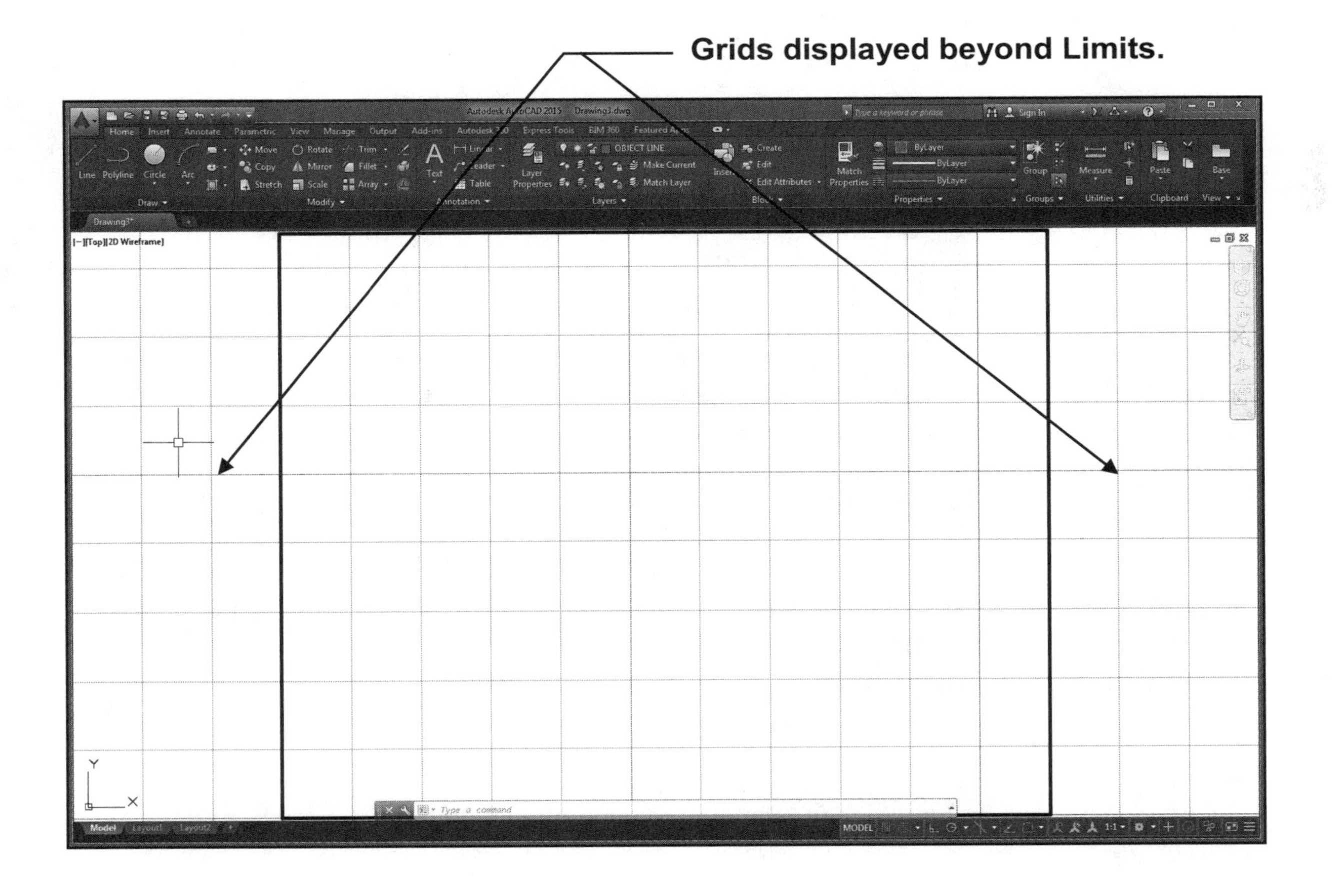

DRAWING SETUP....continued

UNITS AND PRECISION
You now need to select what **unit of measurement** with which you want to work.
Such as: Decimal (0.000) or Architectural (0'-0").
Next you should select how precise you want the measurements. For example, do you want the measurement limited to a 3 place decimal or the nearest 1/8".

HOW TO SET THE UNITS AND PRECISION.

1. Select the **UNITS** command using one of the following:

   Application Menu = Drawing Utilities / Units
   or
   Keyboard = Units <enter>

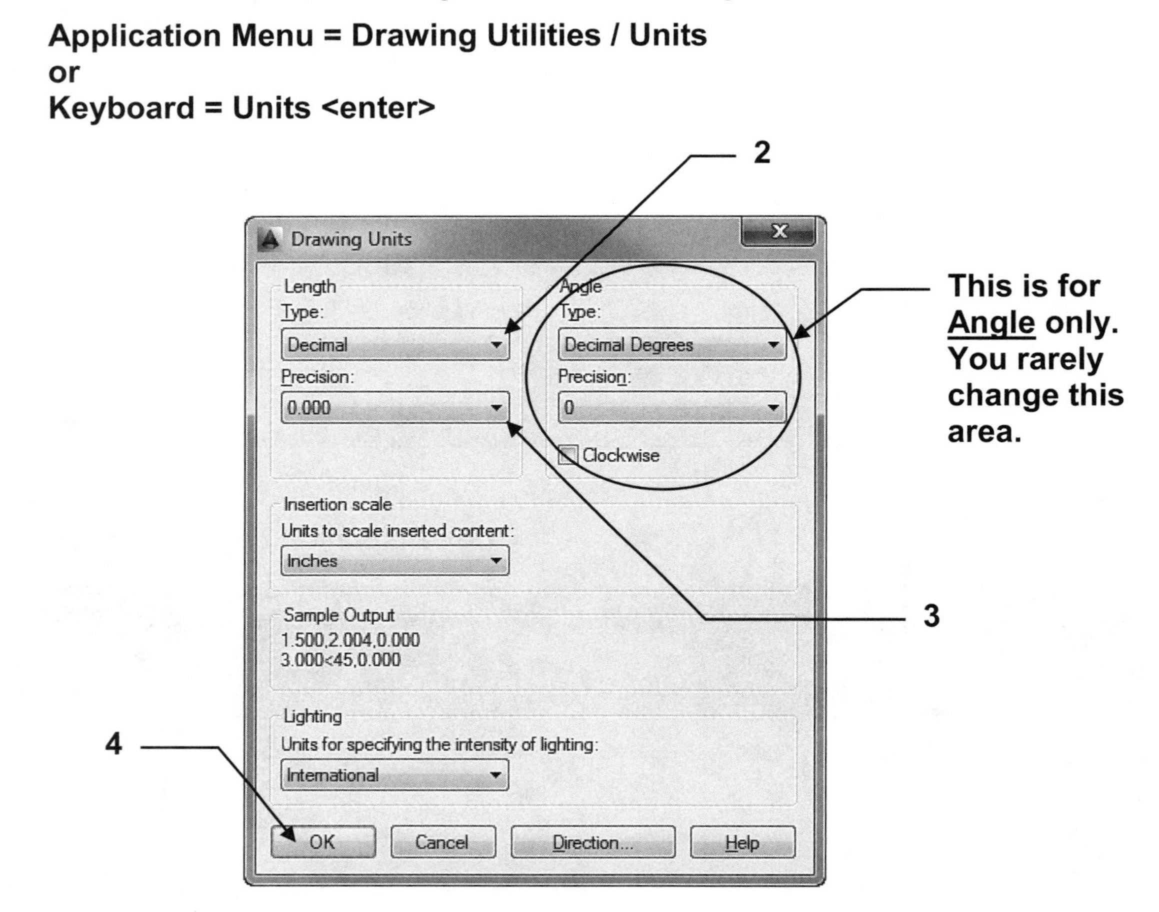

2. **Type:** Select the down arrow and select: **Decimals** or **Architectural**.

3. **Precision:** Select the down arrow and select the appropriate **Precision** associated with the "type".
   Examples: 0.000 for Decimals or 1/16" for Architectural.

4. Select the **OK** button to save your selections.

   Easy, yes?

EXERCISE 4A

INSTRUCTIONS:

1. Start a **NEW** file using **2015-Workbook Helper.dwt.**
2. Set **Units** and **Precision:**

   Units = Fractional **Precision =** 1/2”
3. Set **Drawing Limits:**

   Lower Left corner = 0,0 Upper Right Corner = 20, 15
4. Make sure you use **Zoom / All** after setting Drawing Limits
5. Erase the Rectangle that appears with the template, it will appear too small.
6. Turn **OFF** the **Snap** and **Ortho.**

   (Your cursor should move freely)
7. Draw the objects shown below using:

   Circle, center radius and Line (Use Layer = Object Line)

   Object snap = Center and Tangent
8. **Save** the drawing as: **EX4A**

Note: Use the Tangent object snap at each end of the line.

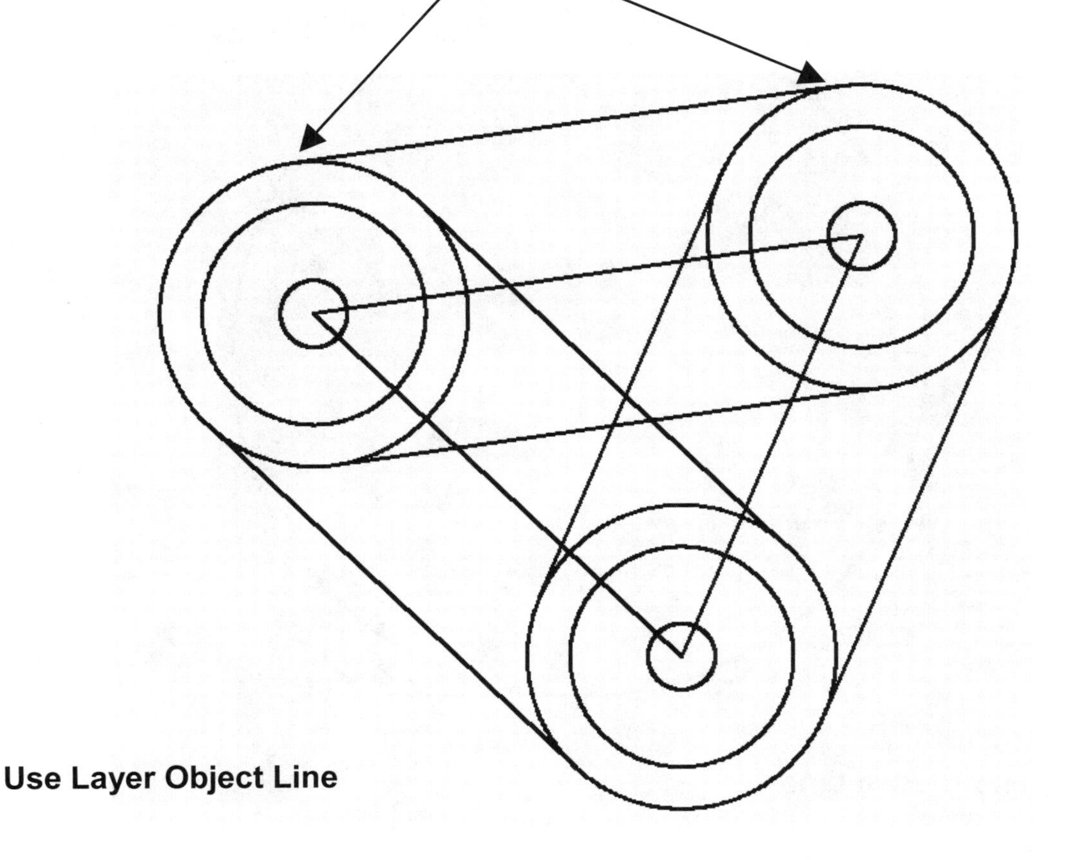

Use Layer Object Line

EXERCISE 4B

INSTRUCTIONS:

1. Start a **NEW** file using **2015-Workbook Helper.dwt.**
2. Set **Units** and **Precision:**

   Units = Fractional　　　　**Precision =** 1/4"
3. Set **Drawing Limits:**

   Lower Left corner = 0,0　　Upper Right Corner = 12, 9
4. Make sure you use **Zoom / All** after setting Drawing Limits
5. Turn **OFF** the **Snap** and **Ortho.**

   (Your cursor should move freely)
6. Draw the objects shown below using:

   Circle, center radius (Use Layer = Object Line)

   Line (Use Layer = Hidden Line)

   Object snap = Quadrant
7. **Save** the drawing as: **EX4B**

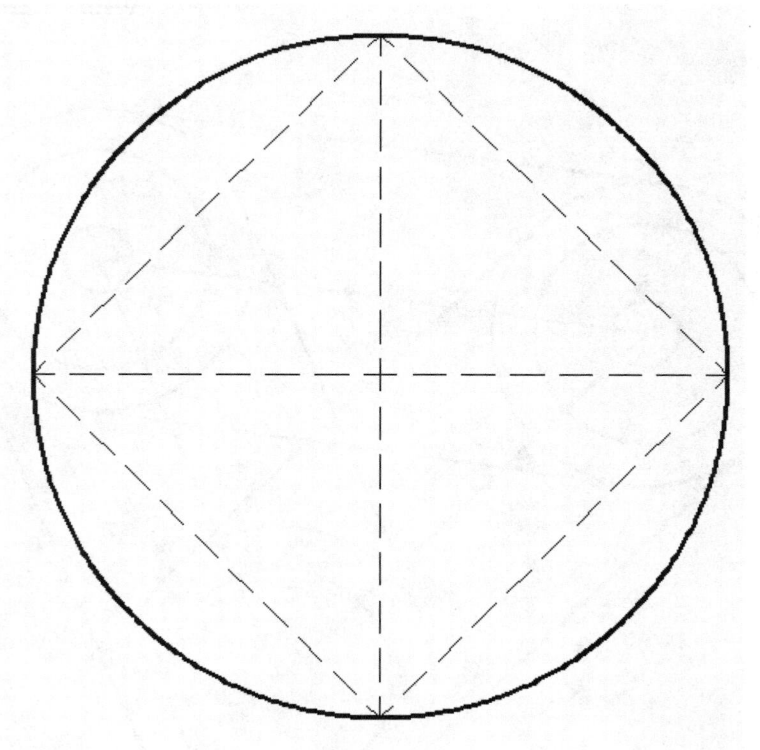

Use Layers:
Object Line and Hidden Line

EXERCISE 4C

INSTRUCTIONS:

1. Start a **NEW** file using **2015-Workbook Helper.dwt.**
2. Set **Units** and **Precision:**

   Units = Architectural **Precision =** 1/2"

   Note: A warning may appear asking you if you "are you sure you want to change the units? Select the OK button.
3. Set **Drawing Limits:**

   Lower Left corner = 0,0 Upper Right Corner = 25, 20
4. Make sure you use **Zoom / All** after setting Drawing Limits
5. Erase the Rectangle that appears with the template
6. Turn **OFF** the **Snap** and **Ortho.**

   (Your cursor should move freely)
7. Draw the objects shown below using:

   Line (Use Layer = Object Line)

   Object snap = Perpendicular
8. **Save** the drawing as: **EX4C**

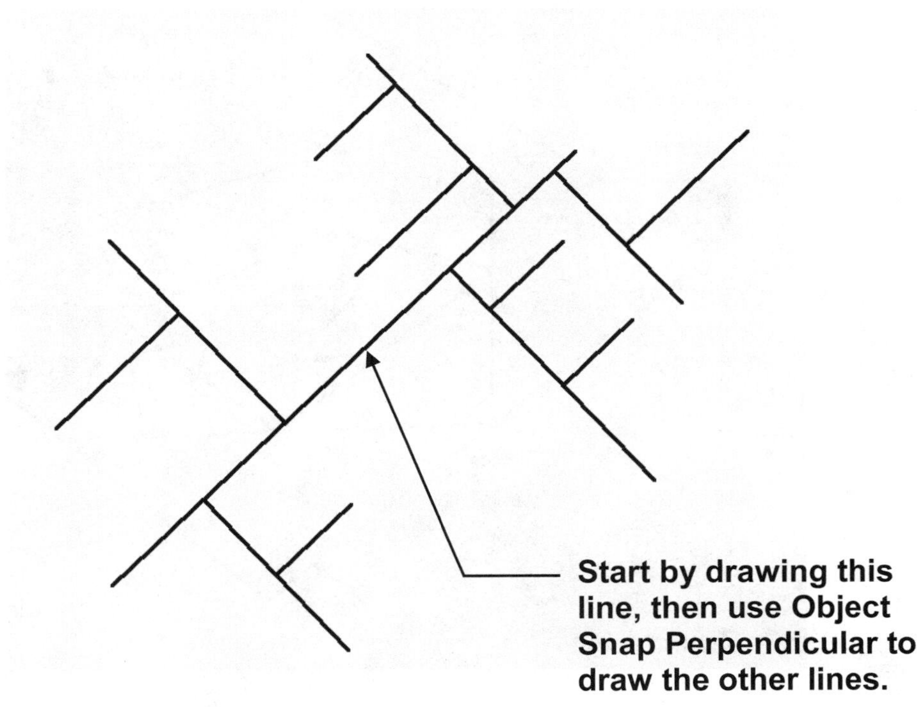

Use Layer = Object Line

EXERCISE 4D

INSTRUCTIONS:

1. Start a **NEW** file using **2015-Workbook Helper.dwt.**
2. Set **Units** and **Precision:**

   Units = Decimals **Precision =** 0.00

   Note: A warning may appear asking you if you "are you sure you want to change the units? Select the OK button.
3. Set **Drawing Limits:**

   Lower Left corner = 0,0 Upper Right Corner = 12,9
4. Make sure you use **Zoom / All** after setting Drawing Limits
5. Turn **OFF Snap** and **Ortho.**

   (Your cursor should move freely)
6. Draw the objects shown below using:

   Line (Use Layer = Object Line)

   Object snap = Midpoint
7. **Save** the drawing as: **EX4D**

Start here

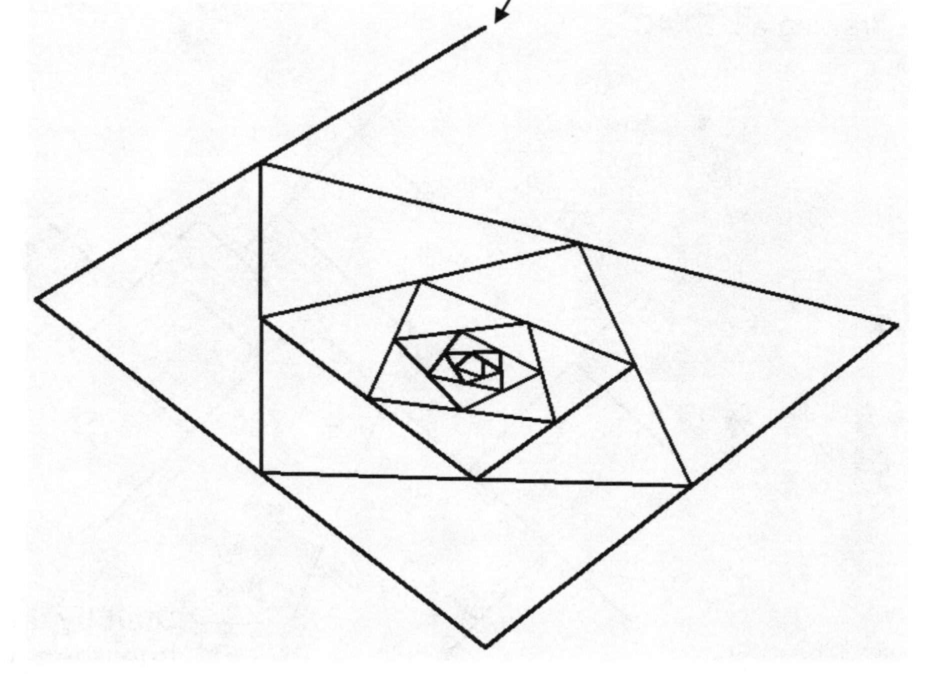

Use Layer = Object Line

EXERCISE 4E

INSTRUCTIONS:

1. Start a **NEW** file using **2015-Workbook Helper.dwt.**
2. Turn **OFF Snap** and **Ortho.**
3. Draw the objects shown below using:

   Line (Use Layer = Object Line)

   Object snap = Intersection
4. **Save** the drawing as: **EX4F**

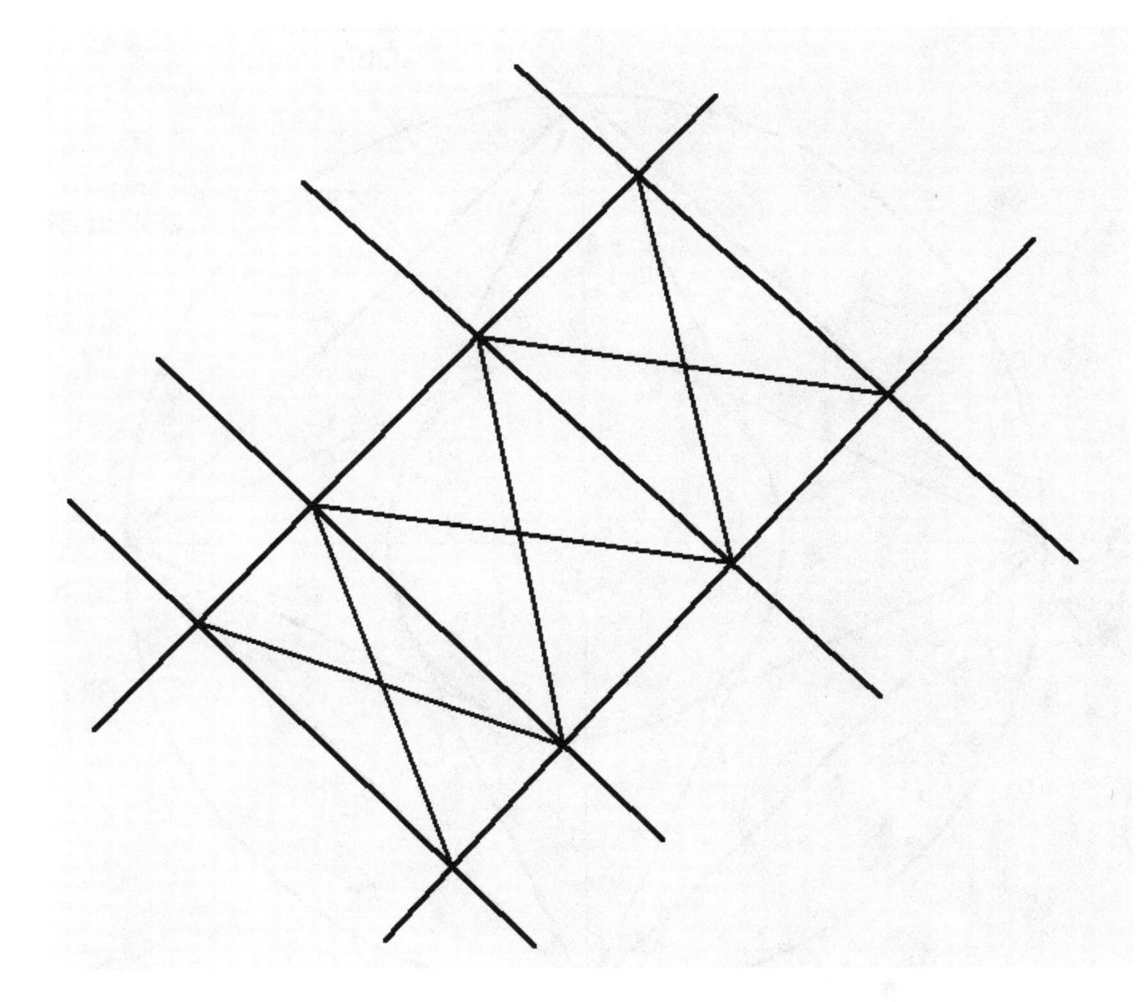

Use Layer = Object Line

EXERCISE 4F

INSTRUCTIONS:

1. Start a **NEW** file using **2015-Workbook Helper.dwt.**
2. Turn **OFF Snap** and **Ortho.**
3. Draw the 2 Circles on layer Object with the following Radii : 1.5 and 3.5.

   (Use Object Snap: Center so both circles will have the same center)
4. Draw the Lines using **Layers**: Object Line and Centerline
5. Use **Object Snap** Quadrant and Tangent
6. **Save** the drawing as: **EX4G**

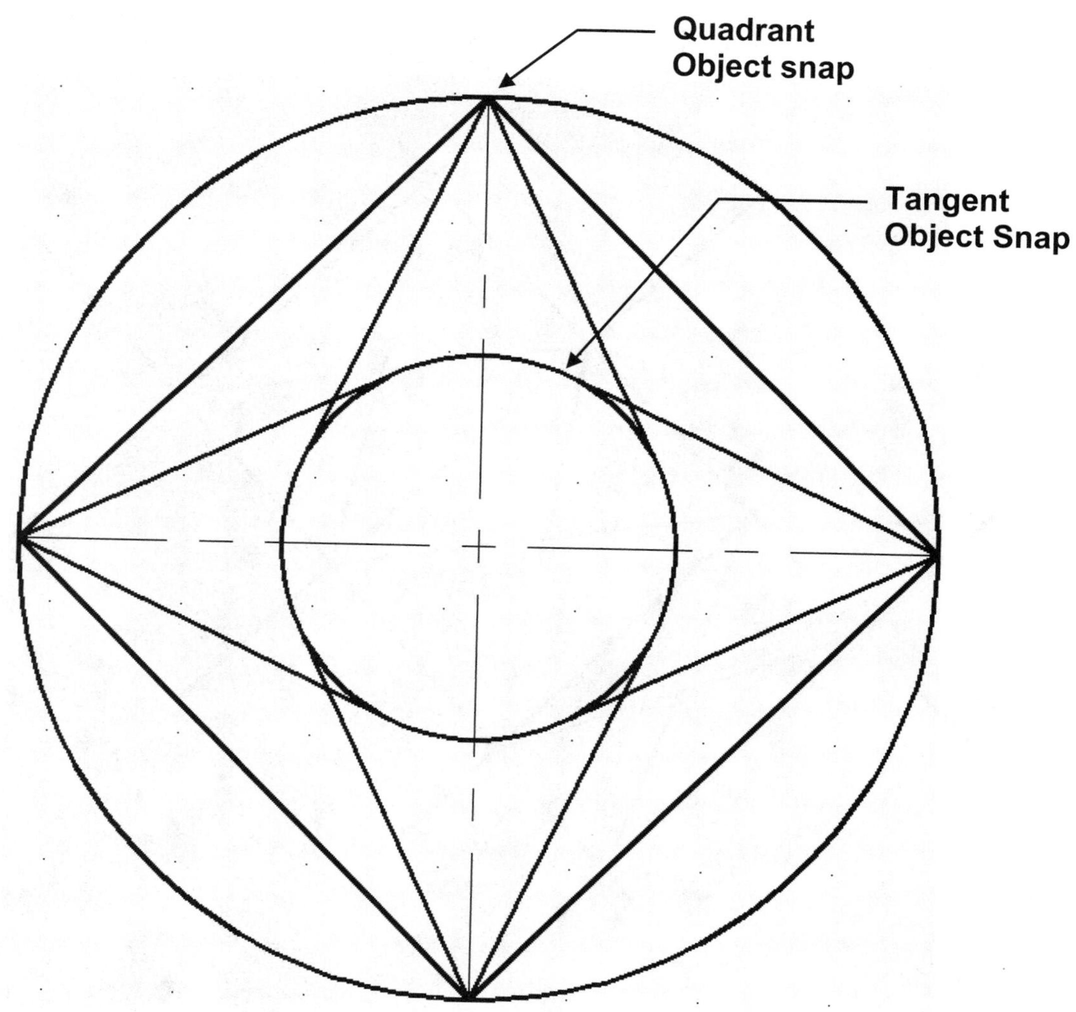

Use Layers = Object Line and Centerline

LEARNING OBJECTIVES

After completing this lesson, you will be able to:

1. Draw an Inscribed or Circumscribed Polygon
2. Create an Ellipse using two different methods
3. Define an Elliptical Arc
4. Create Donuts
5. Define a Location with a Point
6. Select various Point Styles
7. Use 3 new Object Snap modes

LESSON 5

POLYGON

A polygon is an object with multiple edges (flat sides) of equal length. You may specify from 3 to 1024 sides. A polygon appears to be multiple lines but in fact it is one object. You can specify the center and a radius or the edge length. The radius size can be specified Inscribed or Circumscribed.

CENTER, RADIUS METHOD

1. Select the **Polygon** command using one of the following:

Ribbon = Home tab / Draw panel /
or
Keyboard = POL <enter>

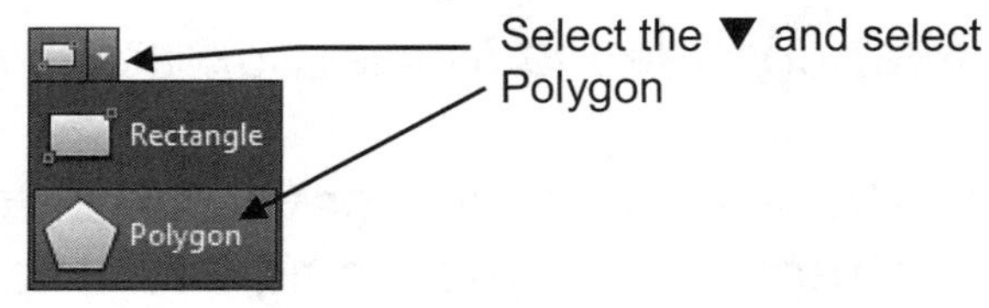

2. The following prompts will appear on the command line:

_polygon Enter number of sides <4>: ***type number of sides <enter>***
Specify center of polygon or [Edge]: ***specify the center location (P1)***
Enter an option [**I**nscribed in circle/**C**ircumscribed about circle]<I>:***type I or C <enter>***
Specify radius of circle: ***type radius or locate with cursor. (P2)***

Note:
The dashed circle is shown only as a reference to help you visualize the difference between Inscribed and Circumscribed. Notice that the radius is the same (2") but the Polygons are different sizes. Selecting Inscribed or Circumscribed is important.

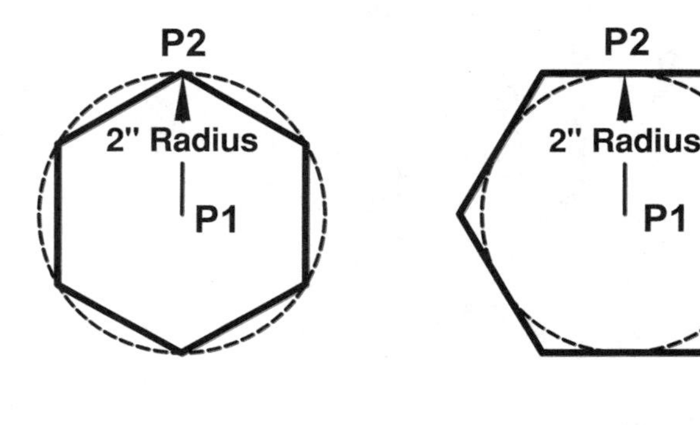

INSCRIBED **CIRCUMSCRIBED**

EDGE METHOD

1. Select the **Polygon** command using one of the options shown above.

2. The following prompts will appear on the command line:

_polygon Enter number of sides <4>: ***type number of sides <enter>***
Specify center of polygon or [Edge]: ***type E <enter>***
Specify first endpoint of edge: ***place first endpoint of edge (P1)***
Specify second endpoint of edge: ***place second endpoint of edge (P2)***

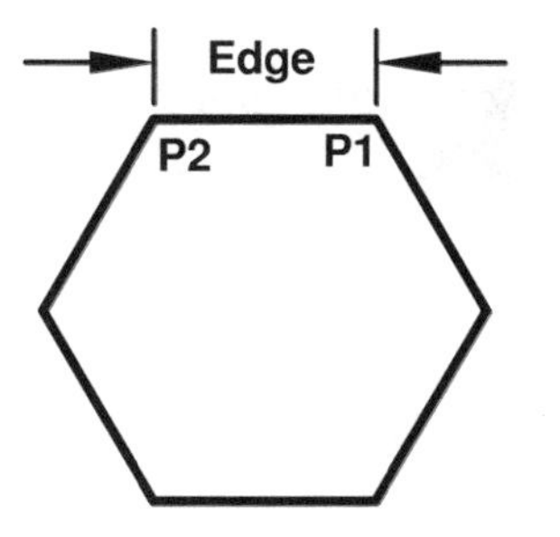

ELLIPSE

There are 3 methods to draw an Ellipse. You may (1) specify 3 points of the axes, (2) define the center point and the axis points or (3) define an elliptical Arc.
The following 3 pages illustrates each of the methods.

AXIS END METHOD

1. Select the **ELLIPSE** command using one of the following:

Ribbon = Home tab / Draw panel
or
Keyboard = EL <enter>

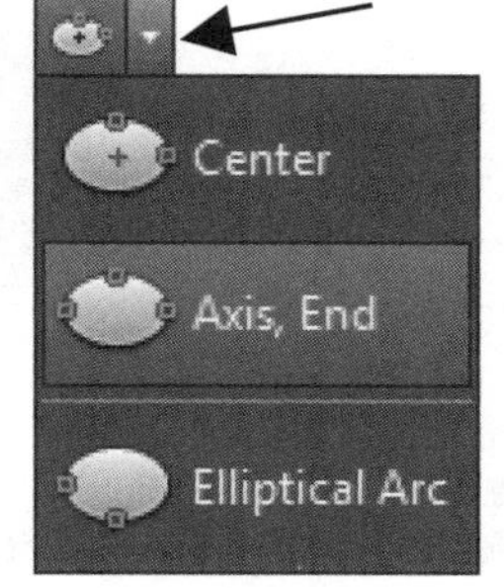

2. The following prompts will appear on the command line:

Command: _ellipse
Specify axis endpoint of ellipse or [Arc/Center]: ***place the first point of either the major or minor axis (P1).***
Specify other endpoint of axis: ***place the other point of the first axis (P2)***
Specify distance to other axis or [Rotation]: ***place the point perpendicular to the first axis (P3).***

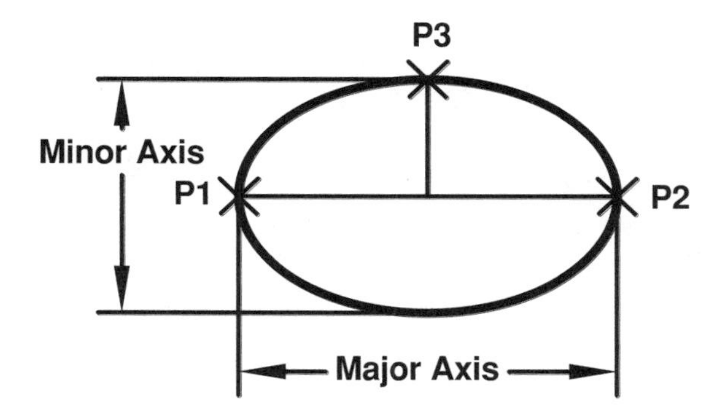

Specifying <u>Major Axis first</u> (P1/P2), then Minor Axis (P3)

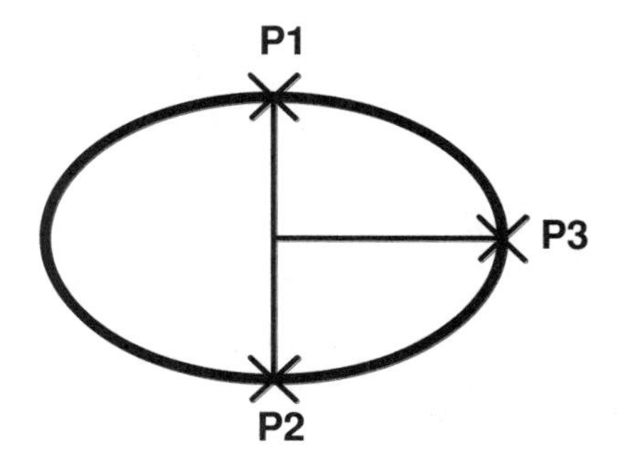

Specifying <u>Minor Axis first</u> (P1/P2), then Major Axis (P3)

Continued on the next page...

ELLIPSE....continued

CENTER METHOD

1. Select the **ELLIPSE** command using one of the following:

Ribbon = Home tab / Draw panel
or
Keyboard = EL <enter> C <enter>

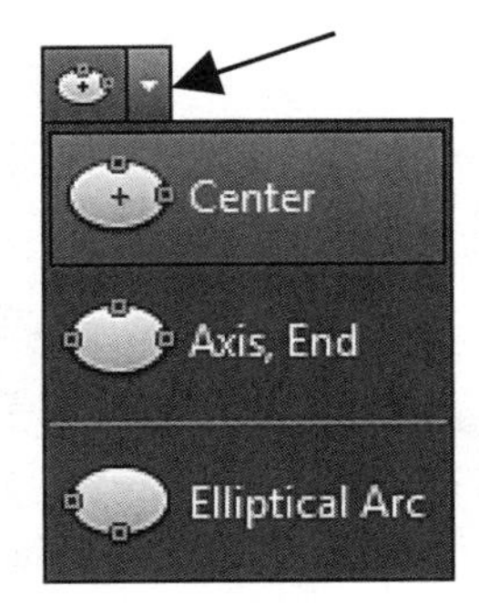

2. The following prompts will appear on the command line:

Command: _ellipse
Specify center of ellipse: ***place center of ellipse (P1)***
Specify endpoint of axis: ***place first axis endpoint (either axis) (P2)***
Specify distance to other axis or [Rotation]: ***place the point perpendicular to the first axis (P3)***

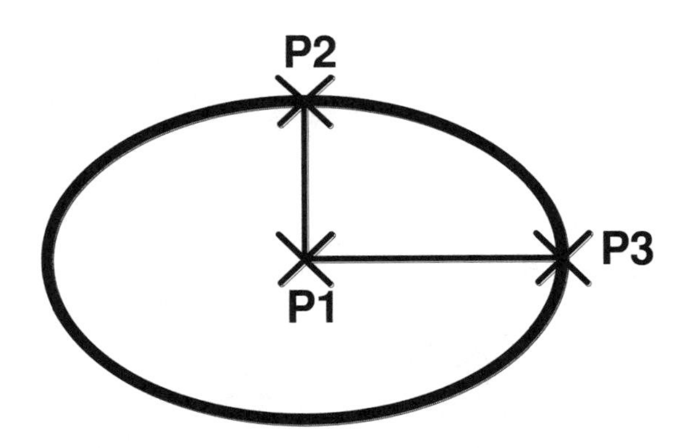

Continued on the next page...

ELLIPSE....continued

ELLIPTICAL ARC METHOD

1. Select the **ELLIPSE** command using one of the following:

   Ribbon = Home tab / Draw panel
   or
   Keyboard = EL <enter> A <enter>

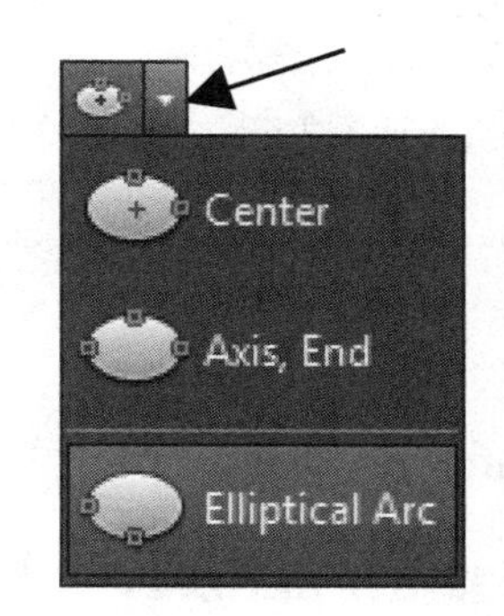

2. The following prompts will appear on the command line:
   Command: _ellipse
   Specify axis endpoint of elliptical arc or [center]: ***type C <enter>***
   Specify center of axis: ***place the center of the elliptical arc (P1)***
   Specify endpoint of axis: ***place first axis point (P2)***
   Specify distance to other axis or [Rotation]: ***place the endpoint perpendicular to the first axis (P3)***
   Specify start angle or [Parameter]: ***place the start angle (P4)***

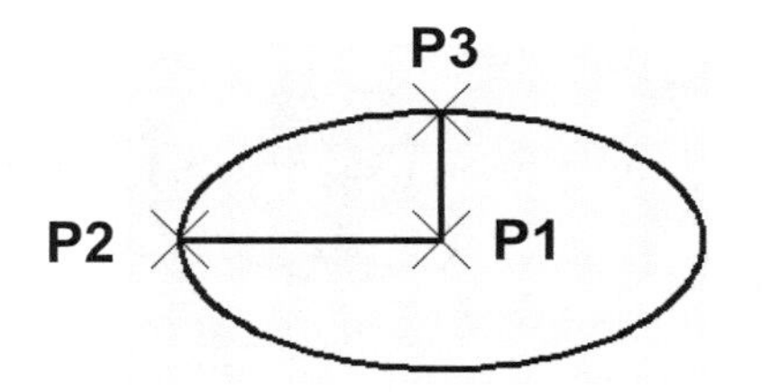

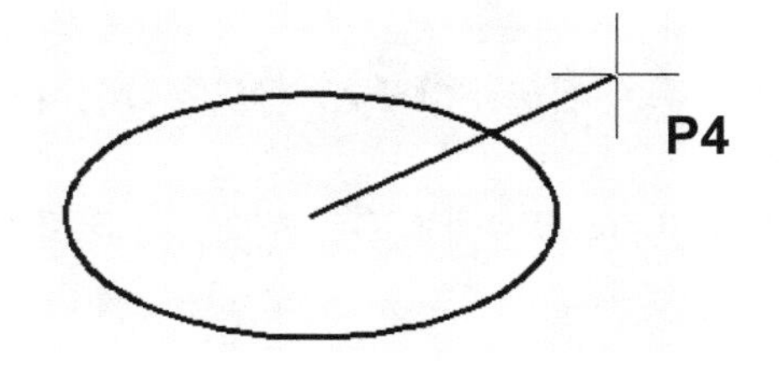

   Specify end angle or [Parameter/Included angle]: ***place end angle (P5)***

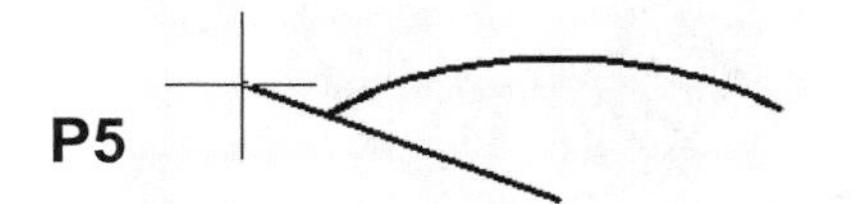

DONUT

A Donut is a circle with ***width.*** You will define the **Inside** and **Outside** diameters.

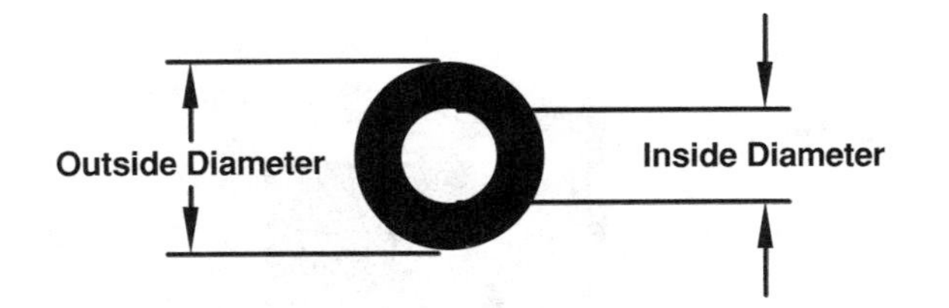

1. Select the **DONUT** command using one of the following:

   Ribbon = Home tab / Draw panel ▼ /
   or
   Keyboard = DO <enter>

2. The following prompts will appear on the command line:

   Command: _donut
   Specify inside diameter of donut: ***type the inside diameter <enter>***
   Specify outside diameter of donut: ***type the outside diameter <enter>***
   Specify center of donut or <exit>: ***place the center of the first donut***
   Specify center of donut or <exit>: ***place the center of the second donut or <enter> to stop***

<u>**Note:**</u>
It will continue to create more donuts until you press **<enter>** to stop the command.

Controlling the "FILL MODE"

1. Command: ***type FILL <enter>***

2. Enter mode [ON / OFF] <OFF>: ***type ON or OFF <enter>***

3. Type ***REGEN* <enter>** to regenerate the drawing to show the latest setting of the ***FILL*** mode.

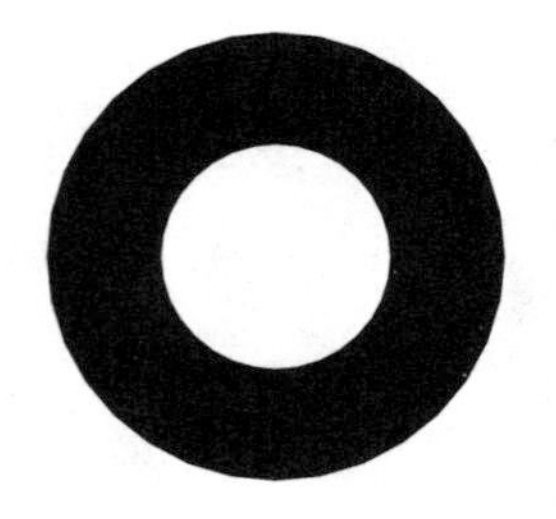

FILL = ON

FILL = OFF

POINT

Points are used to locate a point of reference or location. A **Point** may be represented by one of many **Point Styles** shown below in the Point Style box.

The only object snap option that can be used with Point is **Node**.
(Refer to the next page for more information on Node object snap)

HOW TO USE THE POINT COMMAND

1. Select the **POINT** command using one of the following:

    Ribbon = Home tab / Draw panel ▼ /
    or
    Keyboard = PO <enter>

Note: The Draw panel option creates multiple points until you press the **ESC** key.

The keyboard entry method creates single points.

2. The following prompts will appear on the command line:

    Command: _point
    Current point modes: PDMODE=3 PDSIZE=0.000
    Specify a point: ***place the point location***
    Specify a point: ***place another point or press the "ESC" key to stop***

HOW TO SELECT A "POINT STYLE"

1. Open the Point Style dialog box:

    Ribbon = Home tab / Utilities panel ▼/ Point Style
    or
    Keyboard = ddptype <enter>

2. The Point Style dialog box will appear.

3. Select a point style tile.

4. Select the **OK** button.

Point Size:

Set Size Relative to Screen

Sets the point display size as a percentage of the screen size. The point display does not change when you zoom in or out

Set Size in Absolute Units

Sets the point display size as the actual units you specify under Point Size. Points are displayed larger or smaller when you zoom in or out.

MORE OBJECT SNAPS

3 MORE OBJECT SNAP OPTIONS:

NODe This option snaps to the object "**POINT**" described on the previous page. Select **Node** object snap and place the cursor on the **POINT.** The cursor will snap to the **POINT**.
Note: This is the ONLY object snap that you can use with the object **POINT.**

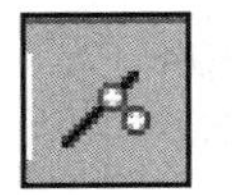

NEArest Snaps to the nearest location on an object.
For example, if you want to attach a Line somewhere on a Circle between quadrants.
Select the Line command then select **Nearest** object snap.
Place the cursor anywhere on the circumference of the Circle and press the left mouse button. The Line will now be accurately attached to the Circle at the location you selected.

M2P **Mid Between 2 Points**
Locates a midpoint between two points

You may select this option from the object snap menu (shift+Rt Click) or you may type ***M2P <enter>*** *when prompted for an endpoint.* *No tool or running object snap option is available.*

HOW TO USE "MTP":

1. Select the Line command and draw 2 parallel lines.
2. Select the **Line** command again.
3. Type **M2P <enter>**
4. Using **Endpoint object snap** select each of the 2 endpoints (P1) and (P2)

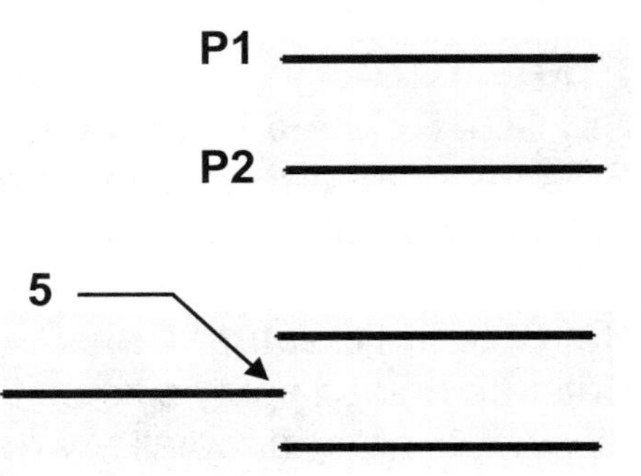

5. The new line's first endpoint should start exactly midpoint between those 2 endpoints.

EXERCISE 5A

INSTRUCTIONS:

1. Start a **NEW** file using **2015-Workbook Helper.dwt.**
2. **Draw** the Circle first using the Center / Radius option.
3. Draw the Circumscribed Polygon next using object snap Center to locate the center of the Polygon at the center of the Circle and Quadrant object snap to locate the radius of the Polygon on the Circle.
4. Draw the Lines last using Object snaps Midpoint and Endpoint.
5. **Ortho** (F8) **ON**
6. **Increment Snap** (F9) **OFF** (It will get in your way)
7. Use layer: Object Line
8. **Save** the drawing as: **EX5A**

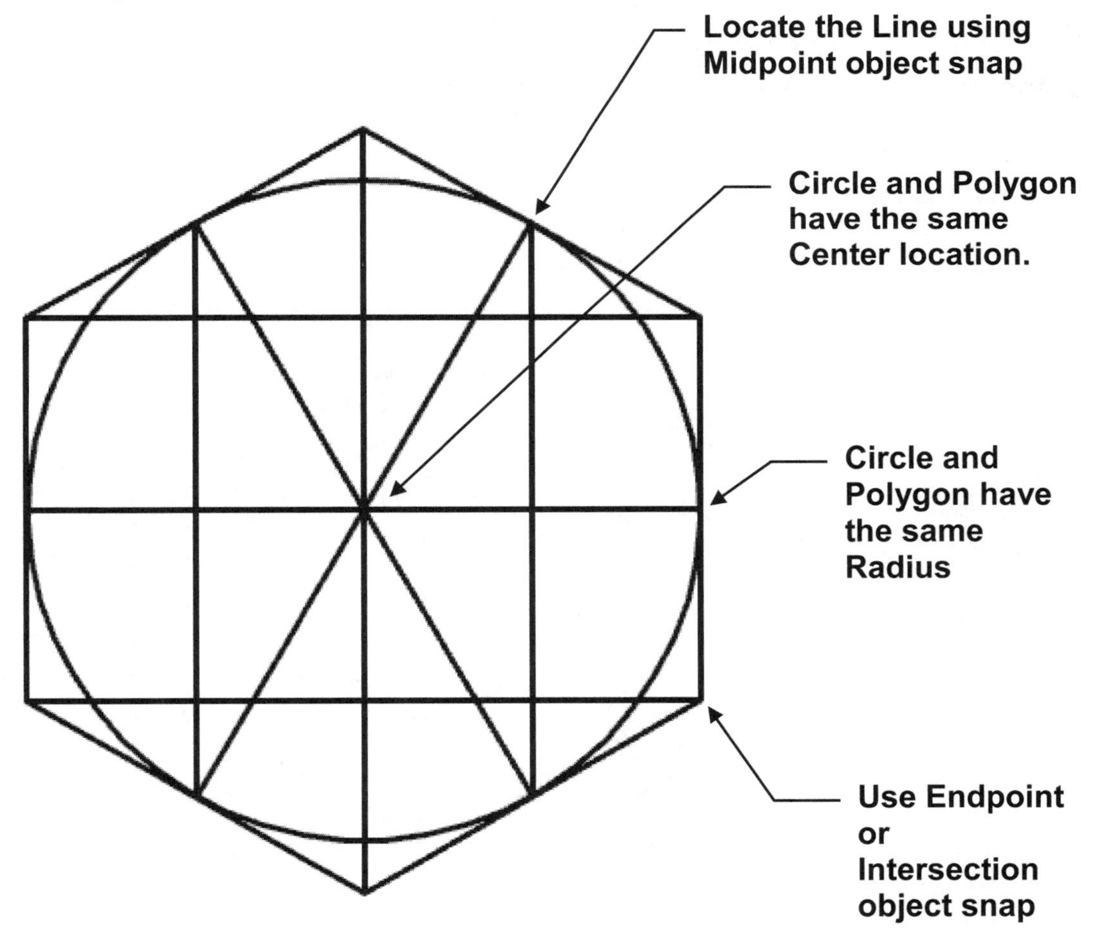

EXERCISE 5B

INSTRUCTIONS:

1. Start a **NEW** file using **2015-Workbook Helper.dwt.**
2. Select the **Point Style** shown below
3. Draw the **Point** first.

Note: Your Point may not appear as large as the one shown below.

4. Draw the Inscribed Polygons
    a. Locate the center of each Polygon using Object Snap **NODE**.
5. **Ortho** (F8) **ON**
6. **Increment Snap** (F9) **OFF**
7. Use layer Object Line

Note:

Place the Polygon points with the cursor. If you type a radius the Polygon will automatically rotate and locate the flat at the bottom.

8. **Save** the drawing as: **EX5B**

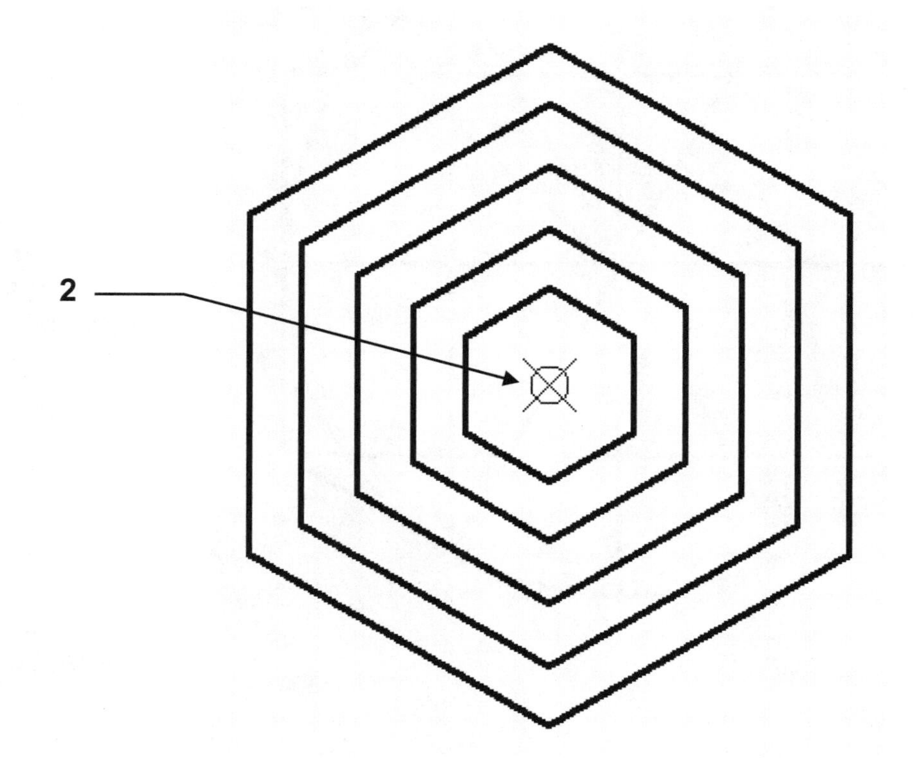

EXERCISE 5C

INSTRUCTIONS:

1. Start a **NEW** file using **2015-Workbook Helper.dwt.**
2. Draw the Objects below using **Ellipse, Line** and **Donut**.
3. Use Object Snap: **Quadrant, Center** and **Nearest**.
4. **Ortho** (F8) **ON**
5. **Increment Snap** (F9) **OFF**
6. Use any layer you like.
7. **Save** the drawing as: **EX5C**

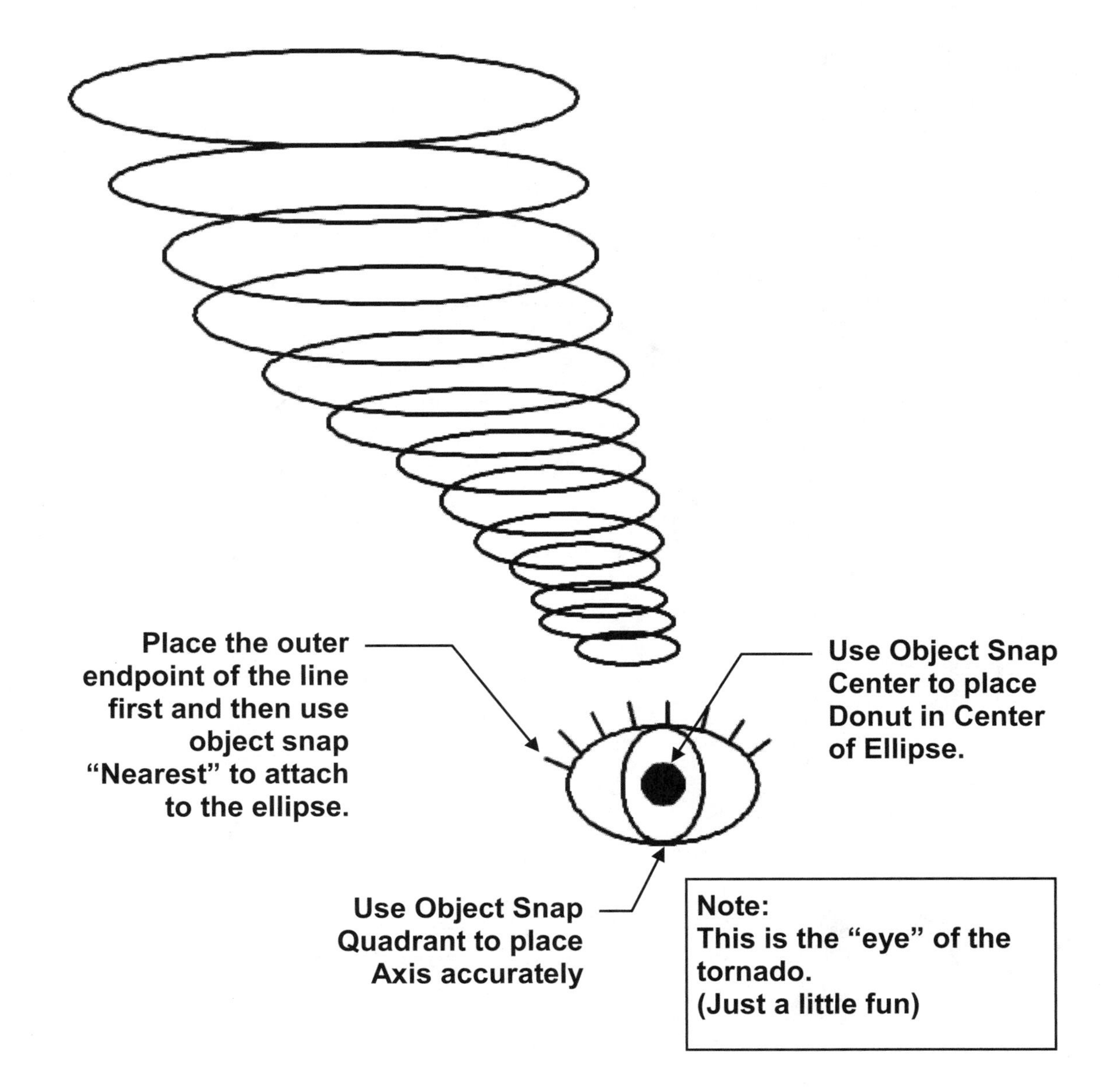

EXERCISE 5D

INSTRUCTIONS:

1. Start a **NEW** file using **2015-Workbook Helper.dwt.**
2. Select the **Point Style** shown below
3. Draw the **Point** first.
4. Draw the **Ellipses:**
    a. Locate the **center** of each Ellipse using Object Snap: **NODE**.
    b. Locate the **Axis** using Object snap: **Quadrant**
5. **Ortho** (F8) **ON**
6. **Increment Snap** (F9) **OFF**
7. Use layer: Object Line
8. **Save** the drawing as: **EX5D**

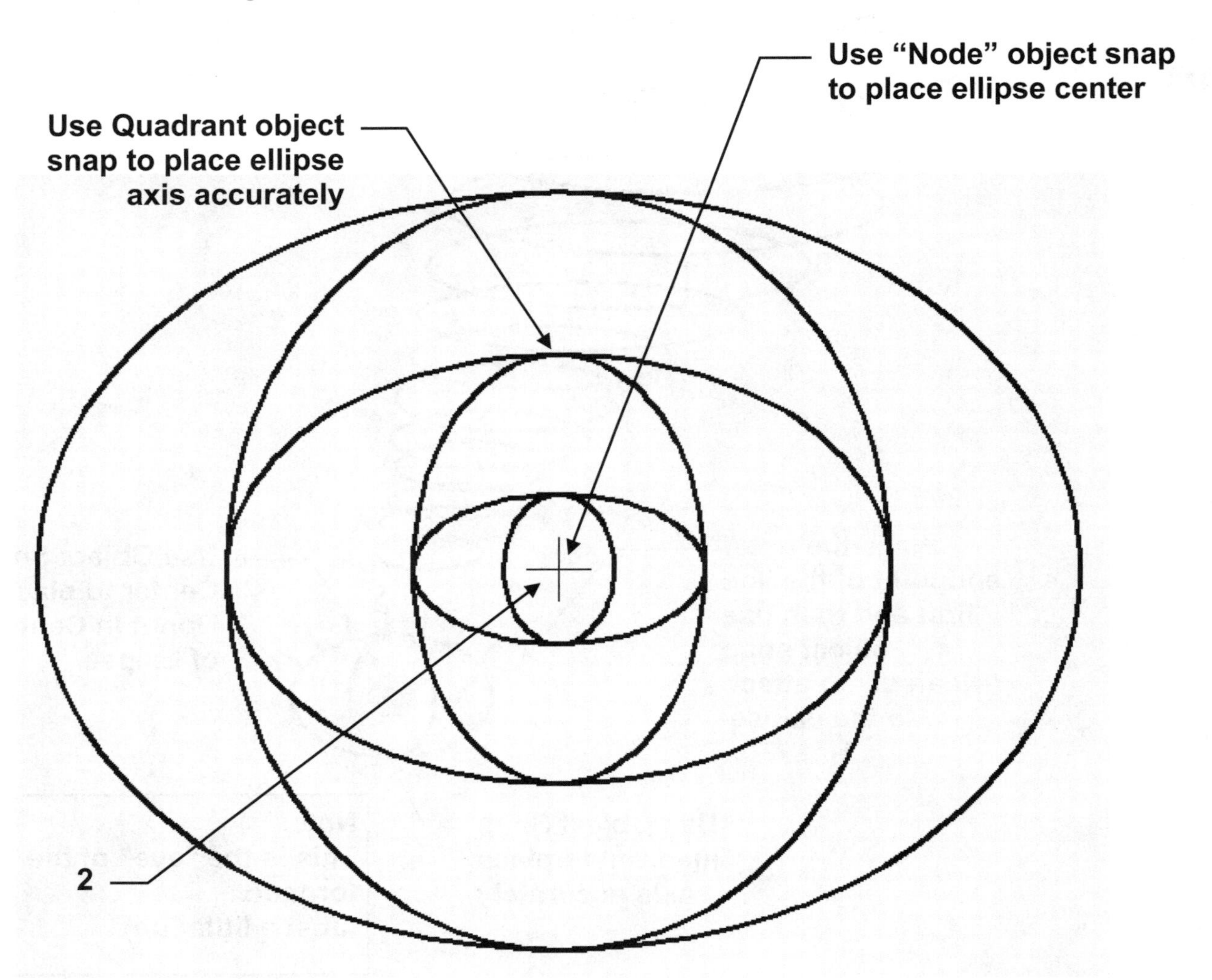

EXERCISE 5E

INSTRUCTIONS:

STEP 1

1. Start a **NEW** file using **2015-Workbook Helper.dwt.**
2. Draw the 3 **DONUTS** shown below.
3. Use Object Snap **Center** to place the centers accurately.
4. **Increment Snap** (F9) **OFF**
5. Use layer: Object Line

Donut sizes

1. ID = 0 OD = 1
2. ID = 1.5 OD = 2
3. ID = 2.5 OD = 3.5

STEP 2

6. Turn the FILL mode **OFF**
7. Type **Regen <enter>**
8. **Save** the drawing as: **EX5E**

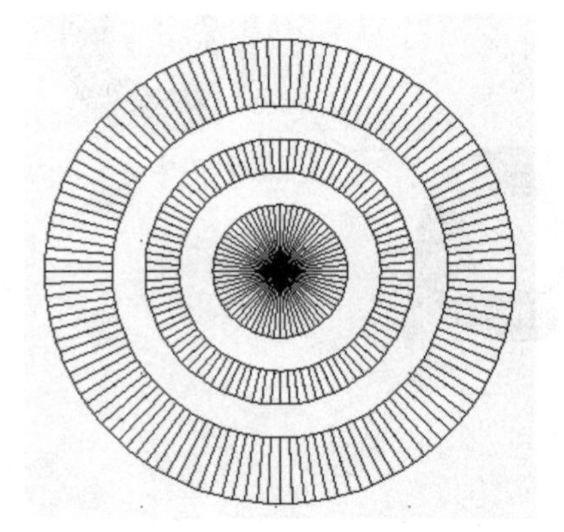

EXERCISE 5F

INSTRUCTIONS:

1. Start a **NEW** file using **2015-Workbook Helper.dwt.**
2. Draw the objects below using: **Point, Polygon, Ellipse** and **Donut**.
3. **Ortho** (F8) **ON**
4. **Increment Snap** (F9) **OFF**
5. Use whatever layers you like
6. **Save** the drawing as: **EX5F**

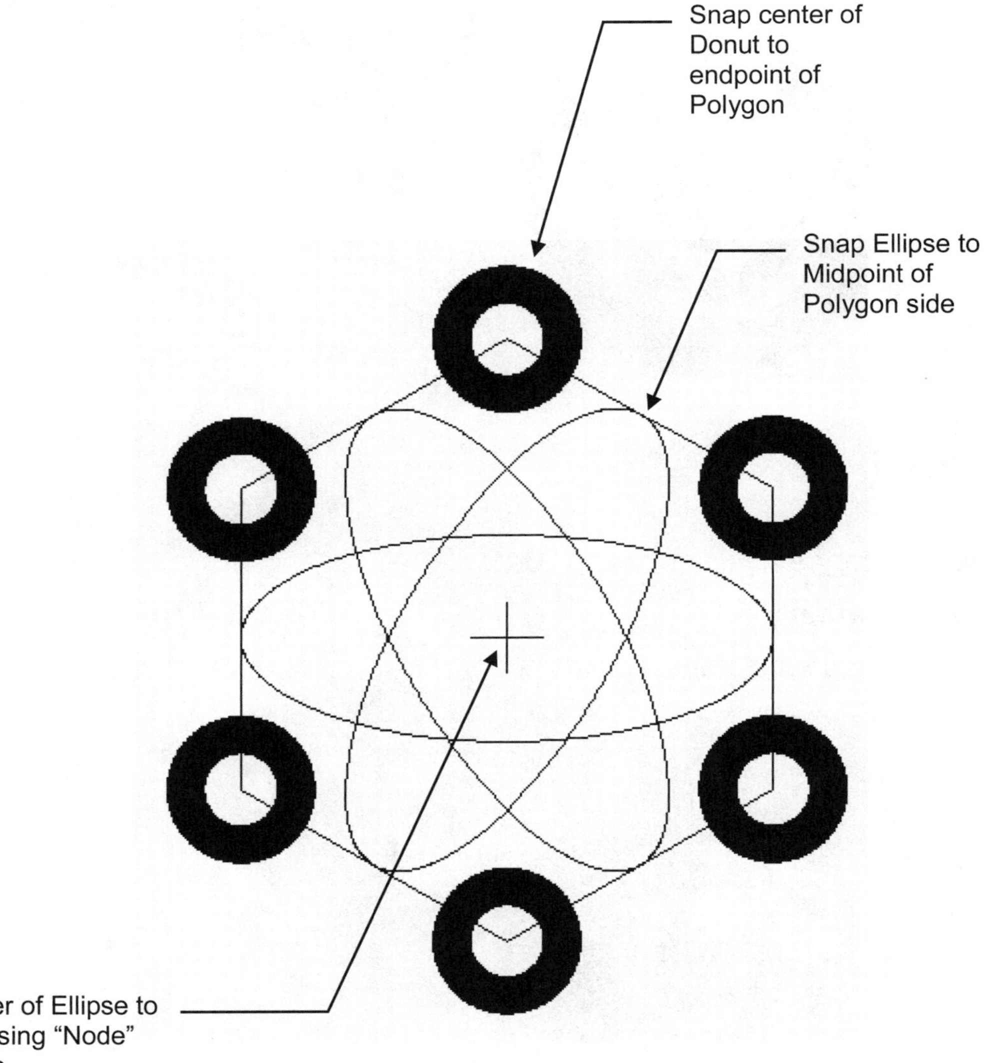

EXERCISE 5G

INSTRUCTIONS:

1. Start a **NEW** file using **2015-Workbook Helper.dwt.**
2. Draw the Zig Zag Lines approximately as shown below.
3. Set Running Object Snap to "**Endpoint**".
4. **Increment Snap** (F9) **OFF**
5. **Ortho** (F8) **= ON**
6. Draw the circles at the "**Midpoint between 2 points**". (P1 and P2)
7. Use layer: Object Line.
8. **Save** the drawing as: **EX5G**

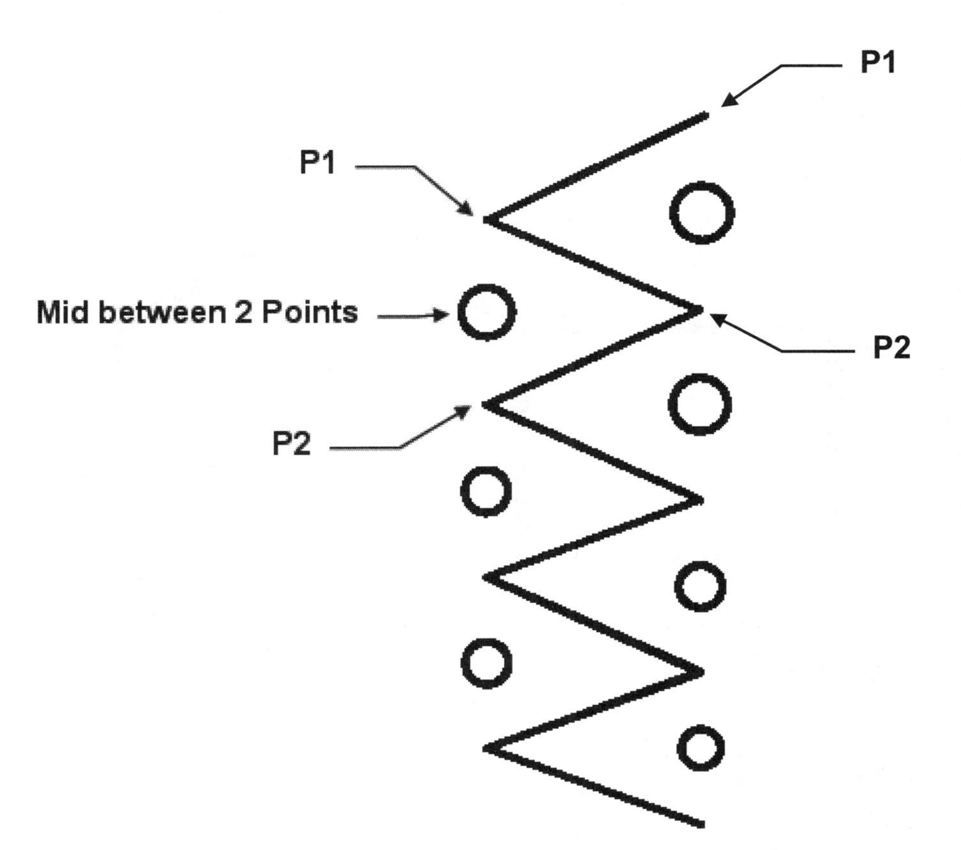

EXERCISE 5H

INSTRUCTIONS:

1. Start a **NEW** file using **2015-Workbook Helper.dwt.**
2. Draw the **Rectangle** on the left first. (any size)
3. Draw the **Polygon** next using the “**EDGE**” option.
    A. Use Object Snap: “**Endpoint**” to place **Edge** points accurately.
4. **Increment Snap** (F9) **OFF**
5. **Ortho** (F8) **= ON**
6. Use Layer: Object Line
7. **Save** the drawing as: **EX5H**

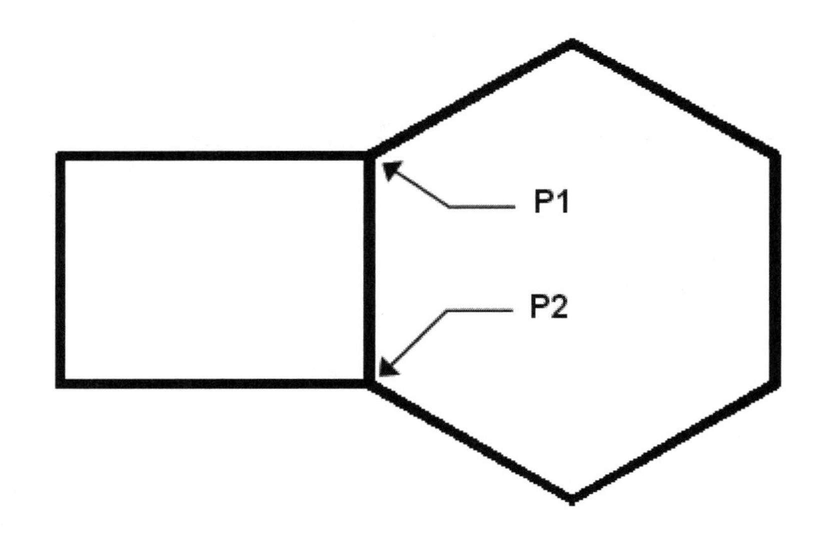

LEARNING OBJECTIVES

After completing this lesson, you will be able to:

1. Use the Break command
2. Trim an object to a cutting edge
3. Extend an object to a boundary
4. Move and object to a new location
5. Explode objects to their primitive entities.

LESSON 6

BREAK

The ***BREAK*** command allows you to break an object at a single point (Break at Point)or between two points (Break). I think of it as breaking a single line segment into two segments or taking a bite out of an object.

<u>METHOD 1 - Break at a Single Point</u>
<u>How to break one Line into two separate objects with no visible space in between</u>.

1. Select the **BREAK AT POINT** command by using::

   Ribbon = Home tab / Modify panel ▼ /

2. _break Select objects: ***select the object to break (P1).***

3. Specify first break point: ***select break location (P2) accurately.***

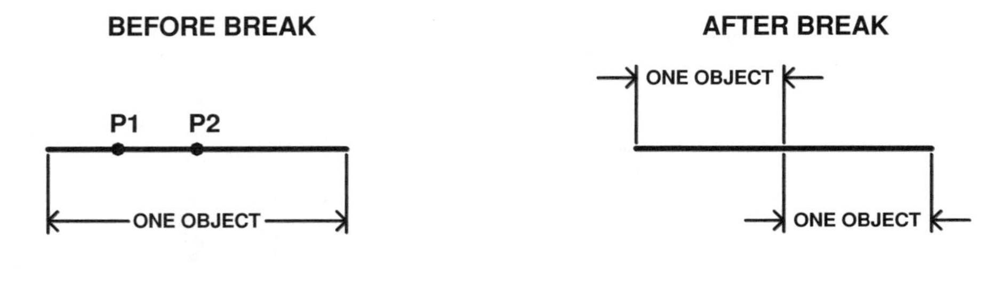

1 Line segment **2 Line segments**

Note:
The single line is now 2 lines but no gap in between the 2 lines.
For example, a 2 inch long line would become two 1 inch lines butted together.

BREAK....continued

METHOD 2
Break between 2 points. (Take a bite out of an object.)
(Use this method if the location of the BREAK is not important.)

1. Select the **BREAK** command using one of the following:

   Ribbon = Home tab / Modify panel ▼ /
   or
   Keyboard = BR <enter>

2. _break Select object: ***pick the first break location (P1).***
3. Specify second break point or [First point]: ***pick the second break location (P2).***

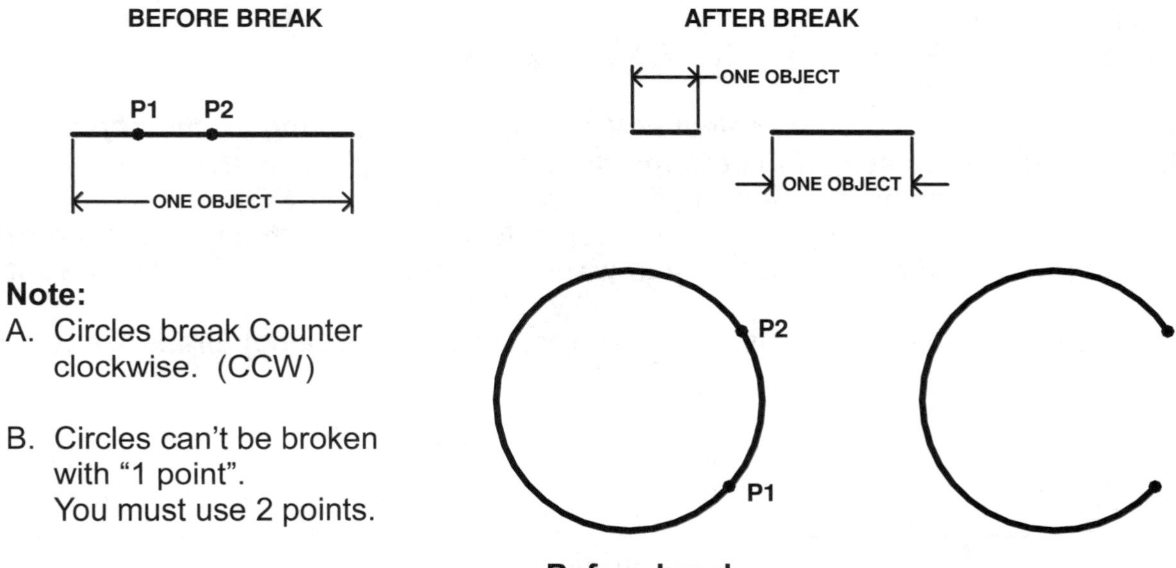

Note:
A. Circles break Counter clockwise. (CCW)

B. Circles can't be broken with "1 point".
You must use 2 points.

The following method is the same as method 2 above; however, use this method if the location of the break is very specific.

1. Select the **BREAK** command.
2. _break Select objects: ***select the object to break (P1) anywhere on the object.***
3. Specify second break point or [First point]: ***type F <enter>.***
4. Specify first break point: ***select the first break location (P2) accurately.***
5. Specify second break point: ***select the second break location (P3) accurately.***

BEFORE BREAK

ONE OBJECT
break
ONE OBJECT

TRIM

The **TRIM** command is used to trim an object to a **cutting edge**. You first select the "Cutting Edge" and then select the part of the object you want to trim. The object to be trimmed must actually intersect the cutting edge or could intersect if the objects were infinite in length.

1. Select the Trim command using one of the following:

   Ribbon = Home tab / Modify panel /
   or
   Keyboard = TR <enter>

2. The following will appear on the command line:

Command: _trim
Current settings: Projection = UCS Edge = Extend
Select cutting edges ...
Select objects or <select all>: ***select cutting edge(s) by clicking on the object (P1)***
Select objects: ***stop selecting cutting edges by pressing the <enter> key***
Select object to trim or shift-select to extend or
[Fence/Crossing/Project/Edge/eRase/Undo]: ***select the object that you want to trim. (P2)***
(Select the part of the object that you want to disappear, not the part you want to remain)
Select object to trim or [Fence/Crossing/Project/Edge/eRase/Undo]: ***press <enter> to stop***

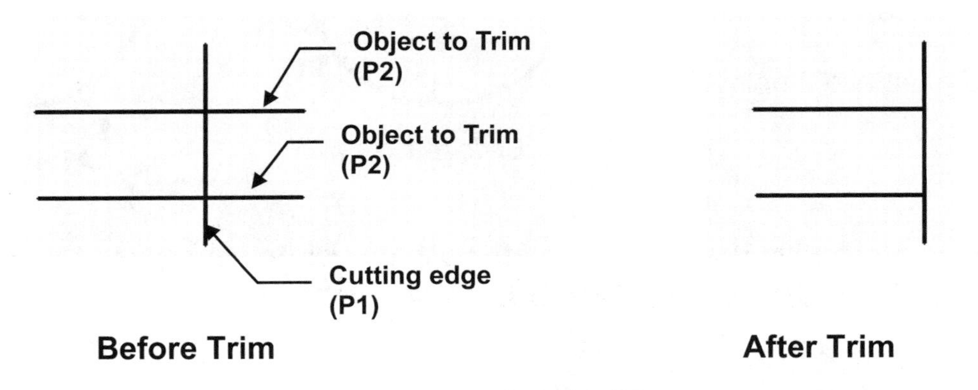

Note: You may toggle between Trim and Extend (page 6-5). Hold down the shift key and the Extend command will activate. Release the shift key and you return to Trim.

Fence See page 6-13
Edge See page 6-5
Project See page 6-5

Crossing You may select objects using a Crossing Window.

eRase You may erase an object instead of trimming while in the Trim command.

Undo You may "undo" the last trimmed object while in the Trim command

EXTEND

The **EXTEND** command is used to extend an object to a **boundary.** The object to be extended must actually or theoretically intersect the boundary.

1. Select the **EXTEND** command using one of the following:

Ribbon = Home tab / Modify panel /
or
Keyboard = EX <enter>

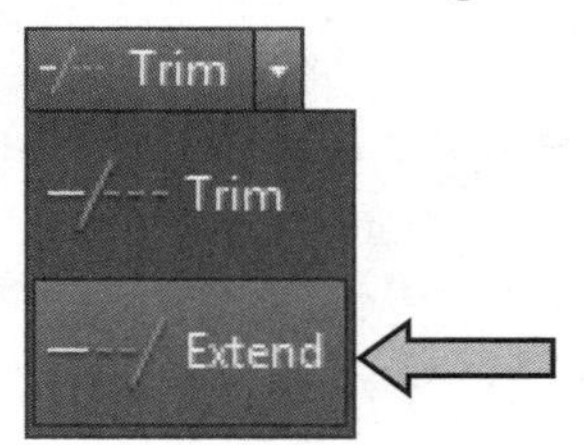

2. The following will appear on the command line:

Command: _extend
Current settings: Projection = UCS Edge = Extend
Select boundary edges ...
Select objects or <select all>: ***select boundary (P1) by clicking on the object.***
Select objects: ***stop selecting boundaries by selecting <enter>.***
Select object to extend or shift-select to Trim or
[Fence/Crossing/Project/Edge/Undo]:***select the object that you want to extend (P2 and P3). (Select the end of the object that you want to extend.)***
Select object to extend or [Fence/Crossing/Project/Edge/Undo]:***stop selecting objects by pressing <enter>.***

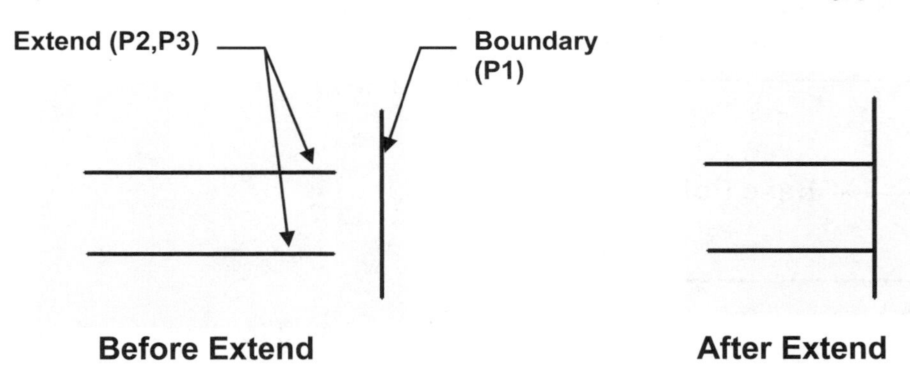

> **Note:** When selecting the object to be extended (P2 and P3 above) click on the end pointing towards the boundary.

Fence See page 6-13
Crossing See page 6-4
Project Same as Edge except used only in "3D".

Edge (Extend or No Extend)
In the **"Extend"** mode, (default mode) the boundary and the Objects to be extended need only intersect if the objects were infinite in length.
In the **"No Extend"** mode the boundary and the objects to be extended must visibly intersect.

Undo See page 6-4

MOVE

The **MOVE** command is used to move object(s) from their current location (base point) to a new location (second displacement point).

1. Select the Move command using one of the following:

   Ribbon = Home tab / Modify panel / 
   or
   Keyboard = M <enter>

2. The following will appear on the command line:

Command: _move
Select objects: ***select the object(s) you want to move (P1).***
Select objects: ***select more objects or stop selecting object(s) by selecting <enter>.***
Specify base point or displacement: ***select a location (P2) (usually on the object).***
Specify second point of displacement or <use first point as displacement>: ***move the object to its <u>new location</u> (P3) and press the left mouse button.***

> ***Warning: If you press <enter> instead of actually picking a new location (P3) with the mouse, Autocad will send it into <u>Outer Space</u>. If this happens just select the undo tool and try again.***

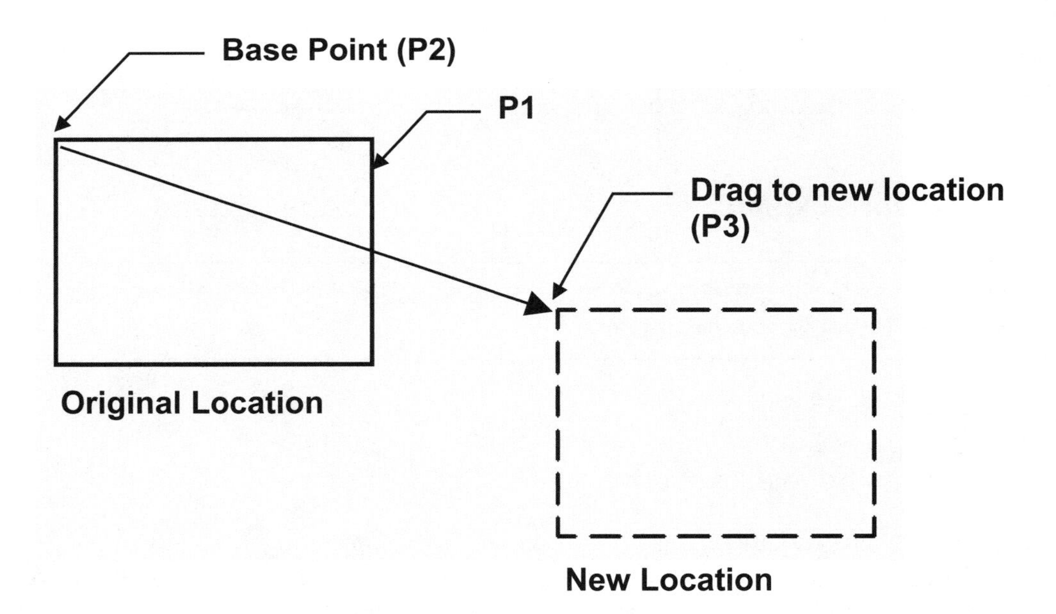

DRAG

The **Drag** option allows you to quickly **move** or **copy** an object(s).

EXAMPLE:

1. Draw a Circle.

2. Select the Circle.
   5 little boxes appear. These are Grips and allow you to edit the object.
   Grips will be discussed more in future lessons.

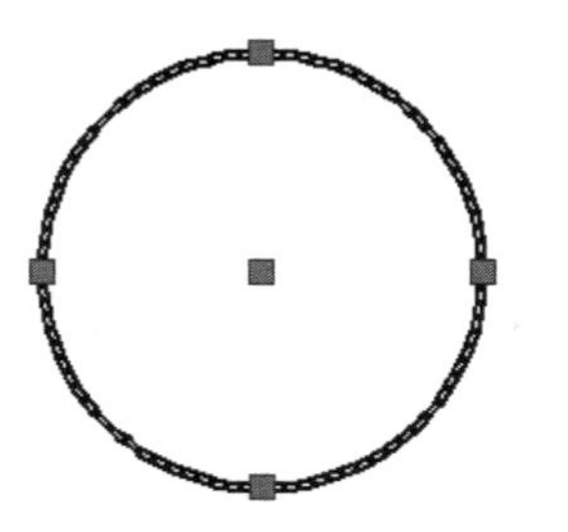

3. Click on the Circle and hold the right hand mouse button down as you drag the Circle to the right.

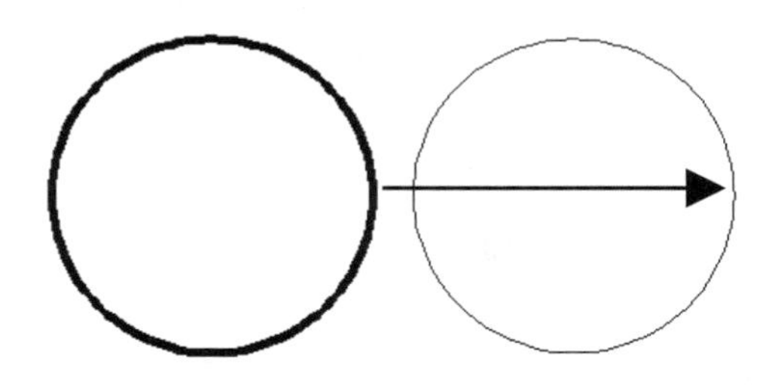

4. When the dragged Circle is in the desired location release the Right Mouse button.

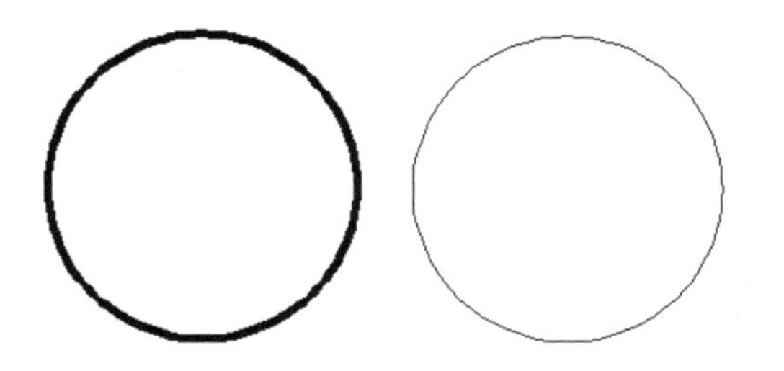

NUDGE

The **Nudge** option allows you to nudge objects in orthogonal increments.

Note:
Snap mode affects the distance and direction in which the objects are nudged.

<u>Nudge objects with **Snap** mode turned OFF</u>:
Objects move two pixels at a time.

<u>Nudge objects with **Snap** mode turned ON:</u>
Objects are moved in increments specified by the current **Snap** spacing.

Note: to set the Increment Snap spacing refer to page 3-7.

EXAMPLE:

1. Draw a Circle.

2. Select the Circle.
   5 little boxes appear. These are Grips and allow you to edit the object. Grips will be discussed more in future lessons.

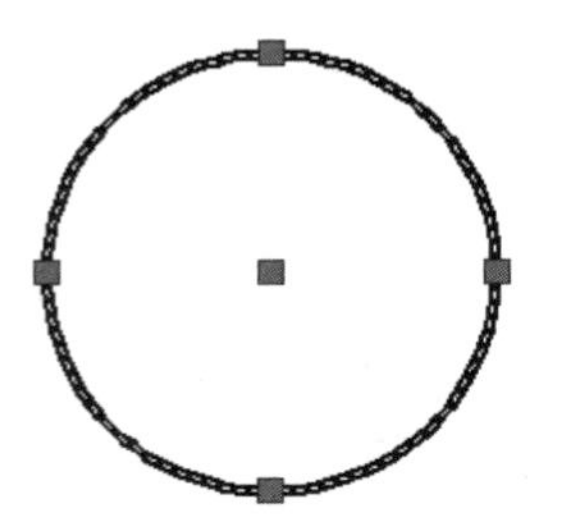

3. Hold down the **Ctrl** key and press one of the **Arrow** keys ←→↑↓

Remember:
The distance the object moves depends on whether you have the **Snap** mode ON or OFF. (Refer to note above)

EXPLODE

The **EXPLODE** command changes (explodes) an object into its primitive objects.
For example: A rectangle is originally one object. If you explode it, it changes into 4 lines. Visually you will not be able to see the change unless you select one of the lines.

1. Select the **Explode** command by using one of the following:

Ribbon = Home tab / Modify panel / 
or
Keyboard = X <enter>

2. The following will appear on the command line:

Command: _explode
Select objects: ***select the object(s) you want to explode.***
Select objects: ***select <enter>.***

Before EXPLODE

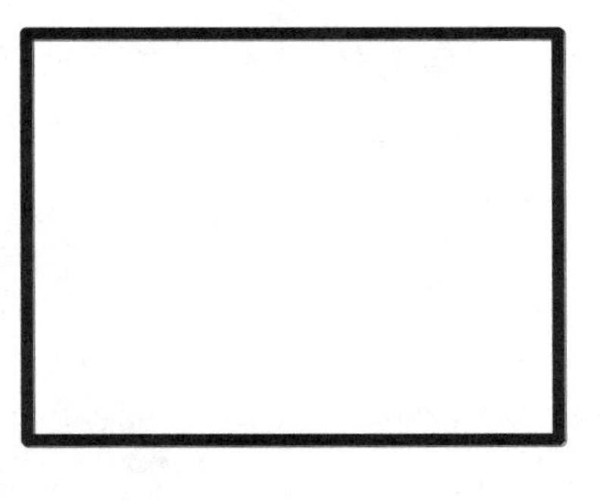

One Object
(Rectangle)

After EXPLODE

4 Objects
(4 Lines)

(Notice there is no visible difference. But now you have 4 lines instead of 1 Rectangle)

Try this:
Draw a rectangle and then click on it. The entire object highlights.
Now explode the rectangle, then click on it again. Only the line you clicked on should be highlighted. Each line that forms the rectangular shape is now an individual object.

EXERCISE 6A

INSTRUCTIONS:

1. Start a **NEW** file using **2015-Workbook Helper.dwt.**
2. **Draw** the objects below:
   A. Use object snap **Midpoint** to locate the center for the Circles.
   B. **Ortho** (F8) **ON** when drawing Horizontal and Vertical lines.
   C. Turn **Increment Snap** (F9) **ON**
   D. Use Layer: Object Line

Center of Circle is the Midpoint of the line

3. Trim the Circles and Rectangle to match the illustration below.
4. **Save** the drawing as: **EX6A**

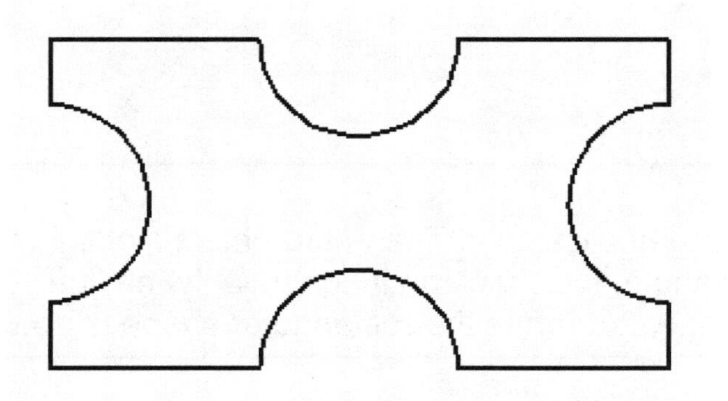

EXERCISE 6B

INSTRUCTIONS:

1. Start a **NEW** file using **2015-Workbook Helper.dwt.**
2. **Draw** the objects below:
   A. **Circles** should have the same center
   B. **Ortho** (F8) **ON** when drawing Horizontal and Vertical lines.
   C. **Ortho** (F8) **OFF** when drawing lines on an Angle.
   D. Turn **Increment Snap** (F9) **OFF**
   E. Use Layer: Object Line

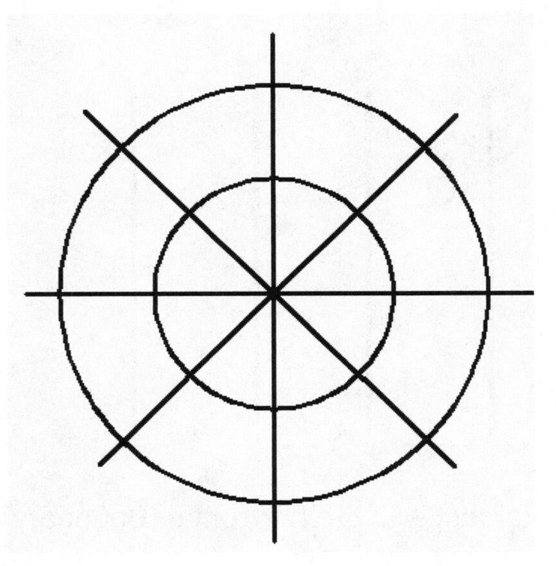

3. **TRIM** the Lines to match the illustration below.
4. **Save** the drawing as: **EX6B**

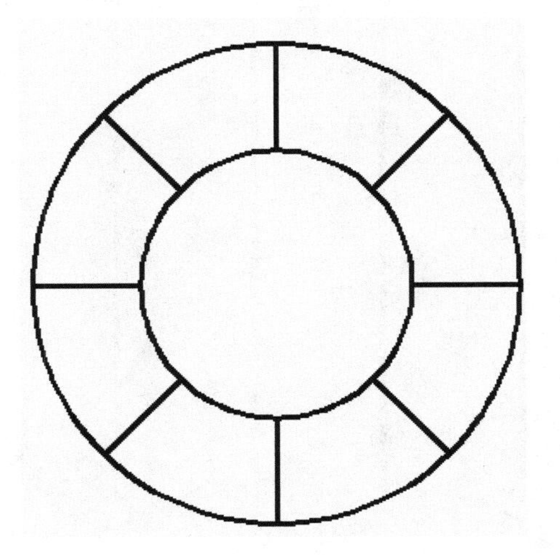

EXERCISE 6C

INSTRUCTIONS:

1. Start a **NEW** file using **2015-Workbook Helper.dwt.**
2. **Draw** the **Lines** below exactly as shown.
   A. **Ortho** (F8) **ON** when drawing Horizontal and Vertical lines.
   B. Turn **Increment Snap** (F9) **ON**
   C. Turn **Osnap** (F3) **OFF**
   D. Use Layer: Object Line

3. **Extend** the vertical lines to intersect with the horizontal (Boundary) Line as shown below.
4. **Save** the drawing as: **EX6C**

EXERCISE 6D

INSTRUCTIONS:

1. Start a **NEW** file using **2015-Workbook Helper.dwt.**
2. Draw the **LINES** below exactly as shown.
3. Select the "**Extend**" command.
4. Select the "**Boundary**" <enter>
5. Now instead of clicking on each vertical line, type **F <enter>.**
6. Place the cursor approximately at location **P1** and click.
7. Move the cursor to approximately location **P2** and click.
8. Press **<enter>** and **<enter>** again.
9. **Save** the drawing as: **EX6D**

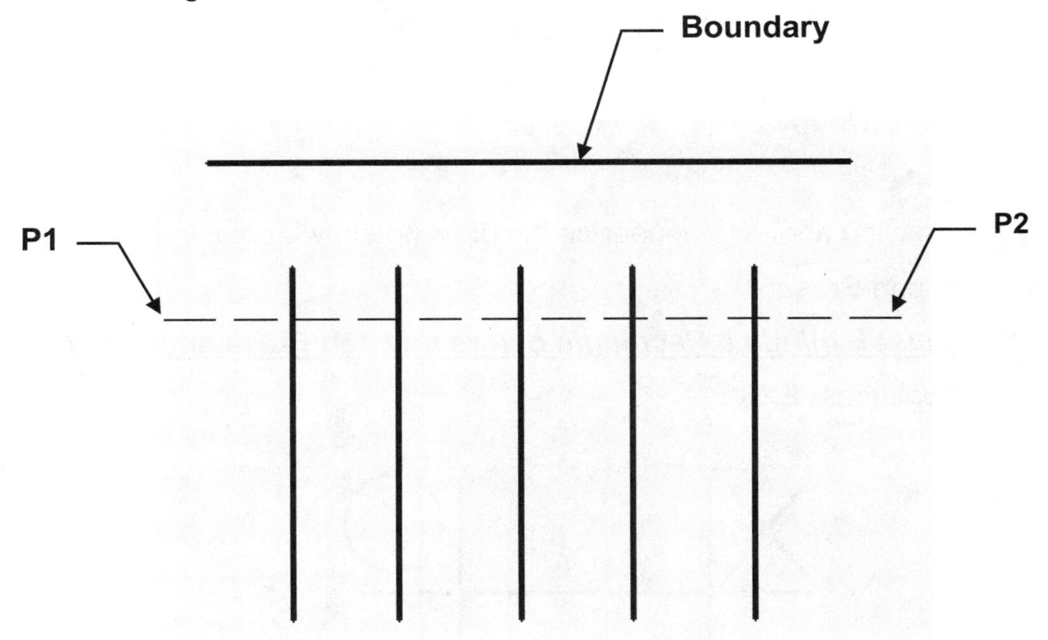

Note:
Be careful to place P1 and P2 above the midpoint of the vertical lines. That instructs AutoCAD to extend the lines in that direction toward the boundary. If you place P1 and P2 below the midpoint of the vertical line, AutoCAD will look for a boundary in that direction. This will confuse AutoCAD because you did not select a boundary below.

EXERCISE 6E

INSTRUCTIONS:

1. Start a **NEW** file using **2015-Workbook Helper.dwt.**
2. **Draw** the objects below using:

   Rectangles, Circles and Lines (for the X's)

   Use Layer: Object Line

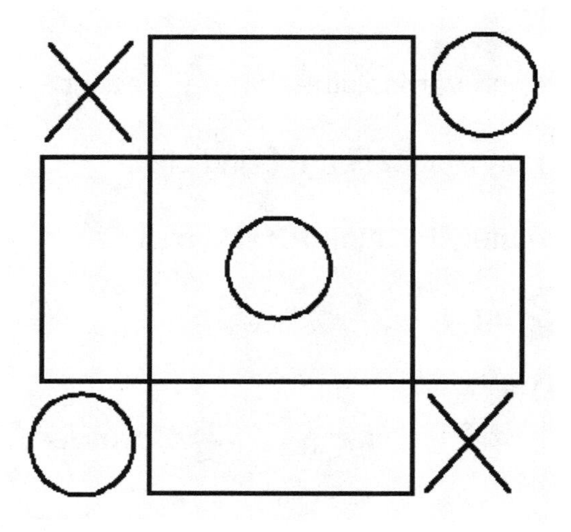

3. Modify the drawing above to appear as the drawing shown below using:

   Explode and **Erase**

   <u>Note: You must Explode a Rectangle before you can Erase an individual line.</u>
4. **Save** the drawing as: **EX6E**

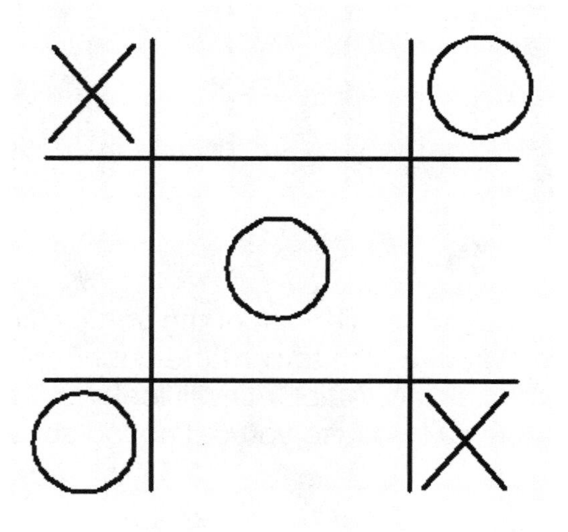

EXERCISE 6F

INSTRUCTIONS:

1. Start a **NEW** file using **2015-Workbook Helper.dwt.**
2. **Draw** the objects below using all of the commands you have learned in the previous lessons.

   Use whatever layers you desire
3. **Save** the drawing as: **EX6F**

Have some fun with this one. But be very accurate.
All objects must intersect exactly. Zoom in and take a look to make sure.

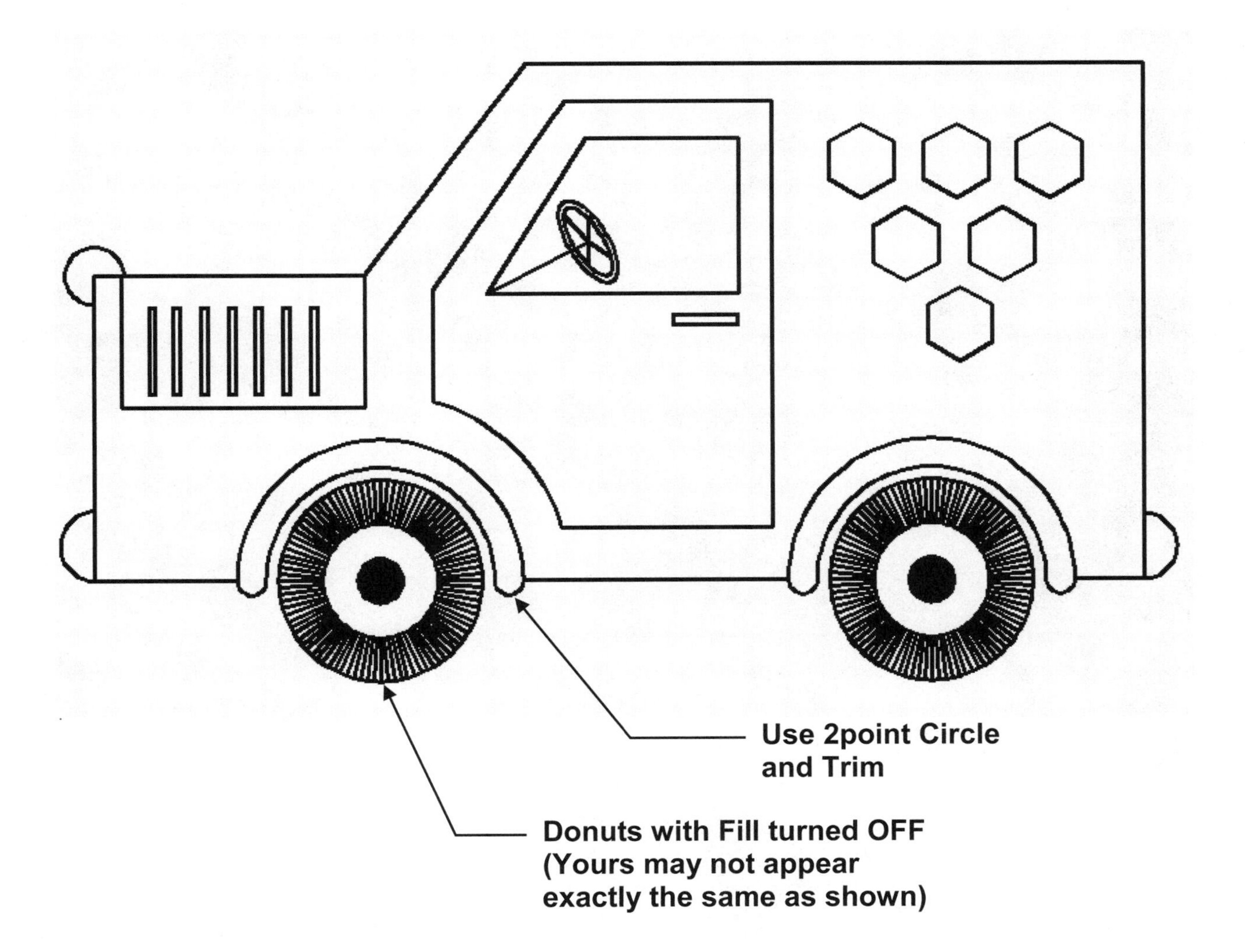

Notes:

LEARNING OBJECTIVES

After completing this lesson, you will be able to:

1. Copy objects
2. Create a mirrored image
3. Add rounded corners
4. Add Chamfered corners

LESSON 7

COPY multiple copies

The **COPY** command creates a duplicate set of the objects selected.
The COPY command is similar to the MOVE command.
(Also refer to page 6-7 for an optional copy method)

The steps required are:

1. Select the objects to be copied,
2. Select a base point
3. Select a New location for the New copy.

The difference between Copy and Move commands:
The Move command merely moves the objects to a new location.
The Copy command makes a copy and you select the location for the new copy.

1. Select the Copy command using one of the following commands:

    Ribbon = Home tab / Modify panel /
    or
    Keyboard = CO <enter>

2. The following will appear on the command line:

Command: _copy
Select objects: ***select the objects you want to copy***

Select objects: ***stop selecting objects by selecting <enter>***

Current settings: Copy mode = Multiple
Specify base point or [Displacement/mOde] <Displacement>: ***select a base point (P1)***

Specify second point or [Array / Exit / Undo] <use first point as displacement>: ***select the new location (P2)***

Specify second point or [Array / Exit / Undo] <Exit>: ***select the new location (P2) for the next copy***

Specify second point or [Array / Exit / Undo] <Exit>: ***select the new location (P2) for the next copy or select Exit to stop.***

Refer to the example on the next page

Continued on the next page...

COPY multiple copies....continued

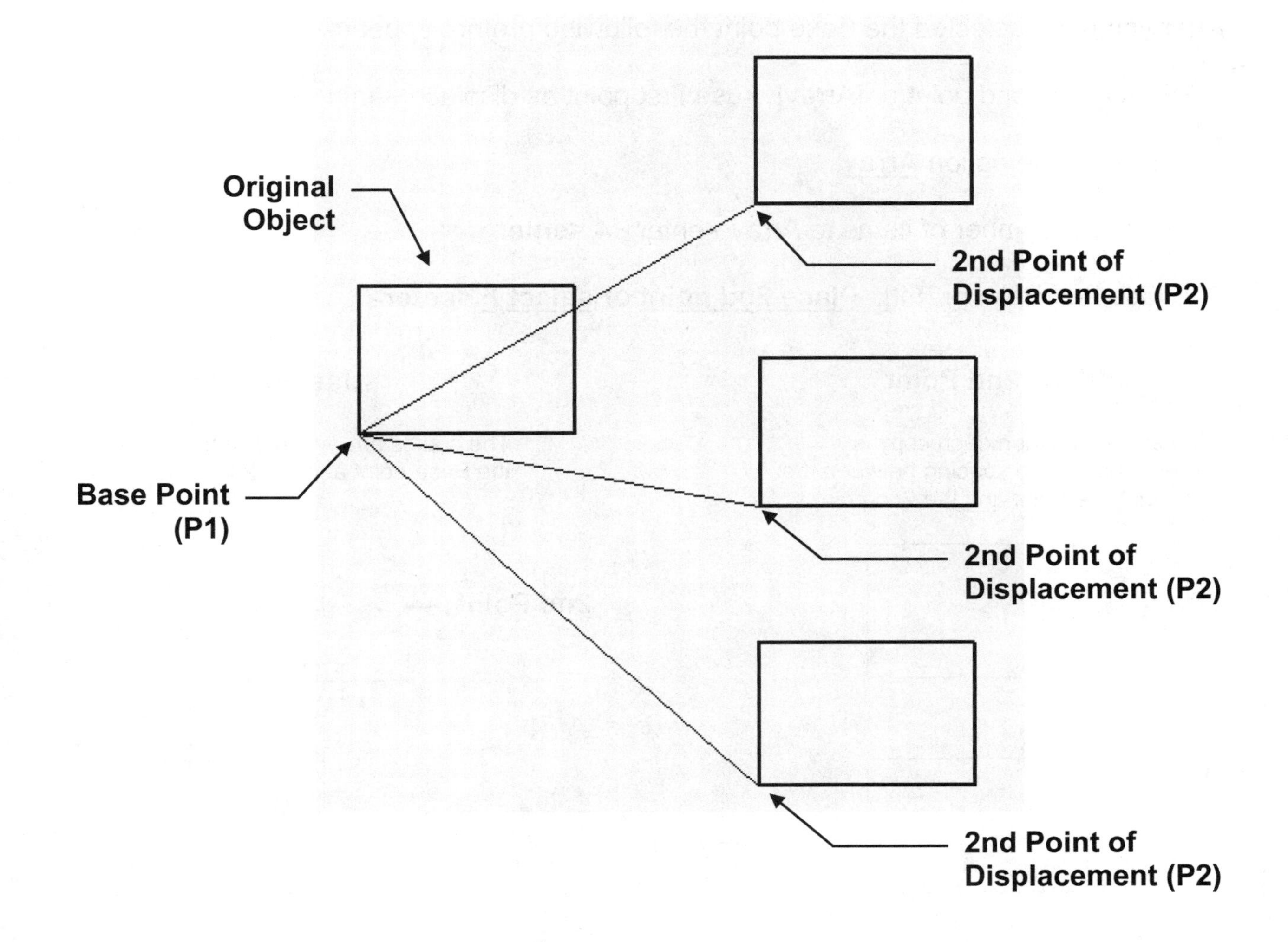

The copy command continues to make copies until you press <enter> to exit.

Changing the "Mode"

You may change the "Mode" to **Single** if you prefer to have AutoCAD stop the Copy command automatically after a single copy.

After you have selected the object(s) to copy the following prompt appears:

Current settings: Copy mode = <u>Single</u>
Specify base point or [Displacement/mOde] <Displacement>:

If you select the option **<u>mOde</u>**, you may select <u>Single</u> or <u>Multiple</u> copy mode.

Continued on the next page...

COPY "Array" option

The Copy command allows you to make an Array of copies.

After you have selected the Base point the following prompt appears:

Specify second point or [Array] <use first point as displacement>:

If you select the option **Array**,

1. Enter the number of items to Array <enter> **4 <enter>**

2. Place 2nd Point or [Fit]: **Place 2nd point or select F <enter>**

Place 2nd Point

The space between each copy is determined by the spacing between the original base point and the 2nd point.

Use Fit option

The copies are evenly spaced between the Base point and the 2nd Point.

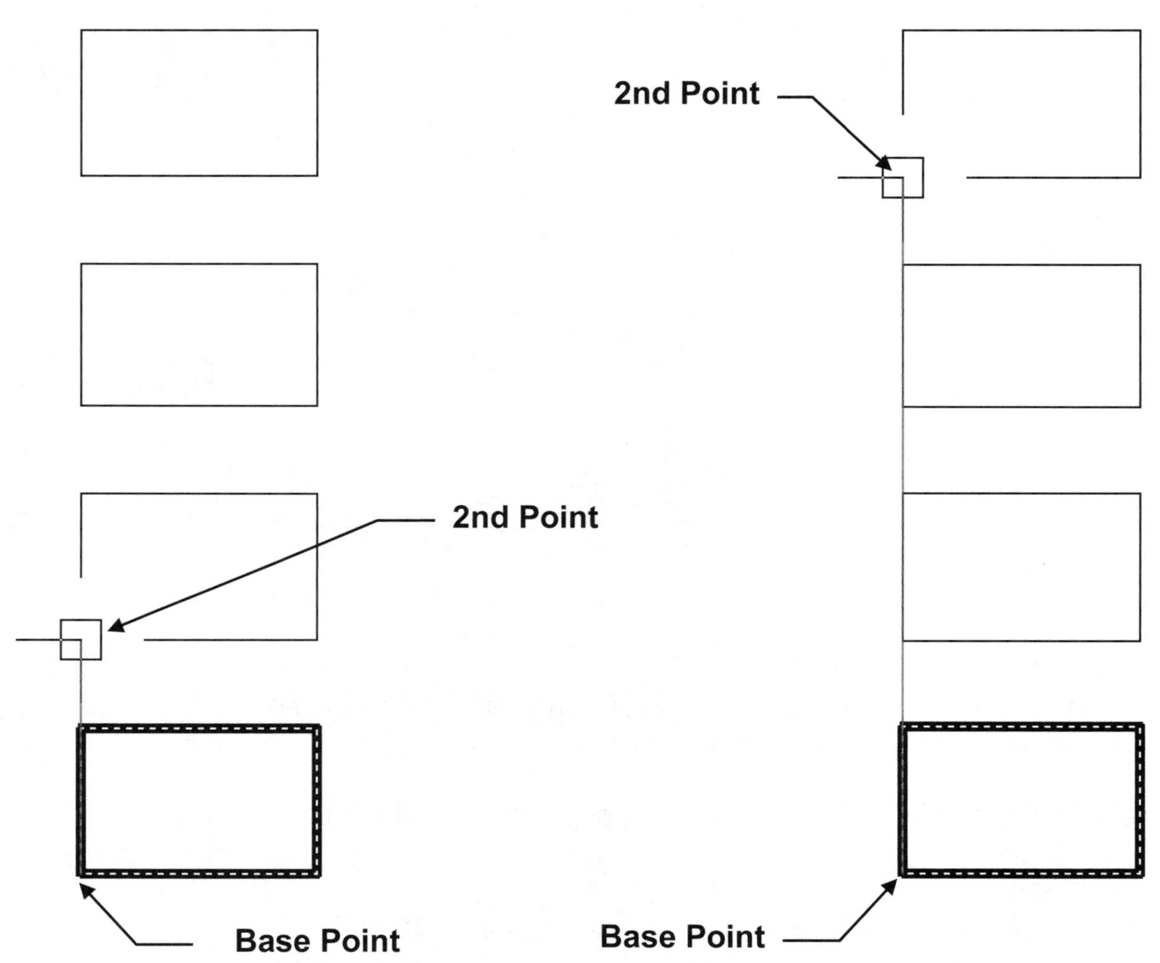

Note: The Array option within the Copy command is a quick method to create multiple copies. But AutoCAD has a more powerful and accurate Array command described in Lesson 13.

MIRROR

The **MIRROR** command allows you to make a mirrored image of any objects you select. You can use this command for creating right / left hand parts or draw half of a symmetrical object and mirror it to save drawing time.

1. Select the **MIRROR** command using one of the following:

   Ribbon = Home tab / Modify panel /
   or
   Keyboard = MI <enter>

2. The following will appear on the command line:

   Command: _mirror
   Select objects: ***select the objects to be mirrored***
   Select objects: ***stop selecting objects by selecting <enter>***
   Specify first point of mirror line: ***select the 1st end of the mirror line***
   Specify second point of mirror line: ***select the 2nd end of the mirror line***
   Erase source objects? [Yes/No] <N>: ***select Y or N***

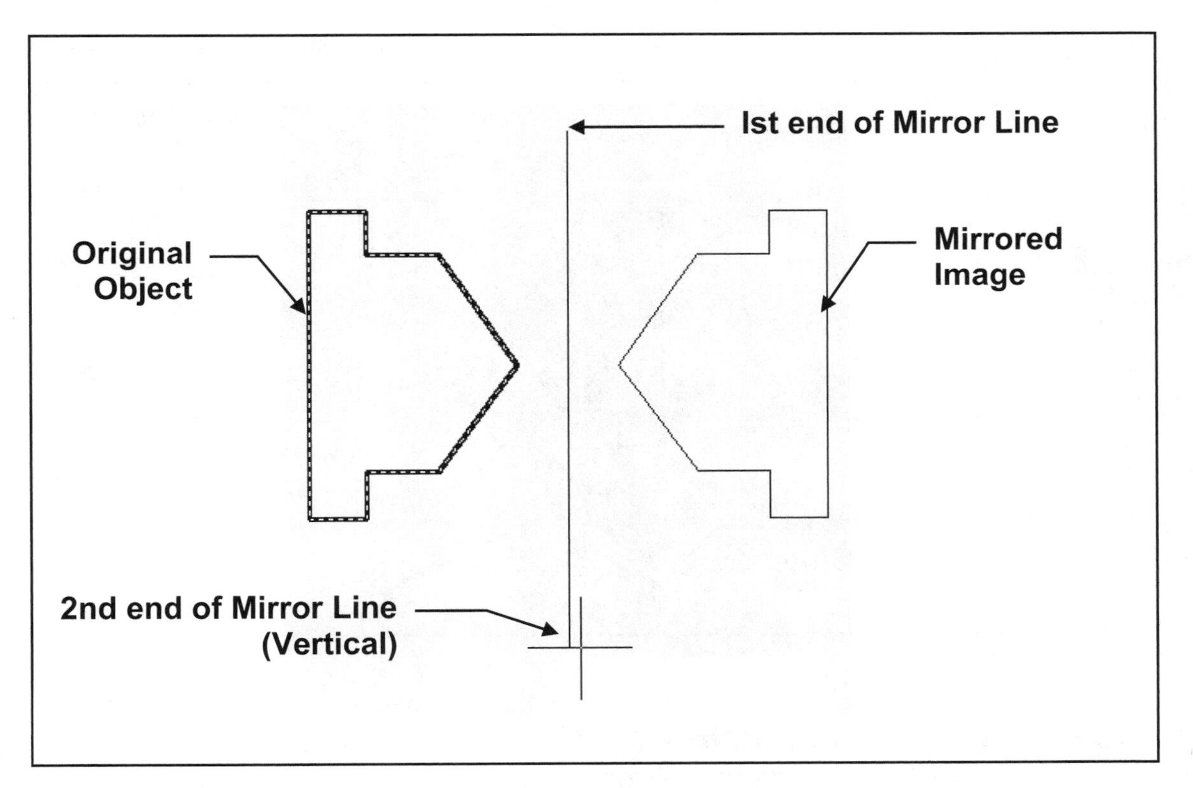

Mirror Line "Vertical"

Continued on the next page...

MIRROR....continued

Note:
The placement of the "Mirror Line" is important. You may make a mirrored copy horizontally, vertically or on an angle. See examples below and on the previous page.

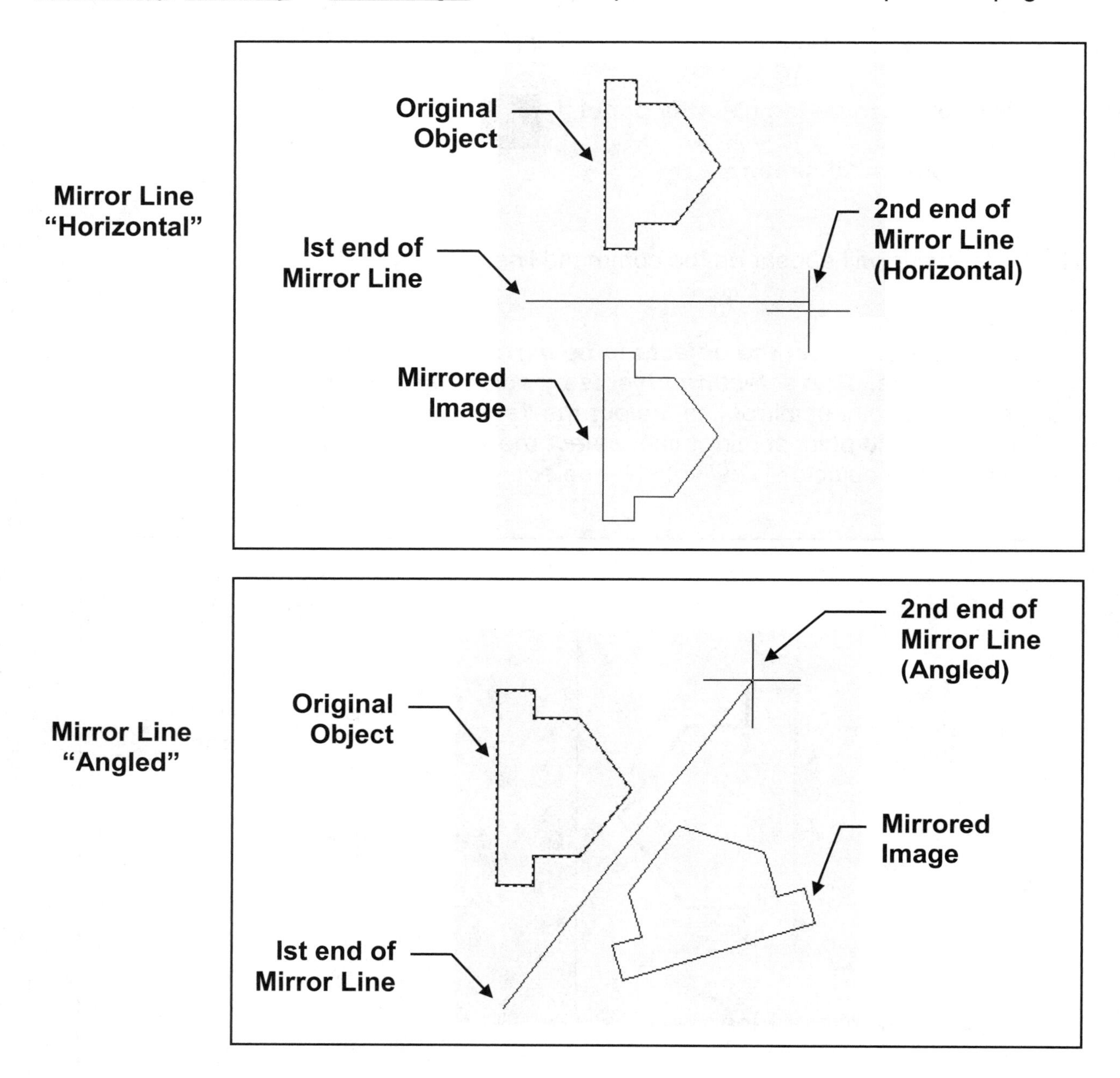

How to control text when using the Mirror command:
(Do the following **before** you use the mirror command)

1. At the command line type **mirrtext <enter>**
2. If you want the text to mirror (reverse reading): type 1 <enter>
   If you do not want the text to mirror: type 0 <enter>

MIRRTEXT SETTING = 1 MIRRTEXT SETTING = 0

FILLET

The **FILLET** command will create a radius between two objects. The objects do not have to be touching. If two parallel lines are selected, it will construct a full radius.

RADIUS A CORNER

1. Select the **FILLET** command using one of the following:

   Ribbon = Home tab / Modify panel /
   or
   Keyboard = F <enter>

2. The following will appear on the command line:

3. **Set the radius of the fillet**
   Command: _fillet
   Current settings: Mode = TRIM, Radius = 0.000
   Select first object or [Undo/Polyline/Radius/Trim/Multiple]: ***type "R" <enter>***
   Specify fillet radius <0.000>: ***type the radius <enter>***

4. **Now fillet the objects**
   Select first object or [Undo/Polyline/Radius/Trim/Multiple]: ***select the 1st object to be filleted***
   Select second object or shift-select to apply corner or [Radius]: ***select the 2nd object to be filleted***

Note: *When you place the Cursor on the second object AutoCAD displays the Fillet and allows you to change the Radius before it is actually drawn. If you choose to change the Radius, select the Radius option, enter a new radius value then select the 2nd object.*

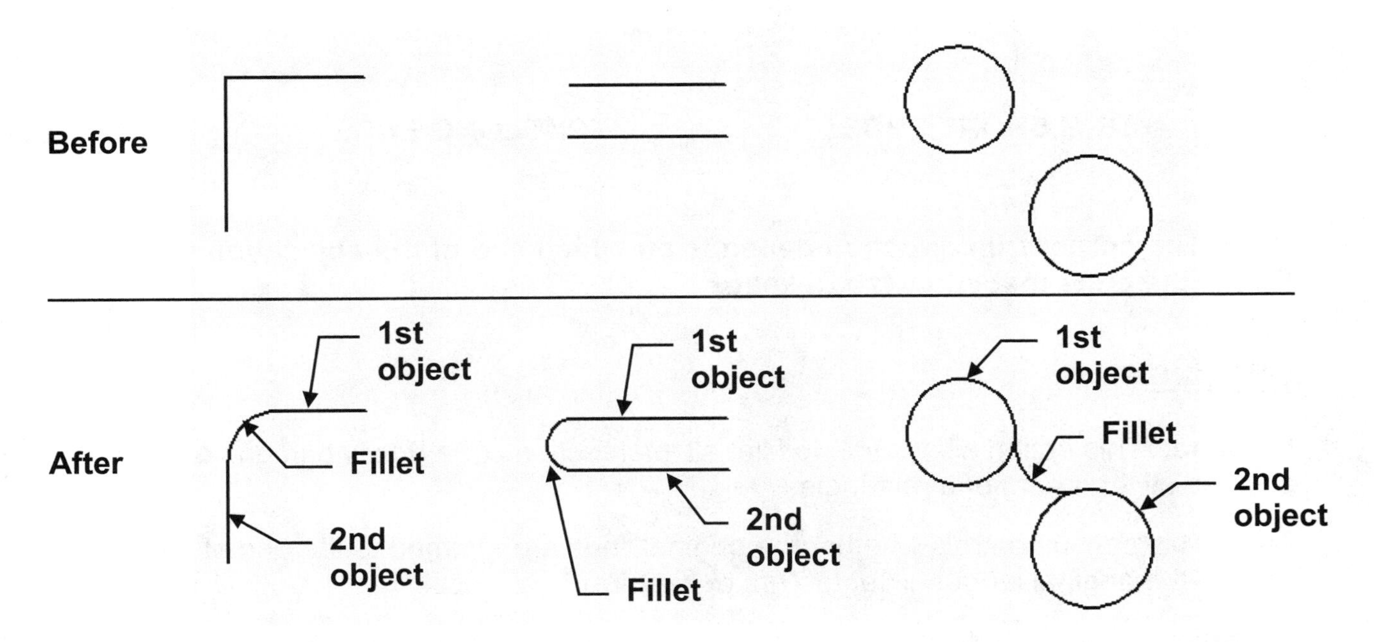

FILLET....continued

The FILLET command may also be used to create a square corner.

SQUARE CORNER

1. Select the **FILLET** command

2. The following will appear on the command line:

   Select first object or [Undo/Polyline/Radius/Trim/Multiple]: ***select the 1st object (P1)***

   Select second object or shift-select to apply corner: ***Hold the shift key down while selecting the 2nd object (P2)***

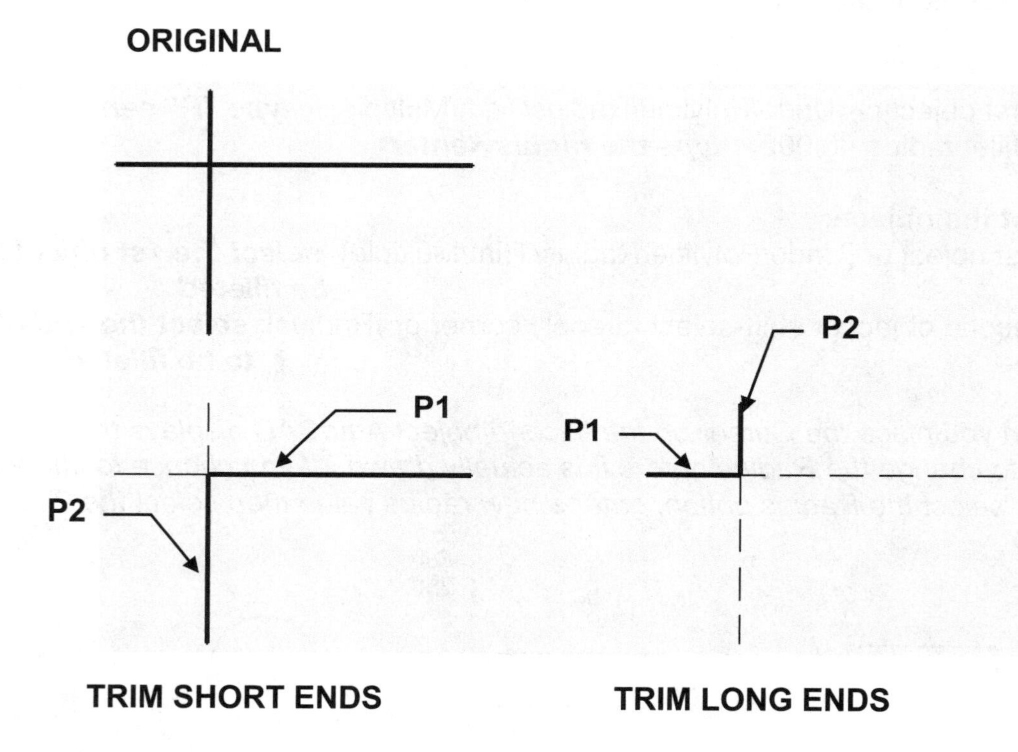

Note: The corner trim direction depends on which end of the object you select. Select the ends that you wish to keep.

OPTIONS:

Polyline: This option allows you to fillet all intersections of a Polyline in one operation, such as all 4 corners of a rectangle.

Trim: This option controls whether the original lines are trimmed to the end of the Arc or remain the original length. (Set to Trim or No trim)

Multiple: Repeats the fillet command until you press <enter> or Esc key.

CHAMFER

The **CHAMFER** command allows you to create a chamfered corner on two lines.
There are two methods: **Distance (below) and Angle (next page)**.

DISTANCE METHOD

Distance Method requires input of a distance for each side of the corner.

1. Select the **CHAMFER** command using one of the following:

   Ribbon = Home tab / Modify panel /
   or
   Keyboard = CHA <enter>

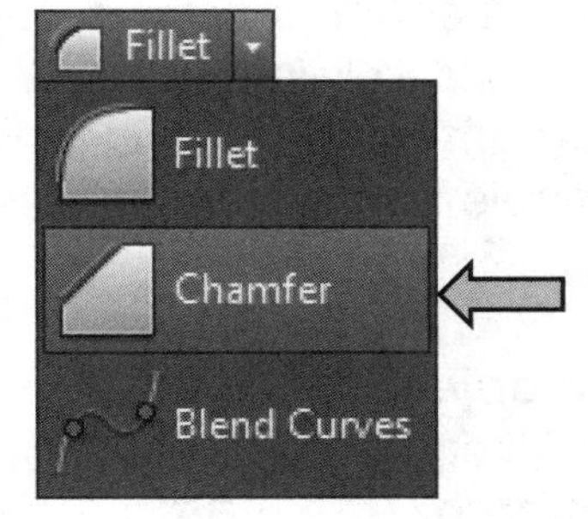

Command: _chamfer
(TRIM mode) Current chamfer Dist1 = 0.000, Dist2 = 0.000
Select first line or [Undo/Polyline/Distance/Angle/Trim/mEthod/Multiple]: ***select "D"<enter>.***
Specify first chamfer distance <0.000>: ***type the distance for first side <enter>.***
Specify second chamfer distance <0.000>: ***type the distance for second side <enter>.***

2. **Now chamfer the object.**
   Select first line or [Undo/Polyline/Distance/Angle/Trim/mEthod/Multiple]: ***select the (First side) to be chamfered (distance 1).***
   Select second line or shift-select to apply corner or [Distance/Angle/Method]: ***select the (Second side) to be chamfered (distance 2).***

Note: When you place the cursor on the second side AutoCAD displays the Chamfer *and allows you to change the Distances before it is actually drawn. If you choose to change the Distance, select the Distance option, enter new distance values then select the 2nd side.*

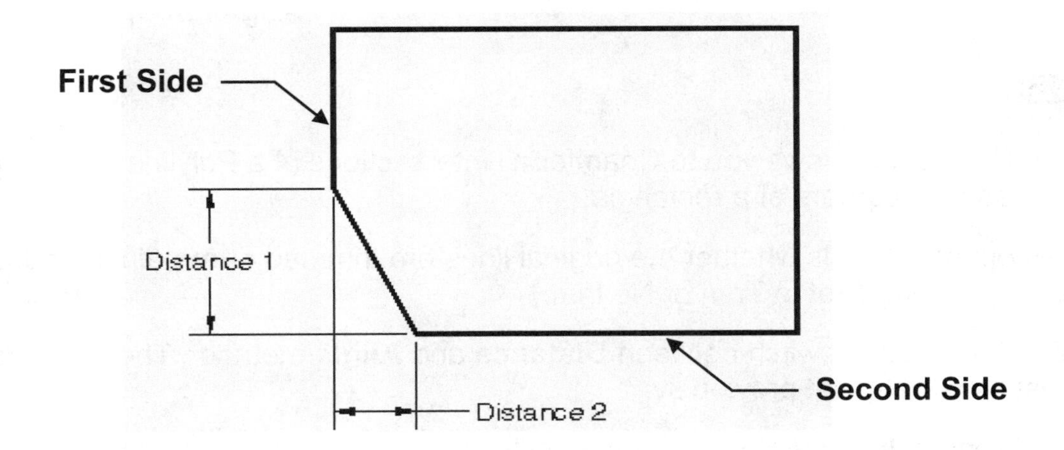

CHAMFER....continued

ANGLE METHOD

Angle method requires input for the length of the line and an angle

1. Select the **CHAMFER** command

   Command: _chamfer
   (TRIM mode) Current chamfer Dist1 = 1.000, Dist2 = 1.000
   Select first line or [Undo/Polyline/Distance/Angle/Trim/method/Multiple]: ***type A <enter>***
   Specify chamfer length on the first line <0.000>: ***type the chamfer length <enter>***
   Specify chamfer angle from the first line <0>: ***type the angle <enter>***

2. **Now Chamfer the object**
   Select first line or [Undo/Polyline/Distance/Angle/Trim/mEthod/Multiple]: ***select the (First Line) to be chamfered. (the length side)***
   Select second line or shift-select to apply corner: ***select the (second line) to be chamfered. (the Angle side)***

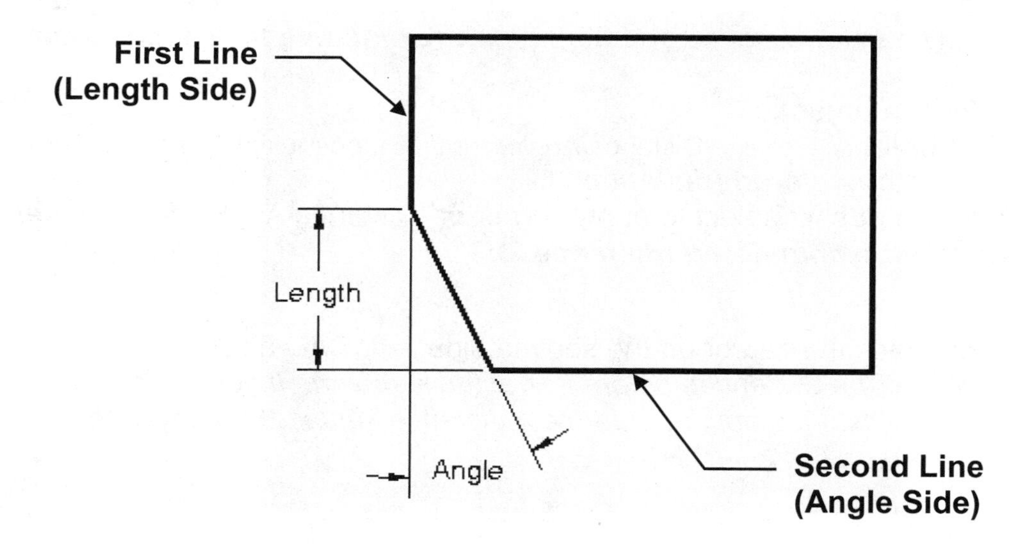

OPTIONS:

Polyline: This option allows you to Chamfer all intersections of a Polyline in one operation. Such as all 4 corners of a rectangle.

Trim: This option controls whether the original lines are trimmed or remain after the corners are chamfered. (Set to Trim or No trim.)

mEthod: Allows you to switch between **Distance** and **Angle** method. The distance or angle must have been set previously.

Multiple: Repeats the Chamfer command until you press <enter> or Esc key.

EXERCISE 7A

INSTRUCTIONS:

1. Start a **NEW** file using **2015-Workbook Helper.dwt.**
2. **Draw** the Lines and one Circle shown below.

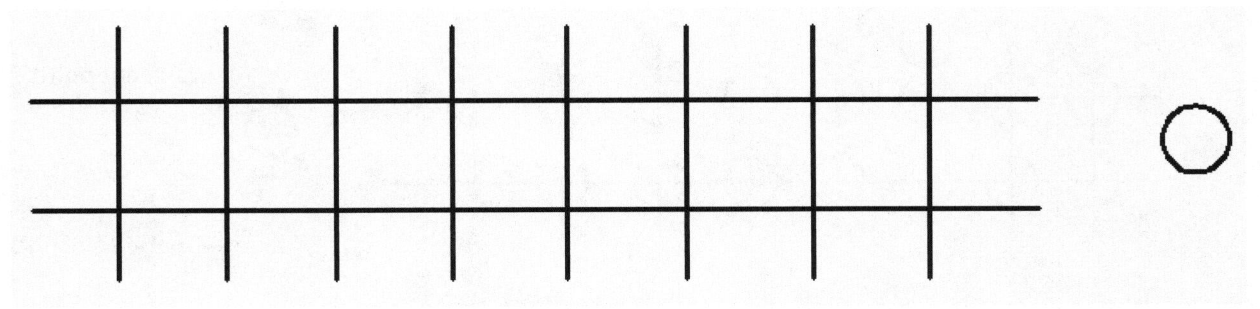

BEFORE COPY

3. Copy the Circle to all of the other locations.
   A. Select the Copy command
   B. Select the Circle
   C. Select the basepoint on the Original Circle.
      Note: the basepoints are different.
   D. Select the New location (2nd point of displacement)
4. **Save** the drawing as: **EX7A**

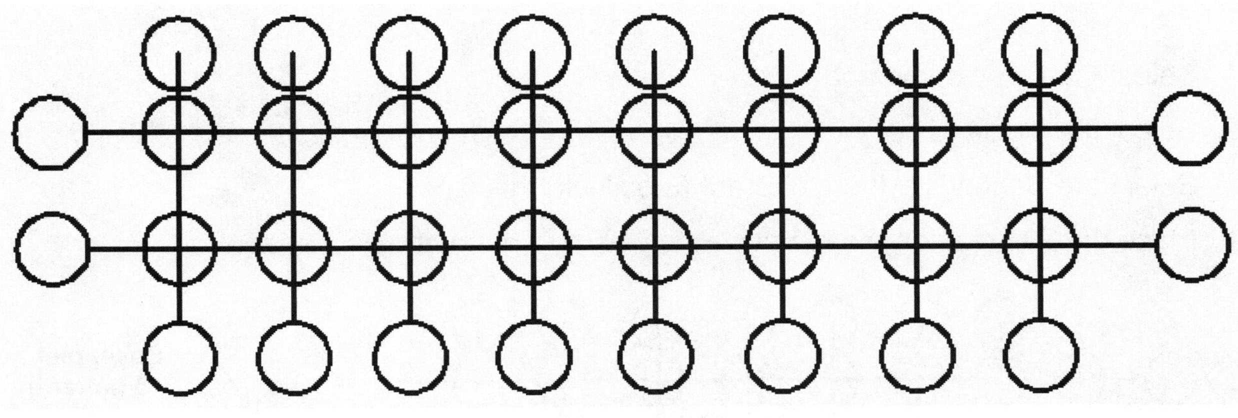

AFTER COPY

Note: Step by step instructions on the next page.

Helpful hints for 7A

1. Select the Copy command
2. Select the Circle
3. Select the Basepoint: the Center of the Circle (use object snap Center)
4. Set Running Object snap to Endpoint and Intersection and place all 24 accurately.

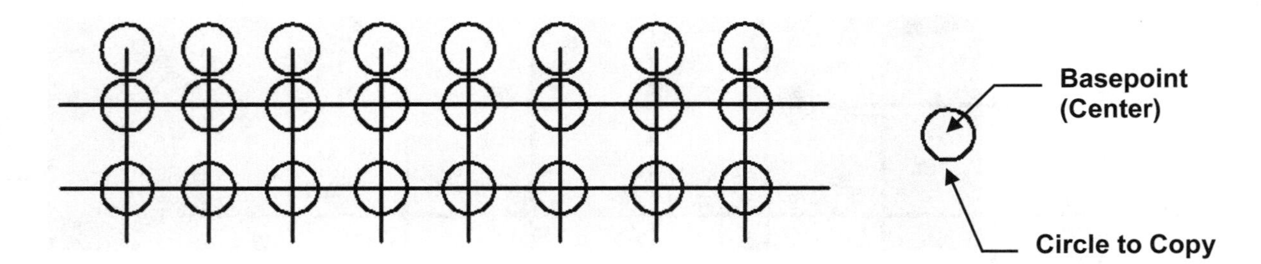

1. Select the Copy command again.
2. Select the Circle to copy
3. Select the Basepoint: the Quadrant (12 o'clock) this time.
4. Place the 8 bottom circles accurately using Endpoint object snap.

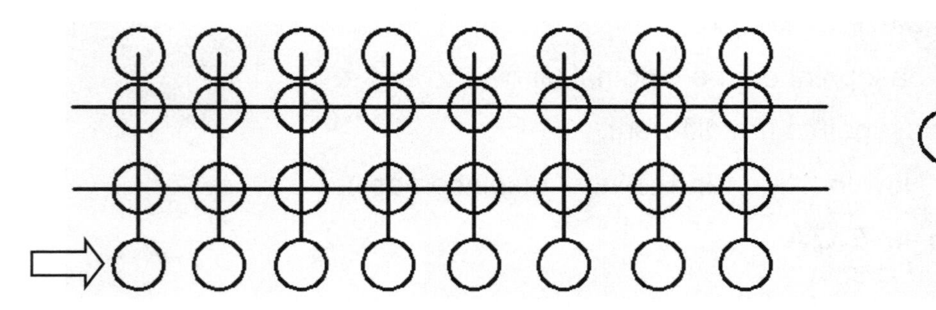

1. Select the Copy command
2. Select the Circle to copy
3. Select the Basepoint: the Quadrant (3 o'clock) this time.
4. Place the 2 circles on the left accurately using Endpoint object snap.

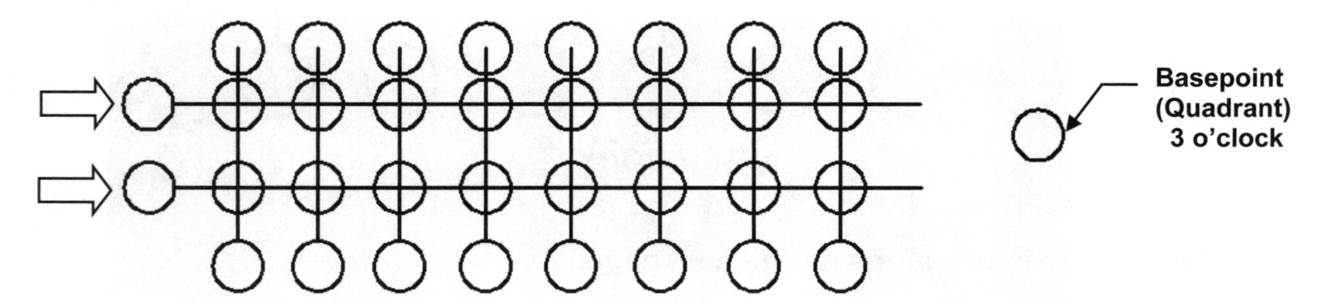

You should be able to figure out the last 2 circles on the right.

EXERCISE 7B

INSTRUCTIONS:

1. Start a **NEW** file using **2015-Workbook Helper.dwt.**
2. **Draw** the Rectangle shown below.
3. Radius the corners using the **Fillet** command.
4. **Save** the drawing as: **EX7B**

BEFORE FILLET

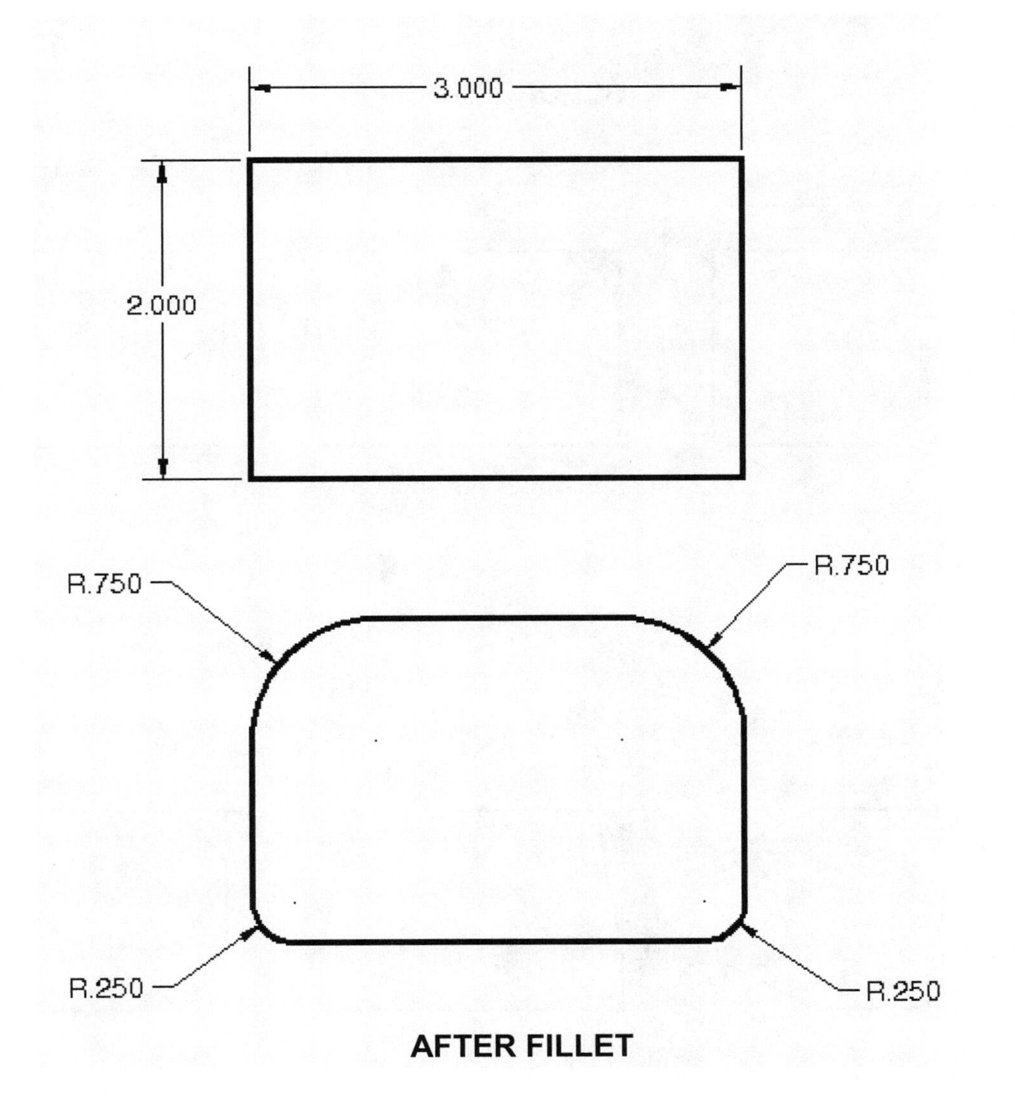

AFTER FILLET

EXERCISE 7C

INSTRUCTIONS:

1. Start a **NEW** file using **2015-Workbook Helper.dwt.**
2. **Draw** the 3 Rectangles shown below.
3. Chamfer the corners using the **Chamfer** command.
4. **Save** the drawing as: **EX7C**

BEFORE CHAMFER

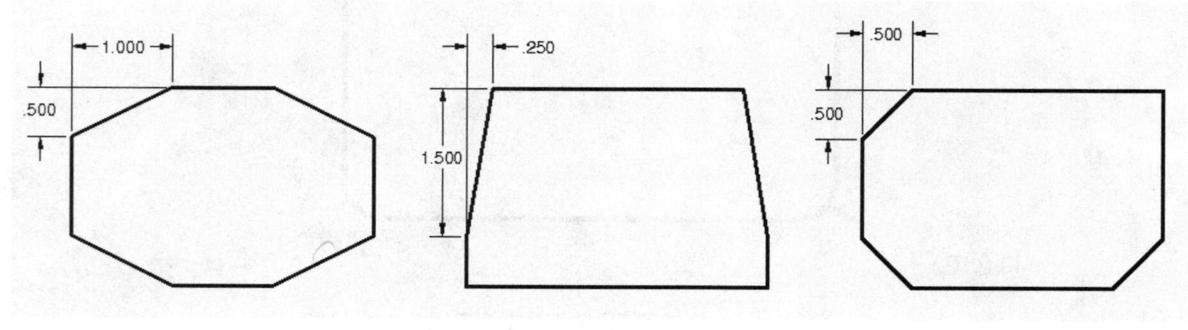

AFTER CHAMFER

EXERCISE 7D

INSTRUCTIONS:

1. Start a **NEW** file using **2015-Workbook Helper.dwt.**
2. **Draw** the Rectangle shown below.
3. Chamfer the corners using the **Chamfer** command.
4. **Save** the drawing as: **EX7D**

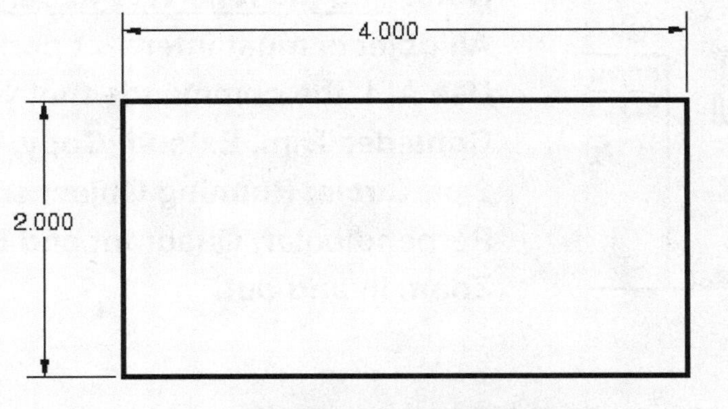

BEFORE CHAMFER

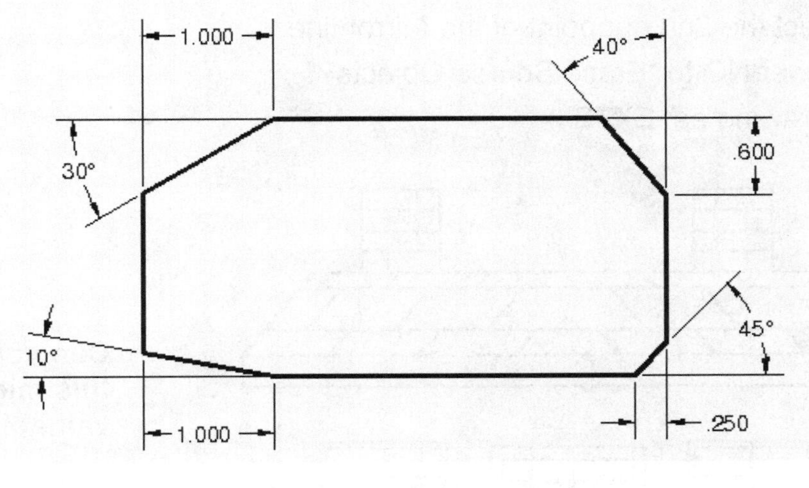

AFTER CHAMFER

EXERCISE 7E

INSTRUCTIONS:

1. Start a **NEW** file using **2015-Workbook Helper.dwt.**
2. **Draw** the "Half house shown below.

   Use Layers: Roof, Wall, Window, Door, Plants and Bricks.

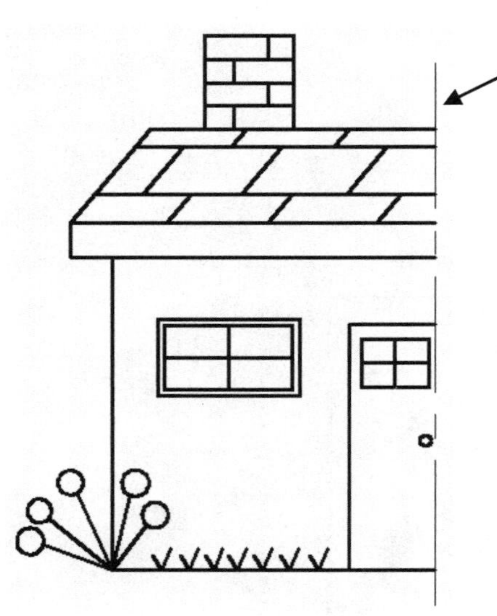

Hint: Draw this line first.
Also snap to it for the Mirror Line.
Your halves should match perfectly.

Note: <u>You must be very accurate</u>.
All objects must intersect perfectly.
Use ALL the commands that you have learned.
Consider Trim, Extend, Copy, Rectangle,
2 pt. Circle, Running Object snap: Nearest,
Perpendicular, Quadrant and Endpoint.
Zoom in and out.

3. Now create a Mirrored image of the half house:
   - A. Select the Mirror command.
   - B. Select the objects to be Mirrored (You could use Window selection).
   - C. Select the 1st endpoint of the Mirror line (Ortho ON).
   - D. Select the 2nd endpoint of the Mirror line.
   - E. Answer NO to "Erase Source Objects?"
4. **Save** the drawing as: **EX7E**

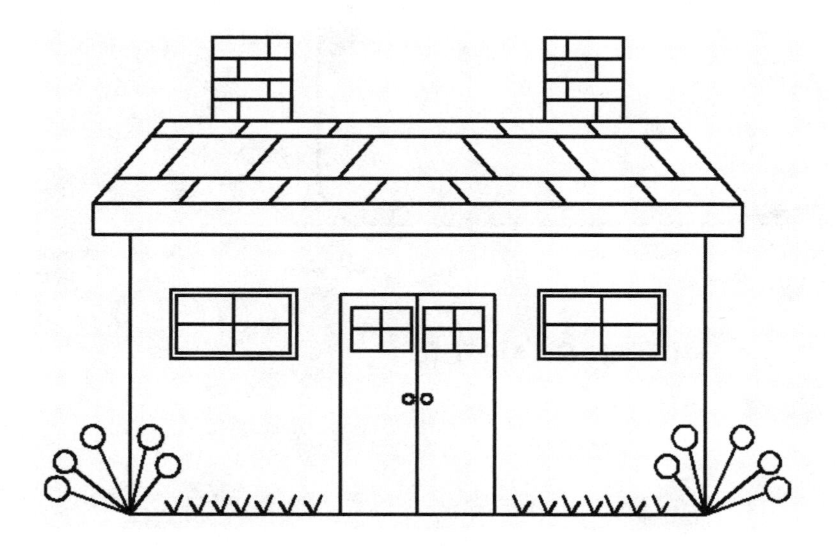

Can you imagine how this might save you valuable time?

LEARNING OBJECTIVES

After completing this lesson, you will be able to:

1. Add a Single Line of text to the drawing
2. Add a paragraph using Multiline text
3. Control tabs, indents and use the Spelling Checker
4. Add Columns
5. Edit existing text

LESSON 8

SINGLE LINE TEXT

SINGLE LINE TEXT allows you to draw one or more lines of text. The text is visible as you type. To place the text in the drawing, you may use the default **START POINT** (the lower left corner of the text), or use one of the many styles of justification described on the next page. Each line of text is treated as a separate object.

USING THE DEFAULT START POINT

1. Select the **SINGLE LINE TEXT** command using one of the following:

Ribbon = Annotate tab / Text panel /
or
Keyboard = DT <enter>

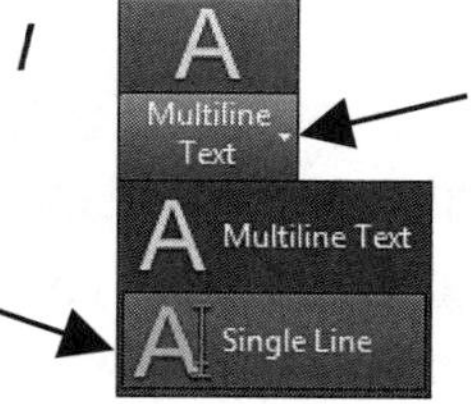

Command: _text
Current text style: "STANDARD" Text height: 0.200 Annotative: No
Specify start point of text or [Justify/Style]: ***Place the cursor where the text should start and left click.***
Specify height <0.200>: ***type the height of your text <enter>***
Specify rotation angle of text <0>: ***type the rotation angle then <enter>***
Enter text: ***type the text string; press enter at the end of the sentence***
Enter text: ***type the text string; press enter at the end of the sentence***
Enter text: ***type the next sentence or press <enter> to stop***

USING JUSTIFICATION

If you need to be very specific, where your text is located, you must use the Justification option. For example if you want your text in the middle of a rectangular box, you would use the justification option "Middle".

The following is an example of Middle justification.

1. Draw a Rectangle 6" wide and 3" high.
2. Draw a Diagonal line from one corner to the other corner.

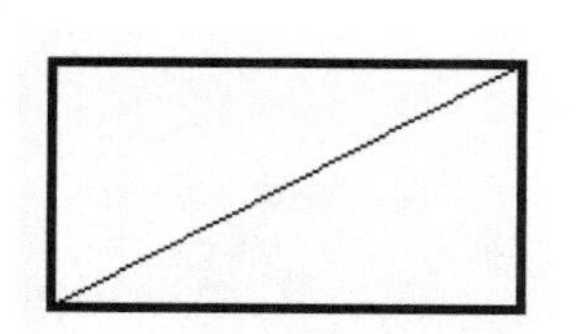

3. Select the SINGLE LINE TEXT command
   Command: _text
   Current text style: "STANDARD" Text height: 0.200
4. Specify start point of text or[Justify/Style]: ***type "J"<enter>***
5. Enter an option [Align/Fit/Center/Middle/Right/TL/ TC/TR/ ML/MC/MR/BL/BC/BR]: ***type M <enter>***
6. Specify middle point of text: ***snap to the midpoint of the diagonal line***
7. Specify height <0.200>: ***1 <enter>***
8. Specify rotation angle of text <0>: ***0 <enter>***
9. Enter text: ***type: HHHH <enter>***
10. Enter text: ***press <enter> to stop***

Also refer to Exercise 8C for "Midpoint between 2 pts" method.

SINGLE LINE TEXT....continued

OTHER JUSTIFICATION OPTIONS:

ALIGN
Aligns the line of text between two points specified.
The height is adjusted automatically.

FIT
Fits the text between two points specified.
The height is specified by you and does not change.

CENTER
This is a tricky one. Center is located at the bottom center of Upper Case letters.

MIDDLE
If only uppercase letters are used **MIDDLE** is located in the middle, horizontally and vertically. If both uppercase and lowercase letters are used **MIDDLE** is located in the middle, horizontally and vertically, but considers the lowercase letters as part of the height.

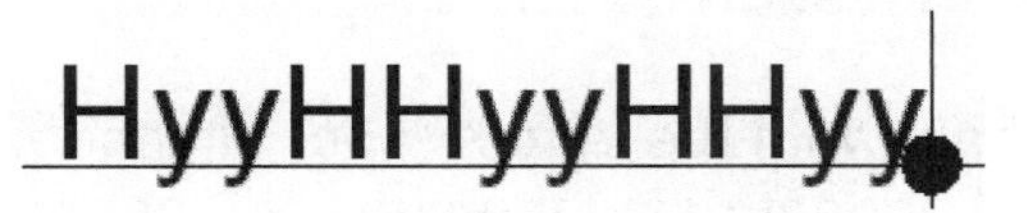

RIGHT
Bottom right of upper case text.

TL, TC, TR
Top left, Top center and Top right of upper and lower case text.

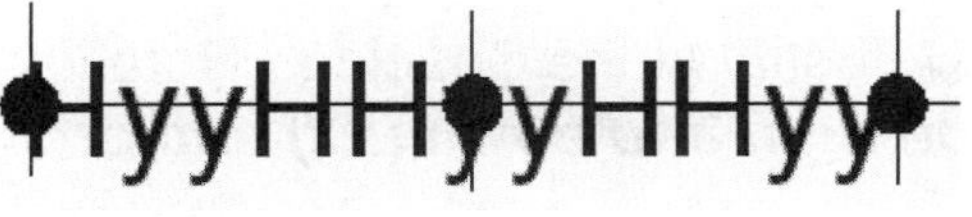

ML, MC, MR
Middle left, Middle center and Middle right of upper case text.
(Notice the difference between “Middle” and “MC”.)

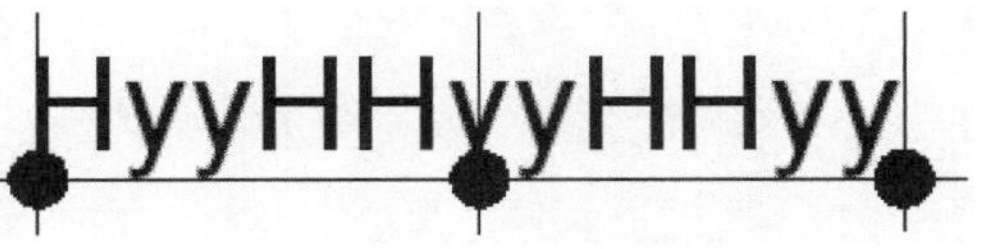

BL, BC, BR
Bottom left, Bottom center and Bottom right of lower case text.
Notice the different location for **BR** and **RIGHT** shown above.
BR considers the lower case letters with tails as part of the height.

MULTILINE TEXT

MULTILINE TEXT command allows you to easily add a sentence, paragraph or tables. The Mtext editor has most of the text editing features of a word processing program. You can underline, bold, italic, add tabs for indenting, change the font, line spacing, and adjust the length and width of the paragraph.

When using Mtext you must first define a text boundary box. The text boundary box is defined by entering where you wish to start the text (first corner) and approximately where you want to end the text (opposite corner). It is very similar to drawing a rectangle. The paragraph text is considered one object rather than several individual sentences.

USING MULTILINE TEXT

1. Select the MULTILINE TEXT command using one of the following:

Ribbon = Annotate tab / Text panel /
or
Keyboard = MT <enter>

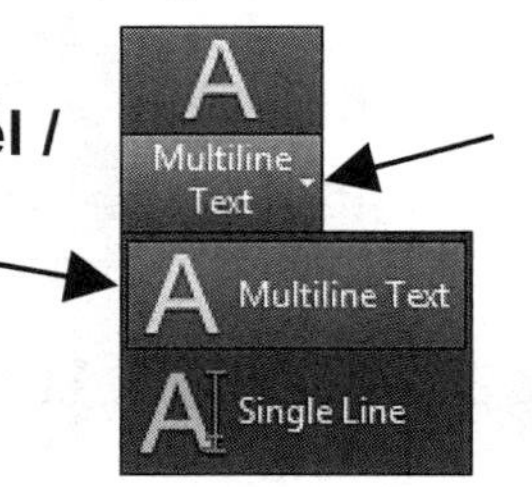

The command line will list the current style, text height and annotative setting.

Mtext Current text style: "STANDARD" Text height: .200 Annotative: No

Note: <u>Annotative</u> will be discussed in lesson 26 and 27.

The cursor will then appear as crosshairs with the letters "abc" attached. These letters indicate how the text will appear using the current font and text height.

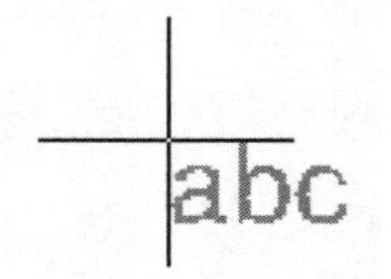

2. Specify first corner: ***Place the cursor at the upper left corner of the area where you want to start the new text boundary box and press the left mouse button. (P1)***

3. Specify opposite corner or [Height / Justify / Line Spacing / Rotation / Style / Width / Columns]: ***Move the cursor to the right and down (P2) and press left mouse button.***

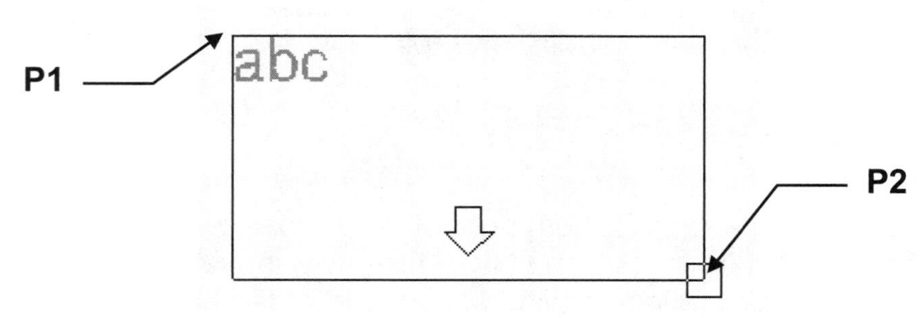

MULTILINE TEXT....continued

The Text Editor Ribbon will appear.

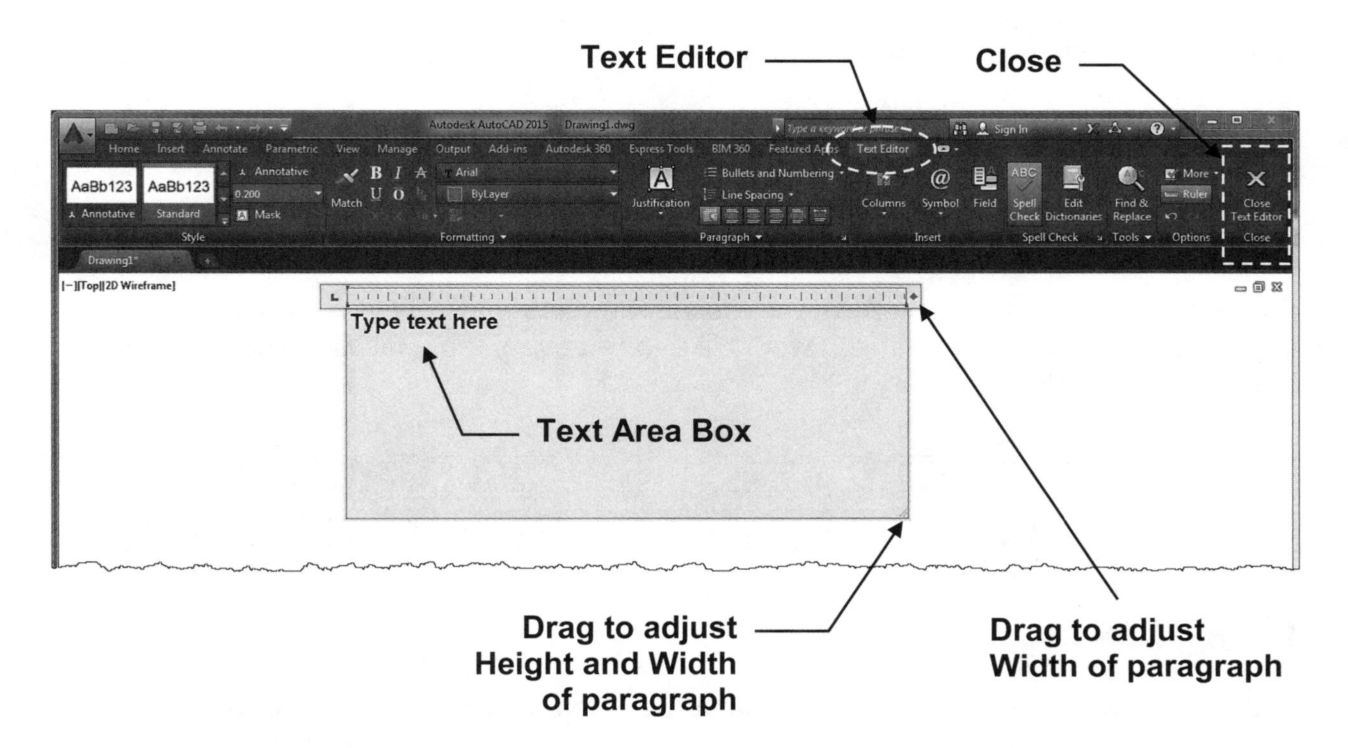

The **Text Editor** allows you to select the Text Style, Font, Height etc. You can add features such as bold, italics, underline and color.

The **Text Area box** allows you to enter the text, add tabs, indent, adjust left hand margins and change the width and height of the paragraph.

4. After you have entered the text in the Text Area box, select the **Close Text Editor** tool.

HOW TO CHANGE THE “abc”, ON THE CROSSHAIRS, TO OTHER LETTERS.

You can personalize the letters that appear attached to the crosshairs using the **MTJIGSTRING** system variable. (10 characters max) The letters will simulate the appearance of the font and height selected but will disappear after you place the lower right corner (P2).

1. Type **MTJIGSTRING** <enter> on the command line.
2. Type the new letters <enter>.

The letters will be saved to the computer, not the drawing. They will appear anytime you use Mtext and will remain until you change them again.

TABS, INDENTS and SPELLING CHECKER

TABS

Setting and removing Tabs is very easy.

The increments are determined by the text height. (For example: If the text height is 1" you may quickly place a tab at any 1" mark on the ruler. To be more specific refer to page 8-9.)

Set or change the stop positions at anytime, using one of the following methods.

Place the cursor on the "Ruler" where you want the tab and left click. A little dark "**L**" will appear. The tab is set. If you would like to remove a tab, just click and drag it off the ruler and it will disappear.

INDENTS

Sliders on the ruler show indention relative to the left side of the text boundary box. The top slider indents the first line of the paragraph, and the bottom slider indents the other lines of the paragraph. (Also refer to page 8-9)

You may change their positions at anytime, using one of the following methods.

Place the cursor on the "Slider" and click and drag it to the new location.

SPELLING CHECKER

If you have Spell check ON you will be alerted as you enter text with a red line under the misspelled word. Right click on the word and AutoCAD will give you some choices.

1. Select the text you wish to Spell check. (Click once on sentence)
2. Select **Annotate tab / Text panel /** 

   The Check Spelling dialog box will appear.
3. Select Start.
   If AutoCAD finds any words misspelled it will suggest a change.
   You may select **Change** or **Ignore**.
   When finished a message will appear stating "**Spelling Check Complete**".

COLUMNS

STATIC COLUMNS

1. Right click in the **Text Box Area** and select **Columns**
2. Select **Column Settings...**

The Column Dialog box appears.

3. Select **Static Columns**
4. Select:
   A. Column Number
   B. Height
   C. Width
   D. Gutter
5. Select the **OK** button
6. The Text Area should appear as shown below with 2 columns divided with a gutter.
7. Start typing in the left hand box. When you fill the left hand box the text will start to spill over into the right hand box.

You may also adjust and make changes to the width and height using the drag tools.

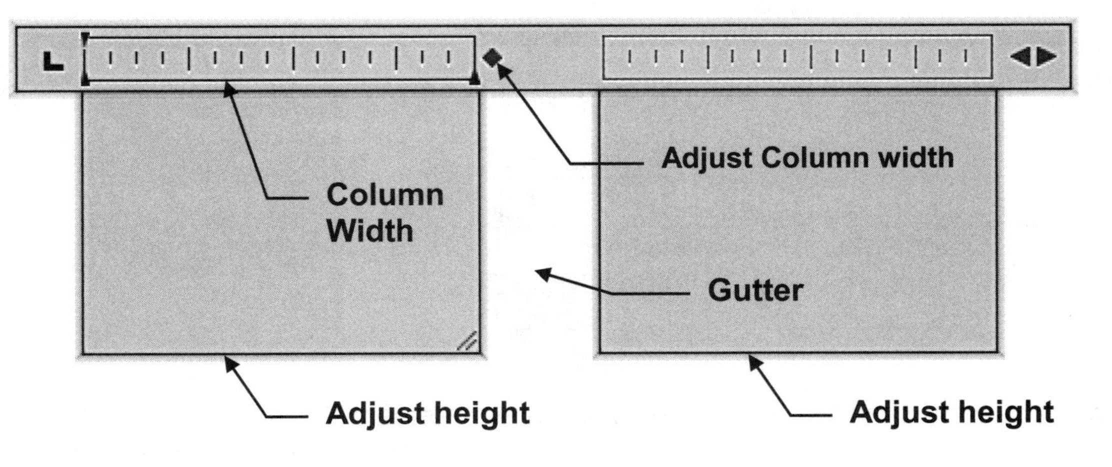

Continued on the next page...

COLUMNS....continued

DYNAMIC COLUMNS

1. Right click in the **Text Box Area** and select **Columns**
2. Select **Column Settings...**

The Column Dialog box appears.

3. Select **Dynamic Columns**
   Auto Height
4. Select:
   A. Height
   B. Width
   C. Gutter
5. Select the **OK** button

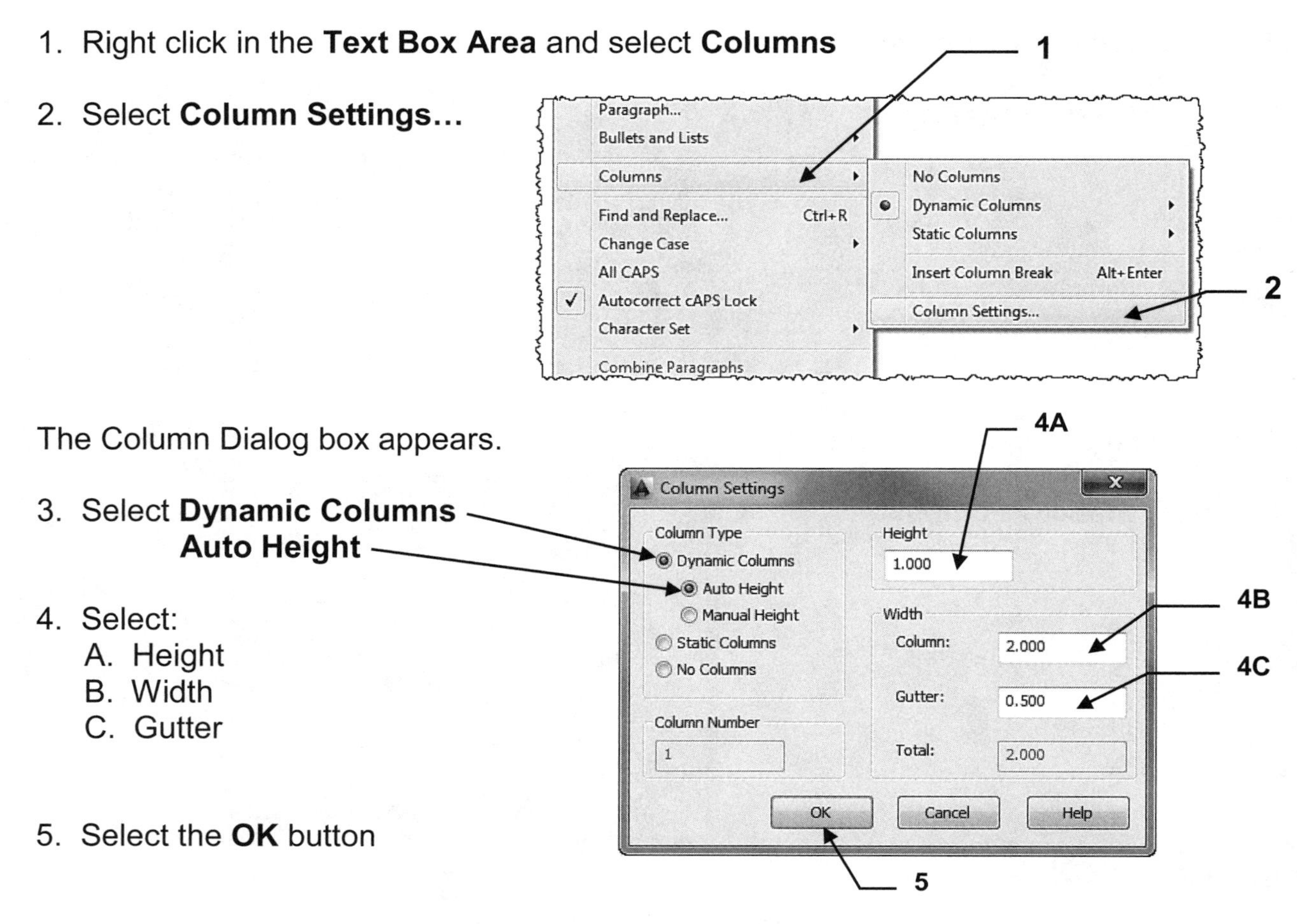

6. The Text Area will first appear with one column with the width and height you set.

7. When the text fills the first column another column will appear.

Today is the day that we

will learn columns.

When the second column fills another column will appear.
You may also adjust and make changes to width and height using the drag tools.

PARAGRAPH and LINE SPACING

PARAGRAPH and LINE SPACING

You may set the tabs, indent and line spacing for individual paragraphs.

1. Right click in the **Text Box Area** and select **Paragraph**.

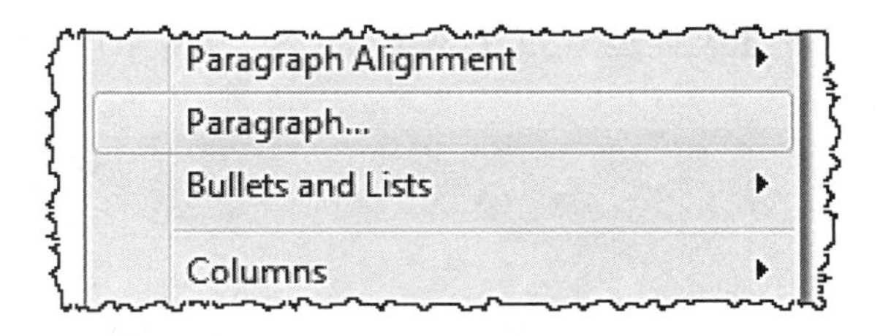

The Paragraph dialog box will appear.

To Add:
1. Type spacing.
2. Select **Add** button.

To Modify:
1. Select the current spacing.
2. Modify the spacing.
3. Select **Modify** button.

To Remove:
1. Select from the list.
2. Select **Remove** button.

You may add or remove tabs here also.

EDITING TEXT

SINGLE LINE TEXT

Editing **Single Line Text** is somewhat limited compared to Multiline Text. In the example below you will learn how to edit the text within a Single Line Text sentence. (In Lesson 12 you will learn additional options for editing Single Line text by using the Properties command.)

1. Double click on the Single Line text you want to edit. The text will highlight.

2. Make the changes in place then press <enter> <enter>.

MULTILINE TEXT

Multiline Text is as easy to edit as it is to input originally. You may use any of the text options shown on the Text Editor tab.

1. Double click on the Multiline text you want to edit.

2. Highlight the text, that you want to change, using click and drag.

3. Make the changes then select the **Close Text Editor** tool.

EDITING TEXT....continued

You may edit many other Multiline Text features.

1. Double click on the Multiline Text you wish to edit.
2. Right click in the **Text Box Area.**

   The menu shown below will appear.

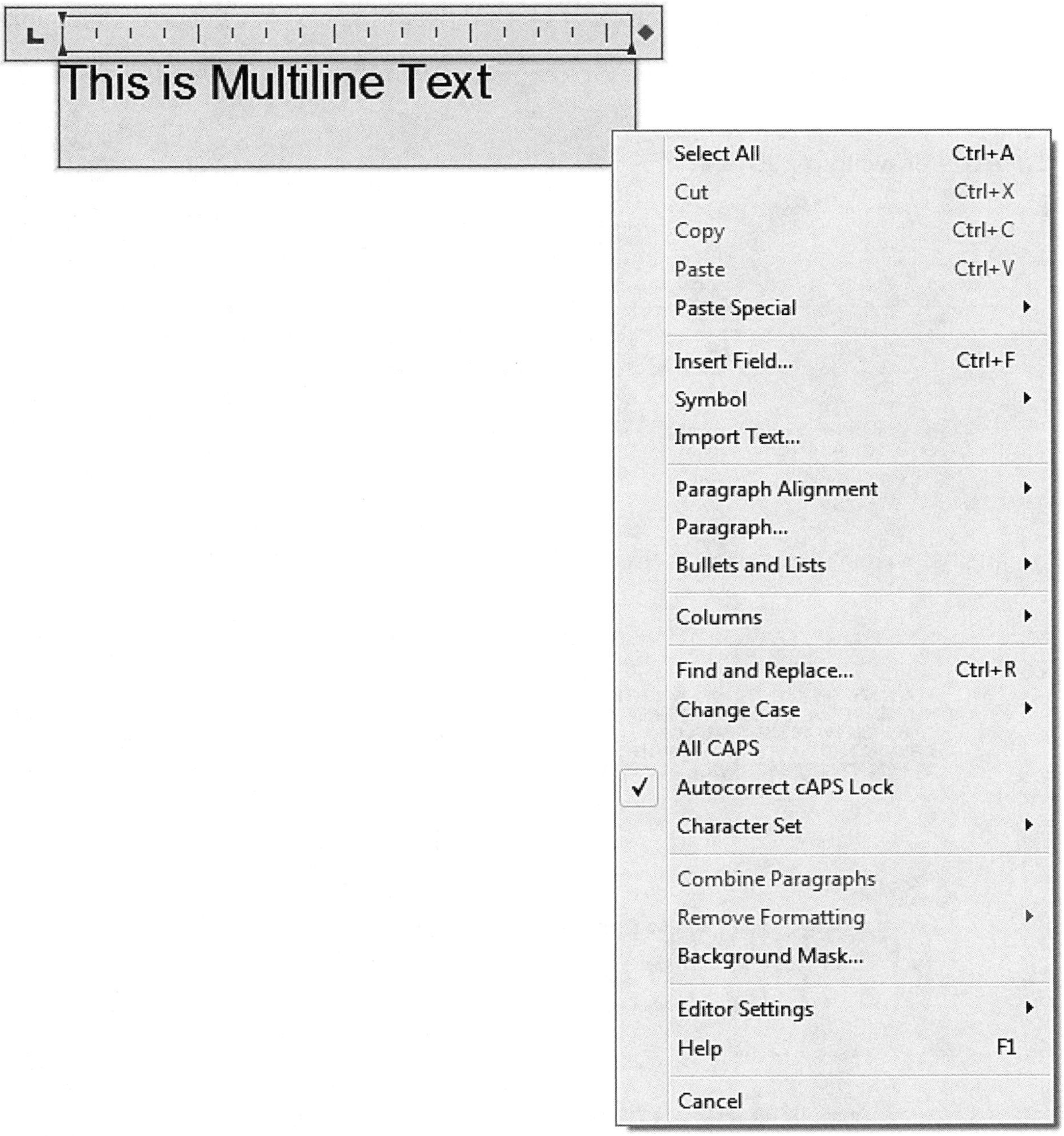

This is Multiline Text

Select All Ctrl+A
Cut Ctrl+X
Copy Ctrl+C
Paste Ctrl+V
Paste Special
Insert Field... Ctrl+F
Symbol
Import Text...
Paragraph Alignment
Paragraph...
Bullets and Lists
Columns
Find and Replace... Ctrl+R
Change Case
All CAPS
✓ Autocorrect cAPS Lock
Character Set
Combine Paragraphs
Remove Formatting
Background Mask...
Editor Settings
Help F1
Cancel

EXERCISE 8A

INSTRUCTIONS:

1. Start a **NEW** file using **2015-Workbook Helper.dwt**
2. Duplicate the text shown below using **Single Line Text**
3. Use Layer Text.
4. Follow the instructions in each block of text. To start the text in the correct location that is stated in each example move the cursor while watching the coordinate display.
5. **Save** the drawing as: **EX8A**

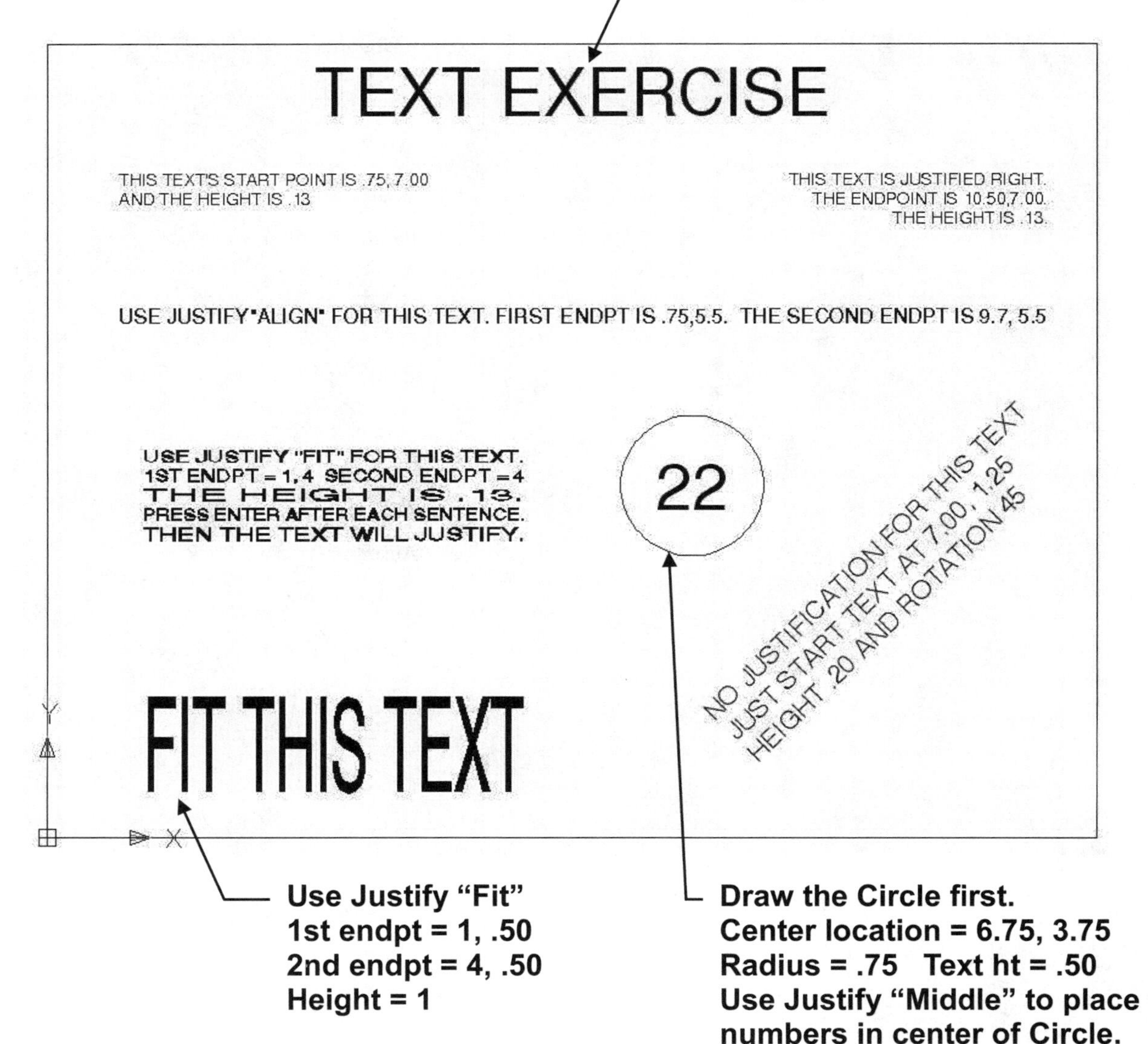

EXERCISE 8B

INSTRUCTIONS:

1. Start a **NEW** file using **2015-Workbook Helper.dwt**
2. Duplicate the text shown below using **MULTILINE** text.
3. Use Layer **Text.**
4. Select **MULTILINE text**
5. Use Text Style: **Standard**
6. Use font: **SansSerif**
7. Text Height: **.250**
8. Follow the directions below. You may make changes to the settings as you type or you may enter all of the text and then go back and edit it. Your choice.
9. Enter all text shown below.
10. Save as: **EX8B**

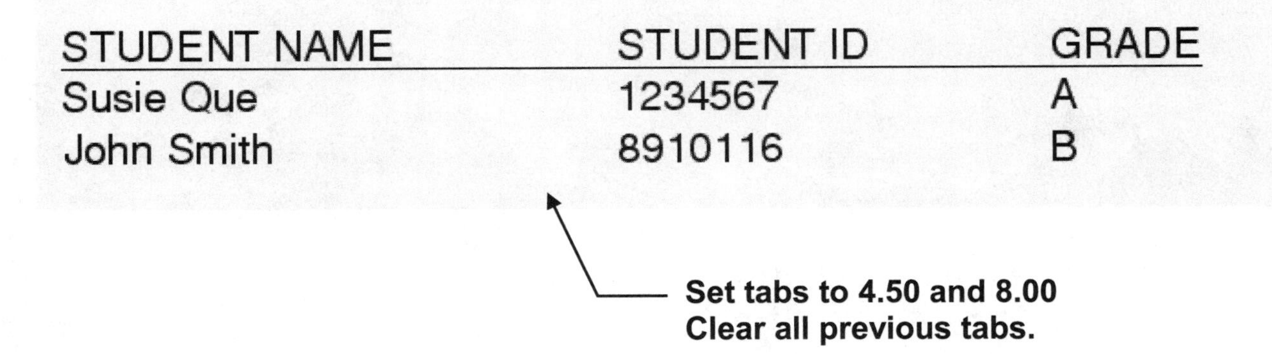

Set tabs to 4.50 and 8.00
Clear all previous tabs.

EXERCISE 8C

INSTRUCTIONS:

1. Start a **NEW** file using **2015-Workbook Helper.dwt**

   The following exercise is designed to teach you how to insert text into the exact middle of a rectangular area using **Single Line Text.**

2. Draw a 6" wide by 3" high rectangle.

3. Select "**SINGLELINE TEXT**"

4. Use Justify: Middle

5. Use "**MTP**" object snap to locate the middle of the rectangle. (Refer to 5-8)
   a. Type **mtp** <enter> on the command line.
   b. Using object snap "Endpoint" snap to **(P1)** corner and then the diagonal corner **(P2)**

6. Use Text Height: 1"

7. Rotation "0"

8. Type the word "**MIDDLE**" and <enter><enter

9. Save as: **EX8C**

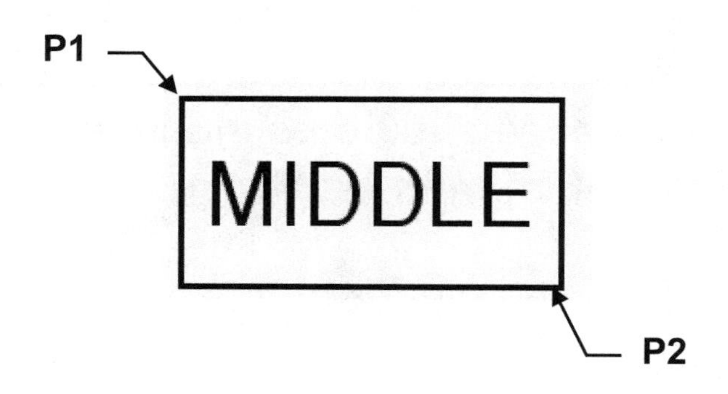

EXERCISE 8D

INSTRUCTIONS:

1. Start a **NEW** file using **2015-Workbook Helper.dwt**

   The following exercise is designed to teach you how to insert text into the exact middle of a rectangular area using **Multiline Text**.

2. Draw a 6" long by 3" wide rectangle.

3. Select "**MULTILINE TEXT**"

4. Start the Text boundary box at the upper left corner **(P1)** of the rectangle. Place the opposite corner at the lower right corner **(P2)** of the rectangle. (Use "Endpoint" object snap to be accurate.)

5. Select **Standard, 1.00, Sans Serif and Middle Center.**

6. Type **MIDDLE** and select **Close Text Editor.**

7. Save as: **EX8D**

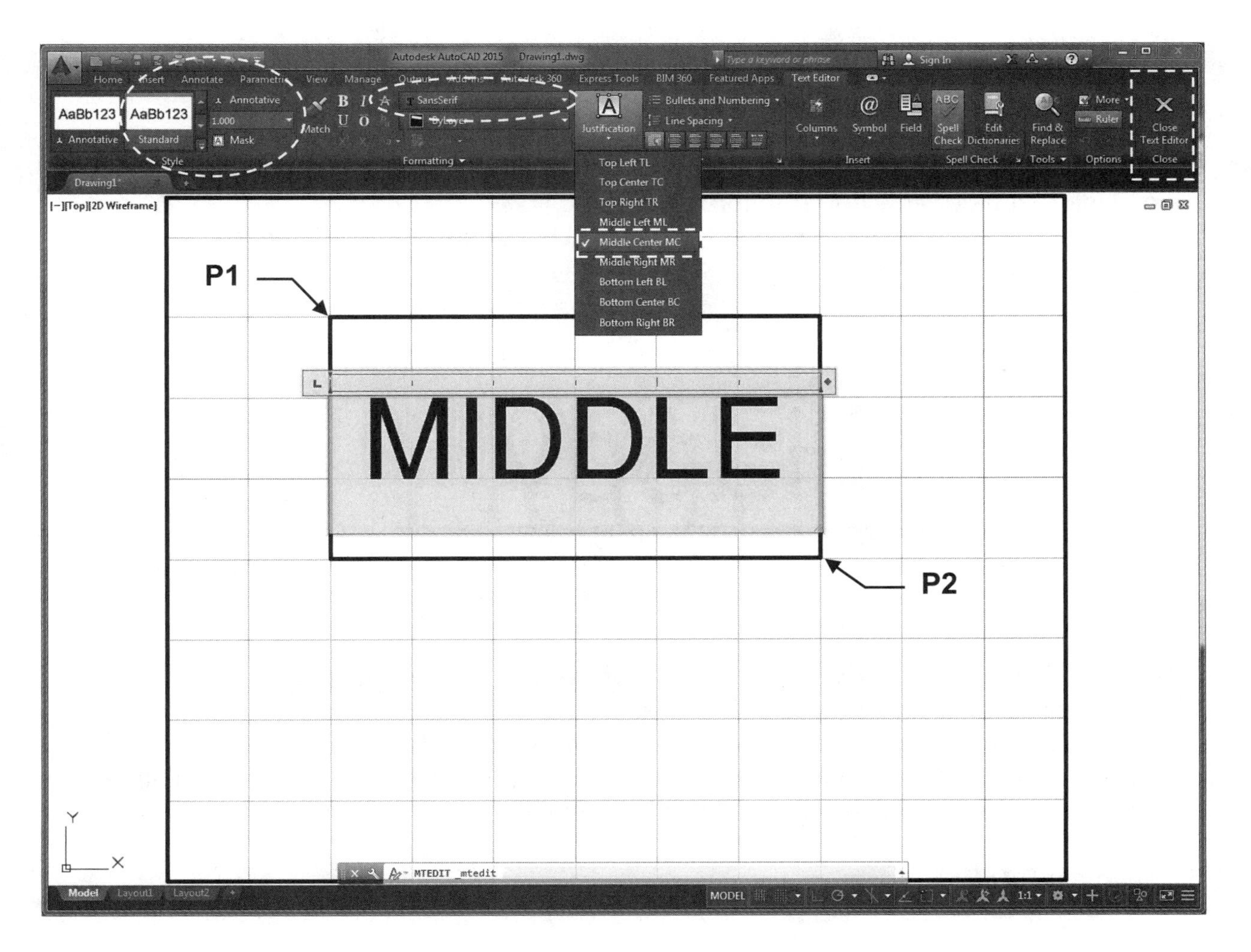

EXERCISE 8E

INSTRUCTIONS:

1. Start a **NEW** file using **2015-Workbook Helper.dwt**
2. Draw **two** 6" long lines as shown.
3. Select Single Line Text
   a. Select **Justify - Center**.
   b. Use Midpoint snap to place the justification point at the midpoint of the line.
   c. Use text height 1" and rotation angle 0.
   d. Type the word "**Happy**" <enter> <enter> **(Use upper and lower case)**

4. Select Single Line Text again.
   a. This time select **Justify - BC**. (bottom center)
   b. Use Midpoint snap to place the justification point at the midpoint of the line.
   c. Use text height 1" and rotation angle 0.
   d. Type the word "Happy" <enter> <enter> (Use upper and lower case)

Happy

Notice the difference between Justify: Center and Bottom Center?
"**Center**" only considers the Upper Case letters when justifying.
"**Bottom Center**" is concerned about those Lower Case letters.
Can you see how you could accidentally place your text too high or too low? Think about the difference between Center and Bottom Center.

5. Save as: **EX8E**

EXERCISE 8F

INSTRUCTIONS:

1. Start a **NEW** file using **2015-Workbook Helper.dwt**
2. Draw the 2 sentences below using **Single Line text**. (Use Layer **Text**)

   Mirror Text setting 1

   Mirror Text setting 0

3. Change the **Mirrtext** setting:

   A. Type **mirrtext <enter>**

   B. Type **1 <enter>**

4. Using the Mirror command, mirror the **first sentence** using a vertical mirror line **(P1 and P2)**

Original | **P1** | **Mirror Copy**

Mirror Text setting 1 | Mirror Text setting 1

P2

5. Now change the **Mirrtext** command setting to **0**
6. Mirror the **second** sentence.

Original | **P1** | **Mirror Copy**

Mirror Text setting 0 | Mirror Text setting 0

P2

Notice the difference in the mirrored copy. Sometimes you will want the mirrored text to be reversed and sometimes you will not. Now you know how to control it.

7. Save as **EX8F**

NOTES:

LEARNING OBJECTIVES

After completing this lesson, you will be able to:

1. Understand the Origin
2. Draw objects accurately using Coordinate Input
3. Input Absolute and Relative coordinates
4. Input using Direct Distance Entry
5. List information about an object
6. Measure the distance between two points
7. Identify a location within the drawing
8. Create your own 8-1/2 x 11 Master Border
9. Print from Model Space

LESSON 9

COORDINATE INPUT

In the previous lessons you have been using the cursor to place objects.
In this lesson you will learn how to place objects in specific locations by entering coordinates. This process is called **Coordinate Input**.

This is not difficult, so do not start to worry.

AutoCAD uses the ***Cartesian Coordinate System.***
The Cartesian Coordinate System has 3 axes, X, Y and Z.

The **X** is the Horizontal axis. (*Right and Left*)
The **Y** is the Vertical axis. (*Up and Down*)
The **Z** is Perpendicular to the X and Y plane.
*(The **Z** axis, which will be discussed in the Advanced workbook.)*

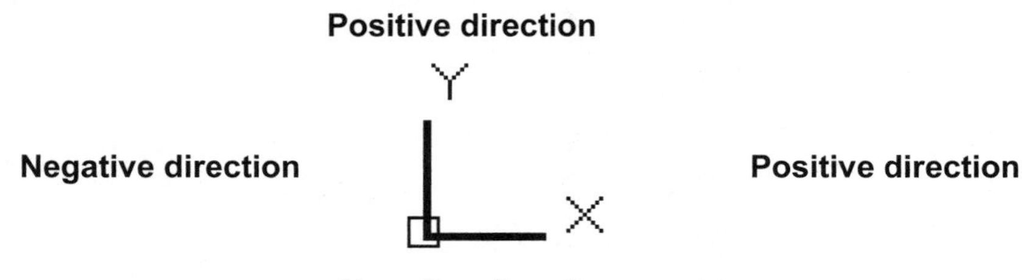

Look at the User Coordinate System (UCS) icon in the lower left corner of your screen. The X and Y are pointing in the positive direction.

The location where the X , Y and Z axes intersect is called the **ORIGIN**.
*The **Origin** always has a coordinate value of X=0, Y=0 and Z=0 (0,0,0)*

When you move the cursor away from the Origin, in the positive direction , the X and Y coordinates are positive.
When you move the cursor in the opposite direction, the X and Y coordinates are negative.

Using this system, every point on the screen can be specified using positive or negative X and Y coordinates.

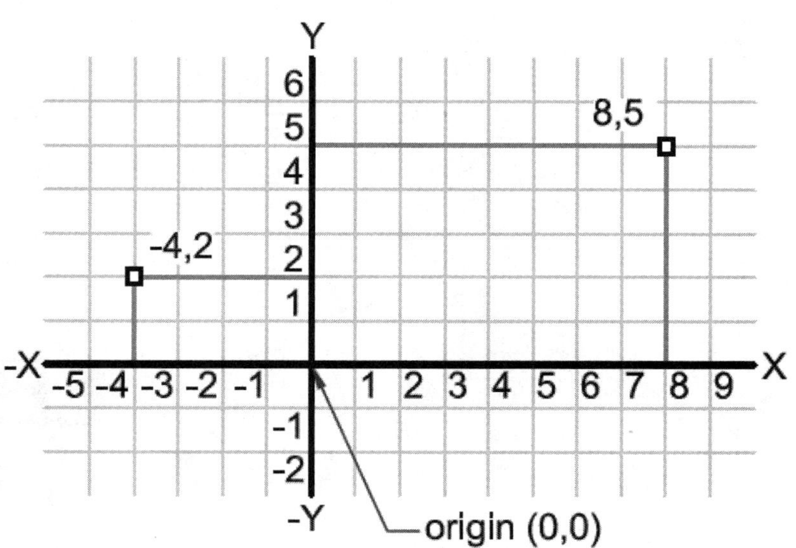

ABSOLUTE COORDINATES

There are 3 types of Coordinate input, **Absolute, Relative** and **Polar**.

(Relative is discussed on the next page and Polar will be discussed in Lesson 11)

ABSOLUTE COORDINATES

When inputting absolute coordinates the input format is: **X, Y** (that is: X comma Y)

Absolute coordinates come ***from the ORIGIN*** and are typed as follows: **8, 5**

The first number (8) represents the **X-axis** (horizontal) distance from the Origin and the second number (5) represents the **Y-axis** (vertical) distance from the Origin.
The two numbers must be separated by a **comma**.

An absolute coordinate of **4, 2** will be **4** units to the right (horizontal) and **2** units up (vertical) from the current location of the Origin.

An absolute coordinate of **-4, -2** will be **4** units to the left (horizontal) and **2** units down (vertical) from the current location of the Origin.

The following are examples of Absolute Coordinate input.
Notice where the Origin is located in each example.

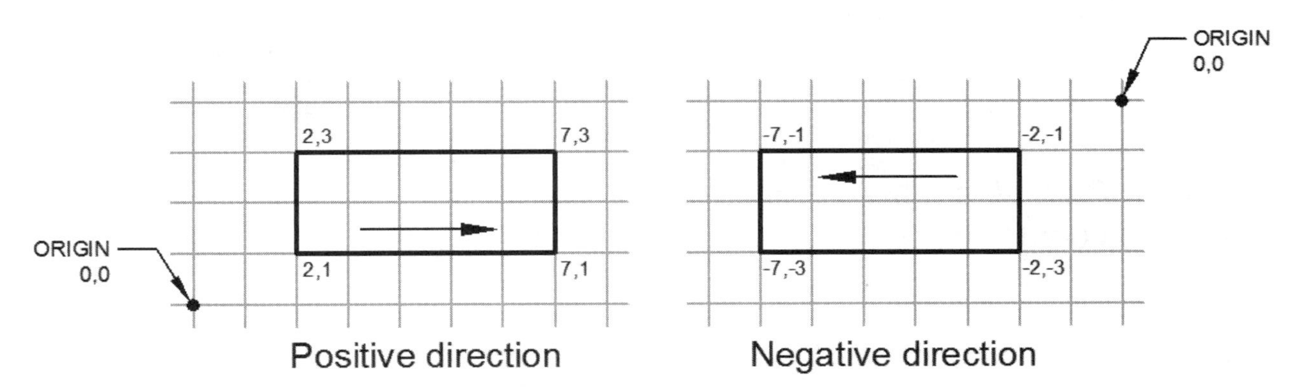

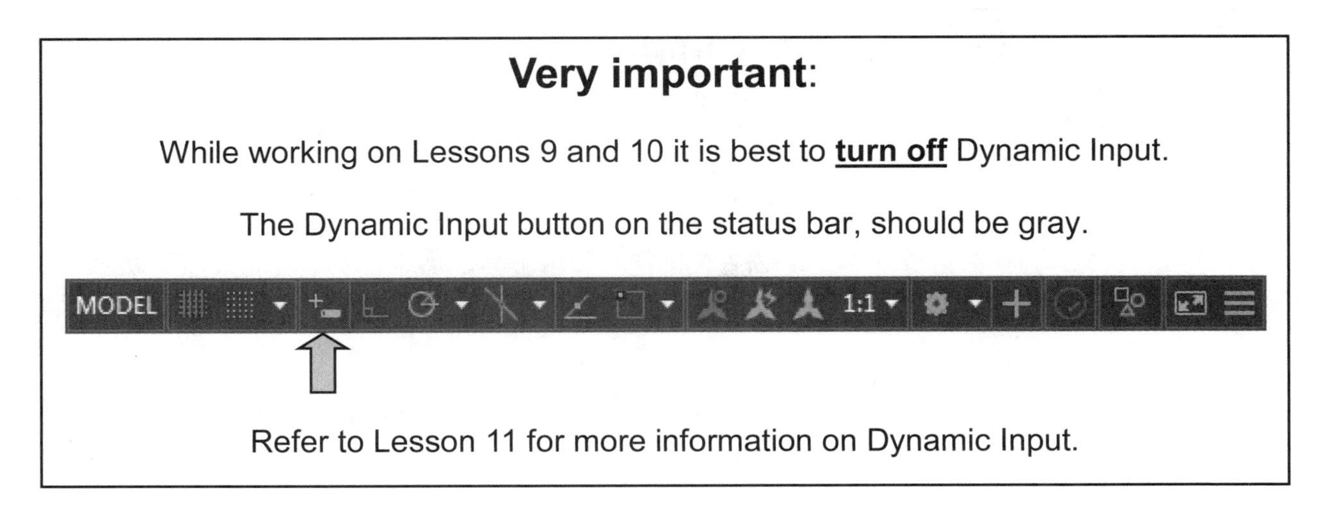

Very important:

While working on Lessons 9 and 10 it is best to **turn off** Dynamic Input.

The Dynamic Input button on the status bar, should be gray.

Refer to Lesson 11 for more information on Dynamic Input.

RELATIVE COORDINATES

RELATIVE COORDINATES

Relative coordinates come ***from the last point entered***. (Not from the Origin)

The first number represents the **X-axis** (horizontal) and the second number represents the **Y-axis** (vertical) just like the absolute coordinates.

To distinguish the relative coordinates from absolute coordinates the two numbers must be preceded by an @ symbol in addition to being separated by a **comma**.

A Relative coordinate of **@5, 2** will go to the **right** 5 units and **up** 2 units from the last point entered.

A Relative coordinate of **@-5, -2** will go to the **left** 5 units and **down** 2 units from the last point entered.

The following is an example of Relative Coordinate input.

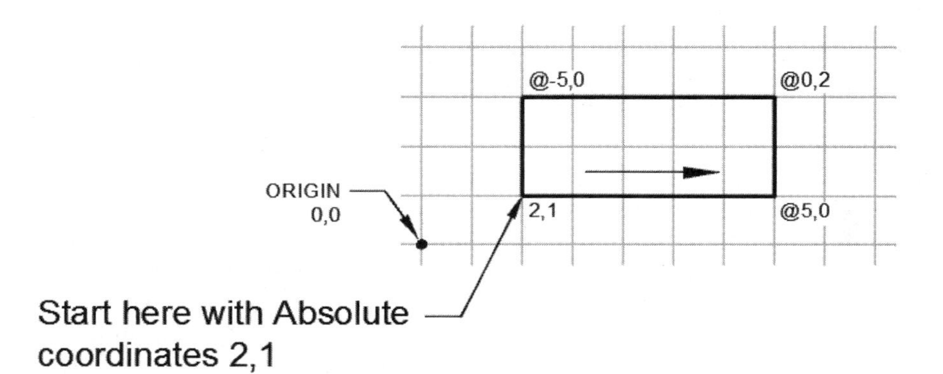

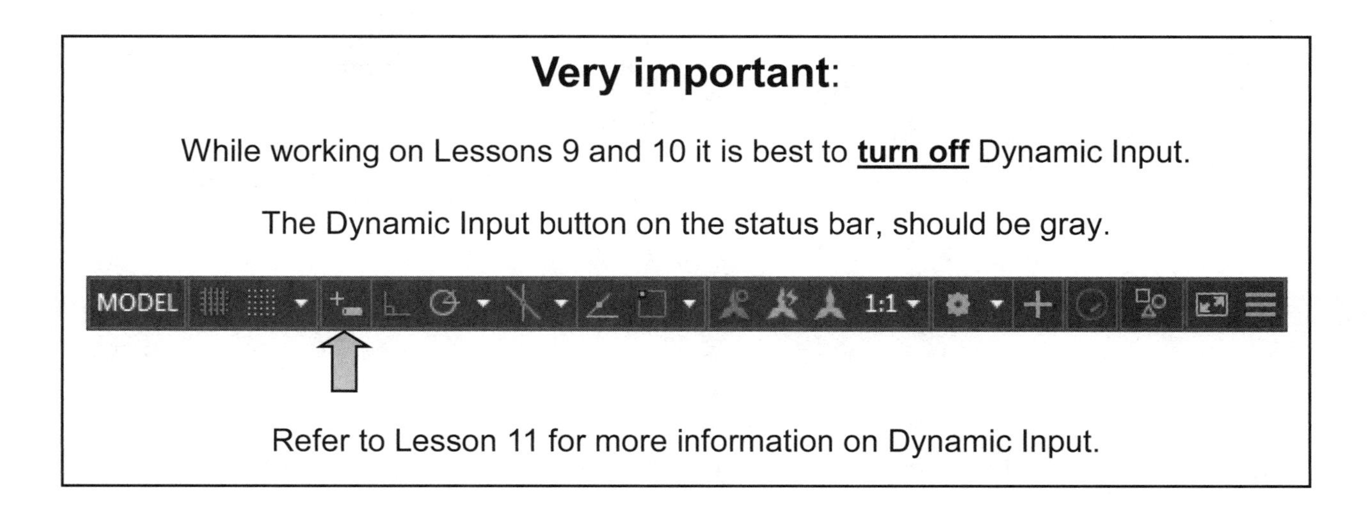

Very important:

While working on Lessons 9 and 10 it is best to **turn off** Dynamic Input.

The Dynamic Input button on the status bar, should be gray.

Refer to Lesson 11 for more information on Dynamic Input.

EXAMPLES OF COORDINATE INPUT

Scenario 1.
If you want to draw a line with the first endpoint "at the Origin" and the second endpoint 3 units in the positive X direction.

0,0 ——————— **3,0**

1. Select the Line command.
2. You are prompted for the first endpoint: **Type 0, 0 <enter>**
3. You are then prompted for the second endpoint: **Type 3, 0 <enter>**

What did you do?
The first endpoint coordinate input, 0,0 means that you do not want to move away from the Origin. You want to start "**ON**" the Origin.

The second endpoint coordinate input, 3, 0 means that you want to move 3 units in the positive X axis. The "0" means you do not want to move in the Y axis. So the line will be exactly horizontal.

Scenario 2.
You want to start a line 1 unit to the right of the origin and 1 unit above and the line will be 4 units in length, perfectly vertical.

@0,4

1. Select the Line command.
2. You are prompted for the first endpoint: **Type 1, 1 <enter>**
3. You are prompted for the second endpoint: **Type @0, 4 <enter>**

1,1

What did you do?
The first endpoint coordinate input, 1, 1 means you want to move 1 unit in the X axis direction and 1 unit in the Y axis direction.

The second endpoint coordinate input @0, 4 means you do not want to move in the X axis "from the last point entered" but you do want to move in the Y axis "from the last point entered. (Remember the @ symbol is only necessary if you are not using DYN)

Scenario 3.
Now try drawing 5 connecting line segments.
(Watch for the negatives)

1. Select the Line command.
2. First endpoint: 2, 4 <enter>
3. Second endpoint: @ 2, -3 <enter>
4. Second endpoint: @ 0, -1 <enter>
5. Second endpoint: @ -1, 0 <enter>
6. Second endpoint: @ -2, 2 <enter>
7. Second endpoint: @ 0, 2 <enter> <enter>

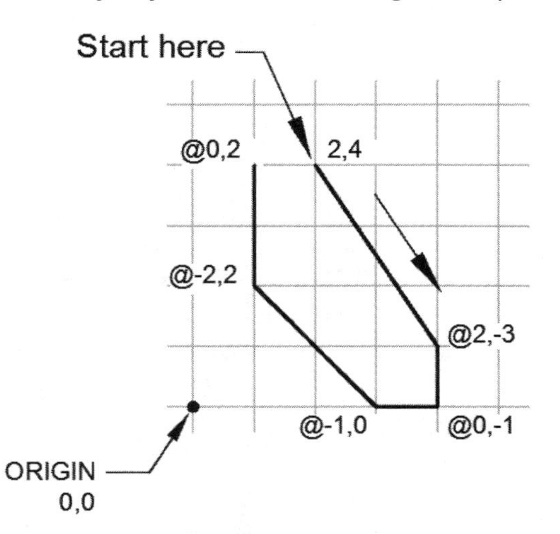

Note: If you enter an incorrect coordinate, just hold down the **Shift key** and press **U** then **<enter>**, the last segment will disappear and you will have another chance at entering the correct coordinate.

DIRECT DISTANCE ENTRY (DDE)

DIRECT DISTANCE ENTRY is a combination of keyboard entry and cursor movement. **DDE** is used to specify distances in the horizontal or vertical axes from the last point entered. **DDE is a *Relative Input*.** Since it is used for Horizontal and Vertical movements, **Orthomode** must be **ON**.

(Note: to specify distances on an angle, refer to Polar Input in Lesson 11)

Using DDE is simple. Just move the cursor and type the distance.
Negative and positive is understood automatically by moving the cursor up (positive), down (negative), right (positive) or left (negative) from the last point entered. No minus or @ sign necessary.

Moving the cursor to the right and typing 5 <enter> tells AutoCAD that the 5 is positive and Horizontal.
Moving the cursor to the left and typing 5 <enter> tells AutoCAD that the 5 is negative and Horizontal.
Moving the cursor up and typing 5 <enter> tells AutoCAD that the 5 is positive and Vertical.
Moving the cursor down and typing 5 <enter> tells AutoCAD that the 5 is negative and Vertical.

EXAMPLE:

1. Orthomode must be ON. Grid OFF
2. Select the Line command.
3. Type: 1, 2 <enter> to enter the first endpoint using Absolute coordinates.
4. Now move your cursor to the right and type: 5 <enter>
5. Now move your cursor up and type: 4 <enter>
6. Now move your cursor to the left and type: 5 <enter> <enter> to stop

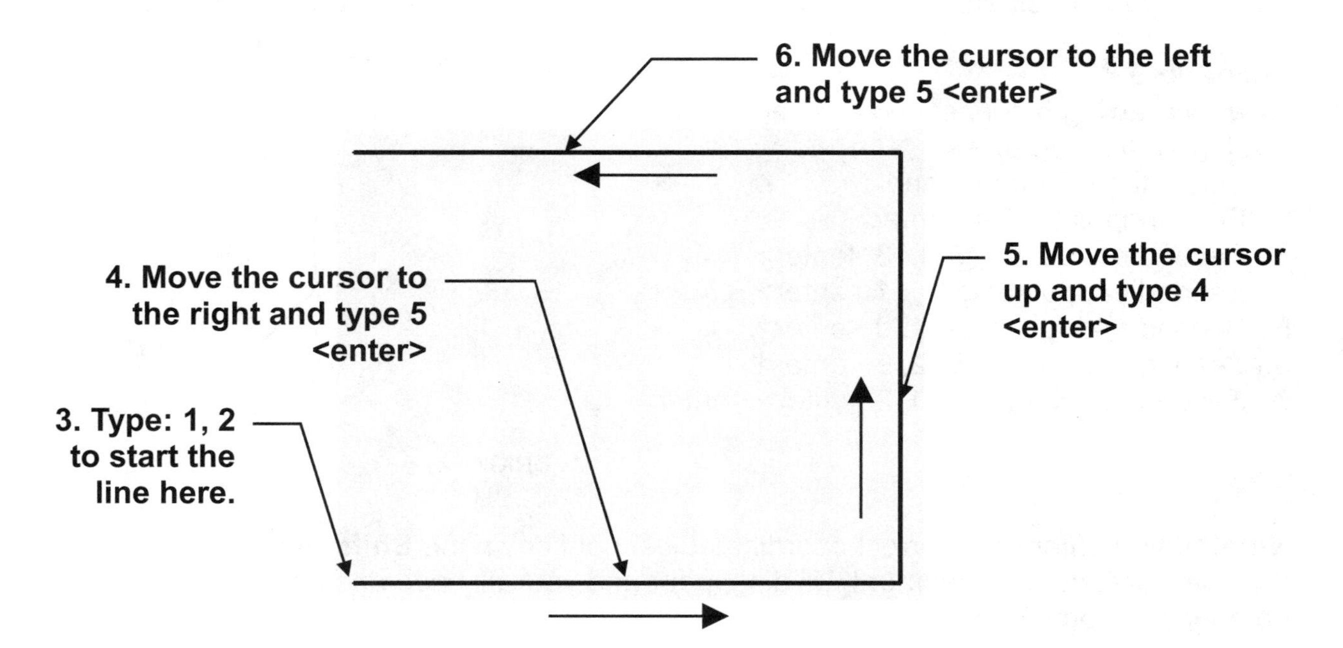

MEASURE TOOLS and ID Point

The following tools are very useful to confirm the location or size of objects.

The Measure tools enables you to measure the **Distance**, **Radius**, **Angle**, **Area**, or **Volume** of a selected object. The default option is Distance.

1..You may access these tools as follows:

Ribbon = Home tab / Utilities Panel / Measure ▼

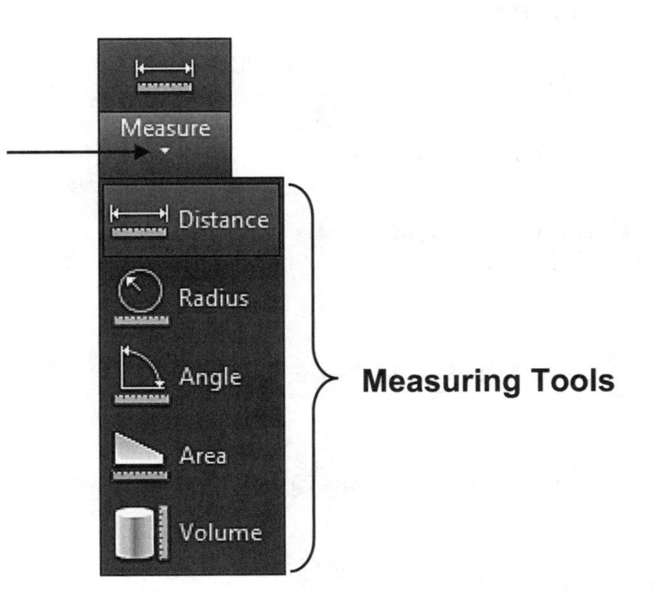

2. Select one of the tools and follow the instructions on the command line.

ID Point

The ID Point command will list the X and Y coordinates of the point that you select. The coordinates listed will be from the Origin.

Ribbon = Home tab / Utilities Panel / ▼
Or
Keyboard = ID <enter>

ID Point
Point Style...
Utilities

1. Select the ID Point command by typing: **ID <enter>**
2. Select a location point, such as the endpoint of a line.
   The X, Y and Z location coordinates for the endpoint will be displayed.

Example:
1. Command: **id <enter>**
2. Snap to the endpoint
3. Coordinates, from the Origin, are displayed.

2. Snap to endpoint

```
Command: '_id Specify point:  X = 5.474     Y = 0.791     Z = 0.000
```

EXERCISE 9A

In this exercise you will create a master template named "**Border A**" to be used for the exercises in Lessons 9 through 25. Follow the easy steps below. Do not skip any.

STEP 1: (Select the Settings)

1. Start a **NEW** file using
   2015-Workbook Helper.dwt

2. Set **Units** (Refer to page 4-14)
   Units = Decimal
   Precision = 0.000

3. Set **Drawing Limits** (Refer to page 4-10)
   Lower Left Corner = 0,0
   Upper Rt. Corner = 11, 8.5

4. Show the new limits using: (Refer to page 4-8)
   Zoom / All

5. **Important**: Change Lineweight settings to **Inches** and **adjust Display scale**.
   Refer to bottom of page 3-12

6. Set **Grids** and **Snap** (Refer to page 3-7, 4-12 and 4-13)
   Snap = .125 **(this is important, it will make it easier to draw the lines)**
   Grids = 1.00

Continued on the next page...

EXERCISE 9A....continued

STEP 2: (Draw the Border Lines)

1. Select Layer **Borderline**

2. Draw the border below using the dimensions shown.

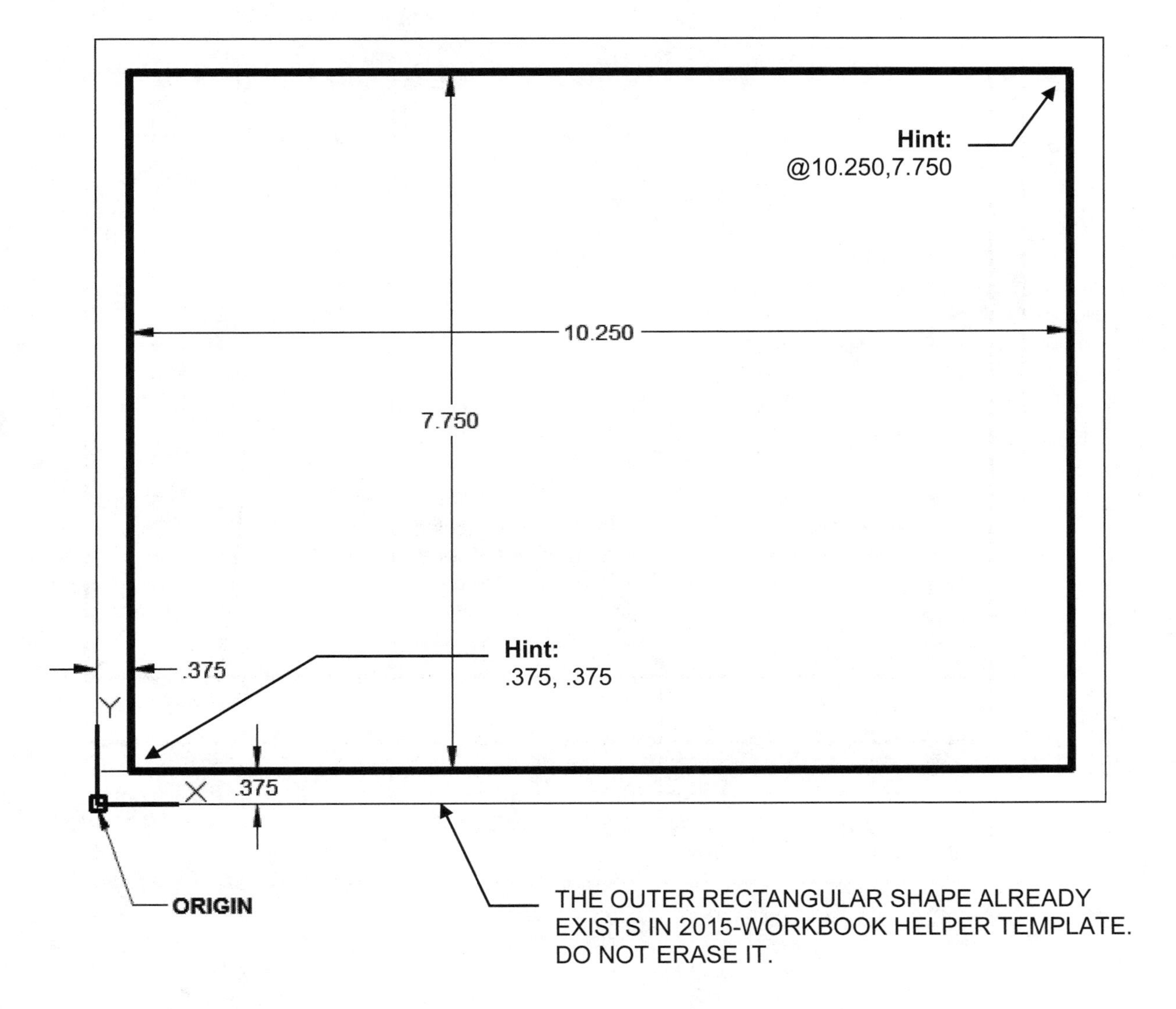

Continued on the next page...

EXERCISE 9A....continued

STEP 3: (Draw the TITLE BLOCK Lines)

1. Select Layer **Borderline**
2. Draw the 3 **TITLE BLOCK** lines as shown below using the dimensions shown.

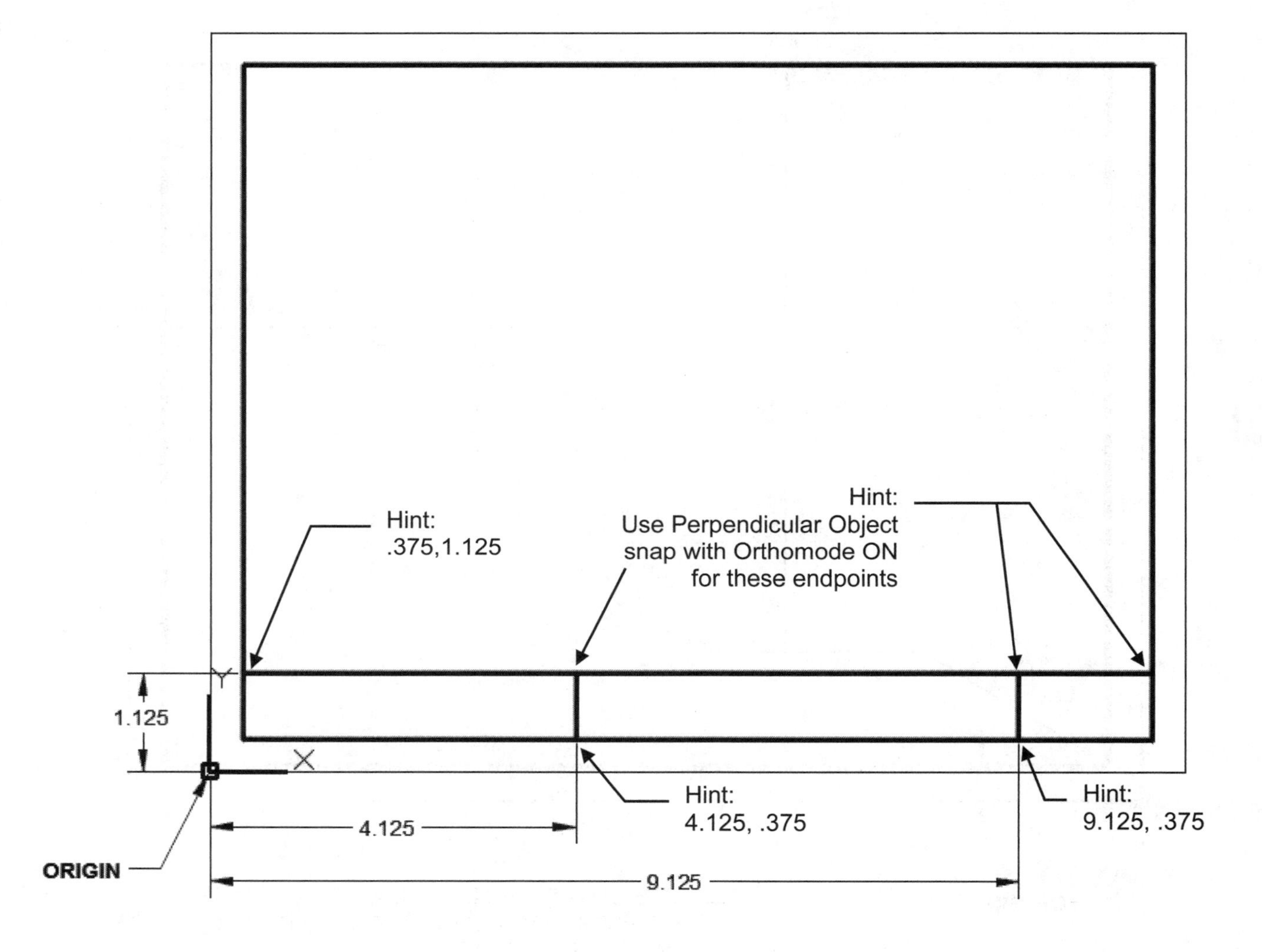

Continued on the next page...

EXERCISE 9A....continued

STEP 4: (Enter the TEXT)

1. Select Layer **Text**

2. Select **Multiline text** command.

3. Select the upper left corner and then the opposite corner.

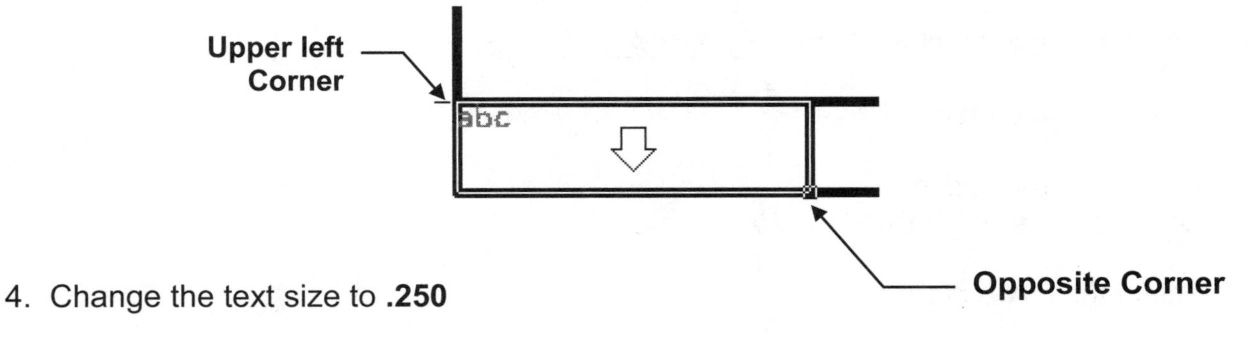

4. Change the text size to **.250**

5. Select **Bold**

6. Select justification **Middle Center**

7. Type the text shown below.

8. Repeat the above for the remaining 2 title boxes

Your name here | Title goes here | Ex-XX

Continued on the next page...

EXERCISE 9A....continued

STEP 5: (Save the border as a Template)

1. Select the **Application Menu** (Refer to page 2-4)

2. Select **Save As** ▶

3. Select **AutoCAD Drawing** **<u>Template</u>**

4. Enter the new template name: **<u>Border A-2015</u>**

5. Select **Save** button

6. Enter the description:
   Use for Lessons 9 through 25

7. Select the **OK** button

You now have a template to use for Lessons 9 through 25.
At the beginning of each exercise you will be instructed to start a **New** drawing using **Border A-2015.dwt**
You will **edit** the **title** and the **Ex-XX** to match the exercise. Editing makes it much easier. You will not need to change the location of the text. Merely edit it.

If you would like to Print this border follow the steps on the following pages.

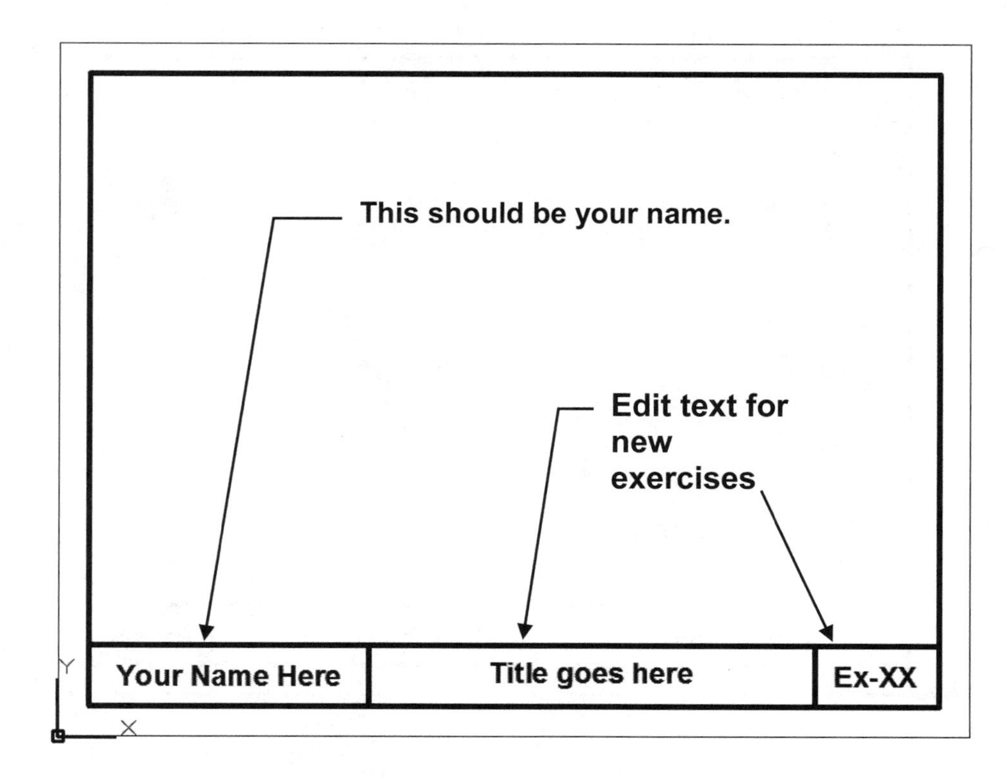

Basic Plotting from Model Space

Note: More Advance plotting methods will be explained in Lessons 26 and 27.

1. **Important:** Open the drawing you want to plot, if it is not already open.
2. Select: **Zoom / All** to center the drawing within the plot area.
3. Select the Plot command using one of the following;

Quick Access toolbar =

Ribbon = Output tab / Plot Panel /

Keyboard = Plot <enter>

The Plot –Model dialog box will appear.

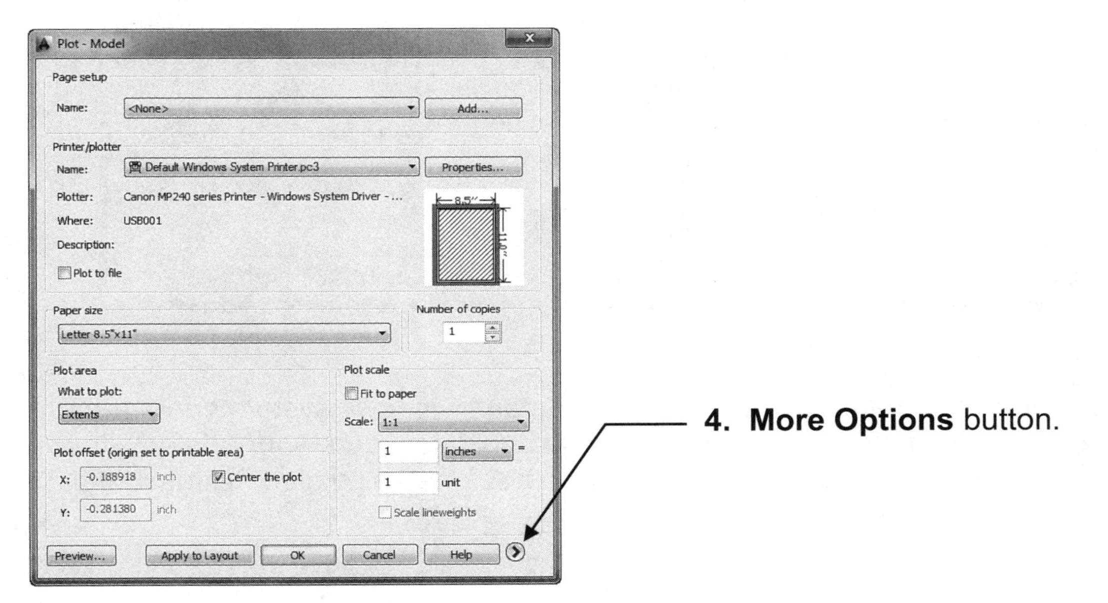

4. Select the **"More Options"** button to expand the dialog box.

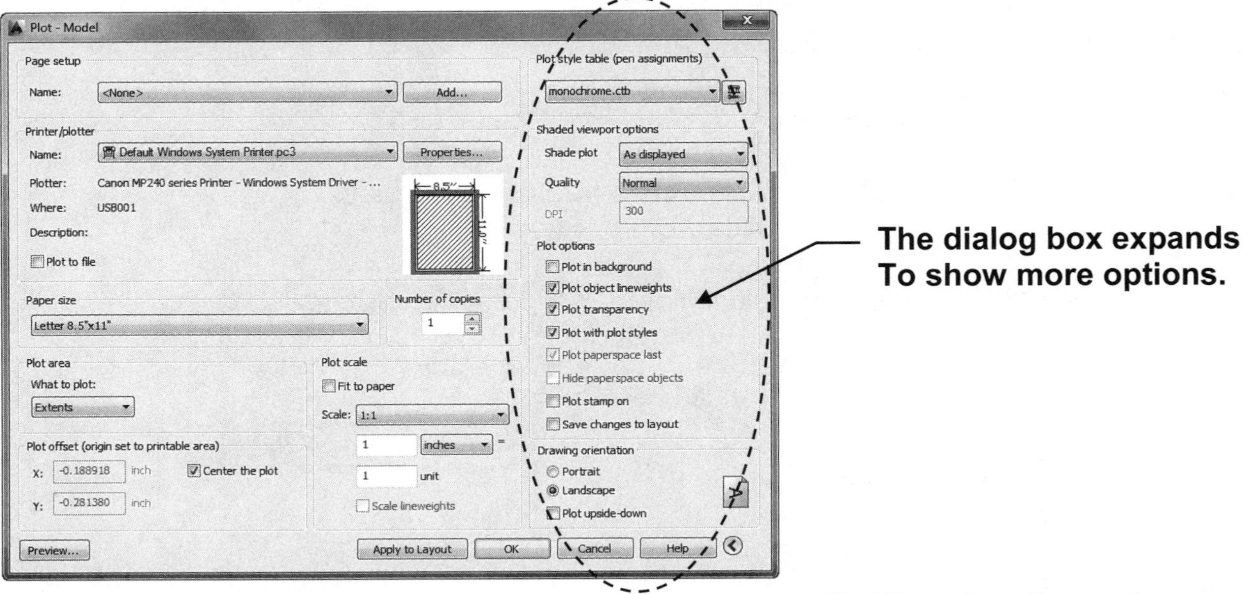

Continued on the next page...

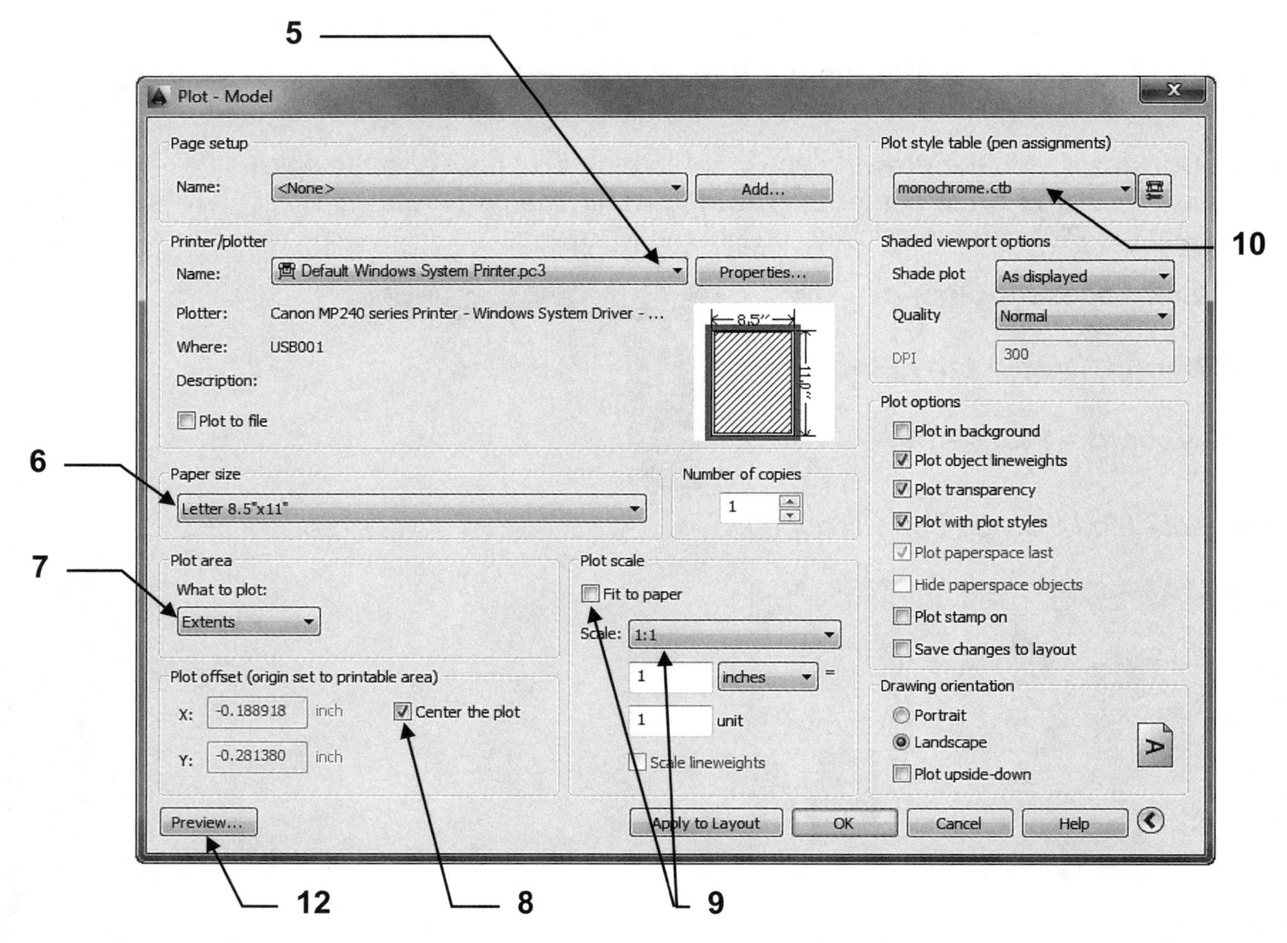

5. Select your printer from the drop down list or “Default windows system printer.pc3”
   Note: If your printer is not shown in the list you should configure your printer within AutoCAD. This is not difficult. Refer to Appendix-A for step by step instructions.

6. Select the Paper size **Letter**.

7. Select the Plot Area **Extents**

8. Select the Plot Offset **Center the Plot**

9. Uncheck the **Fit to** paper box and select Plot Scale **1 : 1**

10. Select the Plot Style table **Monochrome.ctb** (for all black)
    Acad.ctb (for color)

11. If the following box appears, select **Yes**

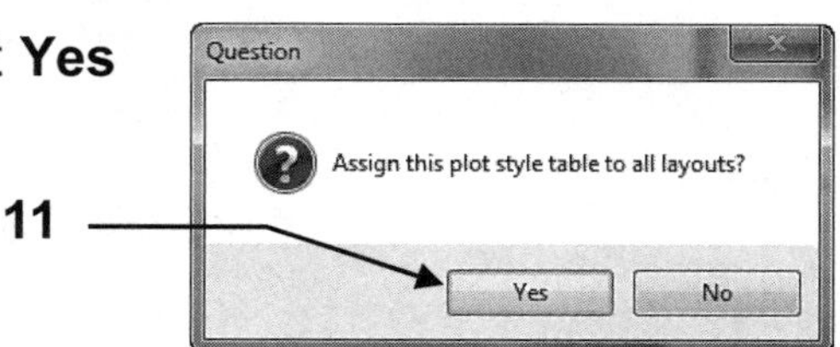

12. Select the **Preview** button.

Continued on the next page...

Note: If you are using a 30 day free trial or student version software, your display may appear with text around the perimeter.

Your Name Here | Titles go here | Ex-XX

Does your display appear as shown above?
If yes, press <enter> and proceed to 13.
If not, recheck 1 through 11.

You have just created a **Page Setup**. All of the settings you have selected can now be saved. You will be able to recall these settings for future plots using this page setup. To save the **Page Setup** you need to **ADD** it to the model tab within this drawing.

13. Select the **ADD** button.

13

Plot - Model
Page setup
Name: <None> Add...
Plot style table (pen assignments)
monochrome.ctb

14. Type the new Page Setup name **Class Model A**

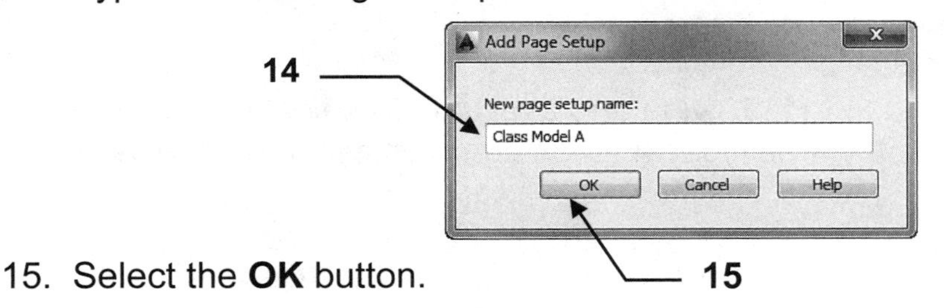

15. Select the **OK** button.

Continued on the next page...

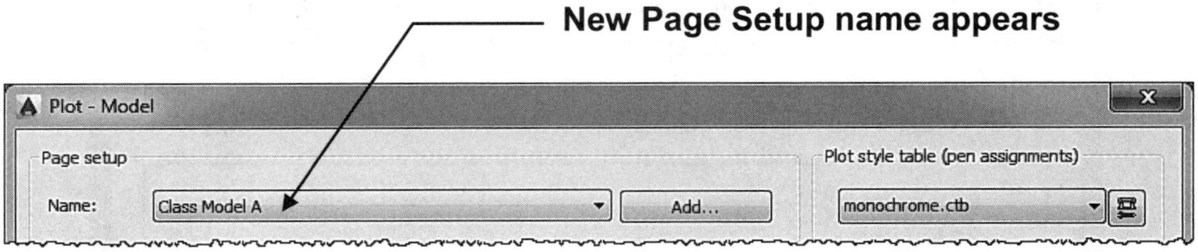

New Page Setup name appears

16. Select the **Apply to Layout** button.

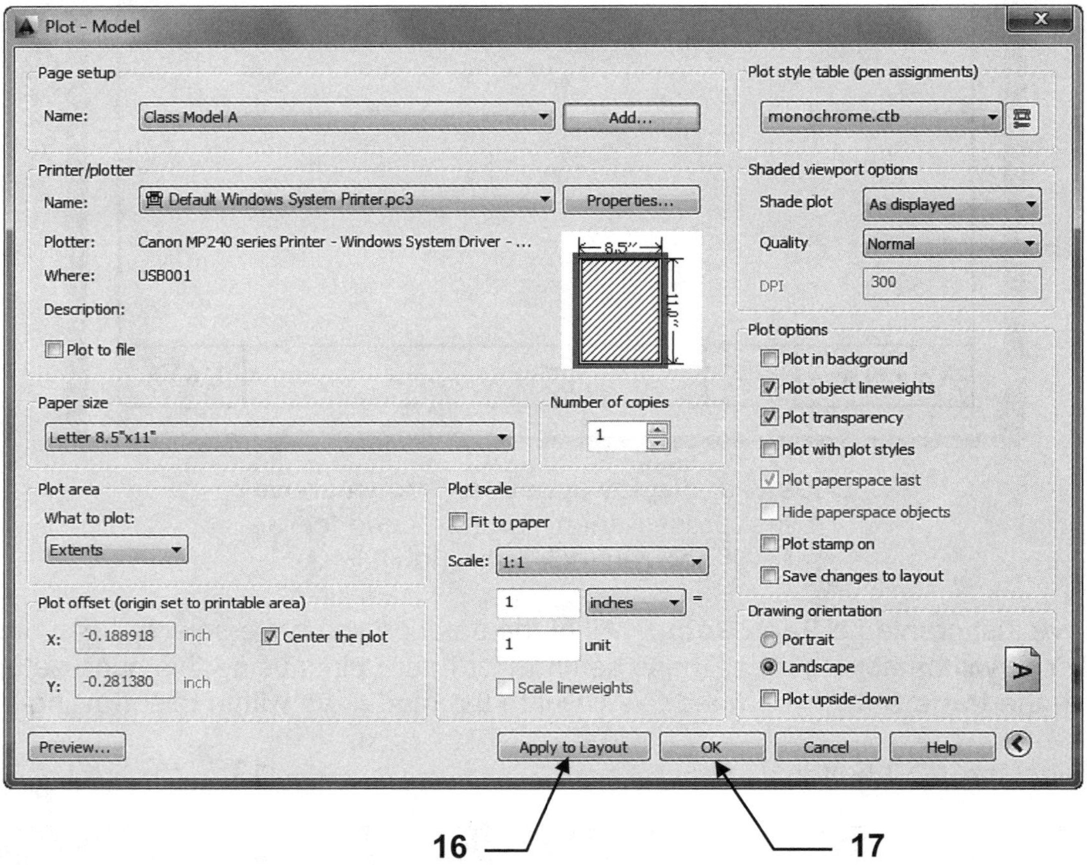

16 **17**

17. Select the **OK** button to send the drawing to the printer or select Cancel if you do not want to print the drawing at this time. The Page Setup will still be saved.

18. Save the entire drawing again as a template using **Save as AutoCAD drawing template: Border A-2015**. (Refer to page 9-12)

    The Page Setup will be saved to the drawing and available to select in the future. You will not have to select all the individual settings again unless you choose to change them.

Now go tape your print out to the refrigerator door.
This is quite an accomplishment.

EXERCISE 9B

INSTRUCTIONS:

1. Start a **NEW** file using **Border A-2015.dwt**
2. Draw the objects below using Absolute and Relative coordinates.
3. Use Layer = Object Line.
4. Edit the Title and Ex-XX by double clicking on the text. Do not erase and replace.
5. Do not dimension (We will get to that soon).
6. Save as **EX9B**
7. Plot using Page Setup **Class Model A** (Refer to page 9-13 through 9-16).

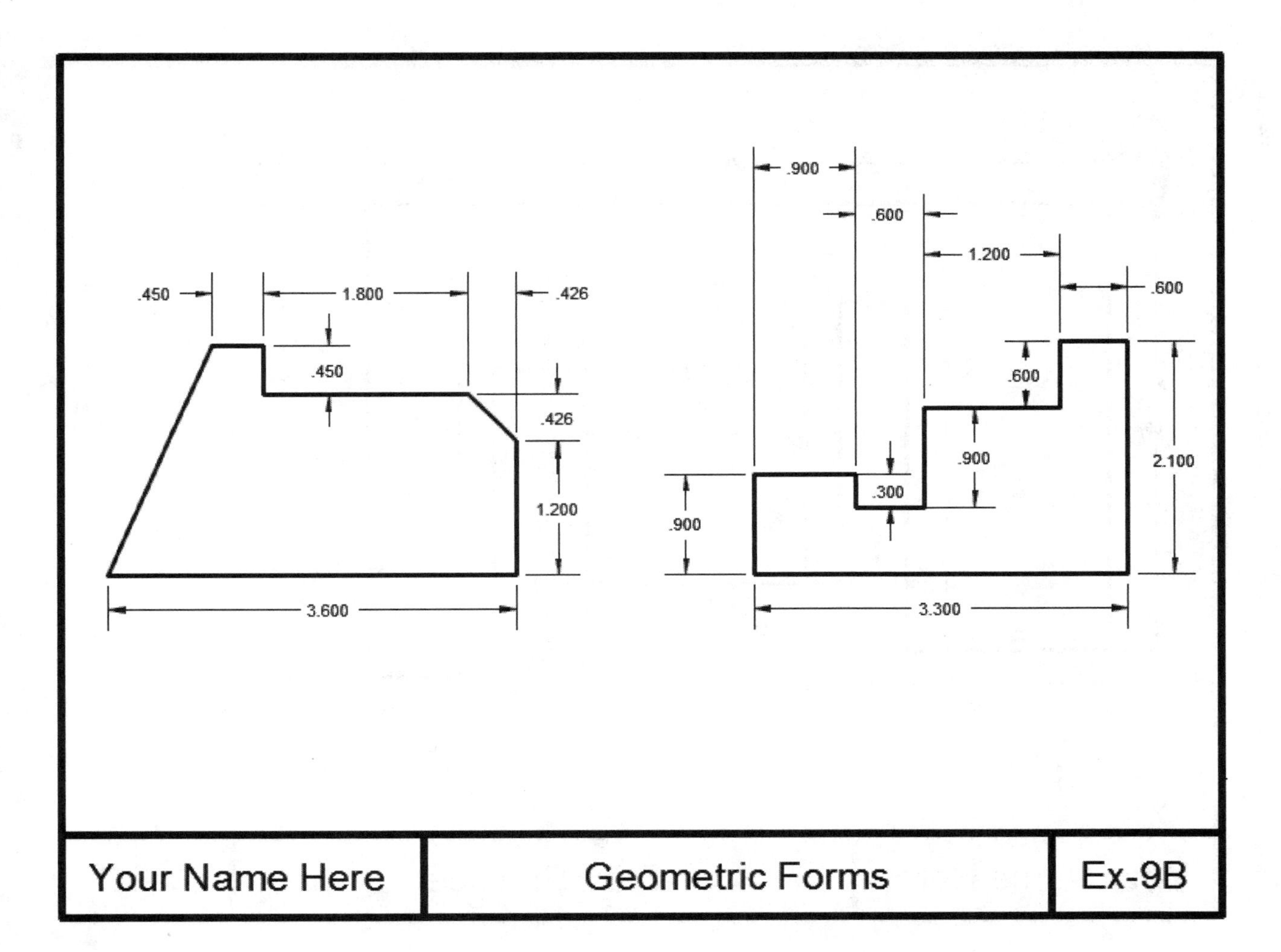

EXERCISE 9C

INSTRUCTIONS:

1. Start a **NEW** file using **Border A-2015.dwt**
2. Draw the objects below using Direct Distance Entry (DDE).
3. Use Layer = Object Line.
4. Edit the Title and Ex-XX by double clicking on the text. Do not erase and replace.
5. Do not dimension.
6. Save as **EX9C**
7. Plot using Page Setup **Class Model A** (Refer to page 9-13 through 9-16).

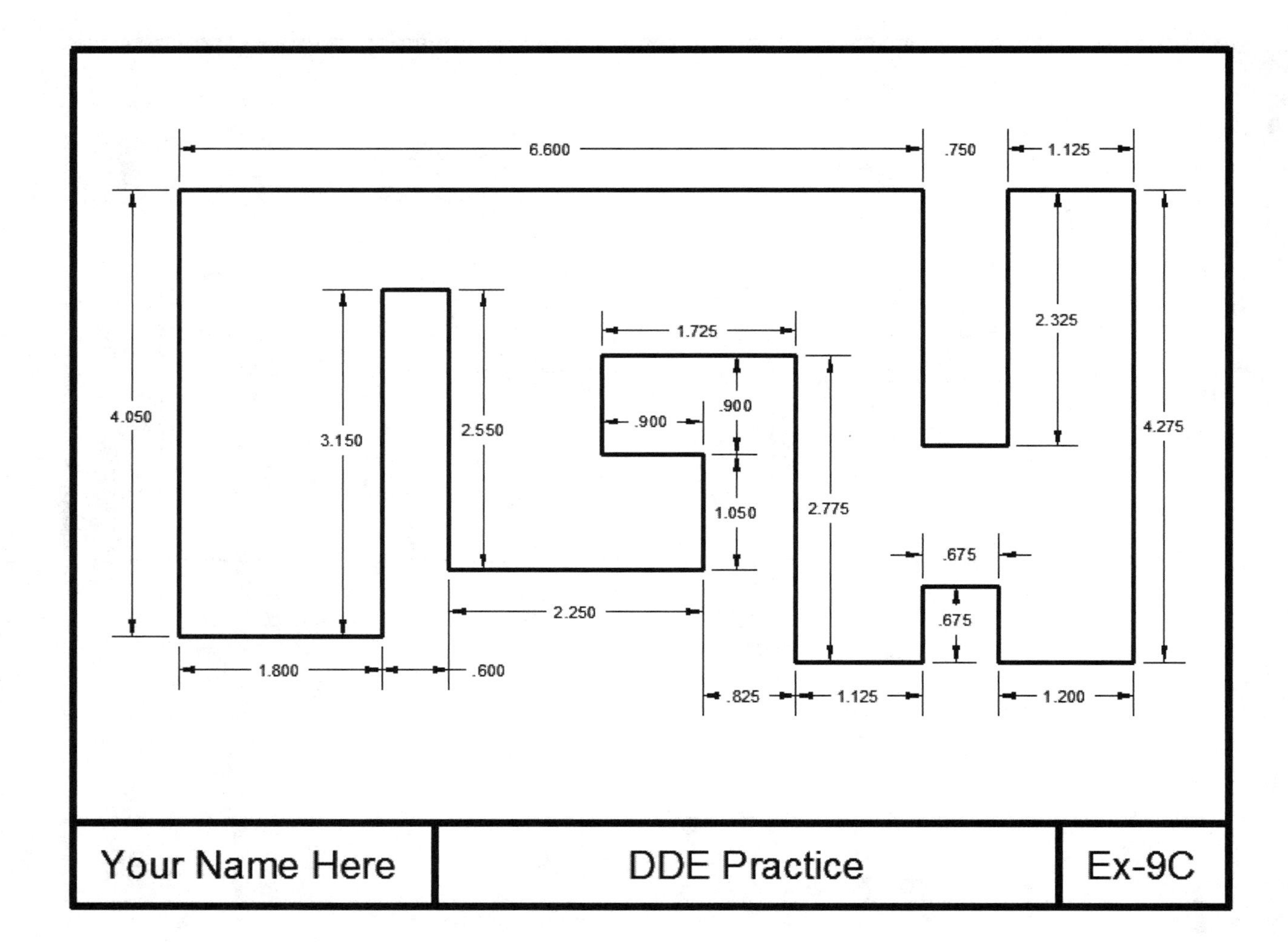

LEARNING OBJECTIVES

After completing this lesson, you will be able to:

1. Move the Origin
2. Control the display of the UCS icon

LESSON 10

MOVING THE ORIGIN

As previously stated in Lesson 9, the **ORIGIN** is where the X, Y, and Z-axes intersect. The Origin's (0,0,0) default location is in the lower left-hand corner of the drawing area. But you can move the Origin anywhere on the screen using the UCS command. (The default location is designated as the "**World**" option or WCS. When it is moved it is UCS, User Coordinate System.)

You may move the Origin many times while creating a drawing. This is not difficult and will make it much easier to draw objects in specific locations.

Refer to the examples on the next page.

To MOVE the Origin:

1. Right click on the Origin icon:

2. Select **Origin** from the shortcut menu

3. Place the new Origin location by entering coordinates or pressing the left mouse button.

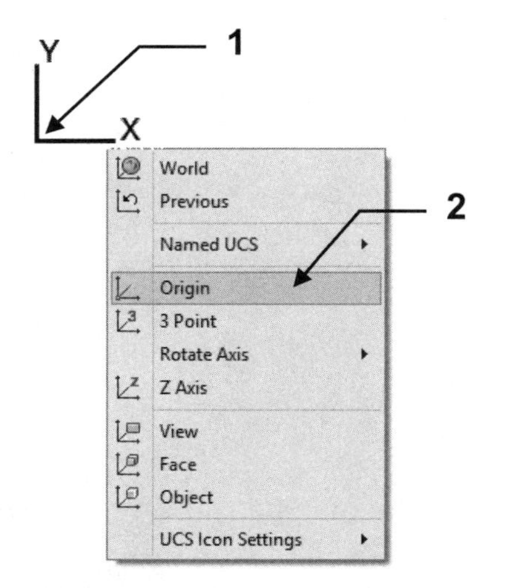

To RETURN the Origin to the default "World" location (the lower left corner):

1. Right click on the Origin icon

2. Select **World** from the shortcut menu.

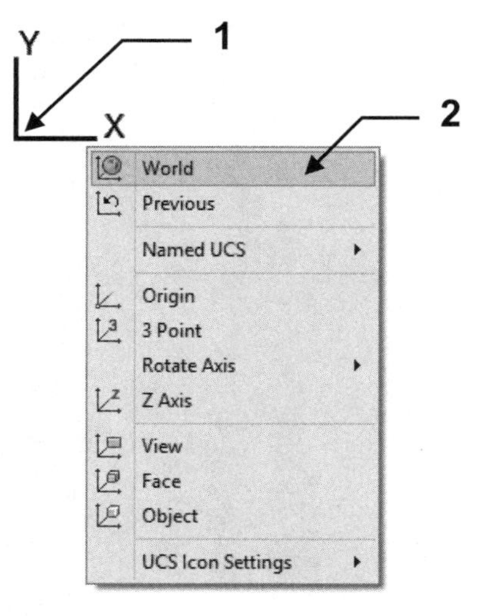

Why move the Origin? Examples on the next page.

MOVING THE ORIGIN....continued

Why move the Origin?

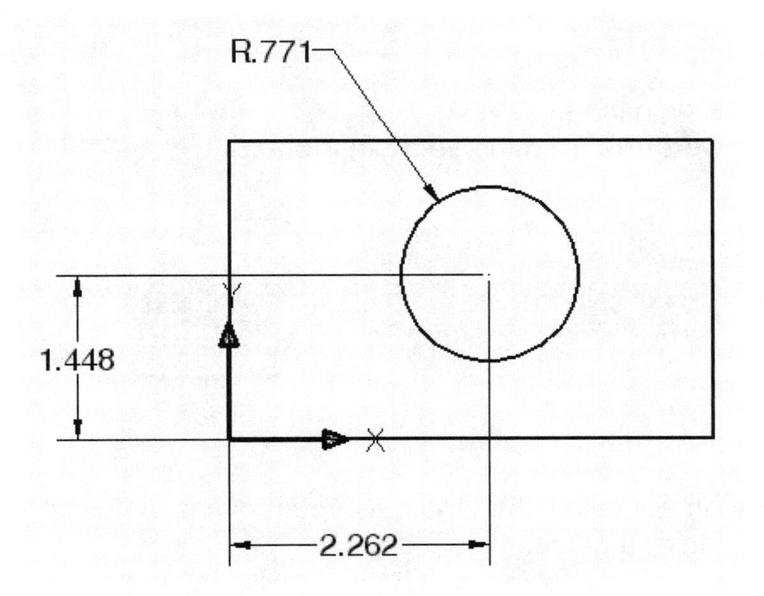

If you move the Origin to the lower left corner of the rectangle it will make it very easy to accurately place the center of the circle.

How to place the Circle accurately:

1. Select **"Origin"** from the shortcut menu as shown on the previous page
2. Snap to the lower left corner of the rectangle using object snap Endpoint. The UCS icon should now be displayed as shown above.
3. Select the Circle command
4. Enter the coordinates to the center of the circle (2.262, 1.448 <enter>)
5. Enter the radius (.771 <enter)
6. Select **"World"** from the shortcut menu to return the UCS icon to it's default location. (Refer to previous page)

More examples of why you would move the Origin

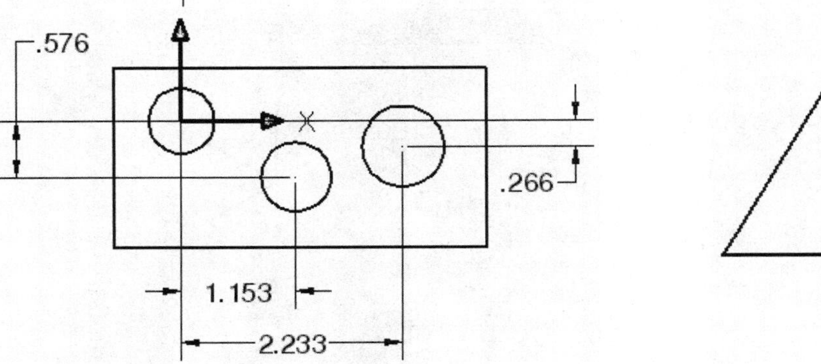

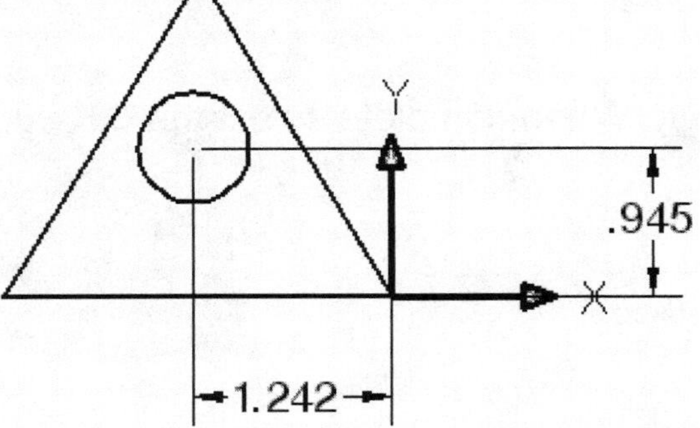

DISPLAYING THE UCS ICON

The **UCS icon** is merely a drawing aid that displays the location of the Origin. It can move with the Origin or stay in the default location. You can even change its appearance.

Show UCS Icon at Origin

1. Right click on the Origin icon.
2. Select **UCS Icon Settings** from the shortcut menu.

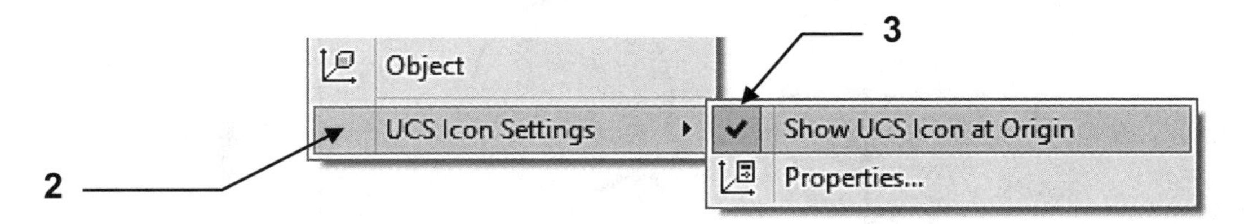

3. **Show UCS Icon at Origin**

☑ The UCS icon will follow the Origin as you move it. You will be able to see the Origin location at a glance.
☐ The UCS icon will not follow the Origin. It will stay in it's default location.

How to change the UCS icon appearance.

1. Right click on the Origin icon.
2. Select **UCS Icon Settings** from the short cut menu.

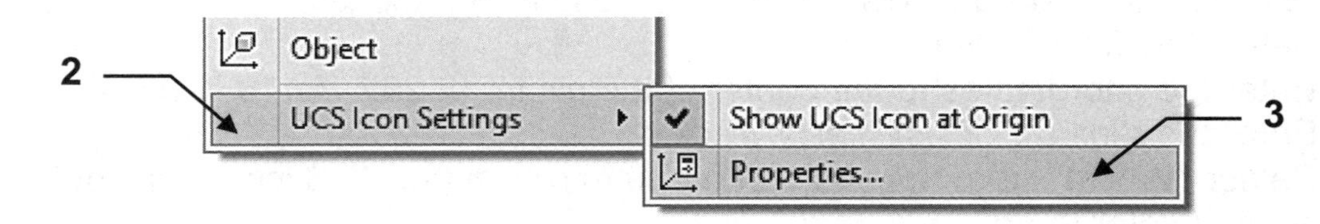

3. Select **Properties**
   When you select this option the dialog box shown below will appear.
   You may change the Style, Size and Color of the icon at any time.
   Changing the appearance is personal preference and will not affect the drawing or commands.

4. When complete select the **OK** button.

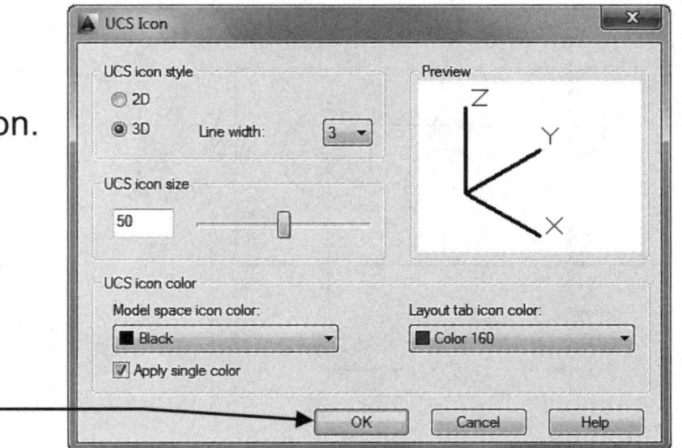

EXERCISE 10A

INSTRUCTIONS:

1. Start a **New** file using **Border A-2015.dwt**
2. Move the Origin to the locations noted before you draw the objects.
3. Use Layer = Object line.
4. Edit the Title and Ex-XX by double clicking on the text. Do not erase and replace.
5. Do not dimension
6. Save as **EX10A**
7. Plot using Page Setup **Class Model A** (Refer to page 9-13 through 9-16)

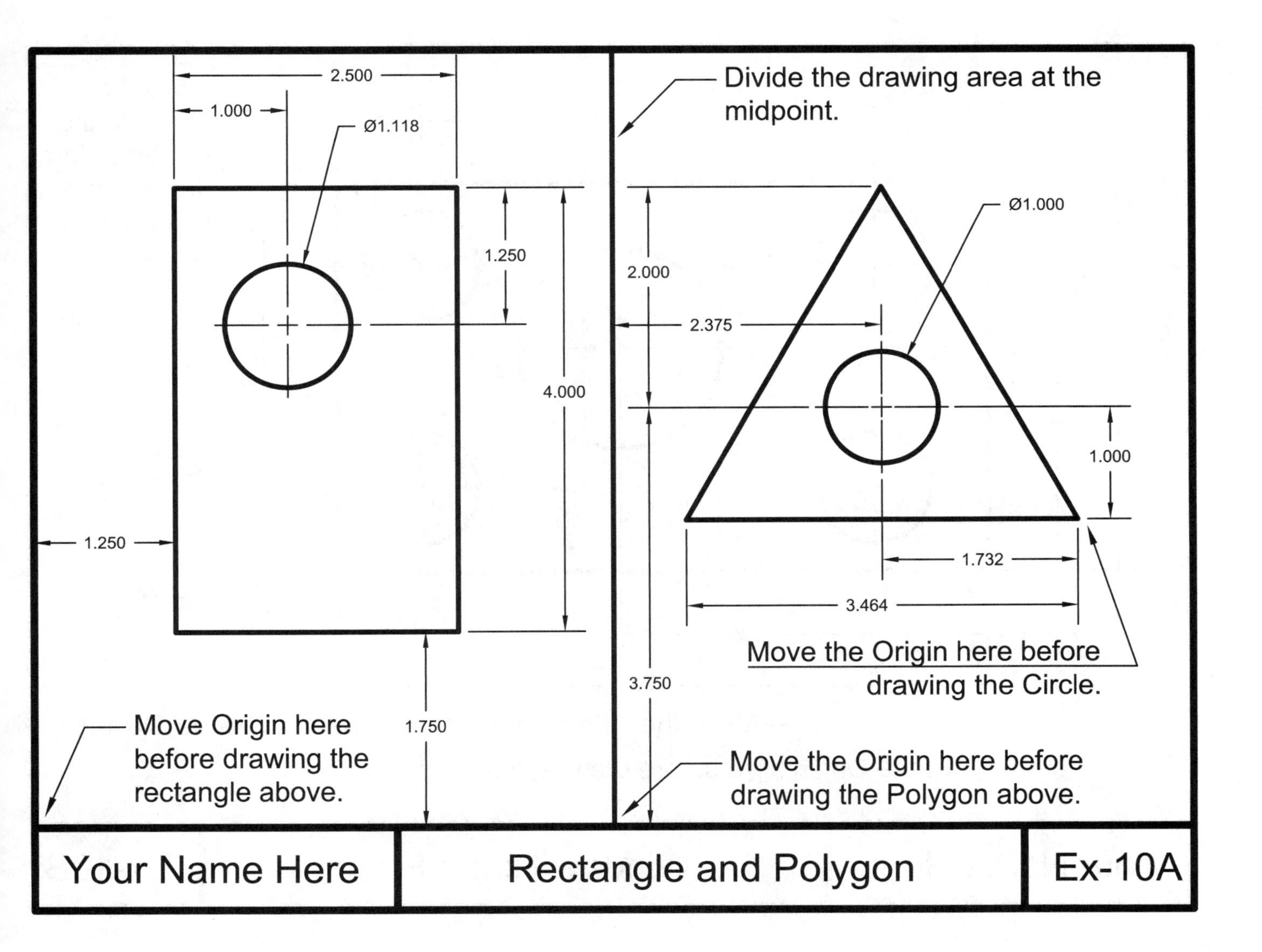

EXERCISE 10B

INSTRUCTIONS:

1. Start a **New** file using **Border A-2015.dwt.**
2. Move the Origin to the locations noted before you draw the objects.
3. Use Layer = Object line.
4. Edit the Title and Ex-XX by double clicking on the text. Do not erase and replace.
5. Do not dimension
6. Save as **EX10B**
7. Plot using Page Setup **Class Model A** (Refer to page 9-13 and 9-16)

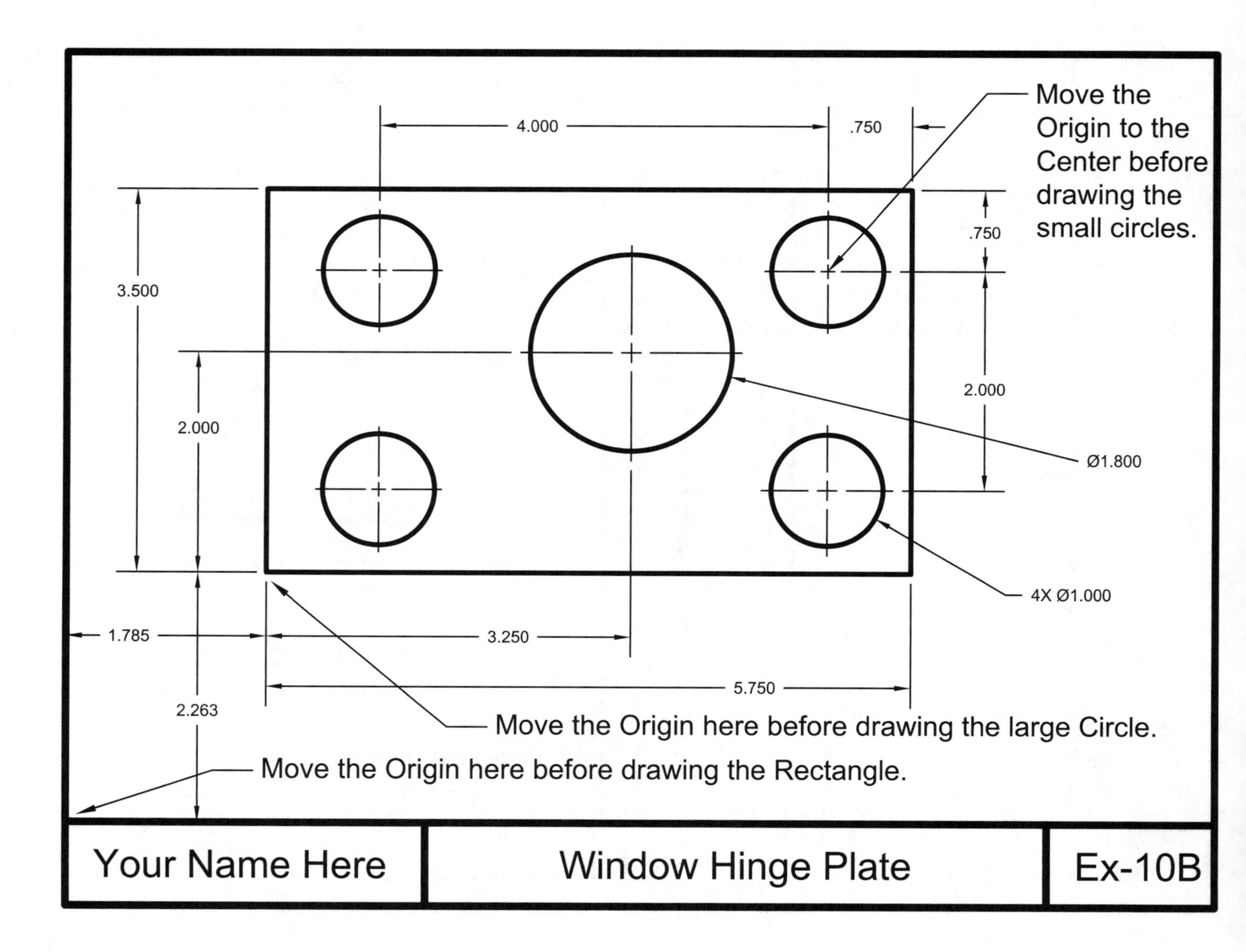

EXERCISE 10C

INSTRUCTIONS:

1. Start a **New** file using **Border A-2015.dwt**
2. You decide where to move the origin.
3. Use Layer = Object line
4. Edit the Title and Ex-XX by double clicking on the text. Do not erase and replace.
5. Do not dimension
6. Save as **EX10C**
7. Plot using Page Setup **Class Model A** (Refer to page 9-13 and 16)

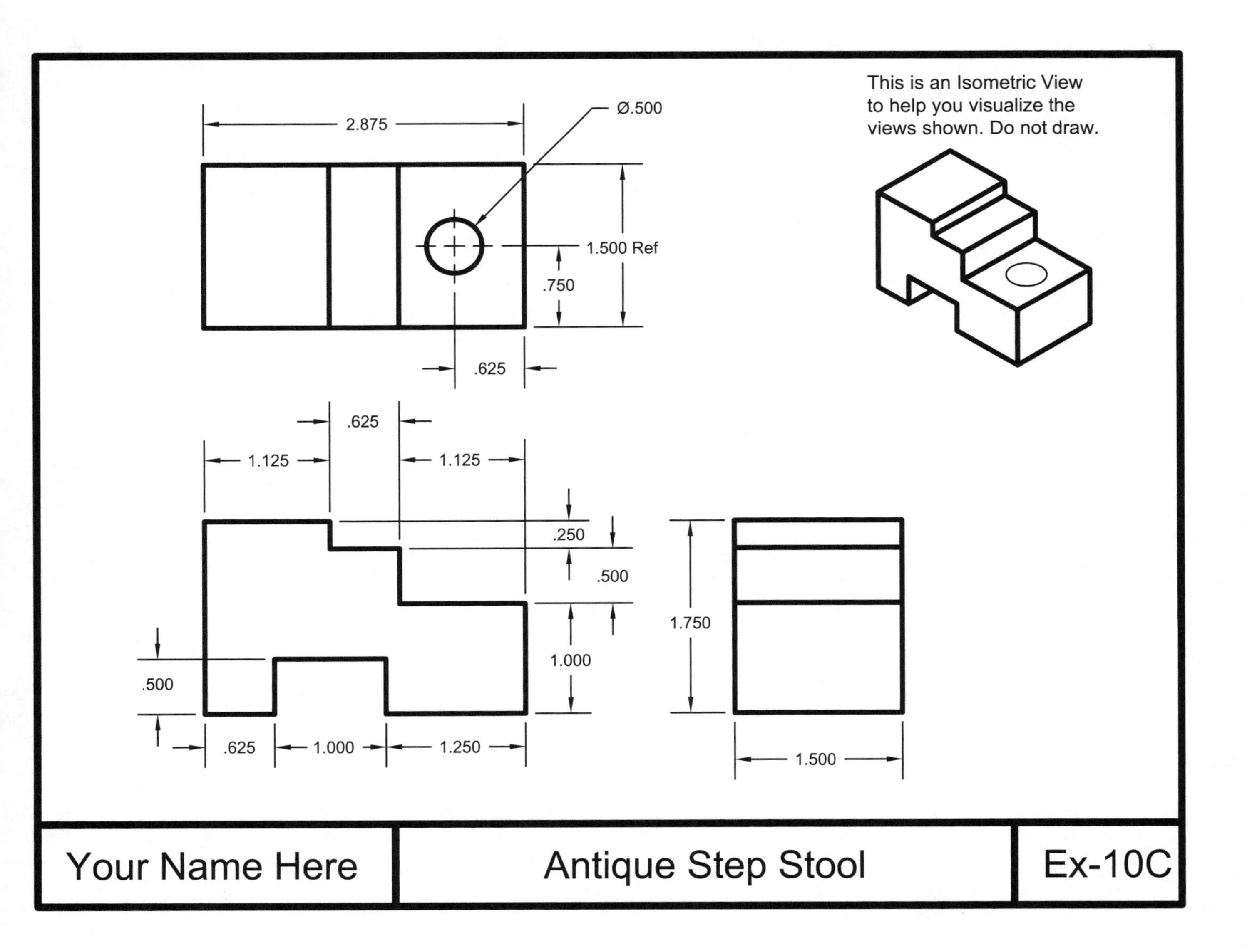

Notes:

LEARNING OBJECTIVES

After completing this lesson, you will be able to:

1. Understand the Polar Degree Clock
2. Draw Lines to a specific length and angle
3. Draw objects using Polar Coordinate Input
4. Use Dynamic Input
5. Use Polar Tracking and Polar Snap

LESSON 11

POLAR COORDINATE INPUT

In Lesson 9 you learned to control the length and direction of horizontal and vertical lines using Relative Input and Direct Distance Entry. Now you will learn how to control the length and **ANGLE** of a line using **POLAR Coordinate Input.**.

UNDERSTANDING THE *"POLAR DEGREE CLOCK"*

Previously when drawing Horizontal and Vertical lines you controlled the direction using a Positive or Negative input. ***Polar Input is different***. The Angle of the line will determine the direction.

For example: If I want to draw a line at a 45 degree angle towards the upper right corner, you would use the angle 45. But if I want to draw a line at a 45 degree angle towards the lower left corner, you would use the angle 225.

You may also use Polar Input for Horizontal and Vertical lines using the angles 0, 90, 180 and 270. No negative input is required.

POLAR DEGREE CLOCK

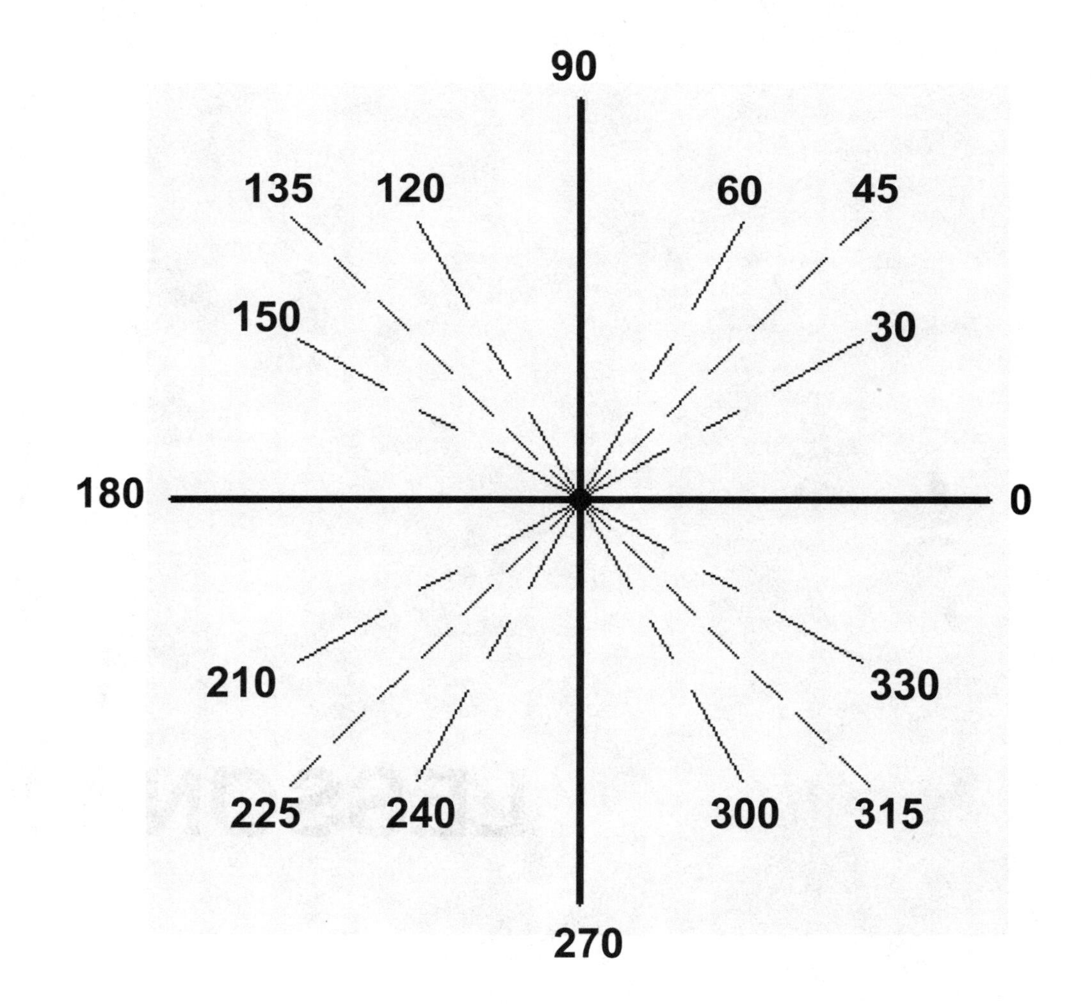

POLAR COORDINATE INPUT....continued

DRAWING WITH *POLAR COORDINATE INPUT*

When entering polar coordinates the first number represents the **Distance** and the second number represents the **Angle**. The two numbers are separated by the **less than (<)** symbol. The input format is: **distance < angle**

Note: If you are using Dynamic Input, refer to the next page.

A Polar coordinate of **@6<45** will be a distance of 6 units and an angle of 45 degrees ***from the last point entered.***

Here is an example of Polar input for 4 line segments.

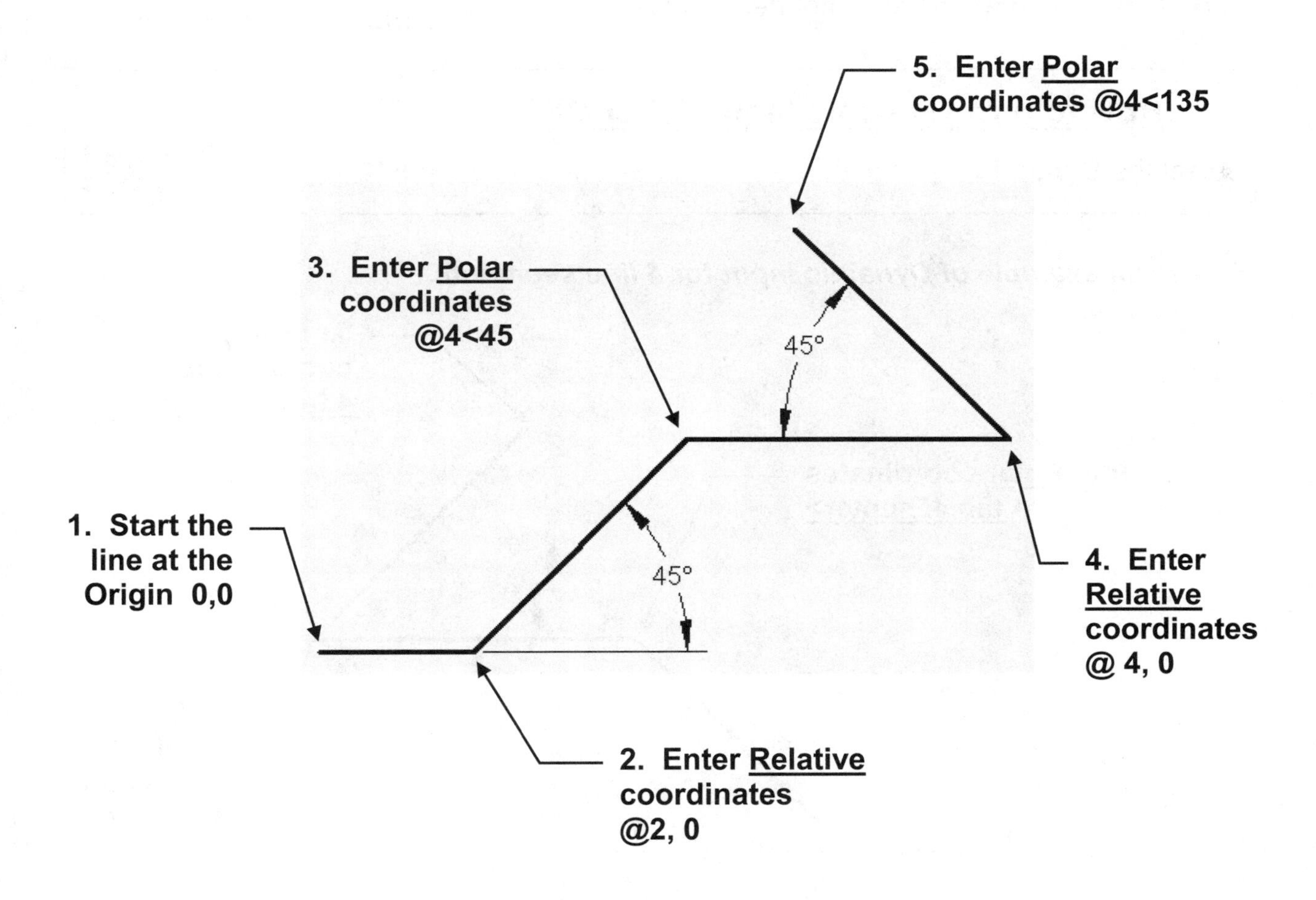

DYNAMIC INPUT

To help you keep your focus in the "drawing area", AutoCAD has provided a command interface called **Dynamic Input**. You may input information within the Dynamic Input tool tip box instead of on the command line.

When AutoCAD prompts you for the **First point** the Dynamic Input tool tip displays the **<u>Absolute: X, Y</u>** distance from the Origin.
Enter the **X** dimension, <u>press the **Tab key**</u>, enter the **Y** dimension then **<enter>.**

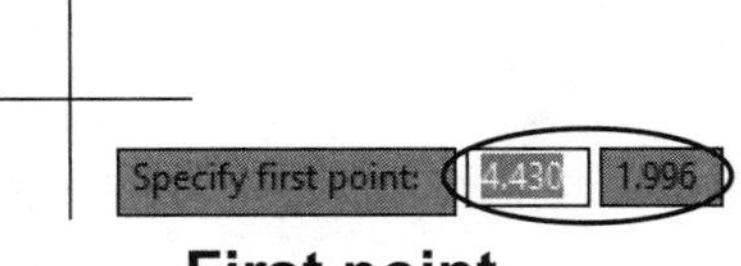

First point

When AutoCAD prompts you for the **Second** and all **Next points** the Dynamic Input tool tip displays the **<u>Relative: Distance and Angle</u>** from the last point entered.
Enter the **distance**, <u>press the **tab key**</u>, move the cursor in the approximate desired angle and enter the **angle** then **<enter>**. (Note: The @ is not necessary)

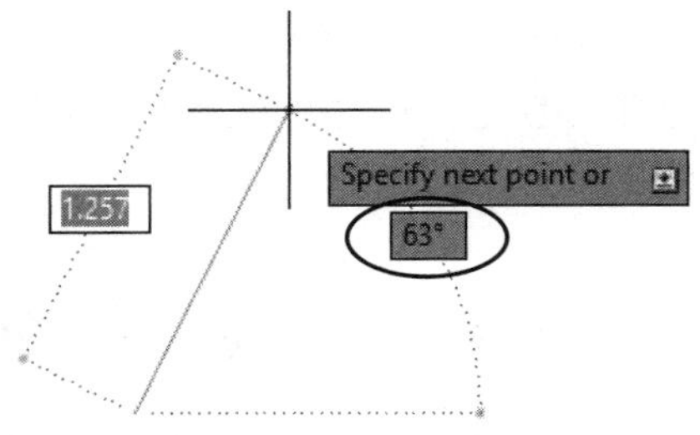

Second and all next points

<u>How to turn Dynamic Input ON or OFF</u>

Select the **Dynamic Input** button on the status bar or use the **F12** key.

Here is an example of Dynamic input for 4 line segments.

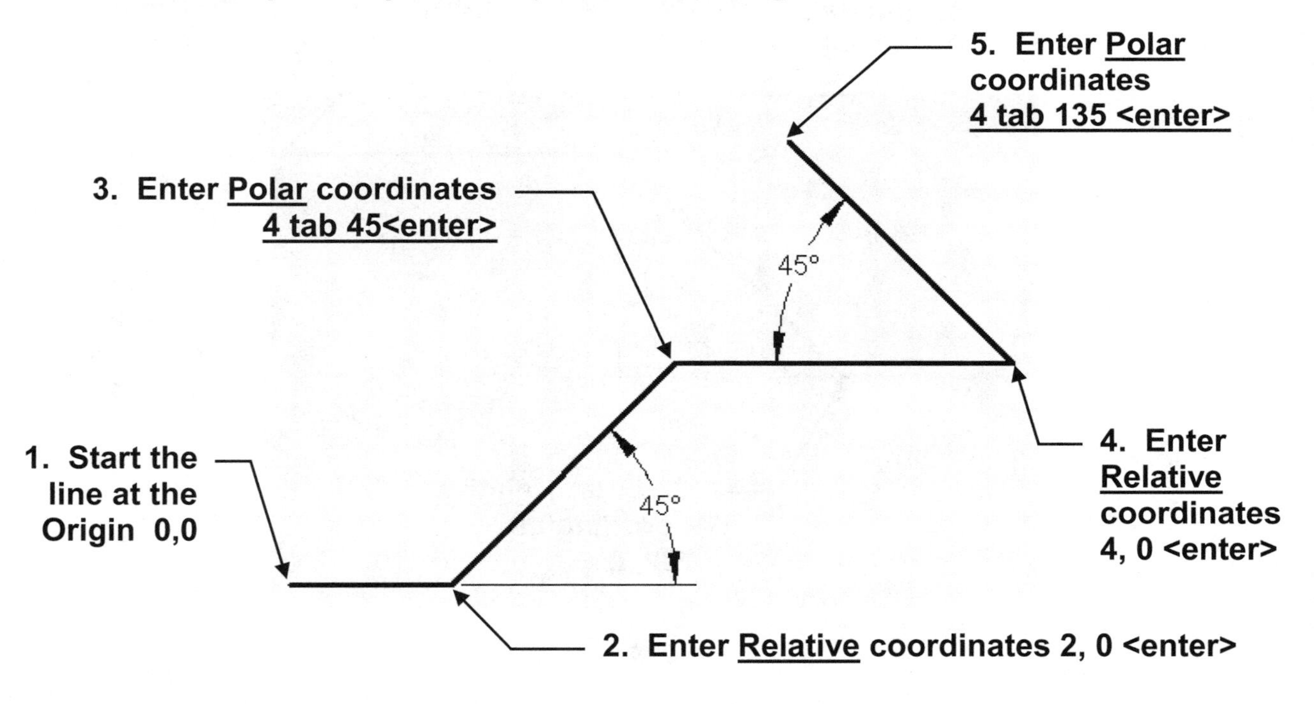

<u>Note</u>: Refer to **Appendix C** for more ways to control the Dynamic Input Tool tip box display.

DYNAMIC INPUT....continued

To enter Cartesian coordinates (X and Y)

1. Enter an "**X**" coordinate value and a **comma**.

2. Enter an "**Y**" coordinate value <enter>.

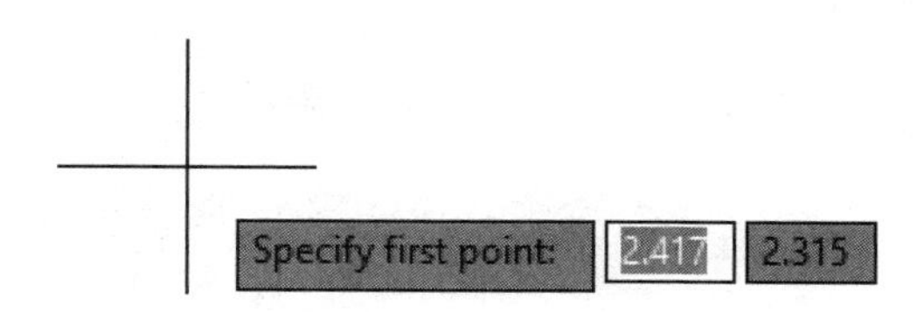

To enter Polar coordinates (from the last point entered)

1. Enter the **distance** value from the last point entered.

2. Press the **Tab** key.

3. Move the cursor in the approximate direction and enter the **angle** value <enter>

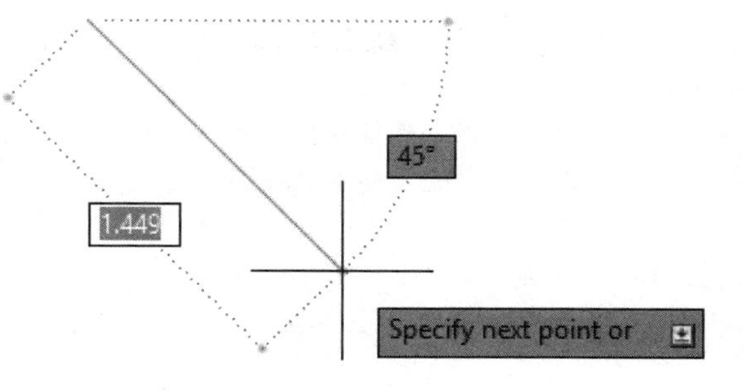

Note: Move the cursor in the approximate direction and enter an angle value of 0-180 **only**
Dynamic Input does not use 360 degrees.
(Refer to example on the next page)

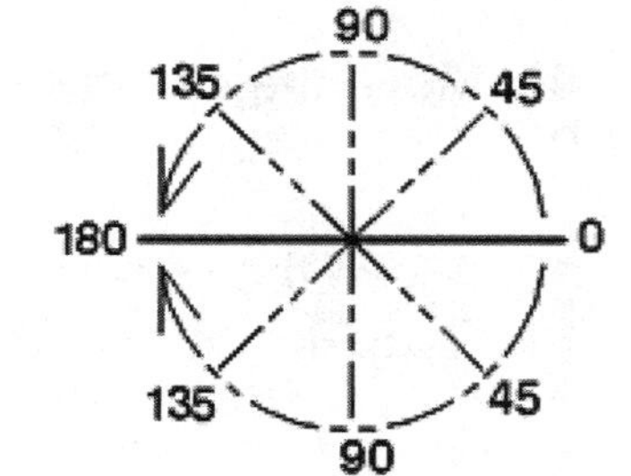

How to specify Absolute or Relative coordinates while using Dynamic Input.

To enter **absolute** coordinates when relative coordinate format is displayed in the tooltip. Enter **#** to temporarily override the setting.

To enter **relative** coordinates when absolute coordinate format is displayed in the tooltip. Enter **@** to temporarily override the setting.

Note about OrthoMode
You may toggle OrthoMode **ON** and **OFF** by holding down the **shift** key.
This is an easy method to use Direct Distance Entry while using Dynamic Input.

USING DYNAMIC INPUT and POLAR COORDINATES

The following is a simple drawing to practice Dynamic Input and Polar coordinates. Think about how this differs from the basic polar input on page 11-2 and 11-3.

Example on the next page.

1 Set the Status Bar as follows:
Dynamic Input = ON   All others = OFF

2. Select the **Line** command:

3. Start the Line near the lower left corner of the drawing area.

Line A 4. Move the cursor to the right.
5. Type 2 <tab>   0 <enter>

Line B 6. Move the cursor up and to the right
7. Type 3 <tab> 45 <enter>

Line C 8. Move the cursor up.
9. Type 2 <tab>   90<enter>

Line D 10. Move the cursor down and to the left.
11. Type 4 <tab>   135<enter>   (Note: 180 – 45 = **135**)

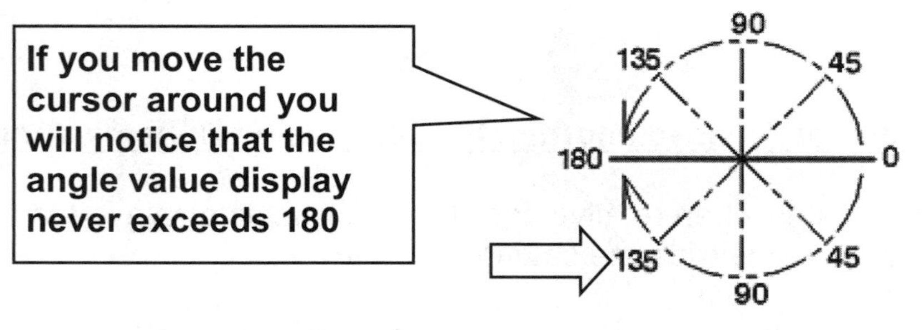

Line E 12. Move the cursor to the left.
13. Type 1.293 <tab> 180 <enter>

Line F 14. Move the cursor down.
15. Type 1.293 <tab> 90 <enter>
16. <enter> to stop

Continued on the next page...

USING DYN INPUT and POLAR COORDS….continued

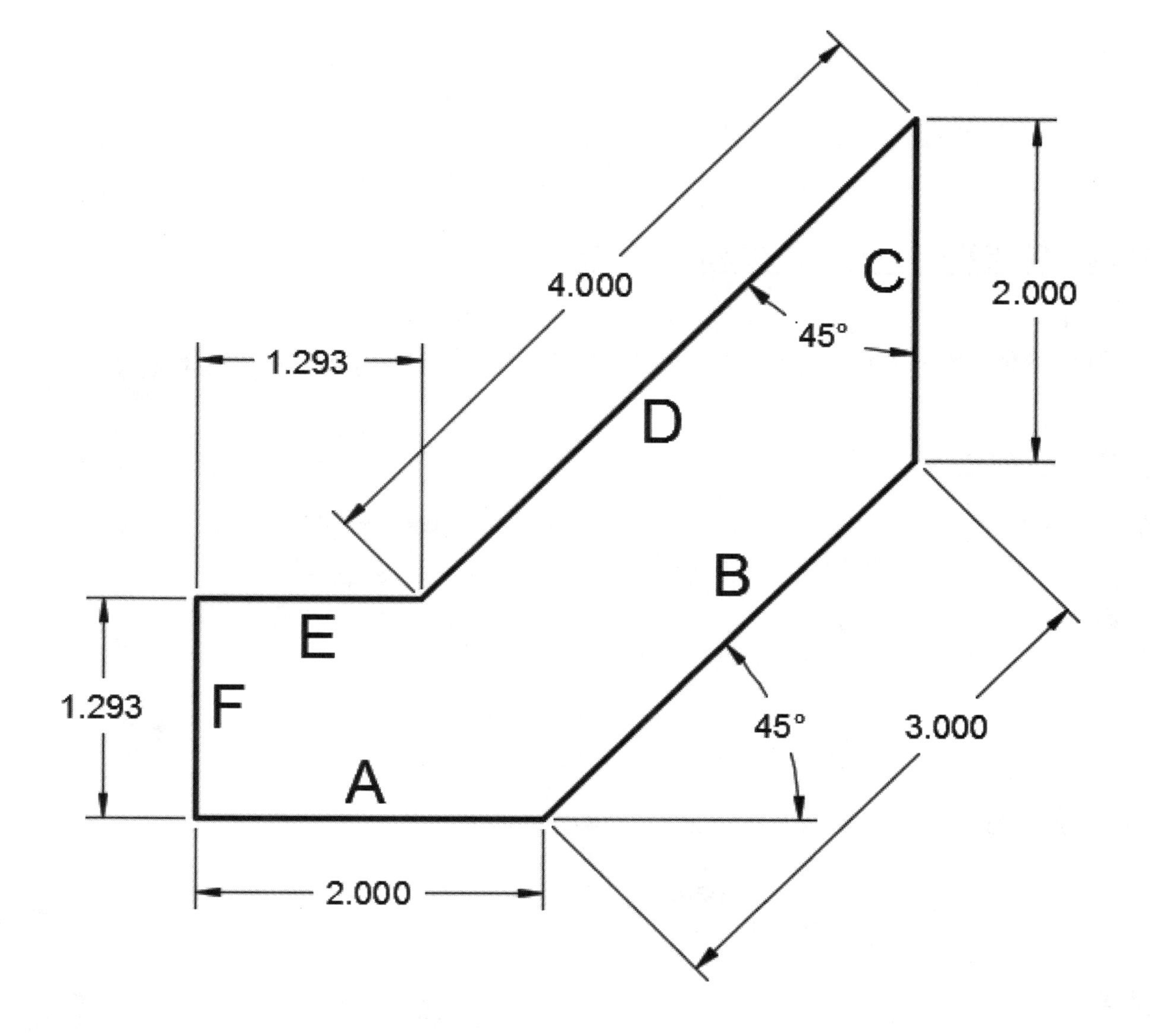

POLAR TRACKING

Polar Tracking can be used <u>instead</u> of **Dynamic Input**. When *Polar Tracking* is "**ON**", a dotted ***"tracking"*** line and a ***"tool tip"*** box appear. The tracking line.... "snaps" to a **preset angle increment** when the cursor approaches one of the preset angles. The word ***"Polar"***, followed by the ***"distance"*** and ***"angle"*** from the last point appears in the box. (A step by step example is described on the next page.)

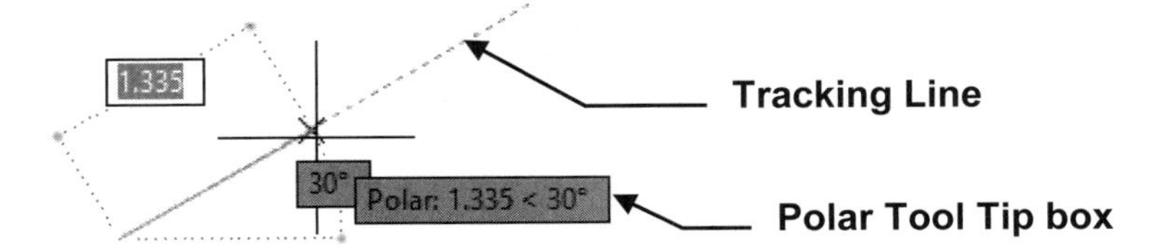

HOW TO SET THE INCREMENT ANGLE

1. Left Click on the **POLAR** button down arrow ▾ on the Status Bar and select "**Tracking Settings**", <u>or select an angle from the list.</u>

Tracking Settings...

POLAR ANGLE SETTINGS

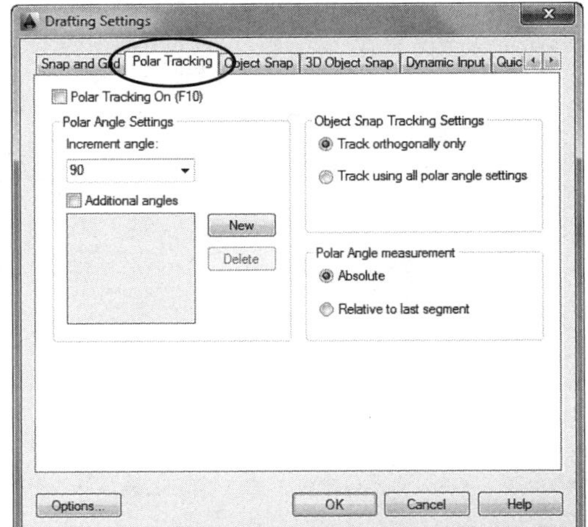

<u>Increment Angle</u> Choose from the Increment Angle list including 90, 45, 30, 22.5, 18, 15,10 and 5. It will also snap to the selected angles multiples. For example: if you choose 30 it will snap to 30, 60, 90, 120, 150, 180, 210, 240, 270, 300, 330 and 0.

<u>Additional Angles</u> Check this box if you would like to use an angle other than one in the Incremental Angle list. For example: 12.5.

<u>New</u> You may add an angle by selecting the "New" button. You will be able to snap to this new angle in addition to the incremental Angle selected. <u>But you will not be able to snap to it's multiple</u>. For example, if you selected 7, you would not be able to snap to 14.

<u>Delete</u> Deletes an Additional Angle. Select the Additional angle to be deleted and then the **Delete** button.

POLAR ANGLE MEASUREMENT

<u>ABSOLUTE</u> Polar tracking <u>angles</u> are relative to the UCS.

<u>RELATIVE TO LAST SEGMENT</u> Polar tracking <u>angles</u> are relative to the last segment.

USING POLAR TRACKING and DDE

1. Set the Polar Tracking Increment Angle to 15.

2. Turn all the Status Bar buttons to Off except POLAR.
   (Note: Dynamic Input should be OFF but you may wish to leave it ON)

3. Select the Line command.

P1 4. Start the Line in the lower left area of the drawing area.

P2 5. Move the cursor in the direction of P2 until the Tool Tip box displays 30 degrees.
6. Type 2 <enter> (for the length).

P3 7. Move the cursor in the direction of P3 until the Tool Tip box displays 90 degrees.
8. Type 2 <enter> (for the length).

P4 9. Move the cursor in the direction of P4 until the Tool Tip box displays 0 degrees.
10. Type 2 <enter> (for the length).

P5 11. Move the cursor in the direction of P5 until the Tool Tip box displays 150 degrees
12. Type 2 <enter> (for the length).

P6 13. Move the cursor in the direction of P6 until the Tool Tip box displays 180 degrees.
14. Type 2 <enter> (for the length).

15. Then type **C** for close.

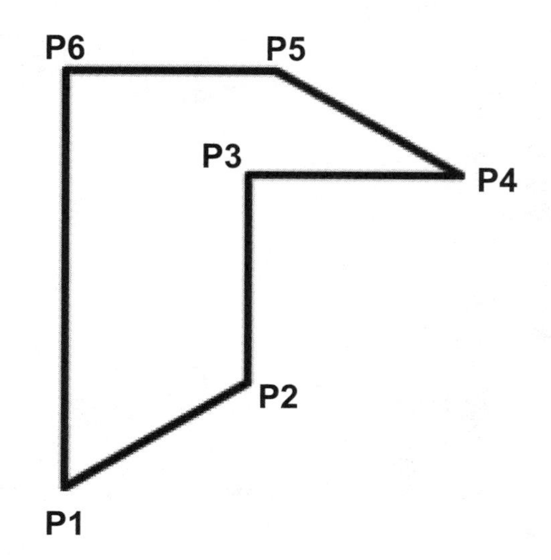

POLAR TRACKING ON or OFF

You may toggle Polar Tracking On or Off using one of the following:
Left click on the **POLAR** button on the Status Bar or Press **F10**

POLAR SNAP

Polar Snap is used with Polar Tracking to make the cursor snap to specific ***distances*** and ***angles***. If you set Polar Snap distance to 1 and Polar Tracking to angle 30 you can draw lines 1, 2, 3 or 4 units long at an angle of 30, 60, 90 etc. without typing anything. You just move the cursor and watch the tool tips.

(A step-by-step example is described on the next page)

SETTING THE POLAR SNAP

1. Set the **Polar Tracking Increment Angle** as shown on page 11-8
2. Left Click on the **SNAP** button down arrow ▾ on the Status Bar and select "**Snap Settings**"

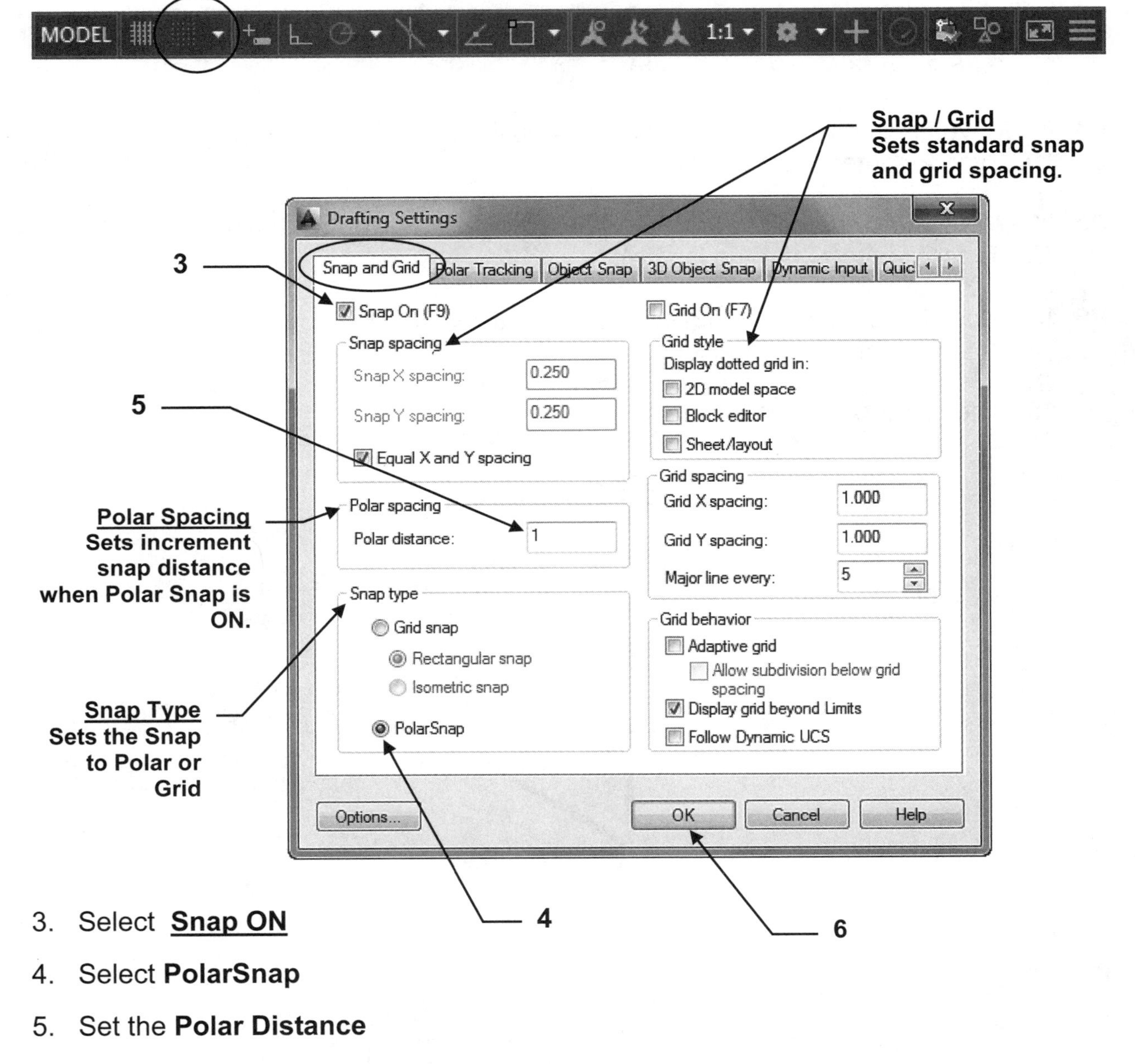

3. Select **Snap ON**
4. Select **PolarSnap**
5. Set the **Polar Distance**
6. Select **OK** button.

USING POLAR TRACKING and POLAR SNAP

Now let's draw the objects below again, but this time with "Polar Snap" .

1. Set **Polar Tracking** Increment Angle to 30 and **Polar Snap** to 1.00.
2. Turn all the Status Bar buttons Off except **SNAP** and **POLAR**.
3. Select the Line command:

P1 4. Start the Line in the lower left area of the drawing area.

P2 5. Move the cursor in the direction of P2 until the Tool Tip box displays
Polar 2.00 <30°

P3 6. Move the cursor in the direction of P3 until the Tool Tip box displays
Polar 2.00 <90°

P4 7. Move the cursor in the direction of P4 until the Tool Tip box displays
Polar 2.00 <0°

P5 8. Move the cursor in the direction of P5 until the Tool Tip box displays
Polar 2.00 <150°

P6 9. Move the cursor in the direction of P6 until the Tool Tip box displays
Polar 2.00 <180°

10 Then type **C** for close.

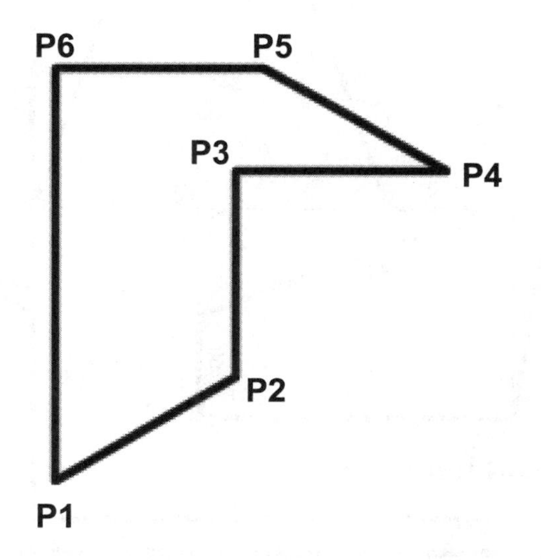

NOTE: You may OVERRIDE the Polar Settings at any time by typing: Polar coordinates (@ Length< Angle) on the Command line.

EXERCISE 11A

INSTRUCTIONS:

1. Start a **New** file using **Border A-2015.dwt**
2. After reviewing the lengths and angles below set the Polar Tracking Increment Angle and Polar Snap distance.

   Note: You may have to "override" a few of the lengths. (See bottom of 11-11)

3. Use Layer = Object line.
4. Edit the Title and Ex-XX by double clicking on the text. Do not erase and replace.
5. Do not dimension
6. Save as **EX11A**
7. Plot using Page Setup **Class Model A**

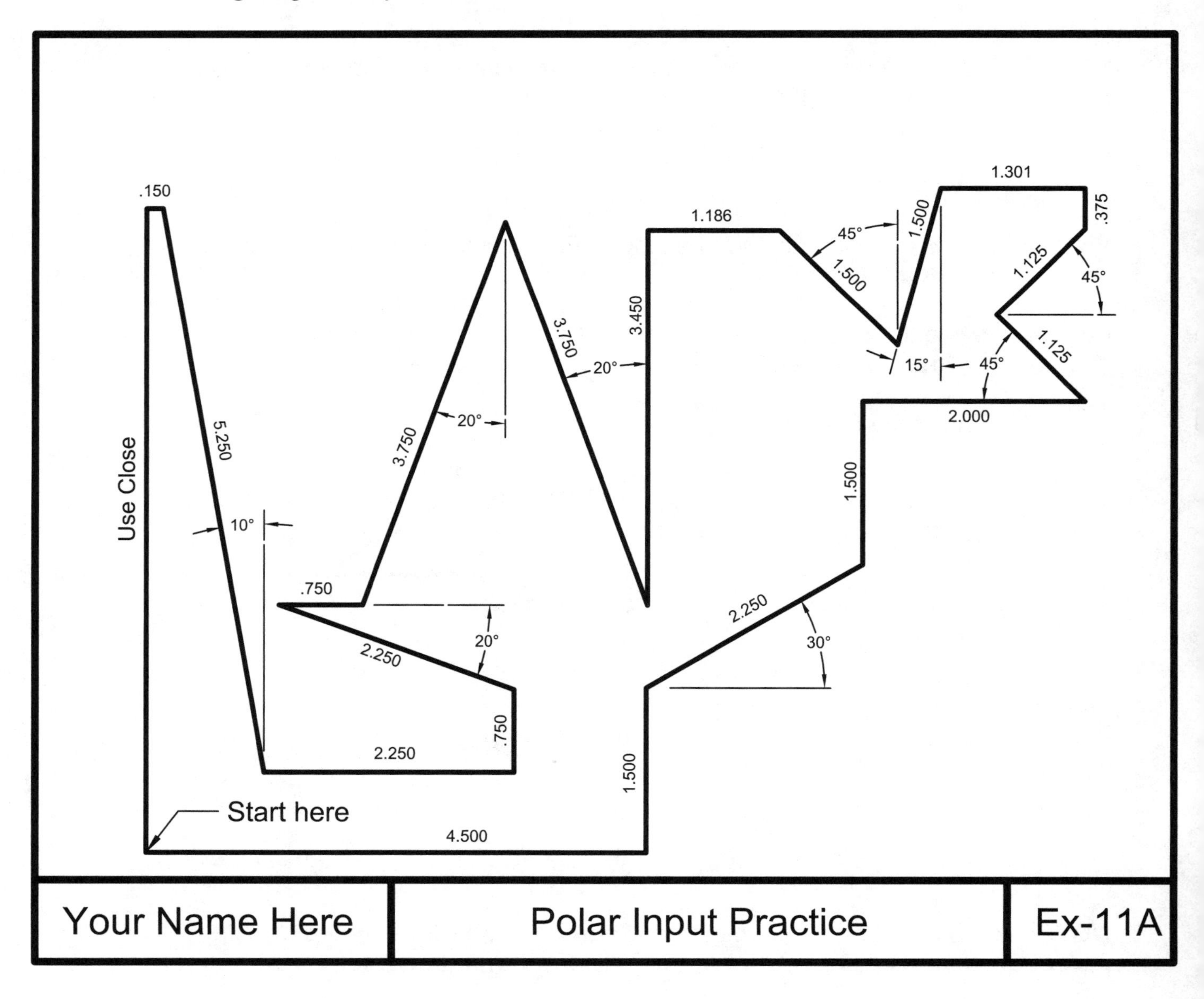

EXERCISE 11B

INSTRUCTIONS:

1. Start a **New** file using **Border A-2015.dwt**
2. After reviewing the lengths and angles below set the
   Polar Tracking Increment Angle to **30**
   Polar Snap distance to **.125**

   Note: the isometric lines are 30, 90, 150, 210, 270 and 330. (Refer to page 11-2)

3. Use Layer = Object line.
4. Edit the Title and Ex-XX by double clicking on the text. Do not erase and replace.
5. Do not dimension
6. Save as **EX11B**
7. Plot using Page Setup **Class Model A**

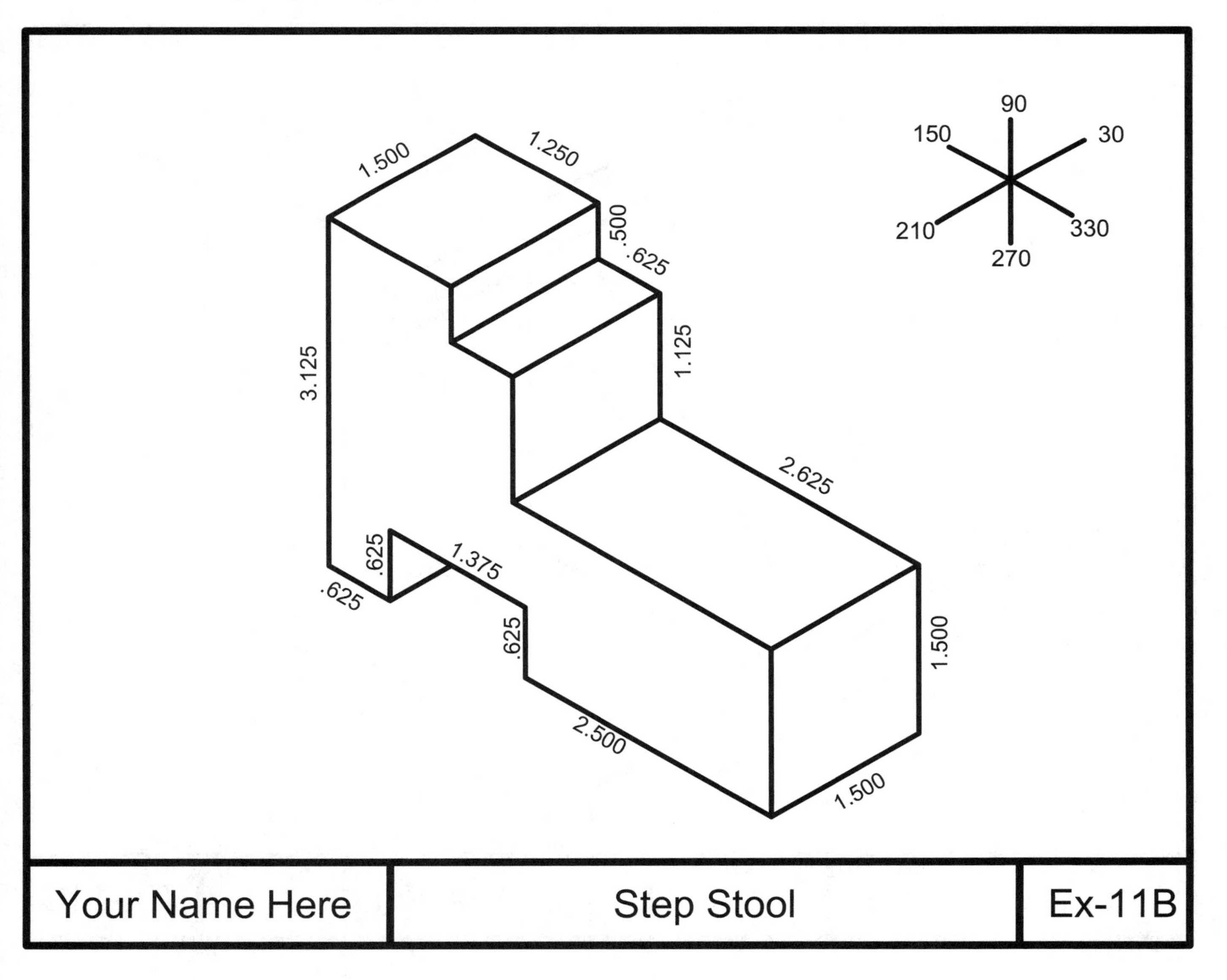

EXERCISE 11C

INSTRUCTIONS:

1. Start a **New** file using **Border A-2015.dwt**
2. After reviewing the lengths and angles below set the Polar Tracking Increment Angle and Polar Snap distance

Note: the isometric lines are 30, 90, 150, 210, 270 and 330. (Refer to page 11-2)

3. Use Layer = Object line.
4. Edit the Title and Ex-XX by double clicking on the text. Do not erase and replace.
5. Do not dimension
6. Save as **EX11C**
7. Plot using Page Setup **Class Model A**

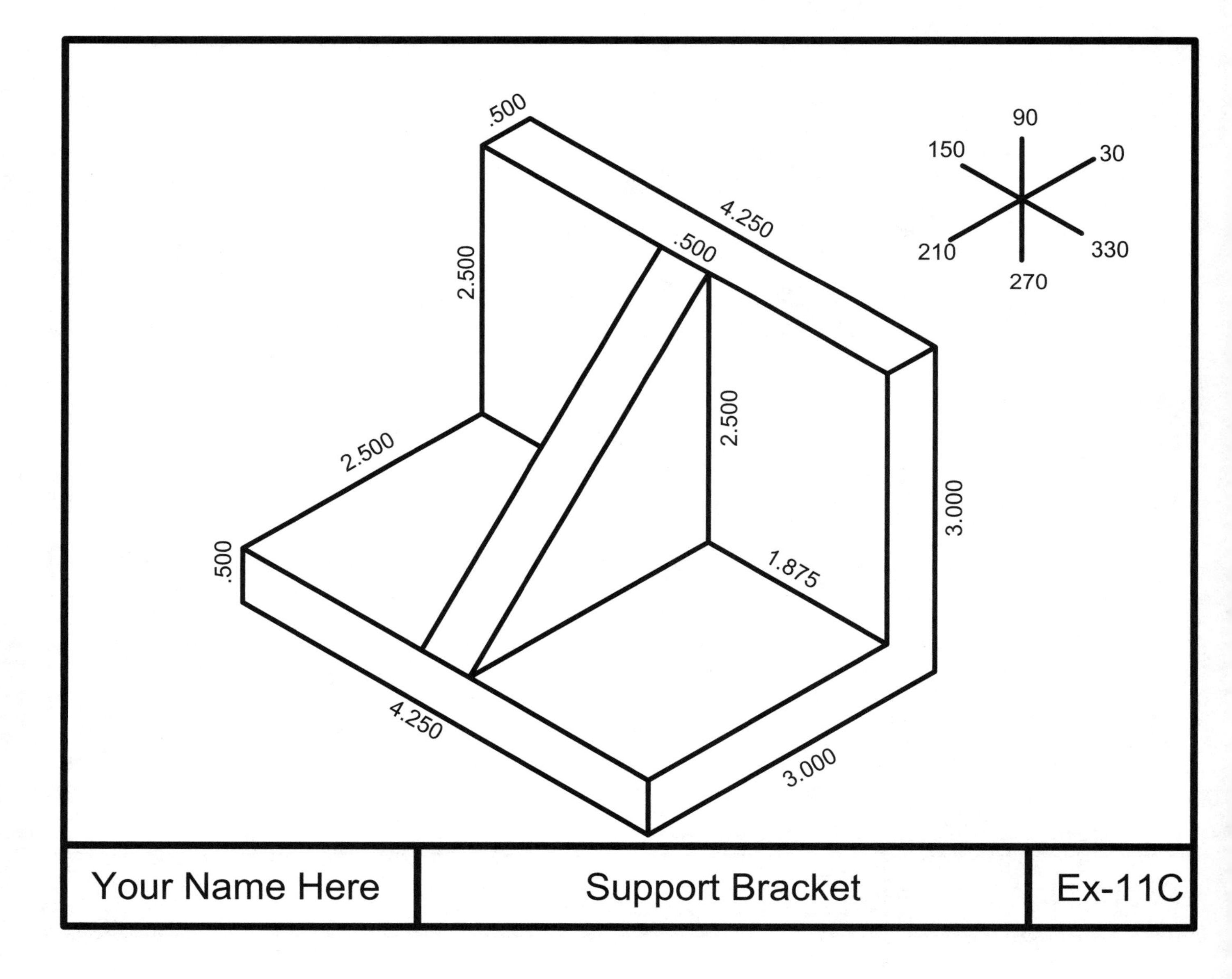

EXERCISE 11D

INSTRUCTIONS:

1. Open **Border A-2015.dwt**
   Note: You will have to do some adding and subtracting on this one.
2. Draw the 5.000 Diameter Circle using Layer Object line.
3. Draw the 4.000 Diameter Circle using Layer Center line.
4. Draw each polar line on layer Centerline using the following example:
   For example: the line marked "X" is drawn as follows:
   a. Place 1st endpoint in the center of the circle.
   b. Enter Polar coordinates for 2nd endpoint (distance 2.750 angle 138)
   (Refer to Polar Clock on page 11-2)
5. Draw the .500 Diameter Circles. Locate their center using object snap Intersection.
6. Edit the Title and Ex-XX by double clicking on the text. Do not erase and replace.
7. Do not dimension
8. Save as **EX11D**
9. Plot using Page Setup **Class Model A**

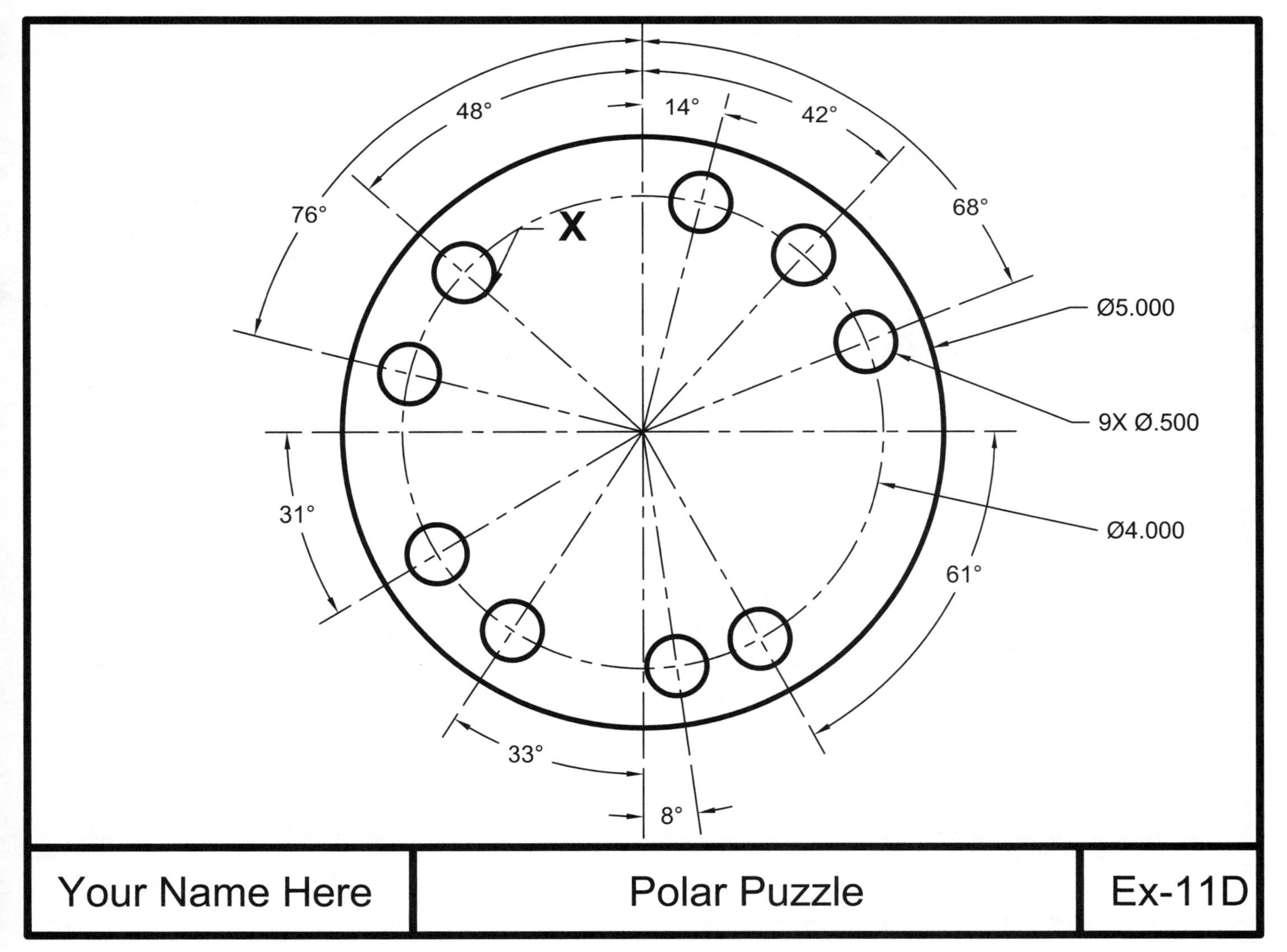

Notes:

LEARNING OBJECTIVES

After completing this lesson, you will be able to:

1. Duplicate an object at a specified distance away
2. Make changes to an objects properties
3. Use the Quick Properties Panel

LESSON 12

OFFSET

The **OFFSET** command duplicates an object parallel to the original object at a specified distance. You can offset Lines, Arcs, Circles, Ellipses, 2D Polylines and Splines. You may duplicate the original object or assign the offset copy to another layer.

Examples of Offset objects.

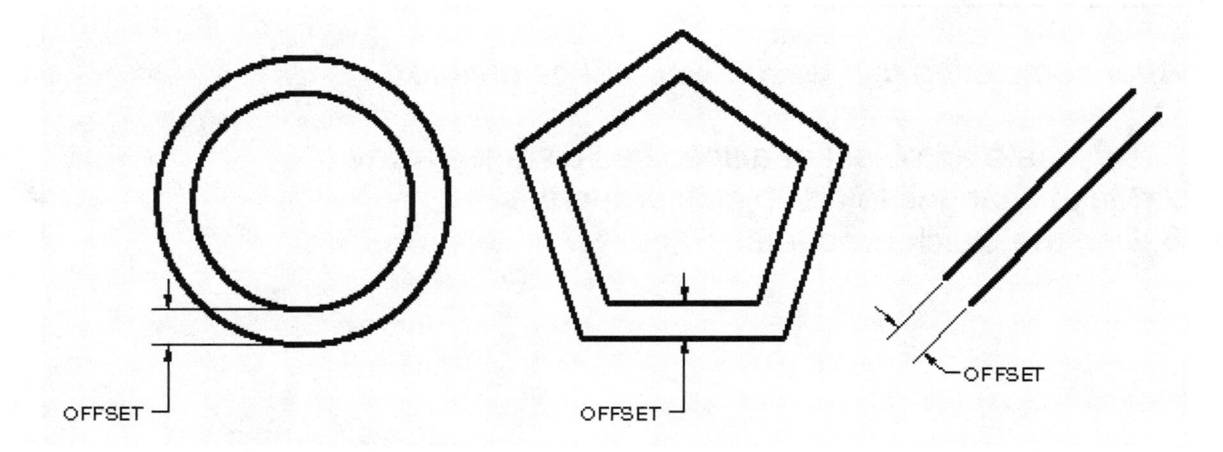

HOW TO USE THE OFFSET COMMAND:

METHOD 1
(Duplicate the Original Object)

1. Select the **OFFSET** command using one of the following:

    Ribbon = Home tab / Modify panel /
    or
    Keyboard = Offset <enter>

2. Command: _offset
   Current settings: Erase source=No Layer=Source OFFSETGAPTYPE=0
   Specify offset distance or [Through/Erase/Layer] <Through>: ***type the offset distance or select Erase or Layer. (see options on the next page)***

3. Select object to offset or <Exit/Undo>: ***select the object to offset.***

4. Specify point on side to offset or [Exit/Multiple/Undo]<Exit>: ***Select which side of the original you want the duplicate to appear by placing your cursor and clicking. (See options on the next page)***

5. Select object to offset or [Exit/Undo]<Exit>: ***Press <enter> to stop.***

OFFSET....continued

METHOD 2

(Duplicate original object but assign the Offset copy to a different layer)

To automatically place the offset copy on a different layer than the original you must first change the "current" layer to the layer you want the offset copy to be placed on.

1. Select the layer that you want the offset copy placed on from the list of layers.

2. Select the OFFSET command (refer to previous page)

3. Command: _offset
   Current settings: Erase source=No Layer=Source OFFSETGAPTYPE=0
   Specify offset distance or [Through/Erase/Layer] <Through>: ***type L <enter>***

4. Enter layer option for offset objects [Current/Source] <Source>: ***select C <enter>***

5. Specify offset distance or [Through/Erase/Layer] <Through>: ***type the offset dist <enter>***

6. Select object to offset or [Exit/Undo] <Exit>: ***select the object to offset.***

7. Specify point on side to offset or [Exit/Multiple/Undo]<Exit>: ***Select which side of the object you want the duplicate to appear by placing your cursor and clicking. (See options below)***

8. Select object to offset or [Exit/Undo]<Exit>: ***Press <enter> to stop.***

OPTIONS:

Through: Creates an object passing through a specified point.

Erase: Erases the source object after it is offset.

Layer: Determines whether offset objects are created on the current layer or on the layer of the source object. Select Layer and then select current or source. (Source is the default)

Multiple: Turns on the multiple offset mode, which allows you to continue creating duplicates of the original without re-selecting the original.

Exit: Exits the Offset command.

Undo: Removes the previous offset copy.

PROPERTIES PALETTE

The **Properties Palette**, shown below, makes it possible to change an object's properties. You simply open the Properties Palette, select an object and you can change any of the properties that are listed.

How to open the Properties Palette:

Ribbon = Home tab / Properties panel / ↘
or
Keyboard = Ctrl + 1

(An example of how to use the Properties Palette is on the next page.)

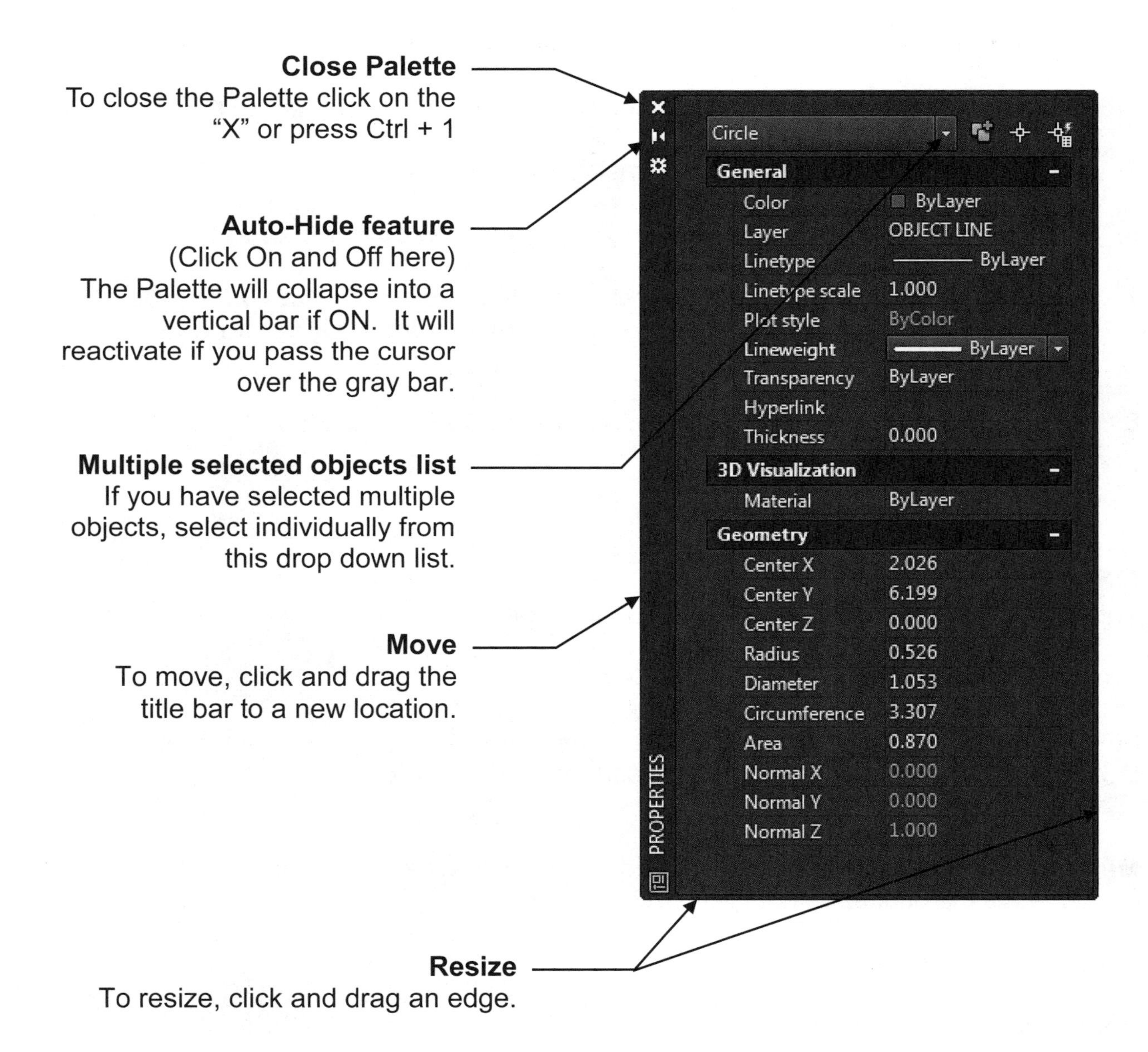

PROPERTIES PALETTE....continued

Example of editing an object using the Properties Palette

1. Draw a 2.00 Radius circle.

2. Open the Properties Palette and select the Circle. (*The Properties for the Circle should appear. You may change any of the properties listed in the Properties Palette for this object. When you press <enter> the circle will change.*)

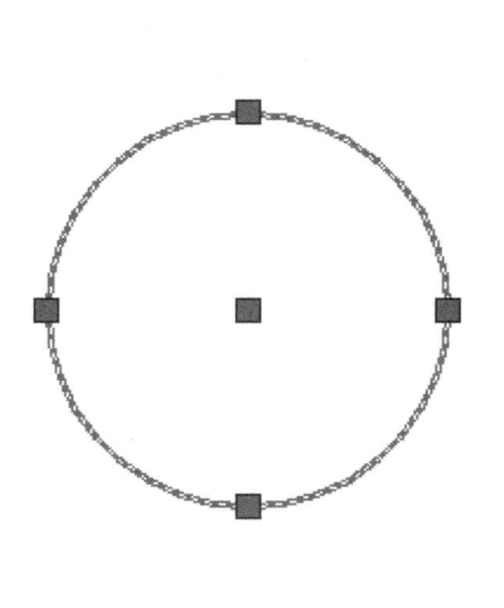

Circle	
General	
Color	ByLayer
Layer	OBJECT LINE
Linetype	ByLayer
Linetype scale	1.000
Plot style	ByColor
Lineweight	ByLayer
Transparency	ByLayer
Hyperlink	
Thickness	0.000
3D Visualization	
Material	ByLayer
Geometry	
Center X	3.398
Center Y	5.468
Center Z	0.000
Radius	2.000
Diameter	4.000
Circumference	12.566
Area	12.566
Normal X	0.000
Normal Y	0.000
Normal Z	1.000

PROPERTIES

3. Highlight and change the "Radius" to 1.00 and the "Layer" to HIDDEN LINE <enter>. *The Circle got smaller and the Layer changed as shown below.*

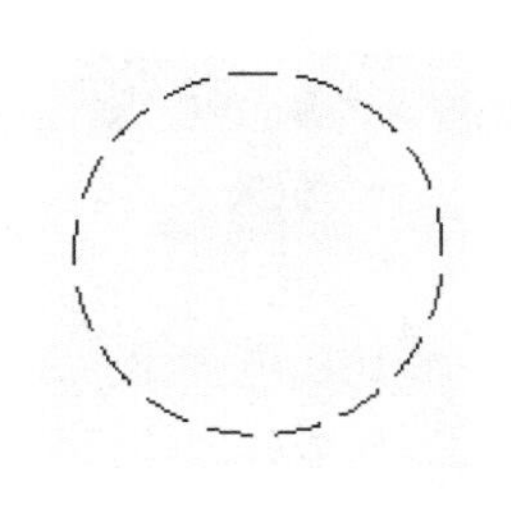

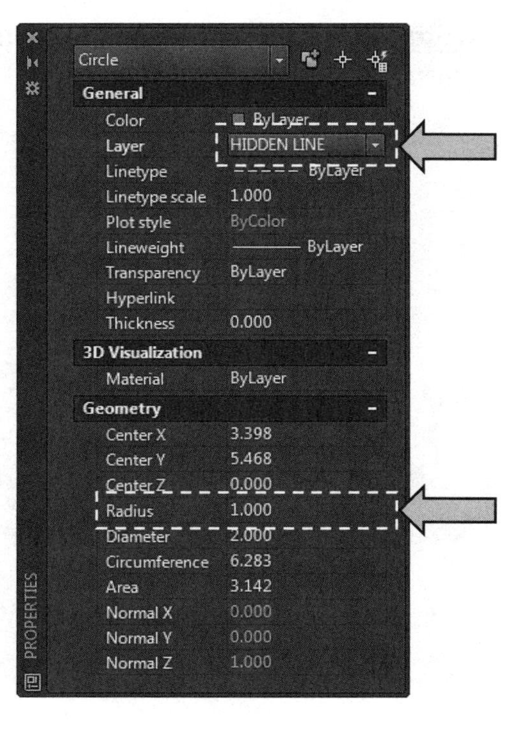

QUICK PROPERTIES PANEL

The Quick Properties Panel, shown below, will only appear if you have it set to ON. The Quick Properties Panel displays fewer properties and appears when you click once on an object. You may make changes to the objects properties using Quick Properties just as you would using the Properties Palette. (AutoCAD is just giving you another option)

How to turn Quick Properties Panel On or OFF.

Select the **Quick Properties** button on the status bar.

MODEL 1:1

1. Select an object

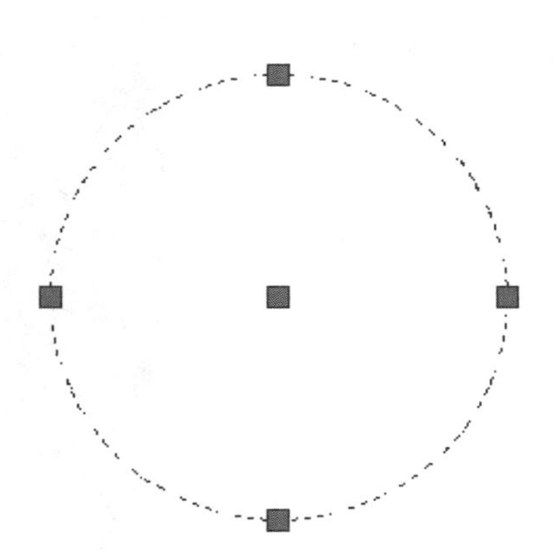

2. The Quick Properties box appears.

   <u>Note</u>: The list depends on the type of object you have selected, such as; Circle, Line, Rectangle, etc.

Circle	
Color	ByLayer
Layer	HIDDEN LINE
Linetype	------ ByLayer
Center X	3.398
Center Y	5.468
Radius	1.000
Diameter	2.000
Circumference	6.283
Area	3.142

Refer to the next page to "Customize" the Quick Properties Panel.

CUSTOMIZING THE QUICK PROPERTIES PANEL

You may add or remove properties from the Quick Properties Panel. And it is easy.

1. Select the **Customize** button.

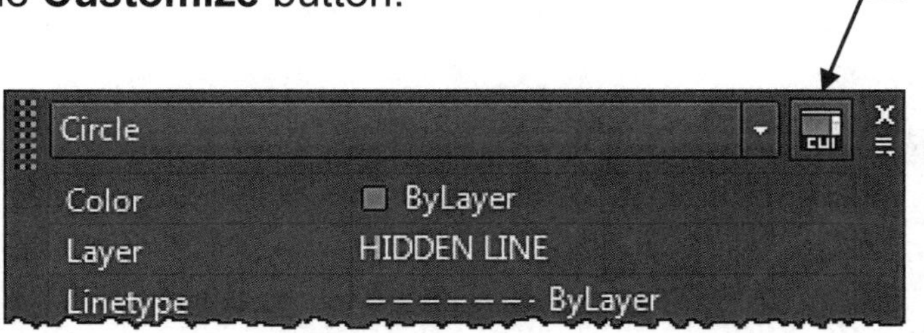

2. Select the Object, from the list, that you would like to customize.

3. Check the boxes for the properties that you want to appear.
   Uncheck the boxes that you do not want to appear.

4. Select **Apply** and **OK**

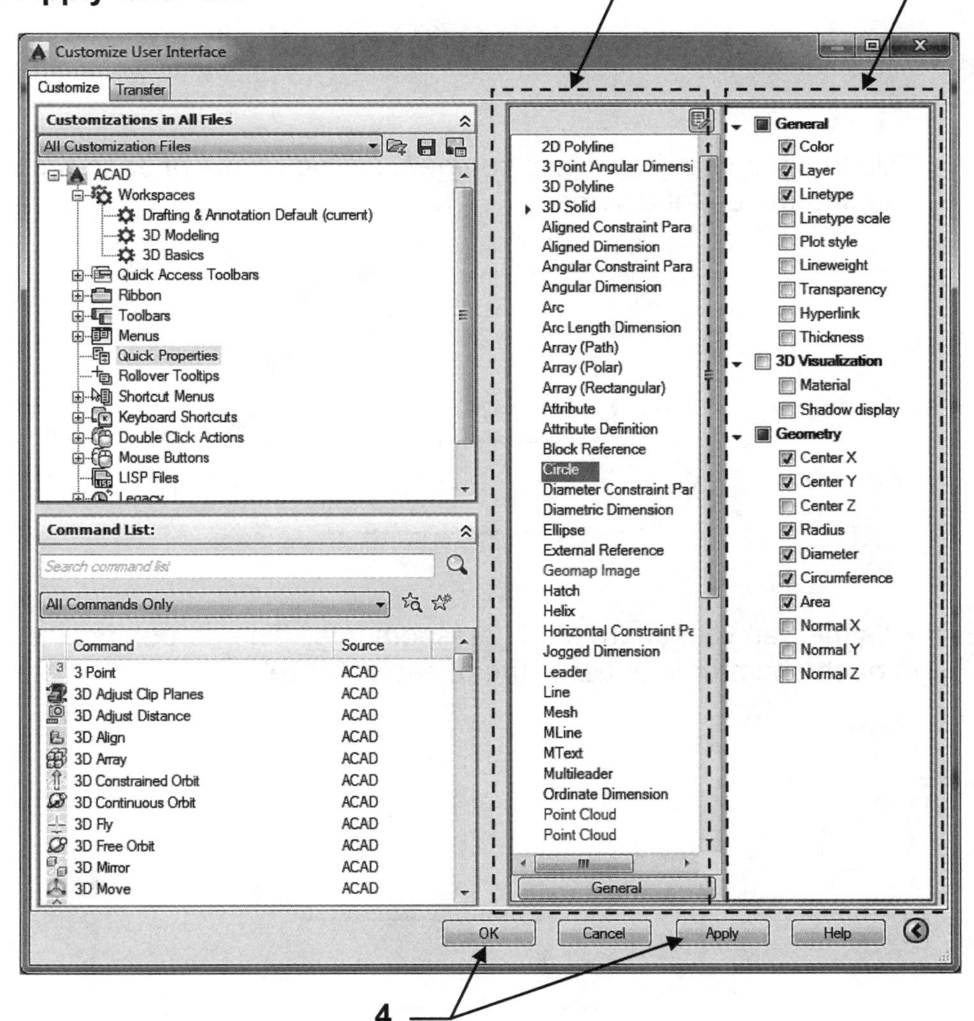

Note: The customizing is saved to the computer not the drawing file.

OFFSETGAPTYPE

When you offset a closed 2D object, such as a rectangle, to create a **larger** object it results in potential gaps between the segments. The **offsetgaptype** system variable controls how these gaps are closed.

To set the offsetgaptype:

Type*:* ***offsetgaptype <enter>***

Enter one of the following:

0 = Fills the gap by extending the polyline segments. (default setting)

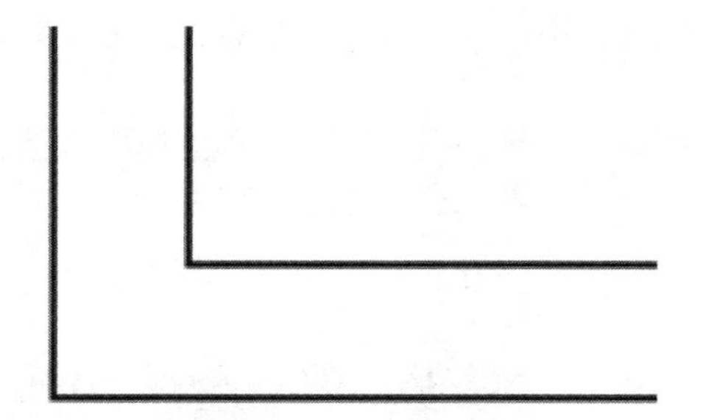

1 = Fills the gap with filleted arc segments. The radius of each arc segment is equal to the offset distance.

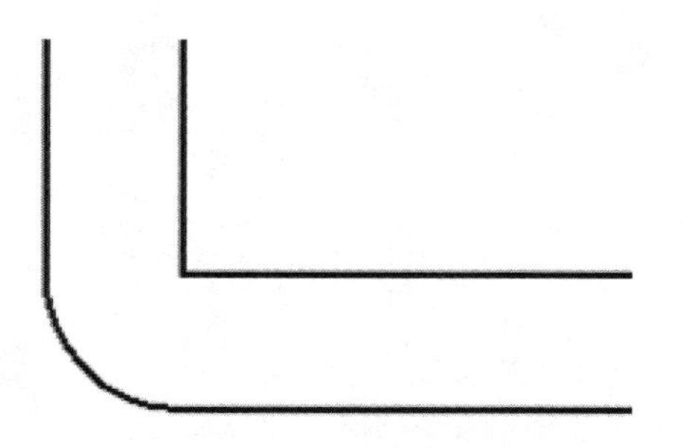

2 = Fills the gap with chamfered line segments. The perpendicular distance to each chamfer is equal to the offset distance.

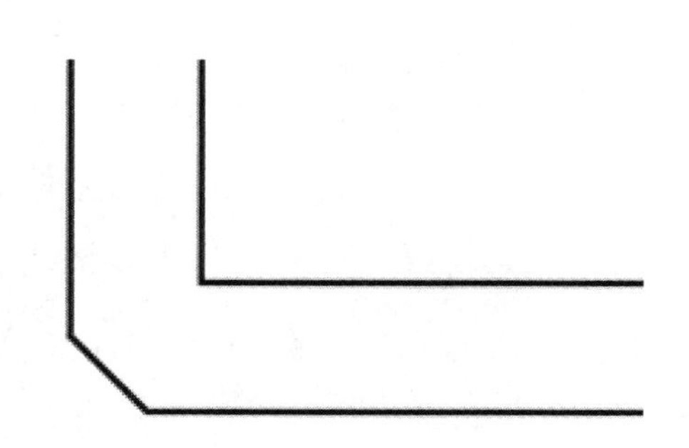

EXERCISE 12A

INSTRUCTIONS:

1. Start a **New** file using **Border A-2015.dwt**
2. Draw the Objects below using Circles, Rectangles and Lines.
3. Use the Offset command to offset the objects.
4. Use Layers = Object line and Hidden line.
5. Edit the Title and Ex-XX by double clicking on the text. Do not erase and replace.
6. Do not dimension
7. Save as **EX12A**
8. Plot using Page Setup **Class Model A**

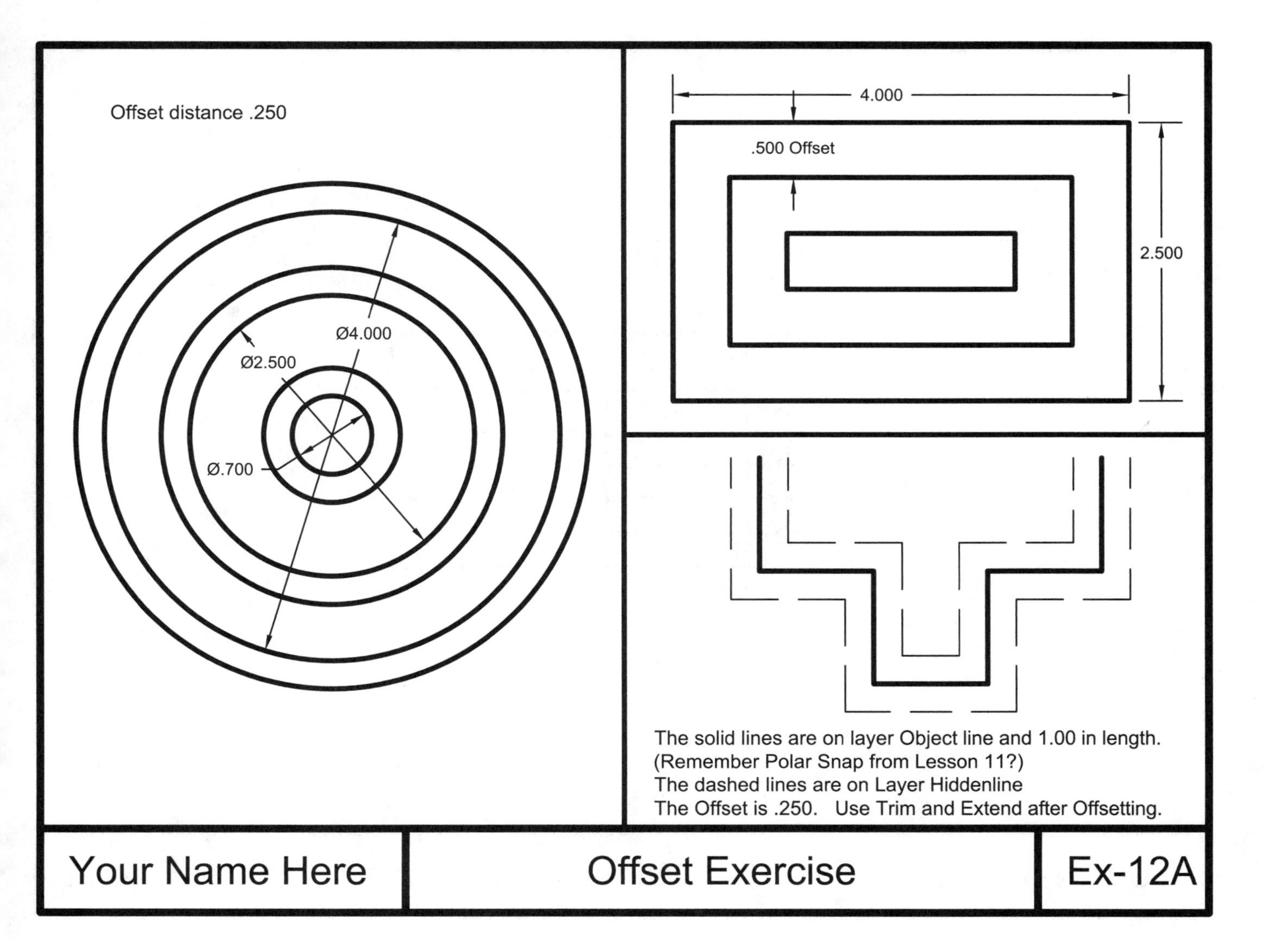

EXERCISE 12B

INSTRUCTIONS:

1. Start a **New** file using **Border A-2015.dwt**
2. Follow the steps shown on the next page to draw the Window Schedule shown below.
3. Use Layers = Object line and Text.
4. Edit the Title and Ex-XX by double clicking on the text. Do not erase and replace.
5. Do not dimension
6. Save as **EX12B**
7. Plot using Page Setup **Class Model A**

Follow instructions on the next page.

WINDOW SCHEDULE		
SYM	SIZE	TYPE
(A)	5'-0" X 4'-0"	WOOD FIXED
(B)	3'-0" X 4'-0"	WOOD FIXED
(C)	3'-0" X 3'-0"	WOOD FIXED
(D)	2'-0" X 3'-0"	ALUMINUM SLIDER

Your Name Here	Window Schedule	Ex-12B

EXERCISE 12B HELPER

Step 1. Draw the window schedule using the offset command.

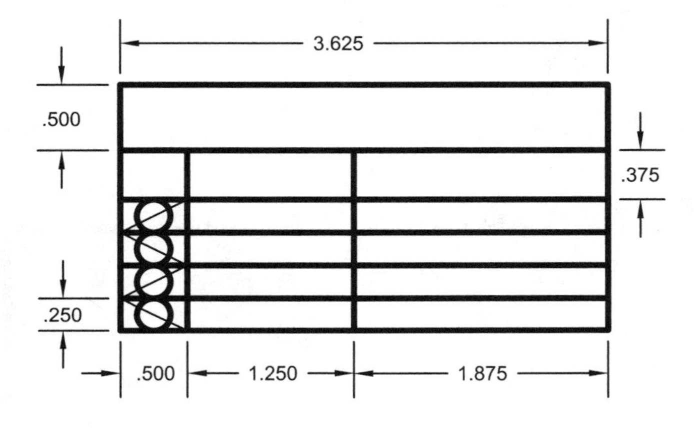

Step 2. Draw the guidelines as shown using object snap "intersection" for the diagonal lines and the offset command for the horizontal and vertical lines.

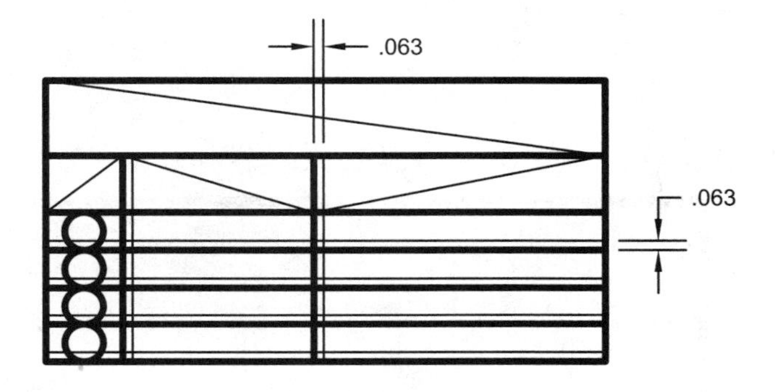

Step 3. Using "Single Line Text" place text as shown.
All text height .125 except title. Title height is .188.

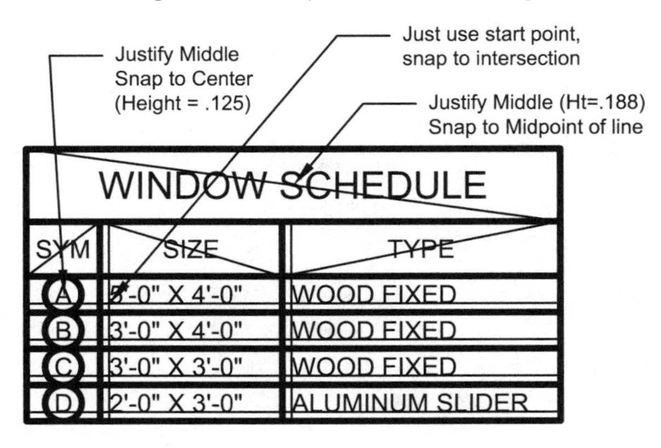

Step 4. Erase the guidelines

WINDOW SCHEDULE		
SYM	SIZE	TYPE
A	5'-0" X 4'-0"	WOOD FIXED
B	3'-0" X 4'-0"	WOOD FIXED
C	3'-0" X 3'-0"	WOOD FIXED
D	2'-0" X 3'-0"	ALUMINUM SLIDER

EXERCISE 12C

INSTRUCTIONS:

1. Start a **New** file using **Border A-2015.dwt**
2. Draw a horizontal and vertical line on Layer Object line.
3. Draw the upper half using the Offset command.
4. MIRROR the upper half. (Do not include the horizontal centerline)
5. Change the horizontal line to layer Center line using Properties Palette or Quick Properties.
6. Edit the Title and Ex-XX by double clicking on the text. Do not erase and replace.
7. Do not dimension
8. Save as **EX12C**
9. Plot using Page Setup **Class Model A**

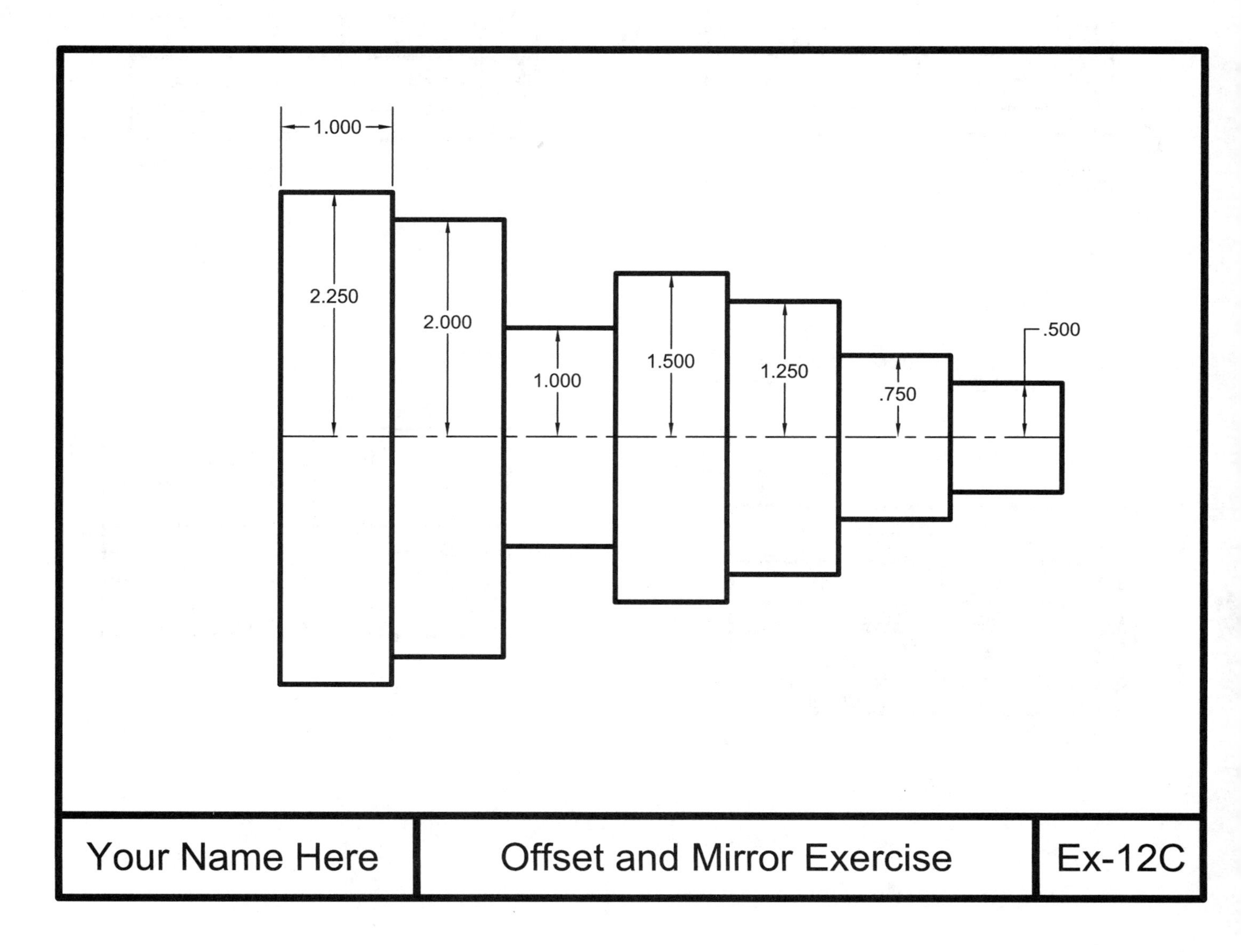

EXERCISE 12D

INSTRUCTIONS:

1. Start a **New** file using **Border A-2015.dwt**
2. Draw the lamp base below.
3. Practice using the "Layer method" when offsetting the hidden lines. Or use the Properties Palette or Quick Properties.
4. Notice that the part is symmetrical. Consider the Mirror command.
5. Use Layers, Object line, Hidden line and Centerline.
6. Edit the Title and Ex-XX by double clicking on the text. Do not erase and replace.
7. Do not dimension
8. Save as **EX12D**
9. Plot using Page Setup **Class Model A**

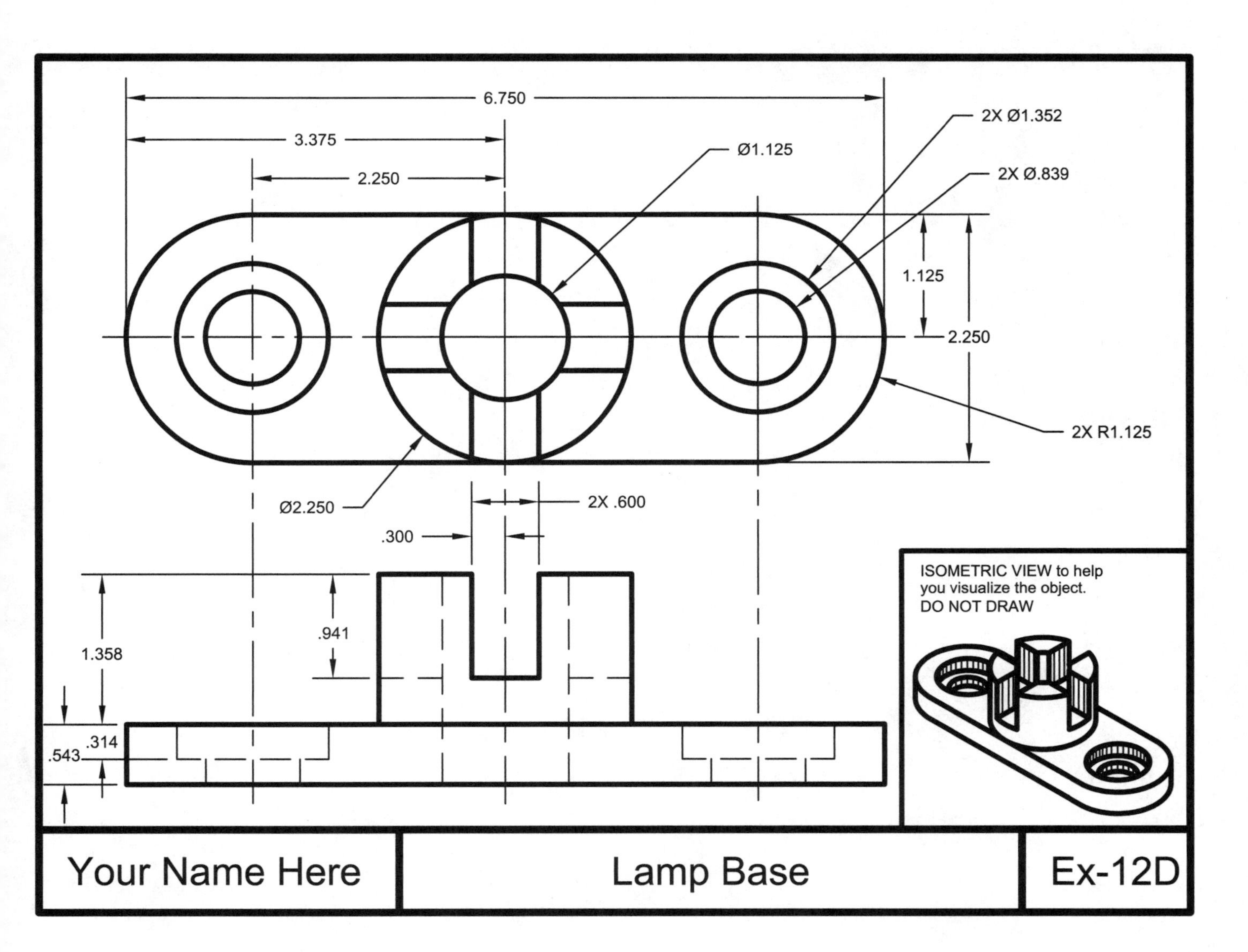

Notes:

LEARNING OBJECTIVES

After completing this lesson, you will be able to:

1. Create multiple copies in a rectangular or circular pattern or Path.
2. Understand how to Array objects

LESSON 13

ARRAY

The ARRAY command allows you to make multiple copies in a **RECTANGULAR** or Circular **(POLAR)** pattern and even on a **PATH**. The maximum limit of copies per array is 100,000. This limit can be changed but should accommodate most users. (Refer to Help menu if you choose to change the limit)

RECTANGULAR ARRAY

This method allows you to make multiple copies of object(s) in a **rectangular pattern**. You specify the number of rows (horizontal), columns (vertical) and the spacing between the rows and columns. The spacing will be equally spaced between copies.

Spacing is sometimes tricky to understand. ***Read this carefully***. The spacing is the distance from a specific location on the original to that same location on the future copy. It is not just the space in between the two. Refer to the example below.

To use the rectangular array command you will select the object(s), specify how many rows and columns desired and the spacing for the rows and the columns.
Refer to step by step instructions on page 13-3.

Example of Rectangular Array:

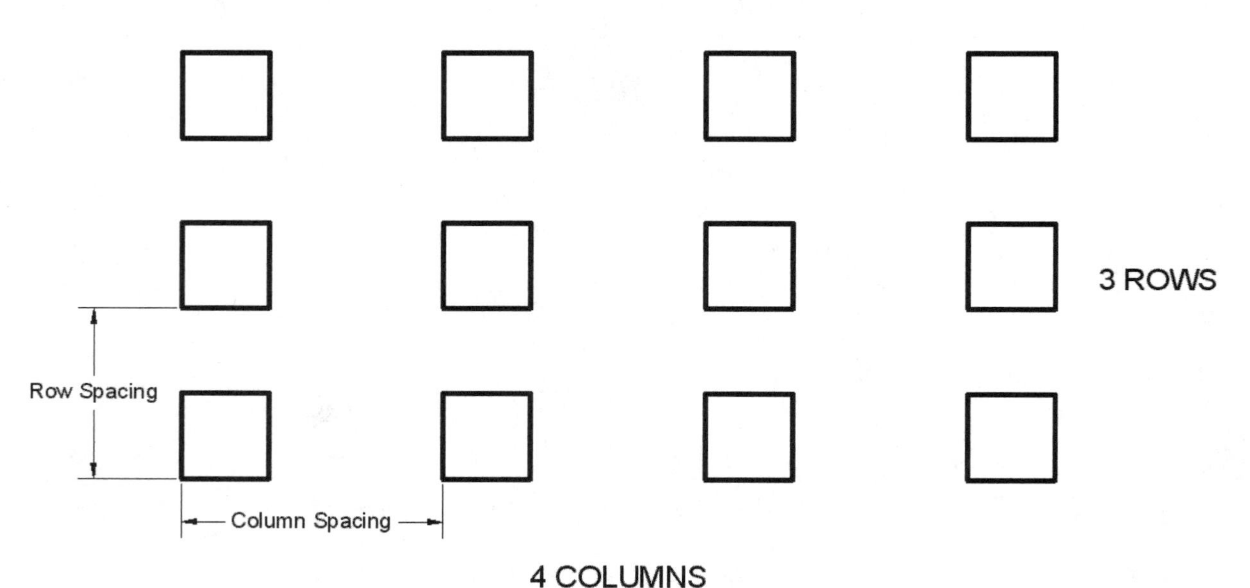

ARRAY....continued

How to create a RECTANGULAR ARRAY

1. Draw a 1" square Rectangle. □

2. Select the **ARRAY** command using one of the following:

   Ribbon = Home tab / Modify panel / Array ▼
   or
   Keyboard = Array <enter>

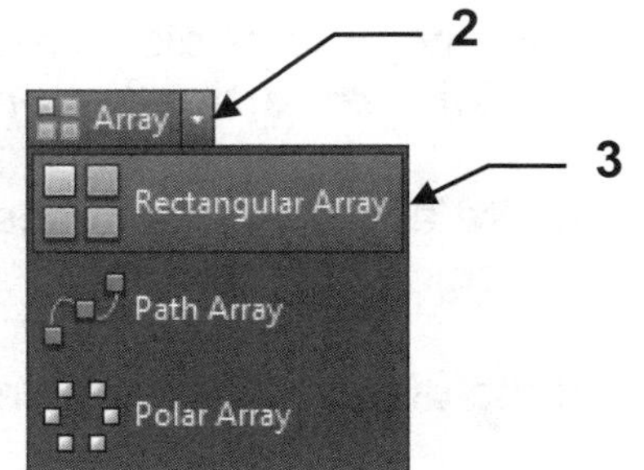

3. Select **Rectangular Array.**

4. Select Objects: ***Select the Object to be Arrayed.***

5. Select Objects: ***Select more objects or <enter> to stop***

The **Array Creation** tab appears and a 3 x 4 default grid array of the object selected.

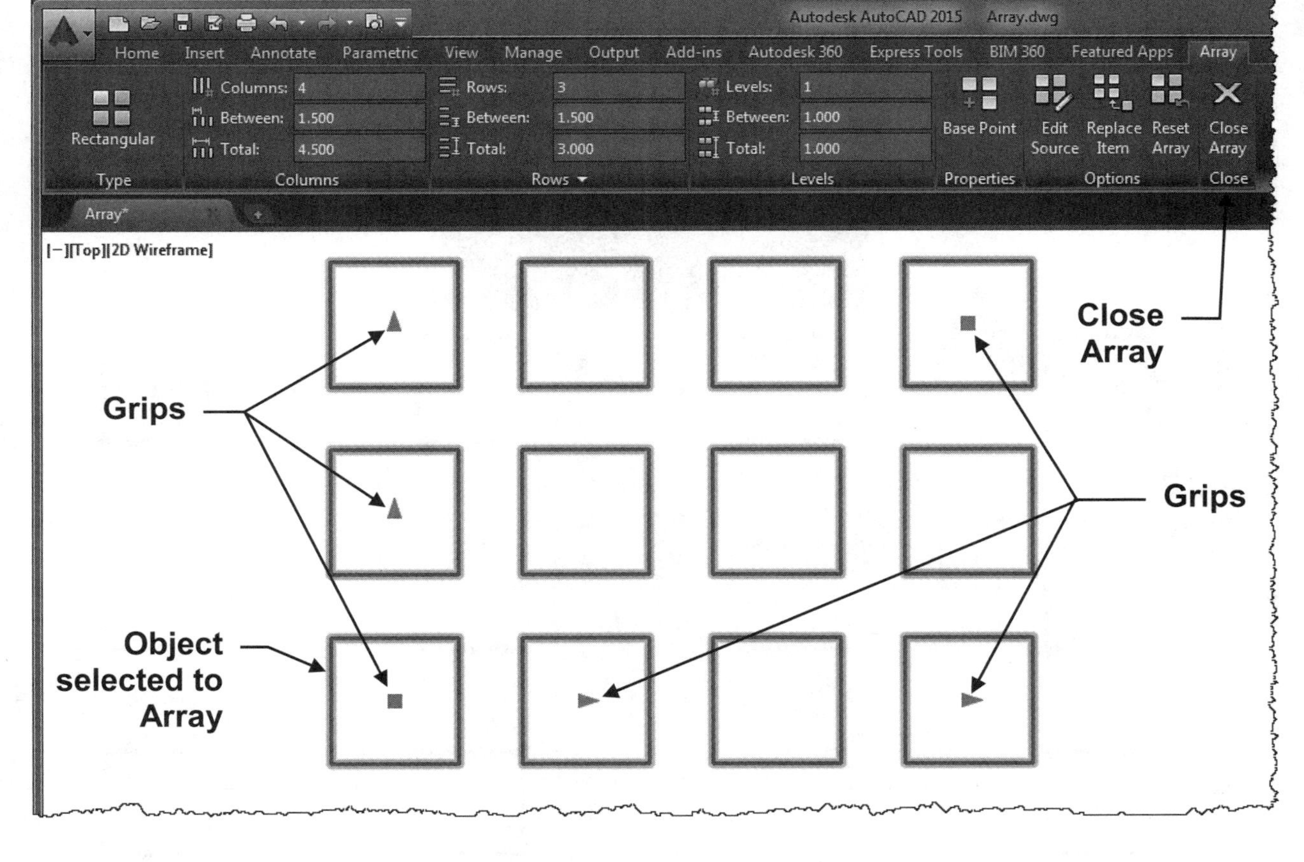

6. Make any changes necessary in the **Array Creation** tab, then press **<enter>** to display any changes.

7. If the display is correct select **Close Array.**

ARRAY....continued

How to edit a RECTANGULAR ARRAY

1. Select the Array to edit.

The **Array** panel is displayed. (The **Quick Properties** will also be displayed if you have the **Quick Properties** button ON in the Status bar.)

2. Make any changes necessary in the **Array** tab, then press **<enter>** to display any changes.

3. If the display is correct select **Close Array.**

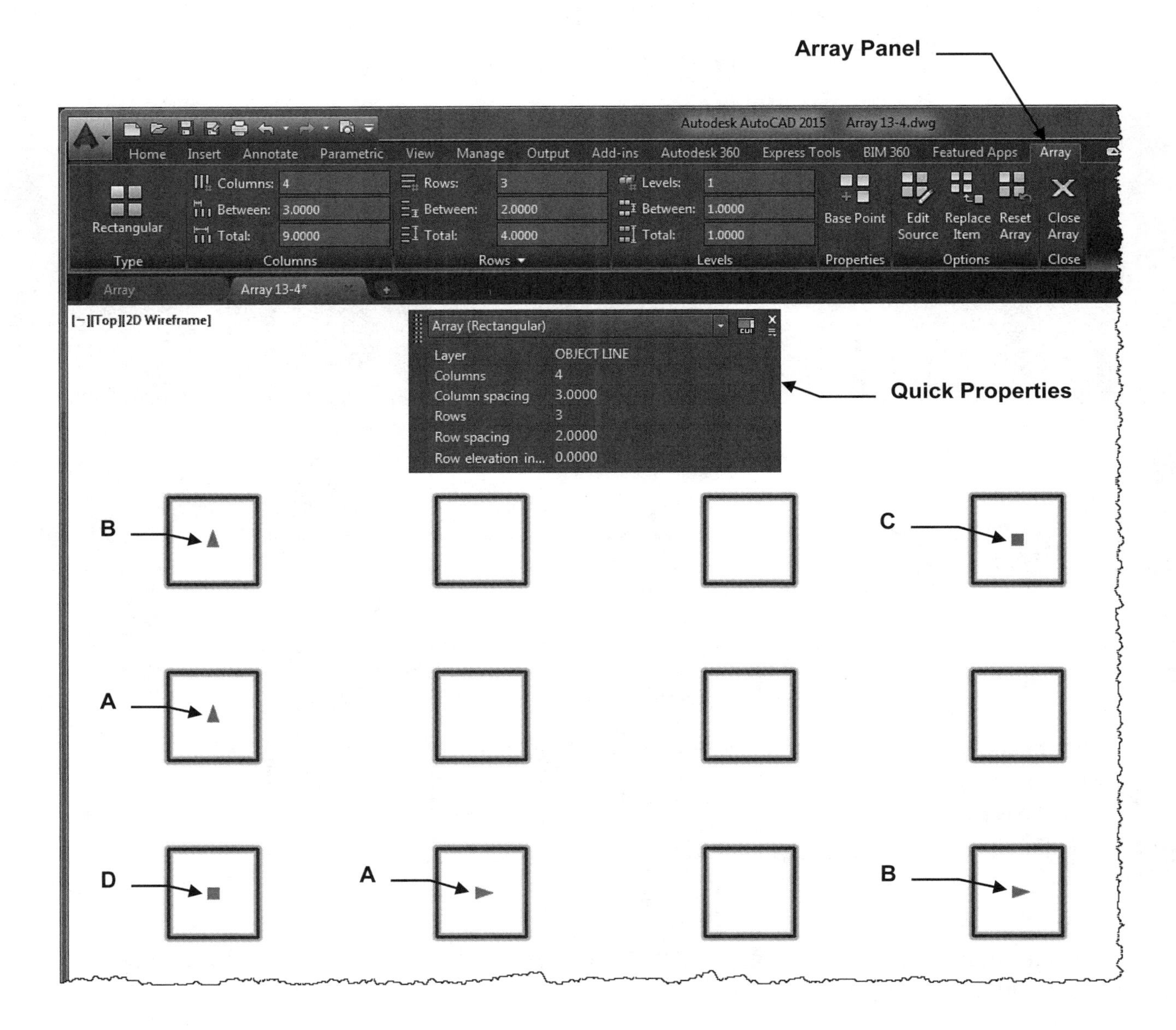

Continued on the next page...

ARRAY....continued

How to edit a RECTANGULAR ARRAY

Using Grips to edit.

You may also use the Grips to edit the spacing. Just click on a grip and drag.

A. The first ▶ or ▲ allows you to change the spacing between the columns or rows.

B. The last ▶ or ▲ allows you to change the total spacing between the base point and the last ▶ or ▲ and also to add extra columns or rows, or change the axis angle.

C. The ■ allows you to change the total row and column spacing simultaneously, and also to add extra columns and rows simultaneously.

D. Use the Base Point grip ■ to **MOVE** the entire Array.

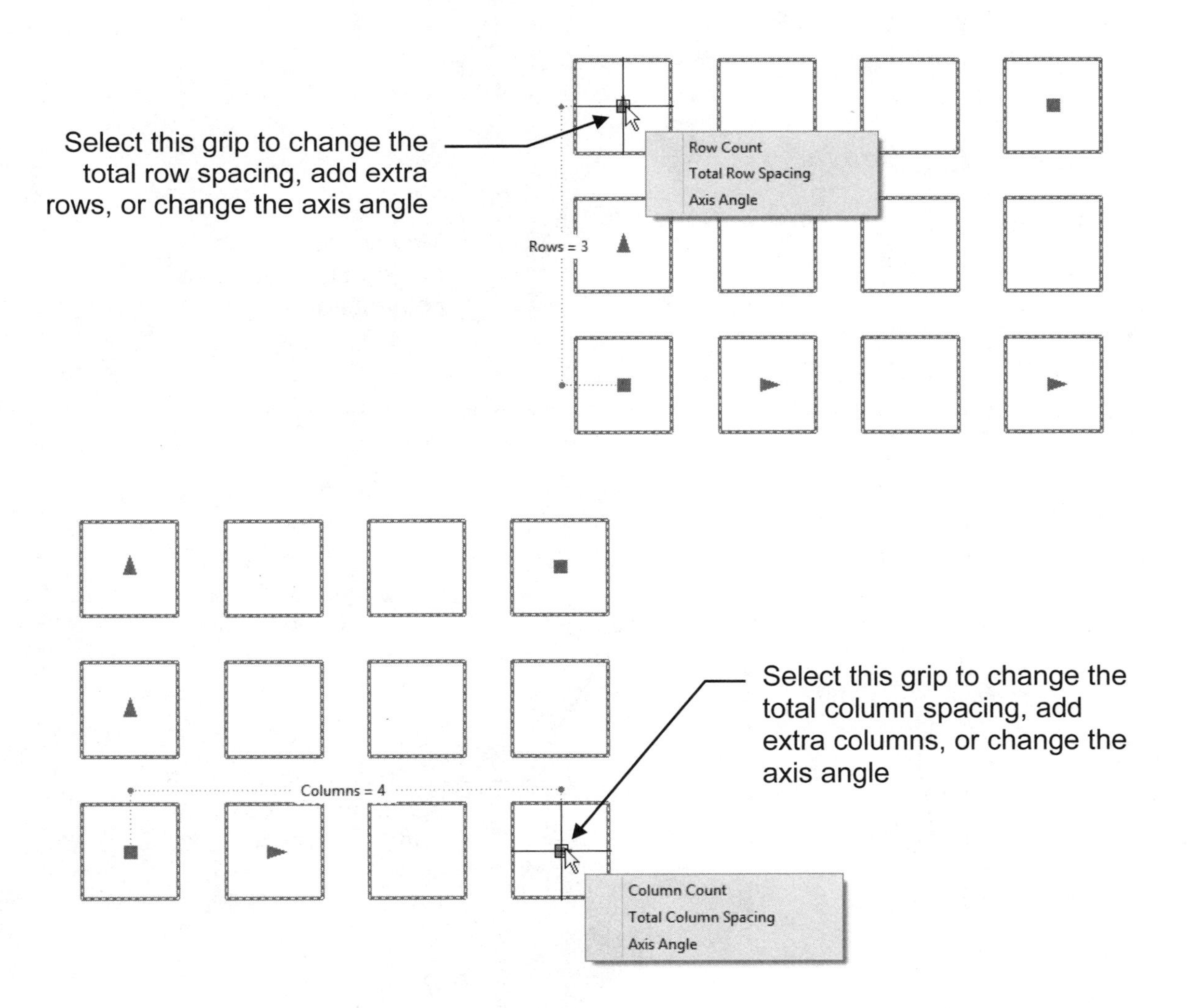

Continued on the next page...

ARRAY….continued

How to edit a RECTANGULAR ARRAY

Using Grips to edit.

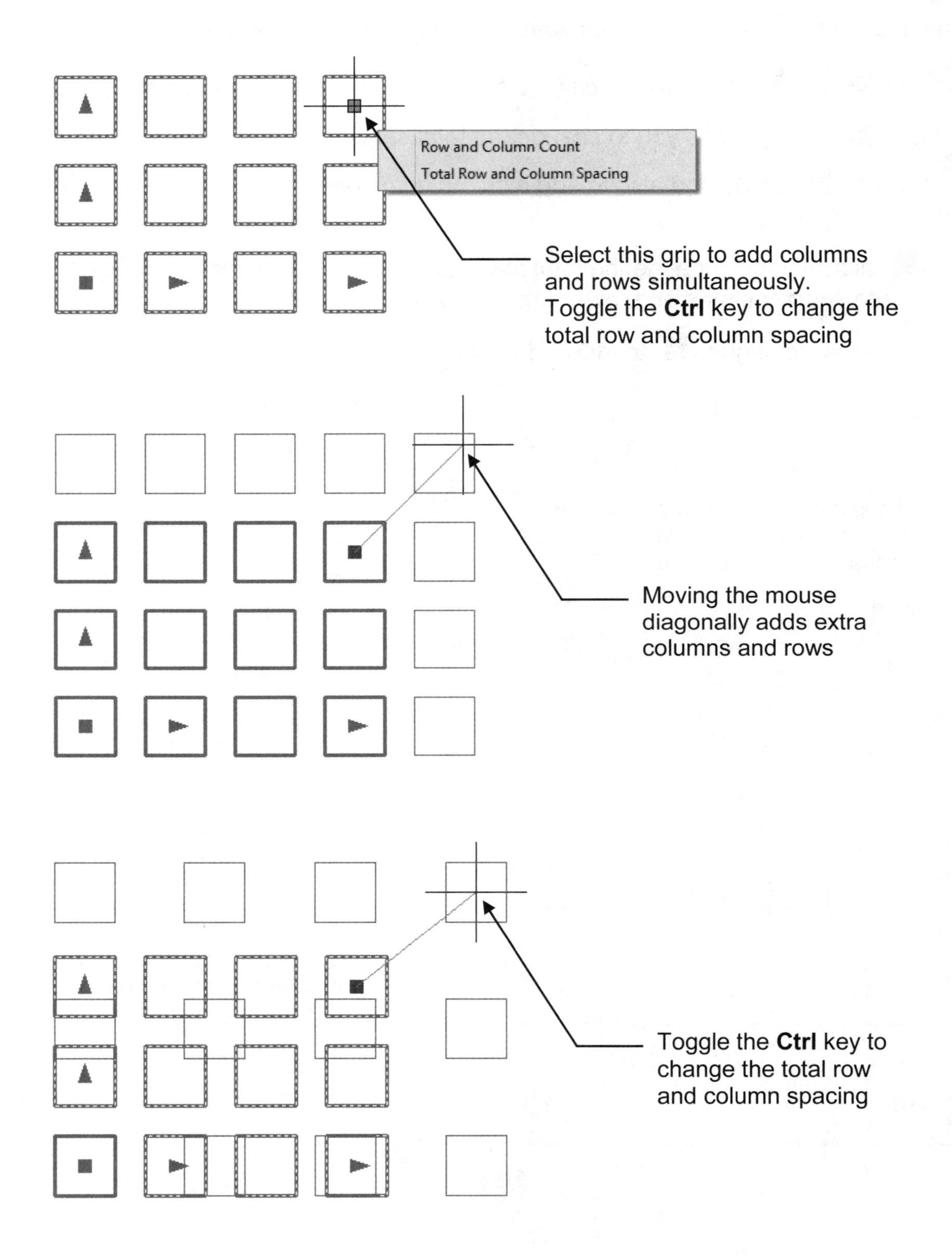

ARRAY....continued

POLAR ARRAY

This method allows you to make multiple copies in a circular pattern. You specify the total number of copies to fill a specific Angle or specify the angle between each copy and angle to fill.

To use the polar array command you select the object(s) to array, specify the center of the array, specify the number of copies or the angle between the copies, the angle to fill and if you would like the copies to rotate as they are copied.

Example of a Polar Array

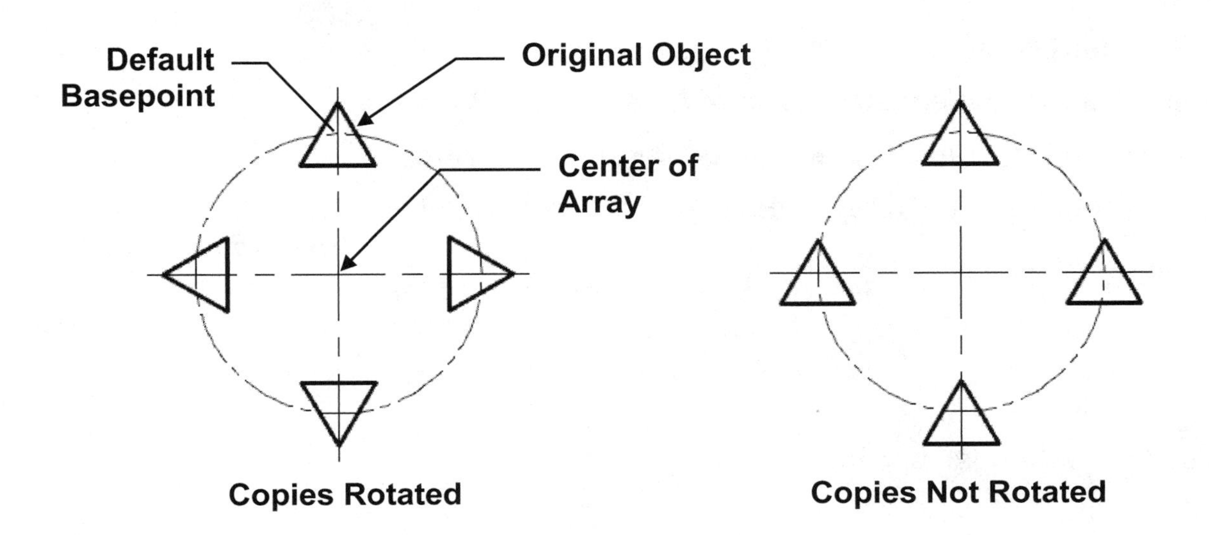

Note: the two examples shown above use the **objects default base point**. The examples below displays what happens if you specify a basepoint.

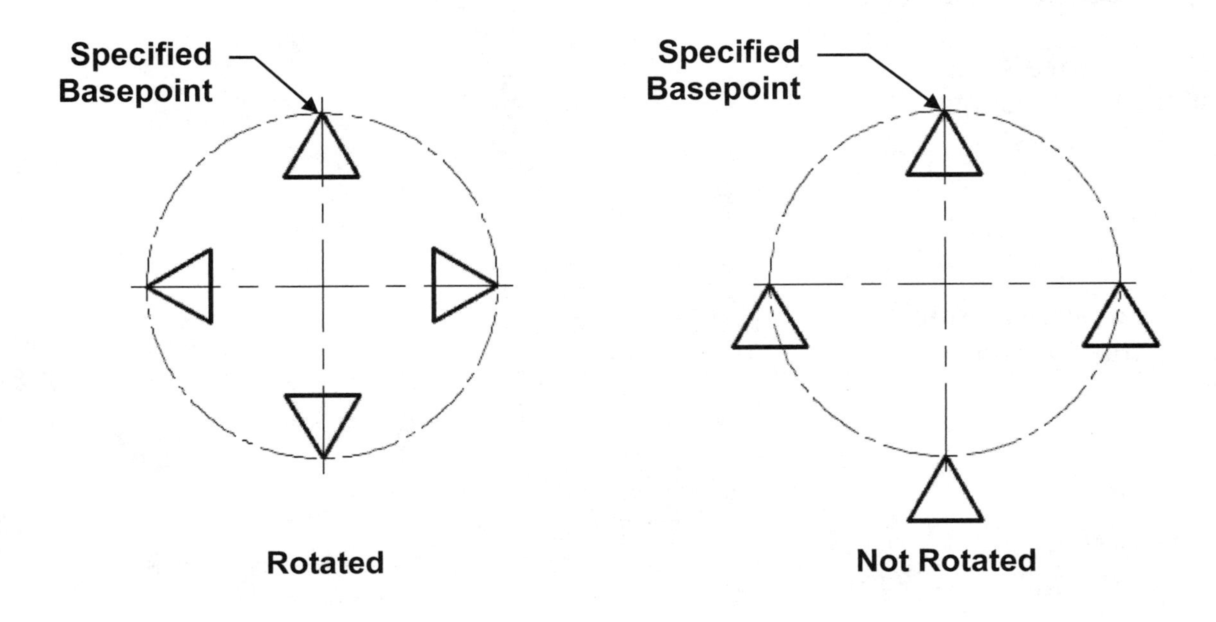

ARRAY....continued

How to create a POLAR ARRAY
Using "Number of Items".

1. Draw a 3" Radius circle.
2. Add a .50 Radius 3 sided Polygon and place as shown.
3. Select the **ARRAY** command using one of the following:

   Ribbon = Home tab / Modify panel / Array ▼
   or
   Keyboard = Array <enter>

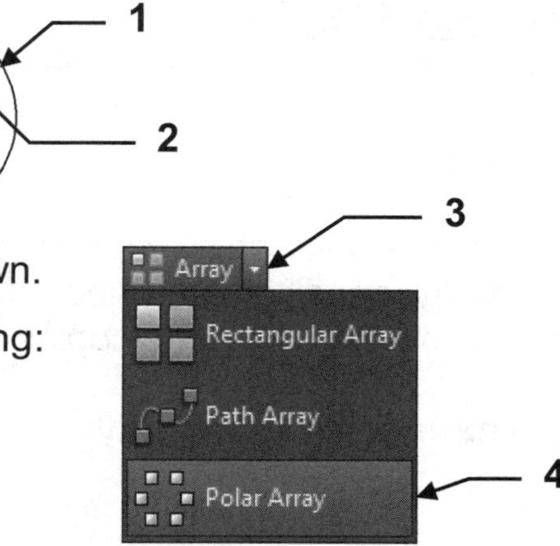

4. Select **Polar Array.**
5. Select Objects: ***Select the Object to be Arrayed. (Polygon)***
6. Select Objects: ***Select more objects or <enter> to stop***
7. Specify center point of array or [Base point / Axis of Rotation] ***Select the Center Point of the Circle***

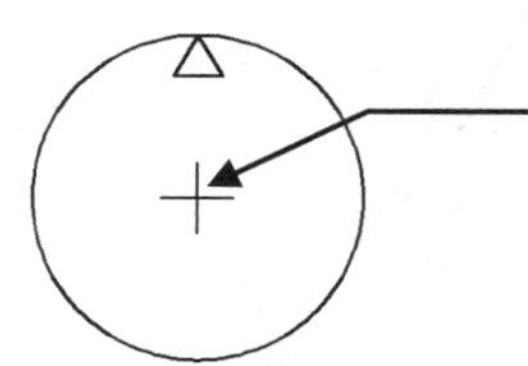

7. Snap to the center of the Circle to select the center of the Array

The **Array Creation** tab appears and the array defaults to 6 items.

8. Enter items: **12**
9. Enter Fill: **360**
10. Press <enter>
    to display the selections.
11. Select **Close Array**
    if display is correct

Note:
12 items were evenly distributed within 360 degrees

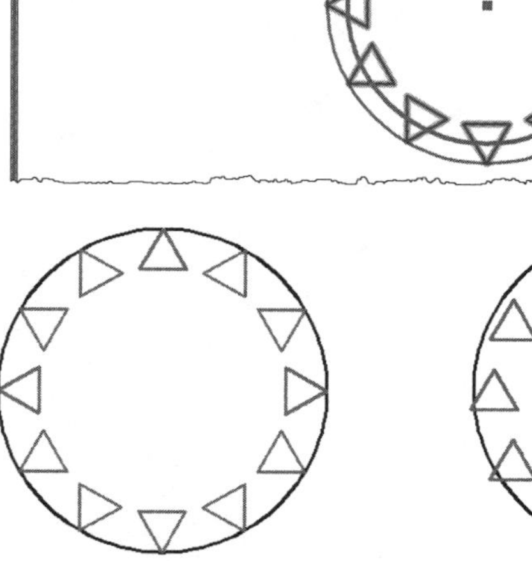

Items Rotated when Polar Arrayed

Items not Rotated when Polar Arrayed (Uncheck)

ARRAY....continued

How to create a POLAR ARRAY

Using "Angle Between".

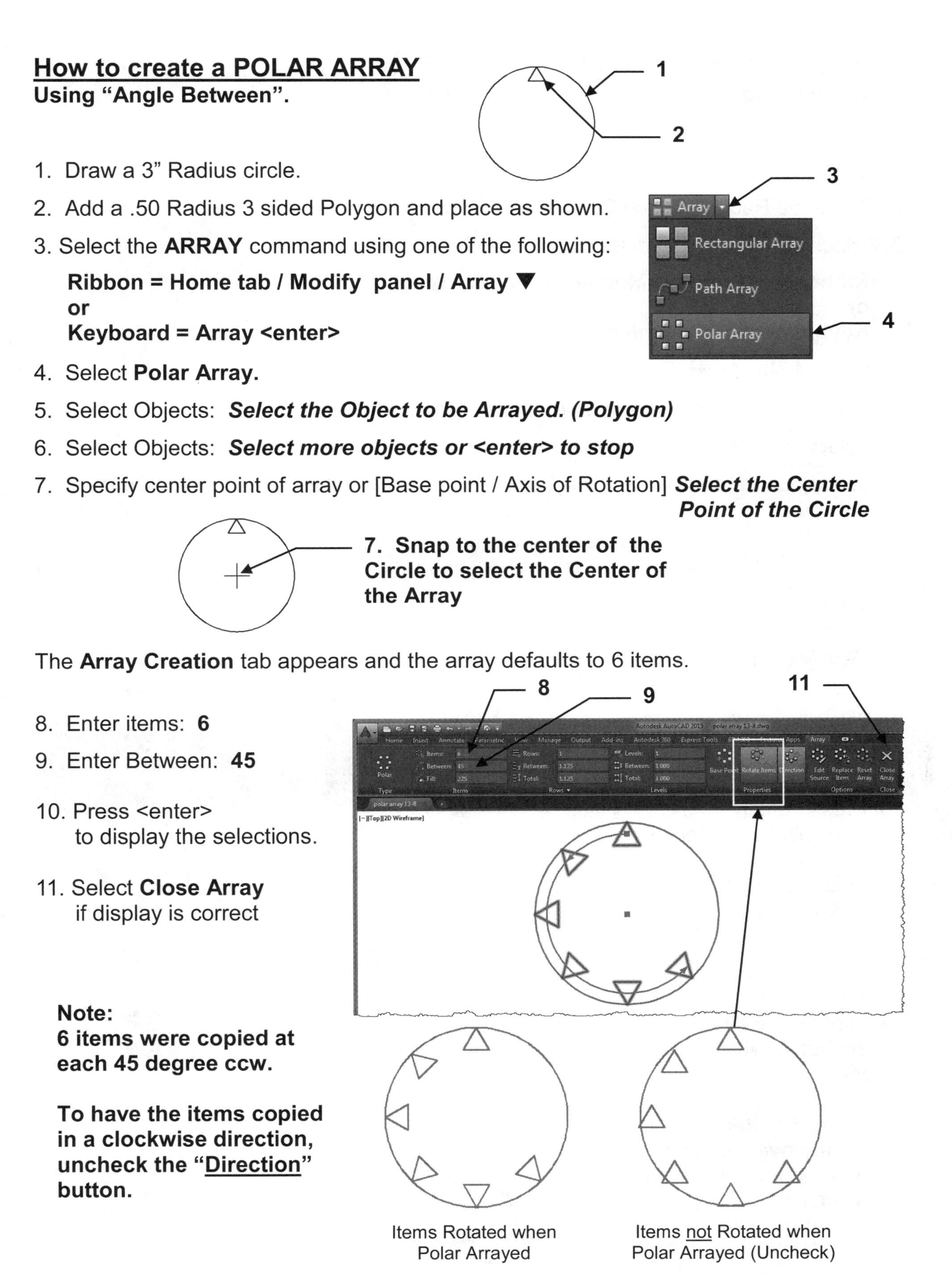

1. Draw a 3" Radius circle.
2. Add a .50 Radius 3 sided Polygon and place as shown.
3. Select the **ARRAY** command using one of the following:

   Ribbon = Home tab / Modify panel / Array ▼
   or
   Keyboard = Array <enter>

4. Select **Polar Array.**
5. Select Objects: ***Select the Object to be Arrayed. (Polygon)***
6. Select Objects: ***Select more objects or <enter> to stop***
7. Specify center point of array or [Base point / Axis of Rotation] ***Select the Center Point of the Circle***

The **Array Creation** tab appears and the array defaults to 6 items.

8. Enter items: **6**
9. Enter Between: **45**
10. Press <enter>
    to display the selections.
11. Select **Close Array**
    if display is correct

Note:
6 items were copied at each 45 degree ccw.

To have the items copied in a clockwise direction, uncheck the "Direction" button.

Items Rotated when Polar Arrayed

Items not Rotated when Polar Arrayed (Uncheck)

ARRAY....continued

How to create a POLAR ARRAY
Using "Fill Angle".

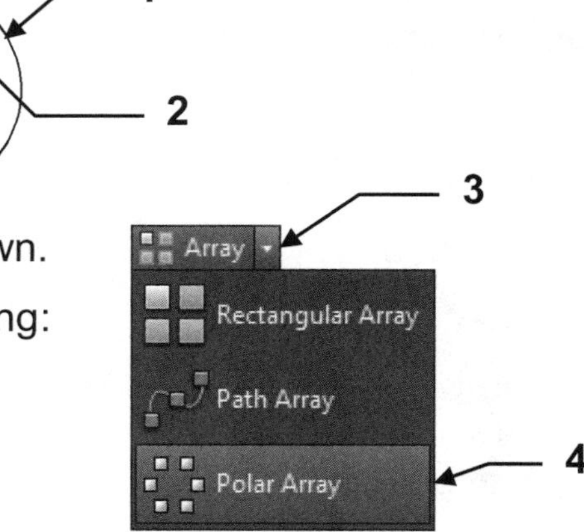

1. Draw a 3" Radius circle.
2. Add a .50 Radius 3 sided Polygon and place as shown.
3. Select the **ARRAY** command using one of the following:

   Ribbon = Home tab / Modify panel / Array ▼
   or
   Keyboard = Array <enter>

4. Select **Polar Array.**
5. Select Objects: ***Select the Object to be Arrayed. (Polygon)***
6. Select Objects: ***Select more objects or <enter> to stop***
7. Specify center point of array or [Base point / Axis of Rotation] ***Select the Center Point of the Circle***

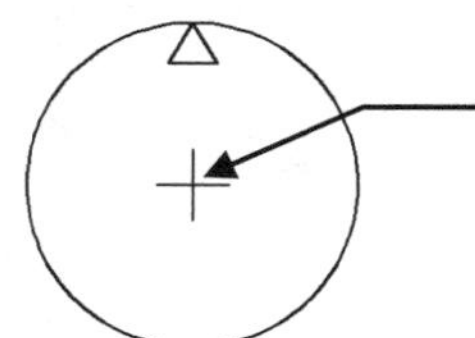

7. Snap to the center of the Circle to select the Center of the Array

The **Array Creation** tab appears and the array defaults to 6 items.

8. Enter items: **8**
9. Enter Fill: **180**
10. Press <enter>
    to display the selections.
11. Select **Close Array**
    if display is correct

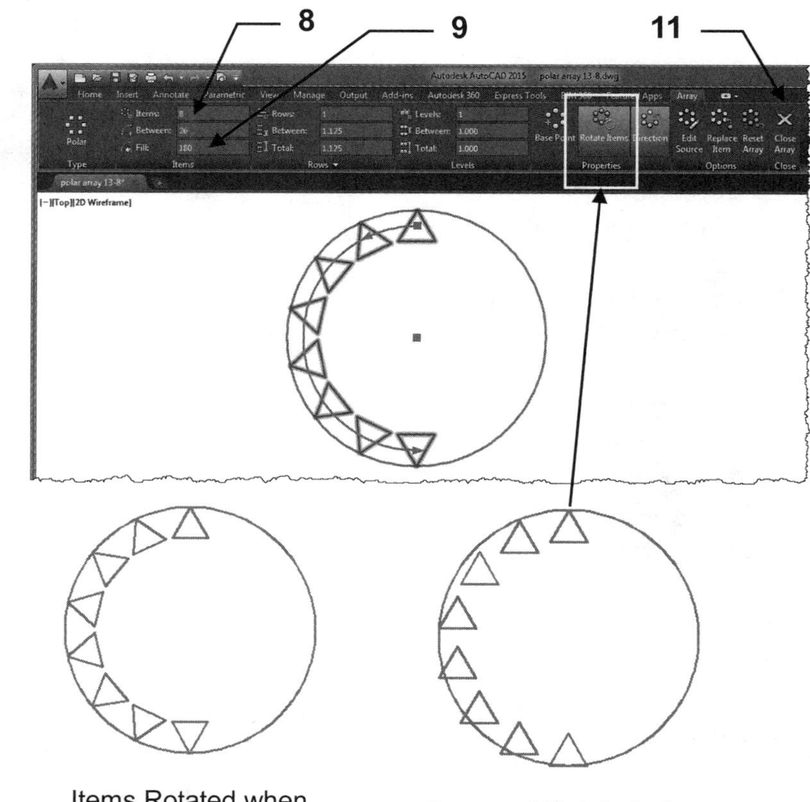

Note:
8 items were evenly distributed within 180 degrees ccw.

To have the items copied in a clockwise direction, uncheck the "Direction" button.

Items Rotated when Polar Arrayed

Items not Rotated when Polar Arrayed (Uncheck)

ARRAY....continued

How to create a PATH ARRAY

1. Draw a Line 6" long at 20 degrees.

2. Add a 0.500" x 0.500" Rectangle as shown.

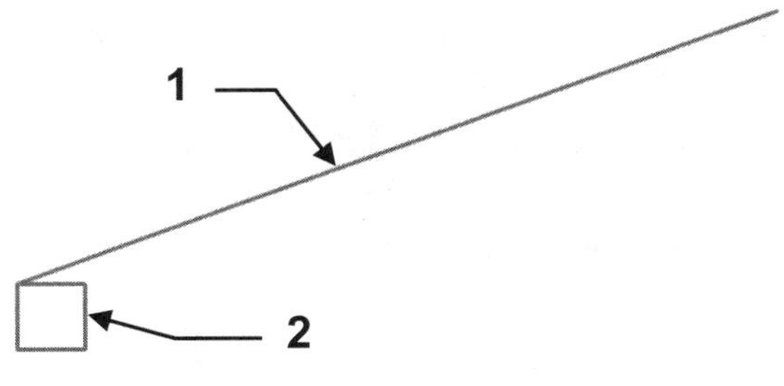

3. Select the **ARRAY** command using one of the following:

   Ribbon = Home tab / Modify panel / Array ▼
   or
   Keyboard = Array <enter>

3
Array
Rectangular Array
Path Array
Polar Array
4

4. Select **Path Array.**

5. Select Objects: ***Select the Object to be Arrayed. (The small Rectangle)***

6. Select Objects: ***Select more objects or <enter> to stop***

7. Select Path Curve: ***Select the Path. (The angled Line)***

> **Note**: The Path can be a line, polyline, spline, helix, arc, circle or ellipse.

The **Array Creation** tab appears and the array defaults to 9 items.

8. Make any alterations and press <enter> to display.

9. If correct select **Close Array**

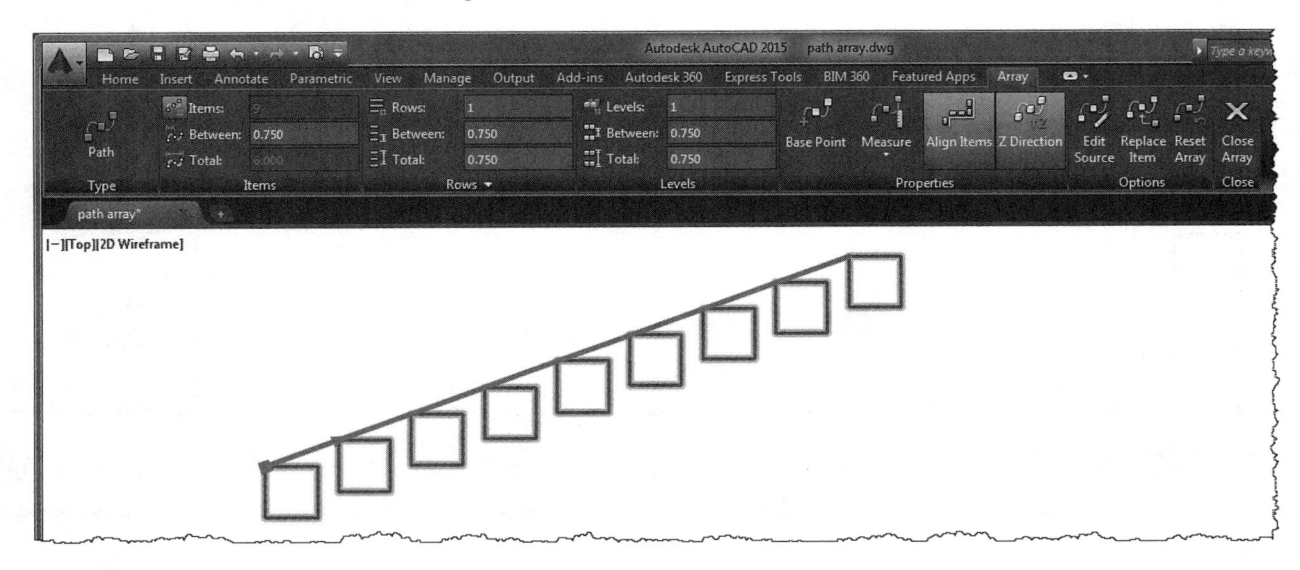

EXERCISE 13A

INSTRUCTIONS:

1. Start a **New** file using **Border A-2015.dwt**
2. Draw the original .500 square on Layer Object.
3. Array the original square as shown.
4. Edit the Title and Ex-XX by double clicking on the text. Do not erase and replace.
5. Do not dimension
6. Save as **EX13A**
7. Plot using Page Setup **Class Model A**

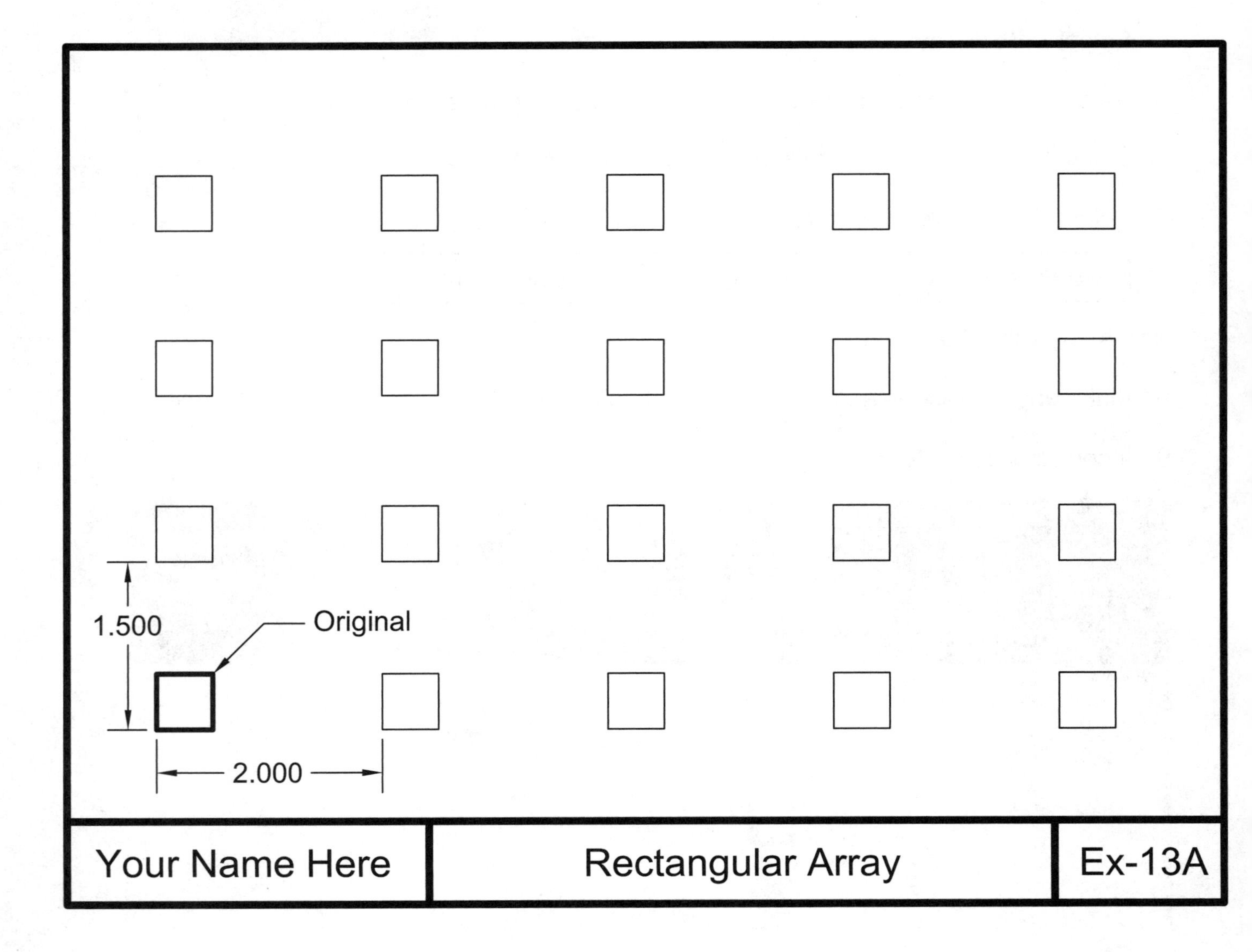

EXERCISE 13B

INSTRUCTIONS:

1. Start a **New** file using **Border A-2015.dwt**
2. Draw the objects shown below using the most efficient methods.
3. **Refer to the next page for more dimensions and helpful hints.**
4. Use Layers = Object line.
5. Edit the Title and Ex-XX by double clicking on the text. Do not erase and replace.
6. Do not dimension
7. Save as **EX13B**
8. Plot using Page Setup **Class Model A**

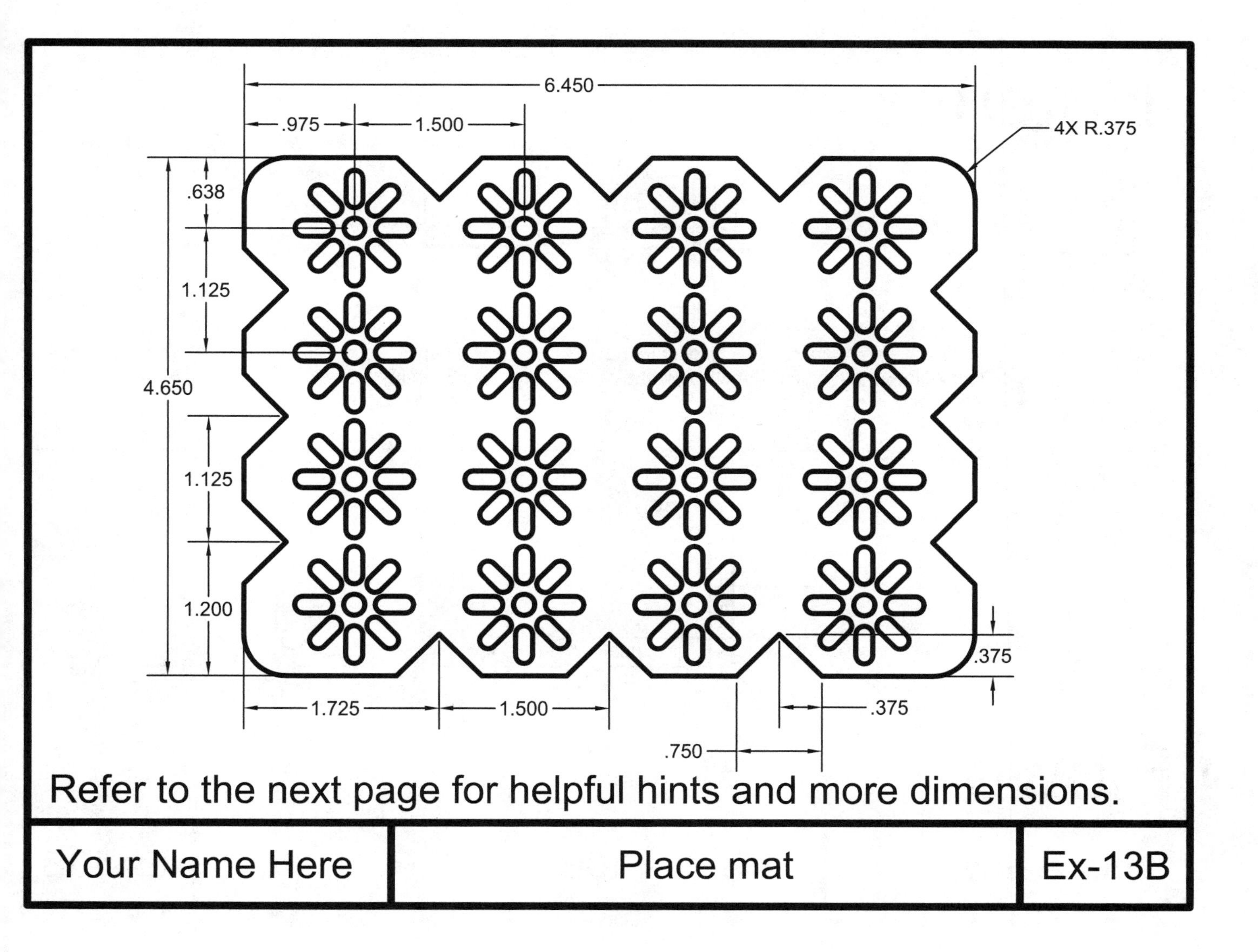

EXERCISE 13B
Helper

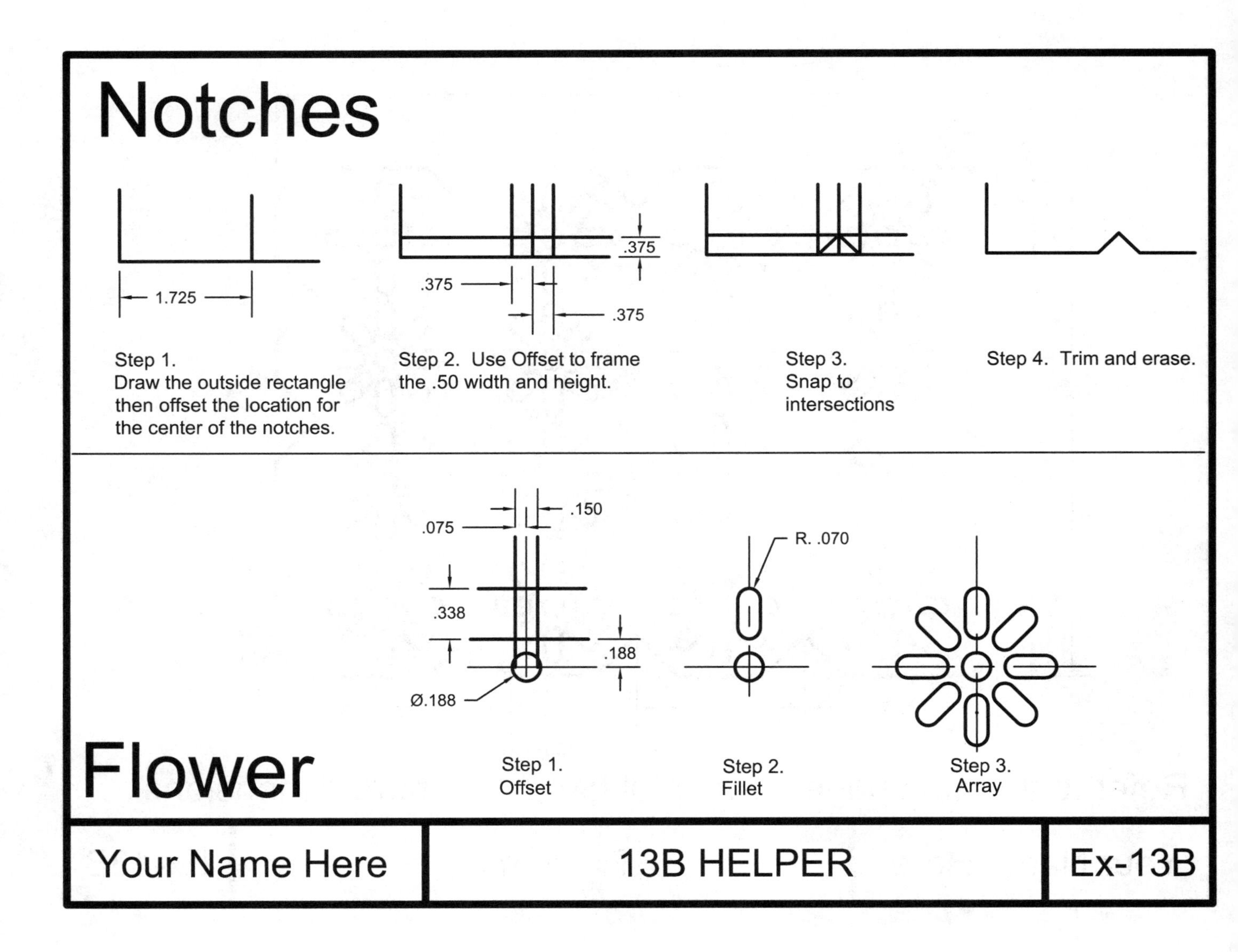

EXERCISE 13C

INSTRUCTIONS:

1. Start a **New** file using **Border A-2015.dwt**
2. Draw the centerlines and circles first.
3. Draw one Polygon at the 12:00 location.
4. Array the original Polygon as shown.
5. Edit the Title and Ex-XX by double clicking on the text. Do not erase and replace.
6. Do not dimension
7. Save as **EX13C**
8. Plot using Page Setup **Class Model A**

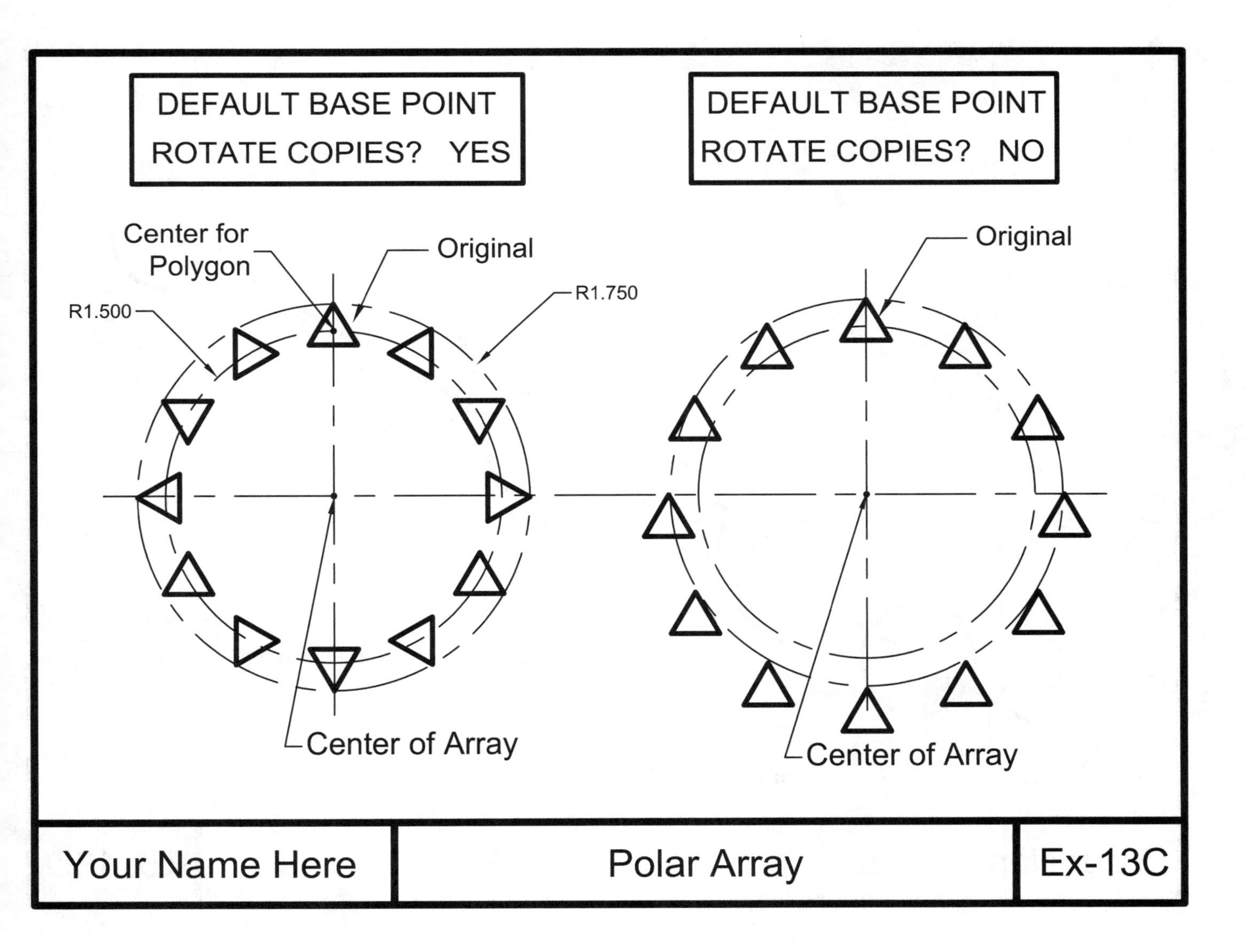

EXERCISE 13D

INSTRUCTIONS:

1. Start a **New** file using **Border A-2015.dwt**
2. Draw Line first. (9.00 Long, 30 degree)
3. Draw one Circle (Radius .50) at lower left end.
4. Array 8 Circles along the Path (Line) evenly divided.
5. Edit the Title and Ex-XX by double clicking on the text. Do not erase and replace.
6. Do not dimension
7. Save as **EX13D**
8. Plot using Page Setup **Class Model A**

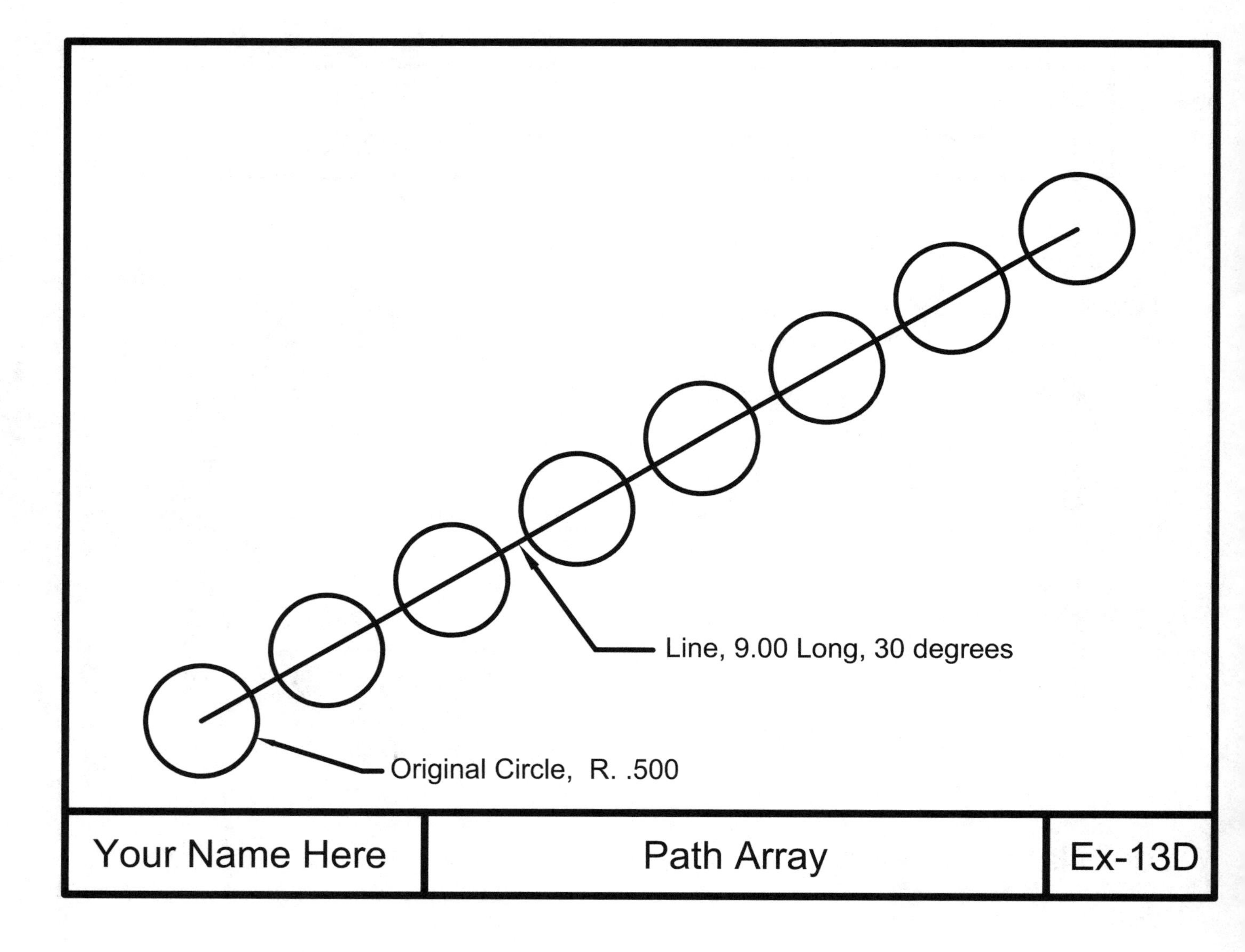

LEARNING OBJECTIVES

After completing this lesson, you will be able to:

1. Make an existing object larger or smaller proportionately
2. Stretch or compress an existing object.
3. Rotate an existing object to a specific angle.

LESSON 14

SCALE

The **SCALE** command is used to make objects larger or smaller proportionately. You may scale using a scale factor or a reference length. You must also specify a base point. The base point is a stationary point from which the objects scale.

1. Select the SCALE command using one of the following:

   Ribbon = Home tab / Modify panel / Scale
   or
   Keyboard = SCALE <enter>

SCALE FACTOR

Command: _scale

2. Select objects: ***select the object(s) to be scaled***
3. Select objects: ***select more object(s) or <enter> to stop***
4. Specify base point: ***select the stationary point on the object***
5. Specify scale factor or [Copy/Reference]: ***type the scale factor <enter>***

If the scale factor is greater than 1, the objects will increase in size.
If the scale factor is less than 1, the objects will decrease in size.

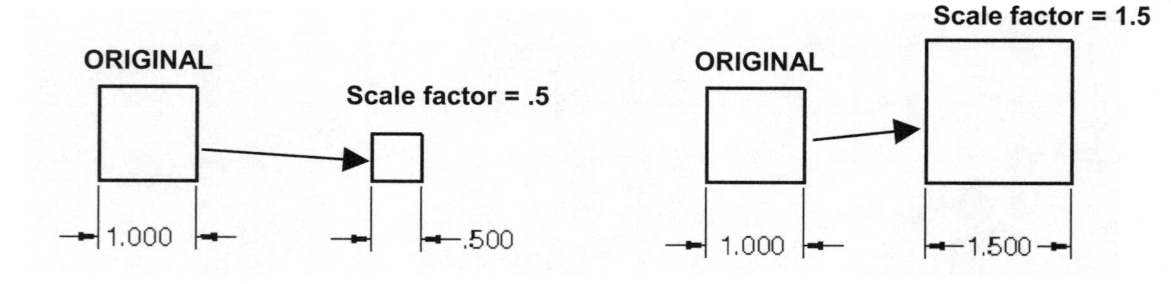

REFERENCE option

Command: _scale

2. Select objects: ***select the object(s) to be scaled***
3. Select objects: ***select more object(s) or <enter> to stop***
4. Specify base point: ***select the stationary point on the object***
5. Specify scale factor or [Copy/Reference]: ***select Reference***
6. Specify reference length <1>: ***specify a reference length***
7. Specify new length: ***specify the new length***

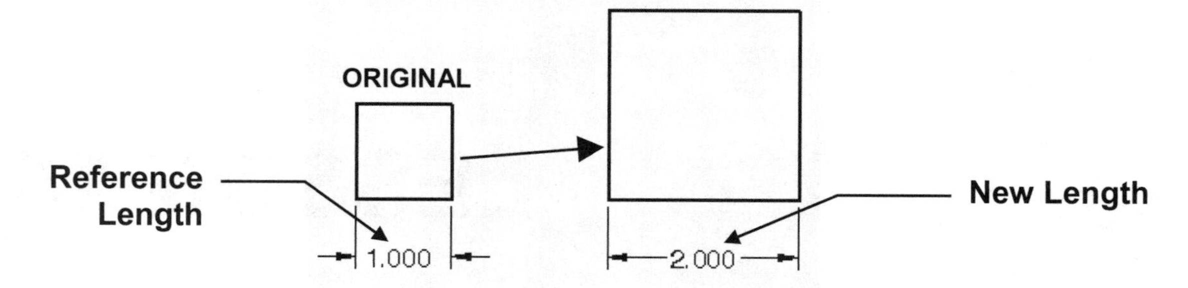

COPY option - creates a duplicate of the selected object. The duplicate is directly on top of the original. The duplicate will be scaled. The Original remains the same.

STRETCH

The **STRETCH** command allows you to stretch or compress object(s). Unlike the Scale command, you can alter an objects proportion with the Stretch command. In other words, you may increase the length without changing the width and vice versa.

Stretch is a very valuable tool. Take some time to really understand this command. It will save you hours when making corrections to drawings.

When selecting the object(s) you must use a **CROSSING** window.
Objects that are crossed, will ***stretch.***
Objects that are totally enclosed, will ***move***.

1. Select the STRETCH command using one of the following:

   Ribbon = Home tab / Modify panel / Stretch
   or
   Keyboard = S <enter>

   Command: _stretch
2. Select objects to stretch by crossing-window or crossing-polygon...
   Select objects: ***select the first corner of the crossing window***
3. Specify opposite corner: ***specify the opposite corner of the crossing window***
4. Select objects: ***<enter>***
5. Specify base point or [Displacement] <Displacement>:
   select a base point (where it stretches from)
6. Specify second point or <use first point as displacement>:
   type coordinates or place location with cursor

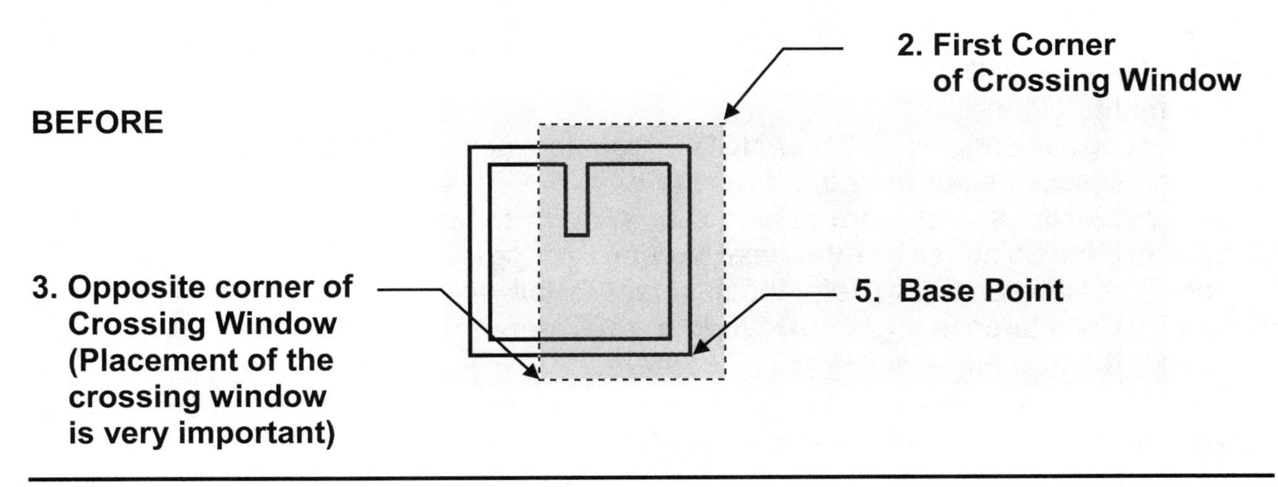

AFTER

(Notice which objects <u>stretched</u> and which objects <u>moved</u>)

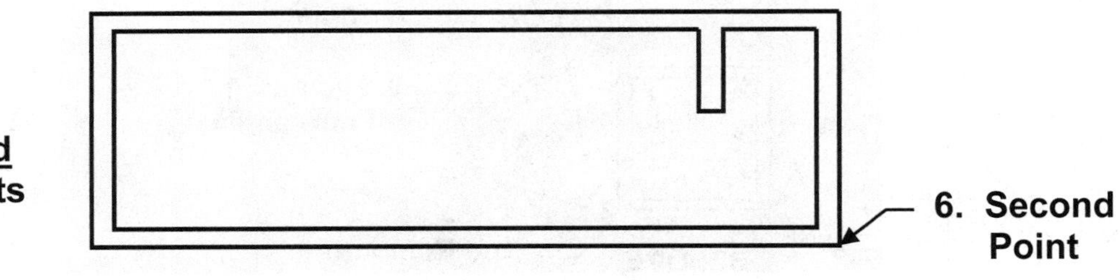

ROTATE

The **ROTATE** command is used to rotate objects around a Base Point. (pivot point)

After selecting the objects and the base point, you will enter the rotation angle from its current rotation angle or select a reference angle followed by the new angle.

A **Positive** rotation angle revolves the objects **Counter- Clockwise**.
A **Negative** rotation angle revolves the objects **Clockwise**.

Select the ROTATE command using one of the following:

> **Ribbon = Home tab / Modify panel /** Rotate
> **or**
> **Keyboard = RO <enter>**

ROTATION ANGLE OPTION

Command: _rotate

1. Current positive angle in UCS: ANGDIR=counterclockwise ANGBASE=0
   Select objects: ***select the object to rotate.***
2. Select objects: ***select more object(s) or <enter> to stop.***
3. Specify base point: ***select the base point (pivot point).***
4. Specify rotation angle or [Copy/Reference]<0>: ***type the angle of rotation.***

REFERENCE OPTION

Command: _rotate

1. Current positive angle in UCS: ANGDIR=counterclockwise ANGBASE=0
   Select objects: ***select the object to rotate.***
2. Select objects: ***select more object(s) or <enter> to stop.***
3. Specify base point: ***select the base point (pivot point).***
4. Specify rotation angle or [Reference]: ***select Reference.***
5. Specify the reference angle <0>: ***Snap to the reference object (1) and (2).***
6. Specify the new angle or [Points]: ***P <enter>.***
7. Specify first point: ***select 1st endpoint of new angle***
8. Specify second point: ***select 2nd endpoint of new angle***

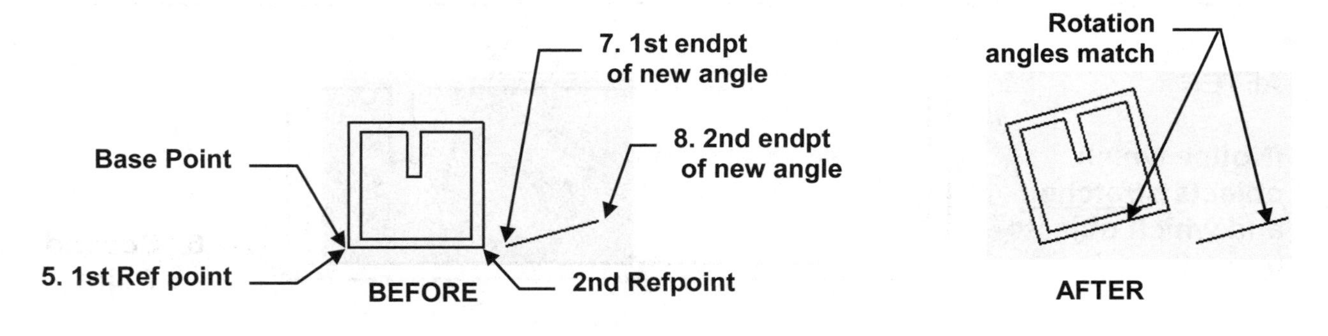

EXERCISE 14A

INSTRUCTIONS:

1. Start a **New** file using **Border A-2015.dwt**
2. Draw both 2" Squares shown below labeled "Original". Use layer Object Line.
3. Scale the original on the left using "scale factor" method.
4. Scale the original on the right using "scale Reference" method.
5. Edit the Title and Ex-XX by double clicking on the text. Do not erase and replace.
6. Do not dimension
7. Save as **EX14A**
8. Plot using Page Setup **Class Model A**

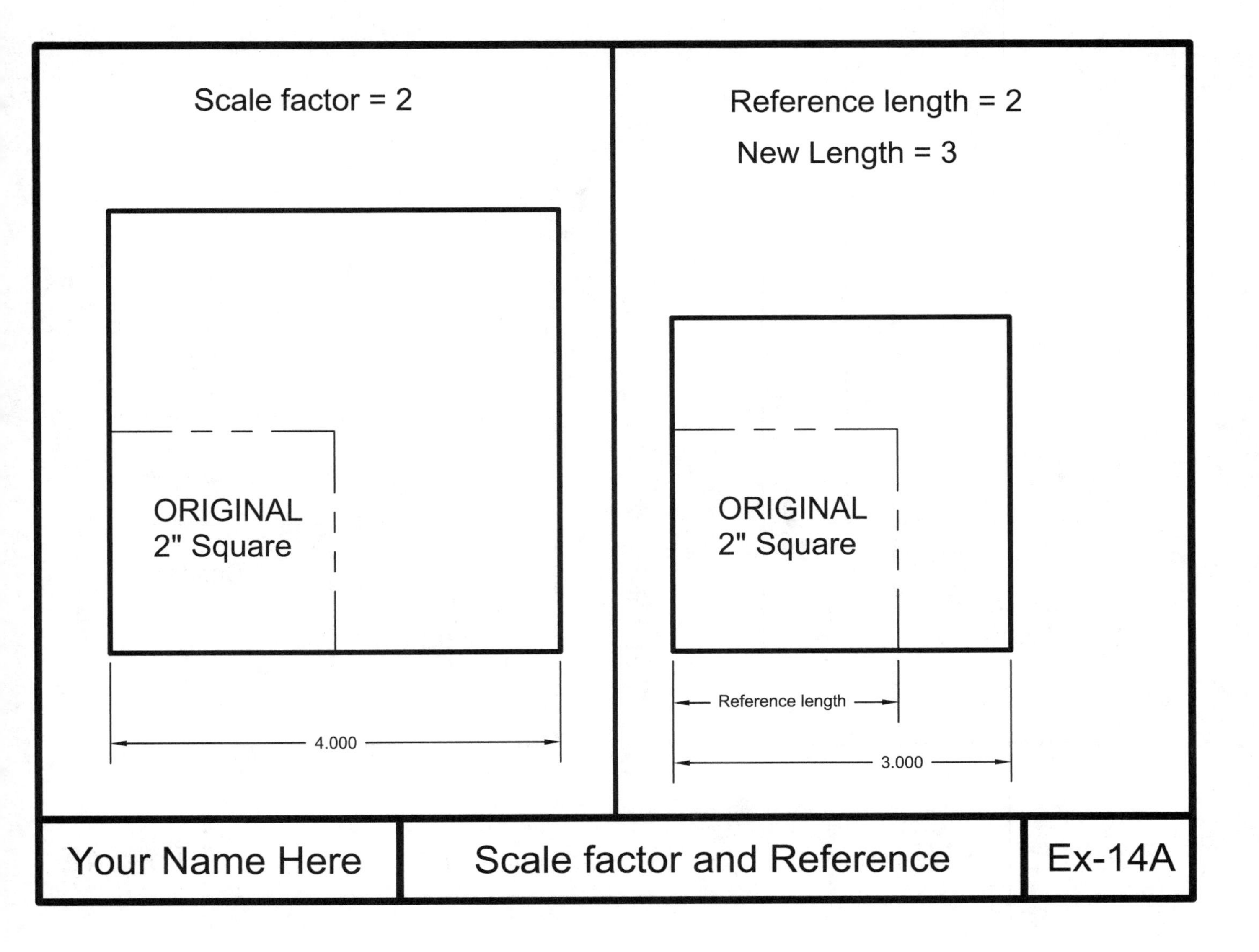

EXERCISE 14B

INSTRUCTIONS:

1. Start a **New** file using **Border A-2015.dwt**
2. Draw both 2" squares as shown below labeled as "Original". Use Layer Object Line.
3. Stretch each as shown.
4. Edit the Title and Ex-XX by double clicking on the text. Do not erase and replace.
5. Do not dimension
6. Save as **EX14B**
7. Plot using Page Setup **Class Model A**

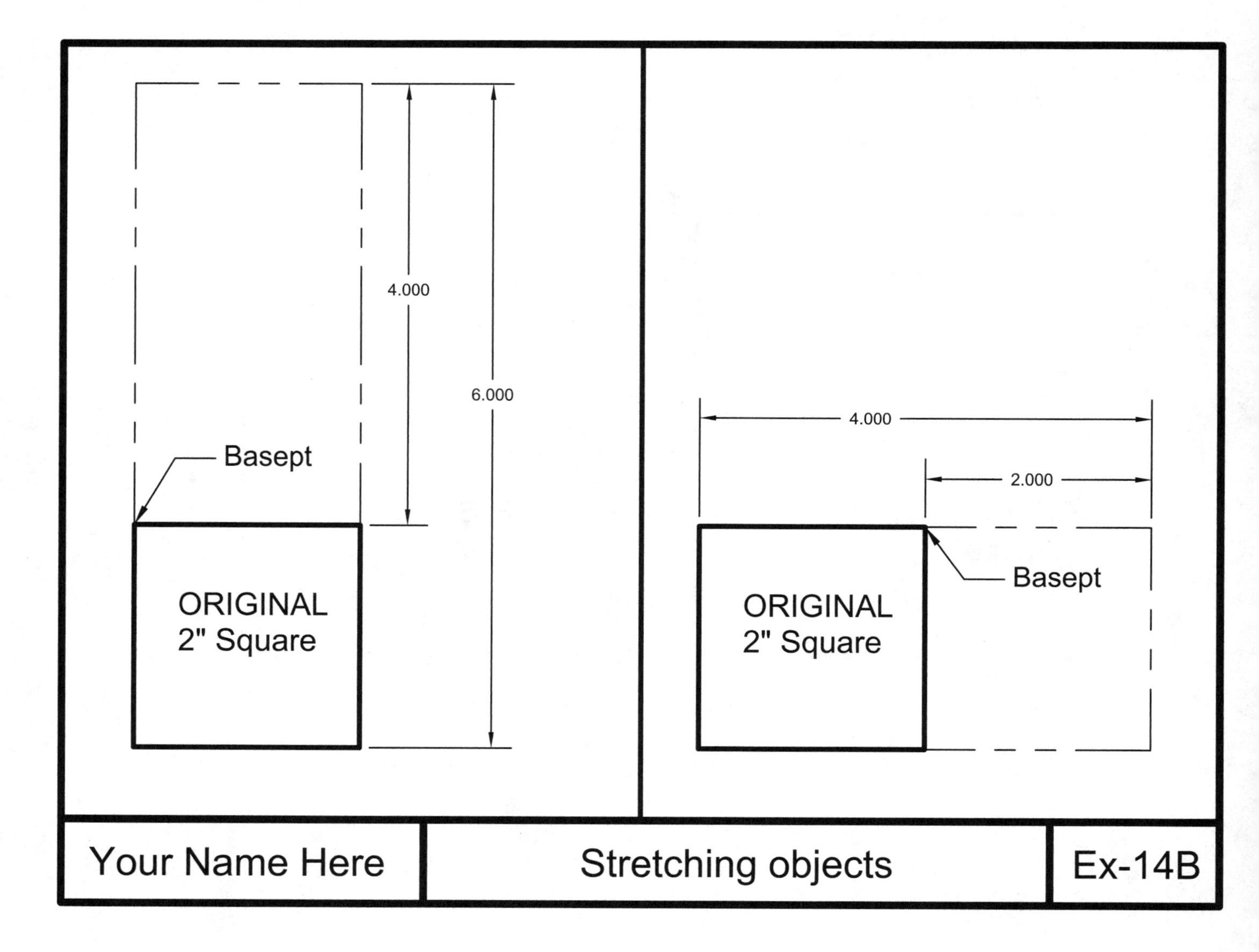

EXERCISE 14C

INSTRUCTIONS:

1. Start a **New** file using **Border A-2015.dwt**
2. Draw the Original shape on the left side of the border. Use Layer Object Line.
3. Stretch the overall length to 8.50 as shown. Notice the placement of the crossing window.

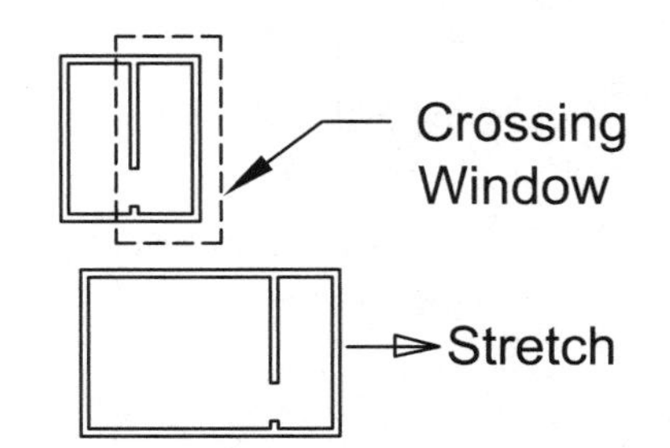

4. Edit the Title and Ex-XX by double clicking on the text. Do not erase and replace.
5. Do not dimension
6. Save as **EX14C**
7. Plot using Page Setup **Class Model A**

Note:
If this was an actual floorplan consider how the stretch command could be very helpful.

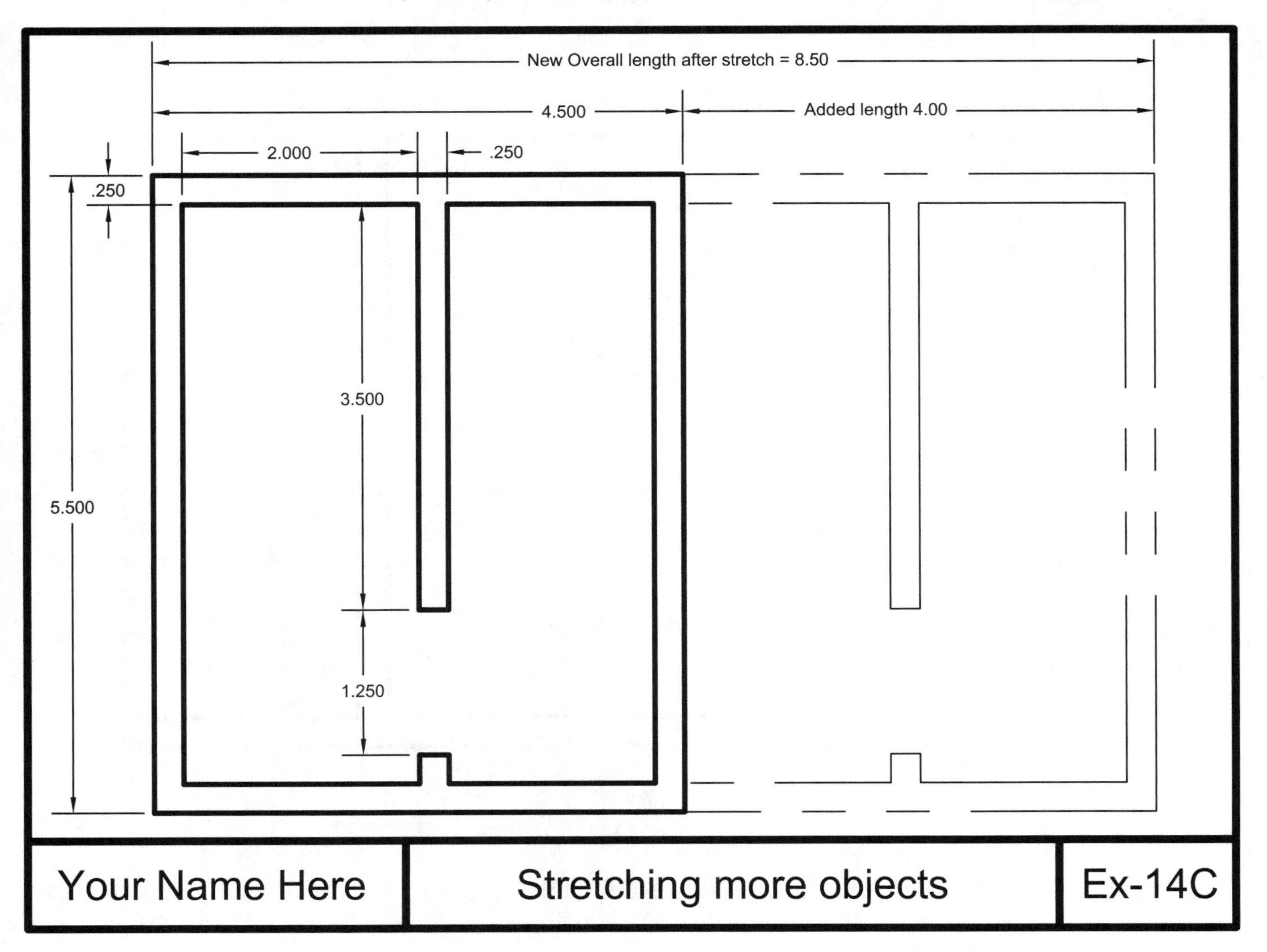

EXERCISE 14D

INSTRUCTIONS:

1. Open **EX14C**

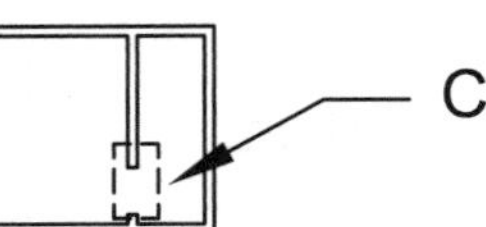

Crossing Window

2. Move the opening to the new location using the Stretch command.
   Notice the placement of the crossing window.

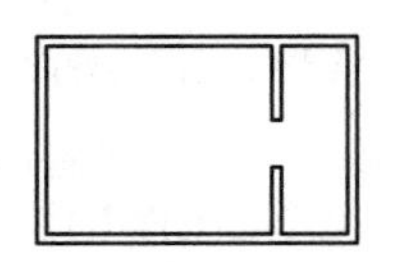

3. Edit the Title and Ex-XX by double clicking on the text. Do not erase and replace.
4. Do not dimension
5. Save as **EX14D**
6. Plot using Page Setup **Class Model A**

Note:
If this was an actual floorplan consider how the stretch command could be very helpful.

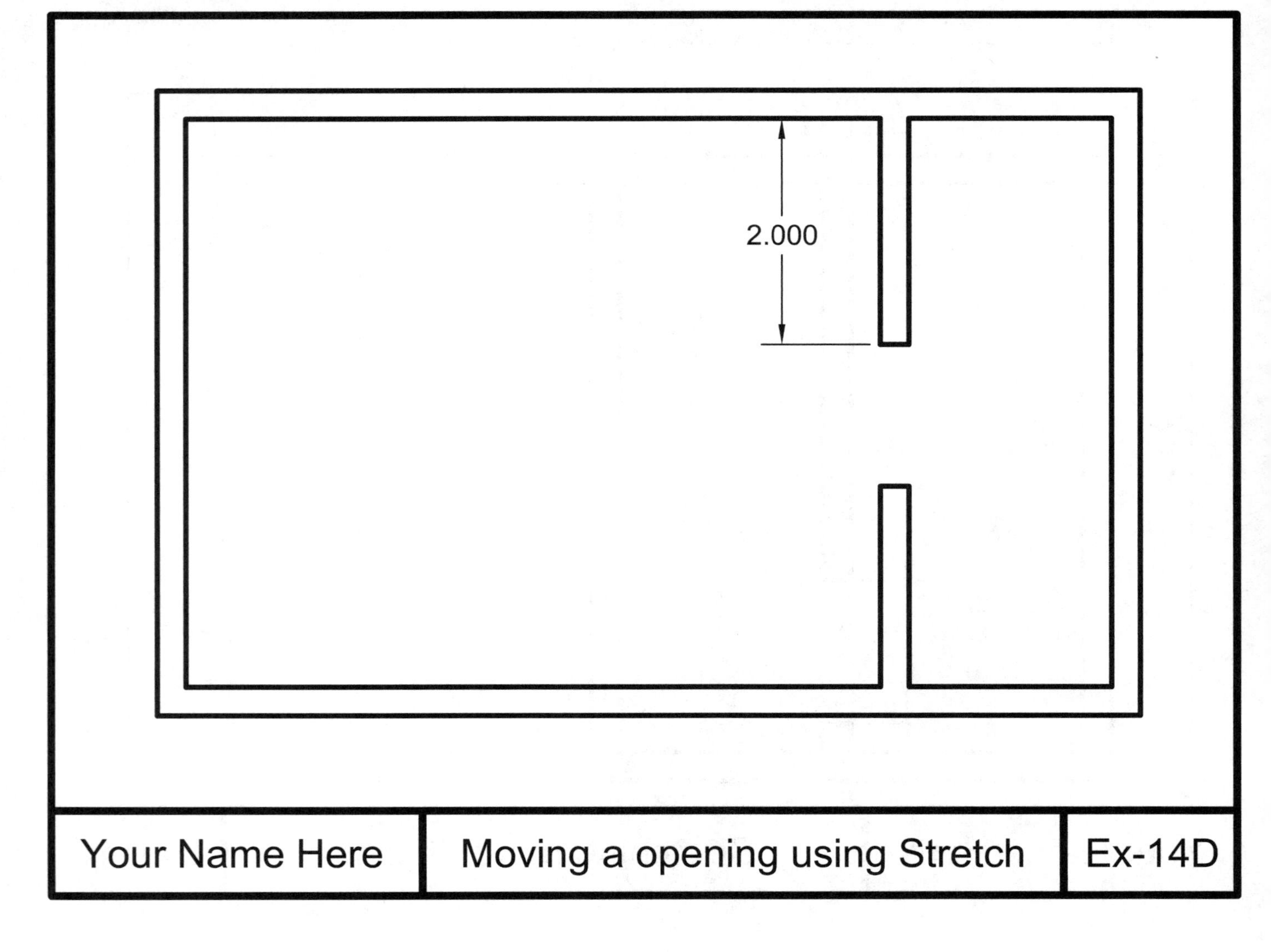

EXERCISE 14E

INSTRUCTIONS:

1. Start a **New** file using **Border A-2015.dwt**
2. Draw the objects at position A. Use Layer Object Line.
3. Copy the objects to position B and C. (Shown as dashed lines)
4. Rotate the copies as shown. **Note**: Base point is important
5. Edit the Title and Ex-XX by double clicking on the text. Do not erase and replace.
6. Do not dimension
7. Save as **EX14E**
8. Plot using Page Setup **Class Model A**

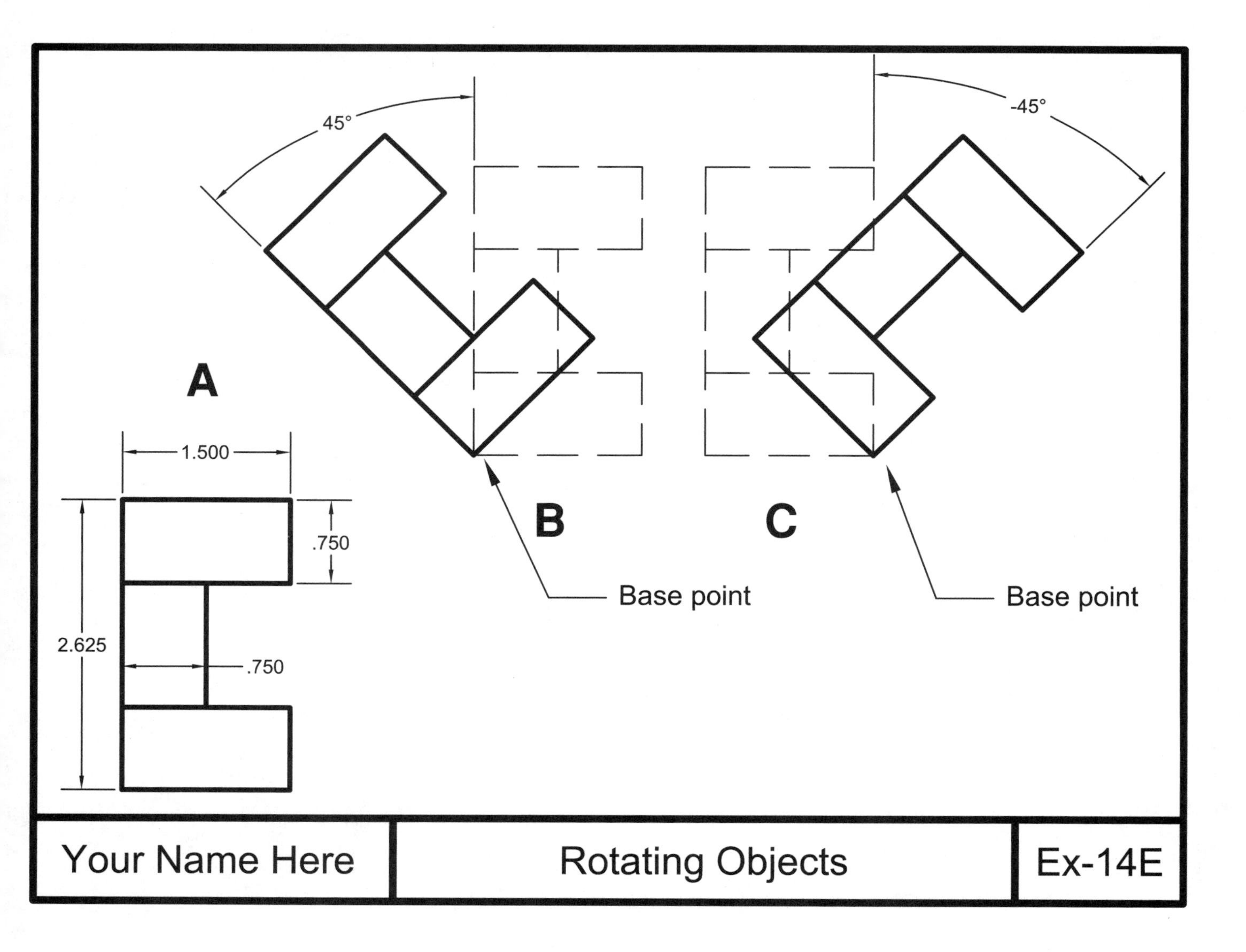

NOTES:

LEARNING OBJECTIVES

After completing this lesson, you will be able to:

1. Place a Cross hatch pattern in a boundary
2. Fill an area with solid fill
3. Add Gradient fills
4. Make changes to existing hatch sets.

LESSON 15

HATCH

The **HATCH** command is used to create hatch lines for section views or filling areas with specific patterns.

To draw **hatch** you must start with a closed boundary. A closed boundary is an area completely enclosed by objects. A rectangle would be a closed boundary. You simply place the cursor inside the closed boundary or select objects.

Note:
A Hatch set is one object.
It is good drawing management to always place Hatch on it's own layer.
Use Layer Hatch. You may also make Hatch appear or disappear with the **FILL** command (pg. 5-6)

HOW TO PLACE HATCH

1. Draw a Rectangle

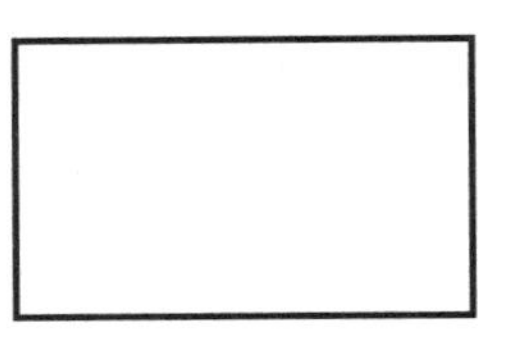

2. Select the Hatch command using one of the following:

   Ribbon = Home tab / Draw panel /
   or
   Keyboard = BH <enter>

   The "Hatch Creation" ribbon tab appears automatically.

3. Place the cursor inside the Rectangle (a closed boundary).

   A hatch pattern <u>preview</u> will appear.

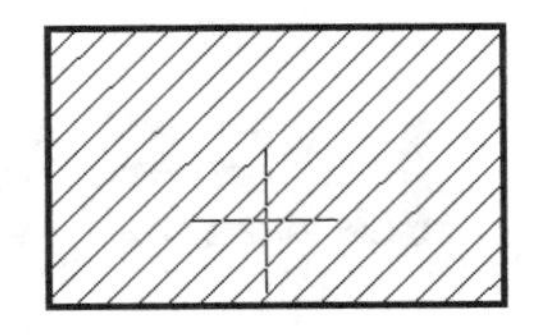

4. Press the left mouse button to accept the Hatch.

5. Select **Close Hatch Creation** or press <enter>

HATCH PROPERTIES

When you select the Hatch command the **Hatch Creation** ribbon tab appears automatically. The panels on this tab help set the properties of the Hatch.
You should set the properties desired previous to placing the hatch set although you can easily edit an existing hatch set.

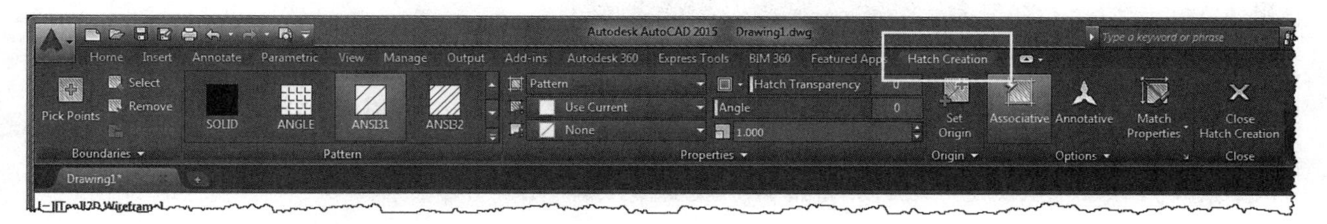

BOUNDARIES Panel

The Boundaries panel allows you to choose what method you will use to select the hatch boundary.

Pick Point:
Pick Point is the default selection. When you select the Hatch command AutoCAD assumes that you want to use the Pick Points method. You merely place the cursor in the closed area to select the boundary. The Hatch set preview will appear. Press the left mouse button to accept.

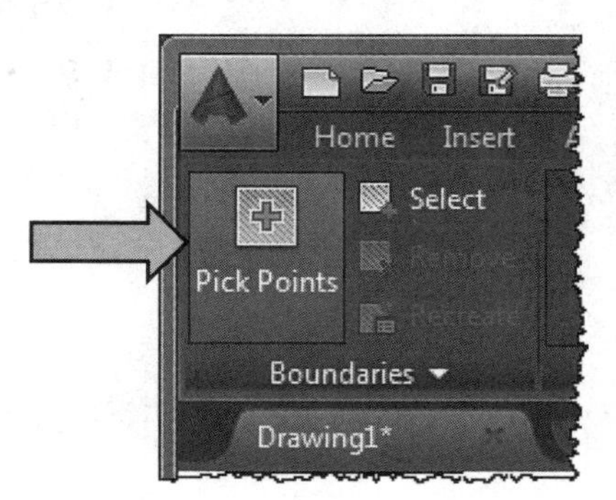

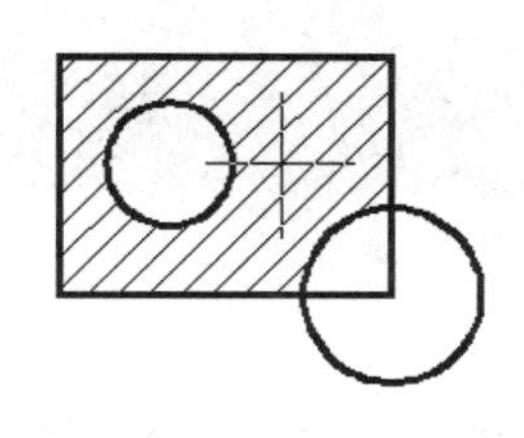

Select and Remove:
You may select or remove objects to a boundary.
Note: **Remove** will not be available unless you click on **Select**.

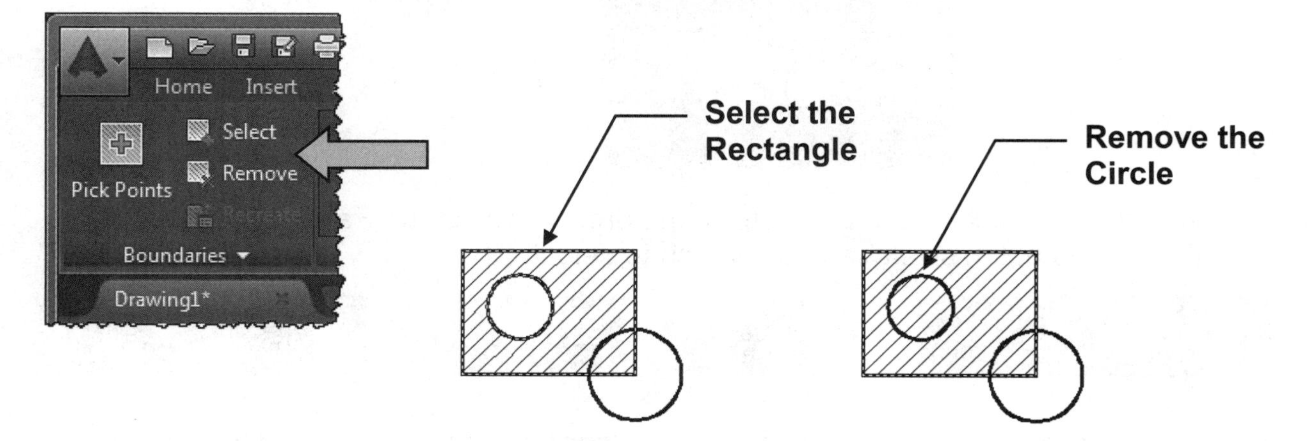

HATCH PROPERTIES....continued

PATTERN Panel

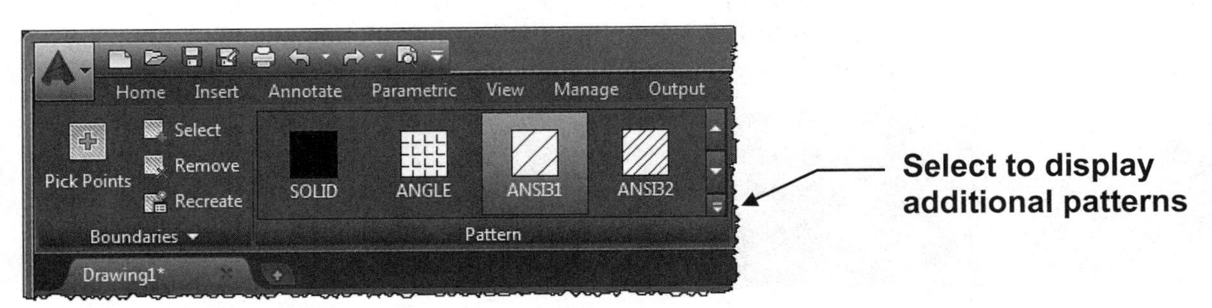

The **PATTERN panel** displays the Hatch swatches that relate to the Hatch Type that has been selected in the Properties panel. Refer to Properties Panel below.

PROPERTIES Panel

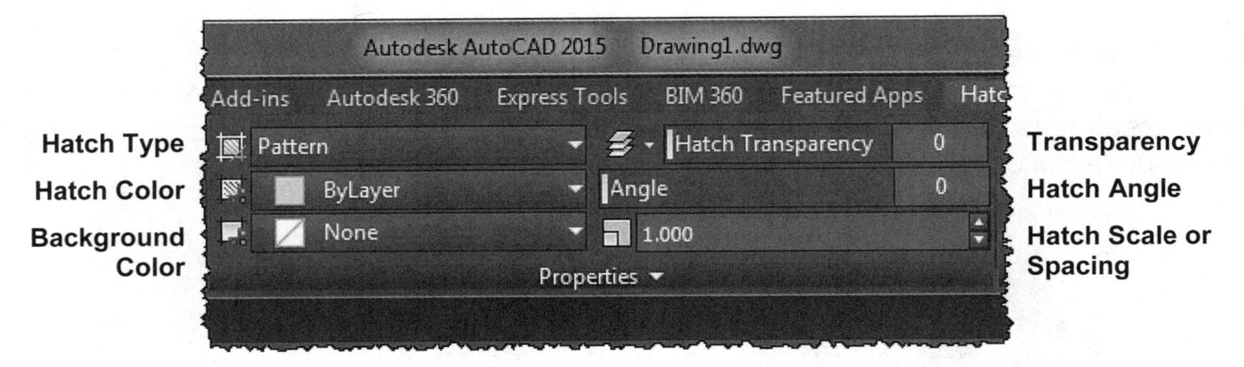

HATCH TYPE

When you select the drop down arrow ▼ you may select one of the Hatch types: **Solid, Gradient, Pattern or User Defined.**

Note: Hatch Types will be explained in more detail on pages 15-8 through 15-11.

Pattern
Solid
Gradient
Pattern
User defined

Select Drop down arrow

When you select a Hatch Type from the drop down list the Pattern panel displays the related hatch swatches from which to select

Pattern **Solid** **User Defined** **Gradient**

HATCH PROPERTIES....continued

PROPERTIES Panel... continued

HATCH COLOR

This color selection is specific to the Hatch and will not affect any other objects.

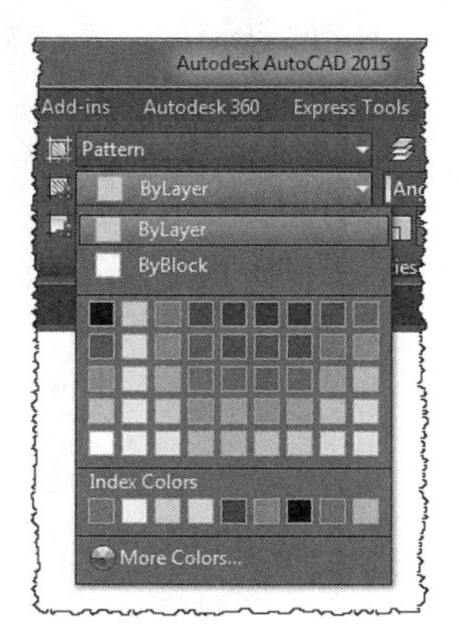

BACKGROUND COLOR

You may select a background color for the hatch area.

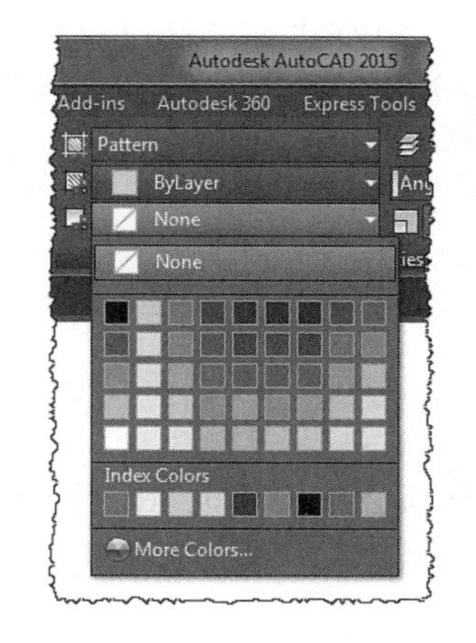

HATCH TRANSPARENCY

Displays the selected transparency setting. You may select to use the current setting for the drawing, the current layer setting or select a specific value.

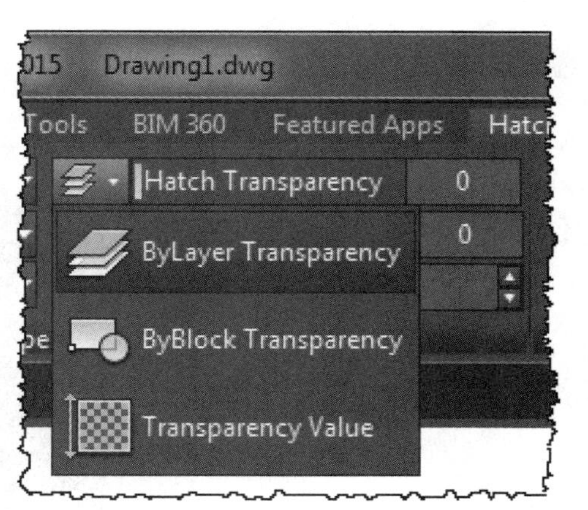

HATCH ANGLE

Pattern: The default angle of a pattern is 0. If you change this angle the pattern will rotate relative to its original design.

User defined: Specify the actual angle of the hatch lines.

HATCH SCALE or SPACING

If Hatch Type **Pattern** is selected this value determines the scale of the Hatch Pattern. A value greater than 1 will increase the scale. A value less than 1 will decrease the scale.

If Hatch Type **User Defined** is selected this value determines the spacing between the hatch lines.

HATCH PROPERTIES....continued

ORIGIN Panel

You may specify where the Hatch will originate. Lower left, lower right, upper left, upper right, center or even the at the UCS Origin.

The Origin locations shown on the right are displayed when Hatch type, Pattern, Solid or User Defined are selected.

The Origin location Centered is displayed only when Hatch type Gradient is selected.

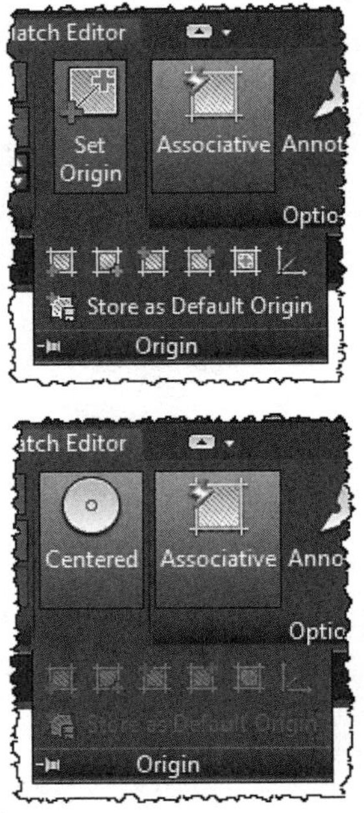

OPTIONS Panel

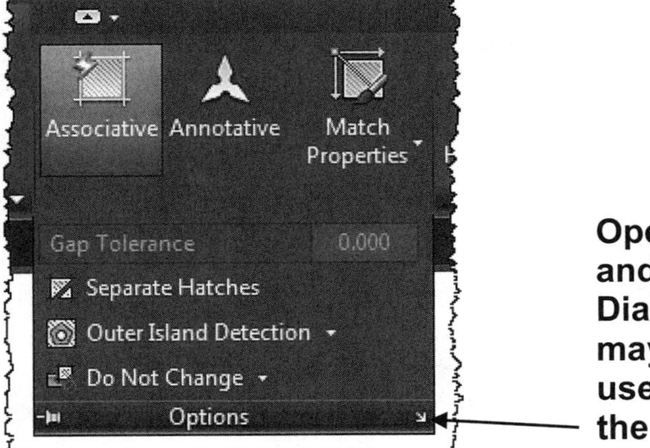

Opens the Hatch and Gradient Dialog Box. You may choose to use it instead of the Ribbon.

Associative: If the Associative option is selected, the hatch set is associated to the boundary. This means if the boundary size is changed the hatch will automatically change to fit the new boundary shape. (Refer to page 15-12 for an example)

Annotative: AutoCAD will automatically adjust the scale to match the current Annotative scale. This option will be discussed in Lesson 27.

HATCH PROPERTIES....continued

OPTIONS Panel....continued

MATCH PROPERTIES

Match Properties allows you to set the properties of the new hatch set by selecting an existing Hatch set. You may choose to "use the current origin" or "use the source hatch origin".

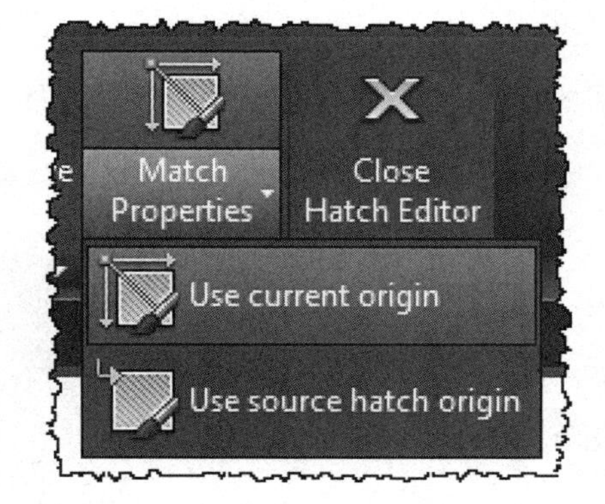

Use current origin: sets all properties <u>except the hatch origin</u>.

Use source hatch origin: sets all properties <u>including the hatch origin</u>.

GAP TOLERANCE

If the area you selected to hatch is not completely closed (gaps) AutoCAD will bridge the gap depending on the Gap tolerance. The Gap tolerance can be set to a value from 0 to 5000. Any gaps equal to or smaller than the value you specify are ignored and the boundary is treated as closed.

CREATE SEPARATE HATCHES

Controls whether HATCH creates a single hatch object or separate hatch objects when selecting several closed boundaries.

OUTER ISLAND DETECTION

These selections determine how Hatch recognizes internal objects.

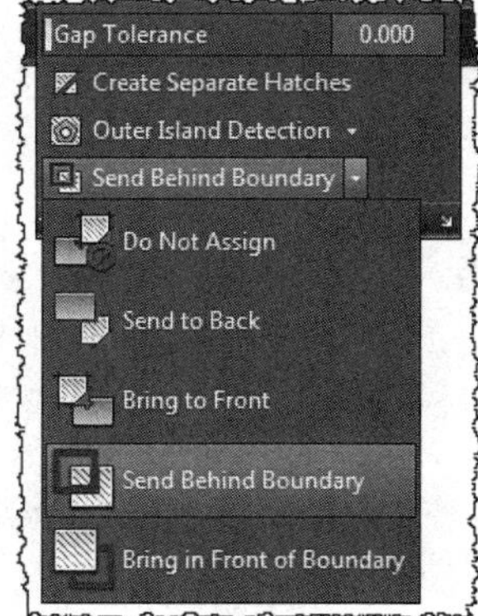

SEND BEHIND BOUNDARIES

These selections determine the draw order of the Hatch set.

HATCH TYPES

PATTERNS

PATTERNS

AutoCAD includes many previously designed Hatch Patterns. **Note:** Using Hatch Patterns will greatly increase the size of the drawing file. So use them conservatively. You may also purchase patterns from other software companies.

1. Select the Hatch Type **PATTERN**.

2. Select a Pattern from the list of hatch patterns displayed in the Pattern panel.

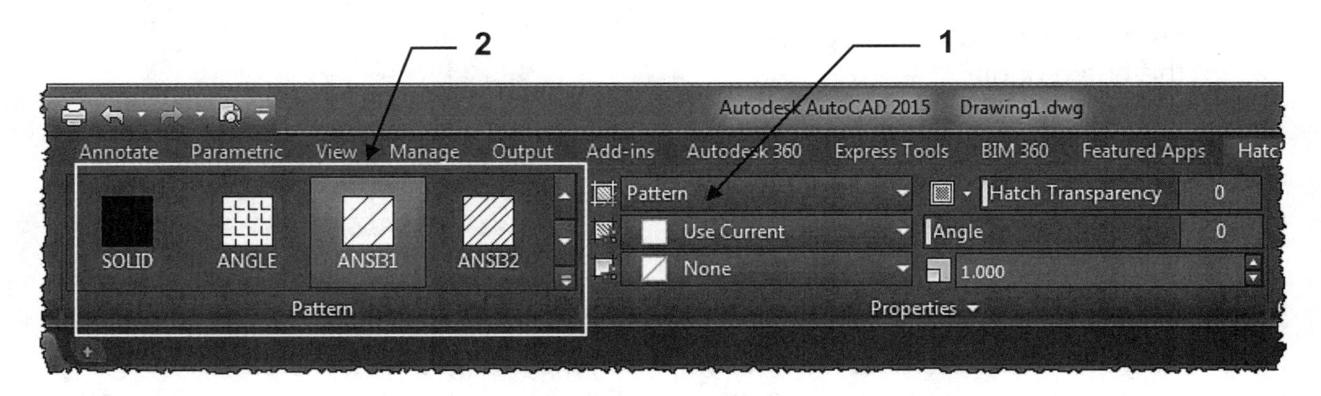

3. Select the Hatch Color, Boundary Background Color, Hatch Transparency.

4. **Angle:** A previously designed pattern has a default angle of 0. If you change this Angle the pattern will rotate the pattern relative to its original design.
   It is important that you understand how to control the angle.

   For example: if **ANSI31** hatch pattern is used and the angle is set to 45 degrees the pattern will rotate relative to its original design and the pattern will appear to be 90 degrees.

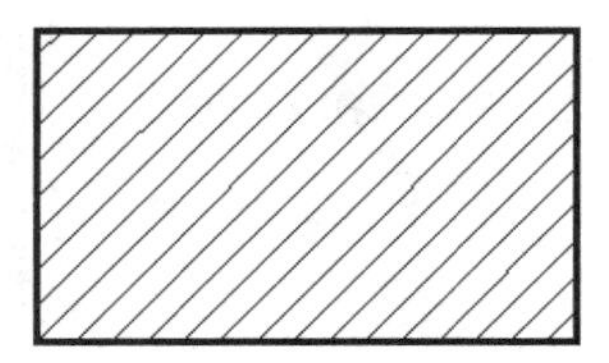

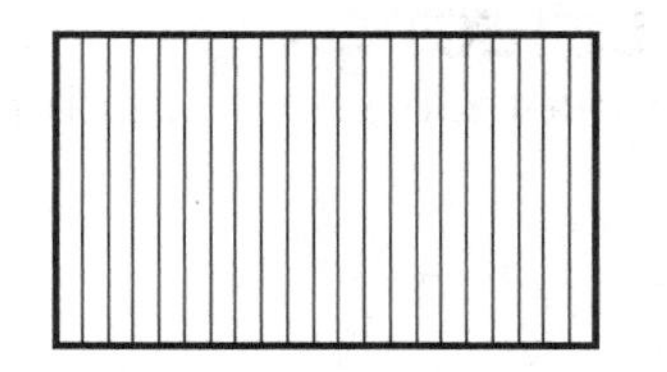

Original Pattern Design
Angle = 0

Pattern Design Rotated
Angle = 45

5. **Scale:** A value greater than 1 will increase the scale. A value less than 1 will decrease the scale. If the Hatch set is Annotative the scale will automatically adjust to the Annotative scale. But you might have to tweak the scale additionally to make it display exactly as you desire.
   (This will be discussed more in Lesson 27. Don't be too concerned about it right now)

HATCH TYPES

USER DEFINED

USER DEFINED
This Hatch Type allows you to simply draw continuous lines. No special pattern.
You specify the **Angle** of the lines and the **Spacing** between the lines.
Note: This Hatch type does not significantly increase the size of the drawing file.

1. Select the Hatch Type **USER DEFINED**.

2. Note the Pattern swatches are not displayed in the Pattern panel. (The Zigzag is only there because alphabetically it was in the same row as User Defined)

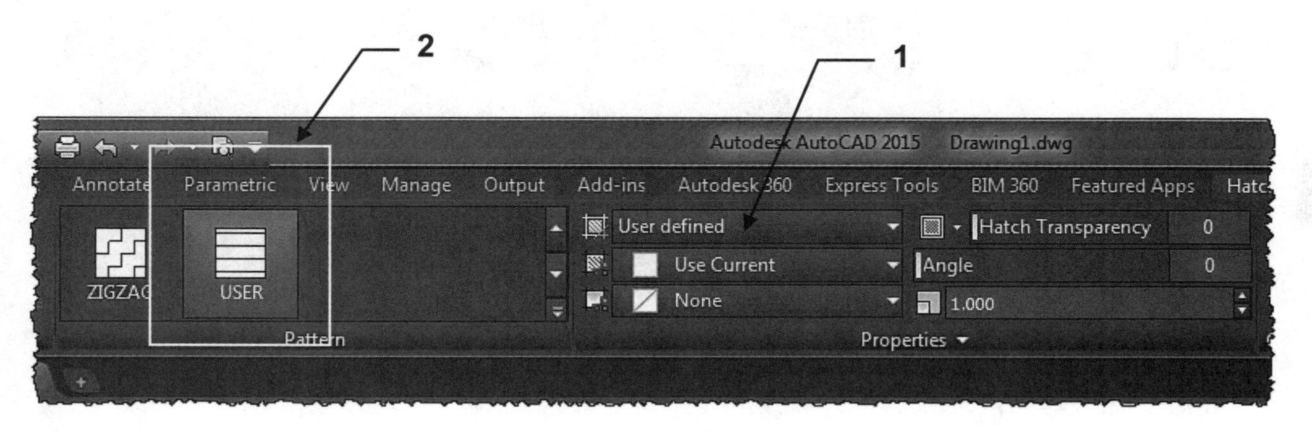

3. Select the Hatch Color, Boundary Background Color, Hatch Transparency.

4. **Angle:** Specify the actual angle that you desire from 0 to 180.

   For example: If you want the lines to be on an angle of 45 degrees you would enter 45. If you want the lines to be on an angle of 90 degrees you would enter 90. (This is different from the angle for Patterns)

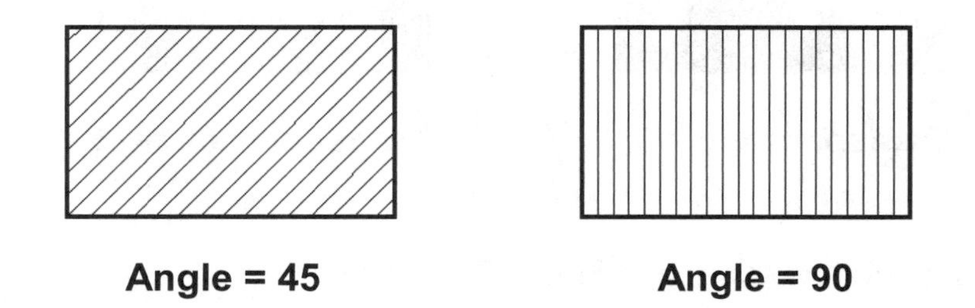

Angle = 45 **Angle = 90**

5. **Spacing:** Specify the actual distance between each of the hatch lines.

HATCH TYPES

SOLID

SOLID
If you would like to fill an area with a solid fill you should use Hatch type **Solid**.

1. Select **Solid** by selecting the Hatch Type **Solid** on the Properties Panel
   or
   select the Solid swatch on the Pattern Panel.

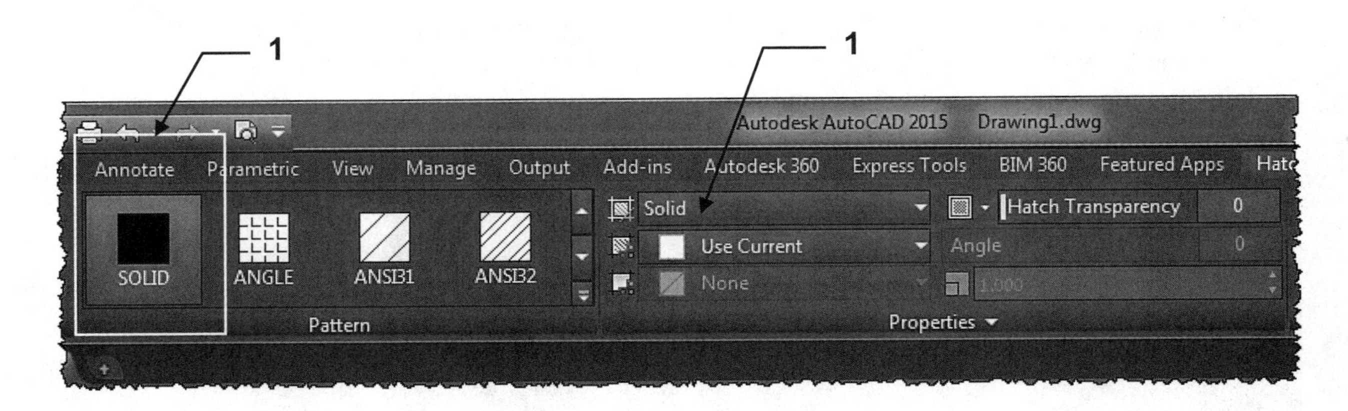

2. Select the Hatch **Color**.
   Note: The Boundary Background Color is not available when using Hatch Type Solid.

3. Select **Transparency**

Transparency = 0

Transparency = 75

4. **Angle:** Not available when using Hatch Type Solid.

5. **Scale:** Not available when using Hatch Type Solid

HATCH TYPES

GRADIENT

GRADIENT

Gradients are fills that gradually change from dark to light or from one color to another. Gradient fills can be used to enhance presentation drawings, giving the appearance of light reflecting on an object, or creating interesting backgrounds for illustrations.

Gradients are definitely fun to experiment with but you will have to practice to achieve complete control. They will also greatly increase the size of the drawing file.

1. Select the Hatch Type **GRADIENT**.

2. Select a Gradient Pattern from the 9 GR_ patterns displayed in the Pattern panel.

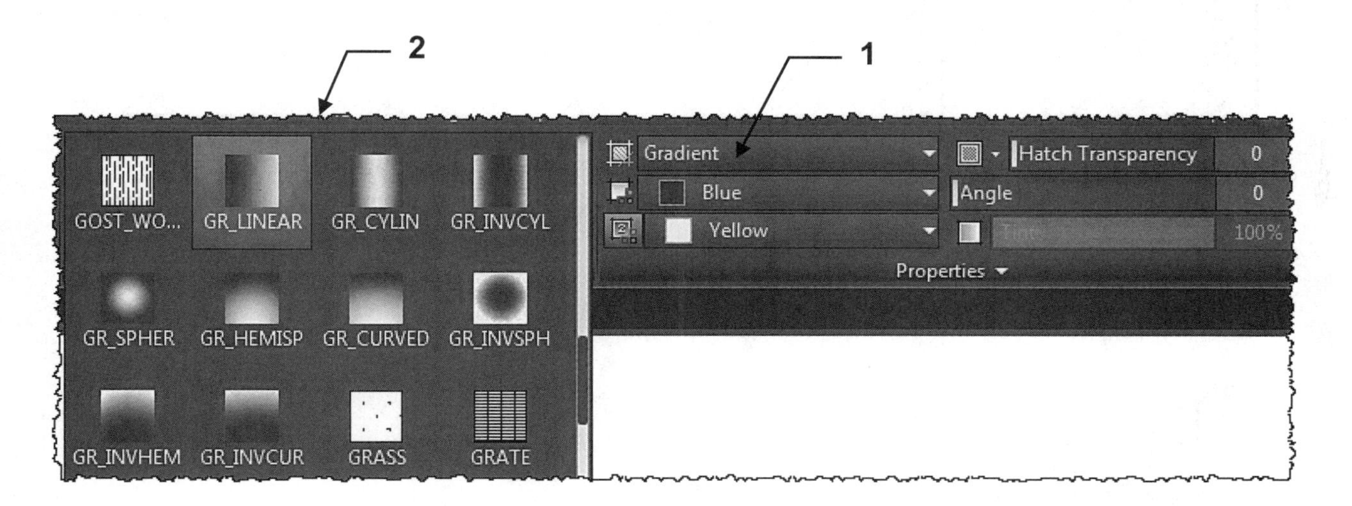

3. Select the Hatch **Color**. The Gradient can be one color or two color. If one color you can select the Tint and Shade of that color. (See step 6 below)

4. Select the Hatch **Transparency**.

5. **Angle:** Specify the actual angle that you desire from 0 to 180. The pattern will rotate the pattern relative to its original design.

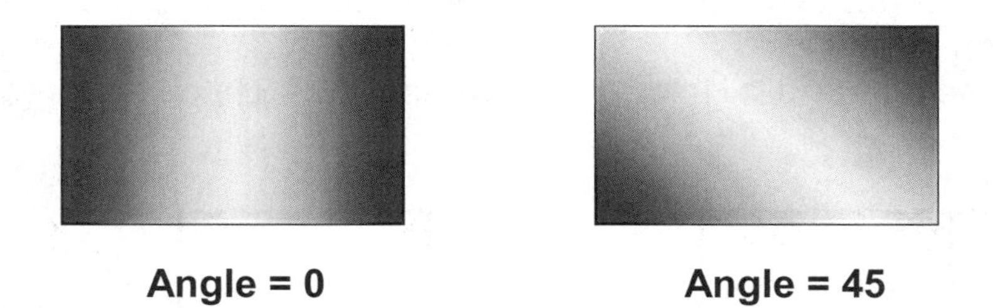

Angle = 0 **Angle = 45**

6. **Tint and Shade:** Tint and Shade is used when you are using only one-color gradient fill. Specify the tint or shade of the color selected in step 3 above.

EDITING HATCH

EDITING THE HATCH SET PROPERTIES

Editing a Hatch set properties is easy.

1. Simply select the Hatch set to edit.
2. The Hatch Editor Ribbon tab will appear.
3. Make new selections. Any changes are applied immediately.

CHANGING THE BOUNDARY

If the Hatch set is **Associative** (see 15-6) you may change the shape of the boundary and the Hatch set will conform to the new shape.
If the Hatch set is **Non-Associative** the Hatch set will not change.

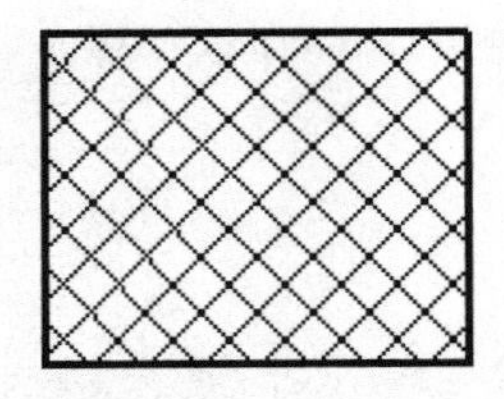

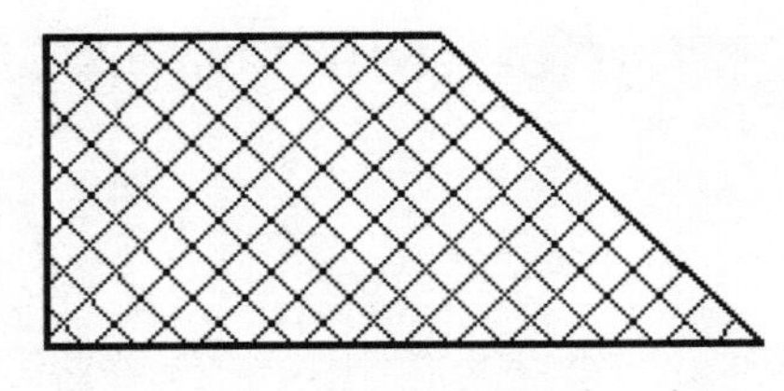

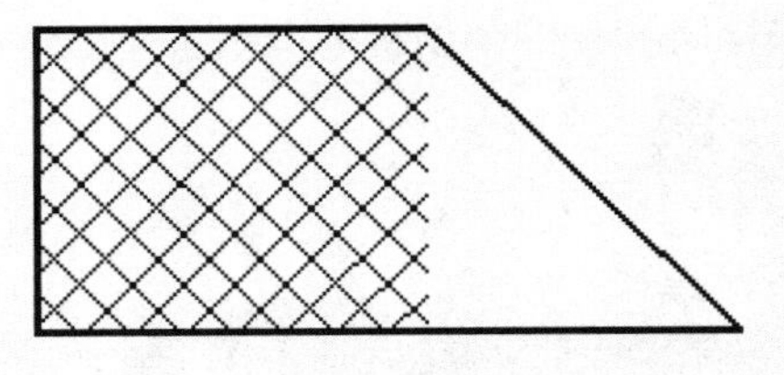

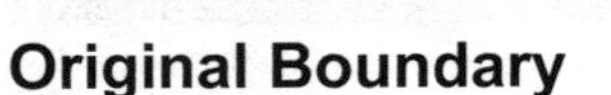

Original Boundary **New Boundary with Associative hatch** **New Boundary with non-Associative hatch**

TRIMMING HATCH

You may trim a hatch set just like any other object.

Before Trim **After Trim**

MIRROR HATCH

An existing Hatch set can be mirrored. The boundary shape will automatically mirror.
But you may control whether the Hatch pattern is mirrored or not.

To control the Hatch pattern mirror: (Note: Set prior to using the Mirror command)

1. Type **mirrhatch <enter>**
2. Enter 0 or 1 <enter> 0 = Hatch not Mirrored 1 = Hatch Mirrored

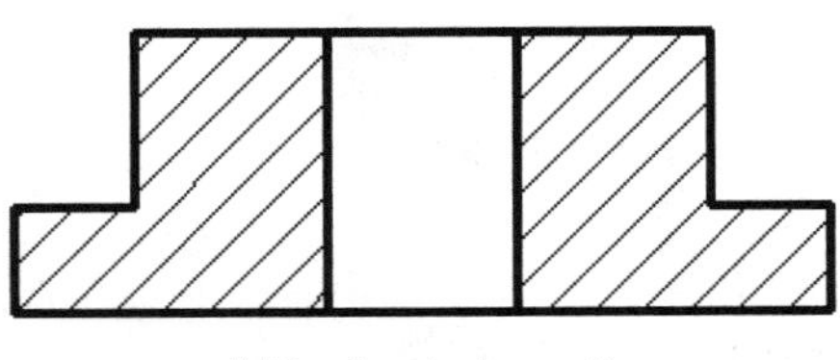

Mirrhatch = 0

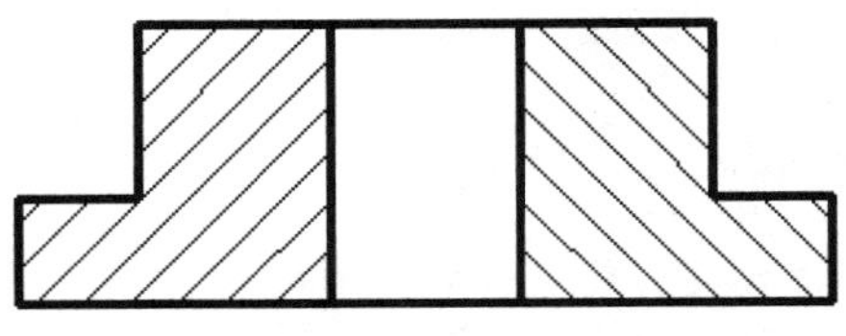

Mirrhatch = 1

EXERCISE 15A

INSTRUCTIONS:

1. Start a **New** file using **Border A-2015.dwt**
2. Draw the Rectangle, Circle and Polygon as shown.
3. Use Layers = Object line and Hatch.
4. Hatch the boundaries shown using the Patterns shown.
5. Edit the Title and Ex-XX by double clicking on the text. Do not erase and replace.
6. Do not dimension
7. Save as **EX15A**
8. Plot using Page Setup **Class Model A**

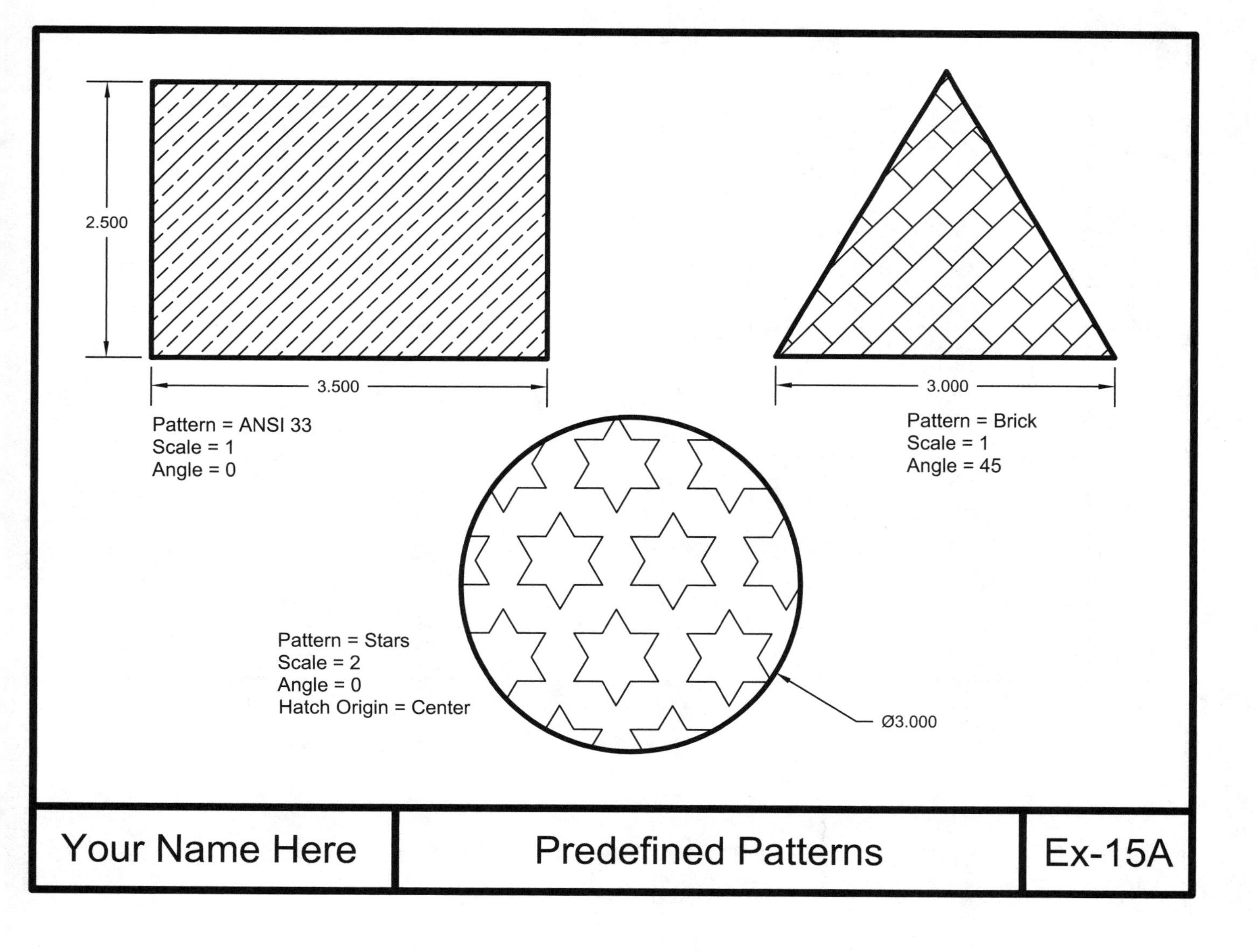

EXERCISE 15B

INSTRUCTIONS:

1. Start a **New** file using **Border A-2015.dwt**
2. Draw the 4 Rectangle as shown.
3. Use Layers = Object line and Hatch.
4. Hatch the boundaries shown using the Patterns shown.
   Notice the angles rotate the hatch pattern from it's original design.
   For example: Angle 45 appears to be rotation 90 because the original design was drawn at a 45 degree.....think about it.

5. Edit the Title and Ex-XX by double clicking on the text. Do not erase and replace.
6. Do not dimension
7. Save as **EX15B**
8. Plot using Page Setup **Class Model A**

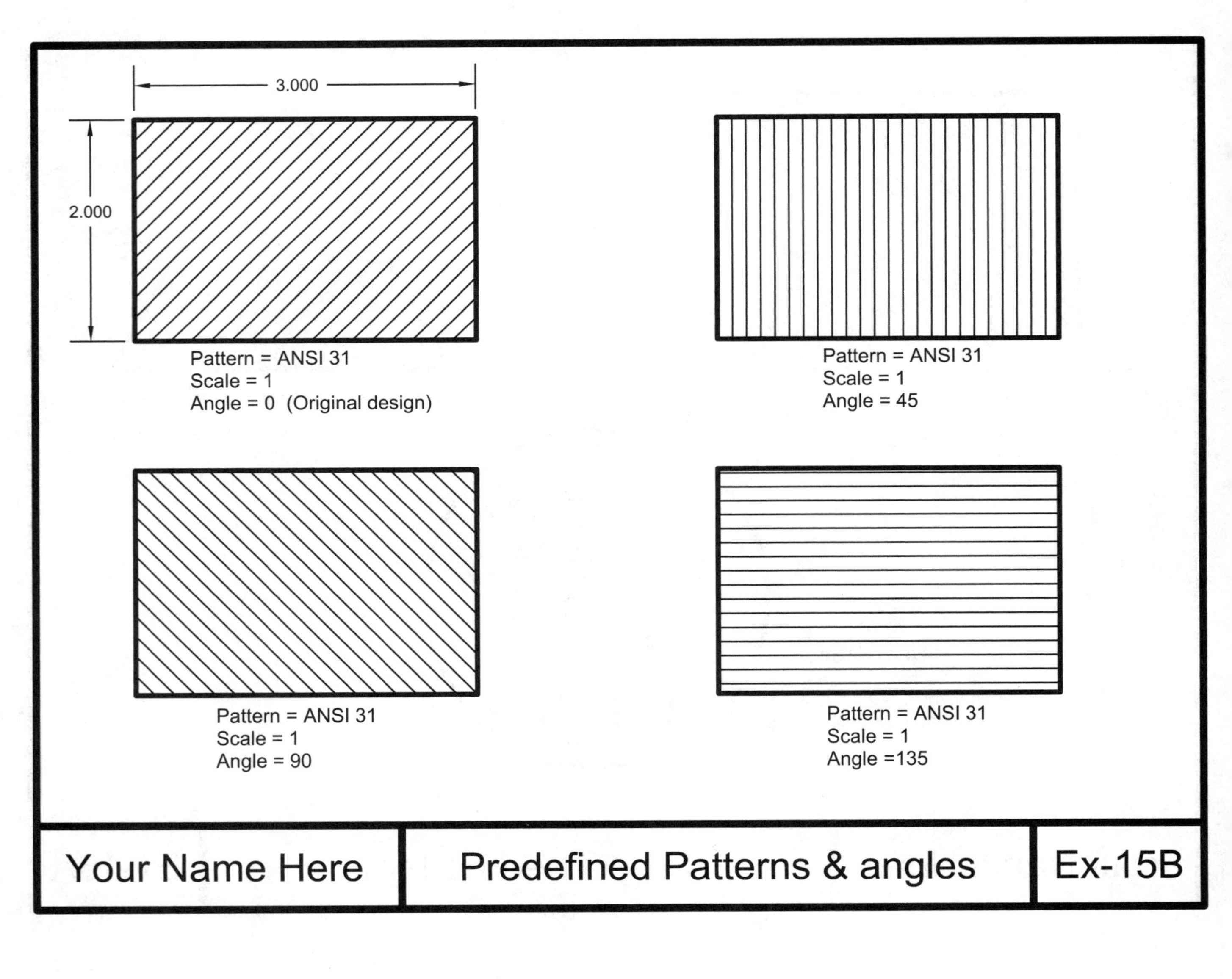

EXERCISE 15C

INSTRUCTIONS:

1. Start a **New** file using **Border A-2015.dwt**
2. Draw the Rectangle as shown using layer Object line.
3. Place the 1" high text in the Middle of the rectangle. Use layer Text.
4. Use Hatch Pattern ANSI31. Use layer Hatch.
5. Select the Boundary using "Select Objects" this time.
6. Edit the Title and Ex-XX by double clicking on the text. Do not erase and replace.
7. Do not dimension
8. Save as **EX15C**
9. Plot using Page Setup **Class Model A**

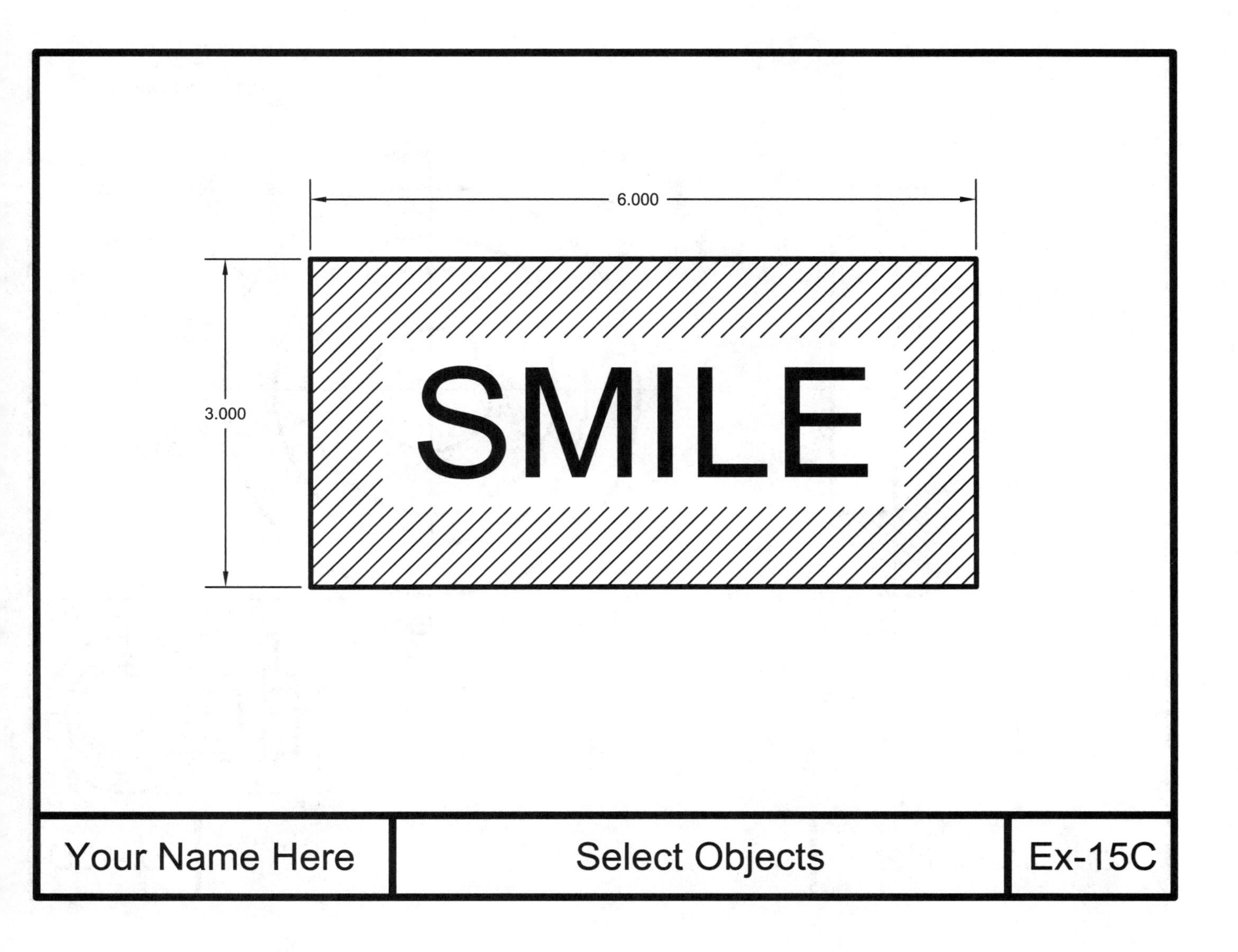

EXERCISE 15D

INSTRUCTIONS:

1. Start a **New** file using **Border A-2015.dwt**
2. Draw the Hub as shown using layer Object line.
3. Use Hatch pattern ANSI 31 Scale = 1 and Angle = 0. Use Layer Hatch
4. Draw the section arrowhead, approximately as shown. Fill with "Solid" hatch.
5. Edit the Title and Ex-XX by double clicking on the text. Do not erase and replace.
6. Do not dimension
7. Save as **EX15D**
8. Plot using Page Setup **Class Model A**

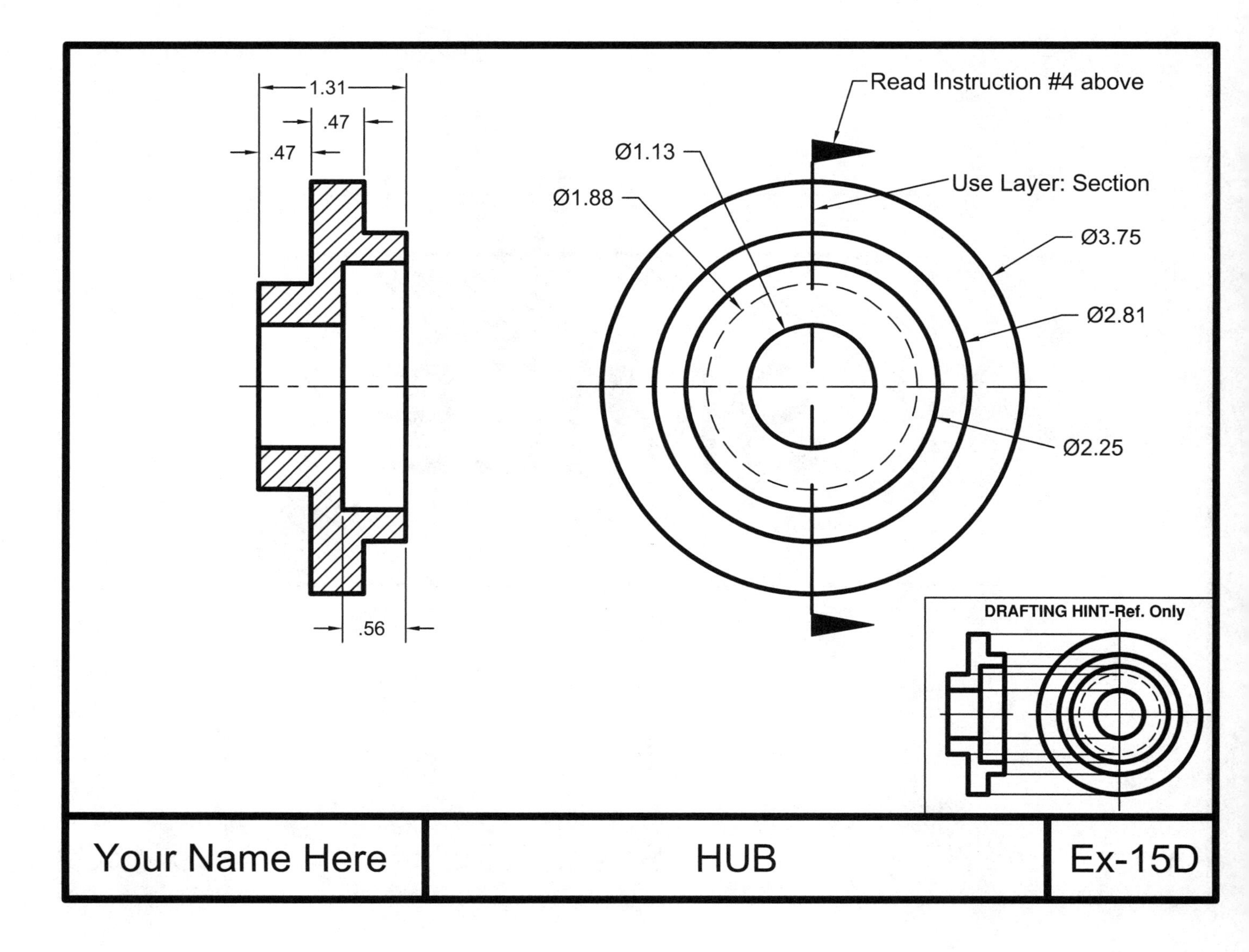

EXERCISE 15E

INSTRUCTIONS:

1. Start a **New** file using **Border A-2015.dwt**
2. Draw the objects shown as Before, using layer Object line.
3. Add Solid Hatch. Use Layer Hatch
4. Edit the Title and Ex-XX by double clicking on the text. Do not erase and replace.
5. Do not dimension
6. Save as **EX15E**
7. Plot using Page Setup **Class Model A**

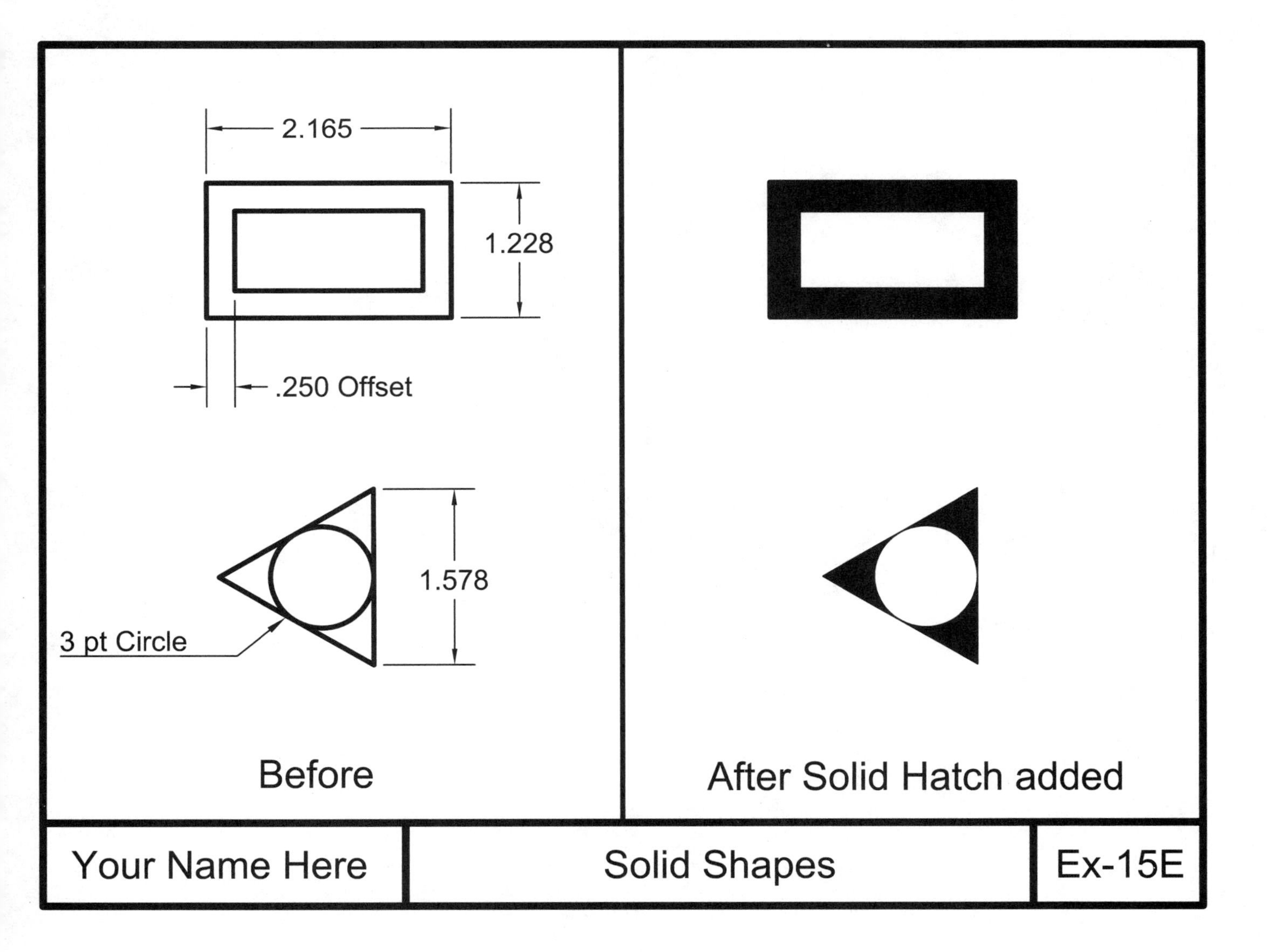

Notes:

LEARNING OBJECTIVES

After completing this lesson, you will be able to:

1. Understand the importance of True Associative dimensioning
2. Create Linear, Continue and Baseline dimensions
3. Draw, Select and Erase objects
4. Use Grips to manipulate dimensions
5. Create a New dimension Style
6. Ignore Hatch Lines when dimensioning

LESSON 16

DIMENSIONING

Dimensioning is basically easy, but as always, there are many options to learn. As a result, I have divided the dimensioning process into multiple lessons.

In addition, Annotative dimensioning will be discussed in Lesson 27. So relax and just take it one lesson at a time.

In this Lesson you will learn how to create a **dimension style** and how to create Linear (horizontal and vertical) dimensions.

Dimensions can be Associative, Non-Associative or Exploded. You need to understand what these are so you may decide which setting you want to use. Most of the time you will use Associative but you may have reasons to also use Non-Associative and Exploded.

Associative
Associative Dimensioning means that the dimension is actually associated to the objects that they dimension. If you move the object, the dimension will move with it. If you change the size of the object, the dimension text value will change also. *(Note: This is not parametric. In other words, you cannot change the dimension text value and expect the object to change. That would be parametric dimensioning)*

Non-Associative
Non-Associative means the dimension is not associated to the objects and will not change if the size of the object changes.

Exploded
Exploded means the dimension will be exploded into lines, text and arrowheads and non-associative.

How to set dimensioning to Associative, Non-Associative or Exploded.

1. On the command line type: ***dimassoc <enter>***

2. Enter the number ***2, 1 or 0 <enter>.***

   2 = Associative

   1 = Non-Associative

   0 = Exploded

DIMENSIONING....continued

How to Re-associate a dimension.

If a dimension is Non-associative, and you would like to make it Associative, you may use the **dimreassociate** command to change.

1. Select **Reassociate** using one of the following:

   Ribbon = Annotate tab / Dimension ▾ panel /
   or
   Keyboard = Dimreassociate <enter>

2. Select objects: ***select the dimension to be reassociated.***

3. Select objects: ***select more dimensions or <enter> to stop.***

4. Specify first extension line origin or [Select object] <next>: (***an "X" will appear to identify which is the first extension); use object snap to select the exact location, on the object, for the extension line point.***

5. Specify second extension line origin <next>: (***the "X" will appear on the second extension) use object snap to select the exact location, on the object, for the extension line point.***

6. ***Continue until all extension line points are selected.***

Note: You must use object snap to specify the exact location for the extension lines.

Regenerating Associative dimensions

Sometimes after panning and zooming, the associative dimensions seem to be floating or not following the object. The **DIMREGEN** command will move the associative dimensions back into their correct location.

Type: ***dimregen <enter>***

LINEAR DIMENSIONING

Linear dimensioning allows you to create horizontal and vertical dimensions.

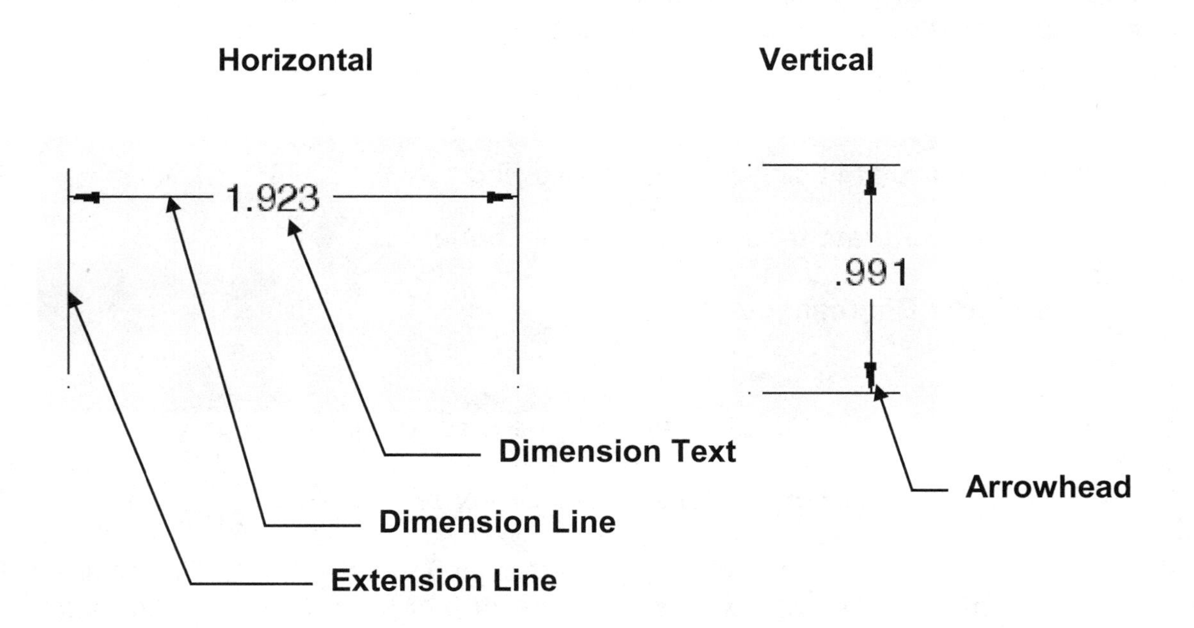

1. Select the **LINEAR** command using one of the following:

   Ribbon = Annotate tab / Dimension panel /
   or
   Keyboard = Dimlinear <enter>

   Command:__dimlinear
2. Specify first extension line origin or <select object>: ***snap to first extension line origin (P1).***
3. Specify second extension line origin: ***snap to second extension line origin (P2).***
4. Specify dimension line location or [Mtext/Text/Angle/Horizontal/Vertical/Rotated]: ***select where you want the dimension line placed (P3).***

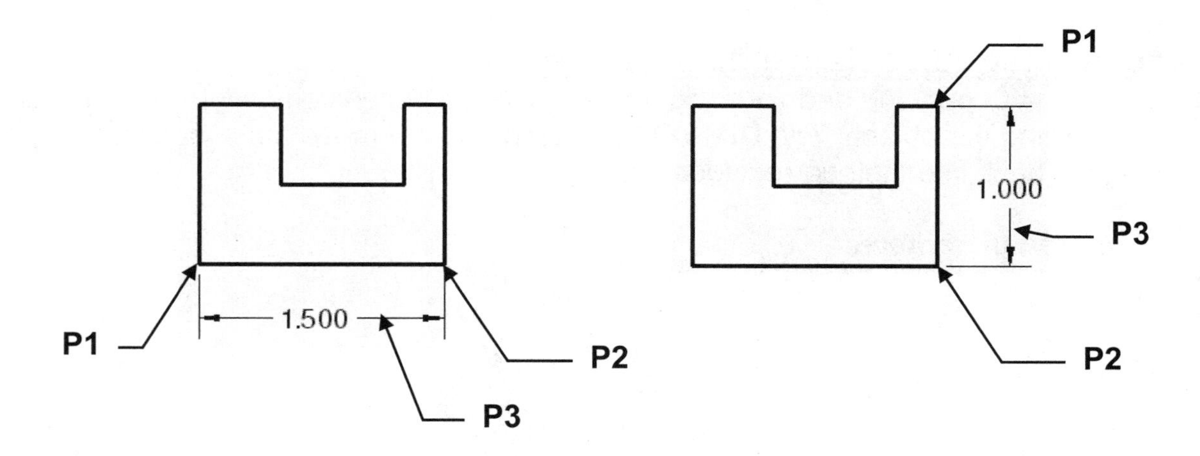

CONTINUE DIMENSIONING

Continue creates a series of dimensions in-line with an existing dimension. If you use the continue dimensioning immediately after a Linear dimension, you do not have to specify the continue extension origin.

1. Create a linear dimension first (**P1 and P2 shown below**)

2. Select the Continue command using one of the following:

Ribbon = Annotate tab / Dimension panel /
or
Keyboard = Dimcontinue <enter>

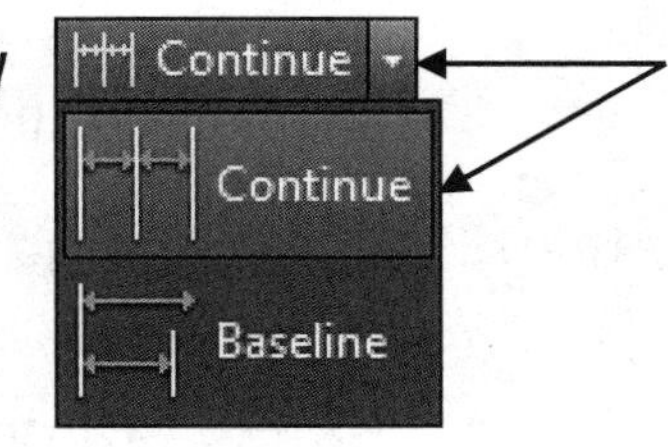

Command: _dimcontinue

3. Specify a second extension line origin or [Undo/Select] <Select>: ***snap to the second extension line origin (P3).***

4. Specify a second extension line origin or [Undo/Select] <Select>: ***snap to the second extension line origin (P4).***

5. Specify a second extension line origin or [Undo/Select] <Select>: ***press <enter> twice to stop.***

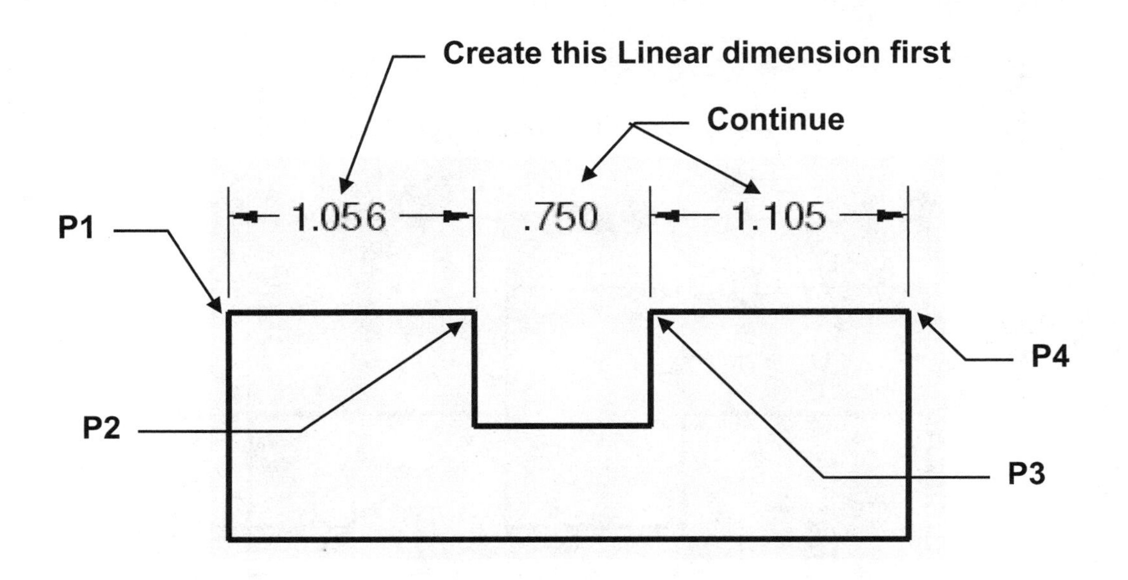

BASELINE DIMENSIONING

Baseline dimensioning allows you to establish a **baseline** for successive dimensions. The spacing between dimensions is automatic and should be set in the dimension style. (See 16-10, Baseline spacing setting)

A Baseline dimension must be used with an existing dimension. If you use Baseline dimensioning immediately after a Linear dimension, you do not have to specify the baseline origin.

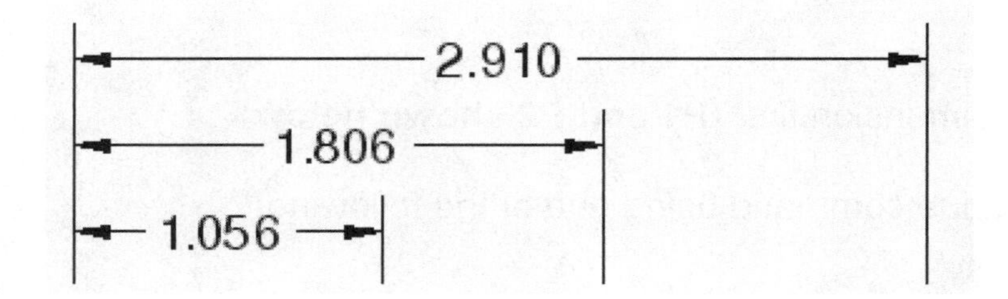

1. Create a linear dimension first **(P1 and P2).**

2. Select the **BASELINE** command using one of the following:

Ribbon = Annotate tab / Dimension panel /
or
Keyboard = Dimbaseline <enter>

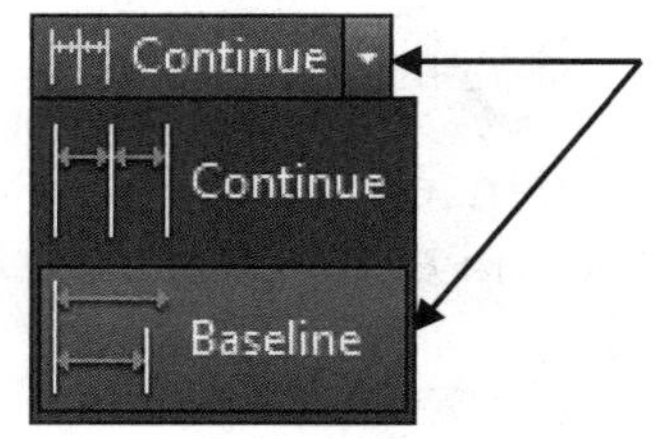

Command: _dimbaseline

3. Specify a second extension line origin or [Undo/Select] <Select>: ***snap to the second extension line origin (P3).***

4. Specify a second extension line origin or [Undo/Select] <Select>: ***snap to (P4).***

5. Specify a second extension line origin or [Undo/Select} <Select>: ***select <enter> twice to stop.***

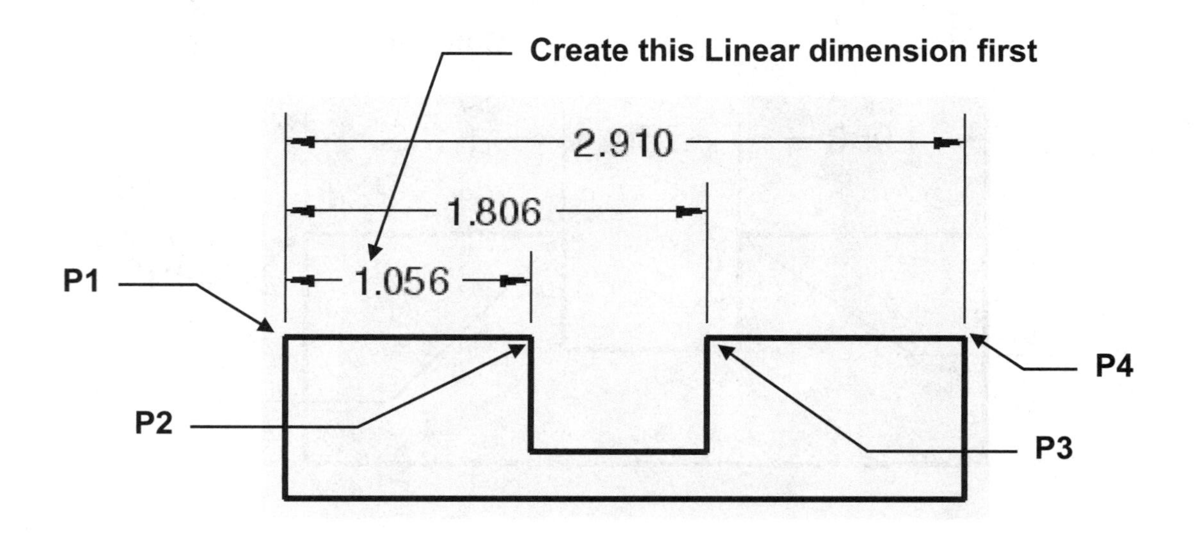

DIMENSION STYLES

Using the "Dimension Style Manager" you can change the appearance of the dimension features, such as length of arrowheads, size of the dimension text, etc. There are over 70 different settings.
You can also Create New, Modify and Override Dimension Styles. All of these are simple, by using the Dimension Style Manager described below.

1. Select the "Dimension Style Manager" using one of the following:

Ribbon = Annotate tab / Dimension panel / ↘
or
Keyboard = Dimstyle <enter>

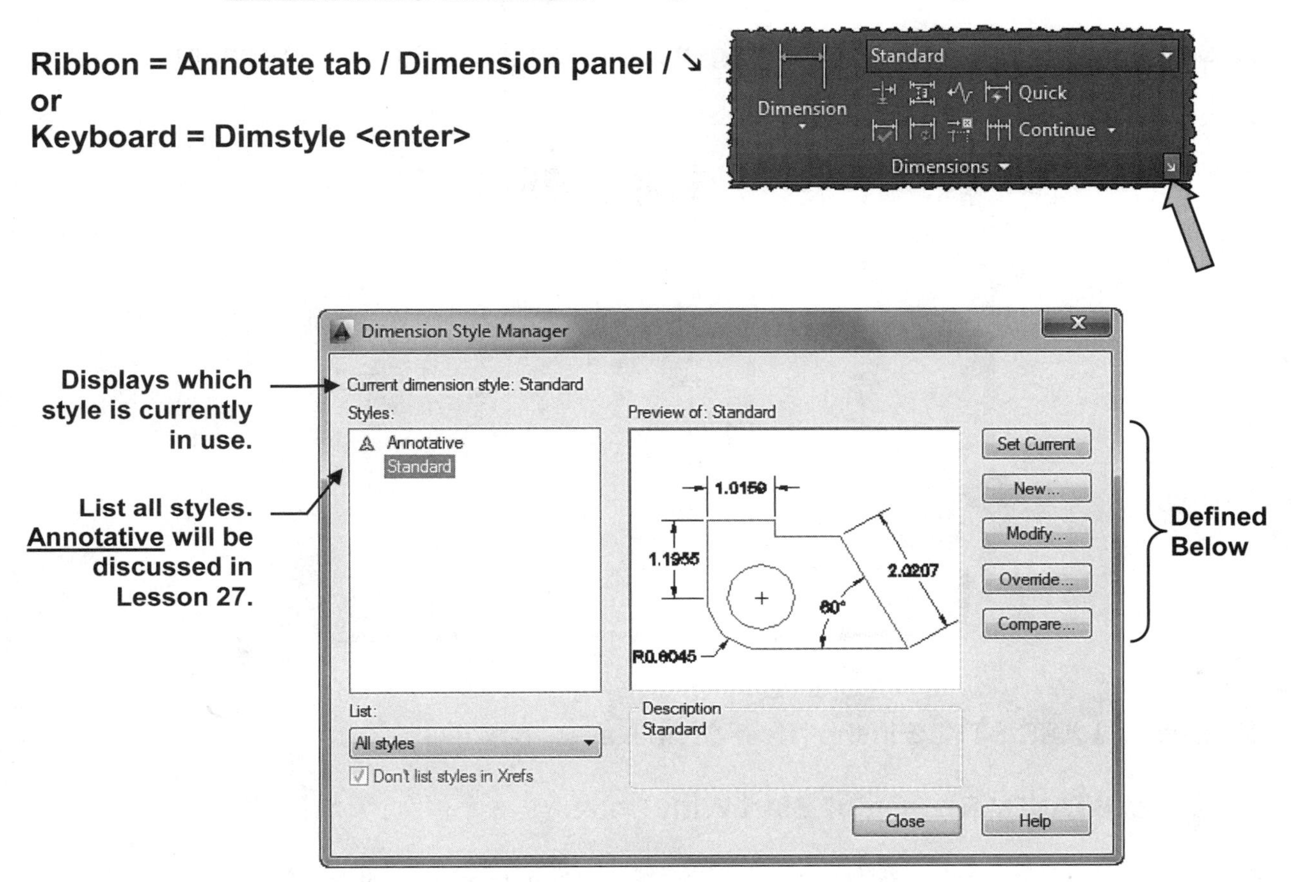

Set Current Select a style from the list of styles and select the **set current** button.

New Select this button to create a new style. When you select this button, the **Create New Dimension Style** dialog box is displayed.

Modify Selecting this button opens the **Modify dimension Style** dialog box which allows you to make changes to the style selected from the "list of styles".

Override An override is a temporary change to the current style. Selecting this button opens the Override Current Style dialog box.

Compare Compares two styles.

CREATING A NEW DIMENSION STYLE

A dimension style is a group of settings that has been saved with a name you assign. When creating a new style you must start with an existing style, such as Standard. Next, assign it a new name, make the desired changes and when you select the OK button the new style will have been successfully created.

How to create a NEW dimension style.

1. Start a **New** file using **Border A-2015.dwt**

2. Select the **Dimension Style Manager** command (Refer to previous page)

3. Select the **NEW** button.

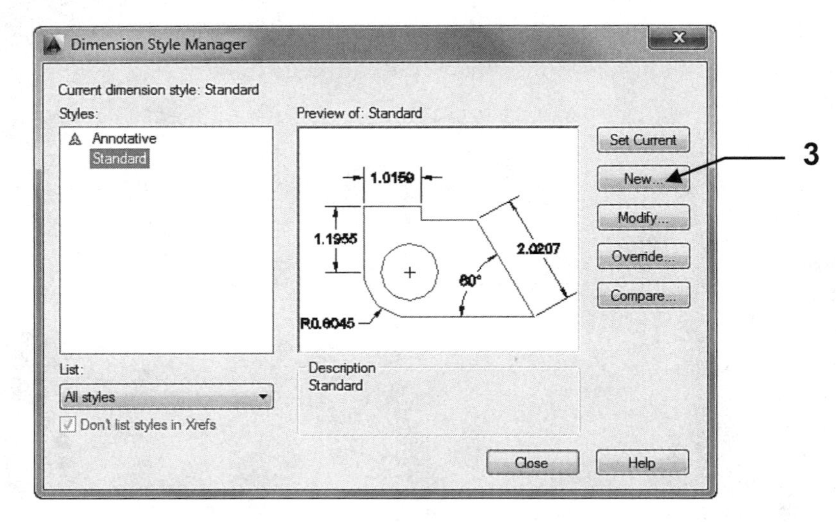

4. Enter **CLASS STYLE** in the "**New Style Name**" box.

5. Select **STANDARD** in the **"Start With:"** box.

6. **"Use For:"** box will be discussed later. For now, leave it set to "**All dimensions**".

7. Select the **CONTINUE** button.

Create New Dimension Style
4
New Style Name:
CLASS STYLE
7
Continue
5
Start With:
Standard
Cancel
Discussed in Lesson 26
Annotative
Help
6
Use for:
All dimensions

CREATING A NEW DIMENSION STYLE....continued

8. Select the **Primary Units** tab and change your setting to match the settings shown below.

8. Primary Units

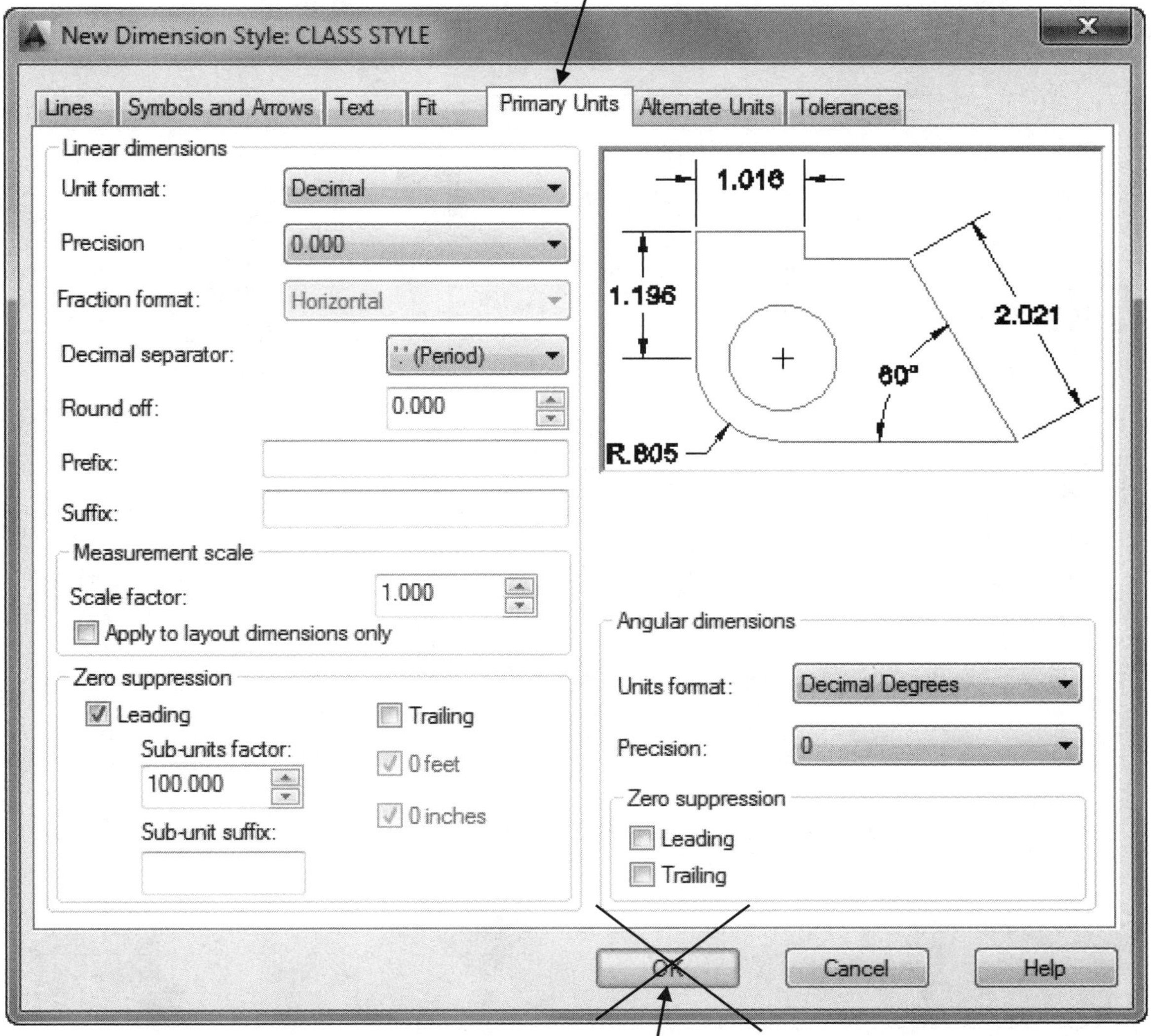

Do not select the OK button yet.

CREATING A NEW DIMENSION STYLE....continued

9. Select the **Lines** tab and change your settings to match the settings shown below.

9. Lines

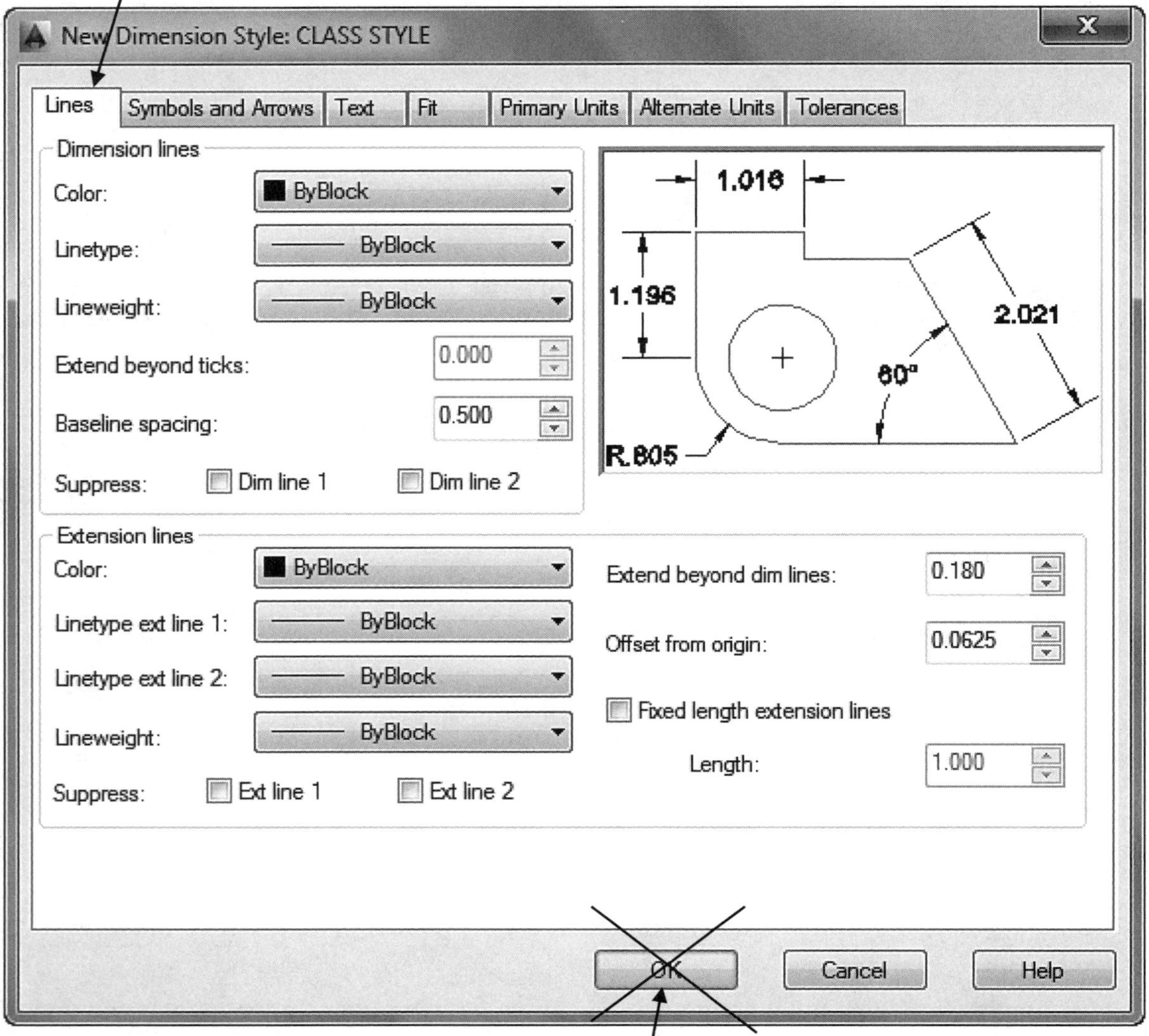

New Dimension Style: CLASS STYLE

Lines | Symbols and Arrows | Text | Fit | Primary Units | Alternate Units | Tolerances

Dimension lines

- Color: ByBlock
- Linetype: ByBlock
- Lineweight: ByBlock
- Extend beyond ticks: 0.000
- Baseline spacing: 0.500
- Suppress: Dim line 1 / Dim line 2

Extension lines

- Color: ByBlock
- Linetype ext line 1: ByBlock
- Linetype ext line 2: ByBlock
- Lineweight: ByBlock
- Suppress: Ext line 1 / Ext line 2
- Extend beyond dim lines: 0.180
- Offset from origin: 0.0625
- Fixed length extension lines
- Length: 1.000

1.016 1.196 2.021 60° R.805

OK | Cancel | Help

Do not select the OK button yet.

CREATING A NEW DIMENSION STYLE....continued

10. Select the **Symbols and Arrows** tab and change your setting to match the settings shown below.

10. Symbols and Arrows

New Dimension Style: CLASS STYLE

Lines | Symbols and Arrows | Text | Fit | Primary Units | Alternate Units | Tolerances

Arrowheads
First: Closed filled
Second: Closed filled
Leader: Closed filled
Arrow size: 0.125

Center marks
None
Mark
Line
0.090

Dimension Break
Break size: 0.125

1.016
1.196
2.021
60°
R.805

Arc length symbol
Preceding dimension text
Above dimension text
None

Radius jog dimension
Jog angle: 45

Linear jog dimension
Jog height factor: 1.500 * Text height

OK | Cancel | Help

Do not select the OK button yet.

CREATING A NEW DIMENSION STYLE....continued

11. Select the **Text** tab and change your settings to match the settings shown below.

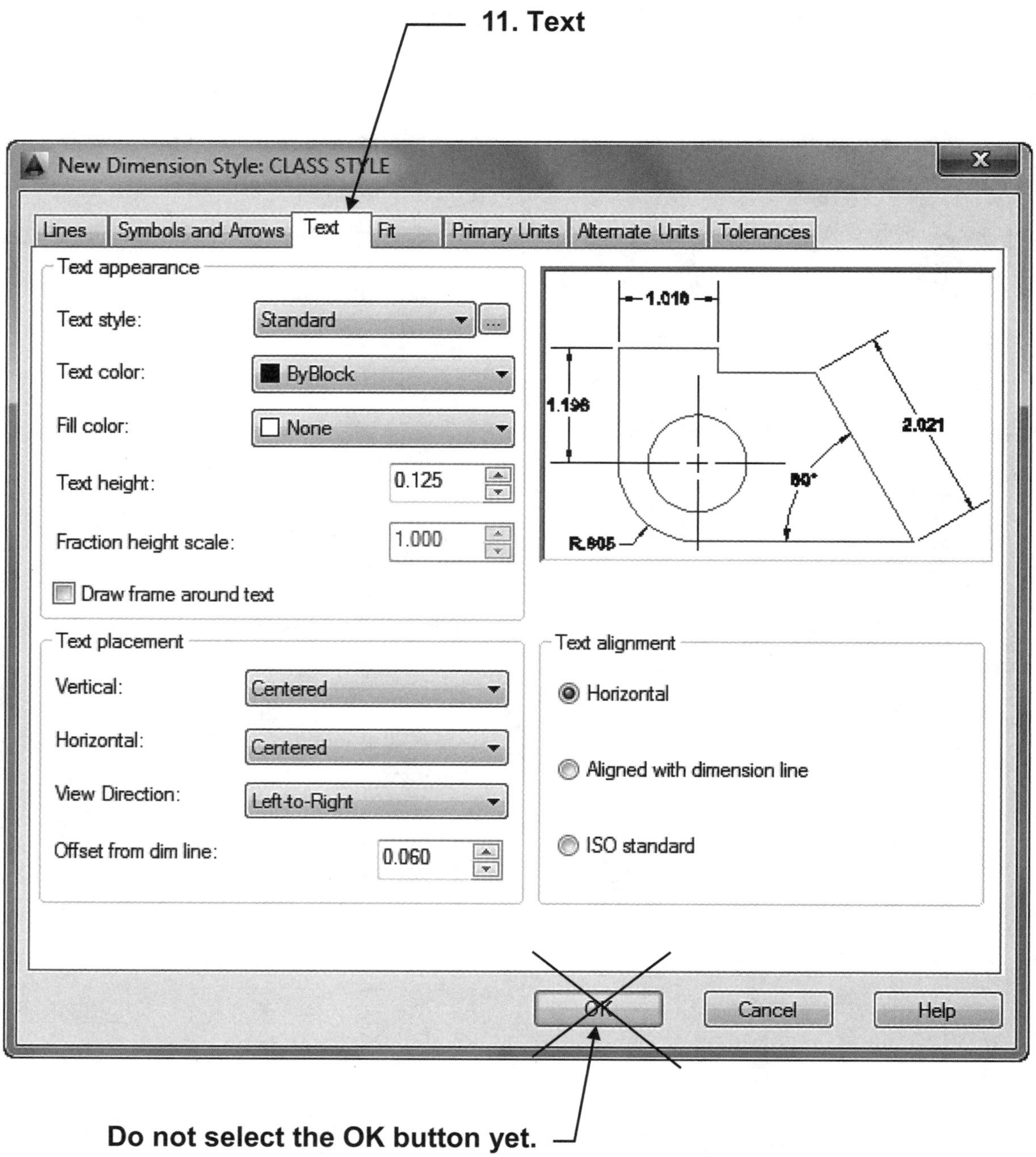

11. Text

Do not select the OK button yet.

CREATING A NEW DIMENSION STYLE....continued

12. Select the **Fit** tab and change your setting to match the settings shown below.

13. Now select the OK button.

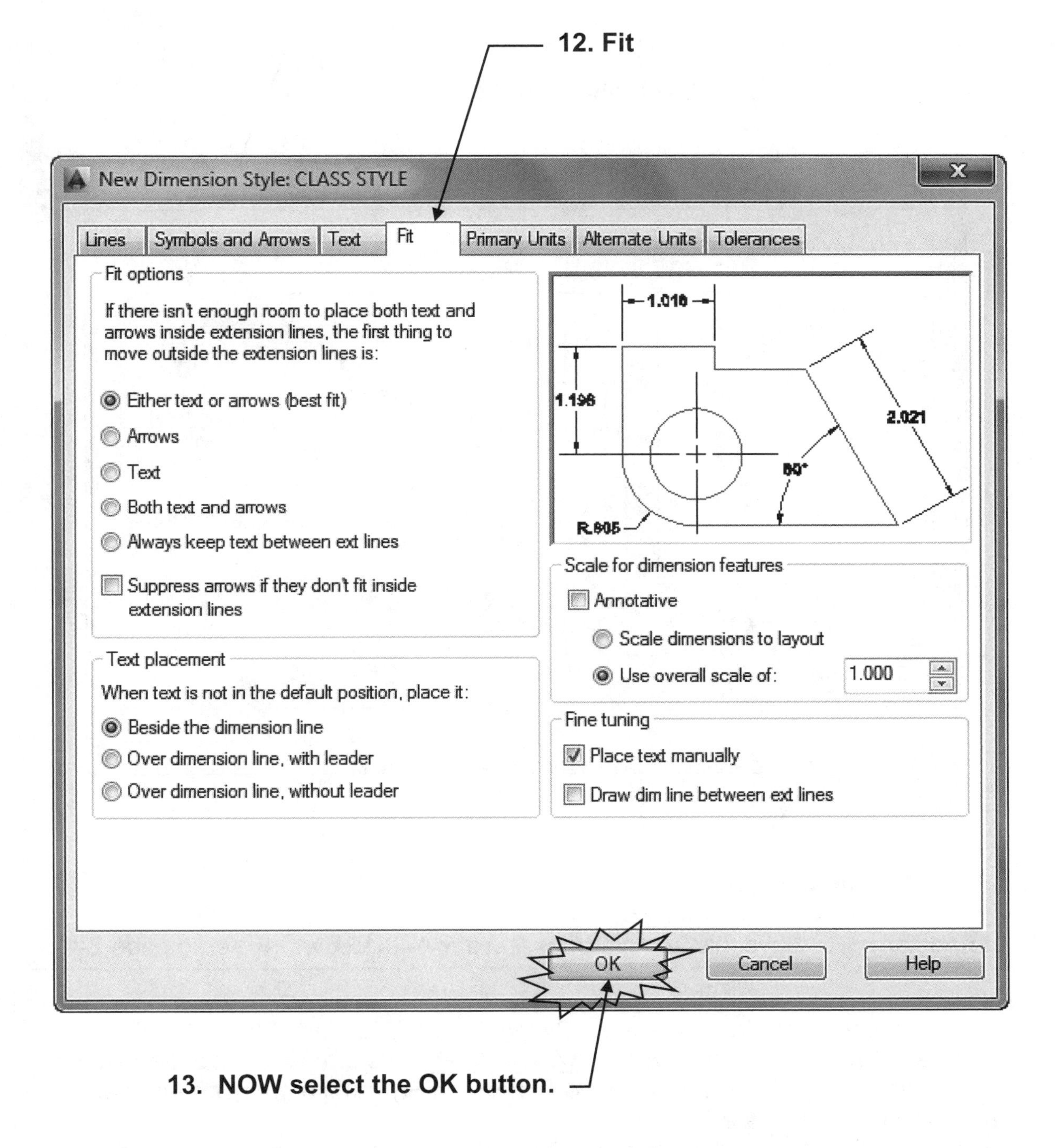

CREATING A NEW DIMENSION STYLE....continued

Your new style **"Class Style"** should be listed.

14. Select the **"Set Current"** button to make your new style "Class Style" the style that will be used.

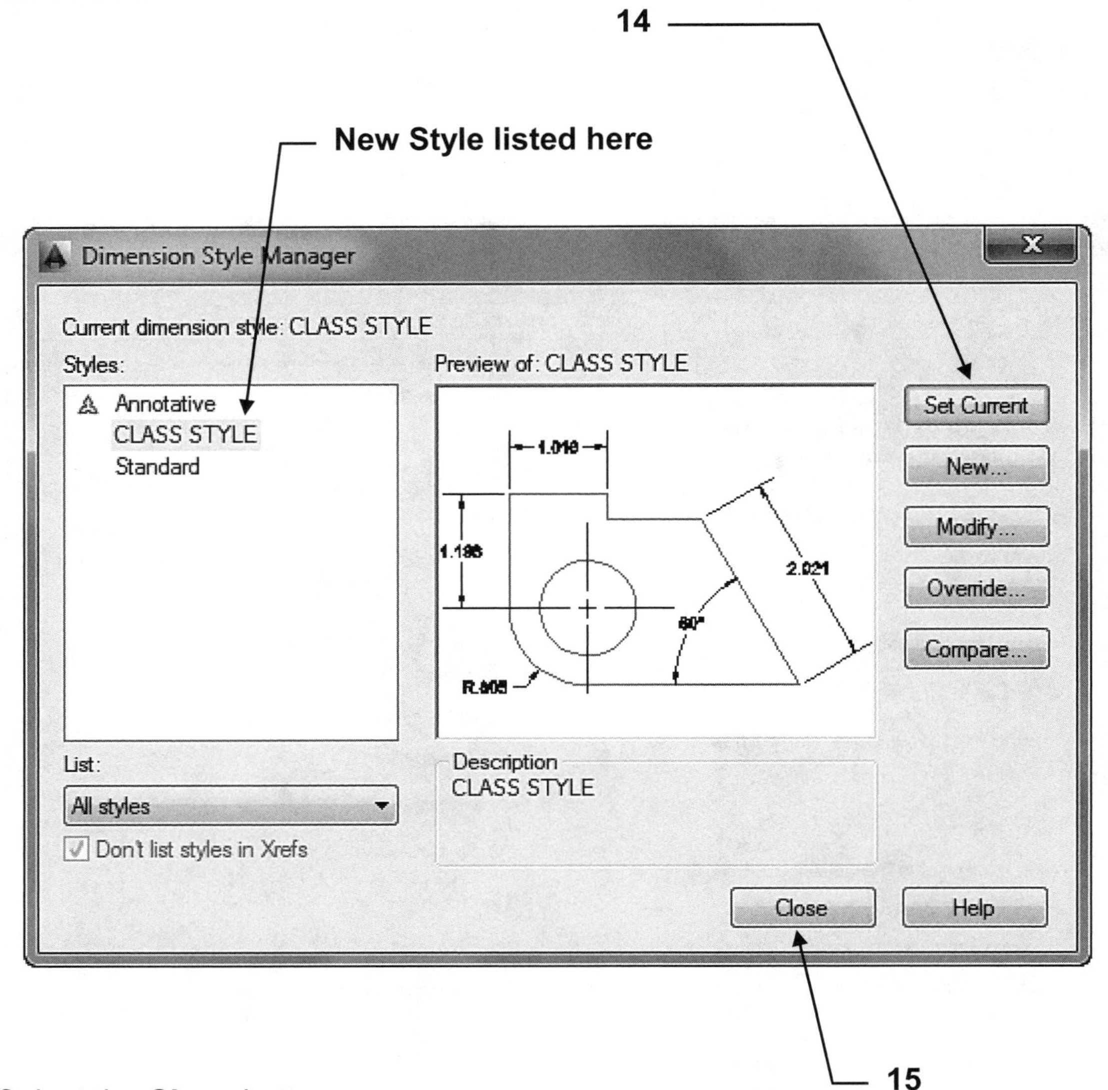

15. Select the **Close** button.

16. **Important:** Re-Save this "template" as **Border A-2015.dwt** (Refer to page 2-4)

> ***Note:***
> *You have successfully created a new "Dimension Style" called "**Class Style**".*
> *This style will be saved in your **Border A -2015** template after you save the template.*
> *So every time you use this template the Dimension Style will already be there and you will not have to create it again.*
> *It is important that you understand that this dimension style resides only in the **Border A-2015** template. If you open another drawing, this dimension style will not be there.*

IGNORING HATCH OBJECTS

Occasionally, when you are dimensioning an object that has "Hatch Lines", your cursor will snap to the Hatch Line instead of the object that you want to dimension.
To prevent this from occurring, select the option **"IGNORE HATCH OBJECTS"**.

EXAMPLE:

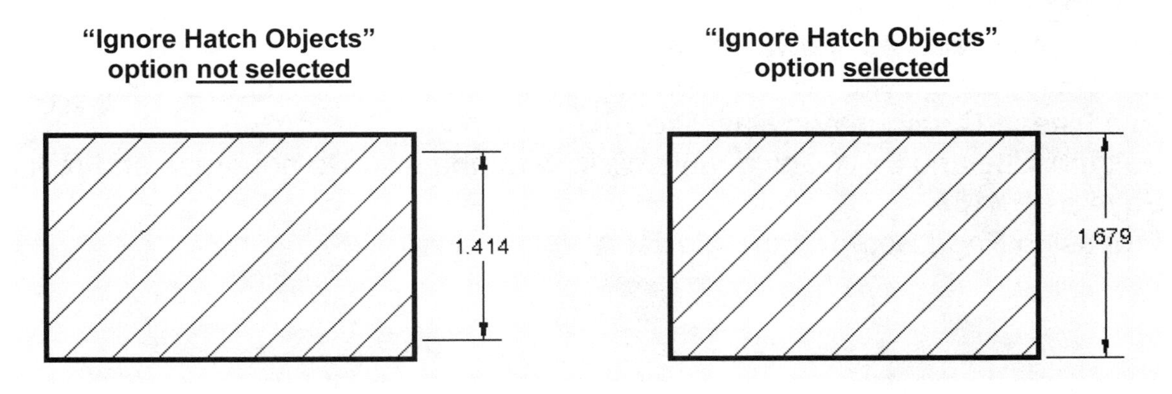

How to select the "IGNORE HATCH OBJECTS" option

1. Type **Options <enter>**
2. Select **Drafting** tab.
3. Check the **Ignore Hatch Objects** box.

EXERCISE 16A

INSTRUCTIONS:

1. Start a **New** file using **Border A-2015.dwt**
2. Create a Dimension Style in your Border A by following the instructions on pages 16-8 through 16-14. Re-Save the template as Border A.dwt before starting the drawing shown below.
3. Draw the Objects as shown using layer Object line.
4. Dimension as shown using Dimension Style "Class Style" on Layer Dimension.
5. Use Linear, Continue and Baseline.
6. Edit the Title and Ex-XX by double clicking on the text. Do not erase and replace.
7. Save as **EX16A**
8. Plot using Page Setup **Class Model A**

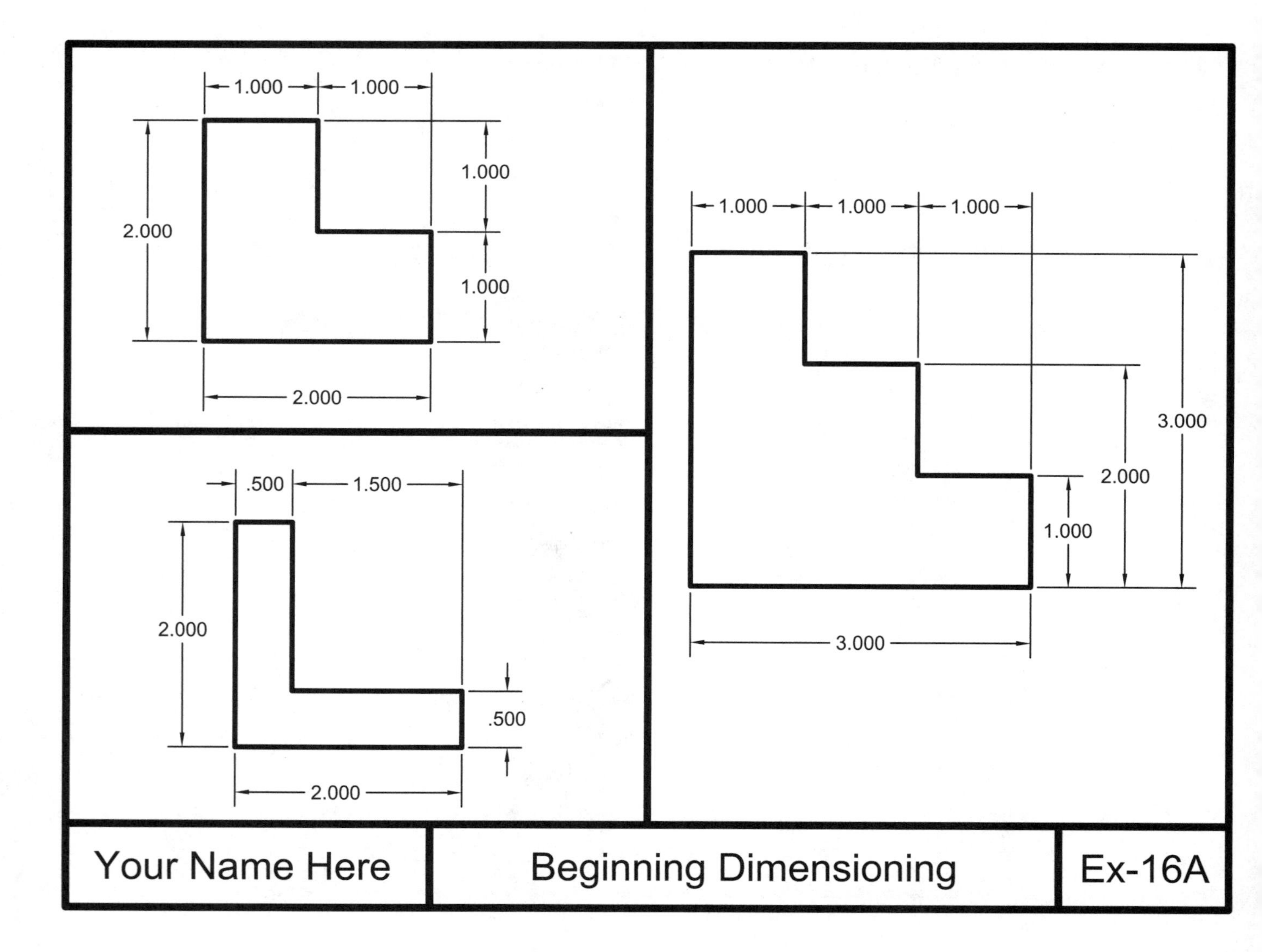

EXERCISE 16B

INSTRUCTIONS:

1. Start a **New** file using **Border A-2015.dwt**
2. Draw the Objects as shown using layer Object line.
3. Dimension as shown using Dimension Style "Class Style" on Layer Dimension.
4. Use Linear, Continue and Baseline.
5. Edit the Title and Ex-XX by double clicking on the text. Do not erase and replace.
6. Save as **EX16B**
7. Plot using Page Setup **Class Model A**

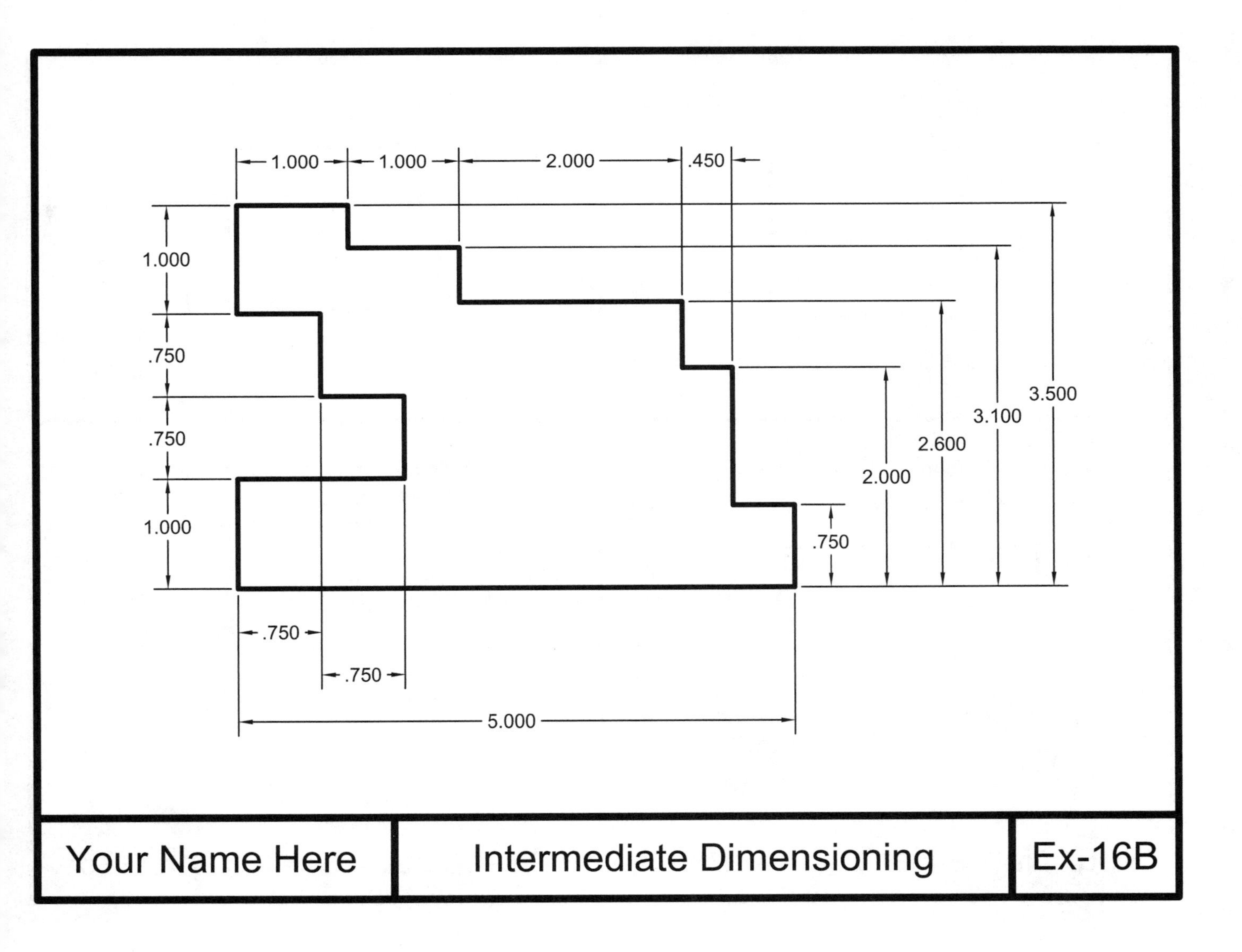

EXERCISE 16C

INSTRUCTIONS:

1. Start a **New** file using **Border A-2015.dwt**
2. Follow the instructions below.
3. Do not divide your drawing into 4 sections. Just draw one rectangle in the middle of your drawing area and follow instructions.
4. Edit the Title and Ex-XX by double clicking on the text. Do not erase and replace.
5. Save as **EX16C**
6. Plot using Page Setup **Class Model A**

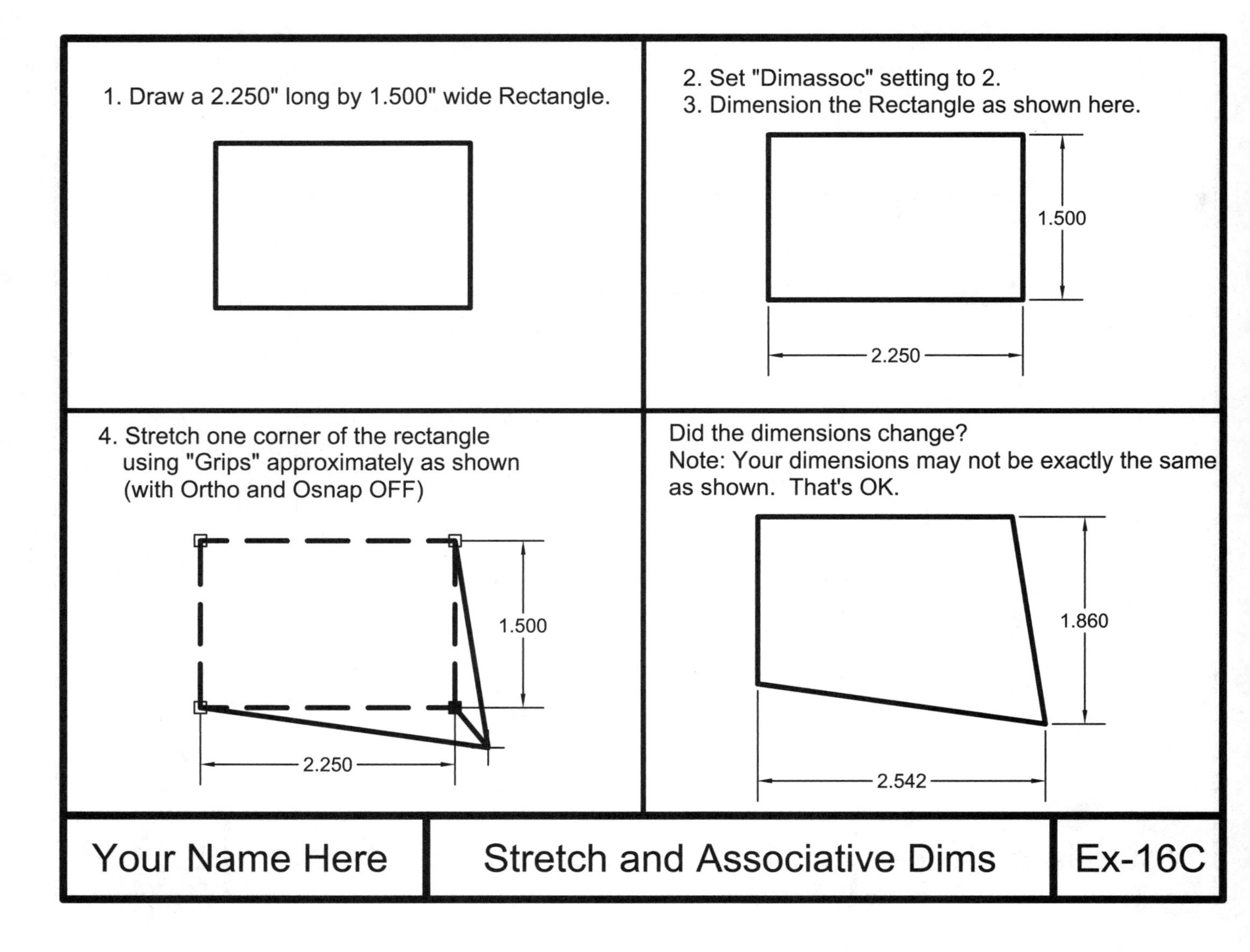

EXERCISE 16D

INSTRUCTIONS:

1. Start a **New** file using **Border A-2015.dwt**
2. Follow the instructions below.
3. Do not divide your drawing into 4 sections. Just draw one rectangle in the middle of your drawing area and follow instructions.
4. Edit the Title and Ex-XX by double clicking on the text. Do not erase and replace.
5. Save as **EX16D**
6. Plot using Page Setup **Class Model A**

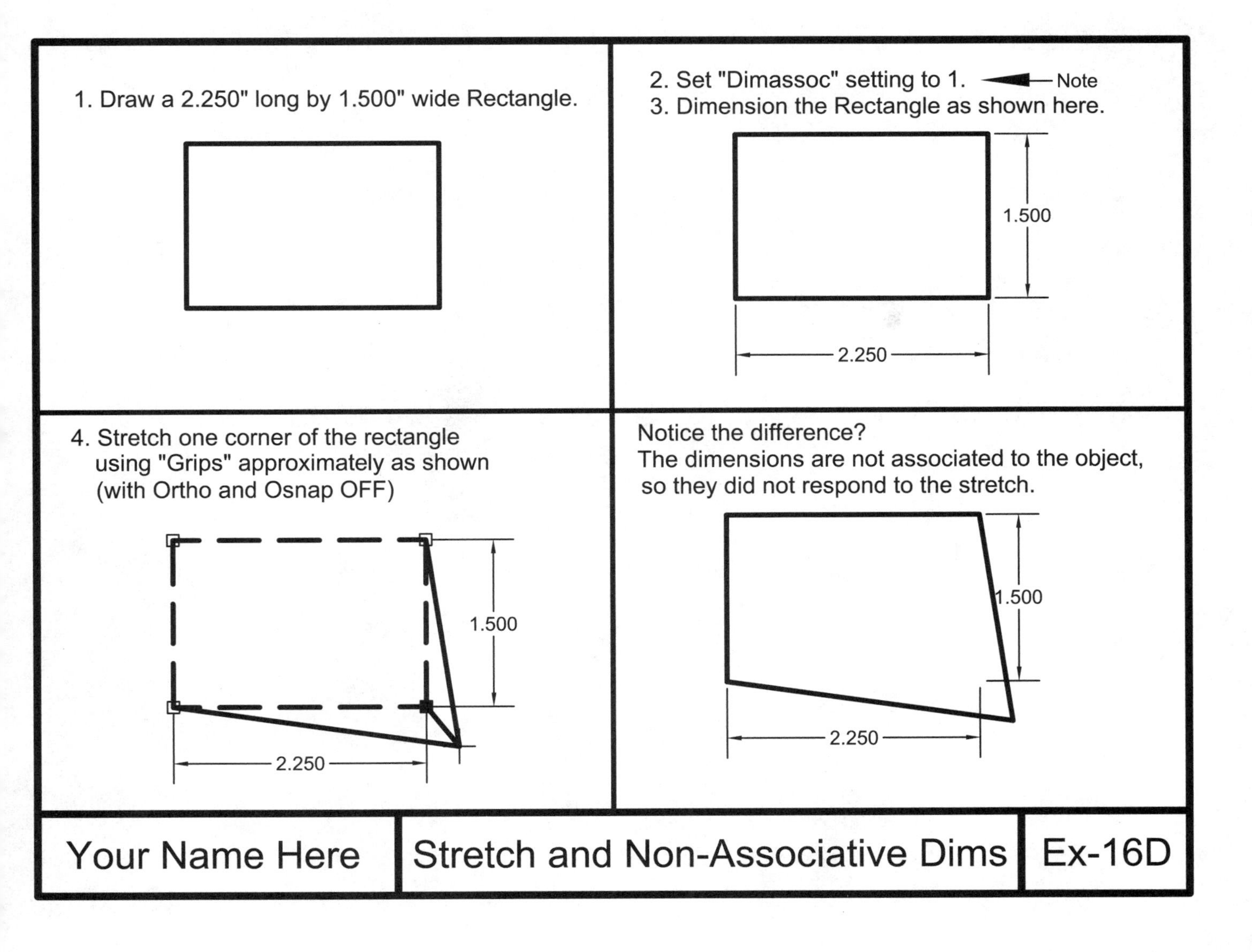

Notes:

LEARNING OBJECTIVES

After completing this lesson, you will be able to:

1. Edit the dimension text without changing the value
2. Move the dimension text within the dimension lines
3. Modify an entire dimension style
4. Override a dimension style
5. Use the Properties Palette to change a dimension
6. Break intersecting extension and dimension lines
7. Add a Jog to a dimension line
8. Adjust the distance between dimensions

LESSON 17

EDITING DIMENSION TEXT VALUES

Sometimes you need to modify the dimension text value. You may add a symbol, a note or even change the text of an existing dimension.
The following describes 2 methods.

Example: Add the word "**Max.**" to the existing dimension value text.

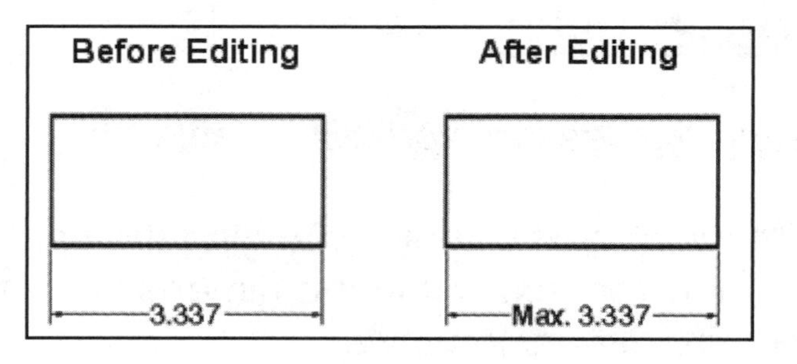

Method 1 (Quick Properties Palette).

1. **Quick Properties** status bar button must be **ON**.
2. Select the dimension you want to override.
3. Scroll down to **Text override**
   (Notice that the actual measurement is directly above it.)
4. Type the new text (**Max**) and **< >** and press <enter>
   (**< >** represents the associative text value 3.337)

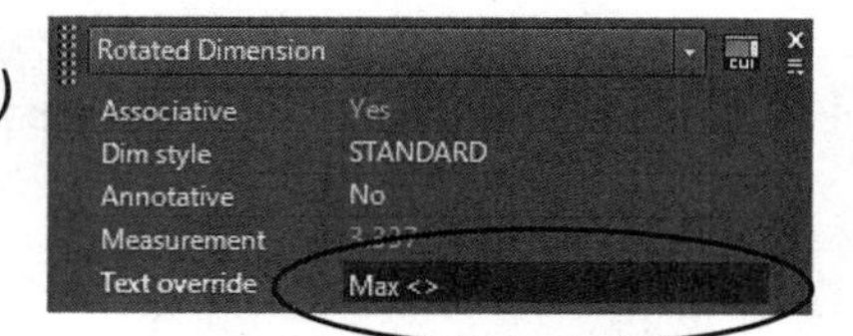

Important:
If you do not use < > the dimension will no longer be Associative.

Method 2 (Text Edit).

1. Type on the Command line: **ed** <enter> (This is the **text edit** command)
2. Select the dimension you want to edit.

Associative Dimension

If the dimension is Associative the dimension text will appear highlighted.

3.337 **Before**

You may add text in front or behind the dimension text and it will remain Associative.
Be careful not to disturb the dimension value text.

After

Non Associative or Exploded Dimension

If the dimension value has been changed or exploded it will appear with a gray background and is not Associative.

4. Make the change.
5. Select the **OK** button.

EDITING THE DIMENSION POSITION

Grips are great tools for repositioning dimensions.
Grips are small, solid-filled squares that are displayed at strategic points on objects. You can drag these grips to stretch, move, rotate, scale, or mirror objects quickly.
Grips may be turned off by typing “grips” <enter> then 0 <enter> .

<u>HOW TO USE GRIPS</u>

1. Select the object (no command can be in use while using grips)

2. Select one of the **blue** grips. It will turn to “**red**”. This indicates that grip is “**hot**”.
   The “**Hot**” grip is the **basepoint**.

3. Move the hot grip to the new location.

4. <u>After editing you must press the **ESC** key to de-activate the grips.</u>

The following is an example of how to use grips to quickly reposition dimensions.

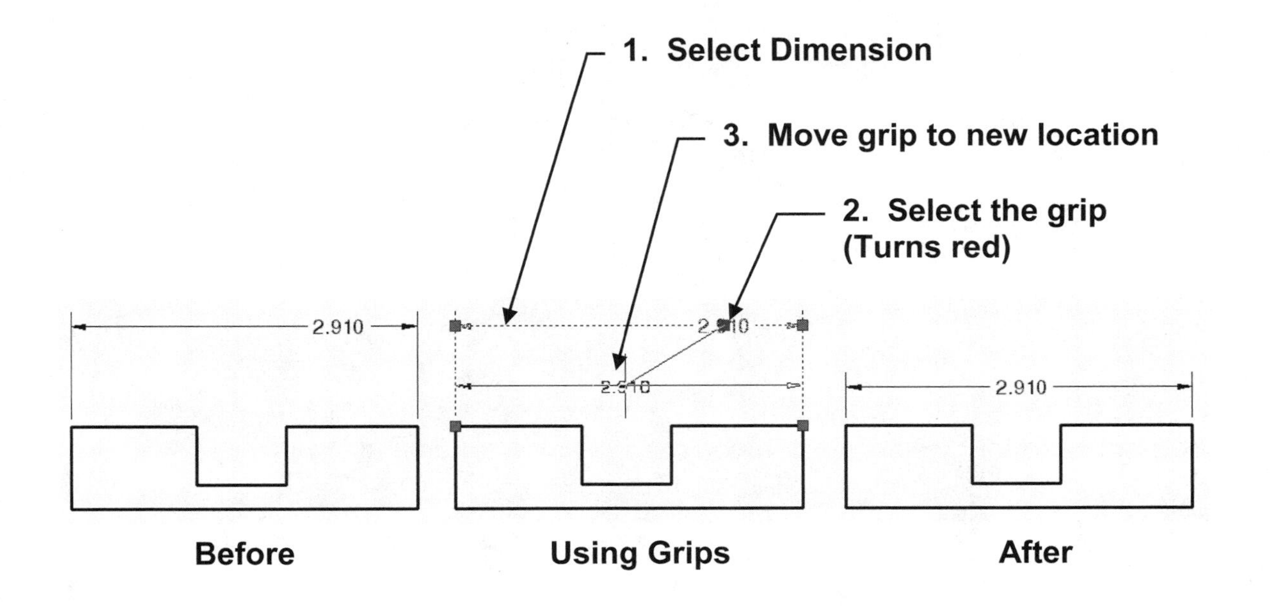

EDITING THE DIMENSION

ADDITIONAL EDITING OPTIONS USING THE SHORTCUT MENU

1. Select the dimension that you want to change.
2. Place the cursor on one of the grips shown below. (Do not press mouse button)
3. Select an option from the short cut menu that appears.

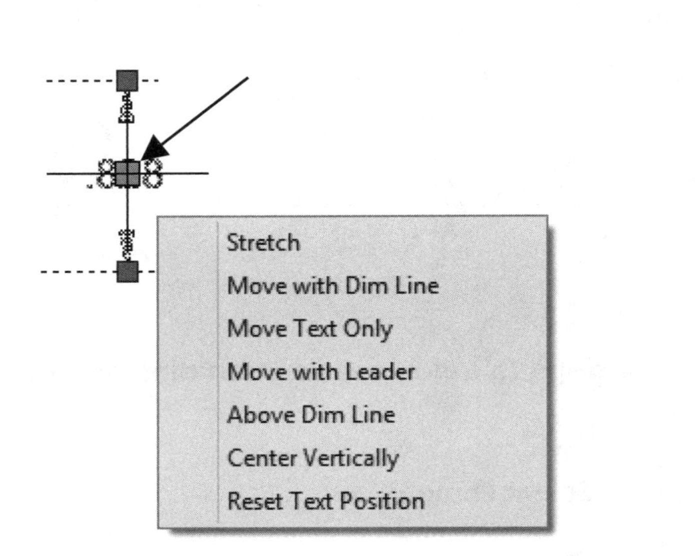

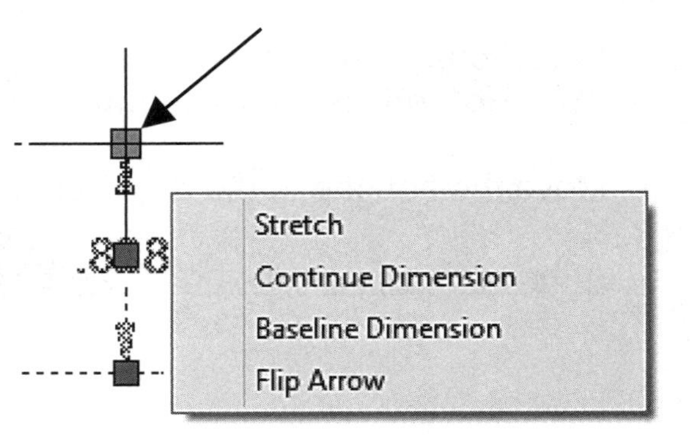

MODIFY AN ENTIRE DIMENSION STYLE

After you have created a Dimension Style, you may find that you have changed your mind about some of the settings. You can easily change the entire Style by using the "Modify" button in the Dimension style Manager dialog box. This will not only change the Style for future use, but it will also update dimensions already in the drawing.

Note: if you do not want to update the dimensions already in the drawing, but want to make a change to a new dimension, refer to Override, page 17-6

1. Select the Dimension Style Manager. (Refer to page 16-8)

2. Select the Dimension Style that you wish to modify.

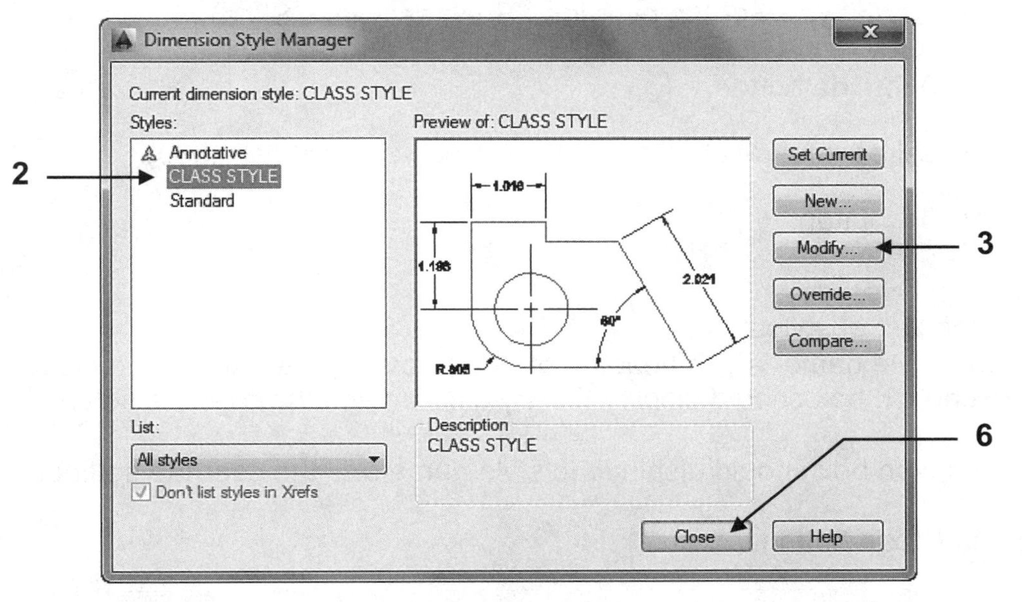

3. Select the **Modify button** from the Dimension Style Manager dialog box.

4. Make the desired changes to the settings.

5. Select the **OK** button.

6. Select the **Close** button.

Now look at your drawing. Have your dimensions updated?

Note:
The method above will not change dimensions that have previously been modified or exploded.

Note: If some of the dimensions have not changed:
1. Type: **-Dimstyle <enter> (notice the (-)dash in front of "dimstyle")**
2. Type: **A <enter>**
3. Select dimensions to update and then <enter>
(Sometimes you have to give them a little nudge.)

OVERRIDE A DIMENSION STYLE

A dimension Override is a **temporary** change to the dimension settings.
An override **will not affect existing dimensions**. It will **only** affect **new dimensions**.
Use this option when you want a **new** dimension just a little bit different but you don't want to create a whole new dimension style and you don't want the existing dimensions to change either.

For example, if you want the new dimension to have a text height of .500 but you want the existing text to remain at .125 ht.

1. Select the **Dimension Style Manager.** (Refer to page 16-8)

2. Select the "**Style**" you want to override. (Such as: Class Style)

3. Select the **Override** button.

4. Make the desired changes to the settings. (Such as: Text ht = .500)

5. Select the **OK** button.

6. Confirm the Override
   Look at the List of styles.
   Under the Style name, a sub heading of **<style overrides>** should be displayed.
   The description box should display the style name and the override settings.

7. The description box should display the style name and the override settings.

8. Select the **Close** button.

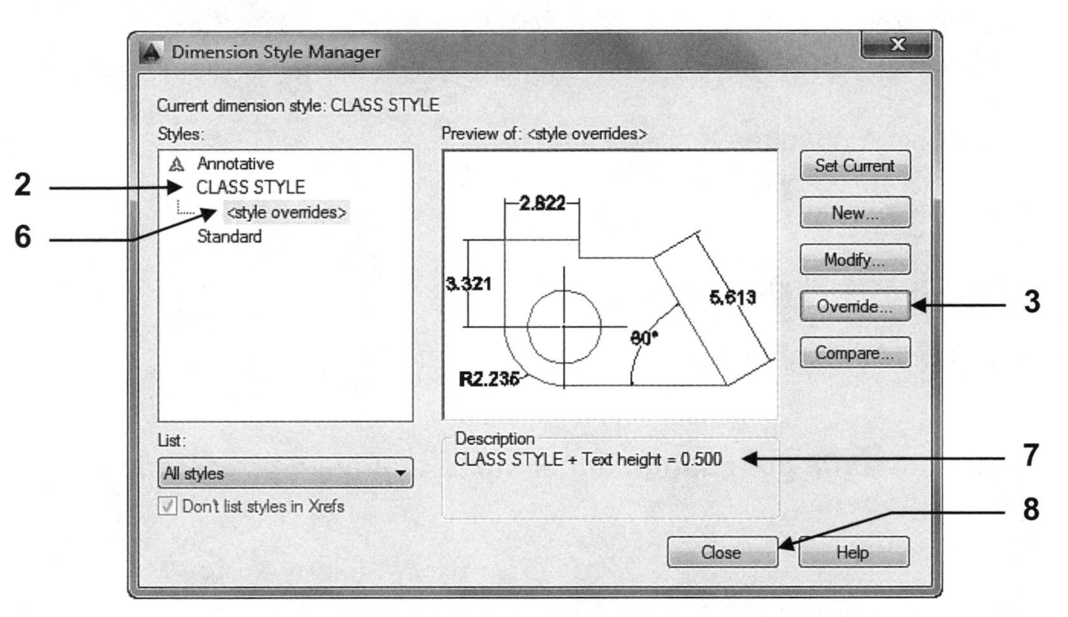

When you want to return to style **Class Style**, select **Class Style** from the styles list then select the **Set Current** button. Each time you select a different style, you must select the **Set Current** button to activate it.

EDIT AN INDIVIDUAL EXISTING DIMENSION

Sometimes you would like to modify the settings of an **individual existing** dimension. This can be achieved using the **Properties palette**.

1. Open the **Properties Palette**. (Refer to page 12-4)
2. Select the dimension to change.
3. Select and change the desired settings.

   Example: Change the dimension text height to .500.

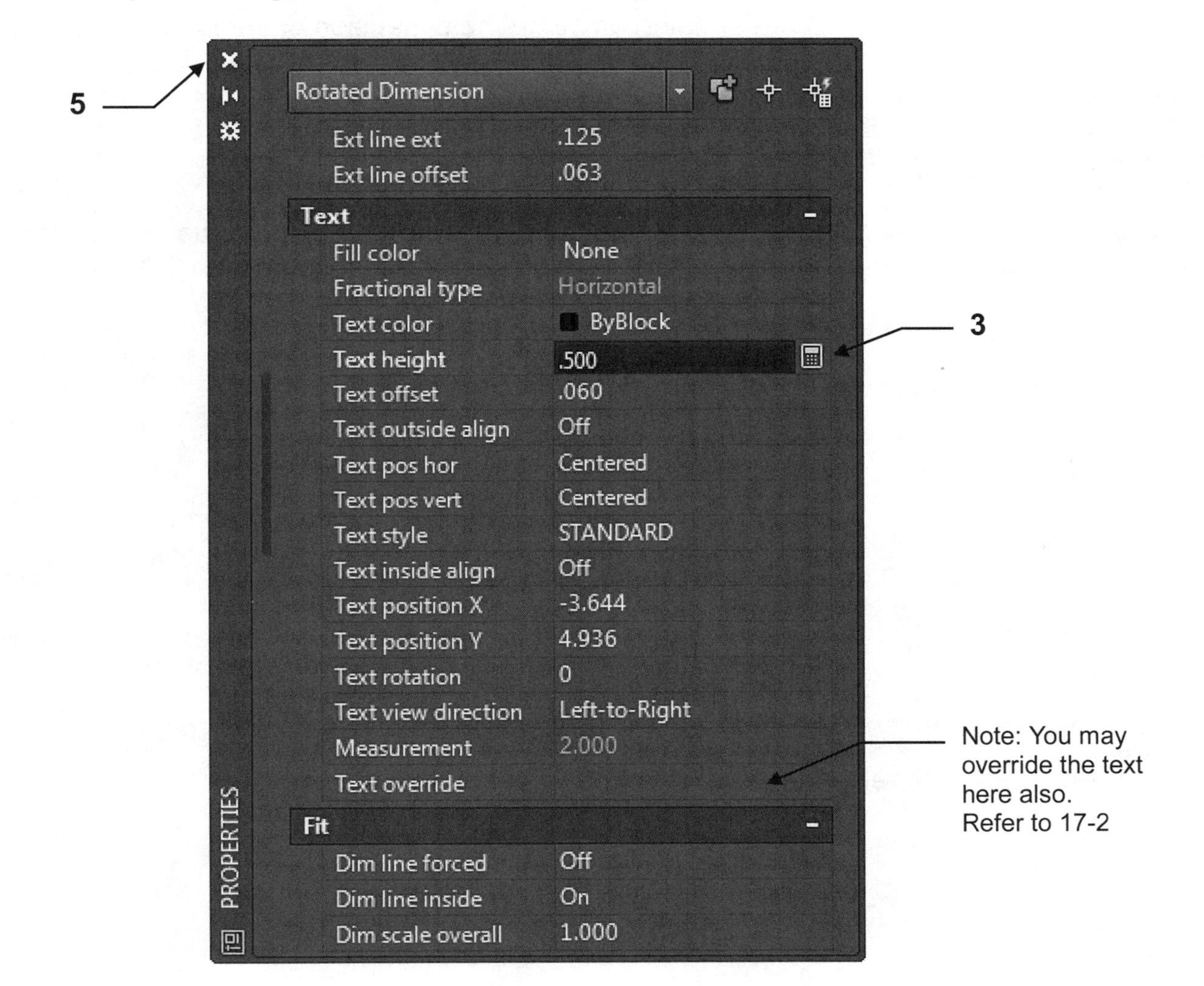

4. Press <enter> . *(The dimension should have changed)*
5. Close the **Properties Palette**.
6. Press the <esc> key to de-activate the grips.

Note: The dimension will remain Associative.

DIMENSION BREAKS

Occasionally extension lines overlap another extension line or even an object. If you do not like this you may use the **Dimbreak** command to break the intersecting lines. **Automatic** (described below) or **Manual** (described on page 17-9) method may be used. You may use the **Remove** (described on page 17-10) option to remove the break.

AUTOMATIC DIMENSION BREAKS

To create an automatically placed dimension break, you select a dimension and then use the Auto option of the DIMBREAK command.

Automatic dimension breaks are updated any time the dimension or intersecting objects are modified.

You control the size of automatically placed dimension breaks on the Symbols and Arrows tab of the Dimension Style dialog box.

The specified size is affected by the dimension break size, dimension scale, and current annotation scale for the current viewport. (Annotation discussed in Lesson 27)

1. Select the Dimbreak command using one of the following:

    Ribbon = Annotate tab / Dimension panel /
    or
    Keyboard = Dimbreak <enter>

    Command: _Dimbreak

2. Select dimension to add / remove break or [Multiple]: ***type M <enter>***
3. Select dimensions: ***select dimension***
4. Select dimensions: ***select another dimension***
5. Select dimensions: ***select another dimension or <enter> to stop selecting***
6. Select object to break dimension or [Auto / Remove] <Auto>: ***<enter>***

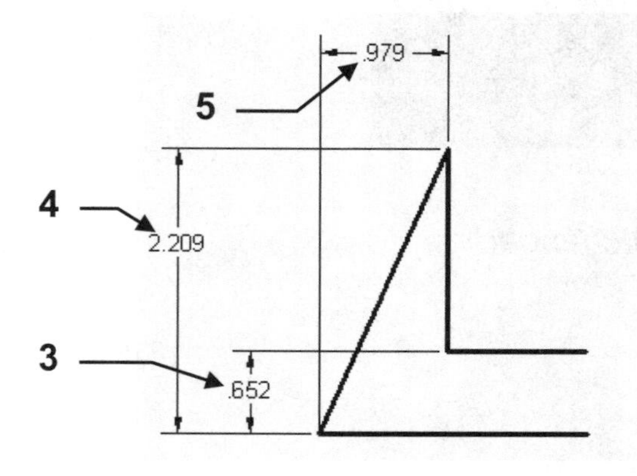

Before Dimbreak

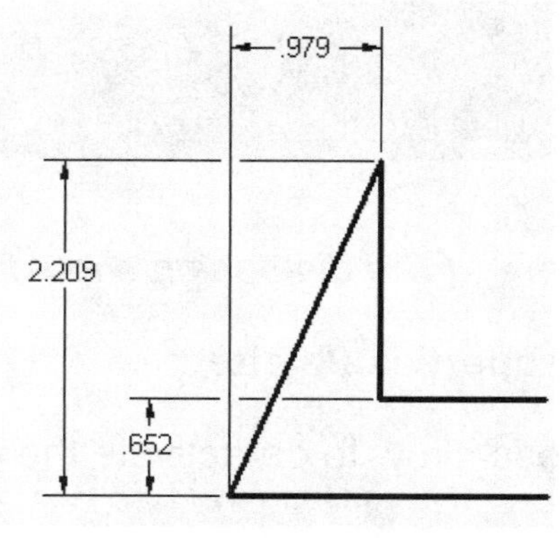

After Dimbreak

MANUAL DIMENSION BREAK

You can place a dimension break by **picking two points** on the dimension, extension, or leader line to determine the size and placement of the break.

Dimension breaks that are added manually by picking two points are not automatically updated if the dimension or intersecting object is modified. So if a dimension with a manually added dimension break is moved or the intersecting object is modified, you might have to restore the dimension and then add the dimension break again.

The size of a dimension break that is created by picking two points is not affected by the current dimension scale or annotation scale value for the current viewport.
(You will learn Annotation in Lesson 27)

1. Select the Dimbreak command using one of the following:

 Ribbon = Annotate tab / Dimension panel /
 or
 Keyboard = Dimbreak <enter>

Command: _Dimbreak
2. Select dimension to add / remove break or [Multiple]: ***select the dimension***
3. Select object to break dimension or [Auto/Manual/Remove] <Auto>: ***type M <enter>***
4. Specify first break point: ***select the first break point location*** (Osnap should be off)
5. Specify second break point: ***select the second break point location***

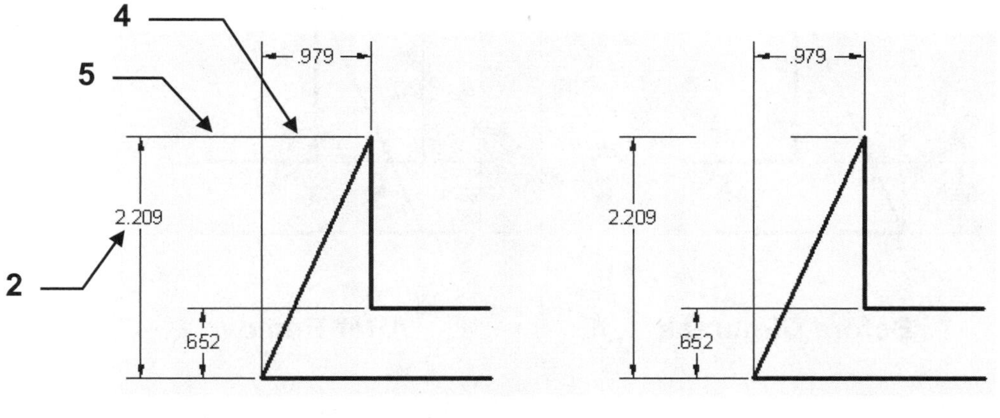

Before Dimbreak **After Dimbreak**

The following objects can be used as cutting edges when adding a dimension break: Dimension, Leader, Line, Circle, Arc, Spline, Ellipse, Polyline, Text, and Multiline text.

DIMENSION BREAK....continued

REMOVE THE BREAK

Removing the break is easy using the Remove option.

1. Select the Dimbreak command using one of the following:

    Ribbon = Annotate tab / Dimension panel /
    or
    Keyboard = Dimbreak <enter>

    Command: _DIMBREAK

2. Select dimension to add / remove break or [Multiple]: ***type M <enter>***

3. Select dimensions: ***select a dimension***

4. Select dimensions: ***select a dimension***

5. Select dimensions: ***select another dimension or <enter> to stop selecting***

6. Select object to break dimension or [Auto/Remove] <Auto>: ***type R <enter>***

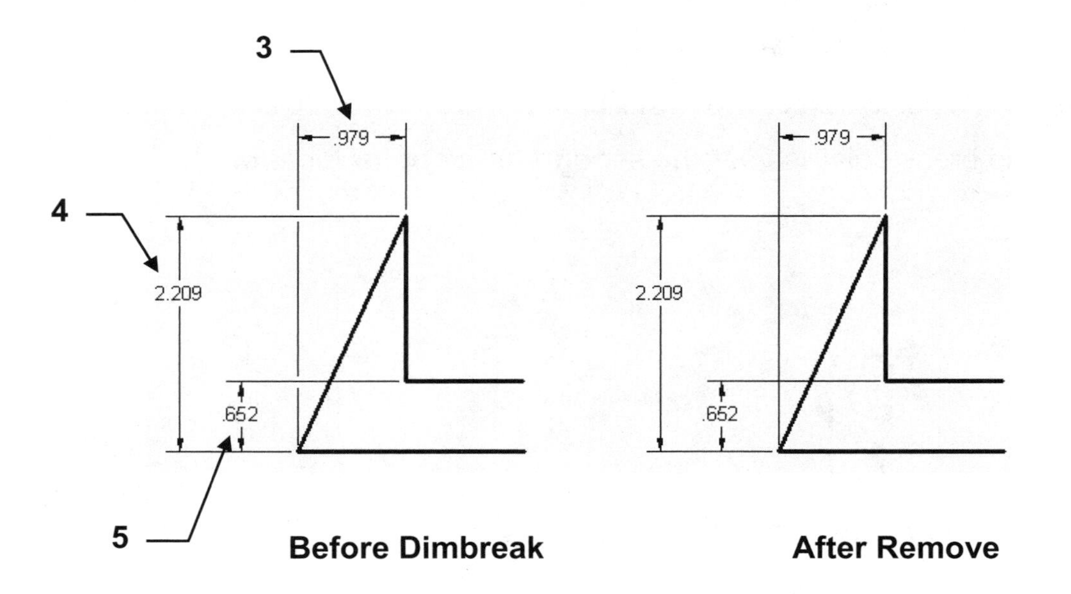

Before Dimbreak **After Remove**

JOG A DIMENSION LINE

Jog lines can be added to linear dimensions. Jog lines are used to represent a dimension value that does not display the actual measurement.

Before you add a jog, the Jog angle and the height factor of the jog should be set in the Symbols & Arrows tab within the Dimension Style.

The height is calculated as a factor of the Text ht.

For example:
if the text height was .250 and the jog height factor was 1.500, the jog would be .375
Formula: (.250 ht. X 1.500 jog ht. factor= .375).

Radius jog dimension
Jog angle: 45
Linear jog dimension
Jog height factor:
1.500 * Text height

1. Select the Jogged Linear command using one of the following:

   Ribbon = Annotate tab / Dimension Panel /
   or
   Keyboard = Dimjogline <enter>

   Command: _DIMJOGLINE
2. Select dimension to add jog or [Remove]: ***Select a dimension***
3. Specify jog location (or press ENTER): ***Press <enter>***
4. After you have added the jog you can re-position it by using grips and adjust the height of the jog symbol using the Properties palette.

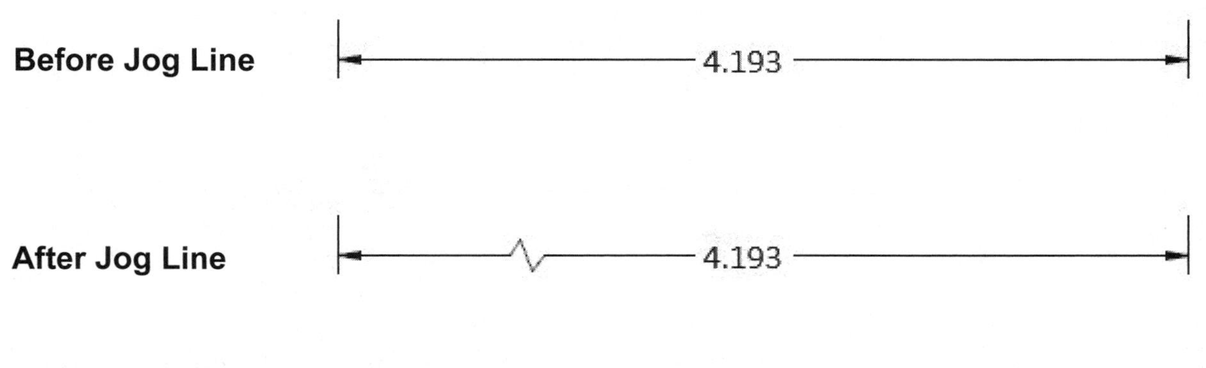

REMOVE A JOG
1. Select the Jogged Linear command
   Command: _DIMJOGLINE
2. Select dimension to add jog or [Remove]: ***type R <enter>***
3. Select jog to remove: ***select the dimension***

ADJUST DISTANCE BETWEEN DIMENSIONS

The Adjust Space command allows you to adjust the distance between existing parallel linear and angular dimensions, so they are equally spaced. You may also align the dimensions to create a string.

1. Select the Dimension Space command using one of the following:

   Ribbon = Annotate tab / Dimension panel /
   or
   Keyboard = Dimspace <enter>

   Command: _DIMSPACE
2. Select base dimension: ***Select the dimension that you want to use as the base dimension when equally spacing dimensions. (P1)***
3. Select dimensions to space: ***select the next dimension to be spaced. (P2)***
4. Select dimensions to space: ***select the next dimension to be spaced. (P3)***
5. Select dimensions to space: ***continue selecting or press <enter> to stop***
6. Enter value or [Auto] <Auto>: ***enter a value or press <enter> for auto***

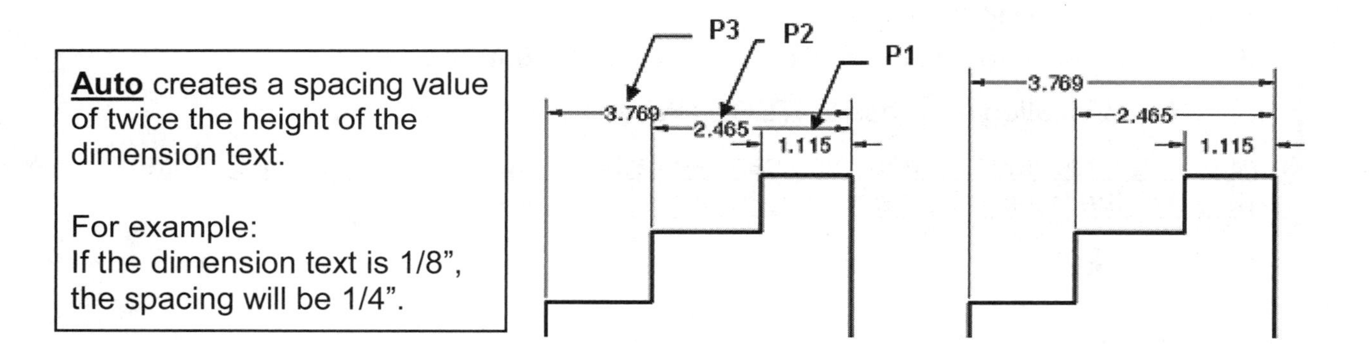

Aligning dimensions

Follow the steps shown above but when asked for the value enter **"0"** <enter>.

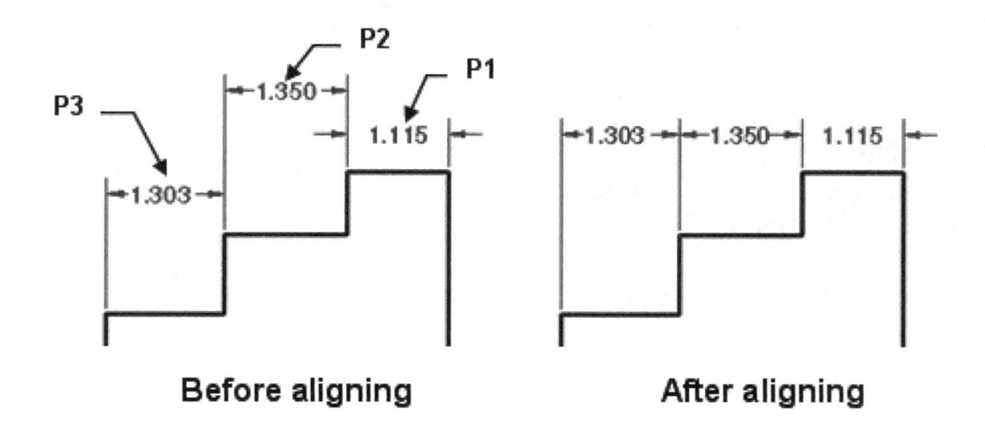

Before aligning **After aligning**

EXERCISE 17A

INSTRUCTIONS:

1. Start a **New** file using **Border A-2015.dwt**
2. Draw the 6" x 4" Rectangle shown below. Use layer Object line.
3. **Upper & Right Dimension:** Use dimension style Class Style. Use Layer Dimension
4. **Lower & Left Dimension:** Use Override. Change the text height setting to .500.
   If the upper & right dimension changed also, you did not use "Override". Try again.
5. Edit the Title and Ex-XX by double clicking on the text. Do not erase and replace.
6. Save as **EX17A**
7. Plot using Page Setup **Class Model A**

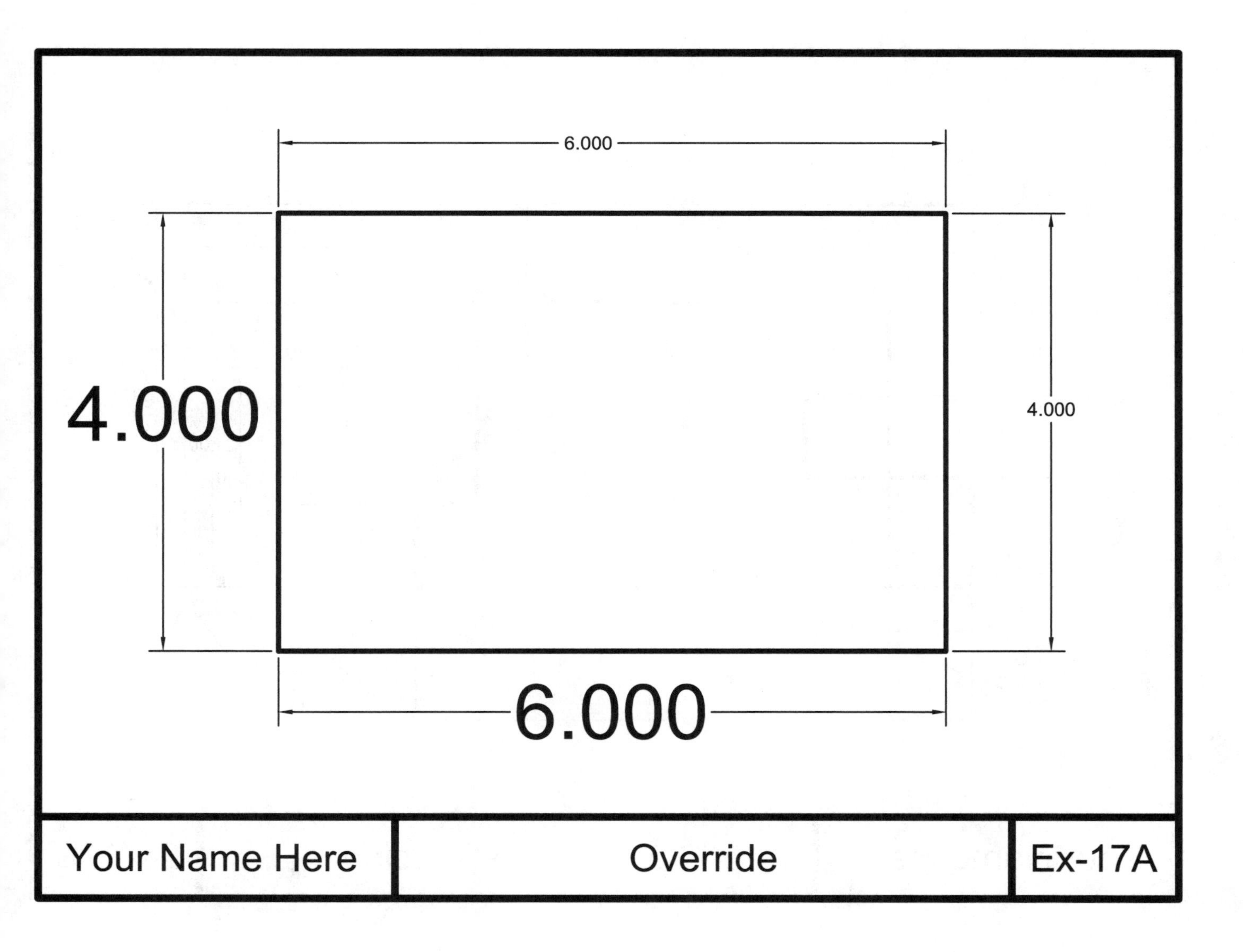

EXERCISE 17B

INSTRUCTIONS:

1. Open **EX16B**
2. Re-space the baseline dimensions on the right side. Use Auto for spacing.
3. Align the (2) .750 dimensions.
4. Edit the Title and Ex-XX by double clicking on the text. Do not erase and replace.
5. Save as **EX17B**
6. Plot using Page Setup **Class Model A**

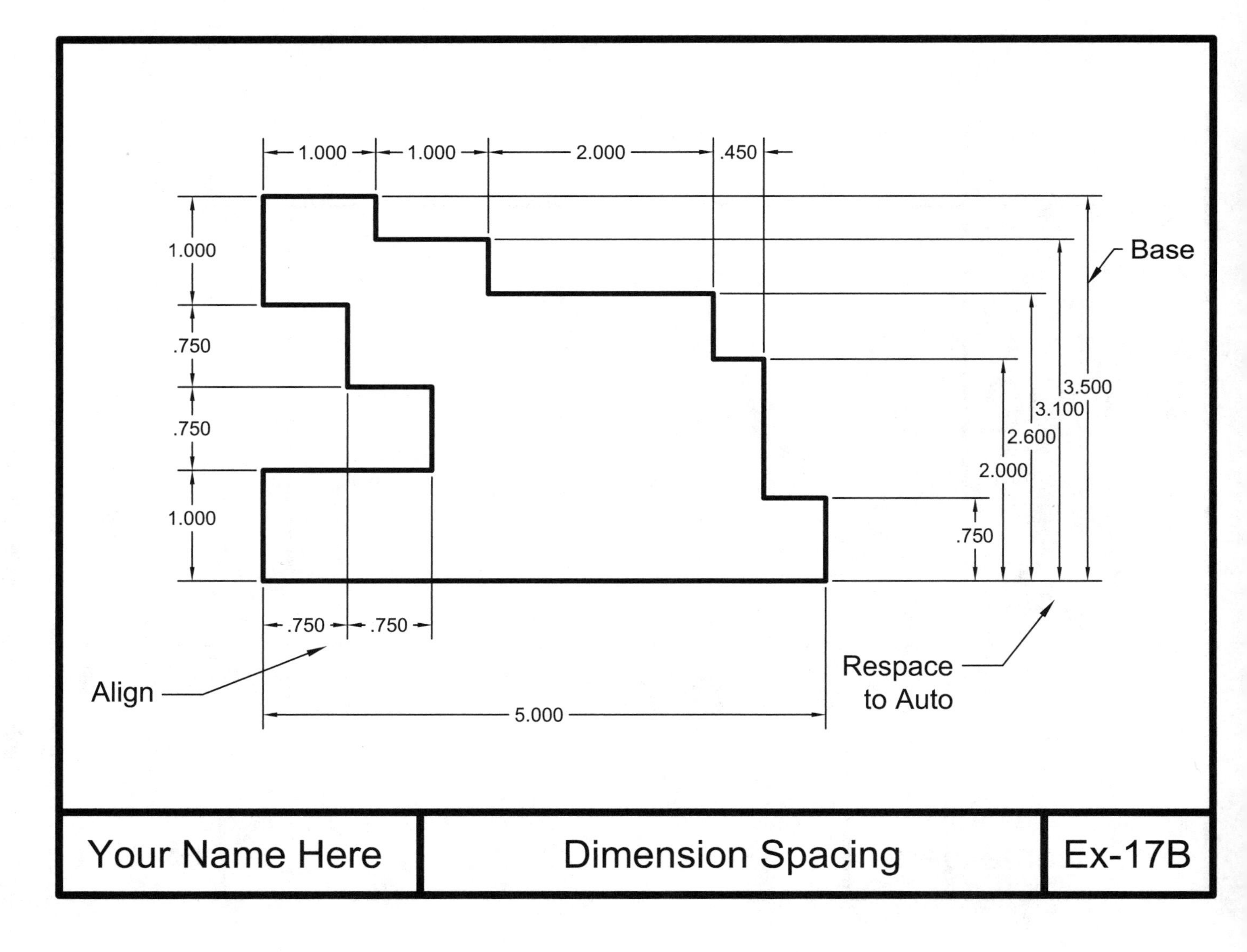

EXERCISE 17C

INSTRUCTIONS:

1. Open **EX17B**
2. Make the changes shown below. Use Grips and the Properties Palette.
3. Notice: Not all dimensions change, so you can't make changes to the Dim. Style.
4. Edit the Title and Ex-XX by double clicking on the text. Do not erase and replace.
5. Save as **EX17C**
6. Plot using Page Setup **Class Model A**

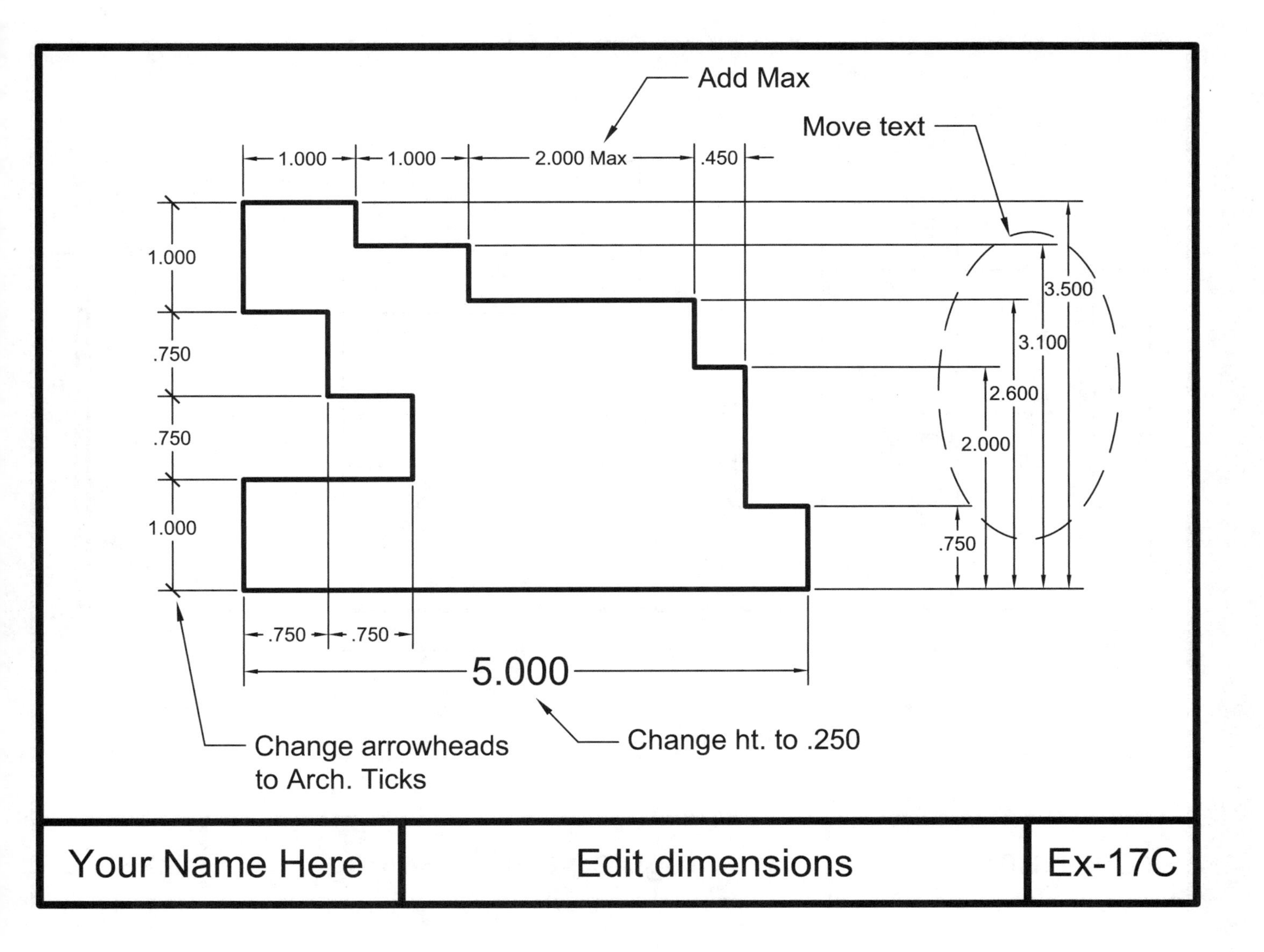

EXERCISE 17D

INSTRUCTIONS:

1. Start a **New** file using **Border A-2015.dwt**
2. Draw the objects shown below.
3. Dimension as shown. Be very careful because there is just enough drawing area.
4. Use Dimension Break to break the extension lines as shown.
5. Use Dimension Jog to create a jog in the lower dimension.
6. Edit the Title and Ex-XX by double clicking on the text. Do not erase and replace.
7. Save as **EX17D**
8. Plot using Page Setup **Class Model A**

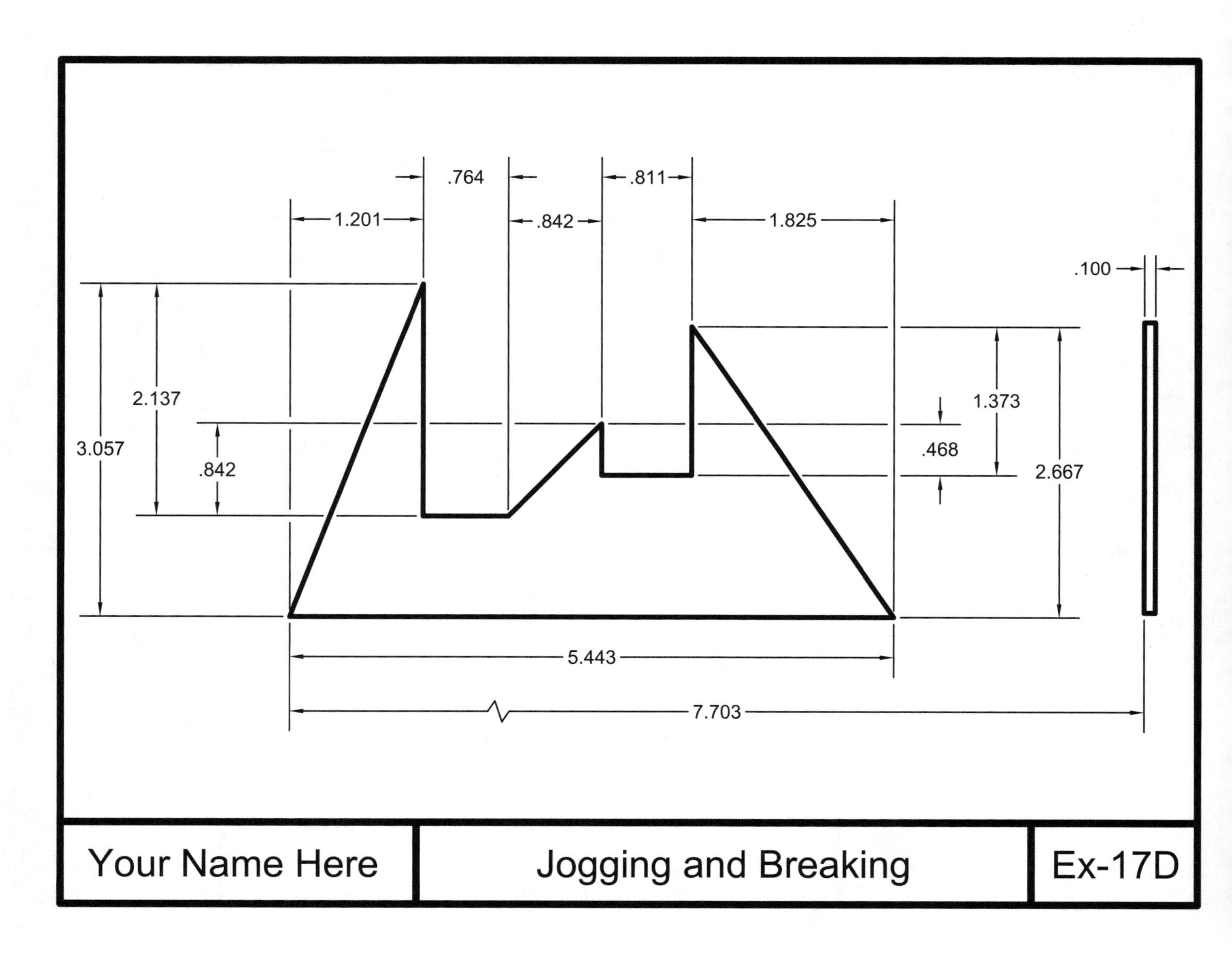

LEARNING OBJECTIVES

After completing this lesson, you will be able to:

1. Add Diameter, Radius and Angular dimensions to a drawing
2. Draw and control Center Marks
3. Flip the direction of an arrowhead
4. Understand the need for Sub-styles
5. Create a sub-style

LESSON 18

DIMENSIONING DIAMETERS

The **DIAMETER** dimensioning command should be used when dimensioning circles and arcs of <u>more than 180 degrees</u>. AutoCAD measures the selected circle or arc and displays the dimension text with the diameter symbol (Ø) in front of it.

1. Select the Diameter command using one of the following:

   Ribbon = Annotate tab / Dimension panel / ▼
   or
   Keyboard = Dimdiameter <enter>

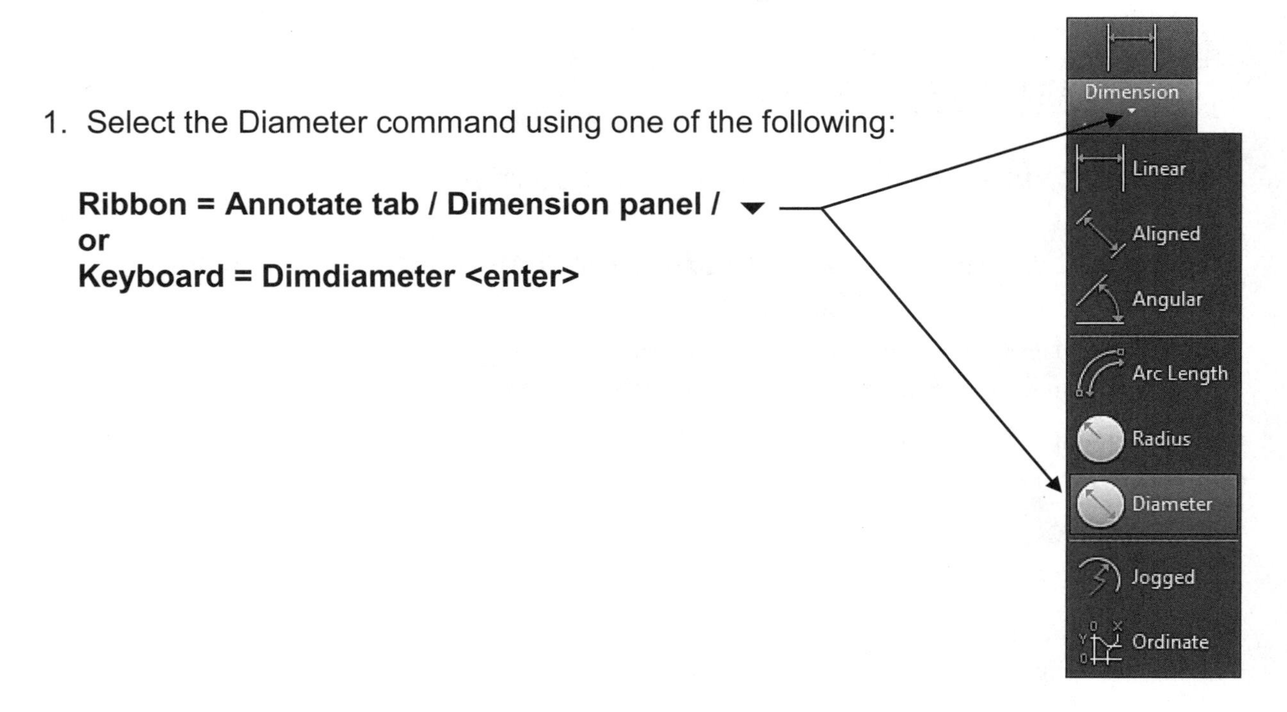

2. Select arc or circle: ***select the arc or circle (P1)*** ***<u>do not use object snap.</u>***
   Dimension text = ***the diameter will be displayed here.***

3. Specify dimension line location or [Mtext/Text/Angle]: ***place dimension text location (P2)***

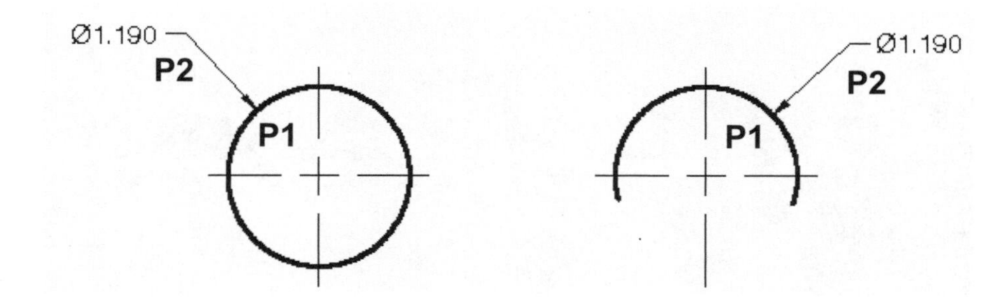

<u>Center marks</u> are automatically drawn as you use the diameter dimensioning command. If the circle already has a center mark or you do not want a center mark, set the center mark setting to <u>**NONE**</u> (Dimension Style / Symbols and Arrows tab) before using Diameter dimensioning. (Refer to 18-7)

DIMENSIONING DIAMETERS....continued

Controlling the diameter dimension appearance.

If you would like your Diameter dimensions to appear as shown in the two examples below, you must change some setting in the Dimension Style **Fit tab**.

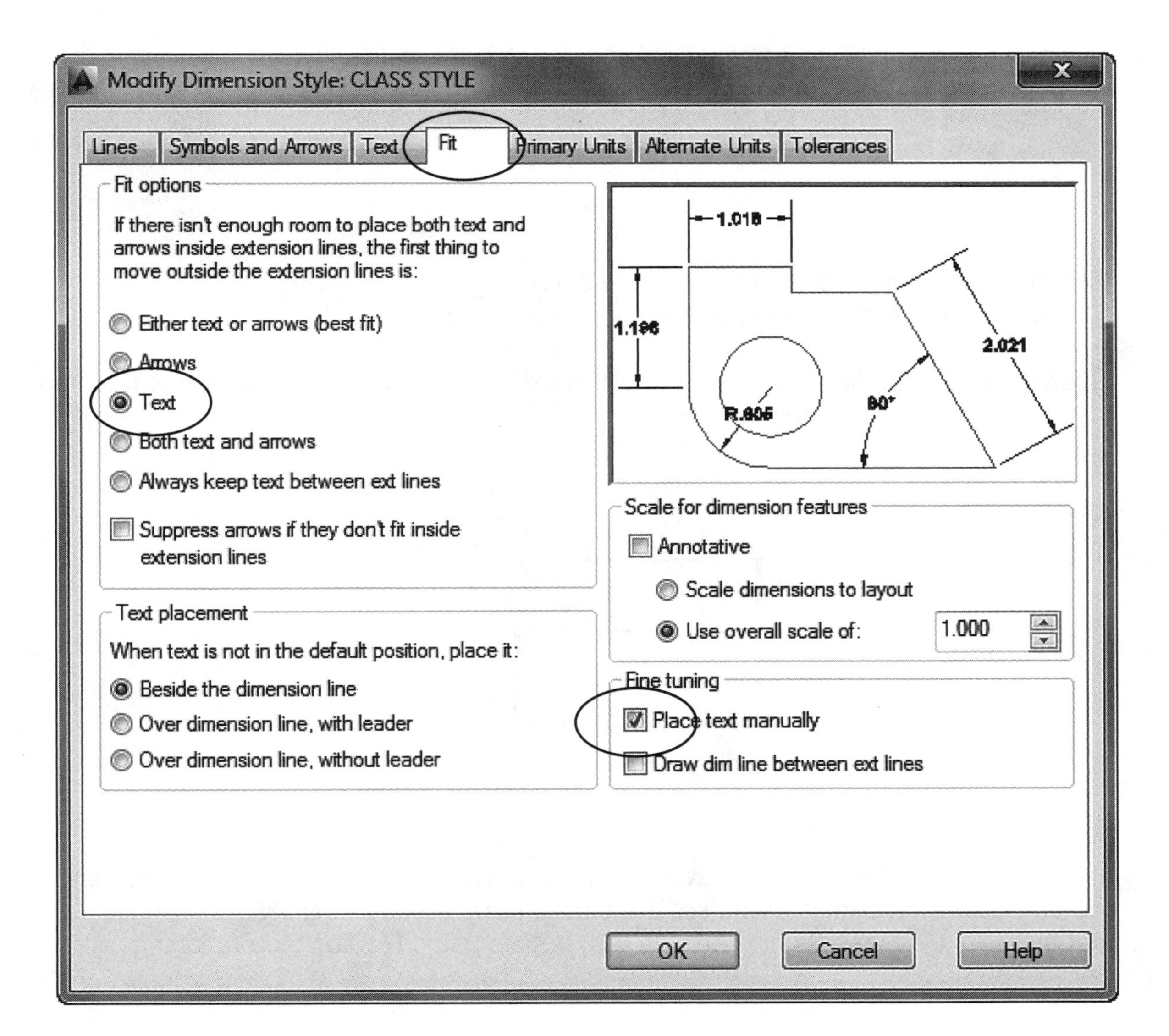

DIMENSIONING RADII

The **Radius** dimensioning command should be used when dimensioning arcs of <u>less than 180 degrees</u>. AutoCAD measures the selected arc and displays the dimension text with the radius symbol (R) in front of it.

1. Select the Radius command using one of the following:

   Ribbon = Annotate tab / Dimension panel / ▼
   or
   Keyboard = Dimradius <enter>

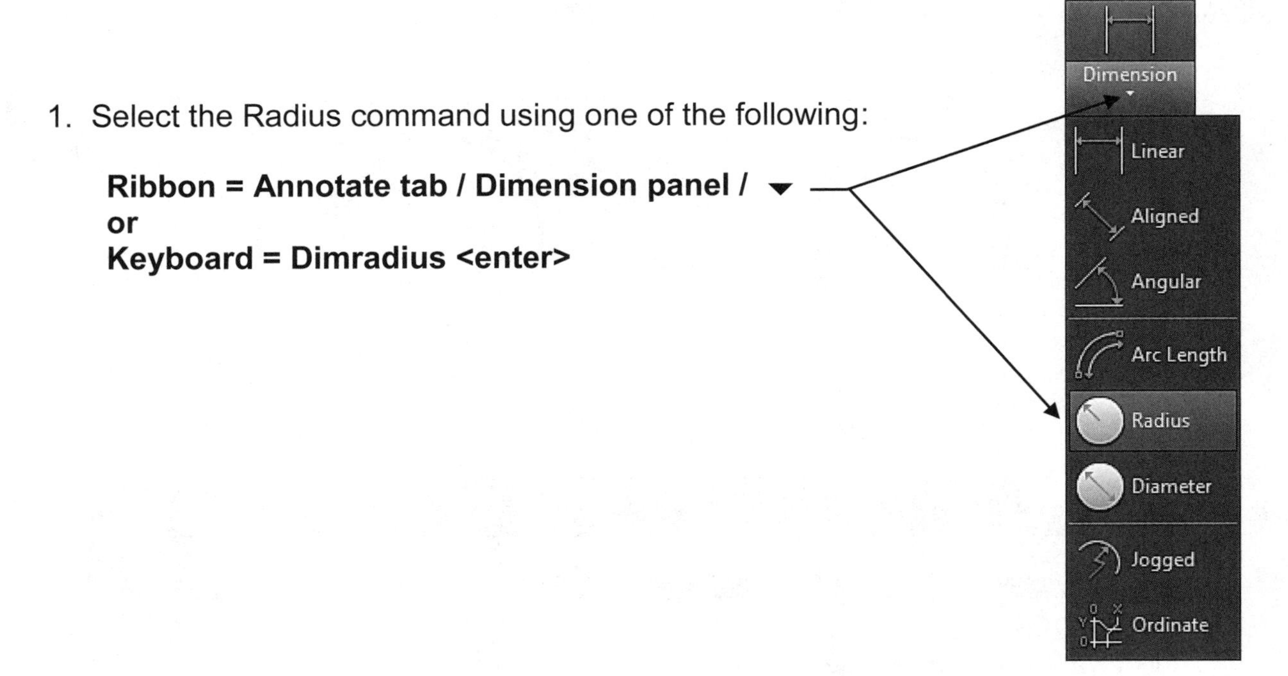

2. Select arc or circle: ***select the arc (P1)*** ***<u>do not use object snap.</u>***
   Dimension text = ***the radius will be displayed here.***

3. Specify dimension line location or [Mtext/Text/Angle]: ***place dimension text location (P2)***

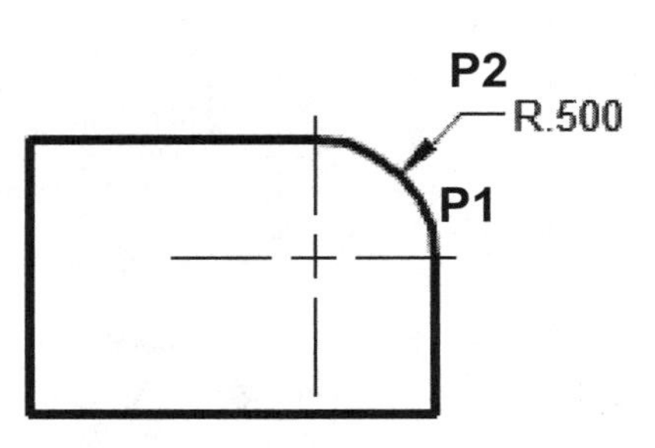

<u>Center marks</u> are automatically drawn as you use the radius dimensioning command. If you do not want a center mark, set the center mark setting to **<u>NONE</u>** (Dimension Style / Symbols and Arrows tab) before using Radius dimensioning. (Refer to 18-7)

DIMENSIONING RADII....continued

Controlling the radius dimension appearance.

If you would like your Radius dimensions to appear as shown in the example below, you must change the "**Fit Options**" and "**Fine Tuning**" in the **Fit tab** in your Dimension Style.

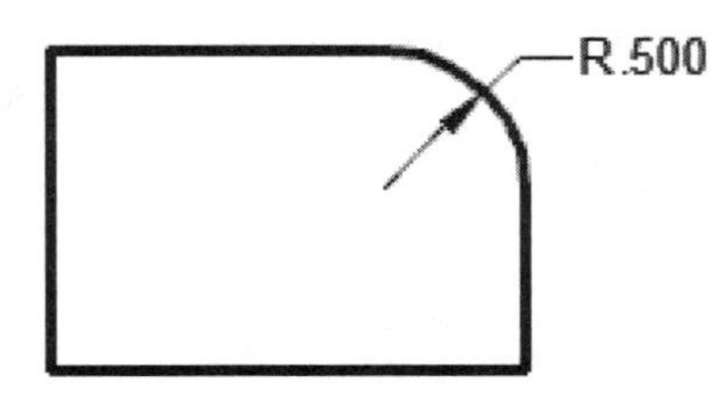

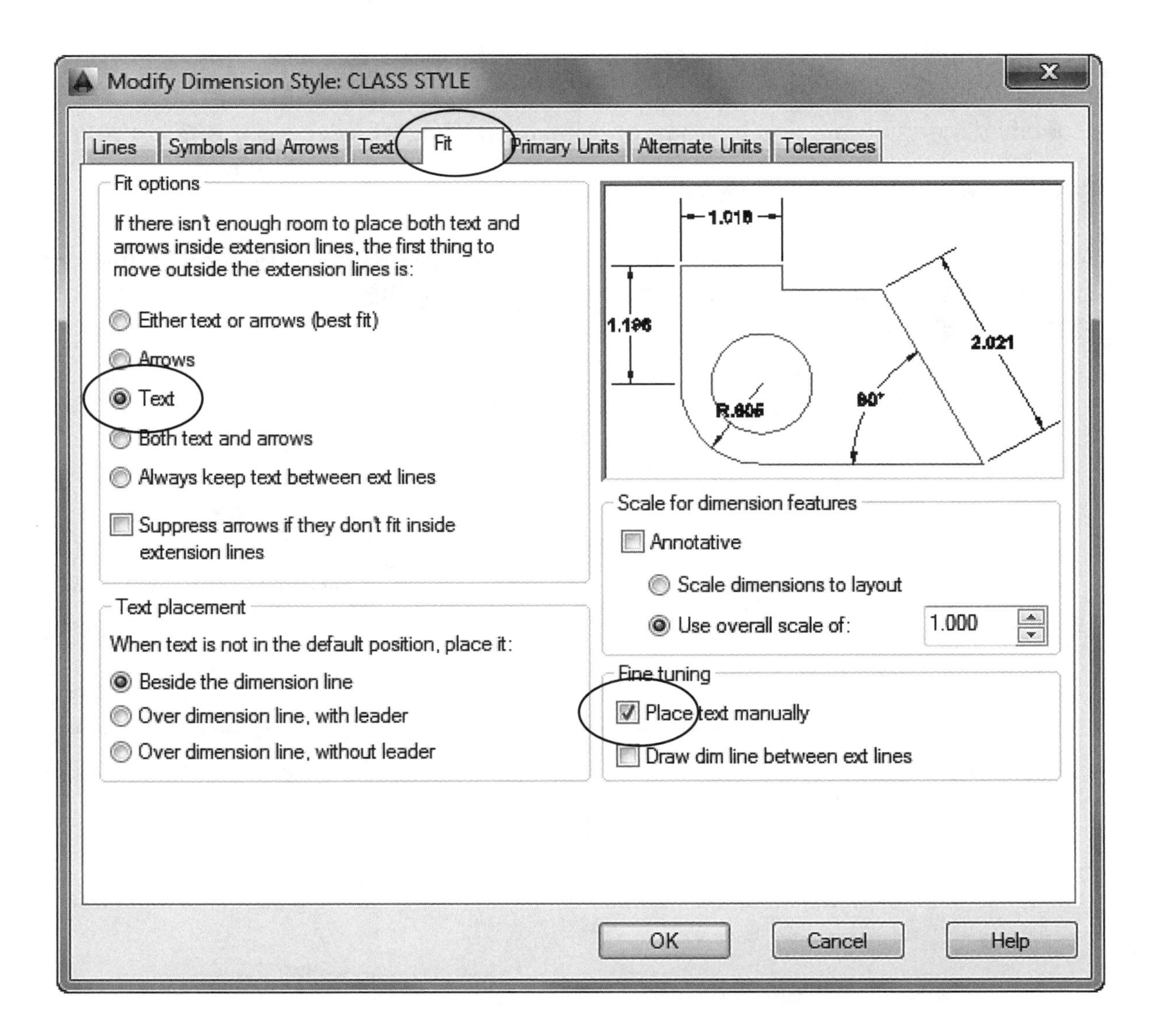

ANGULAR DIMENSIONING

The **ANGULAR** dimension command is used to create an angular dimension between two lines that form an angle. AutoCAD determines the angle between the selected lines and displays the dimension text followed by a degree (°) symbol.

1. Select the **ANGULAR** command.

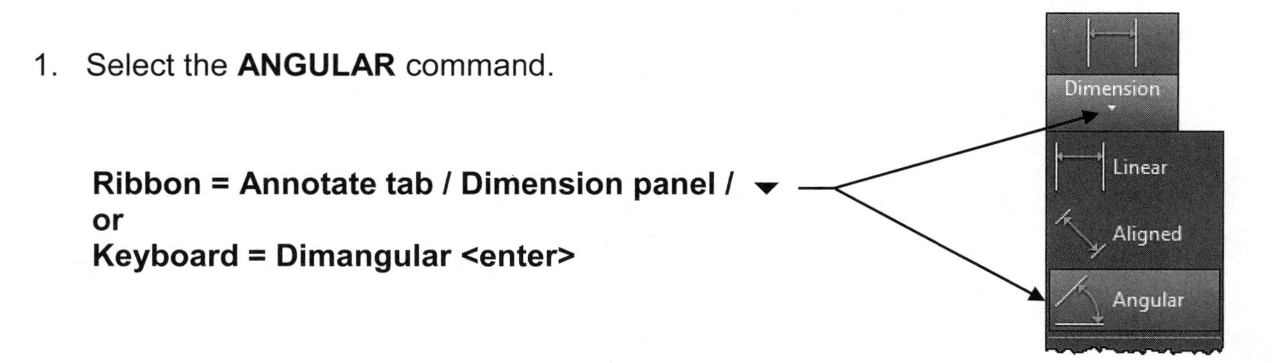

Ribbon = Annotate tab / Dimension panel / ▼
or
Keyboard = Dimangular <enter>

2. Select arc, circle, line, or <specify vertex>: ***select the first line that forms the angle (P1) location is not important, <u>do not use object snap</u>.***

3. Select second line: ***select the second line that forms the angle (P2)***

4. Specify dimension arc line location or [Mtext/Text/Angle]: ***place dimension text location (P3)***

   Dimension text = ***angle will be displayed here***

   Any of the 4 angular dimensions shown below can be displayed by moving the cursor in the direction of the dimension after selecting the 2 lines (**P1** and **P2**) that form the angle.

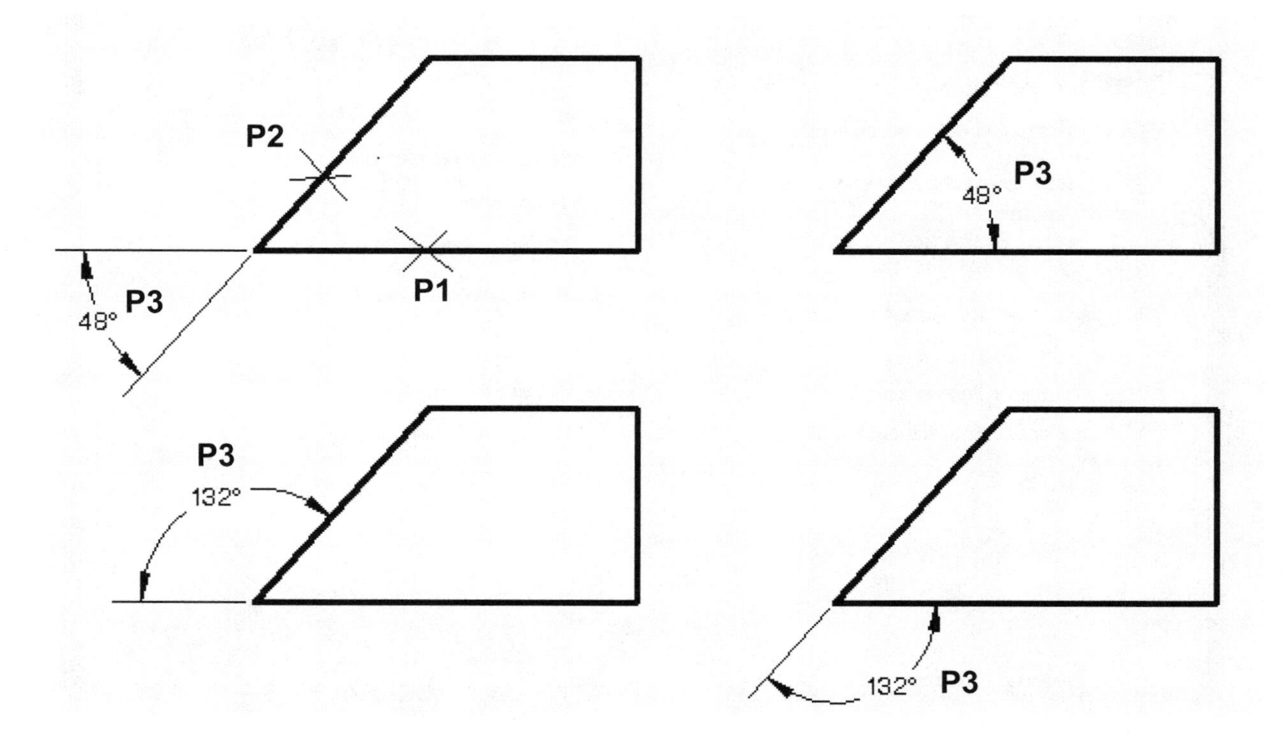

CENTERMARK

CENTERMARKS can ONLY be drawn with circular objects like Circles and Arcs. You set the size and type.

The Center Mark has three types, **None, Mark** and **Line** as shown below.

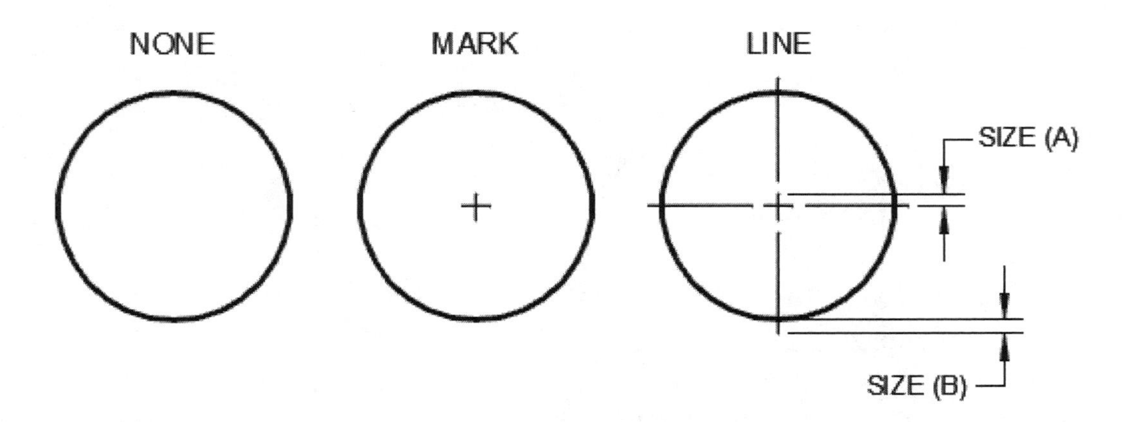

What does "SIZE" mean?

The **size** setting determines both, (A) the length of half of the intersection line and (B) the length extending beyond the circle. (See above right)

Where do you set the CENTERMARK "TYPE" and "SIZE"

1. Select the ***Dimension Style*** command.
2. Select: ***Modify or Override.***
3. Select: ***SYMBOLS and ARROWS*** *tab*
4. Select the ***Center mark type***
5. Set the **Size**.

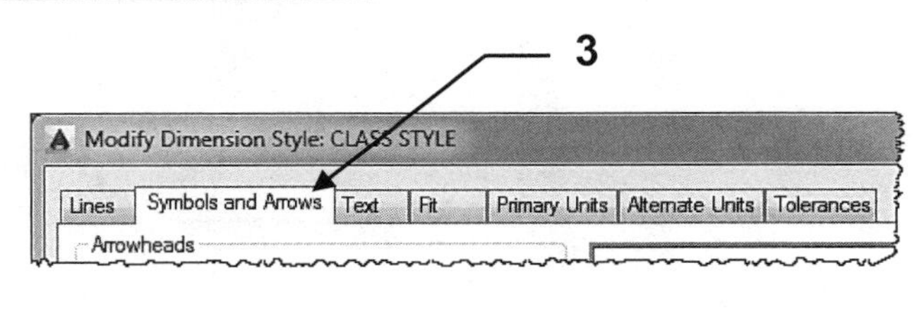

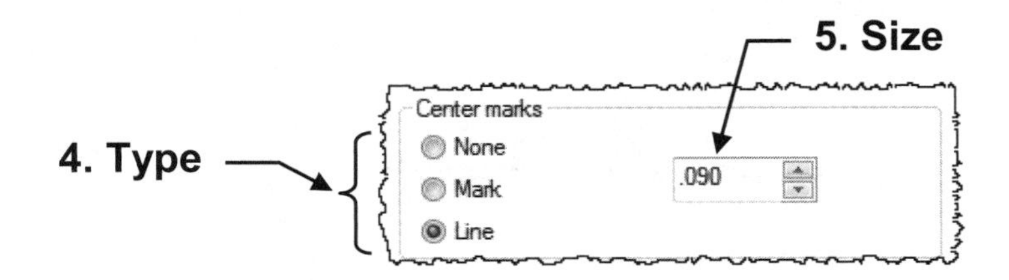

To draw a CENTER MARK

1. Select the **Center mark** command using one of the following:

   Ribbon = Annotation tab / Dimension ▾ panel /
   or
   Keyboard = DCE <enter>

2. Select arc or circle: ***select the arc or circle with the cursor.***

FLIP ARROW

You can easily **flip** the direction of the arrowhead using the **Flip Arrow** option.

How to <u>Flip an arrowhead</u>.

1. Select the dimension that you wish to Flip.

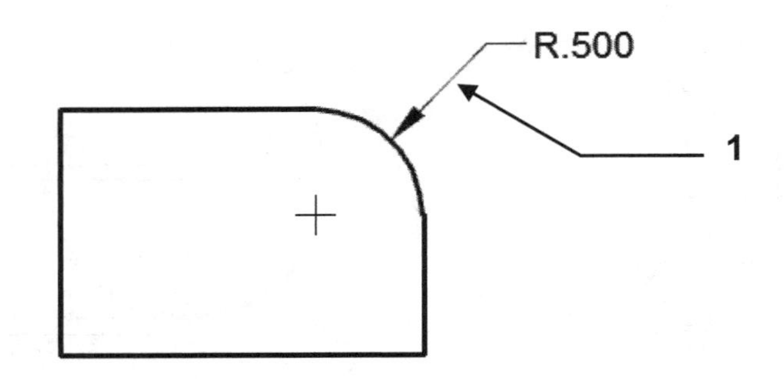

2. Rest your cursor on the arrow grip. (Do not press button, just rest cursor)
3. Select **Flip Arrow** from the shortcut menu.

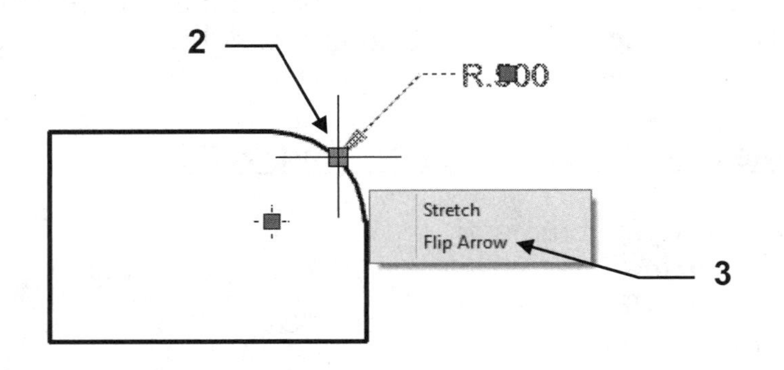

4. The arrowhead Flips.

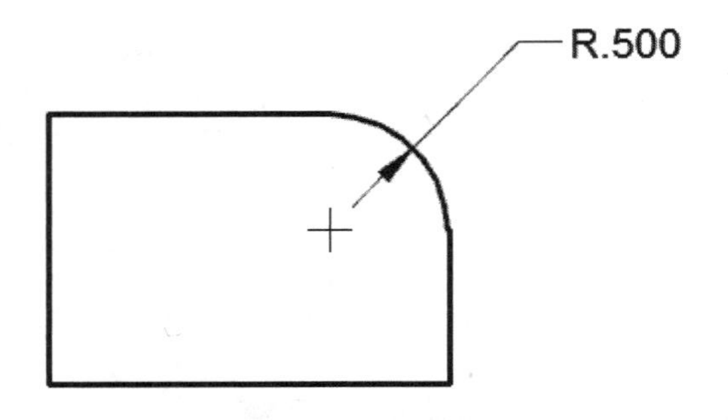

5. Press **Esc** and the grips will disappear. (Do not press <enter>).

CREATING A DIMENSION SUB-STYLE

In Lesson 16 you learned how to create a Dimension Style named Class Style. All of the dimension created with that style appear identical because they have the same settings. Now you are going to learn how to create a "Sub-Style" of the Class Style.

For example:
If you wanted all of the **Diameter** dimensions to have a centerline automatically displayed but you did not want a centerline displayed when using the **Radius** dimension command. To achieve this, you must create a "sub-style" for all **Radius** dimensions.

Sub-styles have also been called "children" of the "Parent" dimension style.
As a result, they form a family.

A Sub-style is permanent, unlike the Override command, which is temporary.

Note: This sounds much more complicated than it is. Just follow the steps below.
It is very easy.

__**How to create a sub-style for Radius dimensions.**__

You will set the center mark to None for the Radius command only.
The Diameter command center mark will not change.

1. Start a **New** file using **Border A-2015.dwt** drawing.

2. Select the **DIMENSION / STYLE** command. (Refer to page 16-8)

3. Select **"Class Style"** from the Style List.

4. Select the **NEW** button.

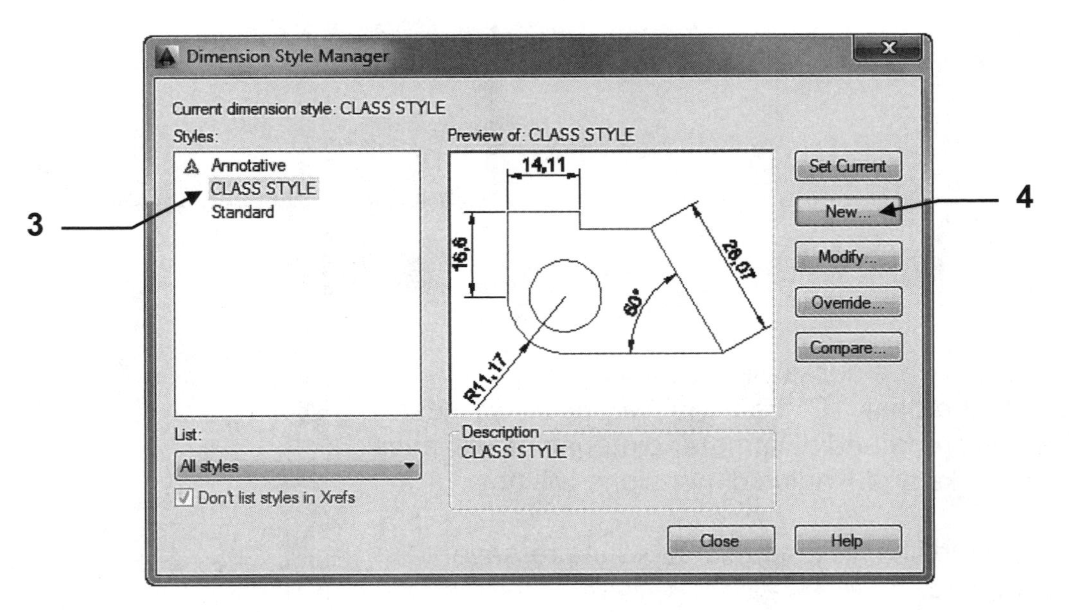

Continued on the next page...

CREATING A DIMENSION SUB-STYLE....continued

5. Change the "Use for" to: **Radius Dimensions**

6. Select **Continue**

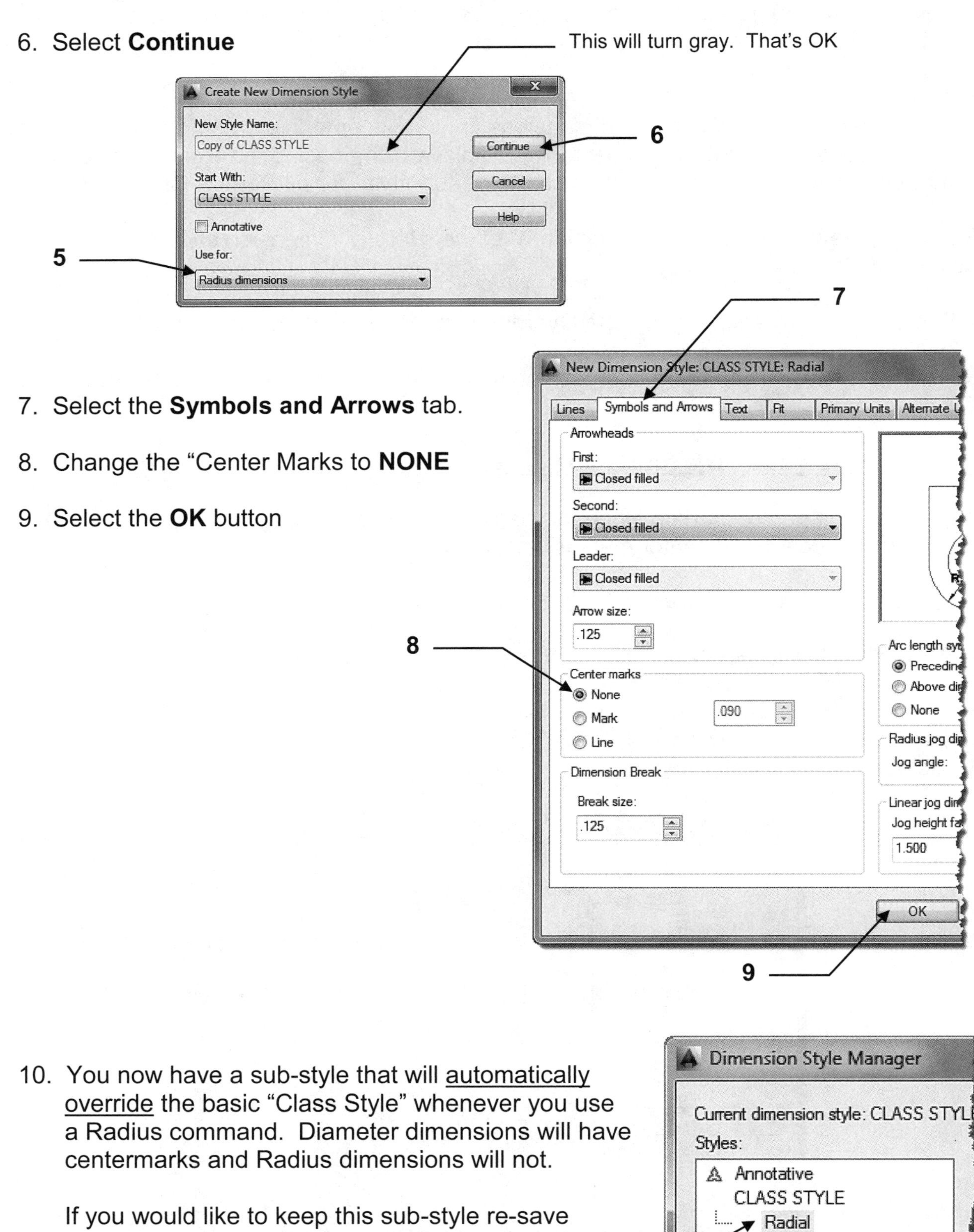

7. Select the **Symbols and Arrows** tab.

8. Change the "Center Marks to **NONE**

9. Select the **OK** button

10. You now have a sub-style that will automatically override the basic "Class Style" whenever you use a Radius command. Diameter dimensions will have centermarks and Radius dimensions will not.

    If you would like to keep this sub-style re-save the template.

EXERCISE 18A

INSTRUCTIONS:

1. Start a **New** file using **Border A-2015.dwt**
2. Draw the objects below using layer Object line.
3. Dimension as shown using Linear, Diameter and Radius. Use Layer Dimension
4. Edit the Title and Ex-XX by double clicking on the text. Do not erase and replace.
5. Save as **EX18A**
6. Plot using Page Setup **Class Model A**

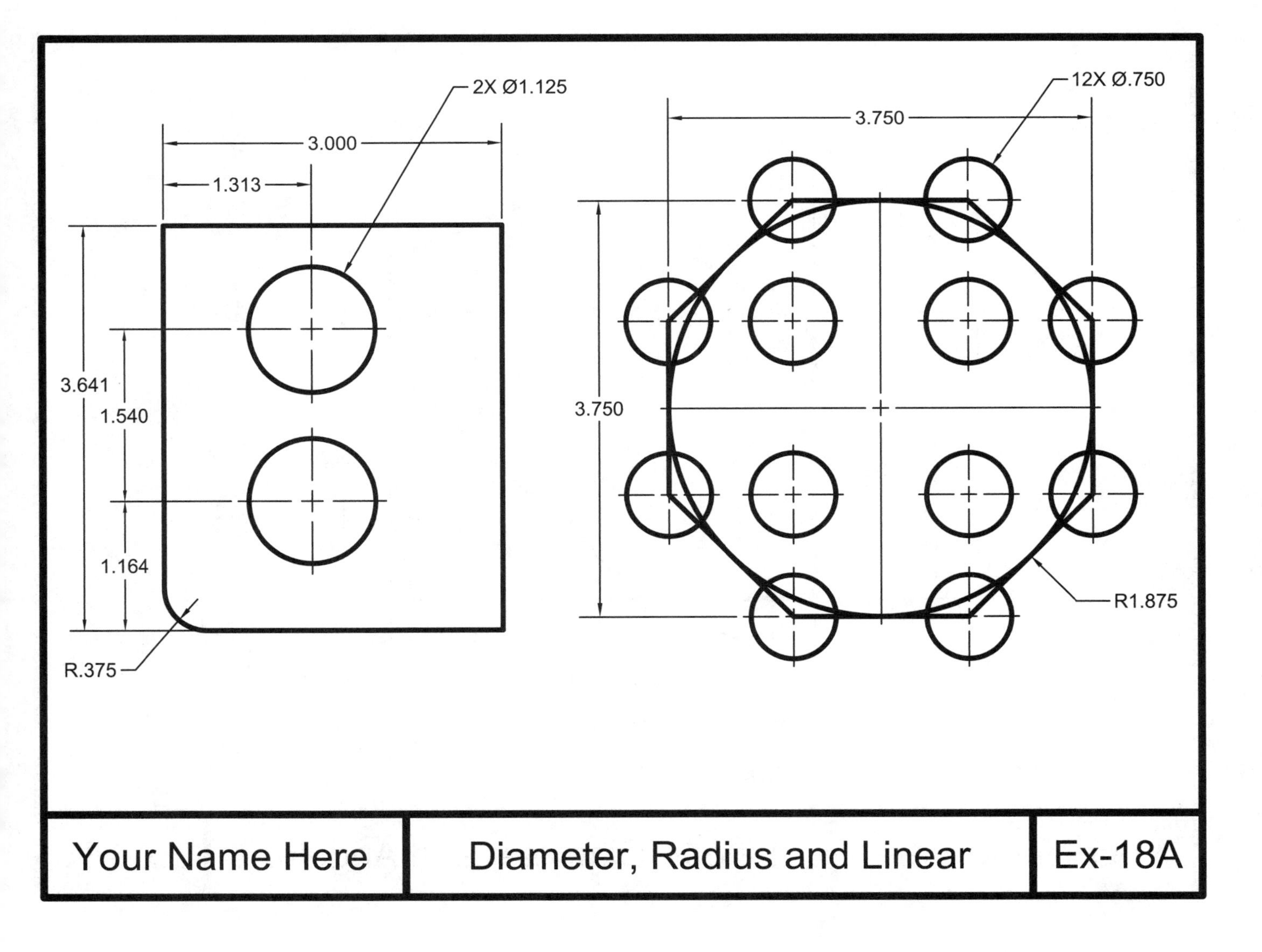

EXERCISE 18B

INSTRUCTIONS:

1. Start a **New** file using **Border A-2015.dwt**
2. Draw the objects below using layer Object line.

   Note: Refer to the next page for drawing hints.

3. Dimension as shown using Linear, Diameter and Angular. Use Layer Dimension
4. Edit the Title and Ex-XX by double clicking on the text. Do not erase and replace.
5. Save as **EX18B**
6. Plot using Page Setup **Class Model A**

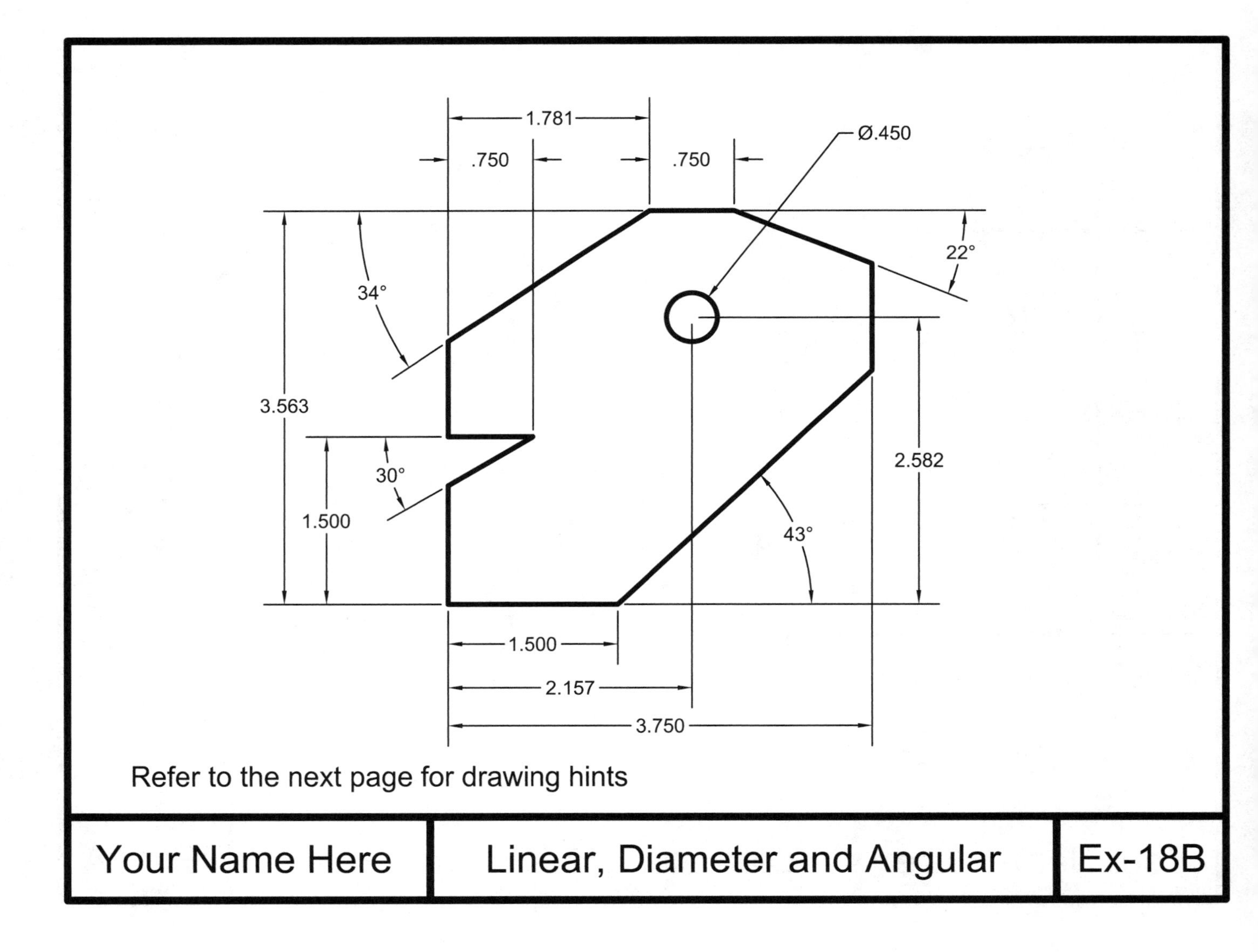

EXERCISE 18B-Helper

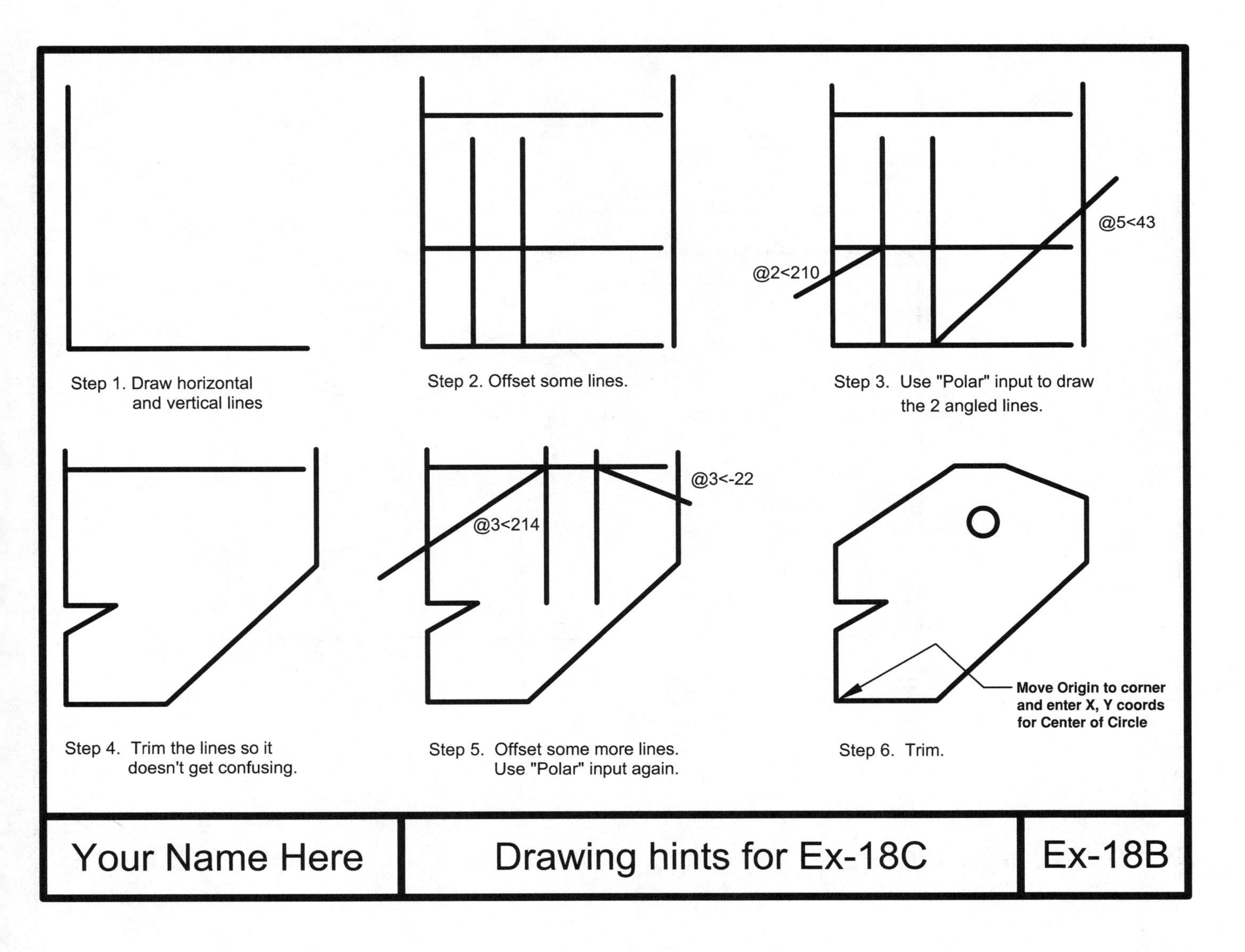

EXERCISE 18C

INSTRUCTIONS:

1. Start a **New** file using **Border A-2015.dwt**
2. Draw the objects below using layer Object line.
3. Dimension as shown using Linear, Diameter and Radius. Use Layer Dimension
4. Edit the Title and Ex-XX by double clicking on the text. Do not erase and replace.
5. Save as **EX18C**
6. Plot using Page Setup **Class Model A**

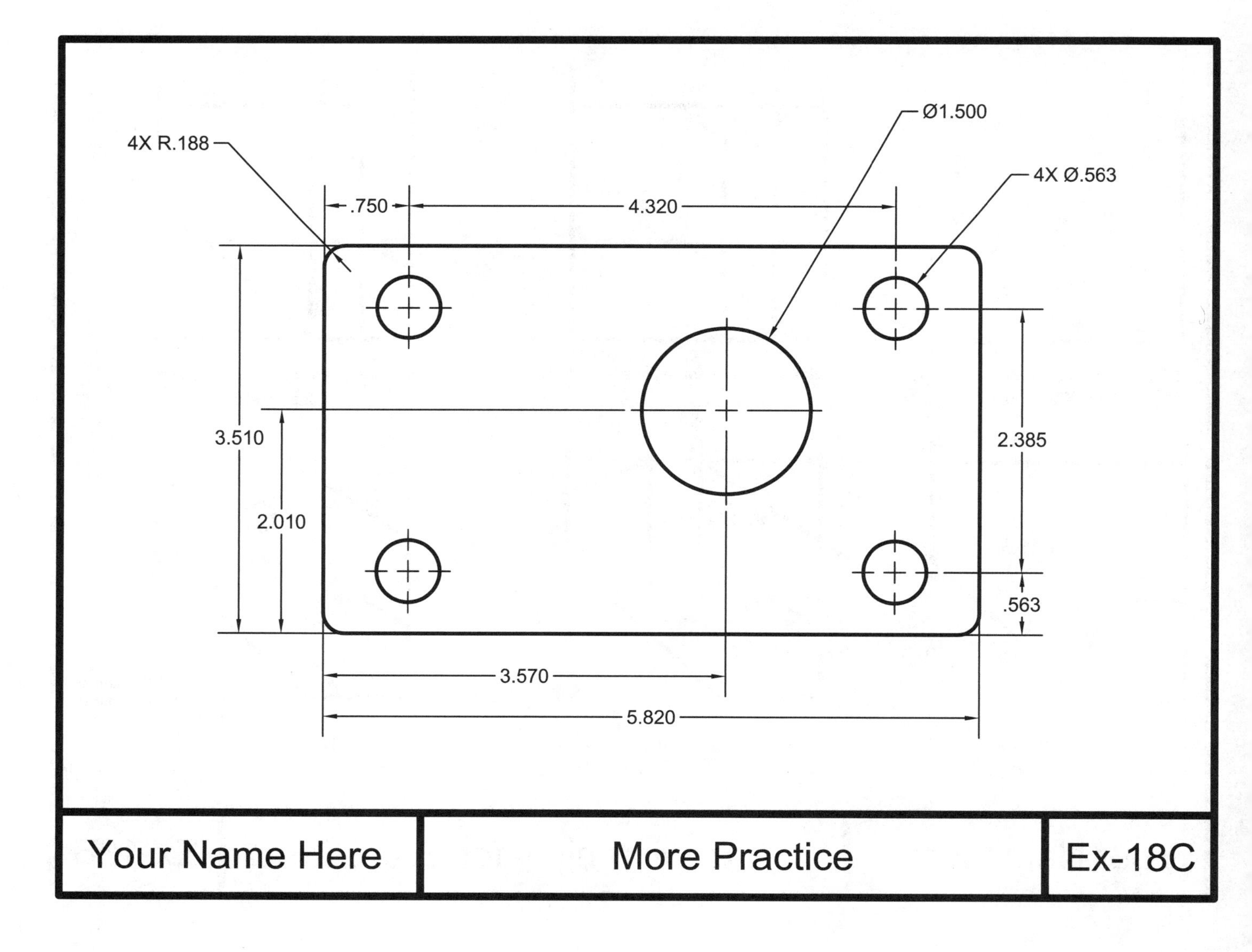

LEARNING OBJECTIVES

After completing this lesson, you will be able to:

1. Create Multileaders
2. Modify Multileader lines
3. Create a New Multileader Style
4. Dimension angular surfaces
5. Add Special Characters
6. Pre-assign a Prefix or Suffix to a dimension.

LESSON 19

MULTILEADER

A multileader is a single object consisting of a **pointer**, **line**, **landing** and **content**. You may draw a multileader, pointer first, tail first or content first.

To select the Multileader use:
Annotate tab / Leaders panel

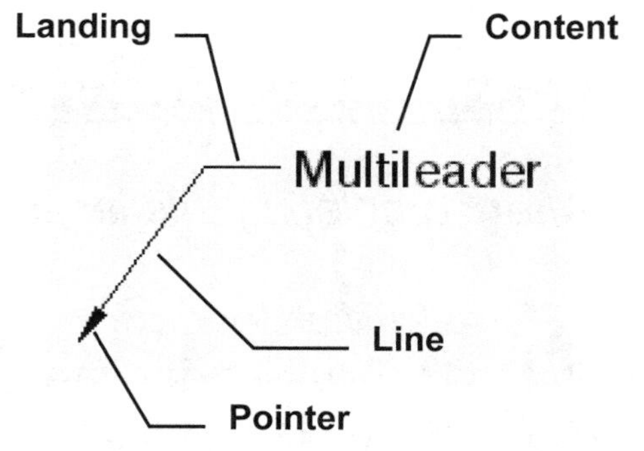

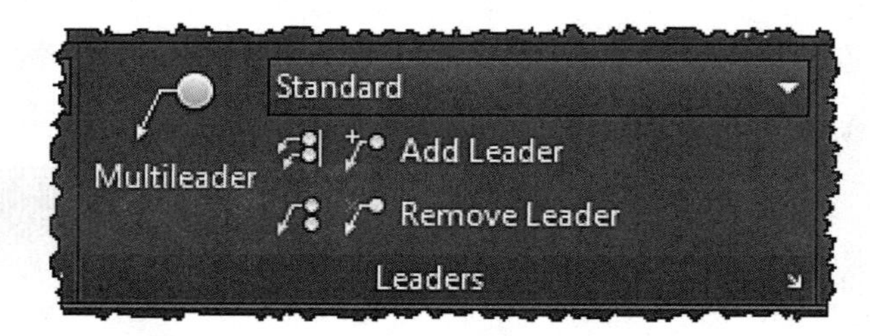

Create a single Multileader

1. Select the **Multileader** tool 

   Command: _mleader
2. Specify leader arrowhead location or [leader Landing first/Content first/Options] <Options>: ***specify arrowhead location (P1)***
3. Specify leader landing location: ***specify the landing location (P2)***
4. The Multitext editor appears. ***Enter text then select "Close Text Editor"***

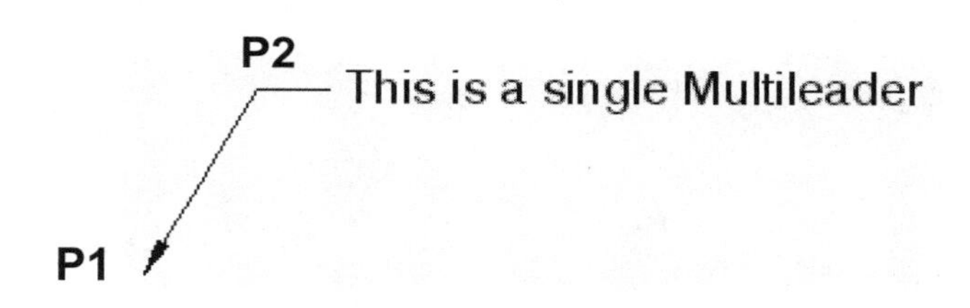

Add leader lines to existing Multileader

1. Select the **Add Leader** tool Add Leader
2. Select a multileader: ***click on the existing multileader line (P1)***
3. Specify leader arrowhead location: ***specify arrowhead location (P2)***
4. Specify leader arrowhead location: ***specify next arrowhead location (P3)***
5. Specify leader arrowhead location: ***specify next arrowhead location (P4)***
6. Specify leader arrowhead location: ***press <enter> to stop***

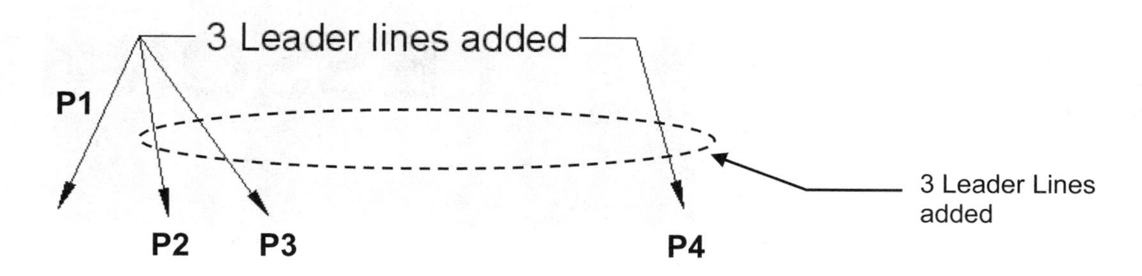

MULTILEADER....continued

Remove a leader line from an existing Multileader

1. Select the **Remove Leader** tool
2. Select a multileader: ***click anywhere on the existing multileader***
3. Specify leaders to remove: ***select the leader line to remove***
4. Specify leaders to remove: ***press <enter> to stop***

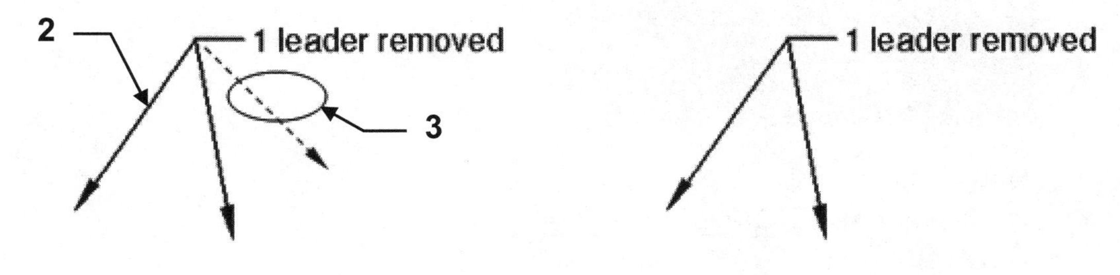

Align Multileaders

1. Select the **Align Multileader** tool 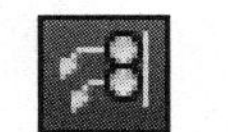

Command: _mleaderalign
2. Select multileaders: ***select the leaders to align (Note 1 and Note 3)***
3. Select multileaders: ***press <enter> to stop selecting***
   Current mode: Use current spacing
4. Select multileader to align to or [Options]: ***select the leader to <u>align</u> <u>to</u> (Note 2)***
5. Specify direction: ***move the cursor and press left mouse button***

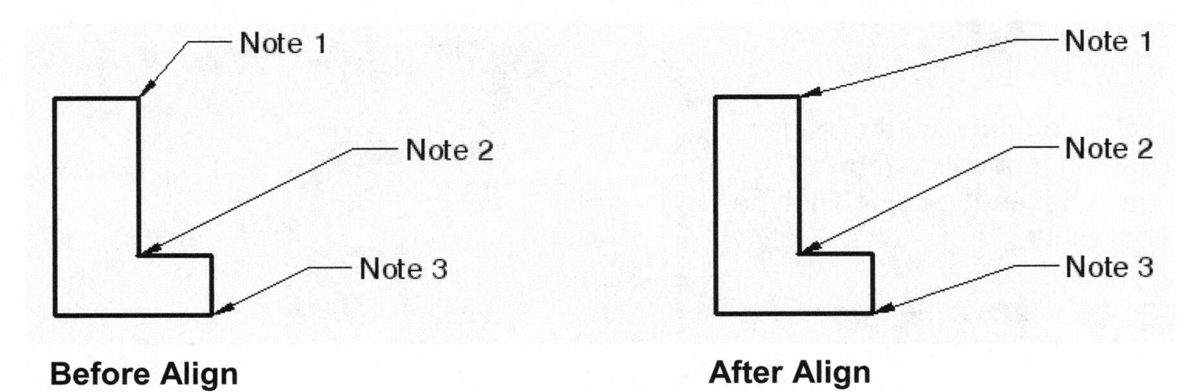

Before Align **After Align**

Note: <u>Collect Multileader</u> works best with Blocks and will be discussed in another lesson.

CREATE A MULTILEADER STYLE

You may create **Multileader styles** to control it's appearance.
This will be similar to creating a dimension style.

1. Start a **New** file using **Border A-2015.dwt**

2. Select the **Multileader Style Manager** by selecting the ↘ on the Leader panel.

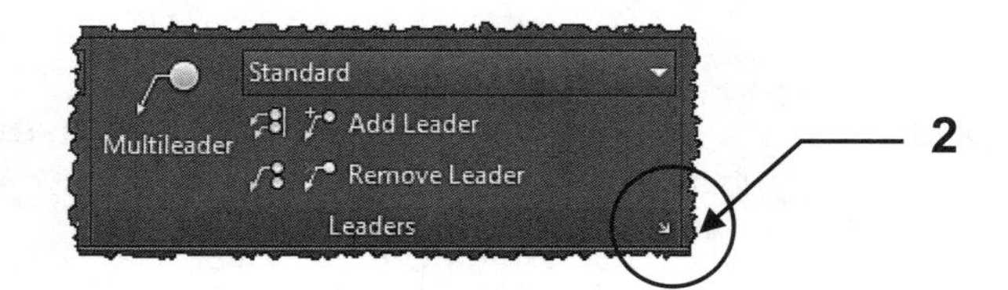

3. Select the **New** button.
4. Enter the New Style name: <u>**Class ML style**</u>
5. Select the **Continue** button.

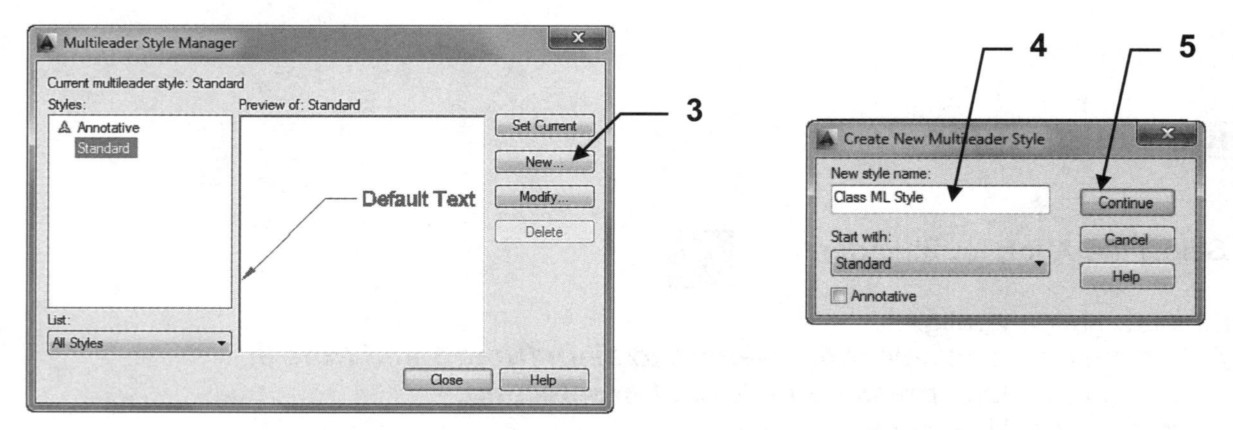

6. Select the <u>**Leader Format**</u> tab and change the settings as shown below.

<u>General</u> – You may set the leader line to straight or spline (curved). You may change the color, linetype and lineweight.

<u>Arrowhead</u> – Select the symbol for the pointer and size of pointer.

<u>Leader break</u> - specify the size of the gap in the leader line if Dimension break is used.
(Refer to page 17-7 for Dimension break reminder)

CREATE A MULTILEADER STYLE....continued

7. Select the **Leader Structure** tab and change the settings as shown below.

Constraints – Specify how many line segments to allow and on what angle.

Landing Settings – Specify the length of the Landing (the horizontal line next to the note) or turn it off by unchecking the "Automatically include landing" box.

Scale – "Annotation" will be discussed in Lesson 27.

8. Select the **Content** tab and change the settings as shown below.

Multileader type – Select what to attach to the landing.

Text Options – Specify the leader note appearance.

Leader Connection – Specify how the note attaches to the landing.

9. Select the **OK** button.

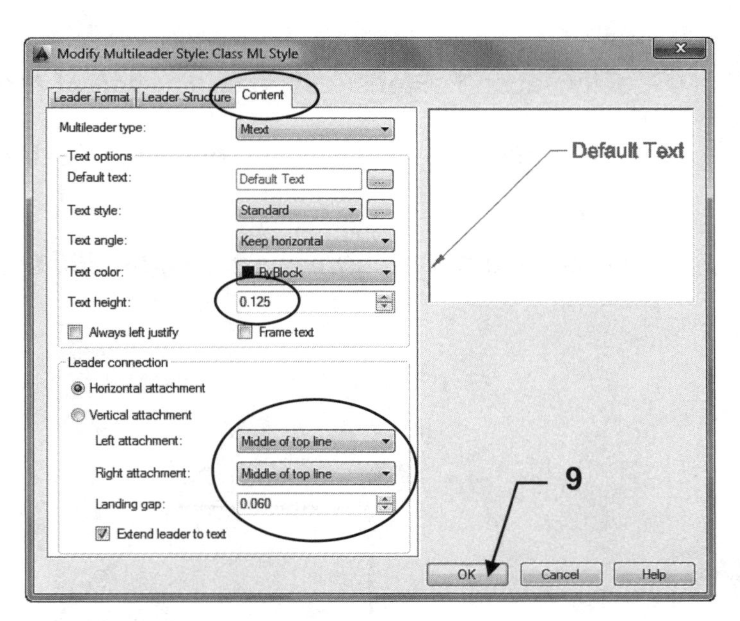

The **Class ML style** should be listed in the "Style" area.

10. Select **Set Current** button.

11. Select **Close** button.

12. **Re-Save** your **Border A-2015.dwt** again.

Now whenever you use **Border A-2015** the Multileader style **Class ML style** will be there.

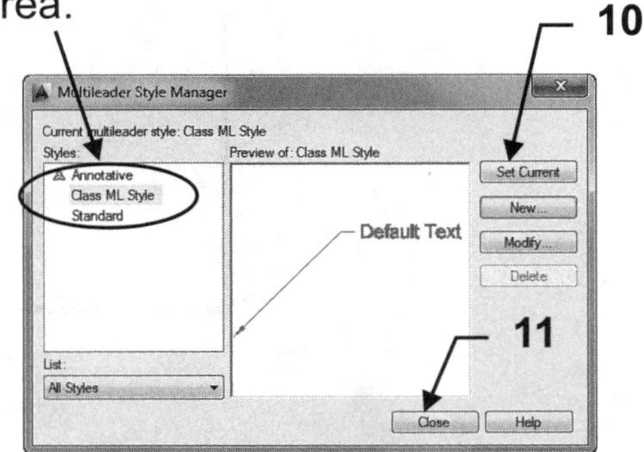

ALIGNED DIMENSIONING

The **ALIGNED** dimension command aligns the dimension with the angle of the object that you are dimensioning. The process is the same as Linear dimensioning. It requires two extension line origins and the placement of the dimension text location. (Example below)

1. Select the **ALIGNED** command using one of the following:

   Ribbon = Annotate tab / Dimension panel / ▾
   or
   Keyboard = Dal <enter>

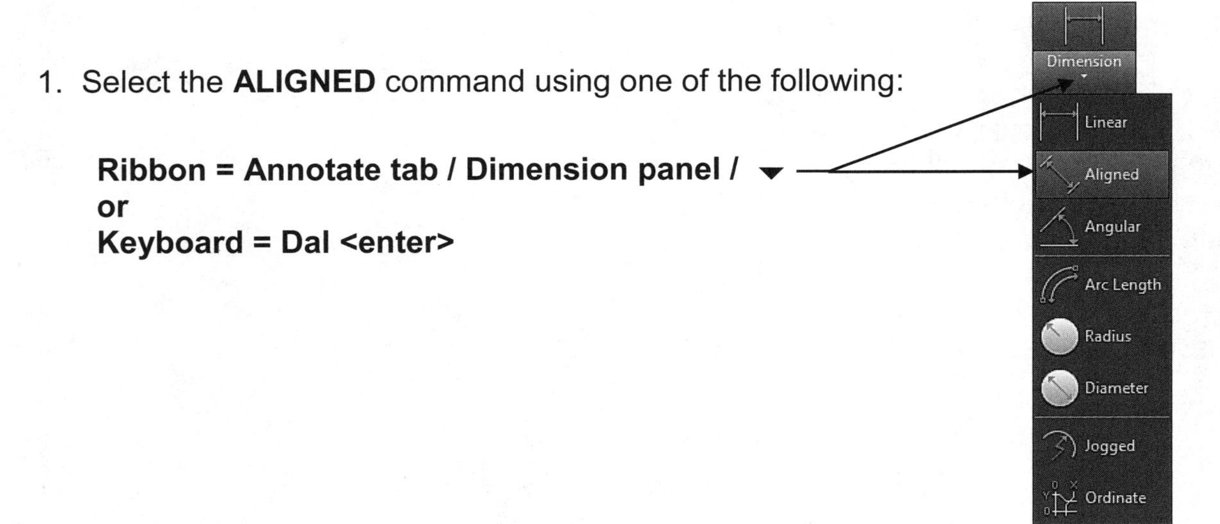

   Command: _dimaligned

2. Specify first extension line origin or <select object>: ***select the first extension line origin (P1)***

3. Specify second extension line origin: ***select the second extension line origin (P2)***

4. Specify dimension line location or [Mtext/Text/Angle]: ***place dimension text location (P3)***

   Dimension text = ***the dimension value will appear here***

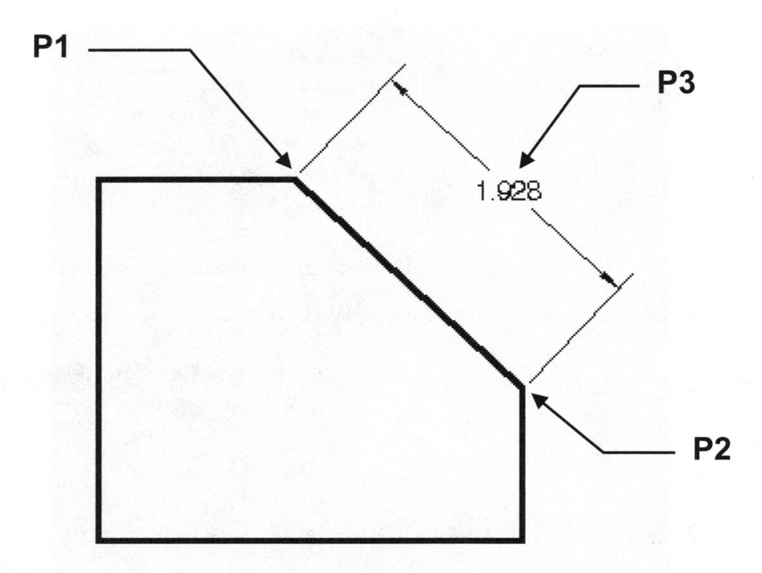

SPECIAL TEXT CHARACTERS

Characters such as the **Degree symbol (°), Diameter symbol (Ø)** and the **Plus / Minus symbols (±)** are created by typing **%%** and then the appropriate "**code**" letter.

SYMBOL		CODE
Ø	Diameter	%%C
°	Degree	%%D
±	Plus / Minus	%%P

SINGLE LINE TEXT

When using "Single Line Text", type the code in the sentence. After you enter the code the symbol will appear.

For example:
Entering 350**%%D** will create: **350°.** The **"D"** is the **"code"** letter for degree**.**

MULTILINE TEXT

When using Multiline Text you may enter the code in the sentence or you may select a symbol using the **Symbol Tool** located on the **Insert panel.**

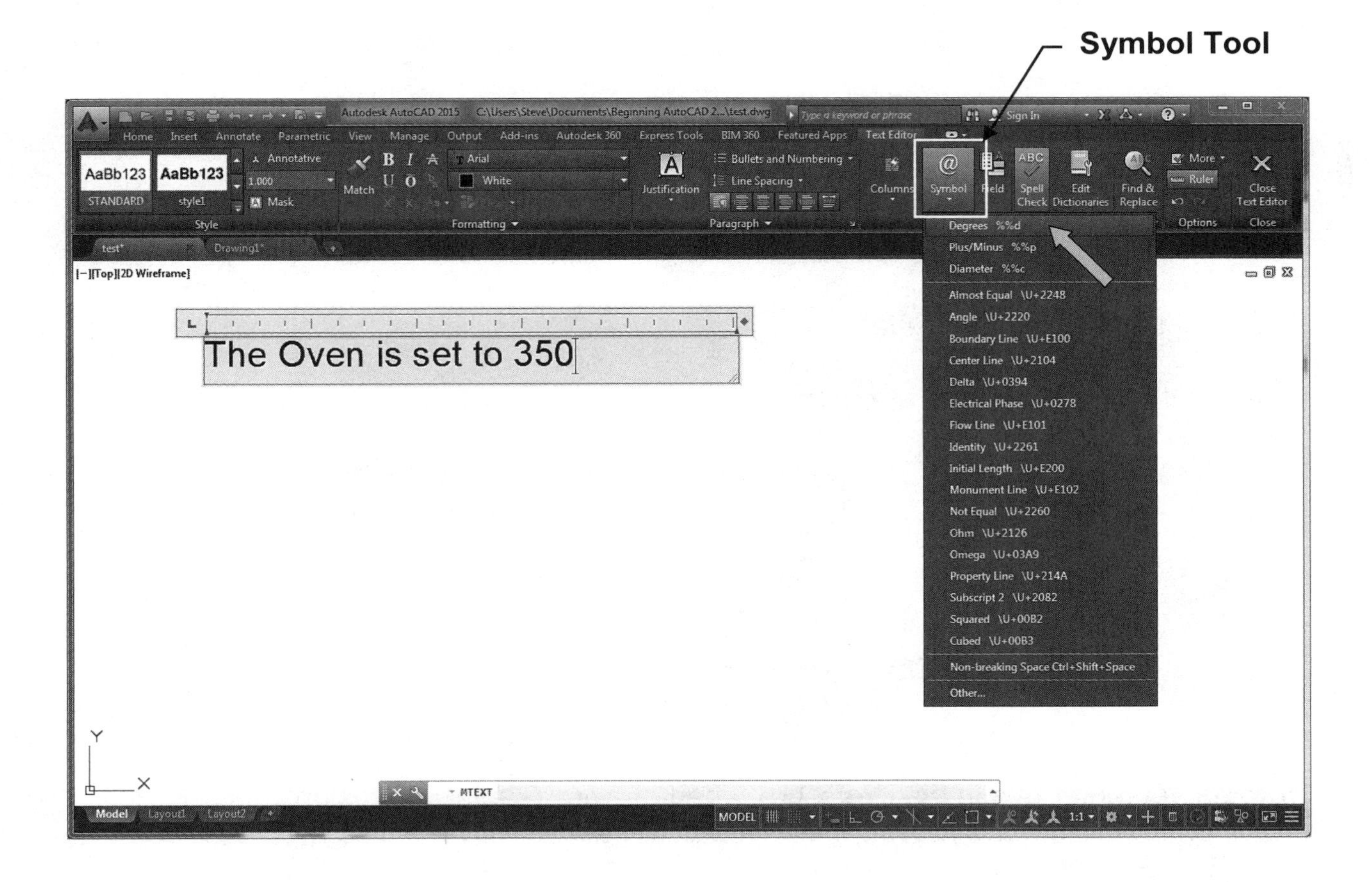

PREFIX and SUFFIX

The **PREFIX** (before) and the **SUFFIX** (after) allows you to preset text to be inserted automatically as you dimension.

Note:
You would use this if you had multiple dimensions that included a prefix or a suffix. If you had only 1 or 2 you would merely edit the dimension text as shown on page 17-2.

1. Select **Dimension Style**.

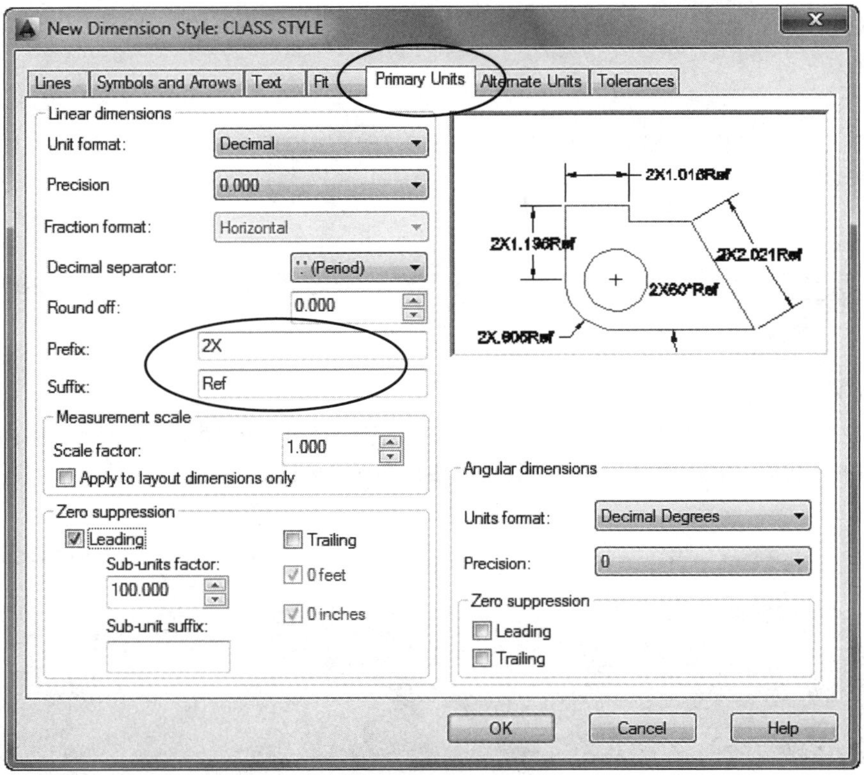

2. Select the **Primary Units** tab.

3. Enter the **Prefix** or **Suffix** content.

   Example:

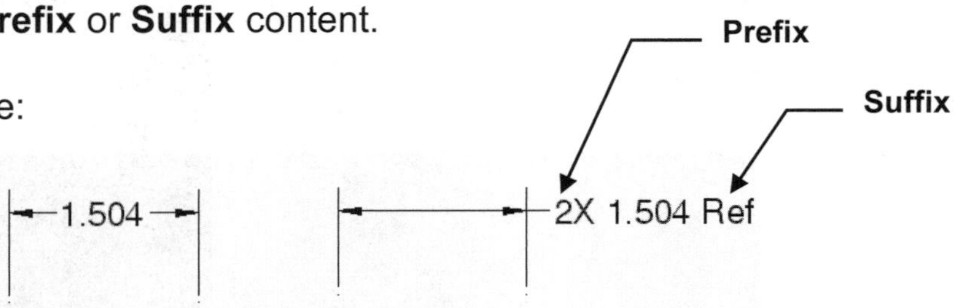

No Prefix/Suffix **Prefix=2X Suffix=Ref**

Note: If you enter text in the Prefix box when drawing Radial dimensions, the **"R"** for radius or the **diameter symbol**, will not be automatically drawn. You must also add the letter R or symbol code to the prefix box. Example: **2X R.** Or **2X %%C**

EXERCISE 19A

INSTRUCTIONS:

1. Start a **New** file using **Border A-2015.dwt**
2. Draw the Lines as shown using Polar coordinates or Dynamic Input. (Lessons 9 &11)
3. Use Layer Object lines.
4. Dimension using Linear, Aligned and Angular.
5. Use Dimension Style "Class Style" and Layer Dimension
6. Edit the Title and Ex-XX by double clicking on the text. Do not erase and replace.
7. Save as **EX19A**
8. Plot using Page Setup **Class Model A**

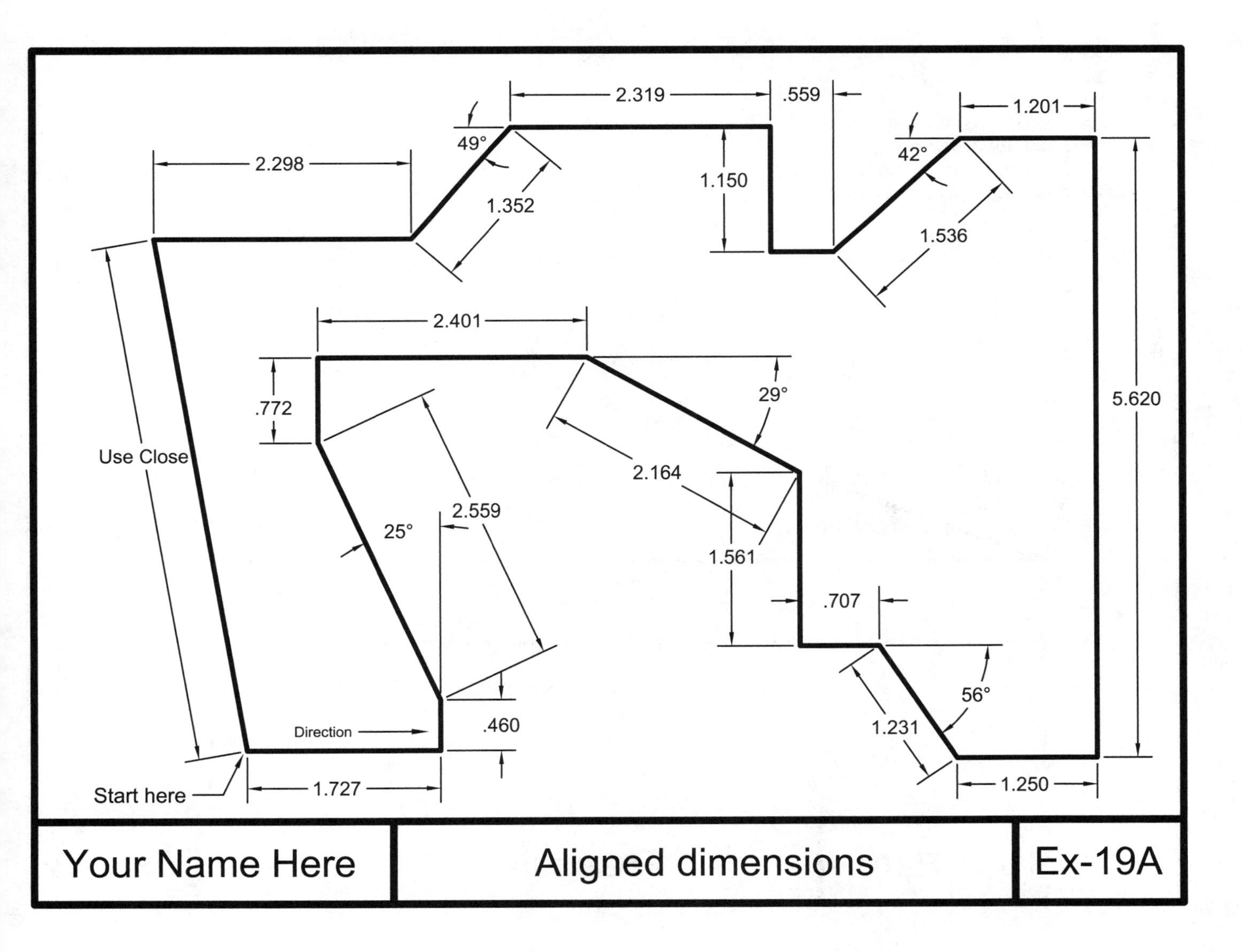

EXERCISE 19B

INSTRUCTIONS:

1. Start a **New** file using **Border A-2015.dwt**
2. Draw the object shown below.
3. Use Layer Object lines.
4. Dimension as shown.
5. Use Dimension Style "Class Style" and Layer Dimension.
6. For the Note: "Special Instructions to Baker" use: Text Ht. = .125 Layer = Text
7. Edit the Title and Ex-XX by double clicking on the text. Do not erase and replace.
8. Save as **EX19B**
9. Plot using Page Setup **Class Model A**

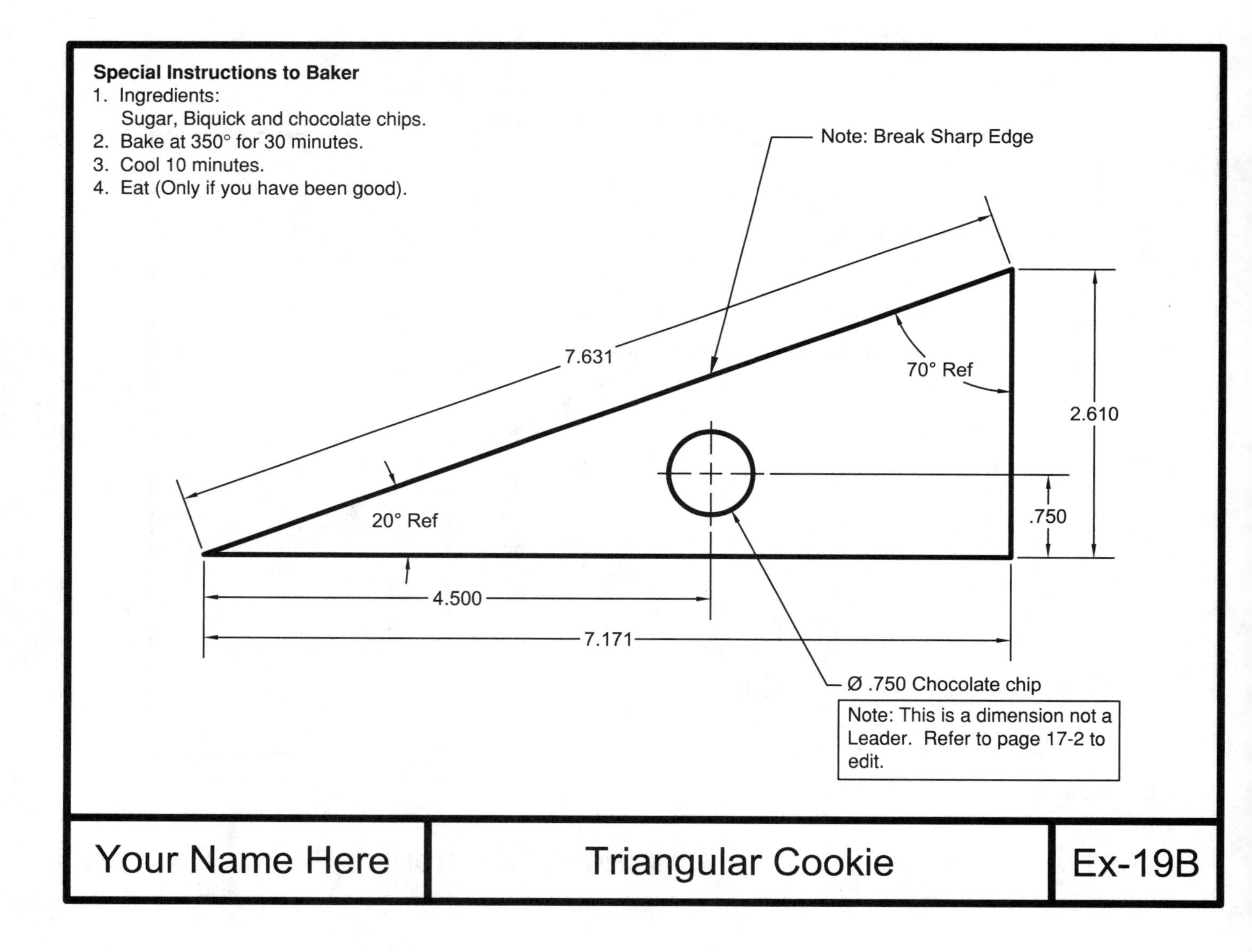

EXERCISE 19C

INSTRUCTIONS:

1. Start a **New** file using **Border A-2015.dwt**
2. Create the Multileader style shown on page 19-4 if you haven't already.
3. Draw the object shown below.
4. Use Layer Object lines.
5. Dimension as shown.
6. Use Dimension Style "Class Style" and "Class ML style" and Layer Dimension.
7. Edit the Title and Ex-XX by double clicking on the text. Do not erase and replace.
8. Save as **EX19C**
9. Plot using Page Setup **Class Model A**

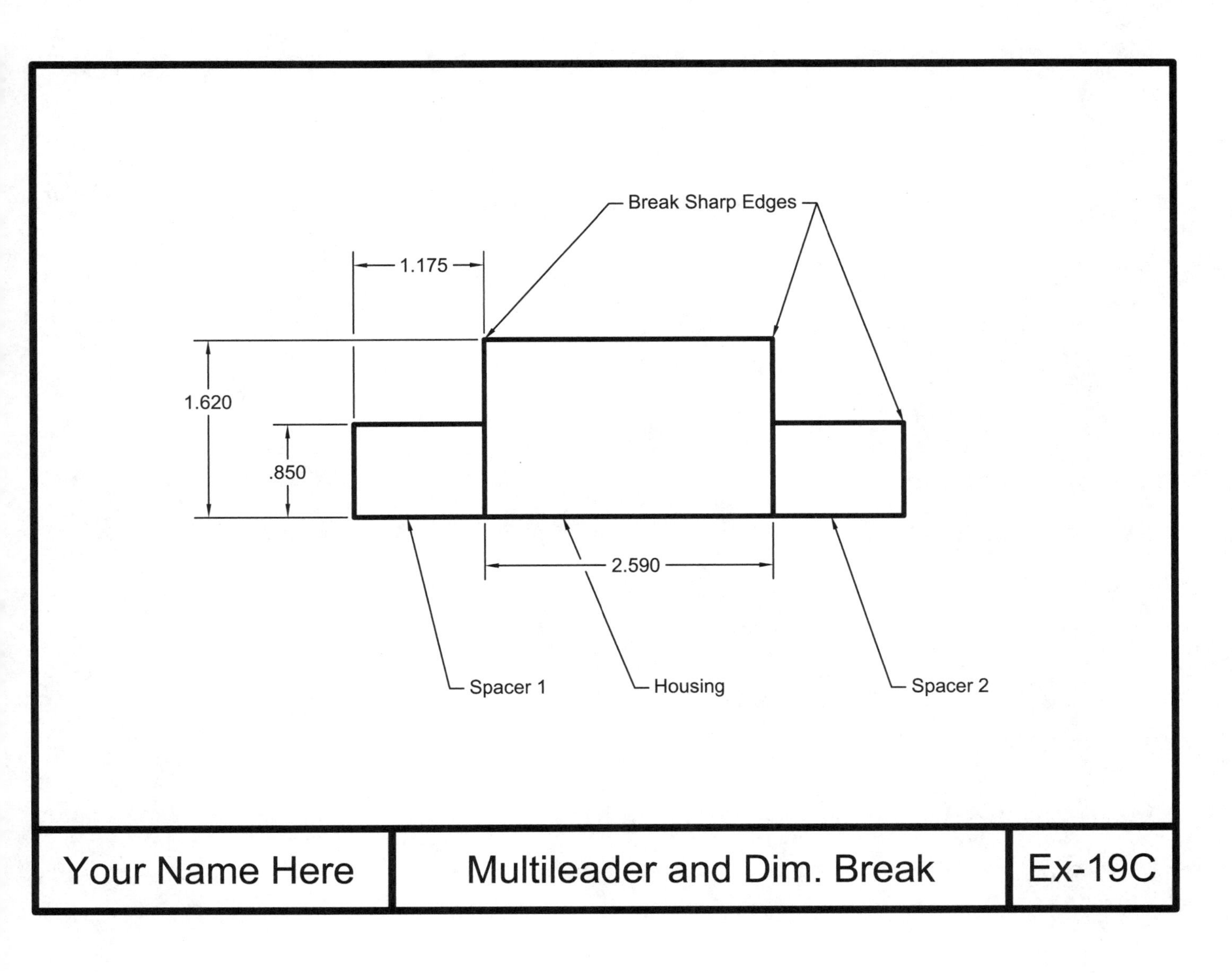

Notes:

LEARNING OBJECTIVES

After completing this lesson, you will be able to:

1. Use Quick Dimension to automatically dimension
2. Edit Multiple Dimensions
3. Edit extension locations using grips

LESSON 20

QUICK DIMENSION

Quick Dimension creates multiple dimensions with one command. Quick Dimension can create Continuous, Staggered, Baseline, Ordinate, Radius and Diameter dimensions.

The following is an example of how to use the "Continuous" option. Staggered, Baseline, Diameter and Radius are explained on the following pages.

1. Select the **Quick Dimension** command using one of the following:

   Ribbon = Annotate tab / Dimensions panel / Quick
   or
   Keyboard = Qdim <enter>

   Command: _qdim
   Associative dimension priority = Endpoint
   Select geometry to dimension

2. Select the objects to be dimensioned with a crossing window or pick each object

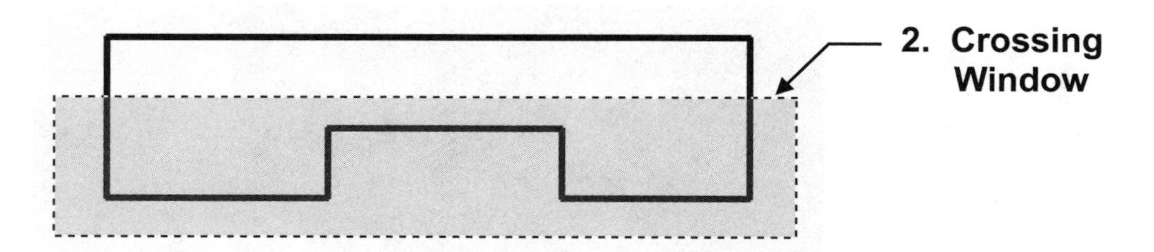

3. Press <enter> to stop selecting objects.

4. Specify dimension line position, or
   [Continuous/Staggered/Baseline/Ordinate/Radius/Diameter/datumPoint/Edit/settings]
   <Continuous>: ***Select "C" <enter> for Continuous.***

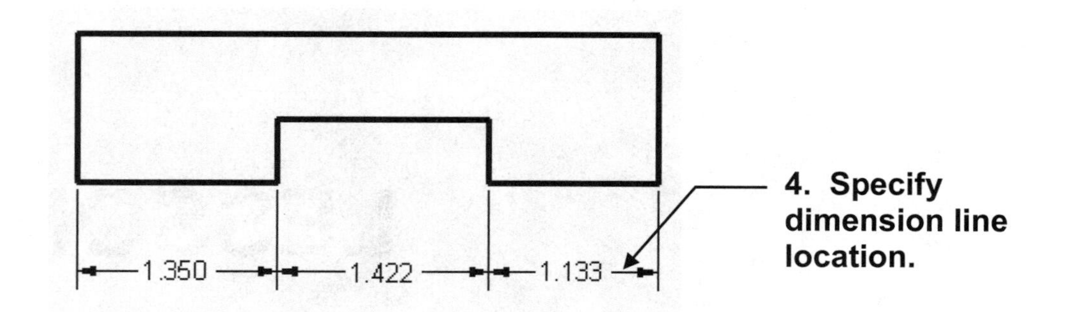

QUICK DIMENSION....continued

STAGGERED

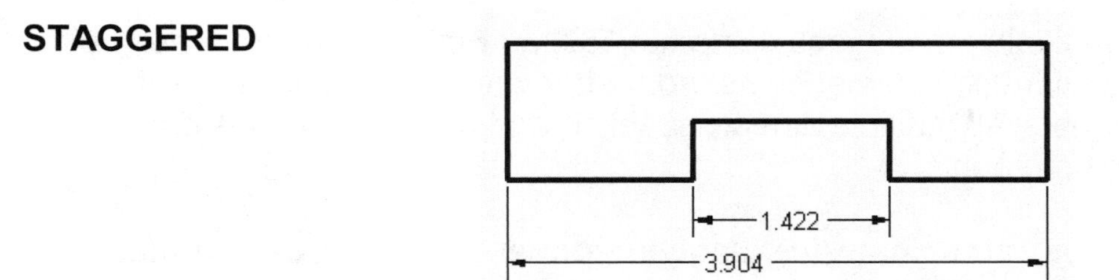

BASELINE

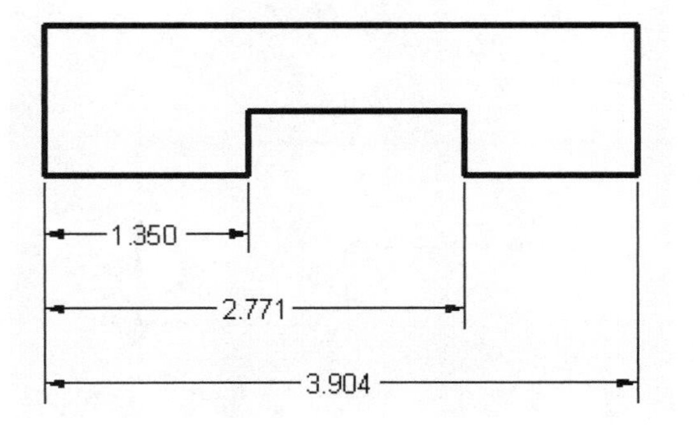

DIAMETER

1. You can use a crossing window. Qdim will automatically filter out any linear dims.
2. Dimension Line Length is determined by the "Baseline Spacing" setting in the Dimension Style. You may stretch it using grips.

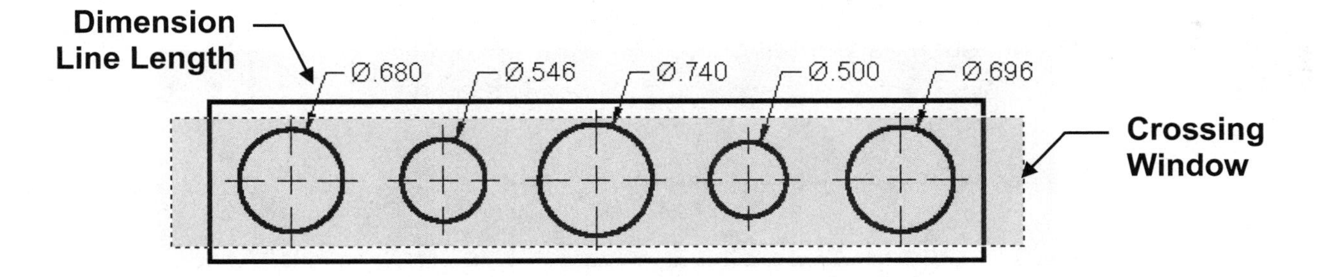

RADIUS

1. You can use a crossing window. Qdim will automatically filter out any linear dims.
2. Dimension Line Length is determined by the "Baseline Spacing" setting in the Dimension Style.

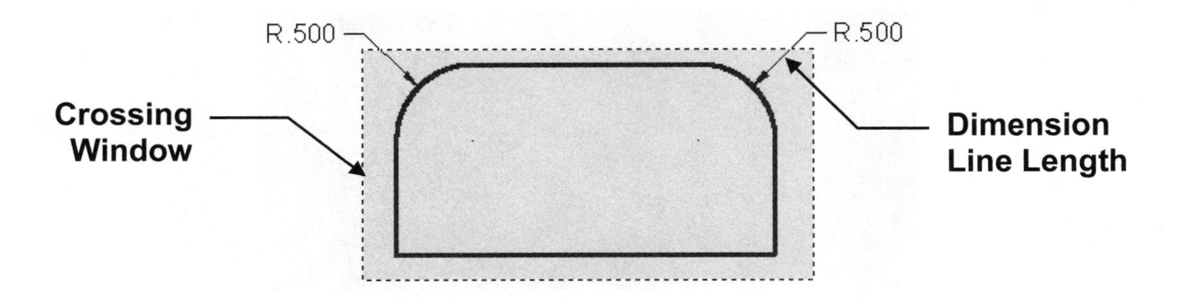

EDITING MULTIPLE DIMENSIONS

You can edit existing multiple dimensions using the QDIM / Edit command. The Qdim, edit command will edit all multiple dimensions, no matter whether they were created originally with Qdim or not. All multiple: linear, baseline and continue dimensions respond to this editing command.

The example below illustrates changing Baseline dimensions to Continuous.

1. Select the **Quick Dimension** command

2. Select the dimensions to edit.

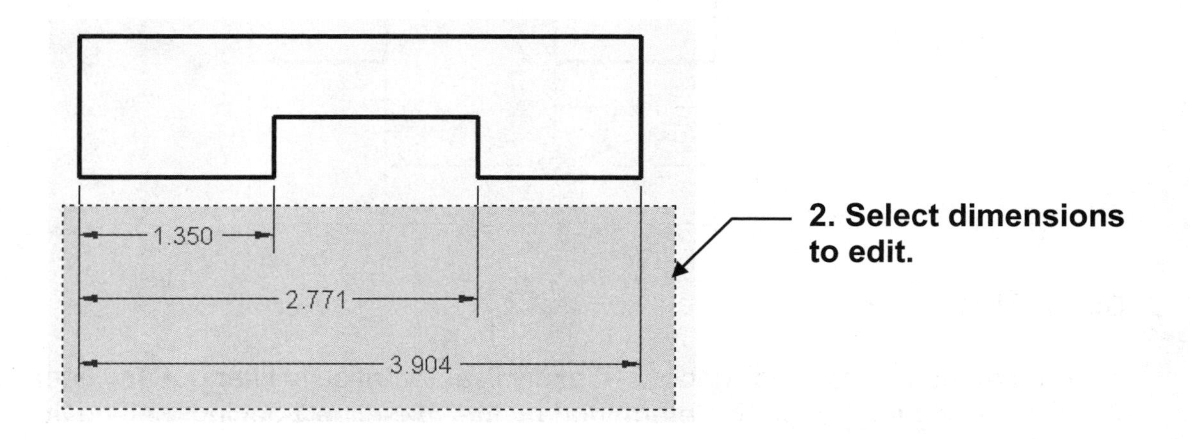

3. Press <enter> to stop selecting dimensions.

4. Select an option, such as Continuous.

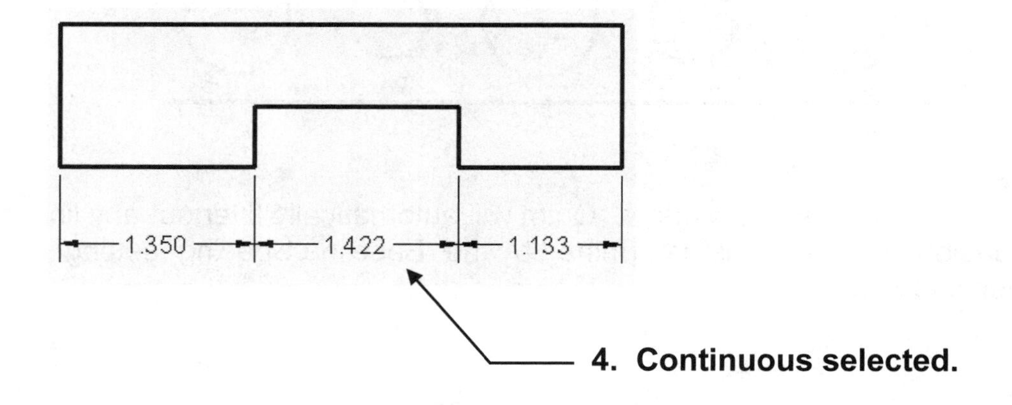

EDITING DIMENSIONS USING GRIPS

Sometimes, when using Quick Dimension, the extension lines point to a location that you do not prefer. If this happens select the dimension, select the grip and drag it to the preferred location. If the dimension is associative (refer to page 16-2) the dimension text will change to reflect the new extension line location.

EXAMPLE:

Before

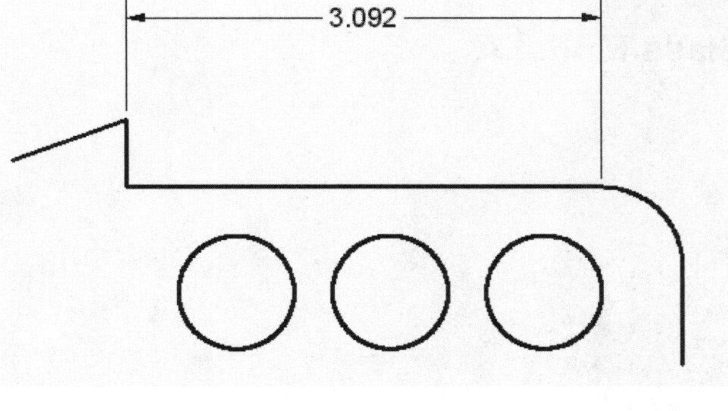

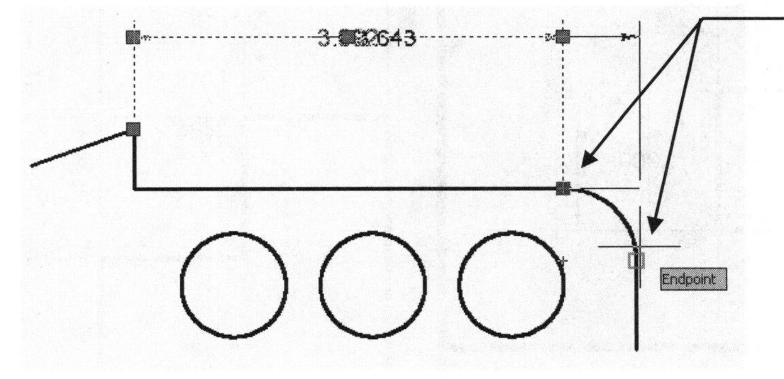

Drag the extension grip to the preferred location

Dimension Text changed to reflect new extension line location

After

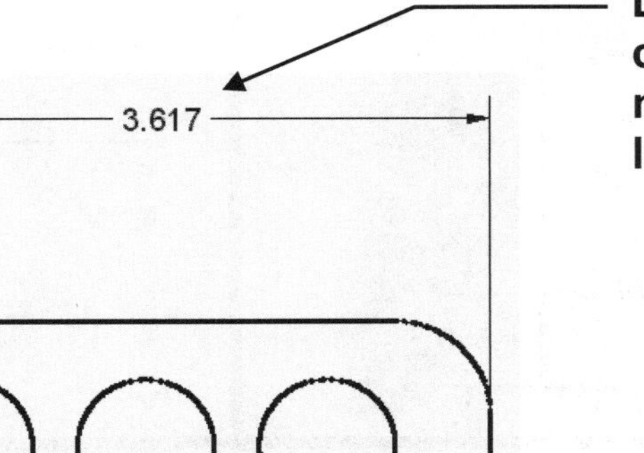

EXERCISE 20A

INSTRUCTIONS:

1. Open **EX16A**
2. Edit the dimensions as shown below using Quick Dimension. (Refer to page 20-4) *Notice: some of the Continuous have changed to Baseline and vice versa. This is a very helpful command.*
3. Edit the Title and Ex-XX by double clicking on the text. Do not erase and replace.
4. Save as **EX20A**
5. Plot using Page Setup **Class Model A**

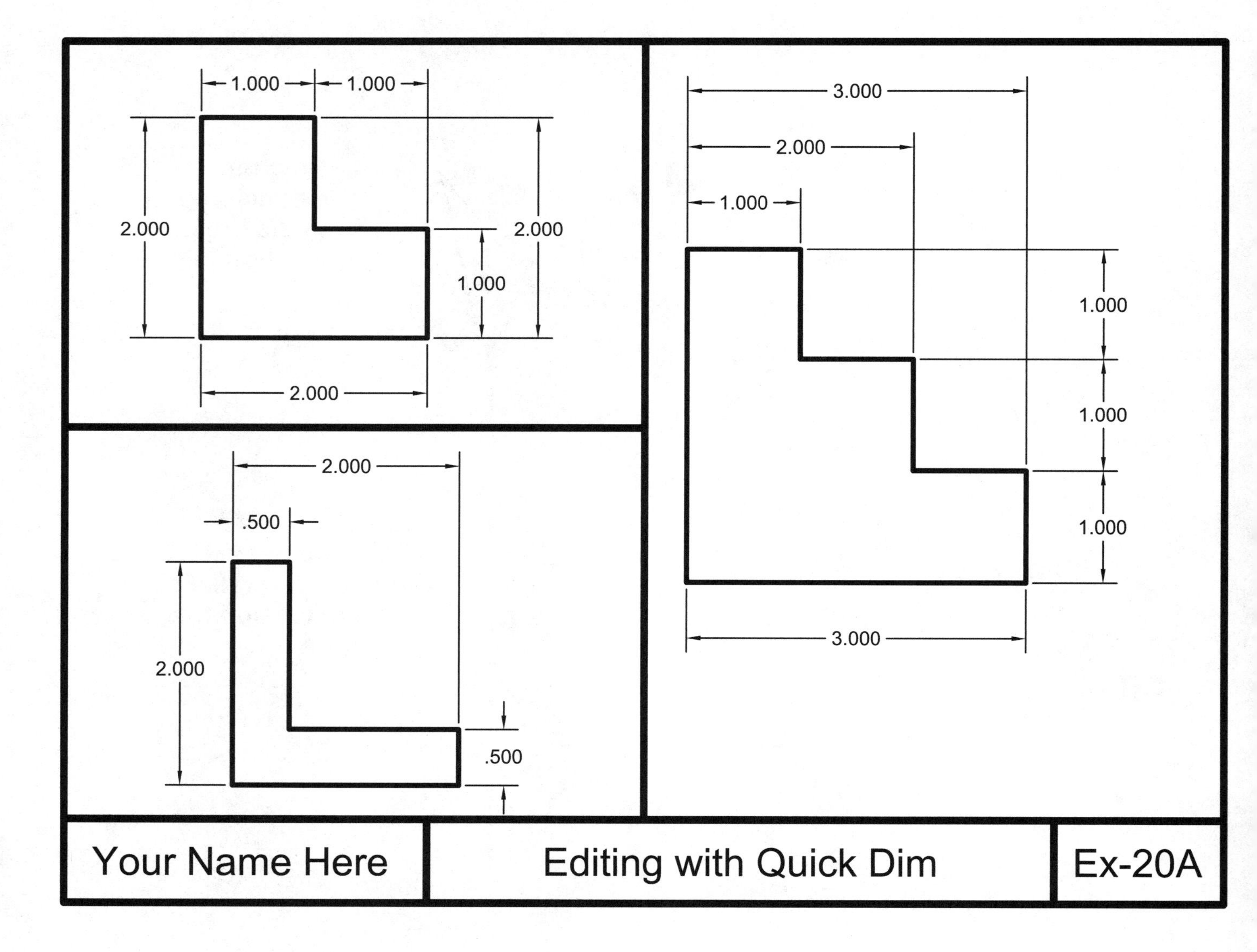

EXERCISE 20B

INSTRUCTIONS:

1. Start a **New** file using **Border A-2015.dwt**
2. Draw the 2 Rectangular forms below.
3. Dimension them using Quick Dimension, Diameter and Radius and a few Linear.

   This time you will have to use "Pick" instead of "Crossing Window" to select the objects to dimension.
   You may also have to move some of the dimension using grips because they may not be exactly where you want them.
   Try it, I think you will find it very interesting.

4. Edit the Title and Ex-XX by double clicking on the text. Do not erase and replace.
5. Save as **EX20B**
6. Plot using Page Setup **Class Model A**

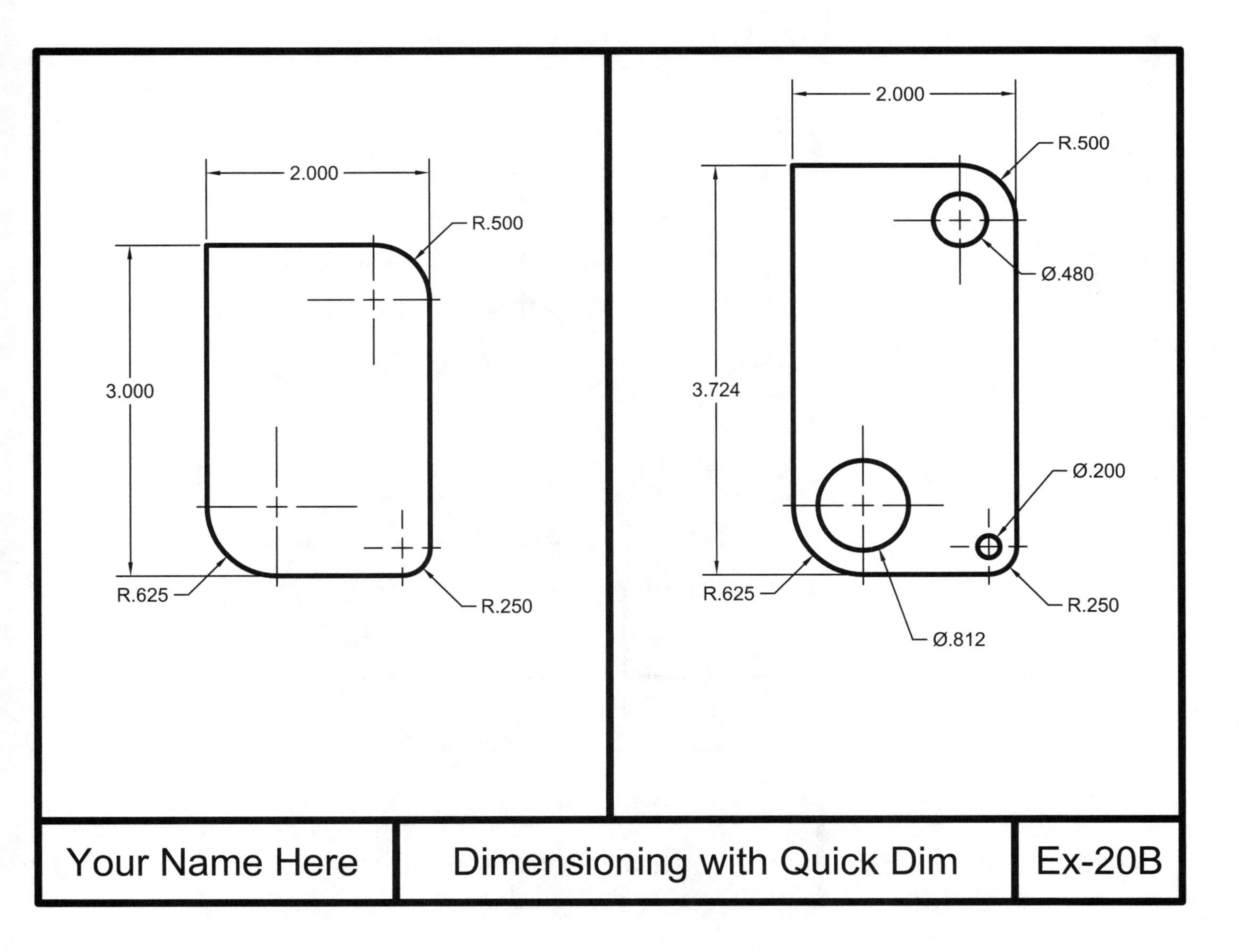

EXERCISE 20C

INSTRUCTIONS:

1. Start a **New** file using **Border A-2015.dwt**
2. Draw the object as shown below using Layer Object line.
3. Dimension using Quick Dimension whenever possible.
   You may have to use "grips" to move the extension lines to the correct endpoint on the object. And you may have to mix in a few Linear dimensions. Experiment and try some new methods.
4. Use Dimension Style "Class Style" and Layer Dimension
5. Edit the Title and Ex-XX by double clicking on the text. Do not erase and replace.
6. Save as **EX20C**
7. Plot using Page Setup **Class Model A**

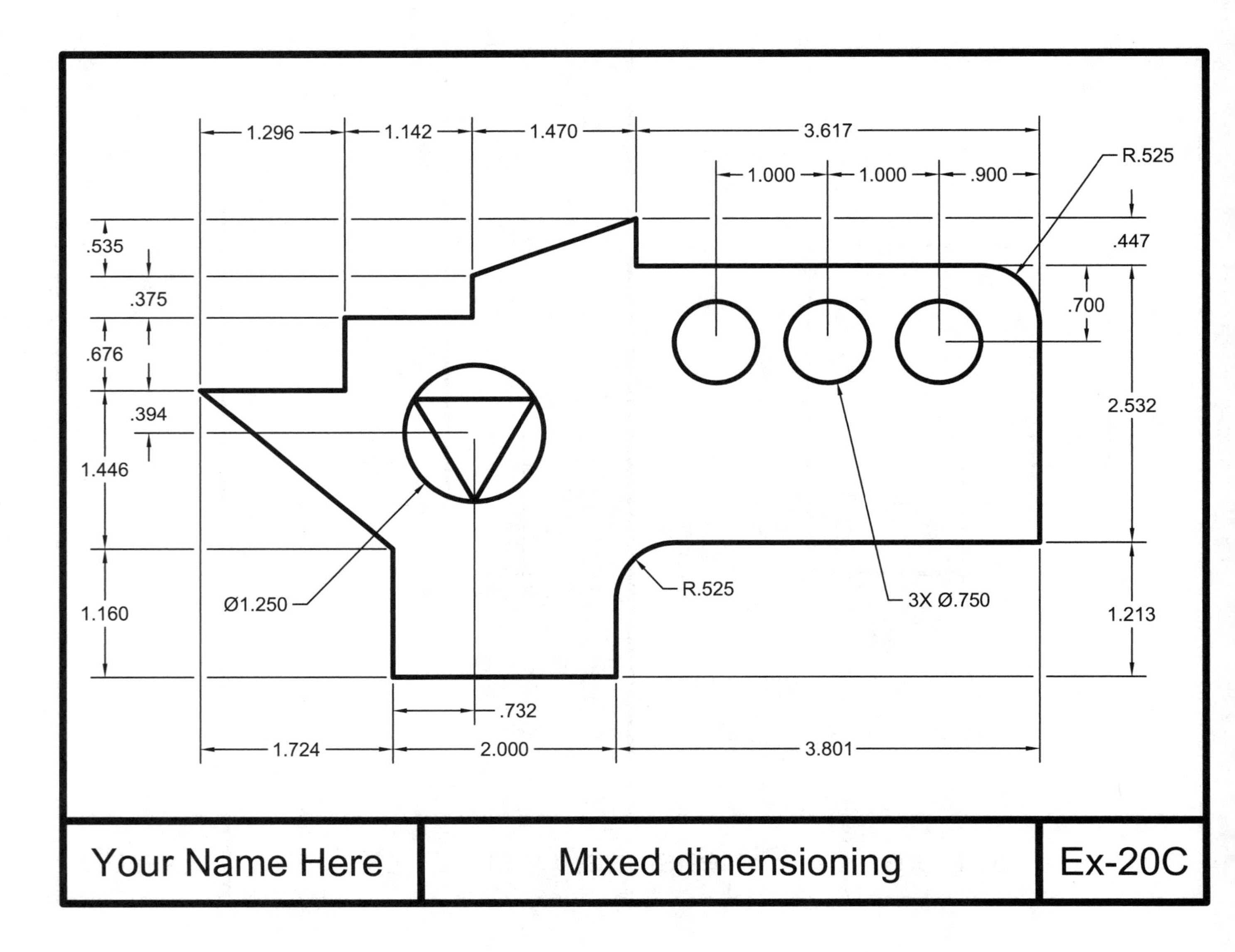

LEARNING OBJECTIVES

After completing this lesson, you will be able to:

1. Match an objects properties with another object
2. Create a Revision Cloud
3. Select a Revision Cloud Style
4. "Wipe Out" an area of an object

LESSON 21

MATCH PROPERTIES

Match Properties is used to "paint" the properties of one object to another. This is a simple and useful command. You first select the object that has the desired properties (the source object) and then select the object you want to "paint" the properties to (destination object).

Only one "source object" can be selected but its properties can be painted to any number of "destination objects".

1. Select the Match Properties command using one of the following:

Ribbon = Home tab / Properties panel /
or
Keyboard = MA <enter>

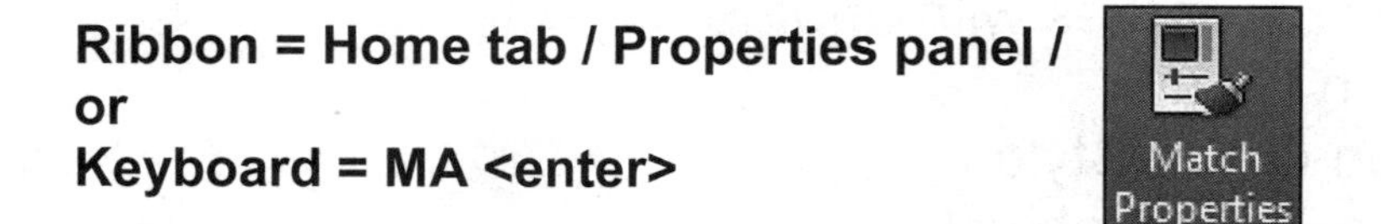

Command: matchprop

2. Select source object: ***select the object with the desired properties to match***

3. Select destination object(s) or [Settings]: ***select the object(s) you want to receive the matching properties.***

4. Select destination object(s) or [Settings]: ***select more objects or <enter> to stop.***

Note: If you do not want to match all of the properties, after you have selected the source object, right click and select "Settings" from the short cut menu, before selecting the destination object. Uncheck all the properties you do not want to match and select the OK button. Then select the destination object(s).

MATCH LAYER

If you draw an object on the wrong layer you can easily change it to the desired layer using Layer Match command. You first select the object that needs to be changed and then select an object that is on the correct layer (object on destination layer).

1. Select the Layer Match command using one of the following:

   Ribbon = Home tab / Layers panel / Match Layer
   or
   Keyboard = Laymch <enter>

   Command: _laymch
2. Select objects to be changed:***select the objects that need to be changed***

3. Select objects: ***select more objects or <enter> to stop selecting***

4. Select object on destination layer or [Name]: ***select the object that is on the layer that you want to change to***

Note:
You may also easily change the layer of an object to another layer as follows:

1. Select the object that you wish to change.

2. Select the Layer drop down menu and select the layer that you wish the object to be placed on.

For example:
If you had a dimension on the **OBJECT LINE** layer by mistake.

1. Select the dimension that is mistakenly on the **OBJECT LINE** layer.

2. Select the **DIMENSION** layer from the drop down menu.

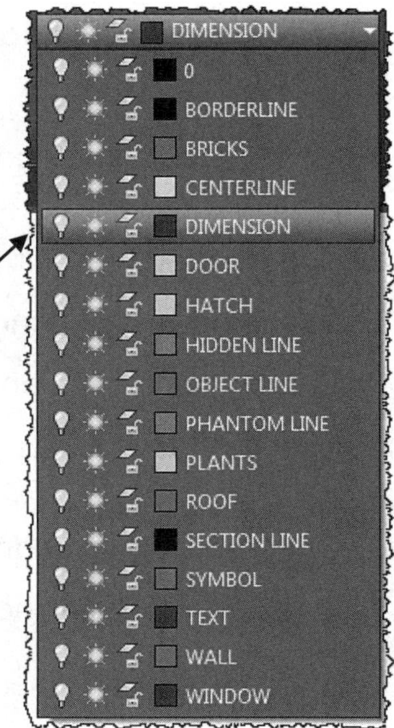

CREATING A REVISION CLOUD

When you make a revision to a drawing it is sometimes helpful to highlight the revision for someone viewing the drawing. A common method to highlight the area is to draw a “Revision Cloud” around the revised area. This can be accomplished easily with the “Revision Cloud” command.

The Revision Cloud command creates a series of sequential arcs to form a cloud shaped object. You set the minimum and maximum arc lengths. (Maximum arc length cannot exceed three times the minimum arc length. Example: Min = 1, Max can be 3 or less) If you set the minimum and maximum different lengths the arcs will vary in size and will display an irregular appearance.

Min & Max same length

Min & Max different length

To draw a Revision Cloud you specify the start point with a left click then drag the cursor to form the outline. AutoCAD automatically draws the arcs. When the cursor gets very close to the start point, AutoCAD snaps the last arc to the first arc and closes the shape.

1. Select the Revision Cloud command using one of the following:

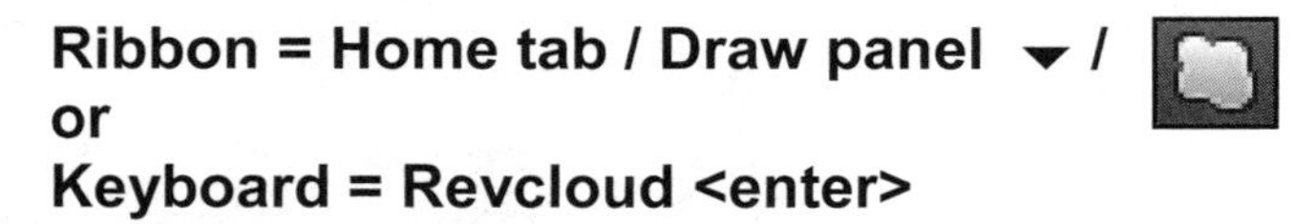

Ribbon = Home tab / Draw panel ▾ /
or
Keyboard = Revcloud <enter>

Command: _revcloud
Minimum arc length: .50 Maximum arc length: .50 Style: Normal

2. Specify start point or [Arc length/Object/Style] <Object>: ***Select “Arc length”***
3. Specify minimum length of arc <.50>: ***Specify the minimum arc length***
4. Specify maximum length of arc <.50>: ***Specify the maximum arc length***
5. Specify start point or [Arc length/Object/Style] <Object>: ***Place cursor at start location & left click.***
6. Guide crosshairs along cloud path...***Move the cursor to create the cloud outline.***
7. Revision cloud finished. ***When the cursor approaches the start point, the cloud closes automatically.***

CONVERT A CLOSED OBJECT TO A REVISION CLOUD

You can convert a closed object, such as a circle, ellipse, rectangle or closed polyline to a revision cloud. The original object is deleted when it is converted.

(If you want the original object to remain, in addition to the new rev cloud, set the variable "delobj" to "0". The default setting is "1".)

1. Draw a closed object such as a circle.

2. Select the Revision Cloud command using one of the following:

   Ribbon = Home tab / Draw panel ▾ /
   or
   Keyboard = Revcloud <enter>

   Command: _revcloud
   Minimum arc length: .50   Maximum arc length: .50   Style: Normal

3. Specify start point or [Arc length/Object/Style] <Object>: ***Select "Arc length"***

4. Specify minimum length of arc <.50>: ***Specify the minimum arc length***

5. Specify maximum length of arc <.50>: ***Specify the maximum arc length***

6. Specify start point or [Arc length/Object/Style] <Object>: ***Select "Object".***

7. Select object: ***Select the object to convert***

8. Select object: Reverse direction [Yes/No] <No>: ***Select Yes or No***
   Revision cloud finished.

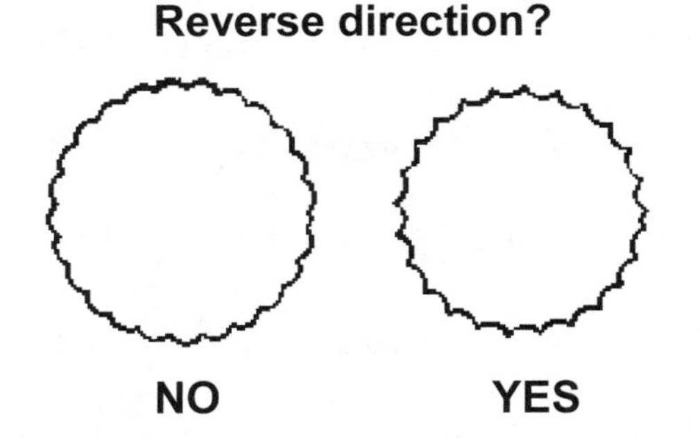

NOTE:
The Match Properties command will not match the arc length from the source cloud to the destination cloud.

REVISION CLOUD STYLE

You may select one of 2 styles for the Revision Cloud; **Normal** or **Calligraphy**.
Normal will draw the cloud with one line width.
Calligraphy will draw the cloud with variable line widths to appear as though you used a chiseled calligraphy pen.

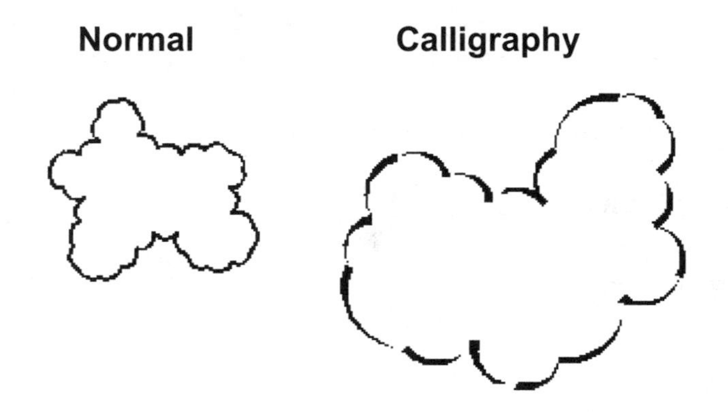

1. Select the Revision Cloud command using one of the following:

   Ribbon = Home tab / Draw panel ▾ /
   or
   Keyboard = Revcloud <enter>

   Command: _revcloud
   Minimum arc length: .50 Maximum arc length: 1.00 Style: Normal

2. Specify start point or [Arc length/Object/Style] <Object>: ***Select "Style"<enter>***

3. Select arc style [Normal/Calligraphy] <Calligraphy>: ***Select "N or C"<enter>***

4. Specify start point or [Arc length/Object/Style] <Object>: ***Select "Arc length"***

5. Specify minimum length of arc <.50>: ***Specify the minimum arc length***

6. Specify maximum length of arc <1.00>: ***Specify the maximum arc length***

7. Specify start point or [Object] <Object>: ***Place cursor at start location & left click.***

8. Guide crosshairs along cloud path...***Move the cursor to create the cloud outline.***

9. Revision cloud finished. ***When the cursor approaches the start point, the cloud closes automatically.***

WIPEOUT

The Wipeout command creates a blank area that covers existing objects. The area has a background that matches the background of the drawing area. This area is bounded by the wipeout frame, which you can turn on or off.

1. Select the Wipeout command using one of the following:

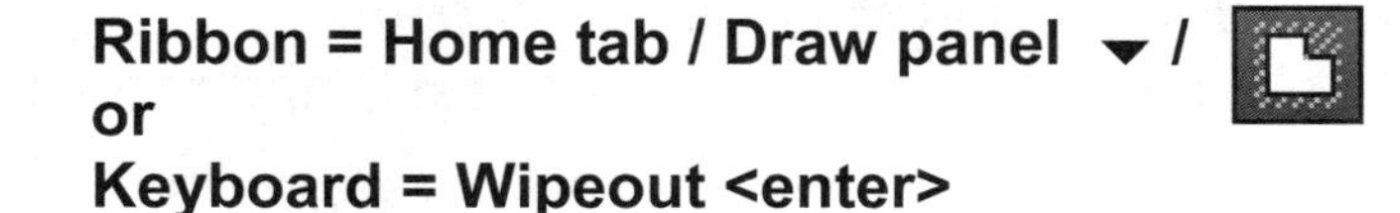

Ribbon = Home tab / Draw panel ▾ /
or
Keyboard = Wipeout <enter>

2. Command: _wipeout Specify first point or [Frames/Polyline] <Polyline>: ***specify the first point of the shape (P1)***
3. Specify next point: ***specify the next point (P2)***
4. Specify next point or [Undo]: ***specify the next point (P3)***
5. Specify next point or [Undo]: ***specify the next point or <enter> to stop***

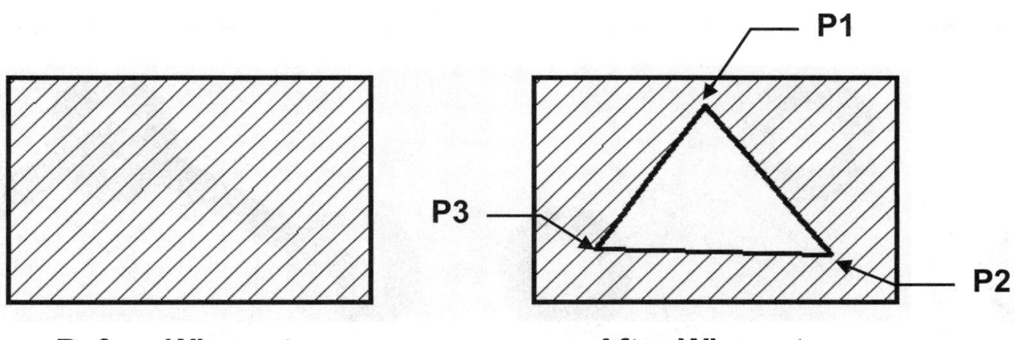

Before Wipeout

After Wipeout

TURNING FRAMES ON OR OFF

1. Select the Wipeout command.
2. Select the “Frames” option.
3. Enter ON or OFF.

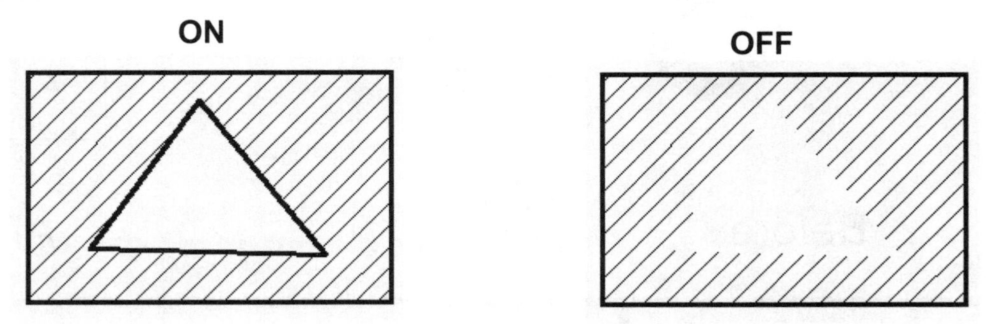

Note: If you want to move the objects and the wipeout area, you must select both and move them at the same time. Do not move them separately.

EXERCISE 21A

INSTRUCTIONS:

1. Open **EX5F.dwg**
2. Using **Match Properties,** change the properties of the Polygon to the same properties as the Donut.
3. Using **Wipeout**, block out the bottom donut approximately as shown.
4. Edit the Title and Ex-XX by double clicking on the text. Do not erase and replace.
5. Save as **EX21A**
6. Plot using Page Setup **Class Model A**

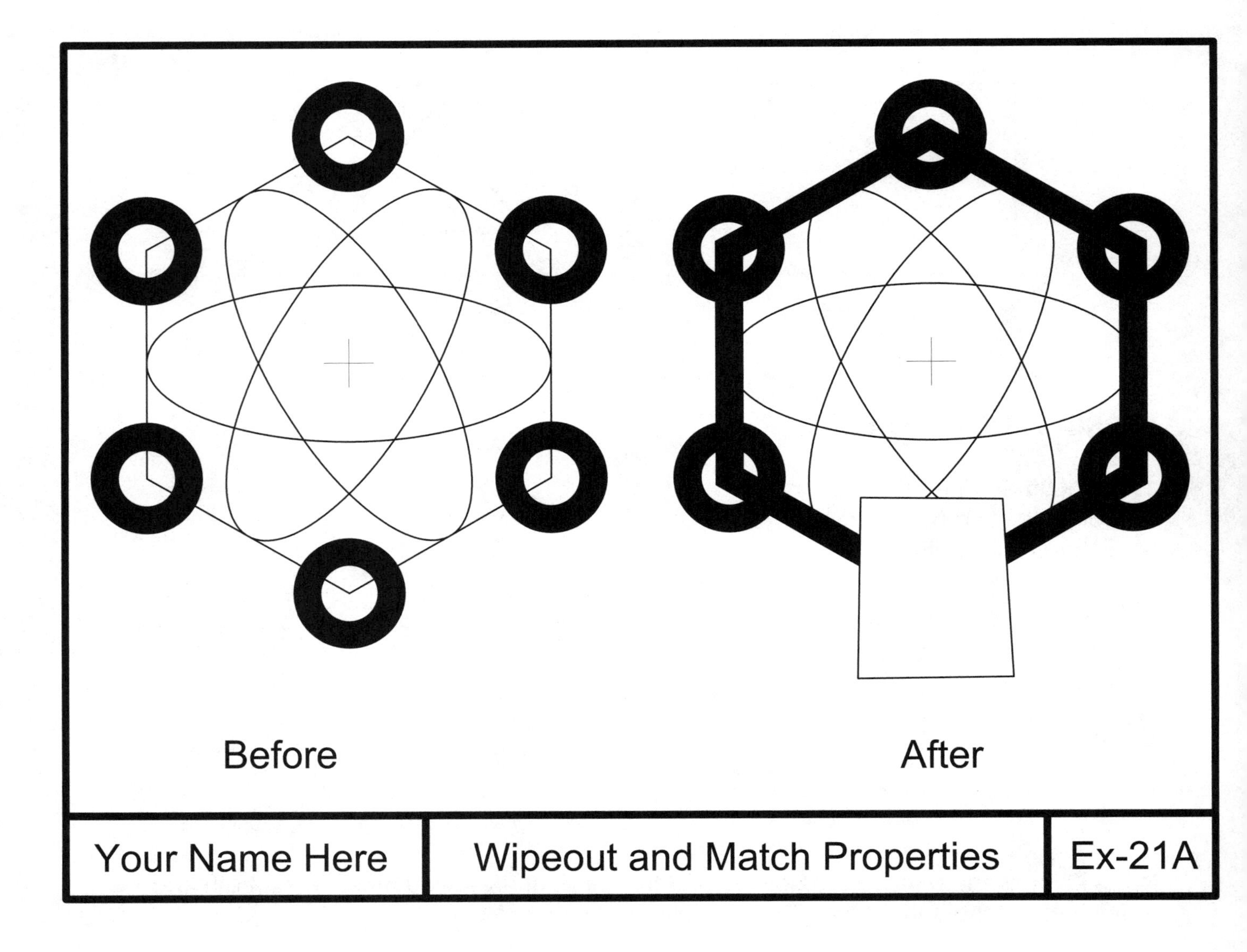

EXERCISE 21B

INSTRUCTIONS:

1. Start a **New** file using **Border A-2015.dwt**
2. Draw the Revision clouds below:
   - #1: Set Min. and Max. Arc Length to .500
   - #2: Set Min. Arc Length to .500 and Max. Arc Length to 1.00
   - #3: Same as #2 but change Style to Calligraphy.
3. Draw (2) 2.50 Diameter Circles and Convert to Revision clouds.
   Set Min and Max Arc Length to .50   Set Style to Normal
   #4 Reverse = No   #5 Reverse = Yes
4. Use Layer Object line.
5. Edit the Title and Ex-XX by double clicking on the text. Do not erase and replace.
6. Save as **EX21B**
7. Plot using Page Setup **Class Model A**

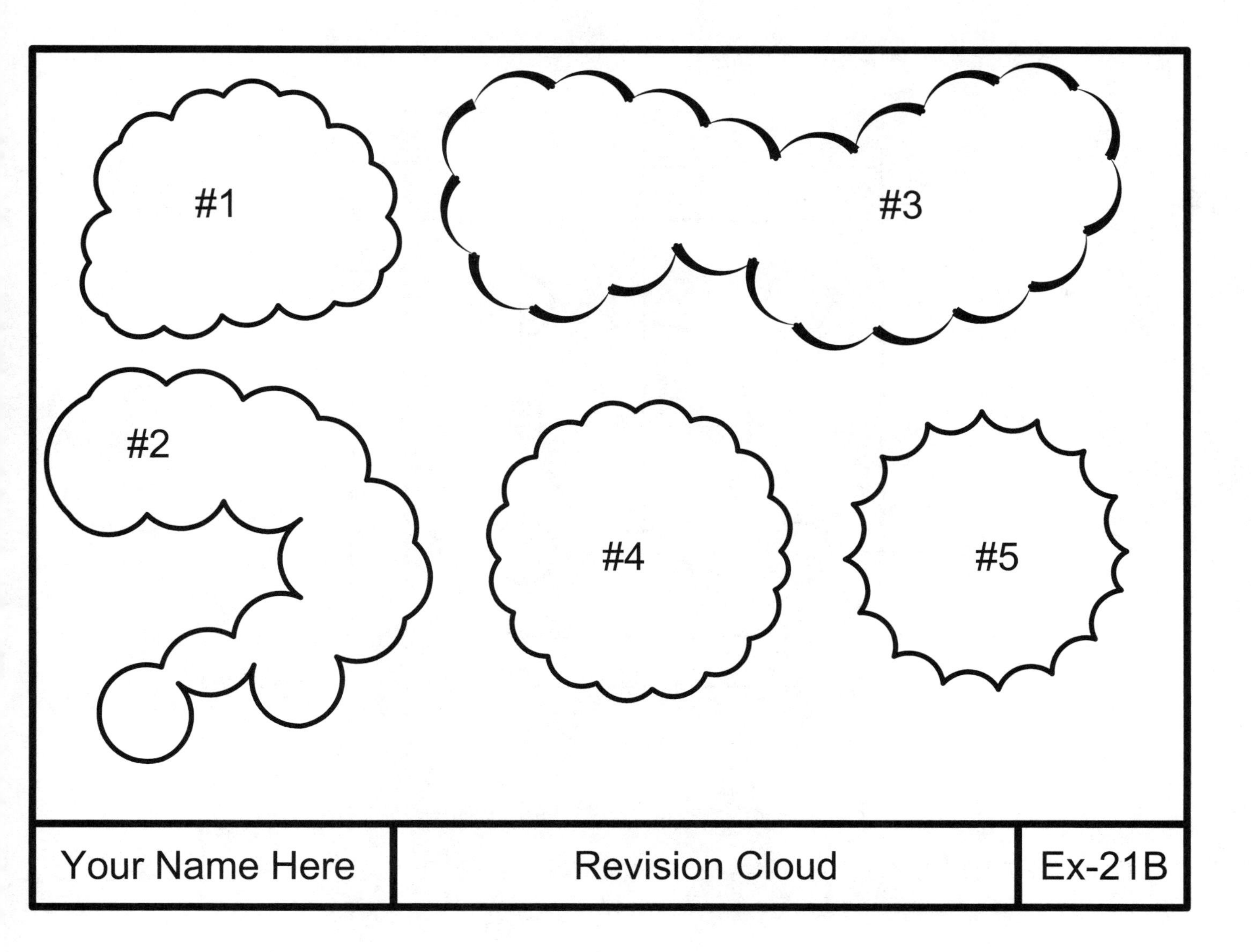

EXERCISE 21C

INSTRUCTIONS:

1. Start a **New** file using **Border A-2015.dwt**
2. The drawing below is designed to give you more practice drawing objects and dimensioning. Use any method that you have learned in the previous lessons.
3. Edit the Title and Ex-XX by double clicking on the text. Do not erase and replace.
4. Save as **EX21C**
5. Plot using Page Setup **Class Model A**

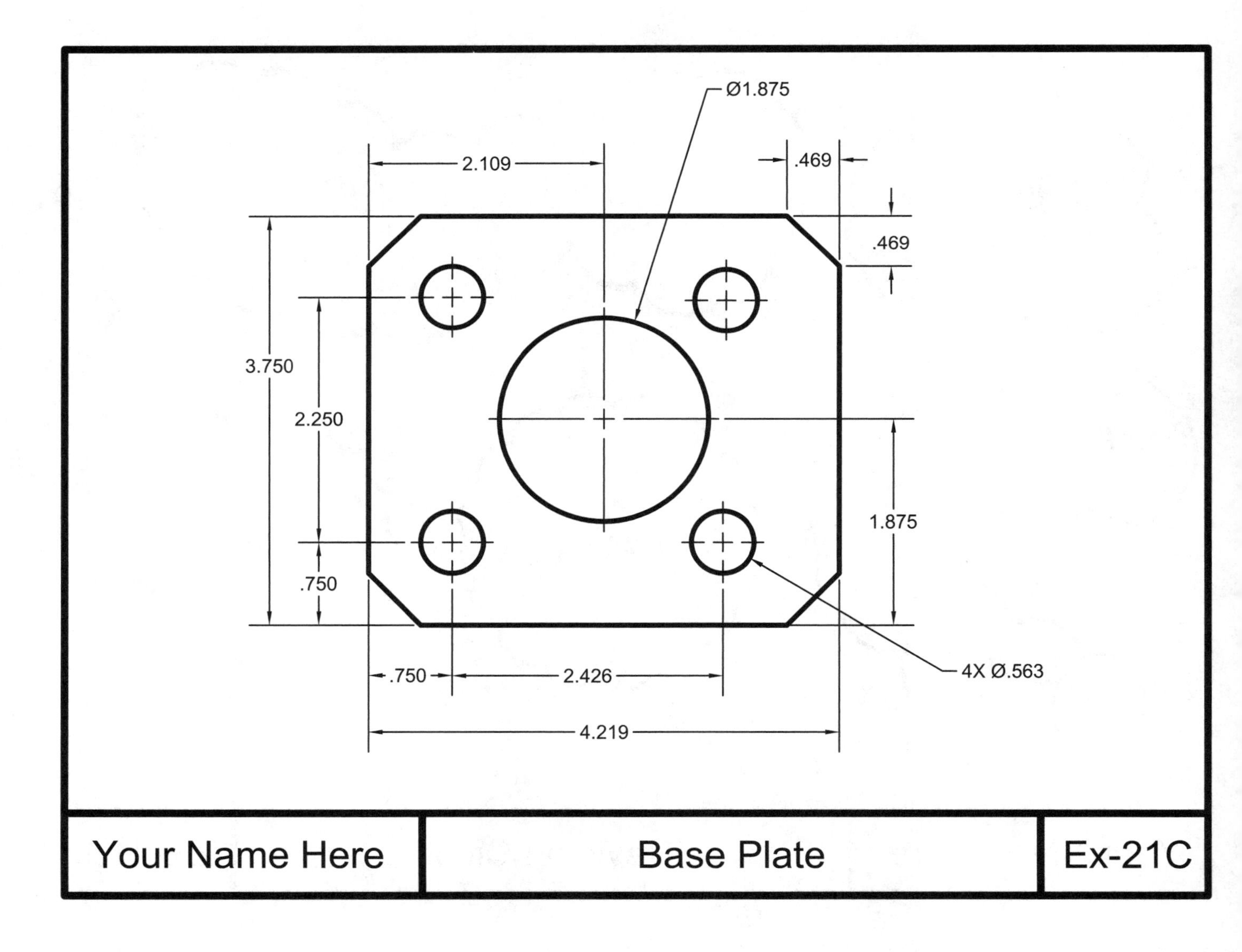

EXERCISE 21D

INSTRUCTIONS:

1. Start a **New** file using **Border A-2015.dwt**
2. The drawing below is designed to give you more practice drawing objects and dimensioning. Use any method that you have learned in the previous lessons.
3. Edit the Title and Ex-XX by double clicking on the text. Do not erase and replace.
4. Save as **EX21D**
5. Plot using Page Setup **Class Model A**

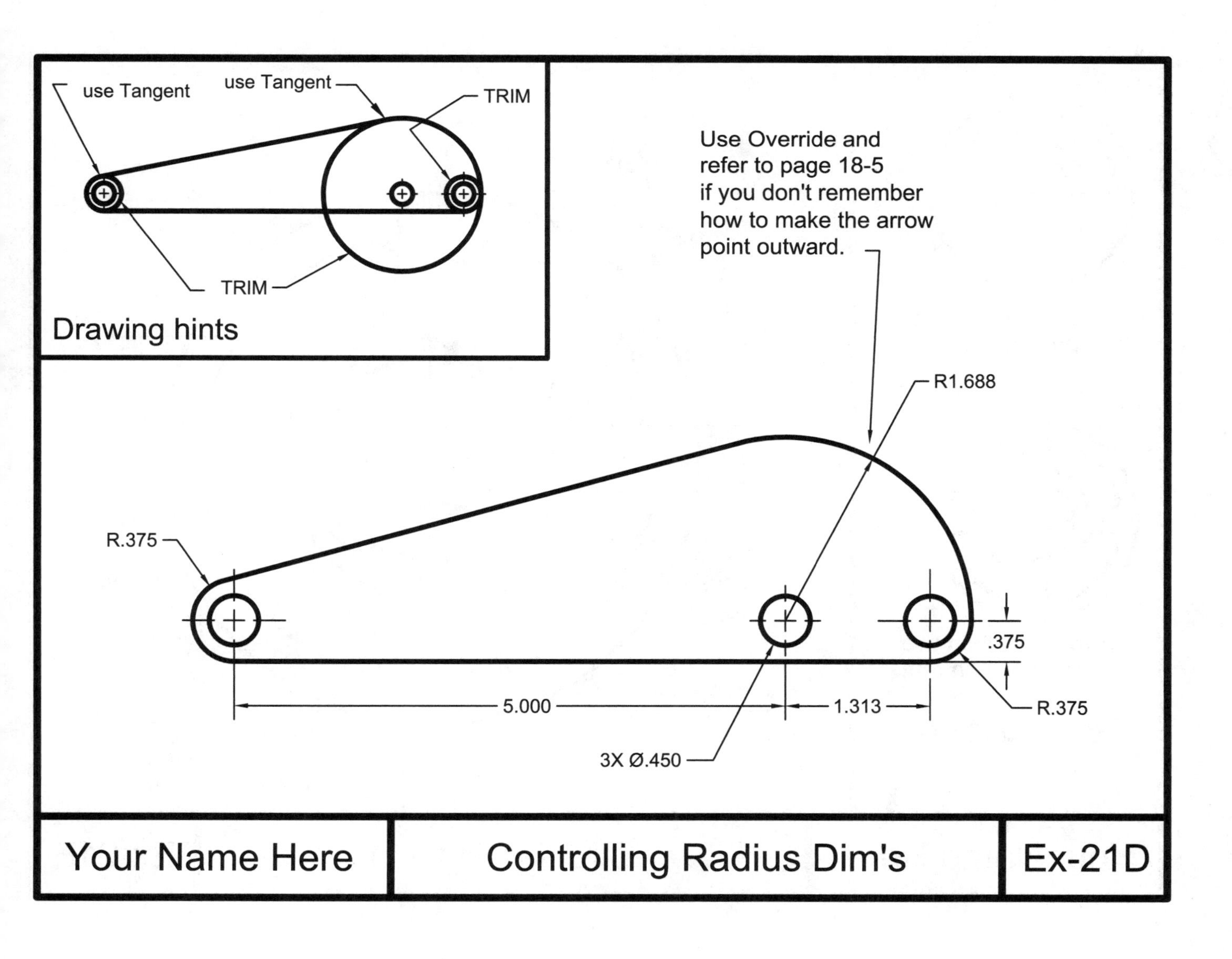

EXERCISE 21E

INSTRUCTIONS:

1. Start a **New** file using **Border A-2015.dwt**
2. The drawing below is designed to give you more practice drawing objects and dimensioning. Use any method that you have learned in the previous lessons.
3. Edit the Title and Ex-XX by double clicking on the text. Do not erase and replace.
4. Save as **EX21E**
5. Plot using Page Setup **Class Model A**

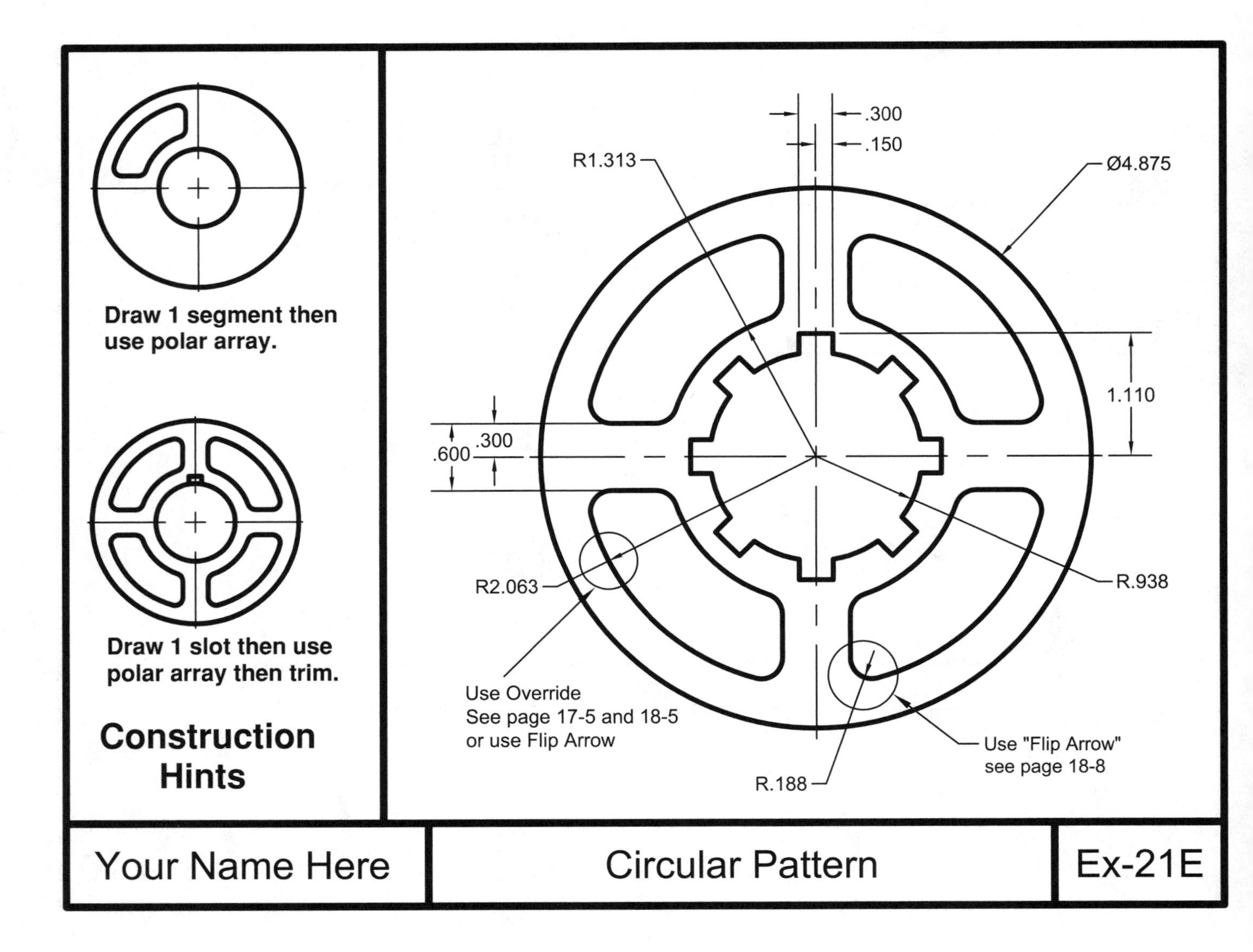

LEARNING OBJECTIVES

After completing this lesson, you will be able to:

1. Draw an Arc using 10 different methods
2. Dimension Arc segments

LESSON 22

ARC

There are 10 ways to draw an ARC in AutoCAD. Not all of the ARCS options are easy to create so you may find it is often easier to **trim a Circle** or use the **Fillet** command.

On the job, you will probably only use 2 of these methods. Which 2 depends on the application.

An **ARC** is a segment of a circle and must be less than 360 degrees.

By default, ARCS are drawn counter-clockwise. You can change the direction by holding down the **Ctrl** key to draw in a clockwise direction. Or in some cases you can enter a negative input to draw in a clockwise direction.

1. Select the Arc Command using one of the following:

   Ribbon = Home tab / Draw panel /
   or
   Keyboard = A <enter>

2. Refer to Exercises 22A through 22J for examples of each method listed below.

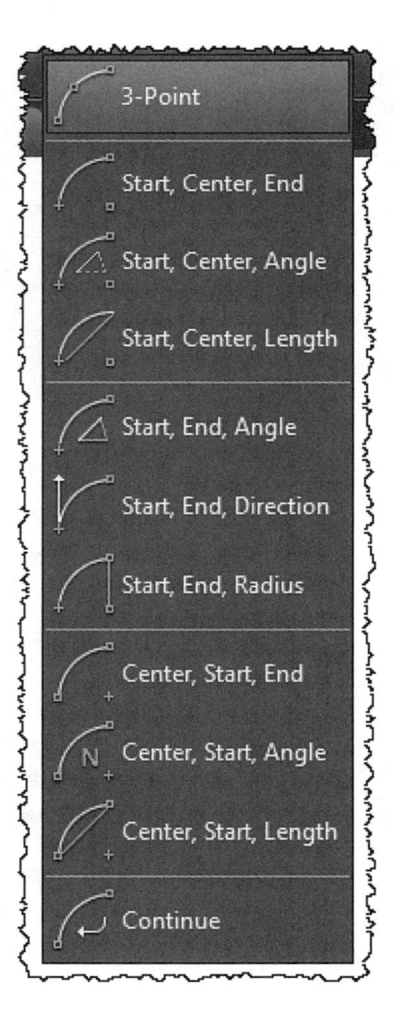

DIMENSIONING ARC LENGTHS

You may dimension the distance along an Arc. This is known as the **Arc length**.
Arc length is an <u>associative</u> dimension.

Example:

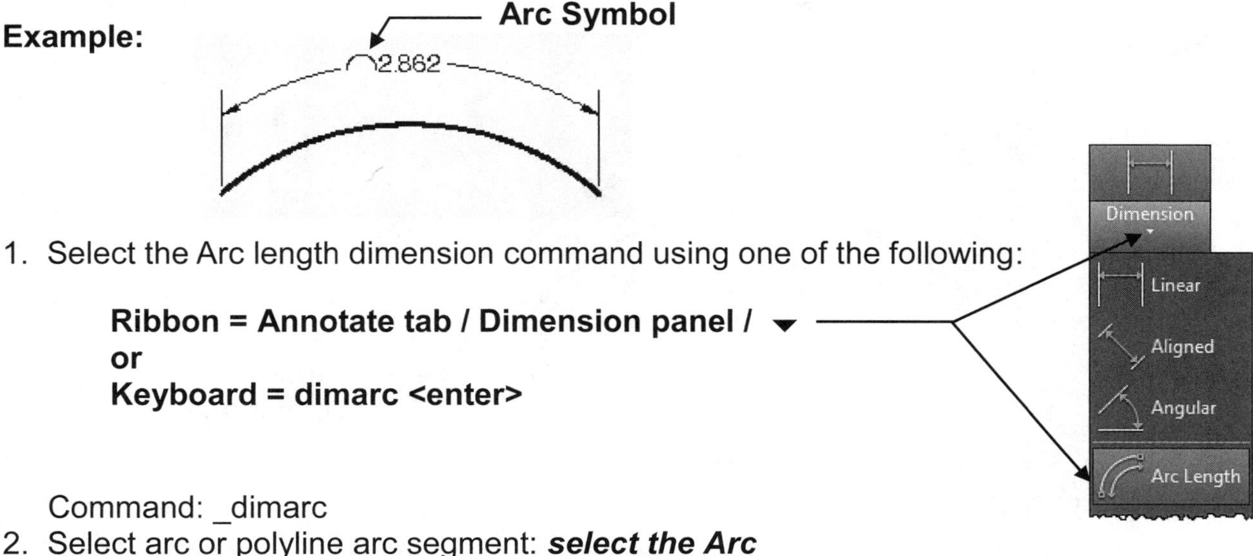

1. Select the Arc length dimension command using one of the following:

   Ribbon = Annotate tab / Dimension panel / ▼
   or
   Keyboard = dimarc <enter>

   Command: _dimarc
2. Select arc or polyline arc segment: ***select the Arc***
3. Specify arc length dimension location, or [Mtext/Text/Angle/Partial/Leader]:
   place the dimension line and text location

   Dimension text = ***dimension value will be shown here***

To differentiate the Arc length dimensions from Linear or Angular dimensions, arc length dimensions display an arc (⌒) symbol by default. (Also called a "hat" or "cap")

The arc symbol may be displayed either <u>above,</u> or <u>preceding</u> the dimension text.
You may also choose not to display the arc symbol.

Example:

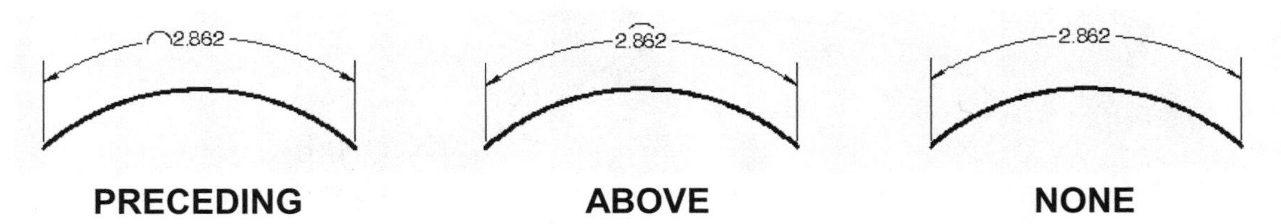

PRECEDING **ABOVE** **NONE**

Specify the placement of the arc symbol in the **Dimension Style / Symbols and Arrows** tab or you may edit its position using the Properties Palette.

Arc length symbol
- Preceding dimension text
- Above dimension text
- None

Continued on the next page...

DIMENSIONING ARC LENGTHS....continued

The extension lines of an Arc length dimension are displayed as radial if the included angle is greater than 90 degrees.

Example:

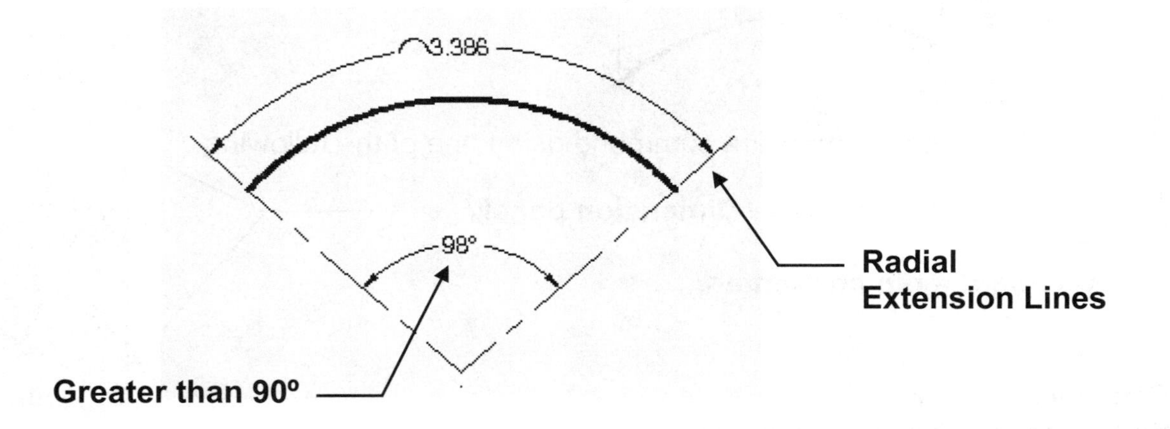

The extension lines of an Arc length dimension are displayed as orthogonal if the included angle is less than 90 degrees.

Example:

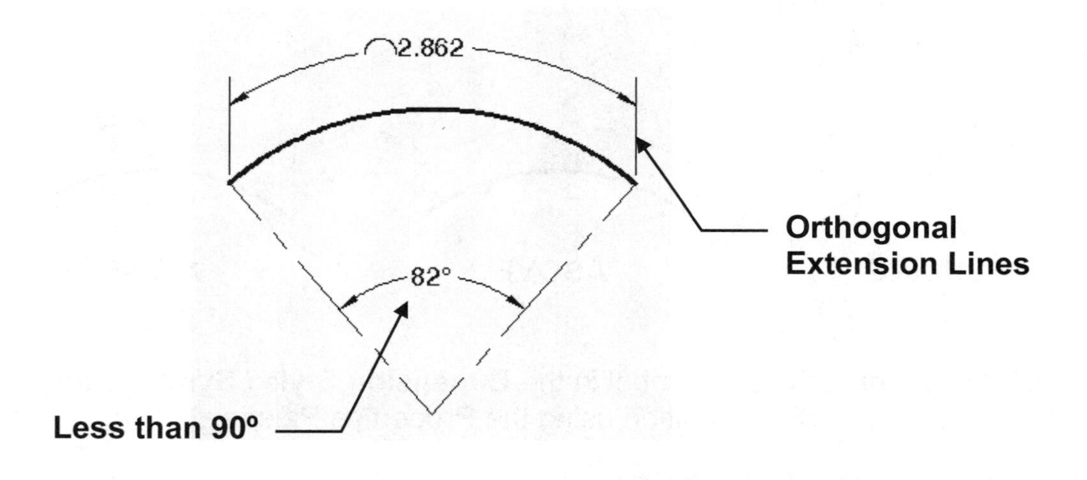

DIMENSIONING A LARGE CURVE

When dimensioning an arc the dimension line should pass through the center of the arc. However, for large curves, the true center of the arc could be very far away, even off the sheet.

When the true center location cannot be displayed you can create a "Jogged" radius dimension.

Example:

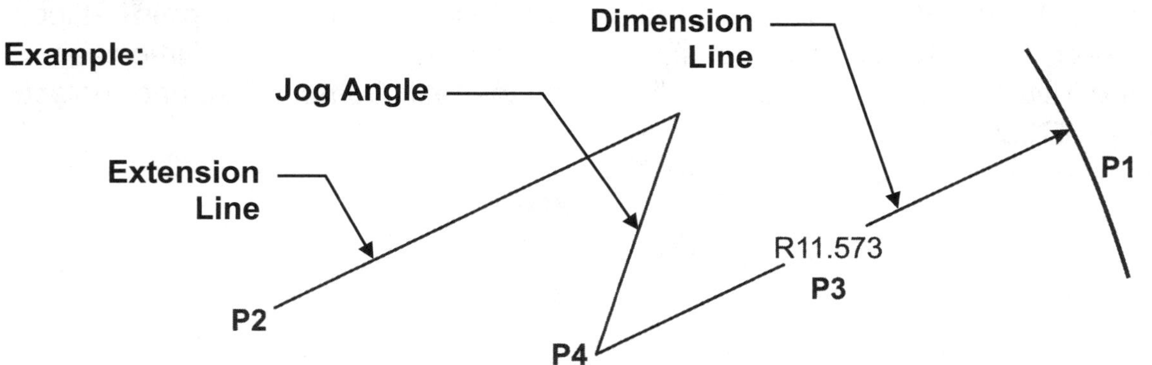

You can specify the jog angle in the **Dimension Style / Symbols and Arrows tab**.

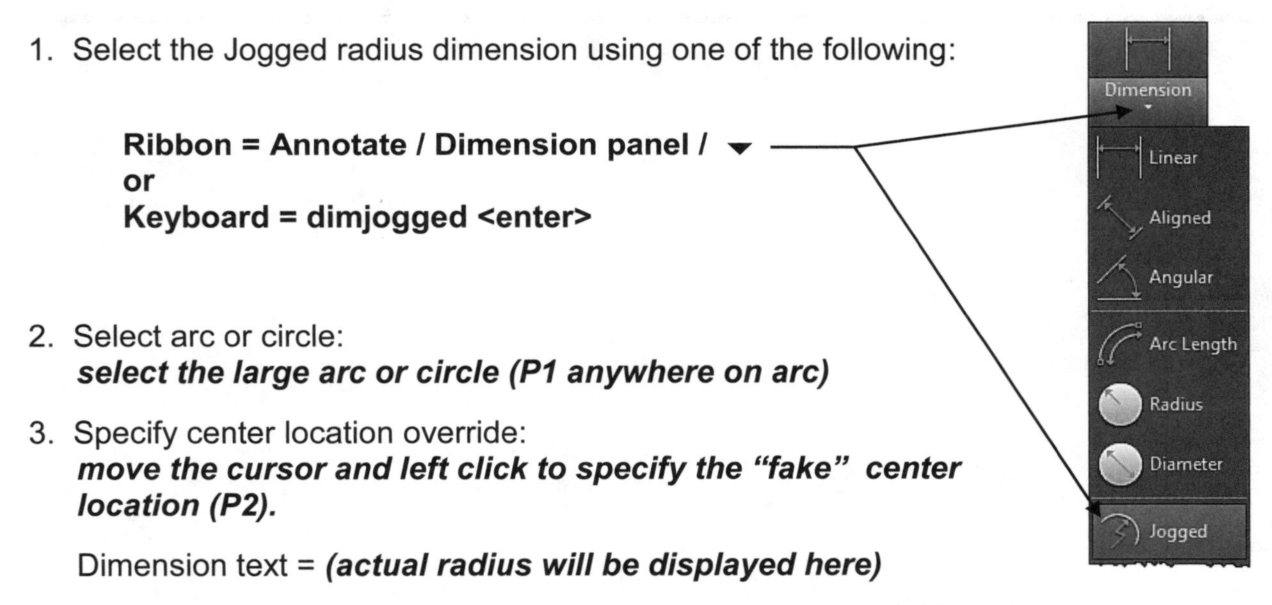

1. Select the Jogged radius dimension using one of the following:

   Ribbon = Annotate / Dimension panel / ▾
   or
   Keyboard = dimjogged <enter>

2. Select arc or circle:
   select the large arc or circle (P1 anywhere on arc)

3. Specify center location override:
   move the cursor and left click to specify the "fake" center location (P2).

   Dimension text = ***(actual radius will be displayed here)***

4. Specify dimension line location or [Mtext/Text/Angle]: ***move the cursor and left click to specify the location for the dimension text (P3).***

5. Specify jog location: ***move the cursor and left click to specify the location for the jog (P4).***

CONTROLLING THE JOG

You can set the Jog Angle and Height factor in:
Dimension Style / Symbols and Arrows tab

Radius jog dimension
Jog angle: 45

Linear jog dimension
Jog height factor:
1.500 * Text height

EXERCISE 22A

INSTRUCTIONS:

1. Start a **New** file using **Border A-2015.dwt**
2. Draw the center lines first on layer Center line.
3. Draw the **3 Point** Arcs using Layer Object Line.
   Place the **1st point** at location (1) **2nd point** at location (2) and **3rd point** at location (3.)
4. Dimension as shown using Dimension Style Class Style and Layer Dimension.
5. Edit the Title and Ex-XX by double clicking on the text. Do not erase and replace.
6. Save as **EX22A**
7. Plot using Page Setup **Class Model A**

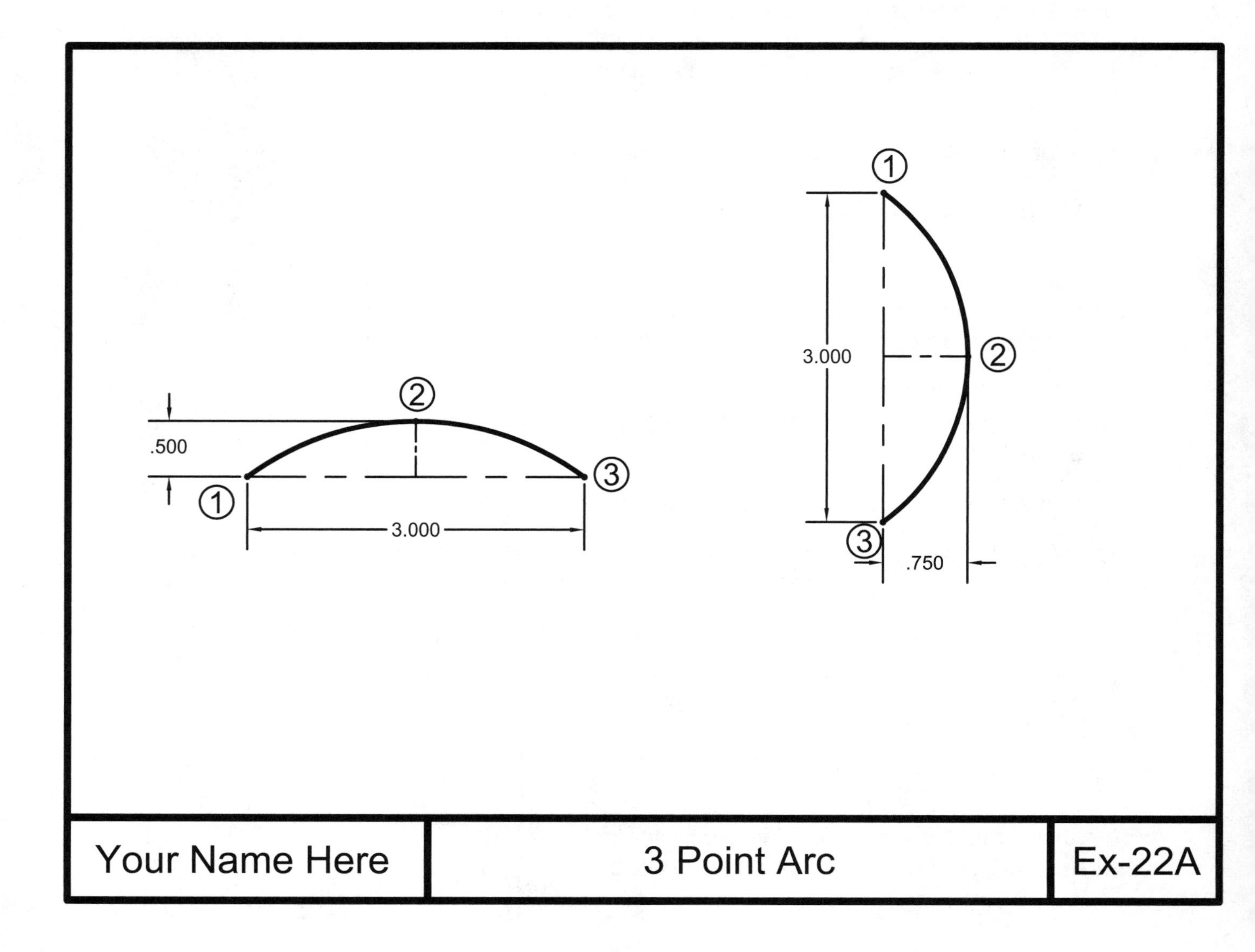

EXERCISE 22B

INSTRUCTIONS:

1. Start a **New** file using **Border A-2015.dwt**
2. Draw the center lines first on layer Center line.
3. Draw the 2 Arcs using method **Start, Center, End** on Layer Object Line.
   Place the **Start** at location (1), **Center** at location (2) and **End** at location (3.)
4. Dimension as shown using Dimension Style Class Style and Layer Dimension.
5. Edit the Title and Ex-XX by double clicking on the text. Do not erase and replace.
6. Save as **EX22B**
7. Plot using Page Setup **Class Model A**

Note: *By default these arcs are always drawn counter-clockwise from the start point. Hold down the* ***Ctrl*** *key to draw in a clockwise direction.*

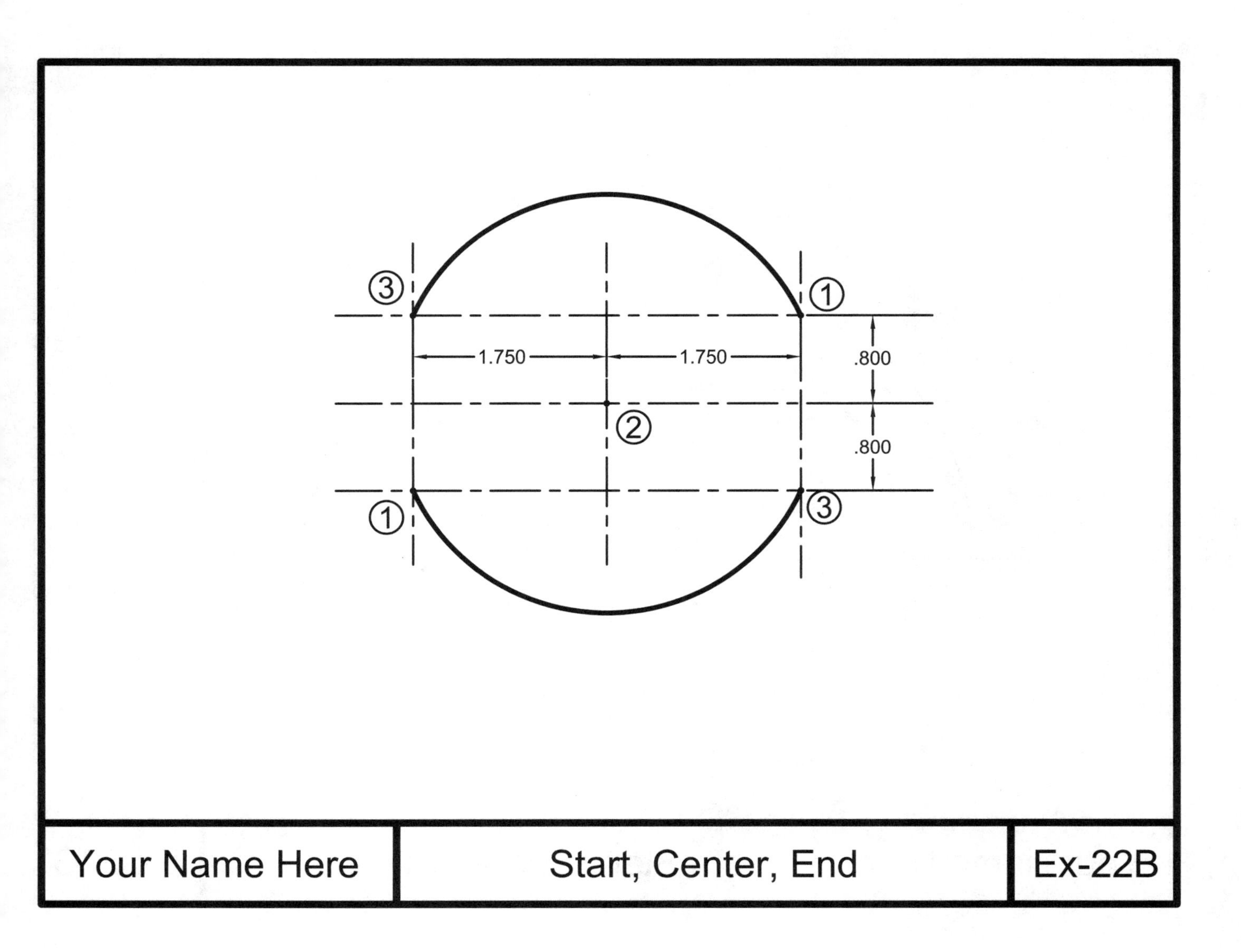

EXERCISE 22C

INSTRUCTIONS:

1. Start a **New** file using **Border A-2015.dwt**
2. Draw the center lines first on layer Center line.
3. Draw the 2 Arcs using method **Start, Center, Angle** on Layer Object Line. Place the **Start** at location (1), **Center** at location (2) and enter **Angle** (3.)
4. Dimension as shown using Dimension Style Class Style and Layer Dimension.
5. Edit the Title and Ex-XX by double clicking on the text. Do not erase and replace.
6. Save as **EX22C**
7. Plot using Page Setup **Class Model A**

Note: *Positive angles are drawn counter-clockwise, negative angles are drawn clockwise. To reverse the directions hold down the* ***Ctrl*** *key.*

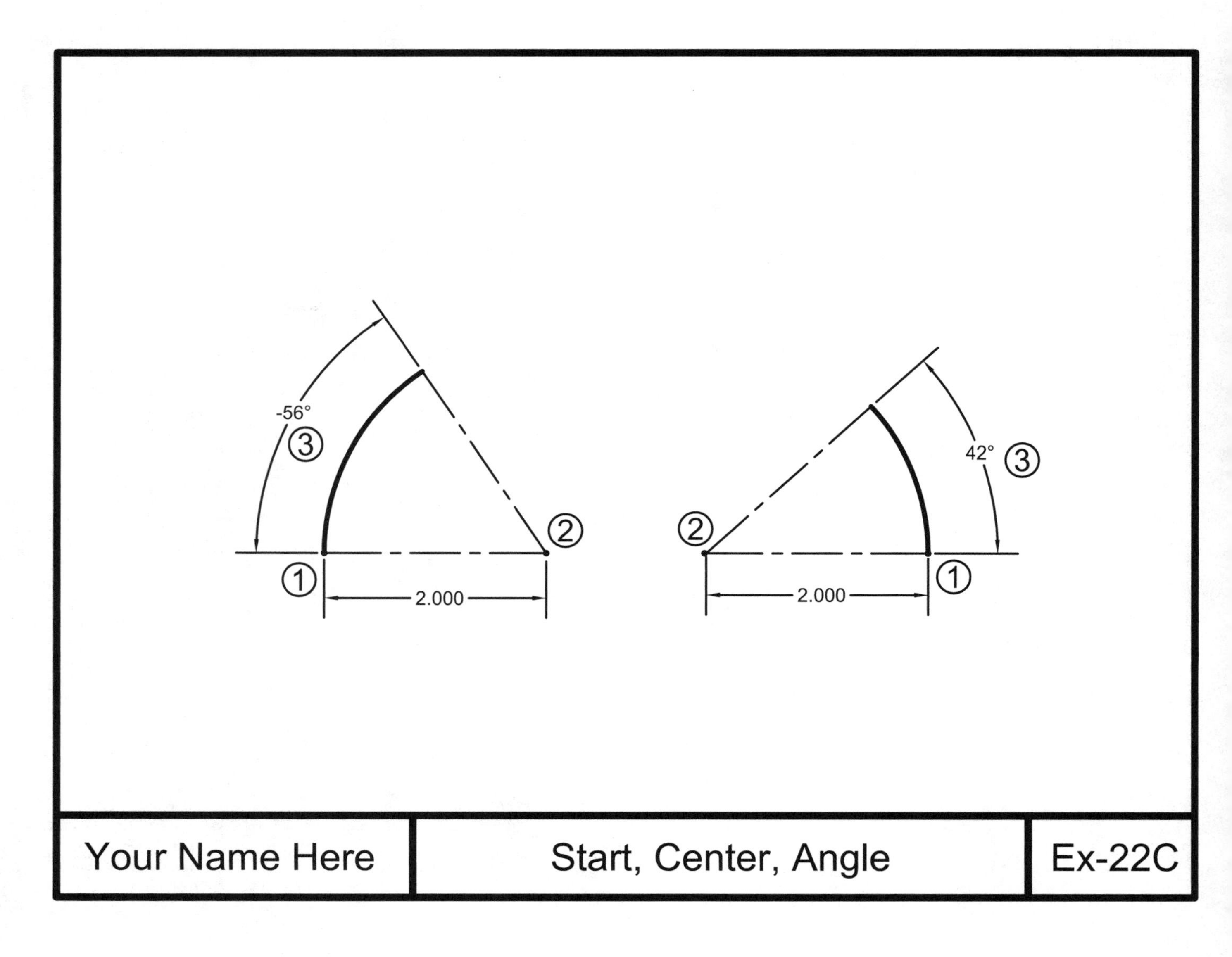

EXERCISE 22D

INSTRUCTIONS:

1. Start a **New** file using **Border A-2015.dwt**
2. Draw the center lines first on layer Center line.
3. Draw the 2 Arcs using method **Start, Center, Length** on Layer Object Line.
   Place the **Start** at location (1), **Center** at location (2) and enter **Length** of Chord (3).
4. Dimension as shown using Dimension Style Class Style and Layer Dimension.
5. Edit the Title and Ex-XX by double clicking on the text. Do not erase and replace.
6. Save as **EX22D**
7. Plot using Page Setup **Class Model A**

Note: *Positive Chord length draws the small segment counter-clockwise, negative Chord length draws the large segment counter-clockwise. To reverse the directions hold down the* ***Ctrl*** *key.*

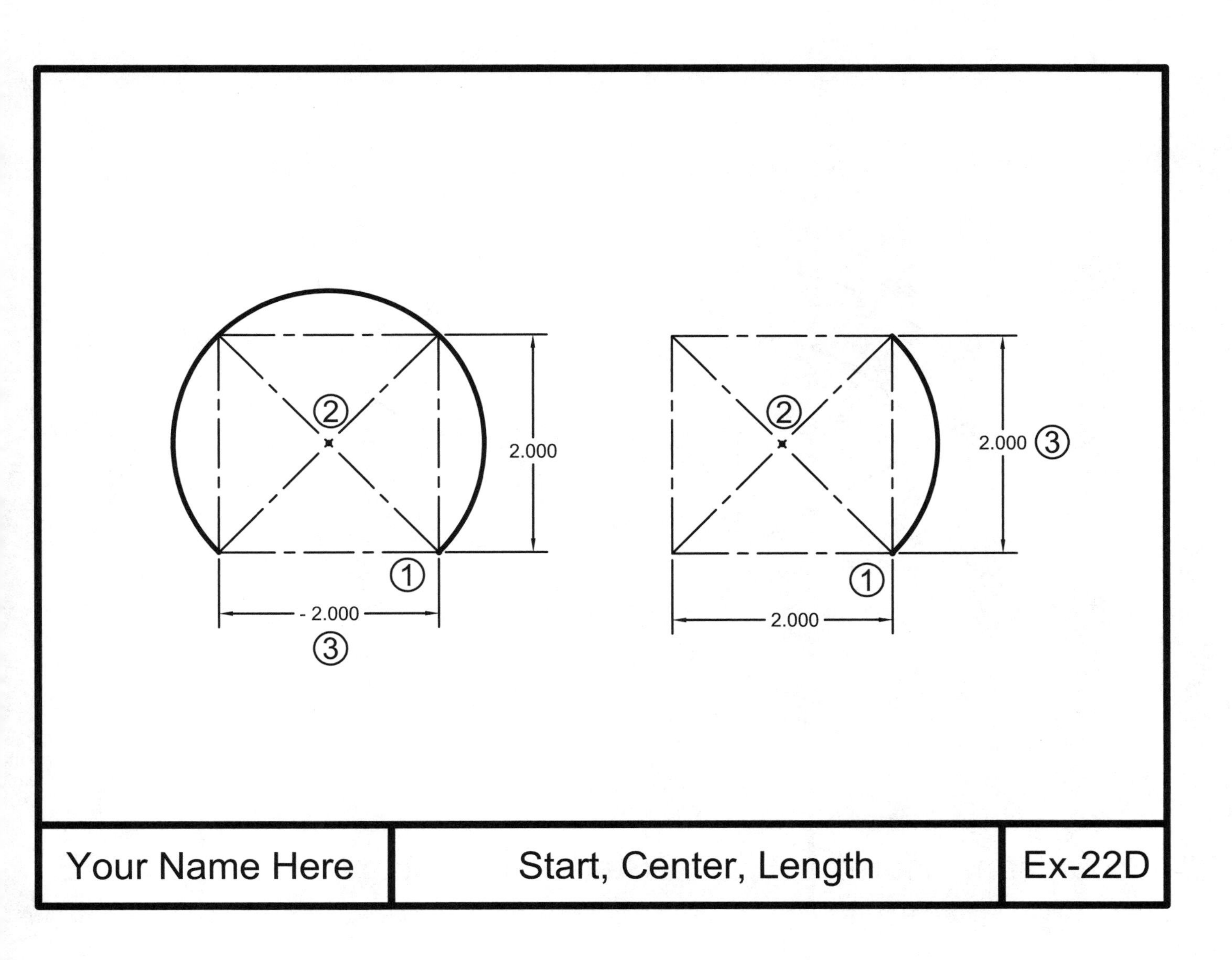

EXERCISE 22E

INSTRUCTIONS:

1. Start a **New** file using **Border A-2015.dwt**
2. Draw the center lines first on layer Center line.
3. Draw the 2 Arcs using method **Start, End, Angle** on Layer Object Line.
   Place the **Start** at location (1), **End** at location (2) and enter **Angle** (3)
4. Dimension as shown using Dimension Style Class Style and Layer Dimension.
5. Edit the Title and Ex-XX by double clicking on the text. Do not erase and replace.
6. Save as **EX22E**
7. Plot using Page Setup **Class Model A**

Note: *Positive Angle draws the arc counter-clockwise, negative Angle draws the arc clockwise. To reverse the directions hold down the* ***Ctrl*** *key.*

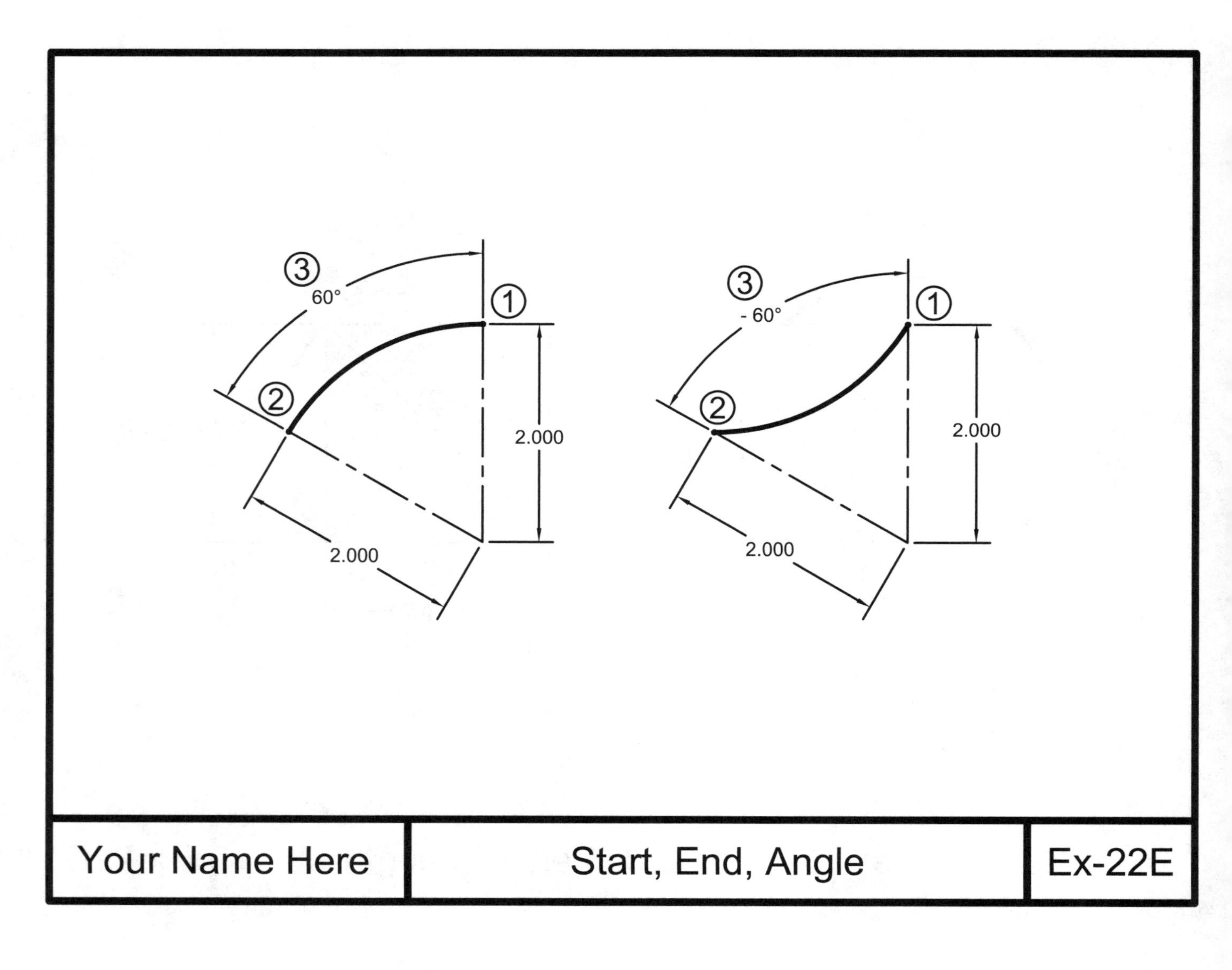

EXERCISE 22F

INSTRUCTIONS:

1. Start a **New** file using **Border A-2015.dwt**
2. Draw the center lines first on layer Center line.
3. Draw the 2 Arcs using method **Start, End, Direction** on Layer Object Line. Place the **Start** at location (1), **End** at location (2) and move your cursor in the **Direction** of (3.)
4. Dimension as shown using Dimension Style Class Style and Layer Dimension.
5. Edit the Title and Ex-XX by double clicking on the text. Do not erase and replace.
6. Save as **EX22F**
7. Plot using Page Setup **Class Model A**

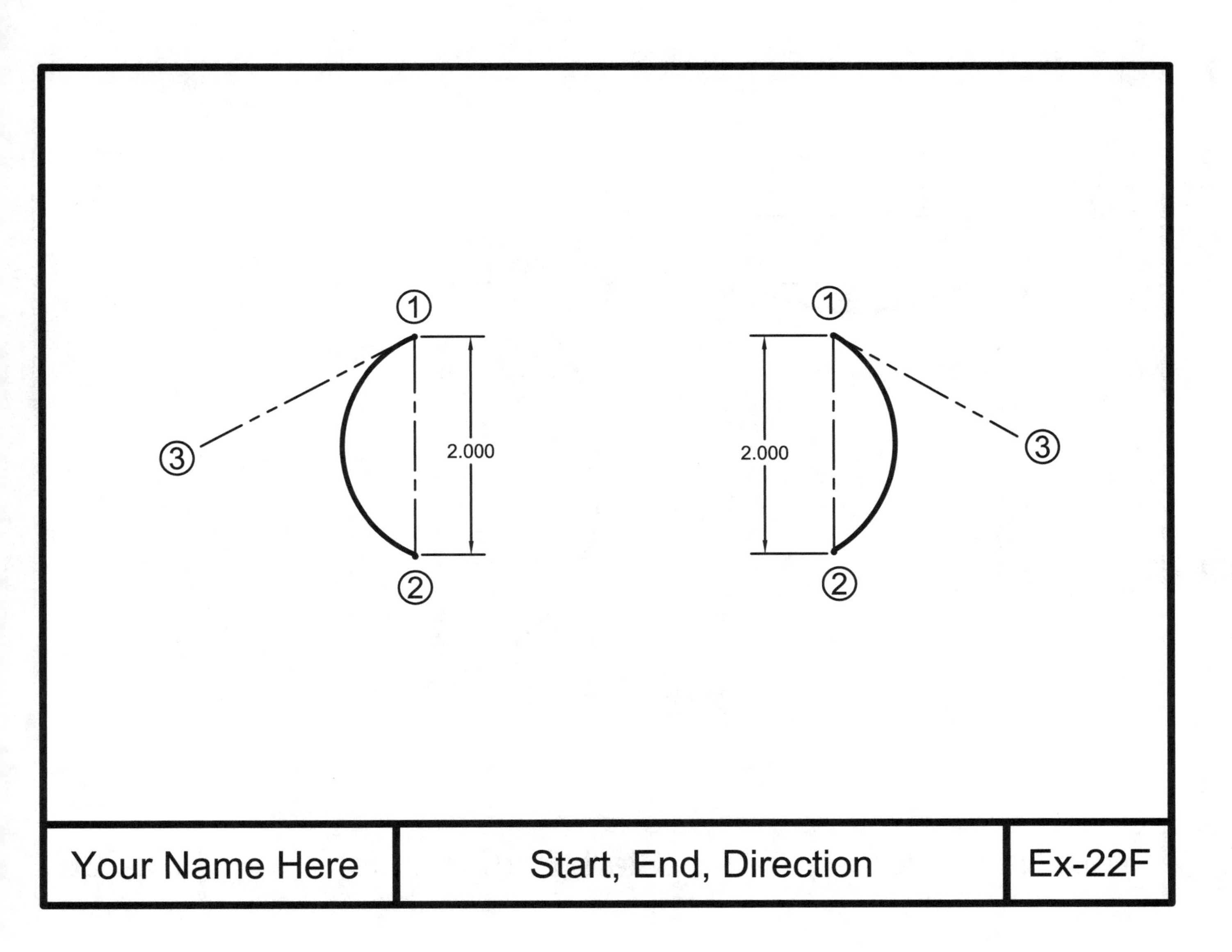

EXERCISE 22G

INSTRUCTIONS:

1. Start a **New** file using **Border A-2015.dwt**
2. Draw the center lines first on layer Center line.
3. Draw the 2 Arcs using method **Start, End, Radius** on Layer Object Line. Place the **Start** at location ①, **End** at location ② and enter the **Radius** of ③.
4. Dimension as shown using Dimension Style Class Style and Layer Dimension.
5. Edit the Title and Ex-XX by double clicking on the text. Do not erase and replace.
6. Save as **EX22G**
7. Plot using Page Setup **Class Model A**

Note: *Positive Radius draws the small segment, negative Radius draws the large segment. To reverse the directions hold down the* ***Ctrl*** *key.*

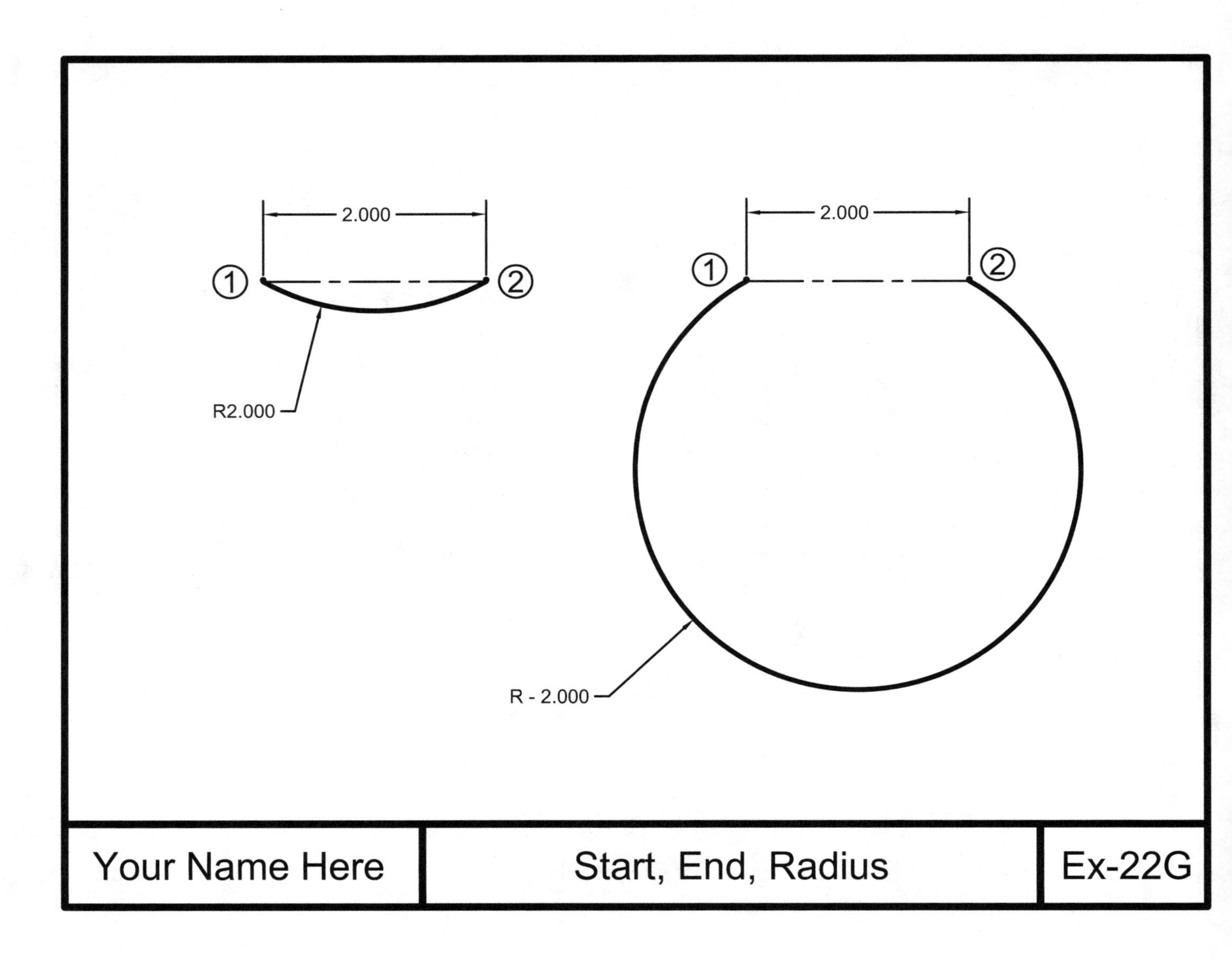

EXERCISE 22H

INSTRUCTIONS:

1. Start a **New** file using **Border A-2015.dwt**
2. Draw the center lines first on layer Center line.
3. Draw the 2 Arcs using method **Center, Start, End** on Layer Object Line.
   Place the **Center** at location ①, **Start** at location ② and **End** at location ③
4. Dimension as shown using Dimension Style Class Style and Layer Dimension.
5. Edit the Title and Ex-XX by double clicking on the text. Do not erase and replace.
6. Save as **EX22H**
7. Plot using Page Setup **Class Model A**

Note: *Draws the Arc Counter Clockwise only. To reverse the directions hold down the **Ctrl** key.*

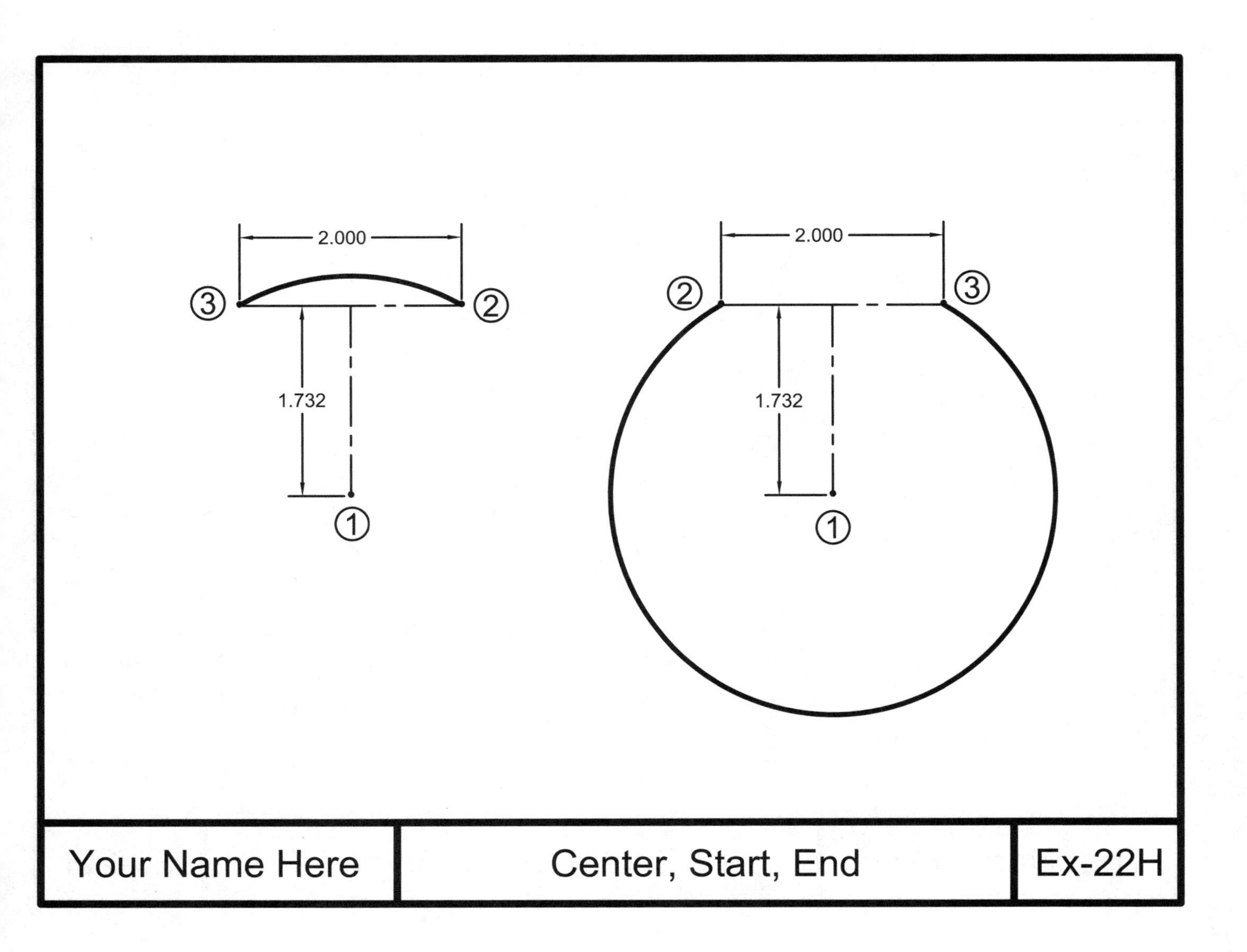

EXERCISE 22I

INSTRUCTIONS:

1. Start a **New** file using **Border A-2015.dwt**
2. Draw the center lines first on layer Center line.
3. Draw the 2 Arcs using method **Center, Start, Angle** on Layer Object Line. Place the **Center** at location (1), **Start** at location (2) and enter **Angle** (3.)
4. Dimension as shown using Dimension Style Class Style and Layer Dimension.
5. Edit the Title and Ex-XX by double clicking on the text. Do not erase and replace.
6. Save as **EX22I**
7. Plot using Page Setup **Class Model A**

Note: *To reverse the arc directions hold down the* ***Ctrl*** *key.*

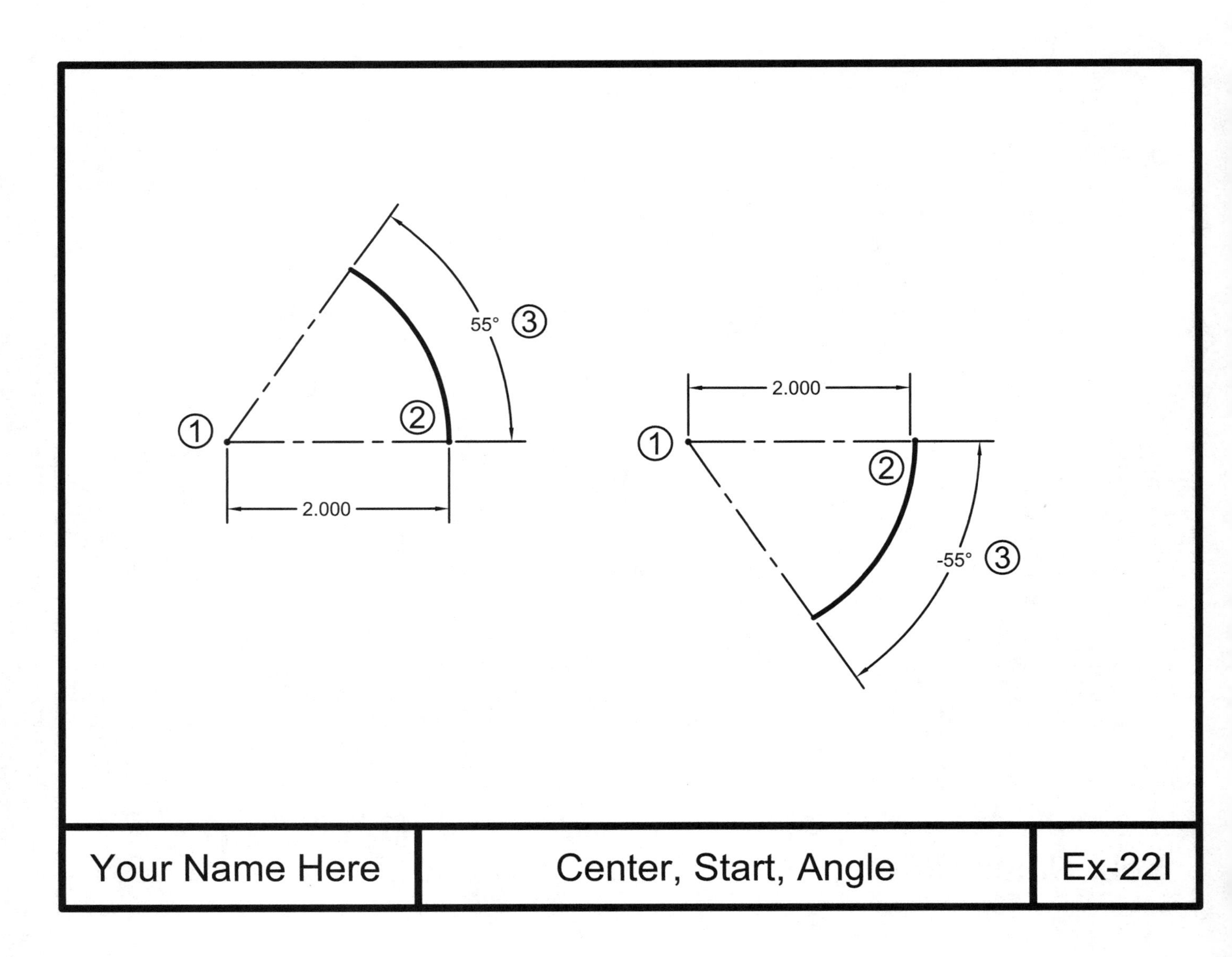

EXERCISE 22J

INSTRUCTIONS:

1. Start a **New** file using **Border A-2015.dwt**
2. Draw the center lines first on layer Center line.
3. Draw the 2 Arcs using method **Center, Start, Length** on Layer Object Line. Place the **Center** at location(1), **Start** at location(2) and enter **Length**(3.)
4. Dimension as shown using Dimension Style Class Style and Layer Dimension.
5. Edit the Title and Ex-XX by double clicking on the text. Do not erase and replace.
6. Save as **EX22J**
7. Plot using Page Setup **Class Model A**

Note: *Positive chord length draws the small segment counter-clockwise, negative chord length draws the large segment counter-clockwise. To reverse the directions hold down the* ***Ctrl*** *key.*

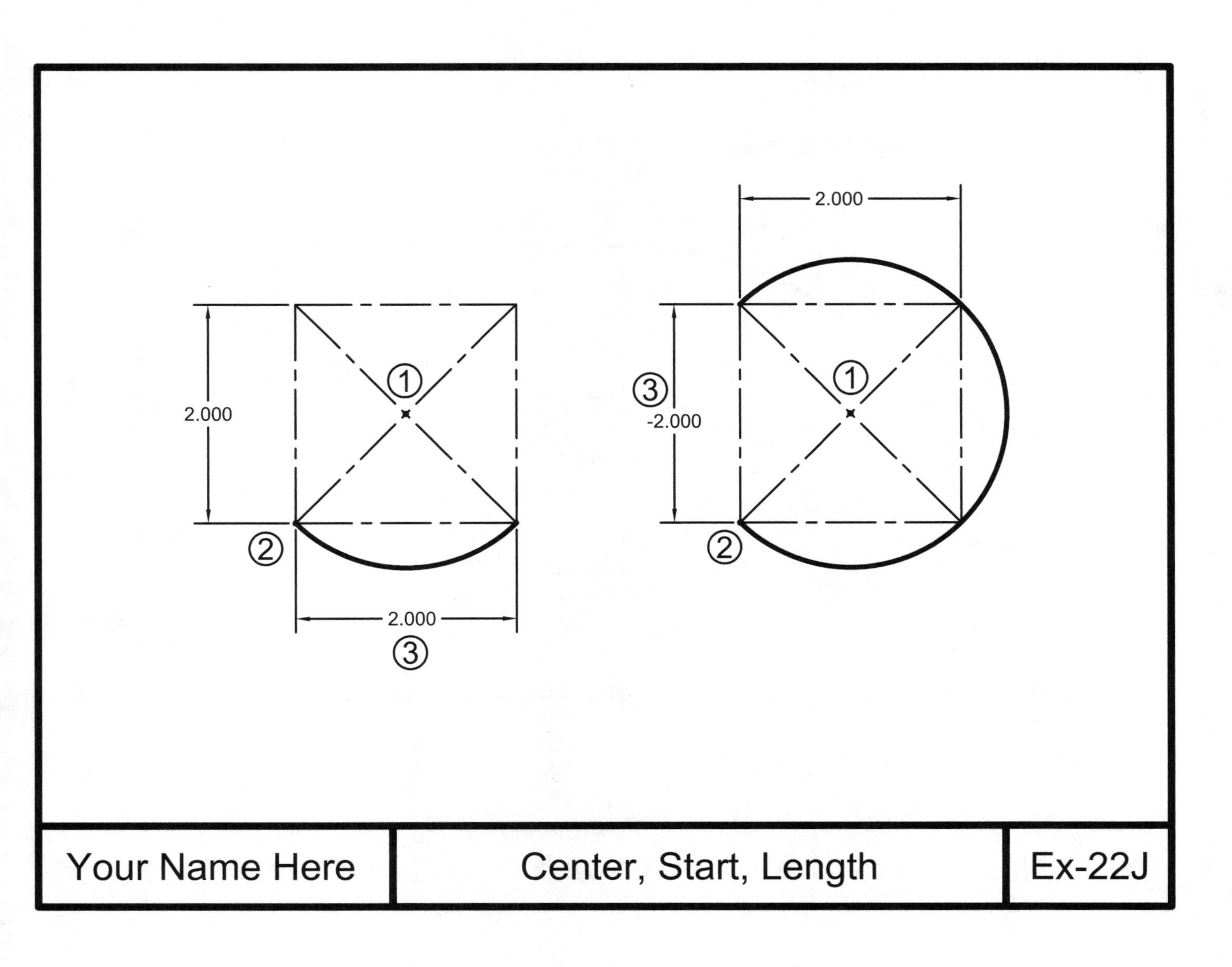

EXERCISE 22K

INSTRUCTIONS:

1. Start a **New** file using **Border A-2015.dwt**
2. Draw the Lines first on layer Center line.
3. Draw the Arc using Layer Object Line. Refer to the previous pages to select method.
4. Dimension as shown using Dimension Style **Class Style** and Layer Dimension.
5. Edit the Title and Ex-XX by double clicking on the text. Do not erase and replace.
6. Save as **EX22K**
7. Plot using Page Setup **Class Model A**

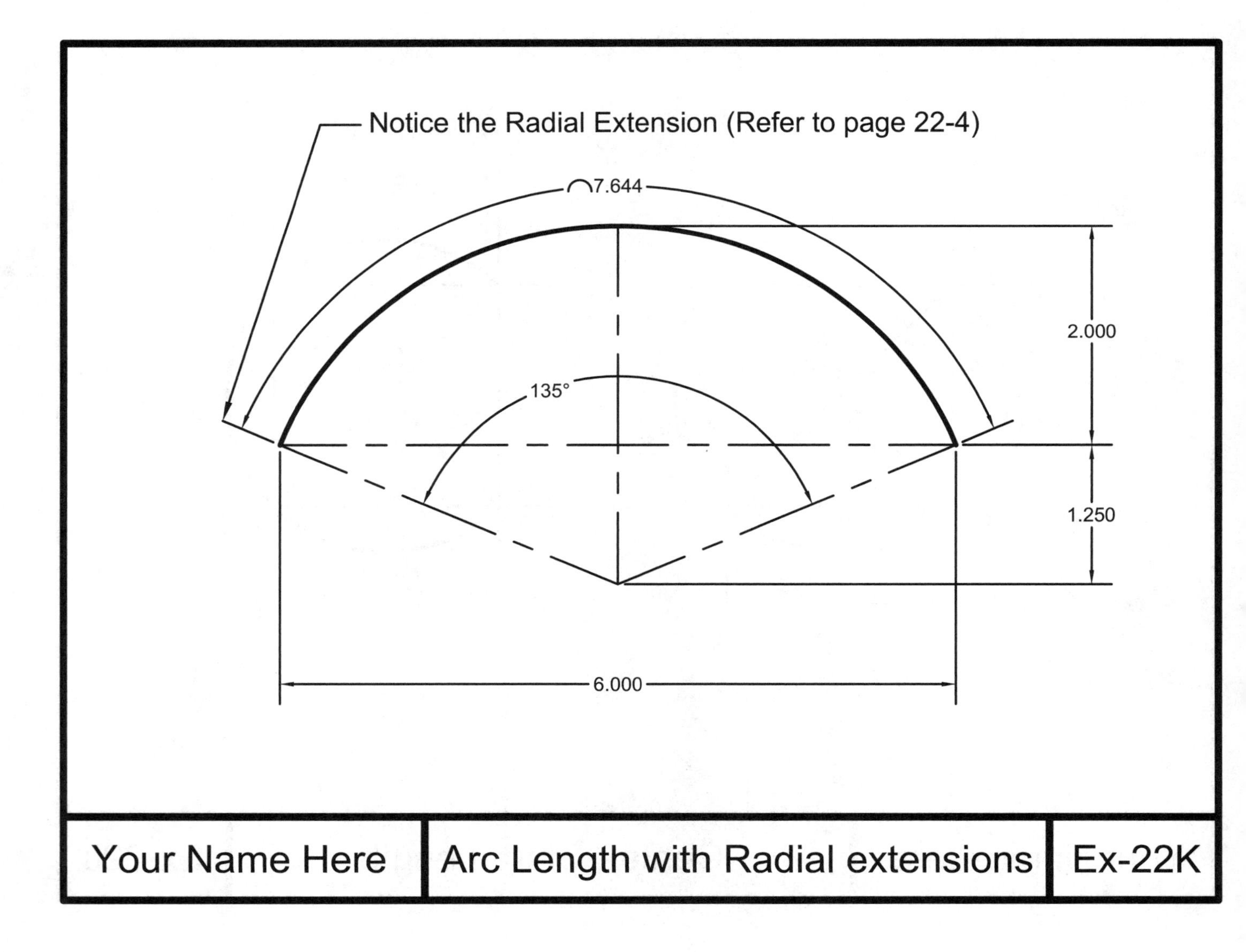

EXERCISE 22L

INSTRUCTIONS:

1. Start a **New** file using **Border A-2015.dwt**
2. Draw the Lines first on layer Center line.
3. Draw the Arc using Layer Object Line. Refer to the previous pages to select method.
4. Dimension as shown using Dimension Style **Class Style** and Layer Dimension.
5. Edit the Title and Ex-XX by double clicking on the text. Do not erase and replace.
6. Save as **EX22L**
7. Plot using Page Setup **Class Model A**

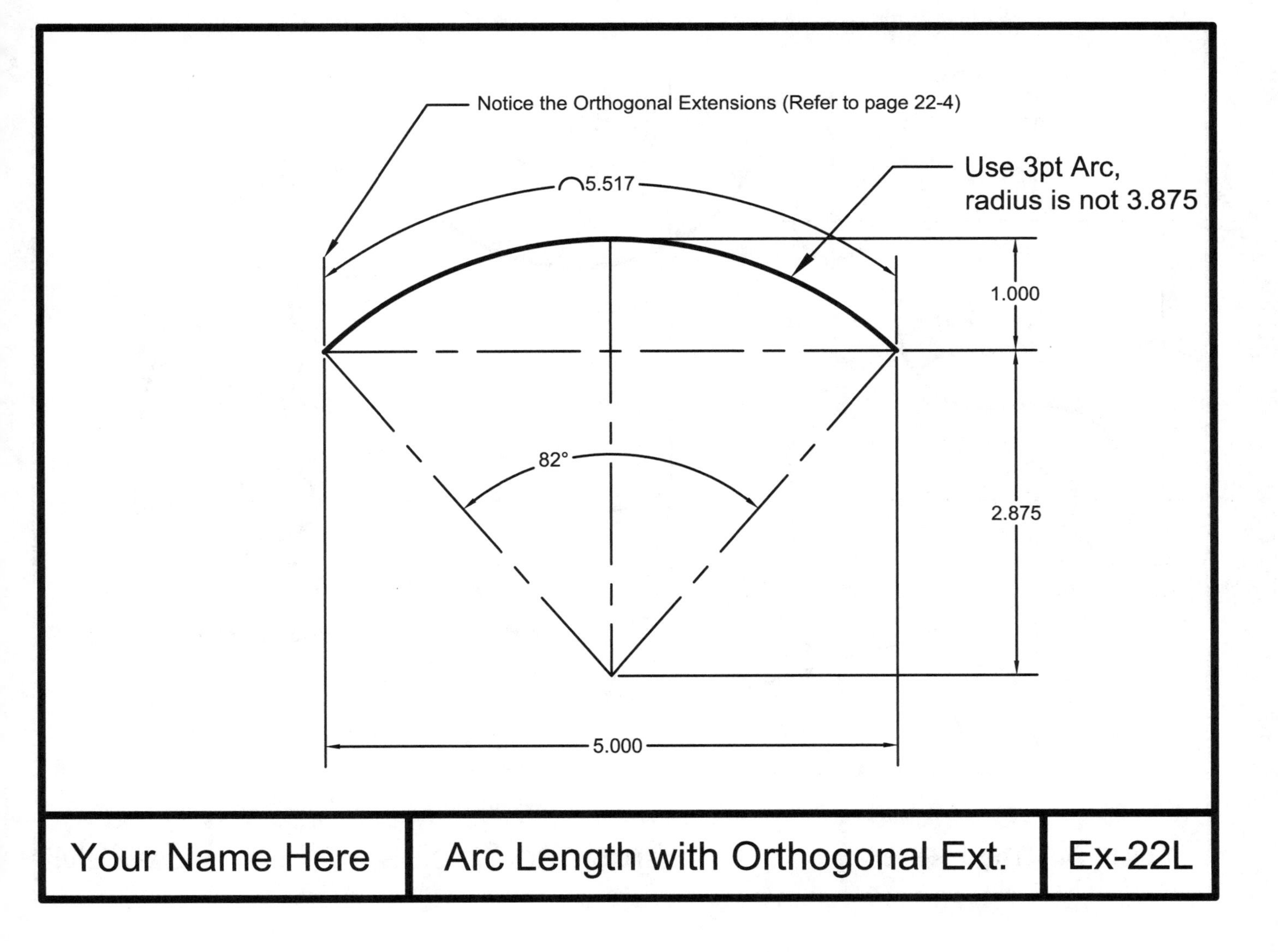

EXERCISE 22M

INSTRUCTIONS:

1. Start a **New** file using **Border A-2015.dwt**
2. Draw objects below using Circles, Arcs, Lines and Fillet on layer Object Line.
3. Dimension as shown using Dimension Style **Class Style** and Layer Dimension. (Refer to page 18-3 and 18-5 for Radius and Diameter dimension settings)
4. Edit the Title and Ex-XX by double clicking on the text. Do not erase and replace.
5. Save as **EX22M**
6. Plot using Page Setup **Class Model A**

Refer to the next page for drawing assistance.

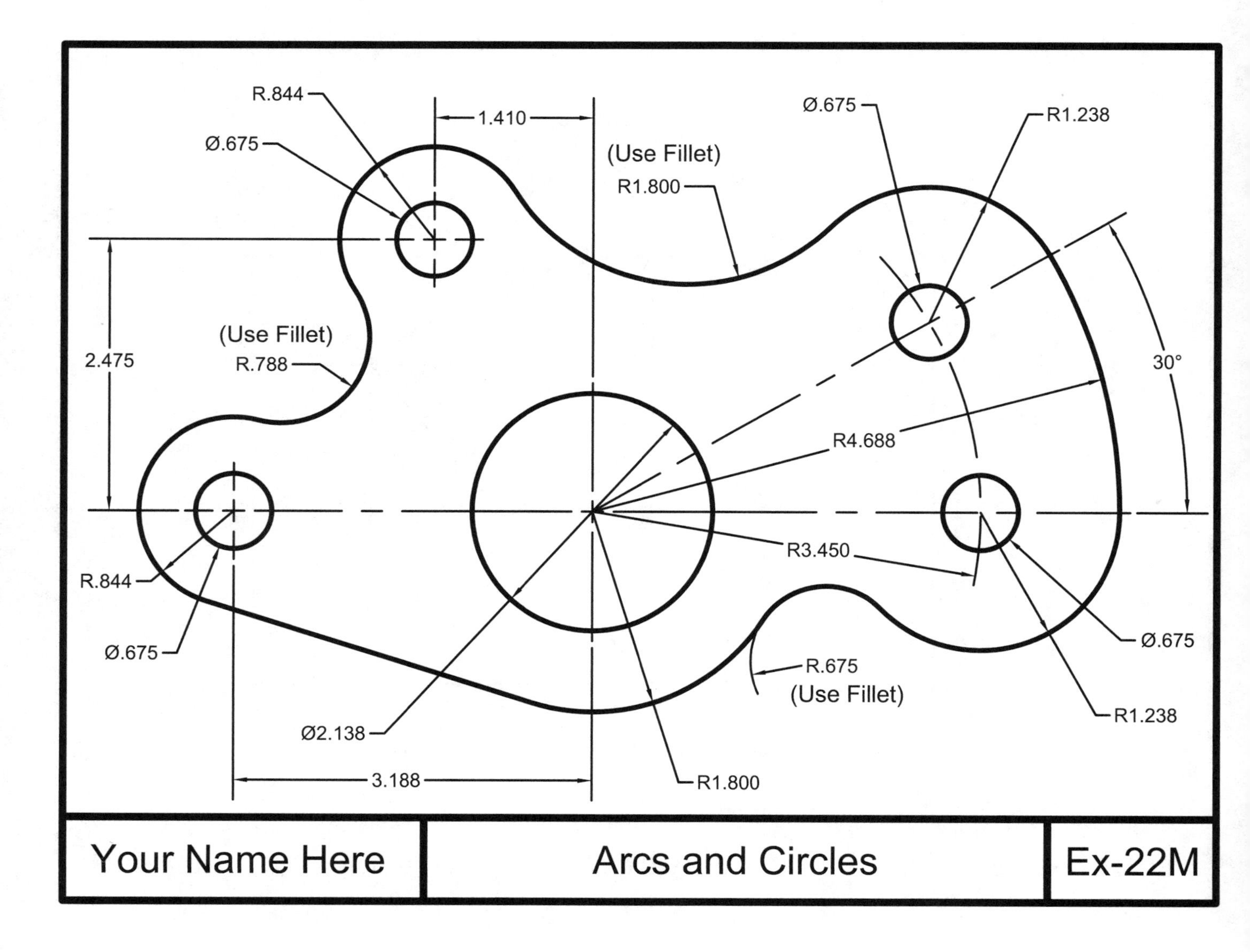

EX 22M Helper

INSTRUCTIONS:

Step 1.

-Draw the horizontal and vertical lines then offset to create intersections for circles.

-Draw the 30 degree line

Step 2.

-Draw the Circles

Step 3.

-Draw the Arc**s** and draw the circles.

Step 4.

-Fillet, draw the tangent line and dimension.

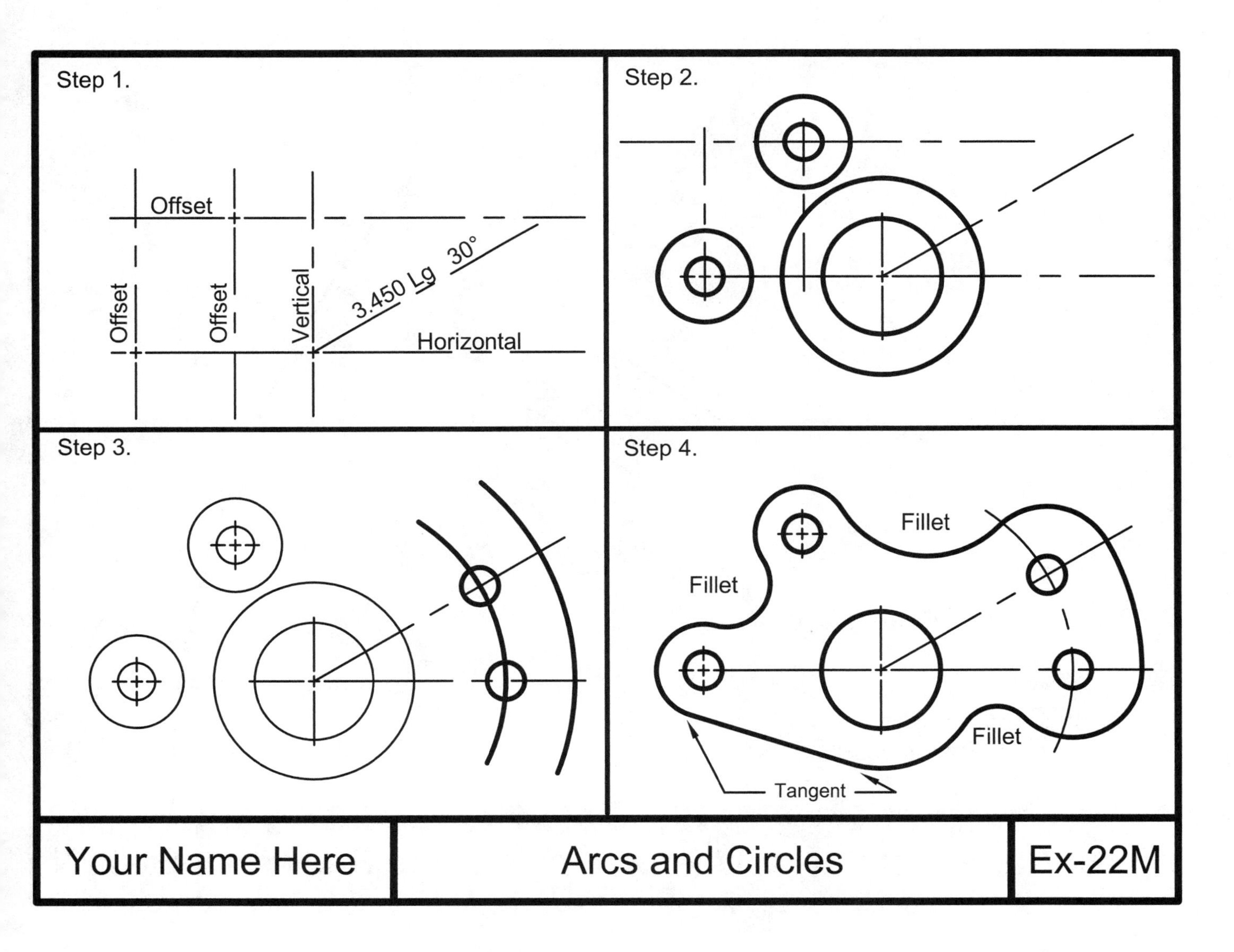

EXERCISE 22N

INSTRUCTIONS:

1. Start a **New** file using **Border A-2015.dwt**
2. Draw a 9" Radius Circle on Layer Object line, as shown below.
3. Trim so it does not extend beyond the the border lines.
4. Dimension as shown using Dimension Style **Class Style** and Layer Dimension. (Refer to page 22-5)
5. Edit the Title and Ex-XX by double clicking on the text. Do not erase and replace.
6. Save as **EX22N**
7. Plot using Page Setup **Class Model A**

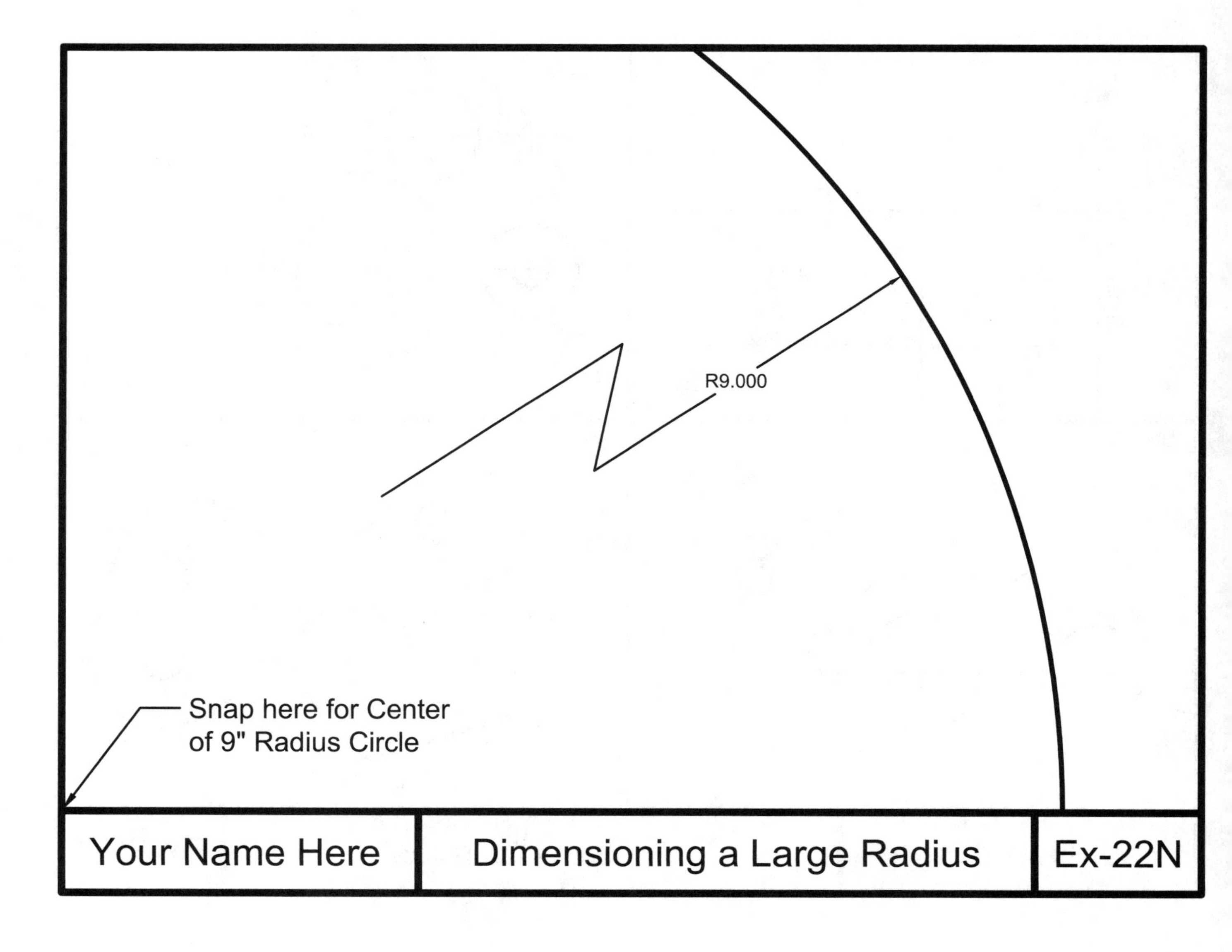

LEARNING OBJECTIVES

After completing this lesson, you will be able to:

1. Understand the Polyline command
2. Draw a Polyline and a Polyarc
3. Assign widths to polylines
4. Set the Fill mode to On or Off.
5. Explode a Polyline

LESSON 23

POLYLINES

A **POLYLINE** is very similar to a LINE. It is created in the same way a line is drawn. It requires first and second endpoints. But a POLYLINE has additional features, as follows:

1. A **POLYLINE** is ONE object, even though it may have many segments.
2. You may specify a specific width to each segment.
3. You may specify a different width to the start and end of a polyline segment.

Select the Polyline Command using one of the following:

Ribbon = Home tab / Draw panel /
or
Keyboard = PL <enter>

THE FOLLOWING ARE EXAMPLES OF POLYLINES WITH WIDTHS ASSIGNED.

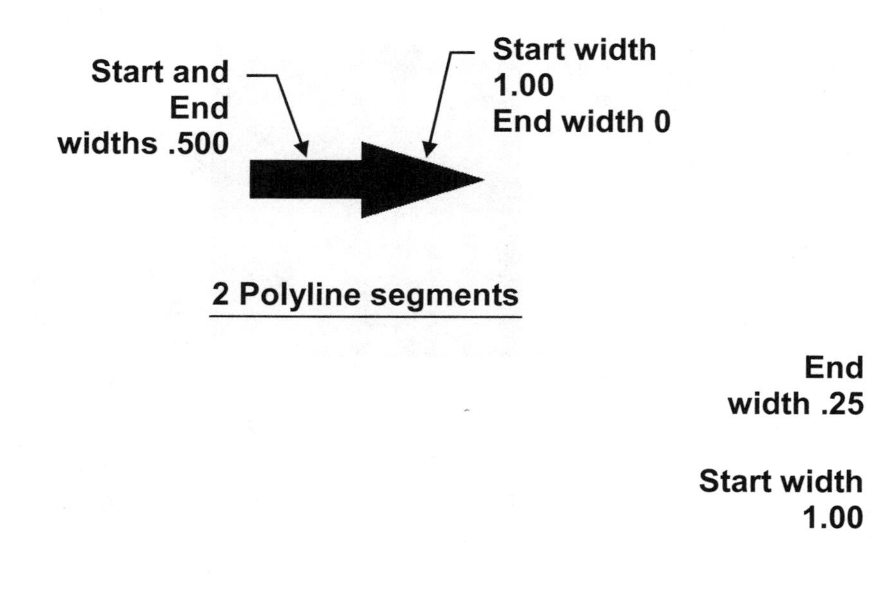

2 Polyline segments

Single Polyline Arc

Refer to the next page for more polyline options

POLYLINES....continued

OPTIONS:

WIDTH
Specify the starting and ending width.

You can create a tapered polyline by specifying different starting and ending widths.

HALFWIDTH
The same as Width except the starting and ending halfwidth specifies half the width rather than the entire width.

ARC
This option allows you to create a circular polyline less than 360 degrees. You may use (2) 180 degree arcs to form a full circular shape.

CLOSE
The close option is the same as in the Line command. Close attaches the last segment to the first segment.

LENGTH
This option allows you to draw a polyline at the same angle as the last polyline drawn. This option is very similar to the OFFSET command. You specify the first endpoint and the length. The new polyline will automatically be drawn at the same angle as the previous polyline.

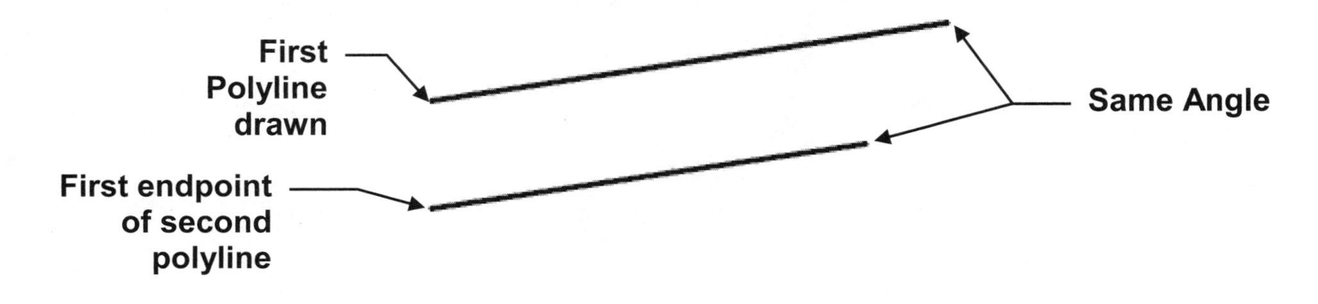

POLYLINES....continued

CONTROLLING THE FILL MODE

If you turn the FILL mode OFF the polylines will appear as shown below.

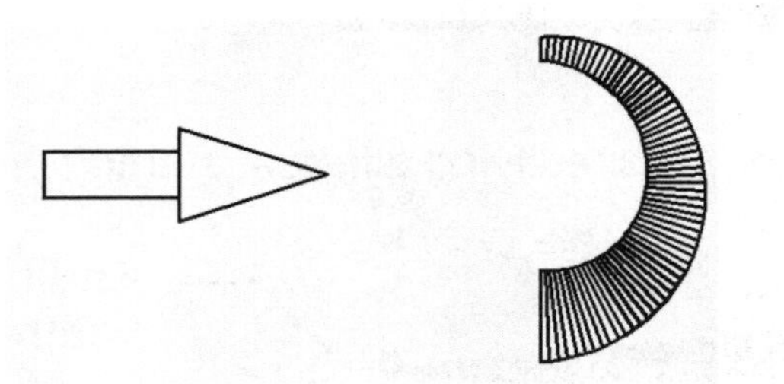

How to turn FILL MODE on or off.

1. Command: ***type* FILL *<enter>***
2. Enter mode [On / Off] <ON>: ***type ON or Off <enter>***
3. Command: ***type REGEN <enter> or select: View / Regen***

EXPLODING A POLYLINE

NOTE: If you explode a POLYLINE it loses its width and turns into a regular line as shown below.

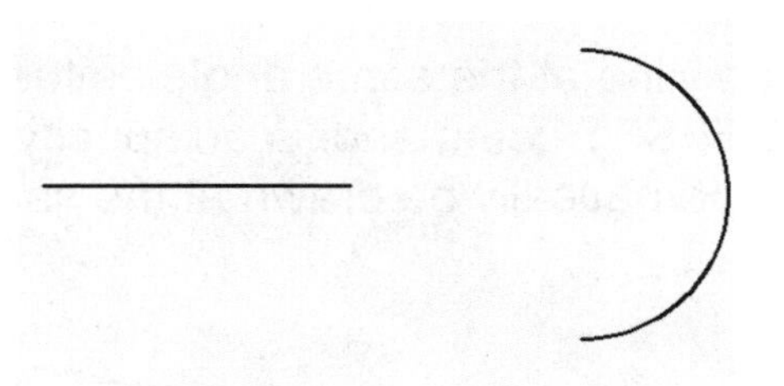

POLYLINES....continued

The following is an example of how to draw a polyline with Width.

1. Select the Polyline Command using one of the following:

   Ribbon = Home tab / Draw panel /
   or
   Keyboard = PL <enter>

   Command: _pline
2. Specify start point: ***place the first endpoint of the line***

   Current line-width is 0.000
3. Specify next point or [Arc/Halfwidth/Length/Undo/Width]: ***select width option***
4. Specify starting width <0.000>: ***1*** <enter>
5. Specify ending width <0.000>: ***1*** <enter>
6. Specify next point or [Arc/Close/Halfwidth/Length/Undo/Width]: ***select Length option***
7. Specify Length: ***3 <enter>***
8. Specify next point or [Arc/Halfwidth/Length/Undo/Width]: ***select width option***
9. Specify starting width <0.000>: ***2*** <enter>
10. Specify ending width <0.000>: ***.5*** <enter>
11. Specify next point or [Arc/Close/Halfwidth/Length/Undo/Width]: ***select Length option***
12. Specify Length: ***2.75 <enter>***
13. Specify next point or [Arc/Halfwidth/Length/Undo/Width]: ***select width option***
14. Specify starting width <0.000>: ***.5*** <enter>
15. Specify ending width <0.000>: ***1*** <enter>
16. Specify next point or [Arc/Close/Halfwidth/Length/Undo/Width]: ***select Length option***
17. Specify Length: ***2.50 <enter>***
18. Specify next point or [Arc/Close/Halfwidth/Length/Undo/Width]: ***press <enter> to stop***

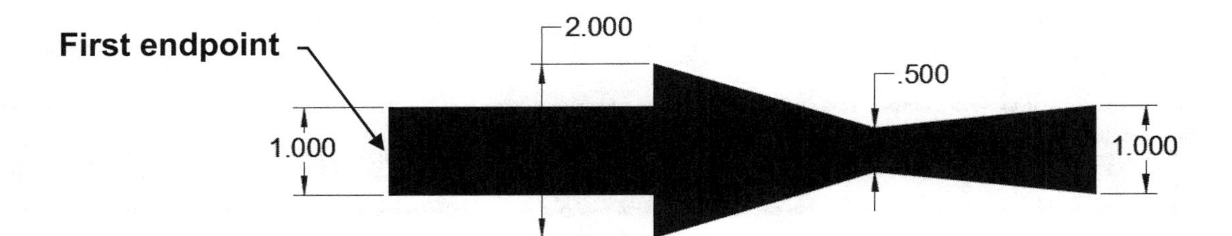

EXERCISE 23A

INSTRUCTIONS:

1. Start a **New** file using **Border A-2015.dwt**
2. Draw the Polylines as shown using layer Object line.
3. Do not dimension
4. Edit the Title and Ex-XX by double clicking on the text. Do not erase and replace.
5. Save as **EX23A**
6. Plot using Page Setup **Class Model A**

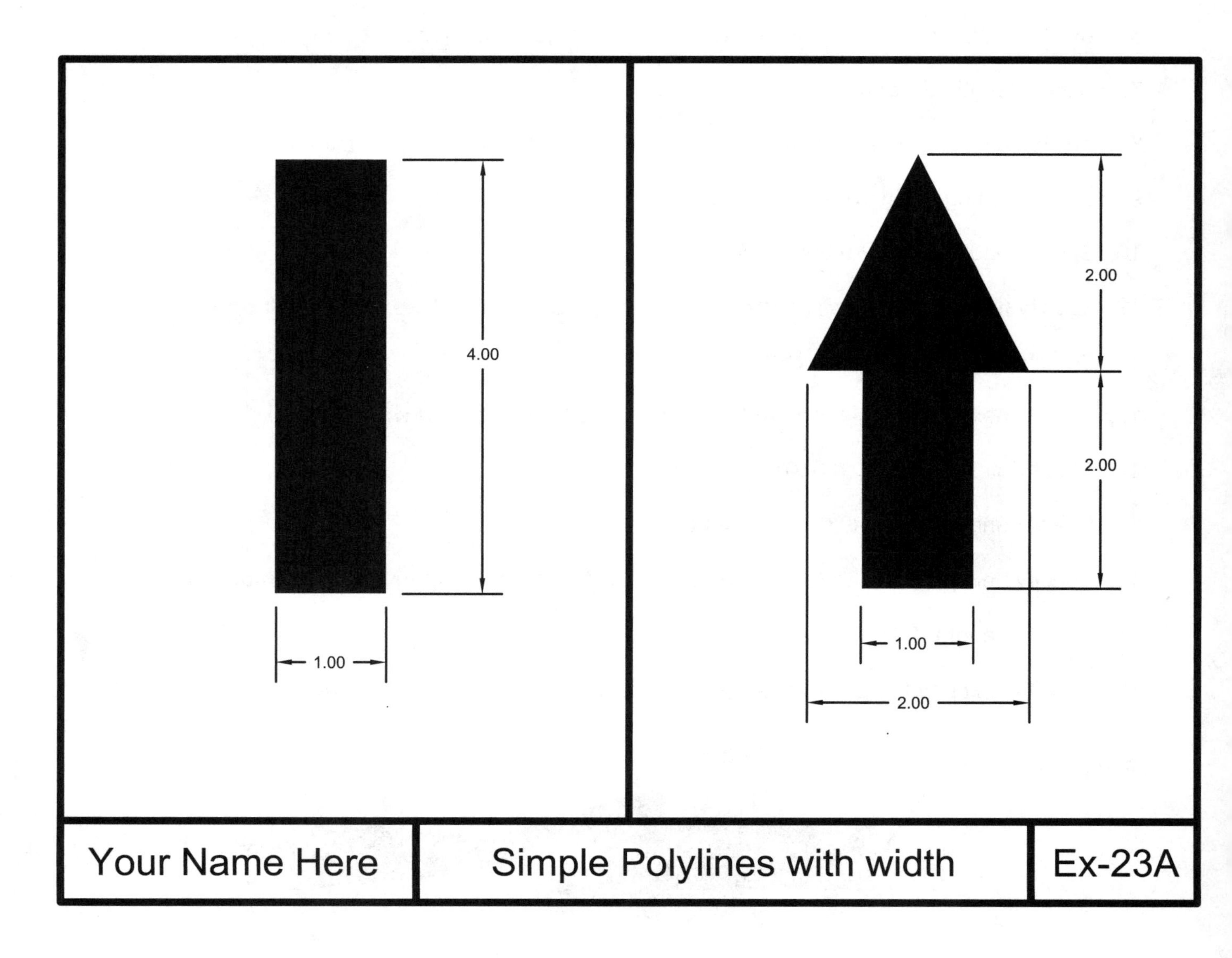

EXERCISE 23B

INSTRUCTIONS:

1. Open **EX23A**
2. Turn off the **Fill** mode. (Refer to page 23-4)
3. Do not dimension
4. Edit the Title and Ex-XX by double clicking on the text. Do not erase and replace.
5. Save as **EX23B**
6. Plot using Page Setup **Class Model A**

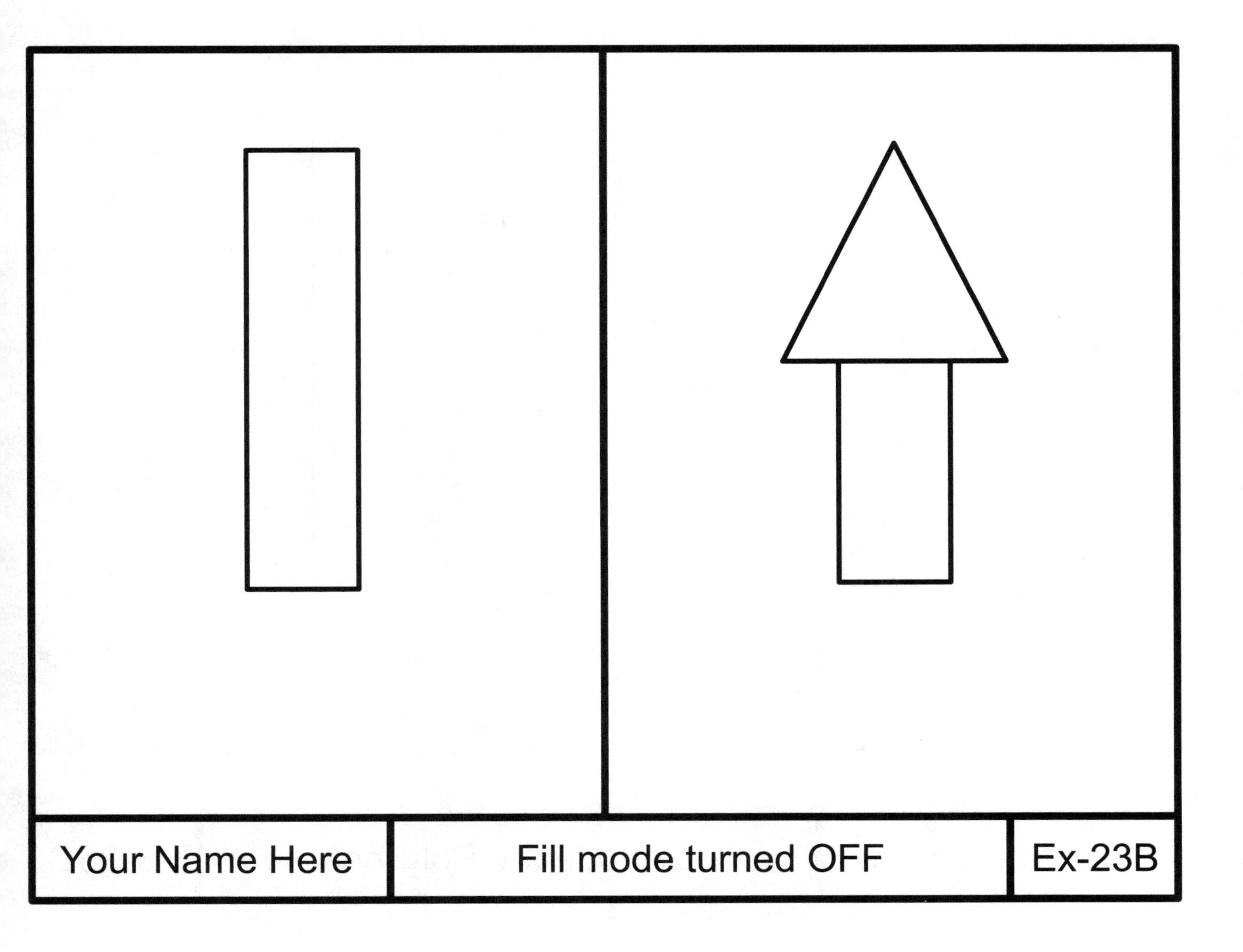

EXERCISE 23C

INSTRUCTIONS:

1. Open **EX23B**
2. Explode each of the polylines.

   Note: I am sure most Architectural students were thinking that the polyline command would be good to use when creating Floor plan walls. But notice what happens if you explode the polylines. Consider Multiline or Double line (LT) commands for Floor plan walls. You can find these commands in the Help Menu

3. Edit the Title and Ex-XX by double clicking on the text. Do not erase and replace.
4. Save as **EX23C**
5. Plot using Page Setup **Class Model A**

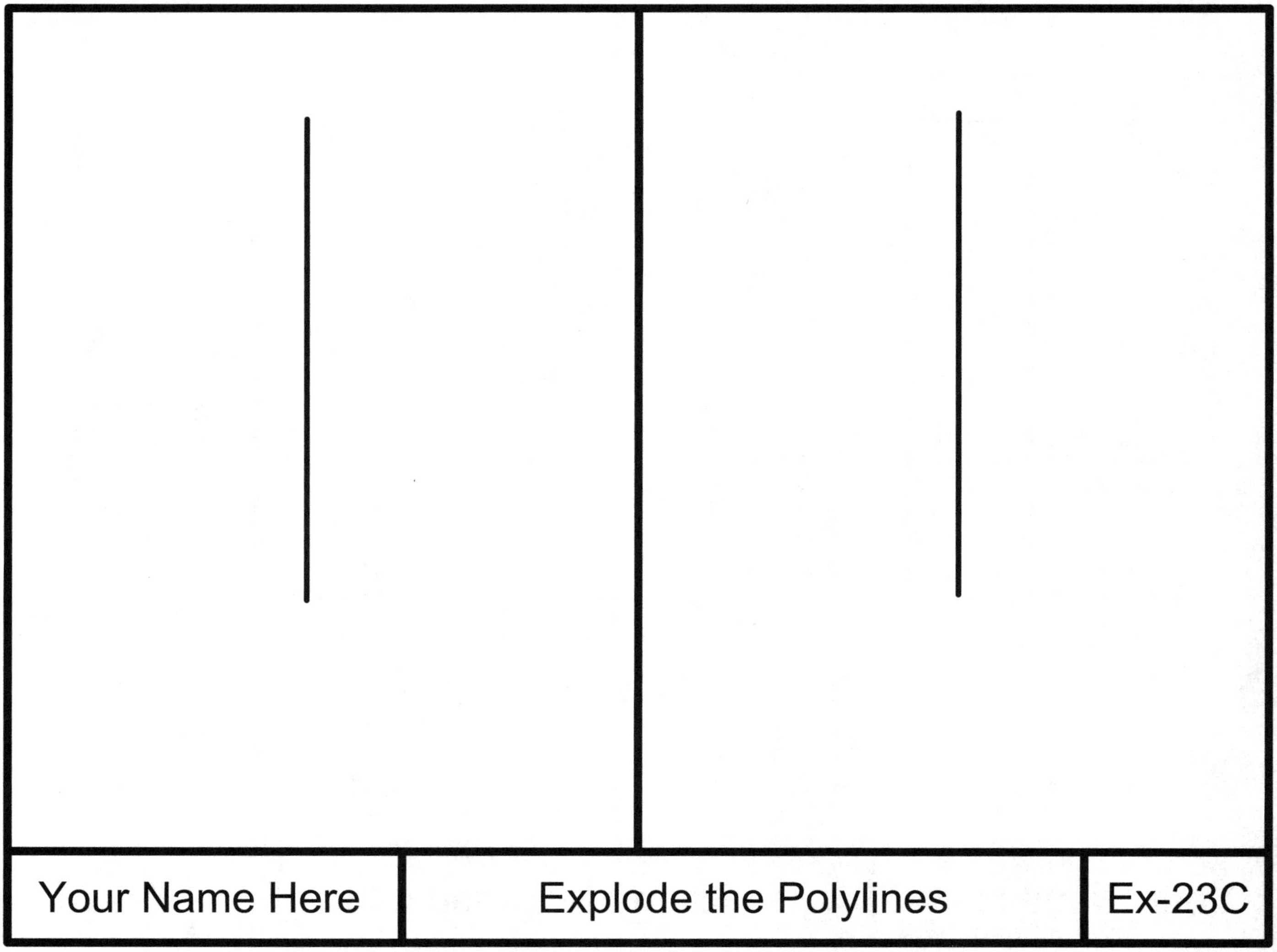

Your Name Here	Explode the Polylines	Ex-23C

EXERCISE 23D

INSTRUCTIONS:

1. Start a **New** file using **Border A-2015.dwt**
2. Draw the Polyline as shown using layer Object Line.
3. Do not Dimension.
4. Edit the Title and Ex-XX by double clicking on the text. Do not erase and replace.
5. Save as **EX23D**
6. Plot using Page Setup **Class Model A**

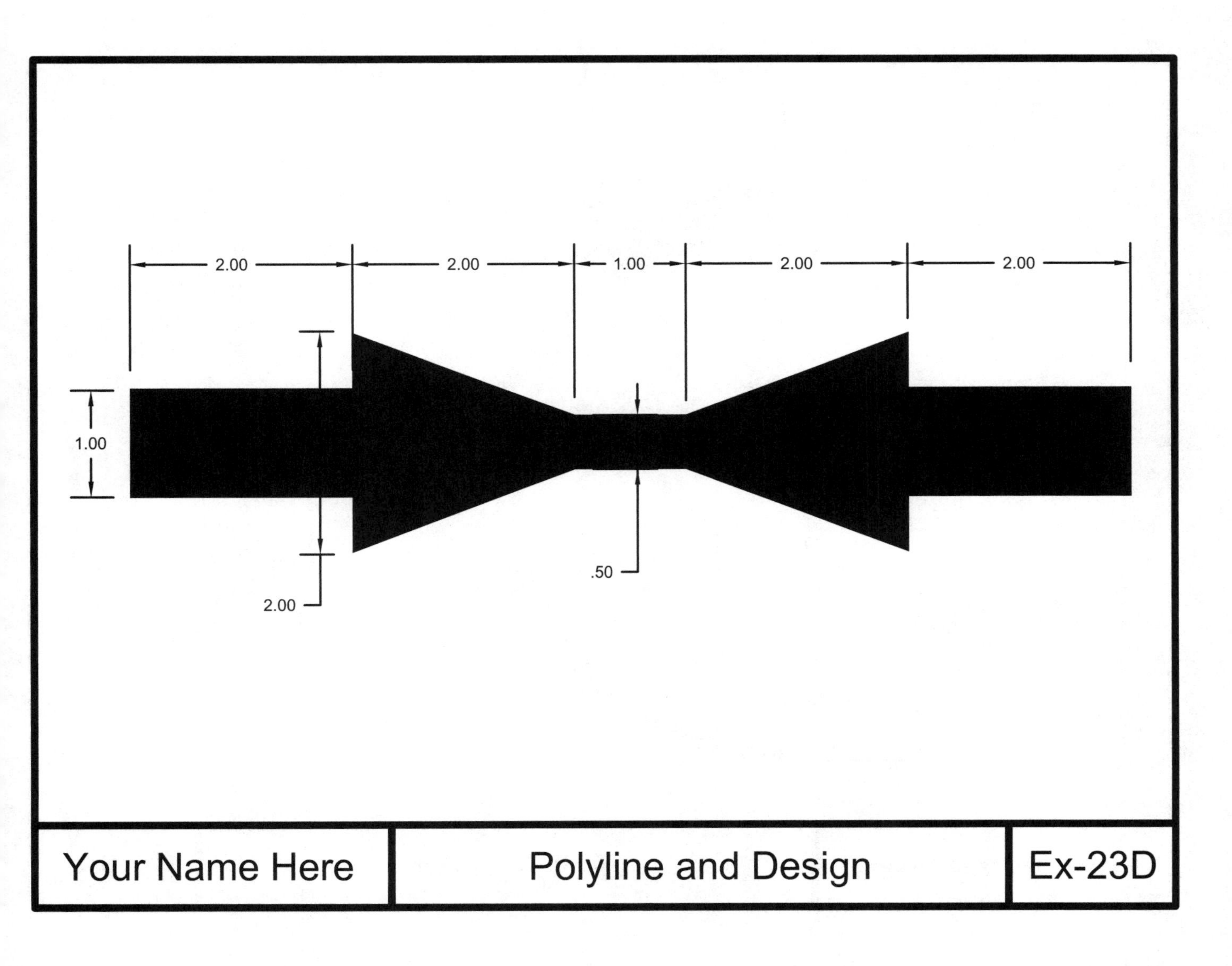

EXERCISE 23E

INSTRUCTIONS:

1. Start a **New** file using **Border A-2015.dwt**
2. Select the Polyline Command.
3. Place the **start point** approximately as shown.
4. Set the **W**idths: Starting width = .10 Ending width = .50
5. Select option **A**rc
6. Select option **R**adius and enter **2.50**
7. Select option **A**ngle and enter **180**
8. Enter **90** for the **Direction of the Chord.**
9. Press **<enter>** to stop
10. Do not Dimension.
11. Edit the Title and Ex-XX by double clicking on the text. Do not erase and replace.
12. Save as **EX23E**
13. Plot using Page Setup **Class Model A**

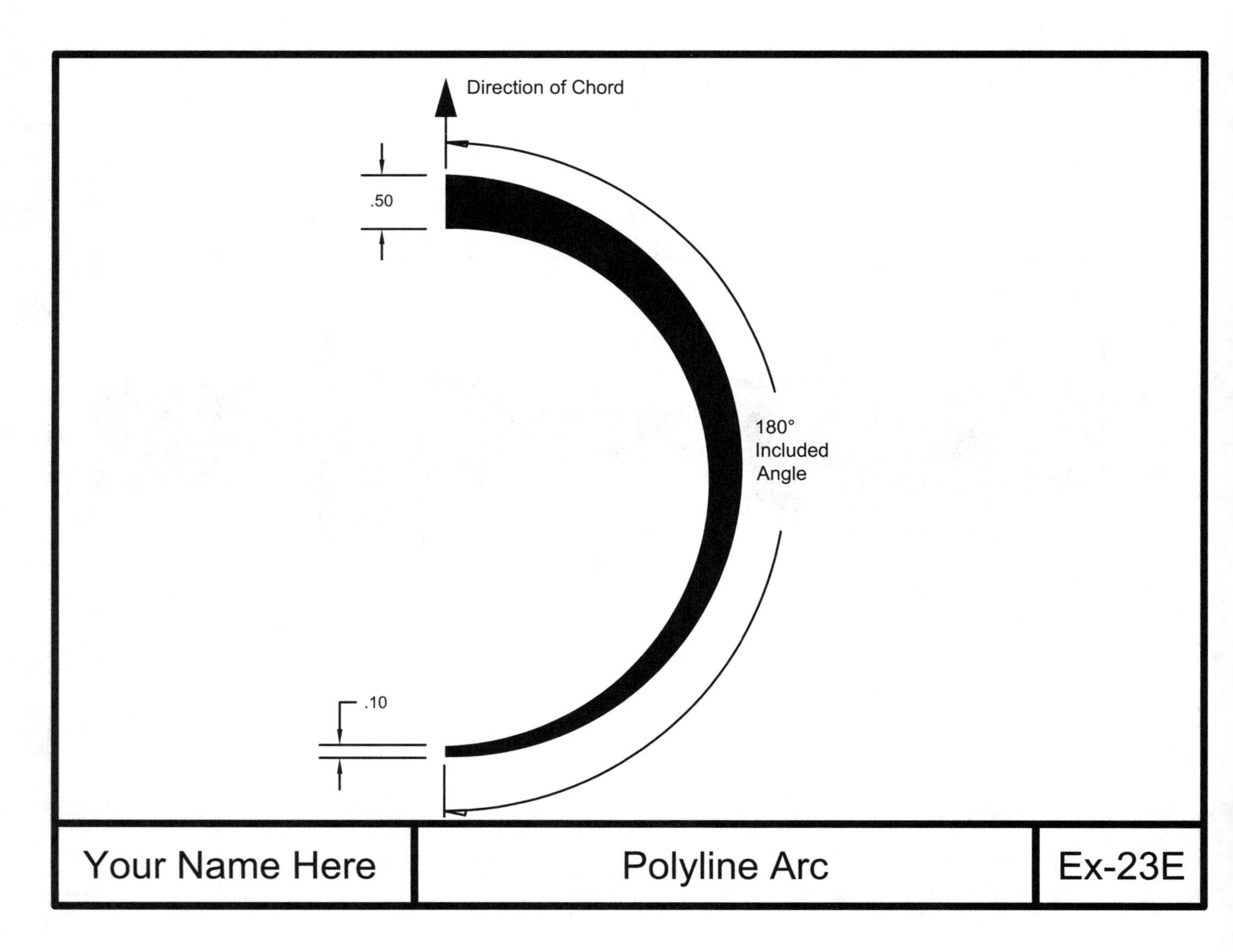

LEARNING OBJECTIVES

After completing this lesson, you will be able to:

1. Edit the width of Polylines
2. Join Polylines
3. Convert Polylines to Curves
4. Convert a basic Line to a Polyline
5. Join Lines, Polylines, Arcs and Splines

LESSON 24

EDITING POLYLINES

The **EDIT POLYLINE** command allows you to make changes to a polyline's option, such as the width. You can also change a regular line into a polyline and JOIN the segments.

<u>**Note:** If you select a line that is **NOT a POLYLINE**, the prompt will ask if you would like to turn it into a POLYLINE.</u>

1. Select the **EDIT POLYLINE** command using one of the following:

    Ribbon = Home tab / Modify ▾ panel /
    or
    Keyboard = PE <enter>

    <u>**Note: You may modify "Multiple" polylines simultaneously**</u>**.**

2. PEDIT Select polyline or [Multiple]: ***select the polyline to be edited or "M"***

3. Enter an option [Close/Join/Width/Edit vertex/Fit/Spline/Decurve/Ltypegen/Undo/ Reverse]: ***select an Option (descriptions of each are listed below.)***

<u>OPTIONS</u>

CLOSE

CLOSE connects the last segment with the first segment of an Open polyline. AutoCAD considers a polyline open unless you use the "Close" option to connect the segments originally.

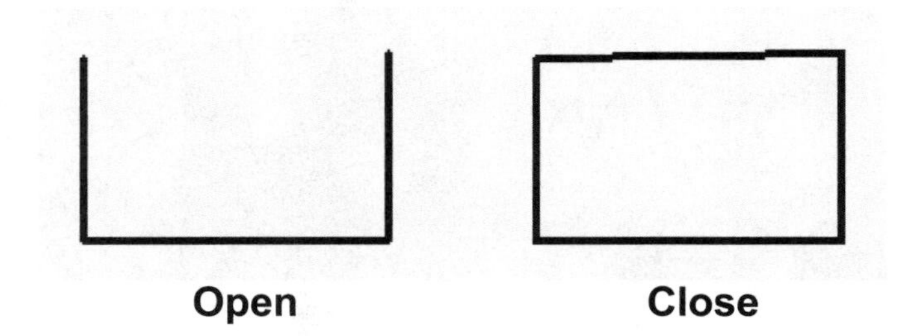

OPEN

OPEN removes the closing segment, but only if the CLOSE option was used to close the polyline originally.

EDITING POLYLINES....continued

OPTIONS:

JOIN
The JOIN option allows you to join individual polyline segments into one polyline.
The segments must have matching endpoints.

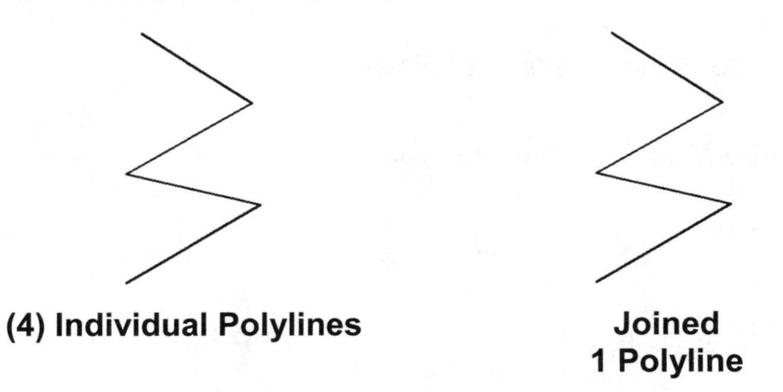

(4) Individual Polylines **Joined 1 Polyline**

WIDTH
The WIDTH option allows you to change the width of the polyline.
But the entire polyline will have the same width.

EDIT VERTEX
This option allows you to change the starting and ending width of each segment individually.

SPLINE
This option allows you to change straight polylines to curves.

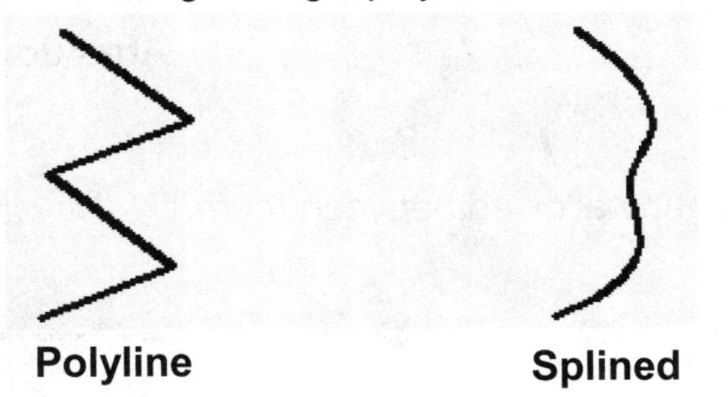

Polyline **Splined**

DECURVE
This option removes the SPLINE curves and returns the polyline to its original straight line segments.

Splined **Decurve**

REVERSE
This option reverses the direction. The start point becomes the end point and vice versa.

JOIN COMMAND

I know this might seem confusing but this Join command is not the same as the Join "option" within the Polyedit tool. The Polyedit option can only be used to join individual polyline segments. This **Join** Command joins similar objects to form a single unbroken object.

1. Select the **JOIN** command using one of the following:

   Ribbon = Home tab / Modify ▾ panel /

   Keyboard = J <enter>

2. Select the **Source** object.

3. Select the **objects to join to the source** object.

Seems easy, and it is, but there are a few rules regarding each type of object you must understand.

LINES
The Lines must be collinear (on the same plane) but can have gaps between them.

Before Join **After Join**

POLYLINES
Same rules as Lines except no gaps allowed between them.

Before Join **After Join**

ARC
The Arcs must lie on the same imaginary circle but can have gaps.

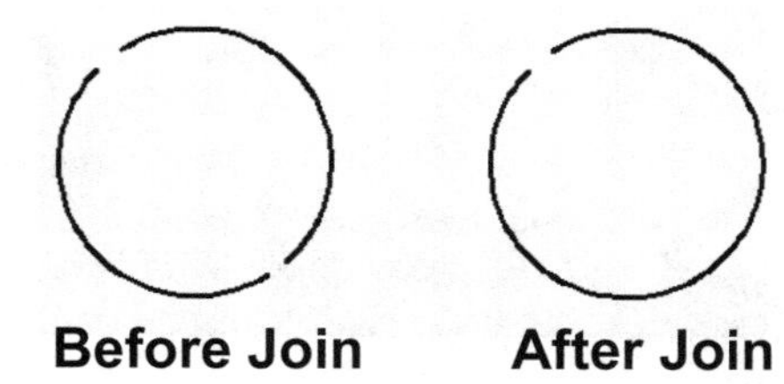

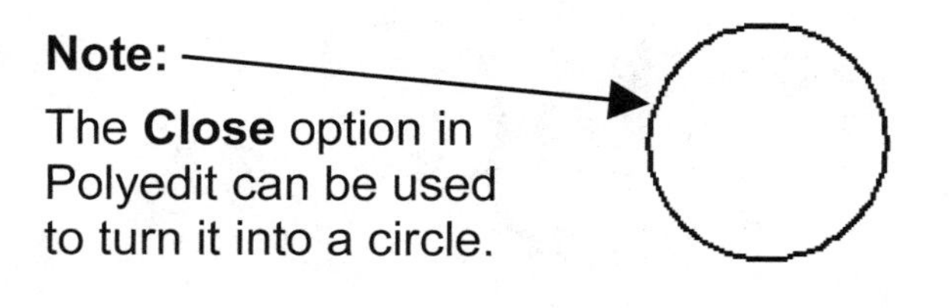

SPLINE
The Spline must lie in the same plane, and contiguous (end to end).
(You haven't learned this command as yet)

EXERCISE 24A

INSTRUCTIONS:

1. Start a **New** file using **Border A-2015.dwt**
2. Draw the Polyline as shown using layer Object line.
3. Using the Polyedit command, Close and Open the polyline.
4. Edit the Title and Ex-XX by double clicking on the text. Do not erase and replace.
5. Do not dimension
6. Save as **EX24A**
7. Not necessary to plot.

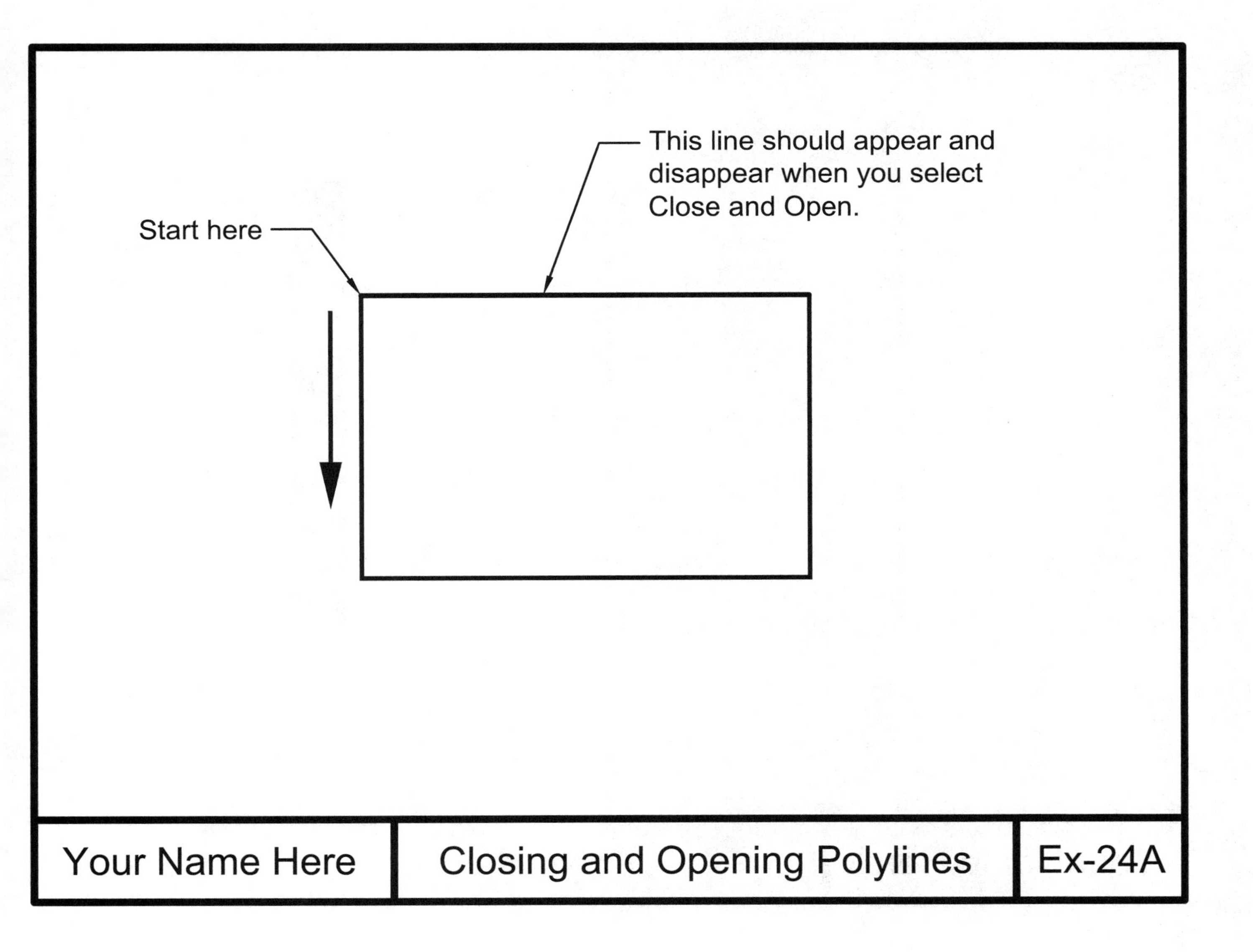

EXERCISE 24B

INSTRUCTIONS:

1. Start a **New** file using **Border A-2015.dwt**

Step 1.

2. Draw the Polyline 4" long and 0 width using layer Object line.

Step 2

3. Change the width of the Polyline to .50
4. Do not dimension
5. Edit the Title and Ex-XX by double clicking on the text. Do not erase and replace.
6. Save as **EX24B**
7. Plot using Page Setup **Class Model A**

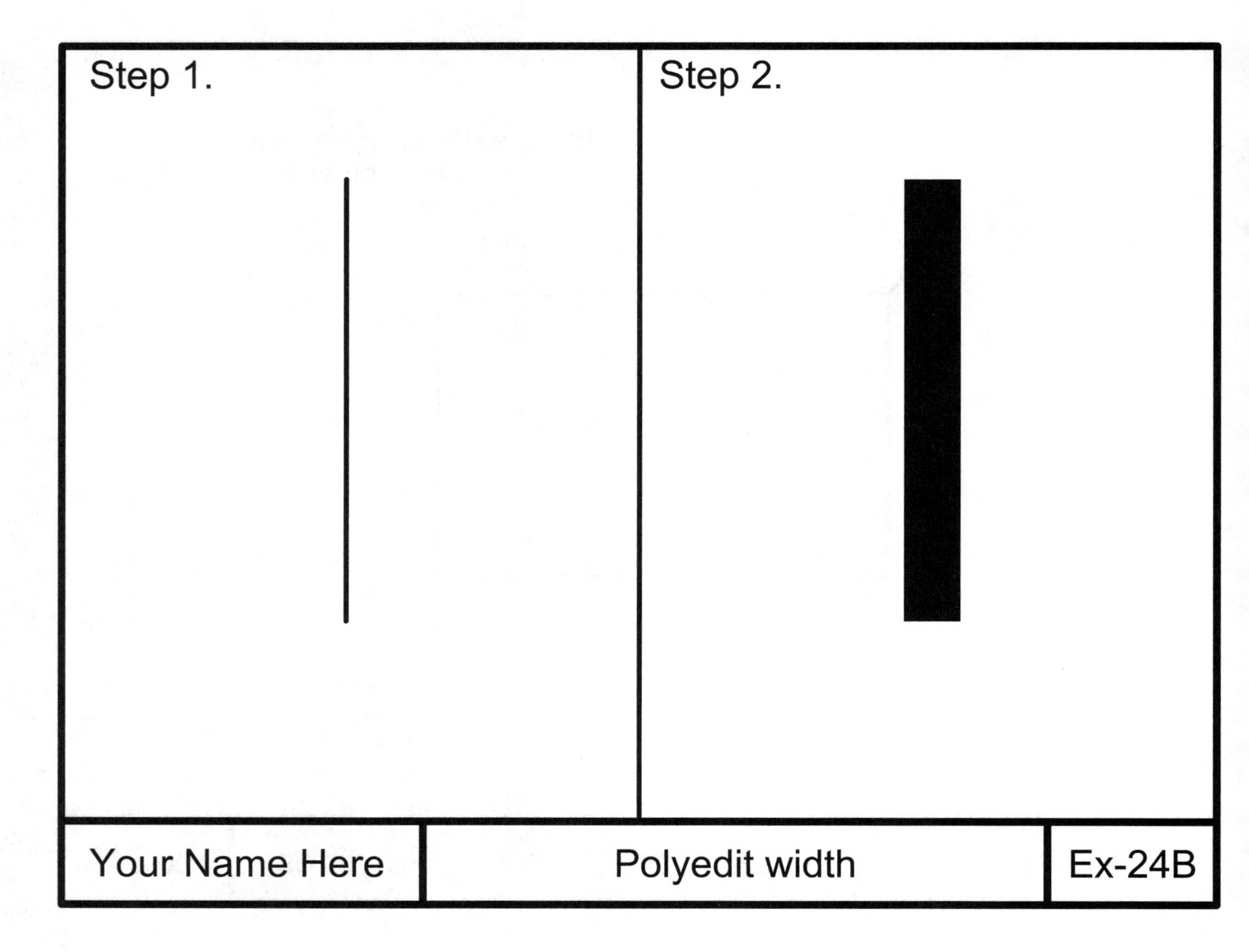

EXERCISE 24C

INSTRUCTIONS:

1. Start a **New** file using **Border A-2015.dwt**

Step 1.

2. Draw 1 continuous polyline with (3) 2" Long segments.

Step 2

3. Select the Polyedit command and select the first segment.
4. Select the Edit Vertex option. (Notice the "X" marking the starting point)
5. Select the Width option and set Start to 1.00 and Ending to .50.
6. Select Next option (The "X" should have moved to the next segment.)
7. Select the Width option and set Start to .50 and Ending to .50
8. Select the Next option (The "X" should have moved to the next segment)
9. Select the Width option and set Start to 2.00 and Ending to 0
10. Select the Exit option and press <enter> to stop
11. Edit the Title and Ex-XX ;   Save as **EX24C;**   Plot using Page Setup **Class Model A**

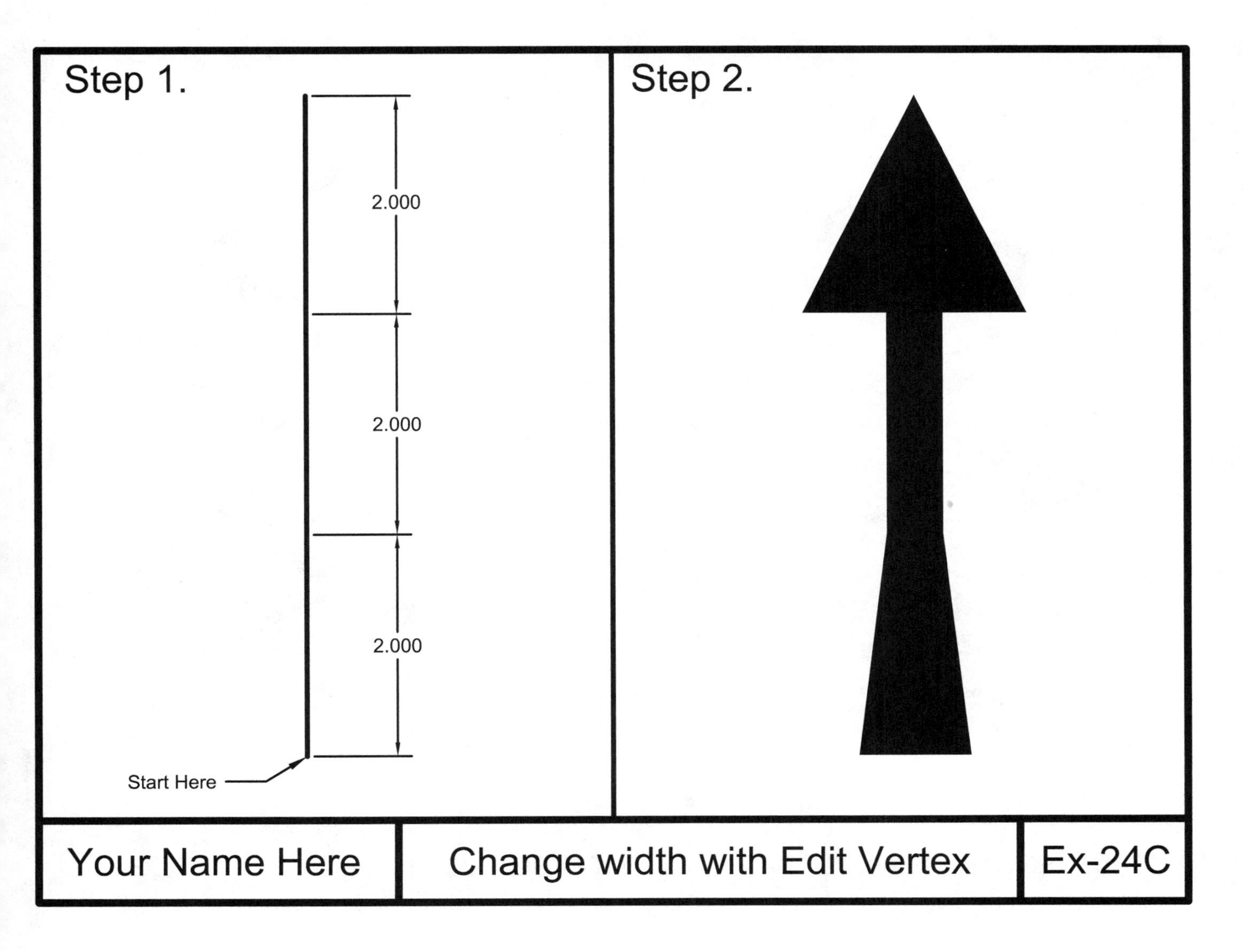

EXERCISE 24D

INSTRUCTIONS:

1. Start a **New** file using **Border A-2015.dwt**

Step 1.

2. Draw the objects below using the **LINE** command **not polyline**.

Step 2

3. Select the Polyedit command and select the first segment drawn.
   (Answer **YES** to "*Not a polyline, do you want to turn it into one*?)
4. Select the Width option and set the width to .10 (Notice only one segment changed)
5. Select the Join option.
6. Select the remaining lines using a crossing window.
   (Now the width should be constant for all lines)
7. Edit the Title and Ex-XX
8. Save as **EX24D**
9. Plot using Page Setup **Class Model A**

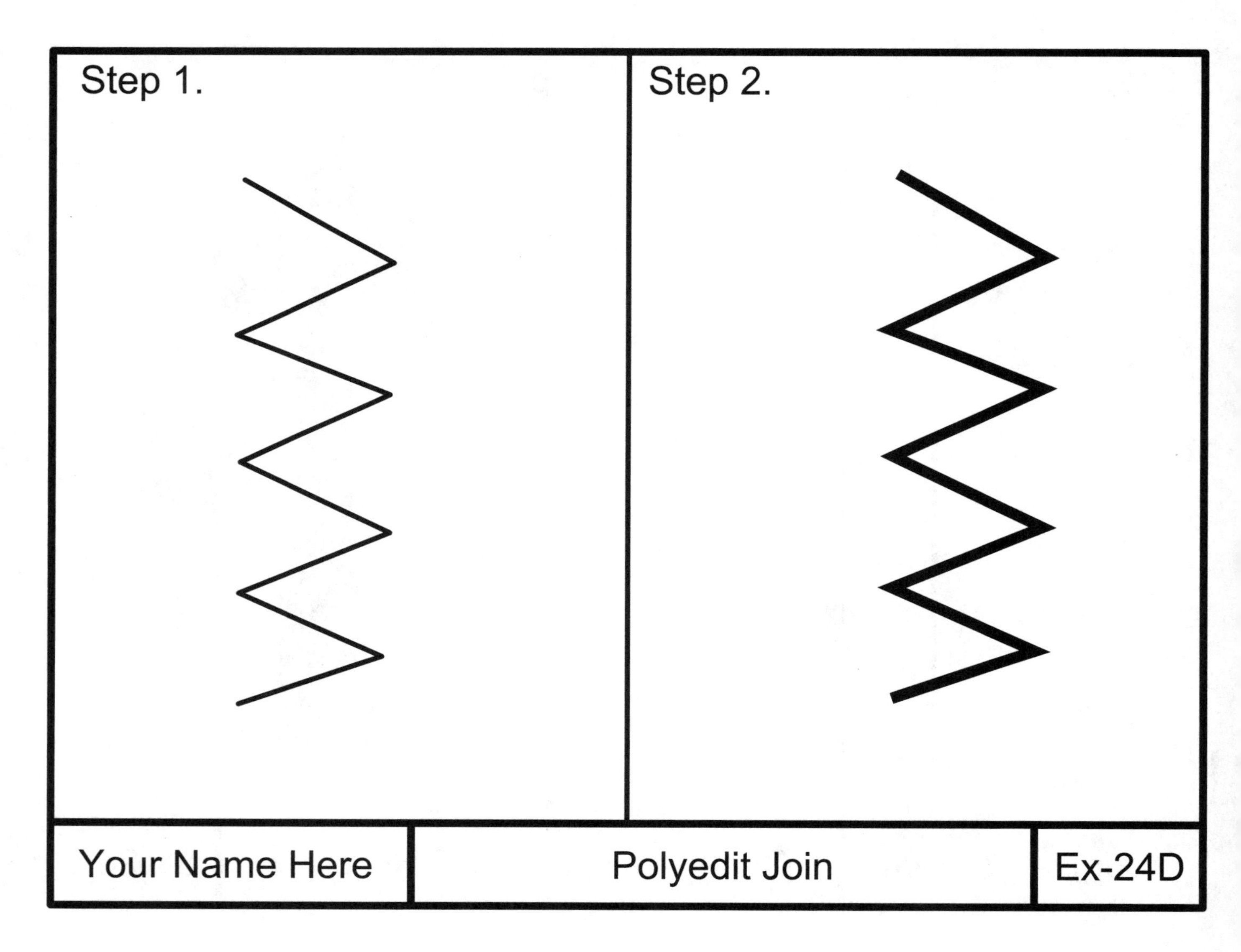

EXERCISE 24E

INSTRUCTIONS:

1. Open **EX24D**
2. Select the Polyedit command
3. Select the Polyline
4. Select the Spline option
5. Edit the Title and Ex-XX
6. Save as **EX24E**
7. Plot using Page Setup **Class Model A**

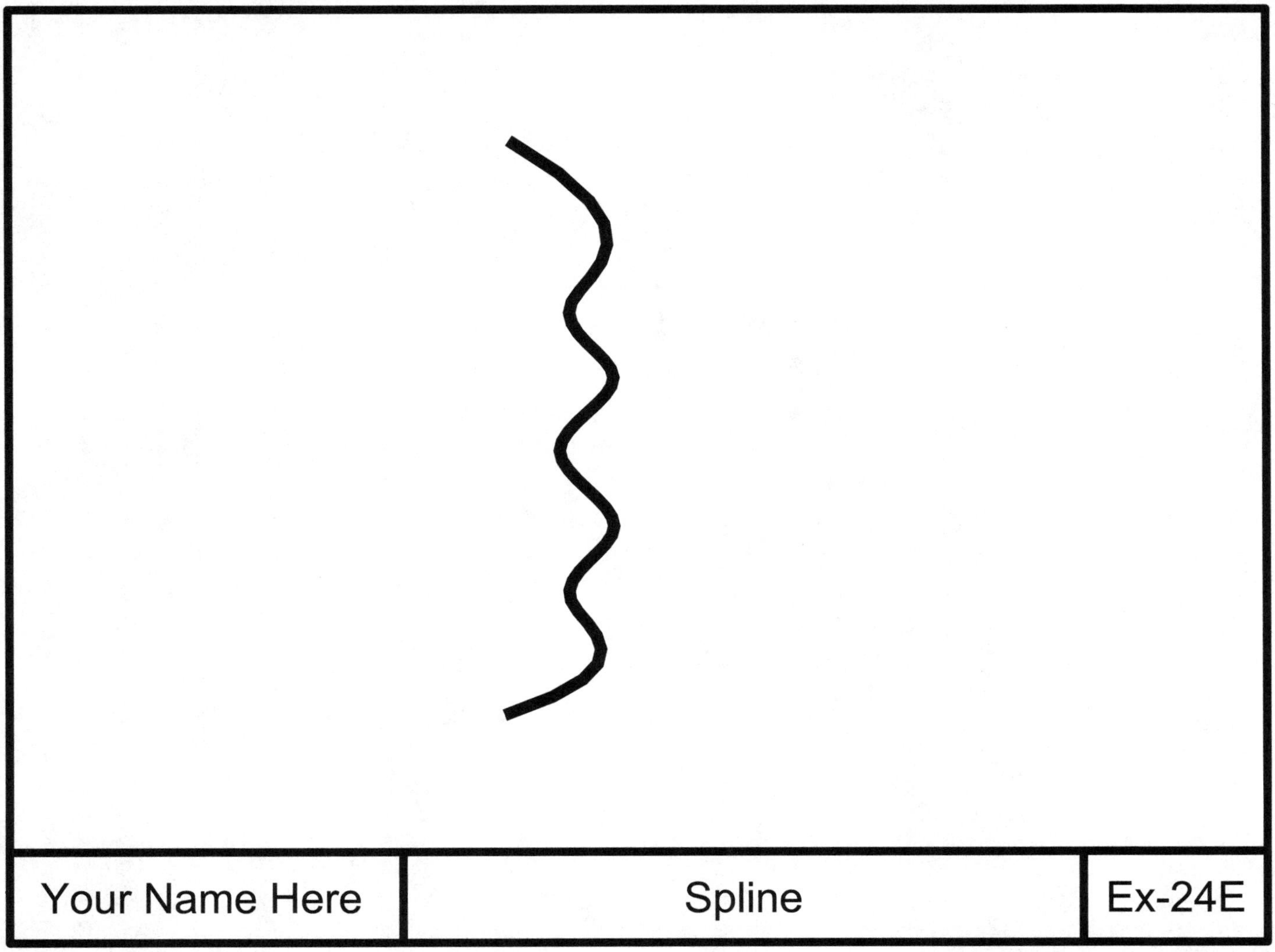

Your Name Here	Spline	Ex-24E

Notes:

LEARNING OBJECTIVES

After completing this lesson, you will be able to:

1. Create a new Text Style
2. Change an existing Text Style
3. Delete a Text Style
4. Select a Point Style
5. Divide an object into specified number of segments
6. Divide an object to a specified measurement

LESSON 25

CREATING NEW TEXT STYLES

AutoCAD provides you with two preset Text Styles named “Standard” and “Annotative”. (Annotative will be discussed in Lesson 27)
You may want to create a new text style with a different font and effects.
The following illustrates how to create a new text style.

1. Select the **TEXT STYLE** command using one of the following:

   Ribbon = Annotate tab / Text panel / ↘
   or
   Keyboard = ST <enter>

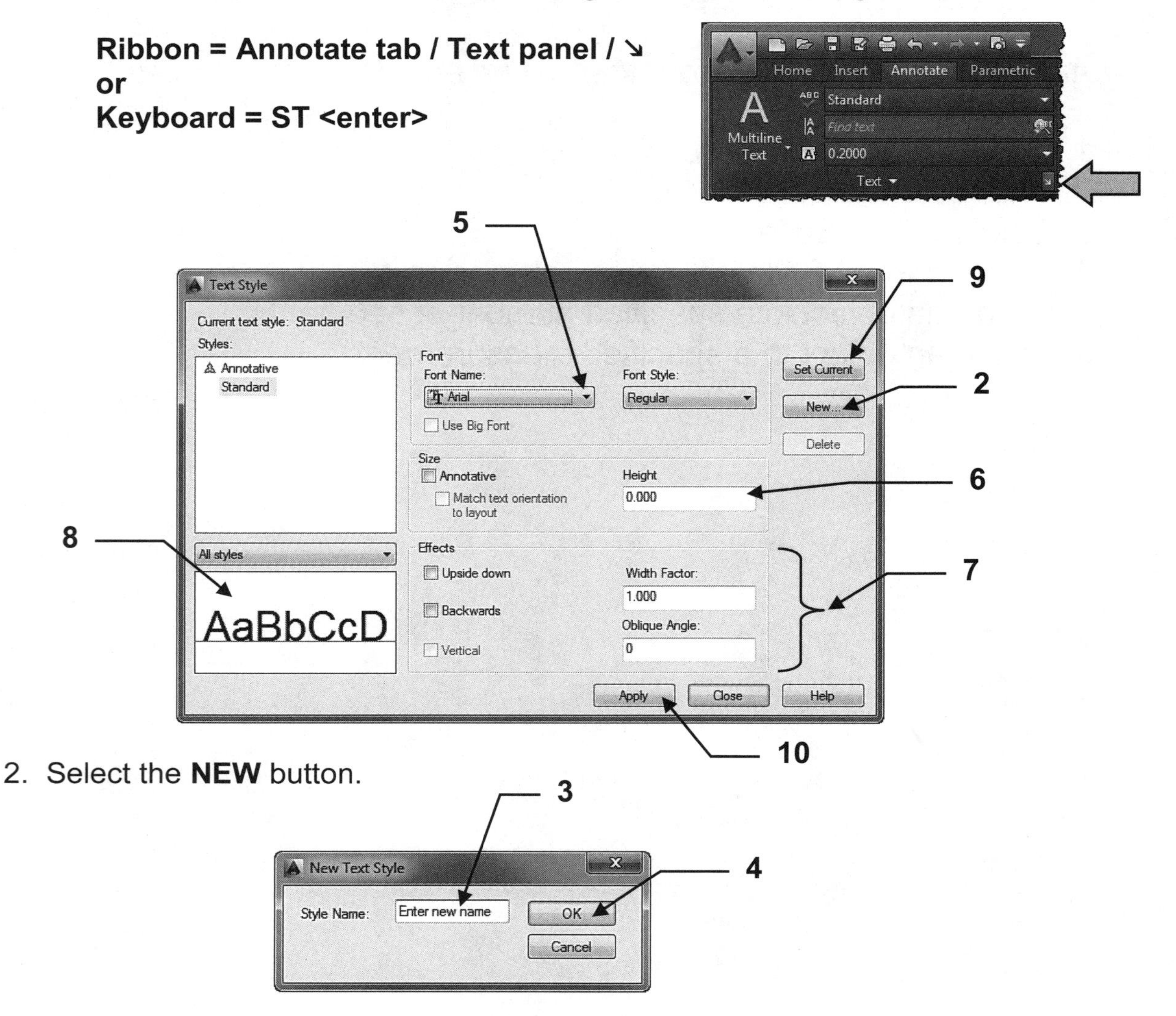

2. Select the **NEW** button.

3. Type the new style name in **STYLE NAME** box.
   Styles names can have a maximum of 31 characters, including letters, numbers, dashes, underlines and dollar signs. You can use Upper or Lower case.

4. Select the **OK** button.

5. Select the **FONT**.

6. Enter the value of the Height.
   Note: If the value is 0, AutoCAD will always prompt you for a height.
   If you enter a number the new text style will have a fixed height and AutoCAD will not prompt you for the height.

CREATING NEW TEXT STYLES....continued

7. Assign **EFFECTS.**

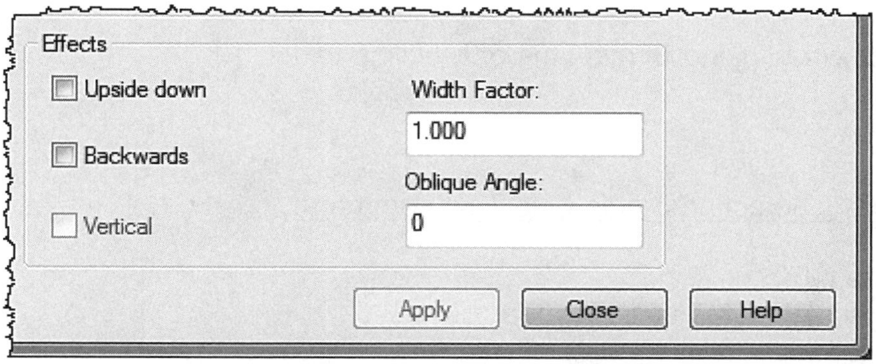

UPSIDE-DOWN
Each letter will be created upside-down in the order in which it was typed. (Note: this is different from rotating text 180 degrees.)

BACKWARDS
The letters will be created backwards as typed.

VERTICAL
Each letter will be inserted directly under the other. Only **.shx** fonts can be used. VERTICAL text will not display in the **PREVIEW** box.

OBLIQUE ANGLE
Creates letter with a slant, like italic. An angle of 0 creates a vertical letter. A positive angle will slant the letter forward. A negative angle will slant the letter backward.

WIDTH FACTOR
This effect compresses or extends the width of each character.
A value less than 1 compresses each character.
A value greater than 1 extends each character.

8. The **PREVIEW** box displays the text with the selected settings

9. Select the **Set Current** button.

10. Select the **Apply** or **Close** button

HOW TO SELECT A TEXT STYLE

After you have created Text Styles you will want to use them. You must select the Text Style before you use it.

Below are the methods when using Single Line or Multiline text.

Single Line Text

Select the style before selecting Single Line Text command.

1. Select the **Annotate** tab.
2. Using the Text panel, select the style down arrow ▼
3. Select the Text Style

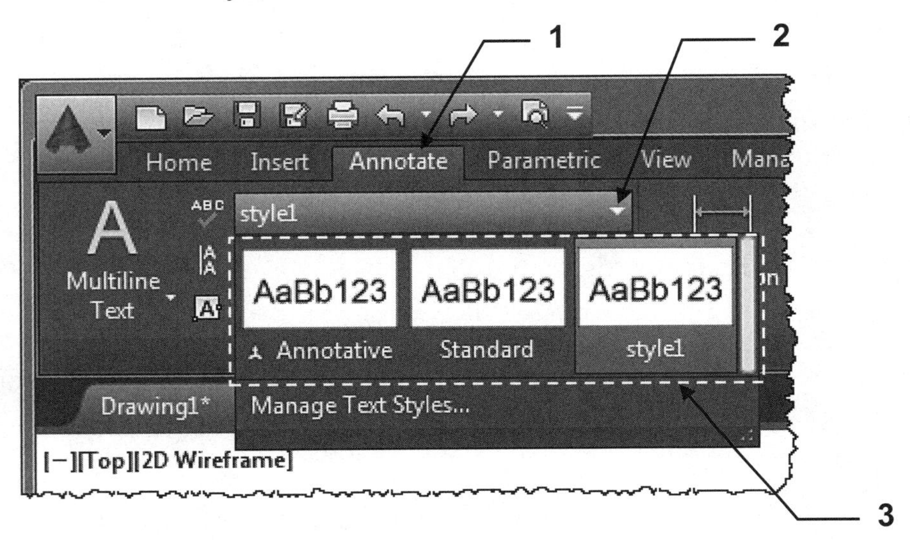

Multiline Text

1. Select **Multiline Text** and place the **first corner** and **opposite corner.**
2. Find the **Style** panel and scroll through the text styles available using the up and down arrows.

Select the Text Style within the Text Editor.

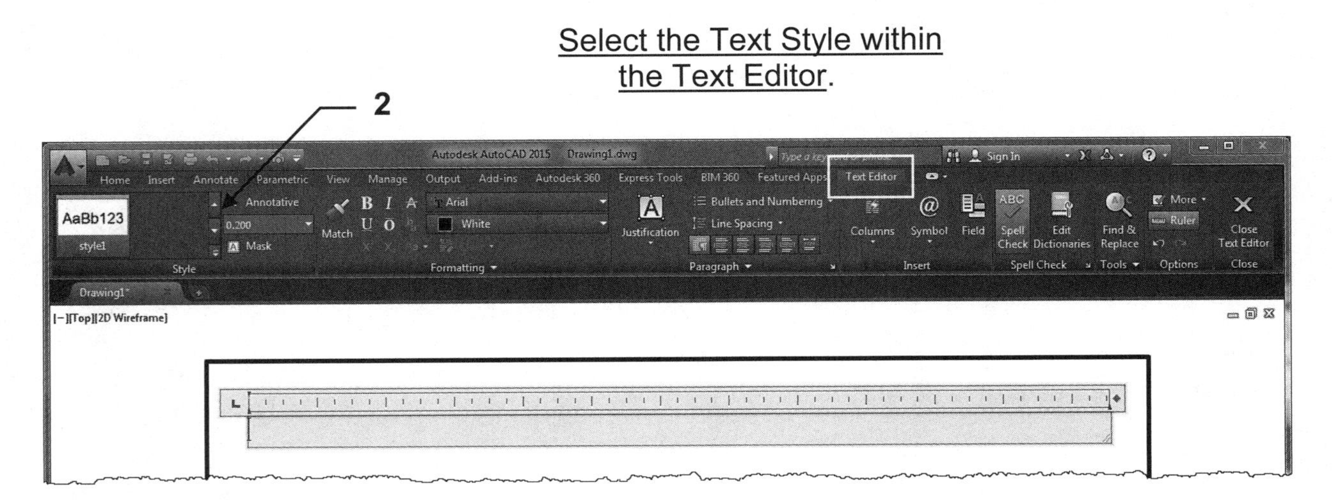

DELETE A TEXT STYLE

1. Select the **TEXT STYLE** command using one of the following:

   Ribbon = Annotate tab / Text panel / ↘
   or
   Keyboard = ST <enter>

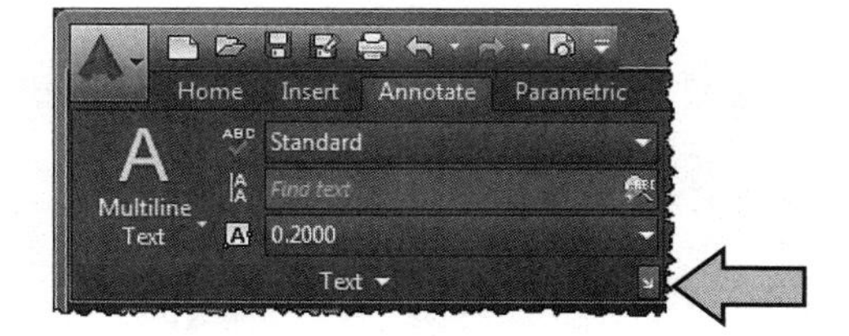

2. First, select a **Text Style** that you <u>do not want to Delete</u> and select **Set Current** button. (You can't Delete a Text Style that is in use.)

3. Select the **Text Style** that you want to Delete.

4. Select the **Delete** button.

5. Warning appears, select **OK** or **Cancel**

6. Select the **Close** button.

Note: Also refer to page 29-9 for the PURGE command. The Purge command will remove any unused text styles, dimension styles, layers and linetypes.

CHANGE EFFECTS OF A TEXT STYLE

1. Select the **TEXT STYLE** command using one of the following:

    Ribbon = Annotate tab / Text panel / ↘
    or
    Keyboard = ST <enter>

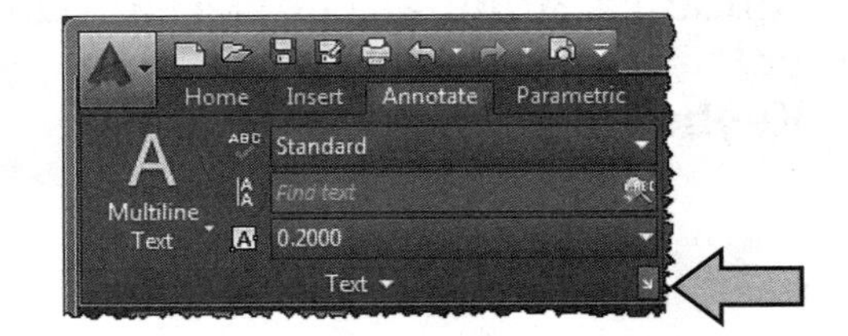

2. Select the **Text Style** you wish to change.

3. Make the changes in the **EFFECTS** boxes.

    Note about Vertical:
    Only **.shx** fonts can be vertical
    Vertical text will not display in the **PREVIEW** box.

4. Select the **Apply** button. (*Apply will stay gray if you did not change a setting.)*

5. Select the **Close** button.

DIVIDE COMMAND

The **DIVIDE** command divides an object mathematically by the NUMBER of segments you specify. A POINT (object) is placed at each interval on the object.

Note: the object selected is **NOT** broken into segments. The **POINTS** are simply drawn **ON** the object.

EXAMPLE:

This LINE has been DIVIDED into 4 EQUAL lengths.
But remember, the line is not broken into segments.
The Points are simply drawn **ON** the object.

1. First open the Point Style box and select the **POINT STYLE** to be placed on the object.

   Ribbon = Home tab / Utilities Panel ▼/ Point Style
   or
   Keyboard = ddptype <enter>

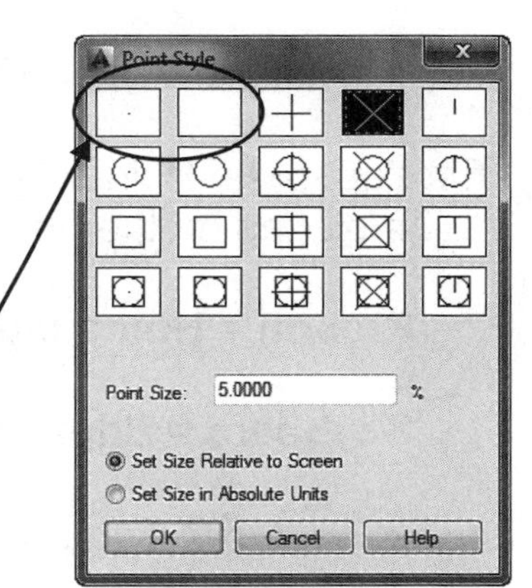

Select either of these ONLY when you want to make the points disappear yet still be there.

(Refer to page 5-7 if you need a refresher on Points)

2. Next select the **DIVIDE** command using one of the following:

   Ribbon = Home tab / Draw panel ▾ /
   or
   Keyboard = DIV <enter>

3. Select Object to divide: ***select the object to divide***

4. Enter the number of segments or [Block]: ***type the number of segments <enter>***

MEASURE COMMAND

The **MEASURE** command is very similar to the **DIVIDE** command because point objects are drawn at intervals on an object. However, the **MEASURE** command allows you to specify the **LENGTH** of the segments rather than the number of segments.

Note: the object selected is **NOT** broken into segments. The **POINTS** are simply drawn **ON** the object.

EXAMPLE:

Select Object here to start the measurement From this end.

The **MEASURE**ment was started at the left endpoint, and ended just short of the right end of the line. The remainder is less than the measurement length specified. You designate which end you want the measurement to start by selecting the end when prompted to select the object.

1. First open the Point Style box and select the **POINT STYLE** to be placed on the object.

   Ribbon = Home tab / Utilities Panel ▼/ Point Style
   or
   Keyboard = ddptype <enter>

(Refer to page 5-7 if you need a refresher on Points)

Point Style

Point Size: 5.0000 %

Set Size Relative to Screen

Set Size in Absolute Units

OK Cancel Help

2. Next select the **MEASURE** command using one of the following:

   Ribbon = Home tab / Draw panel ▼ /
   or
   Keyboard = ME <enter>

3. Select Object to MEASURE: ***select the object to MEASURE***
   (Note: this selection point is also where the MEASUREment will start.)

4. Specify length of segment or [Block]: ***type the length of one segment <enter>***

EXERCISE 25A

INSTRUCTIONS:

1. Start a **New** file using **Border A-2015.dwt**
2. Create the 3 Text Styles shown below and on the next page.
   (Style1, Style2 and Style3)
   Select all of the options shown.
3. **Save** the drawing as: **EX25A**

STYLE1 *Note: Style2 and Style3 on the next page.*

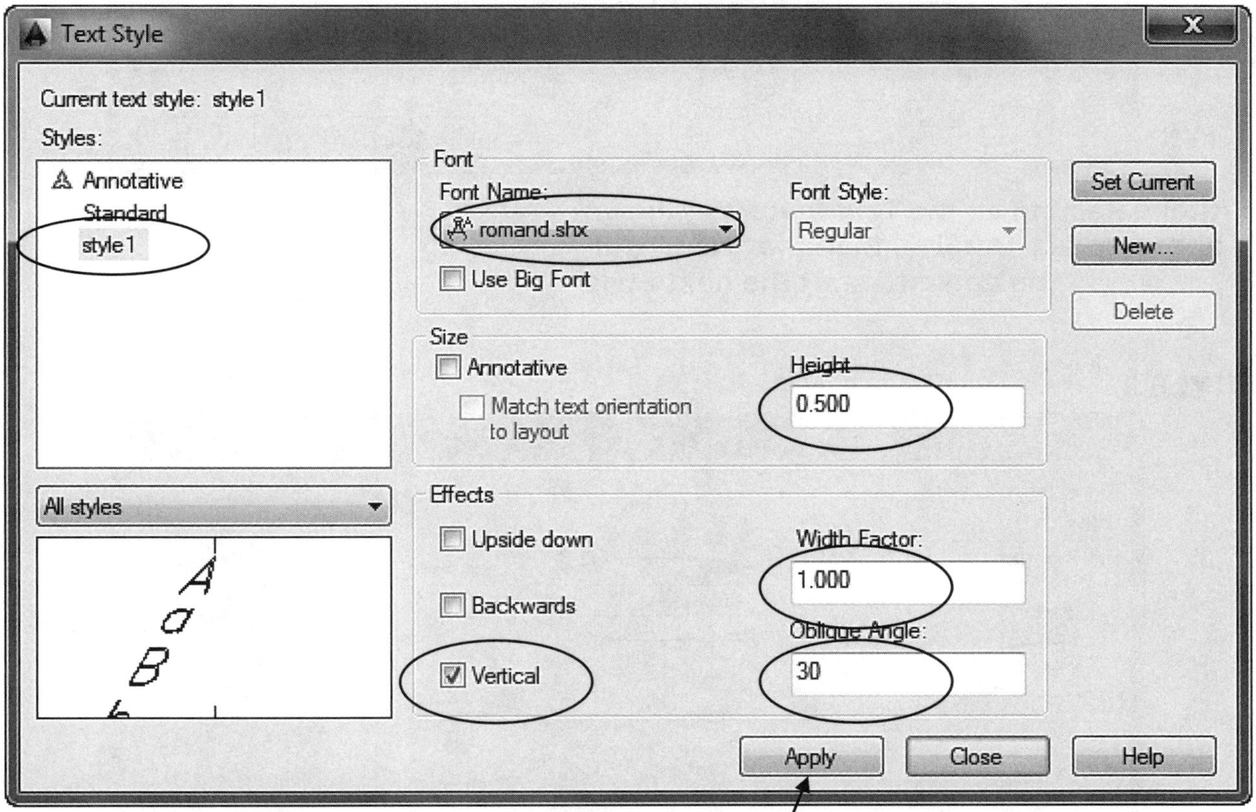

Text Style

Current text style: style1

Styles:

Annotative

Standard

style1

All styles

Font

Font Name: romand.shx

Font Style: Regular

Use Big Font

Size

Annotative

Match text orientation to layout

Height: 0.500

Effects

Upside down

Backwards

Vertical

Width Factor: 1.000

Oblique Angle: 30

Set Current

New...

Delete

Apply

Close

Help

After selecting all the Text features shown above select "Apply", then the "New" button again to start the next Style

EXERCISE 25A....continued

STYLE2

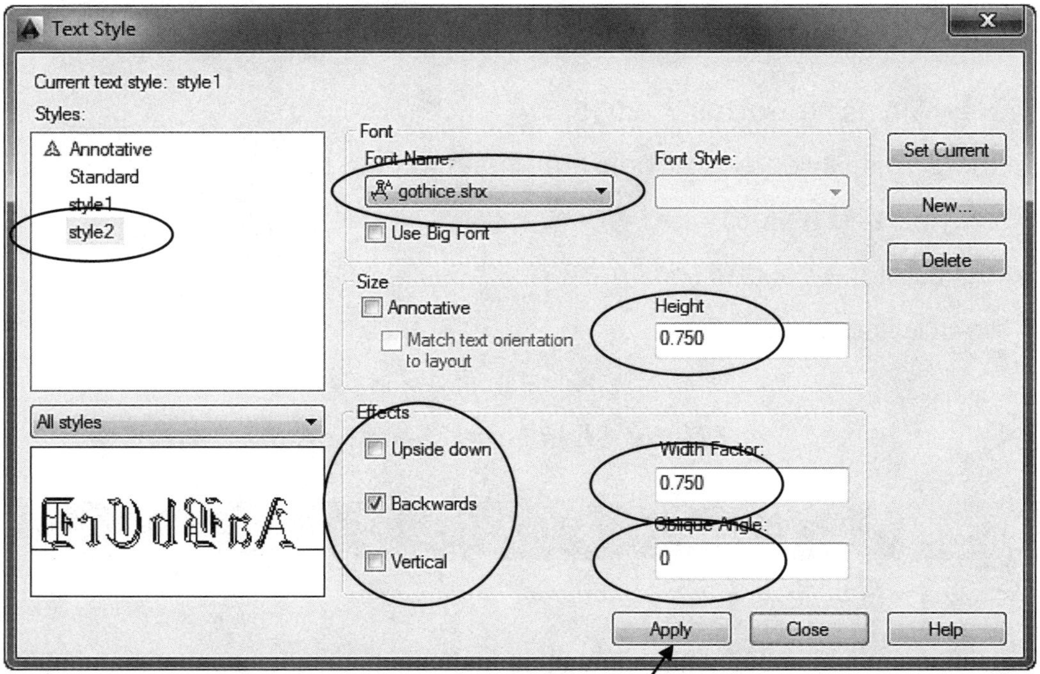

Text Style

Current text style: style1

Styles: Annotative, Standard, style1, style2

Font Name: gothice.shx

Font Style:

Use Big Font

Size: Annotative; Match text orientation to layout

Height: 0.750

Effects: Upside down; Backwards (checked); Vertical

Width Factor: 0.750

Oblique Angle: 0

All styles

Set Current | New... | Delete | Apply | Close | Help

After selecting all the Text features shown above select "Apply", then the "New" button again to start the next Style

STYLE 3

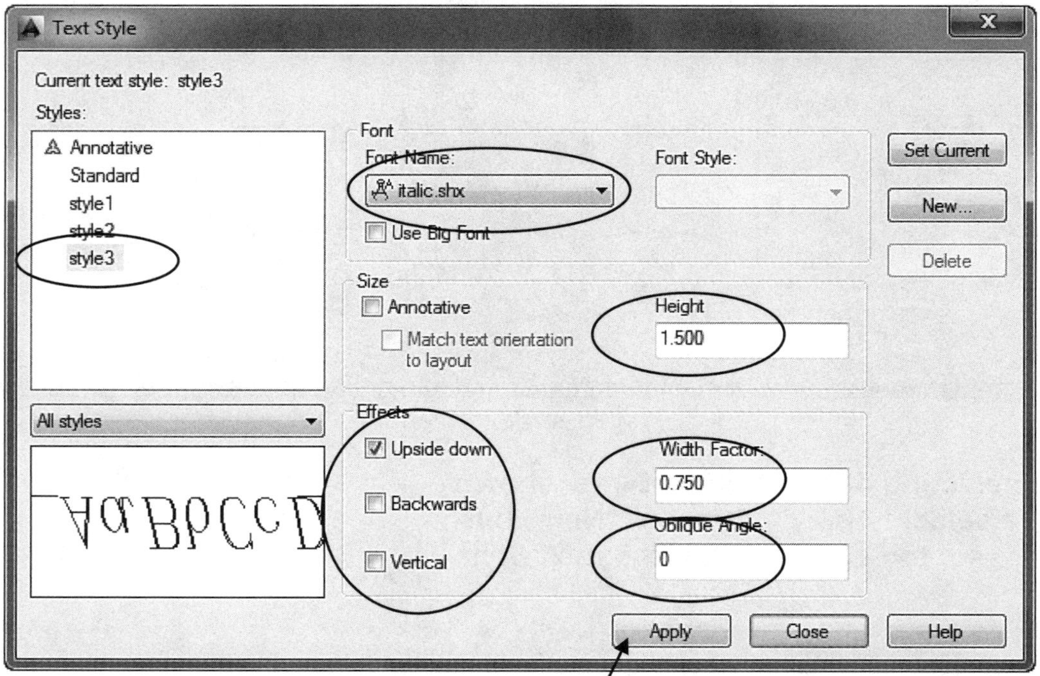

Text Style

Current text style: style3

Styles: Annotative, Standard, style1, style2, style3

Font Name: italic.shx

Font Style:

Use Big Font

Size: Annotative; Match text orientation to layout

Height: 1.500

Effects: Upside down (checked); Backwards; Vertical

Width Factor: 0.750

Oblique Angle: 0

All styles

Set Current | New... | Delete | Apply | Close | Help

Select Apply button and Close button

EXERCISE 25B

INSTRUCTIONS:

1. Open **EX25A** (If not already open)
2. Before entering text, move the Origin to the lower left corner as shown.
3. Draw the 3 Text Styles using "Single Line Text".

   Remember to select the text style before selecting the Single Line text command.

4. Place the start point exactly as shown. (enter coordinates)
5. Use layer Text.
6. Edit the Title and Ex-XX by double clicking on the text. Do not erase and replace.
7. Save as **EX25B**
8. Plot using Page Setup **Class Model A**

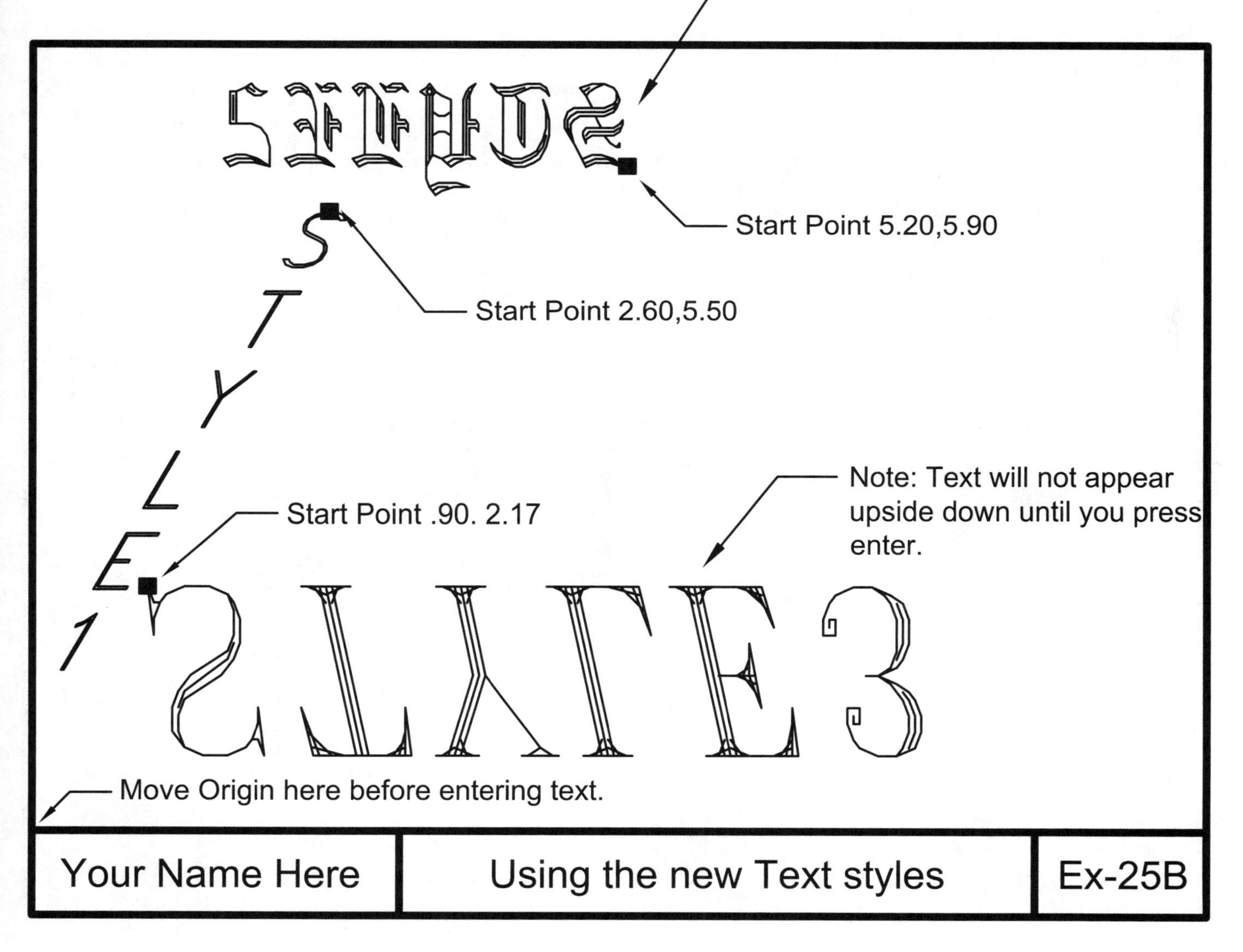

EXERCISE 25C

INSTRUCTIONS:

1. Open **EX25B** (if not already open)
2. Change the Text to Text Style "Standard".
   a. Select all text
   b. Quick Properties will appear
   c. Change Text "varies" to "Standard"

3. Edit the Title and Ex-XX by double clicking on the text.
4. Save as **EX25C**
5. Plot using Page Setup **Class Model A**

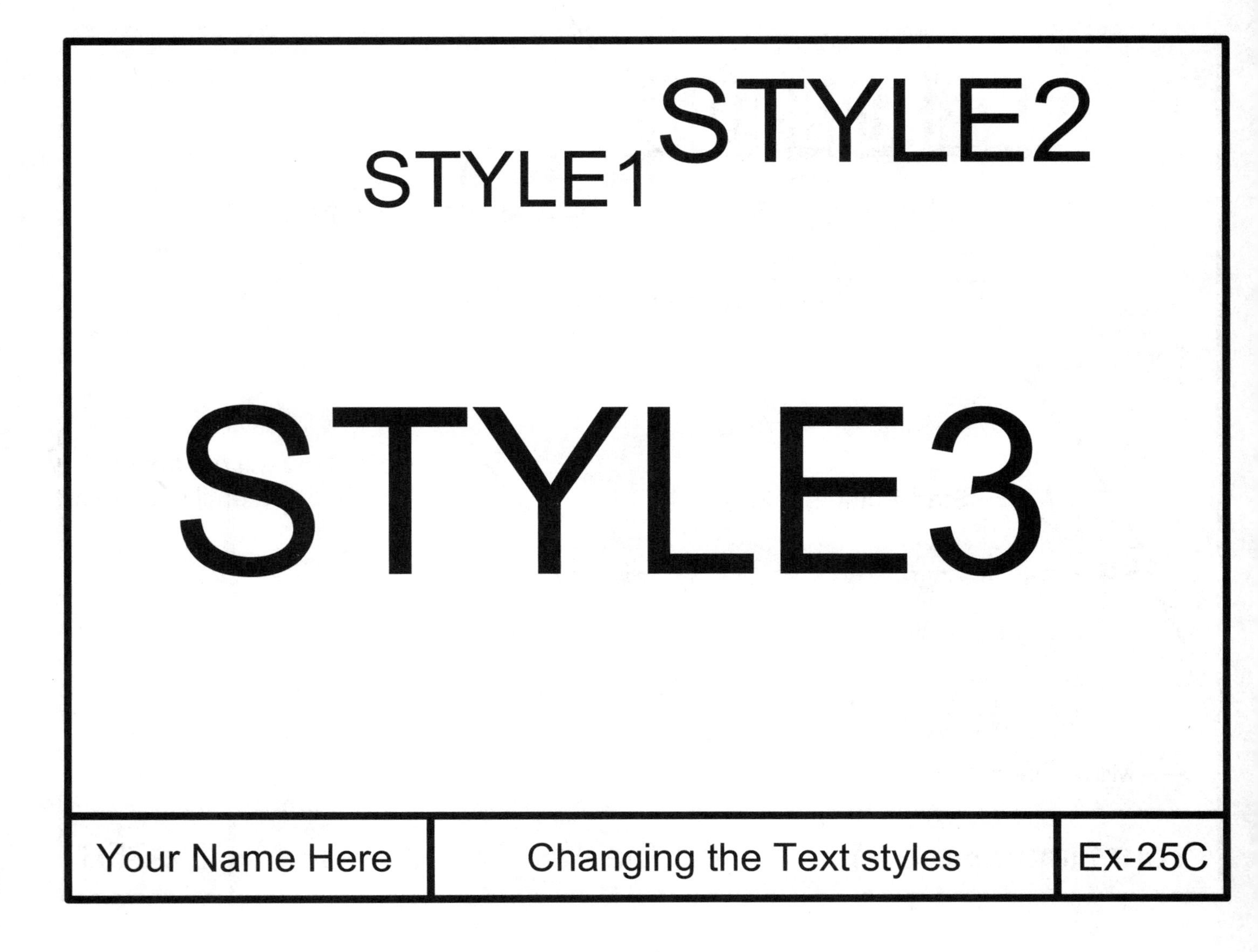

EXERCISE 25D

INSTRUCTIONS:

1. Open **EX25C** (If not already open)
2. Delete Text Styles 1, 2 and 3 as follows:

    a. Select **TEXT STYLE box.**

    b. Select the “Standard” text style.

    c. Select **Set Current** button. (This makes Standard the current text style)

    AutoCAD will not allow you to delete an active or current text style. That is why you must make the “Standard” text style current before you may delete a text style.

    d. Select the style you want to **delete**.

    e. Click on the **DELETE** button.

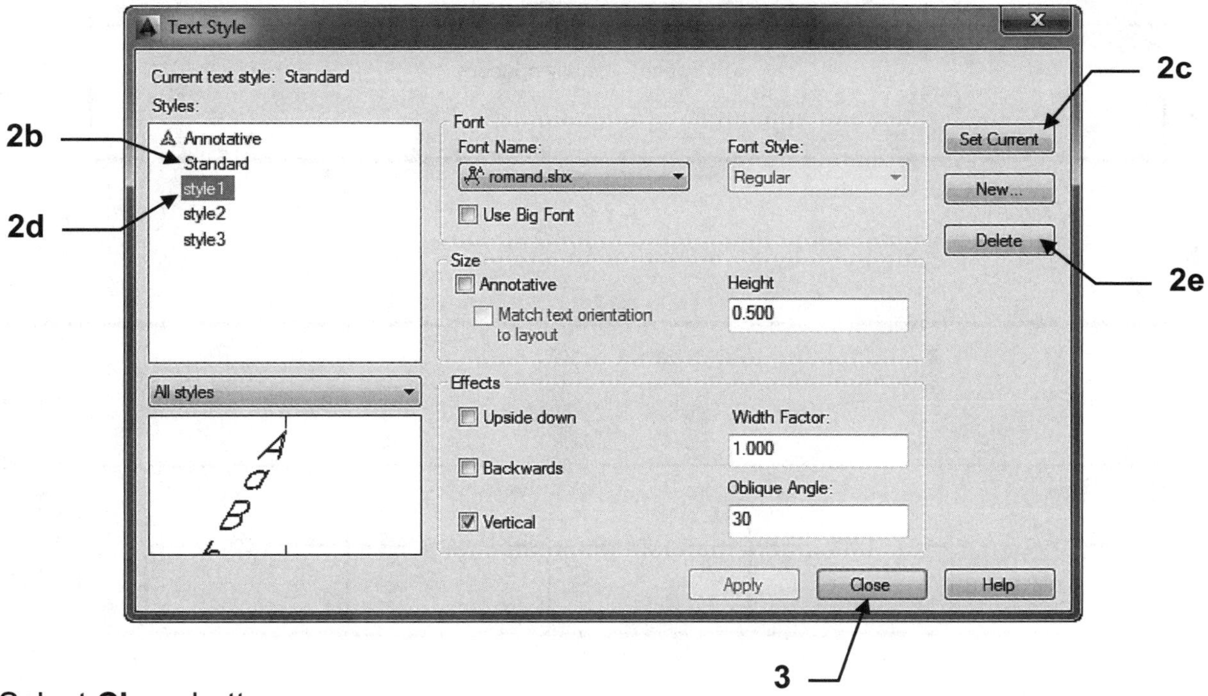

3. Select **Close** button.
4. **Save** as: **EX25D**

EXERCISE 25E

INSTRUCTIONS:

1. Start a **New** file using **Border A-2015.dwt**
2. Draw the (2) 8" long lines shown below on layer Object line.
3. **Divide** the upper line into 10 equal segments.
4. Using **Measure** create (5) 1.50 long segments on the lower line as shown.
5. Edit the Title and Ex-XX by double clicking on the text. Do not erase and replace.
6. Dimension as shown. (Remember "Node" is the object snap to use for Points)
7. Save as **EX25E**
8. Plot using Page Setup **Class Model A**

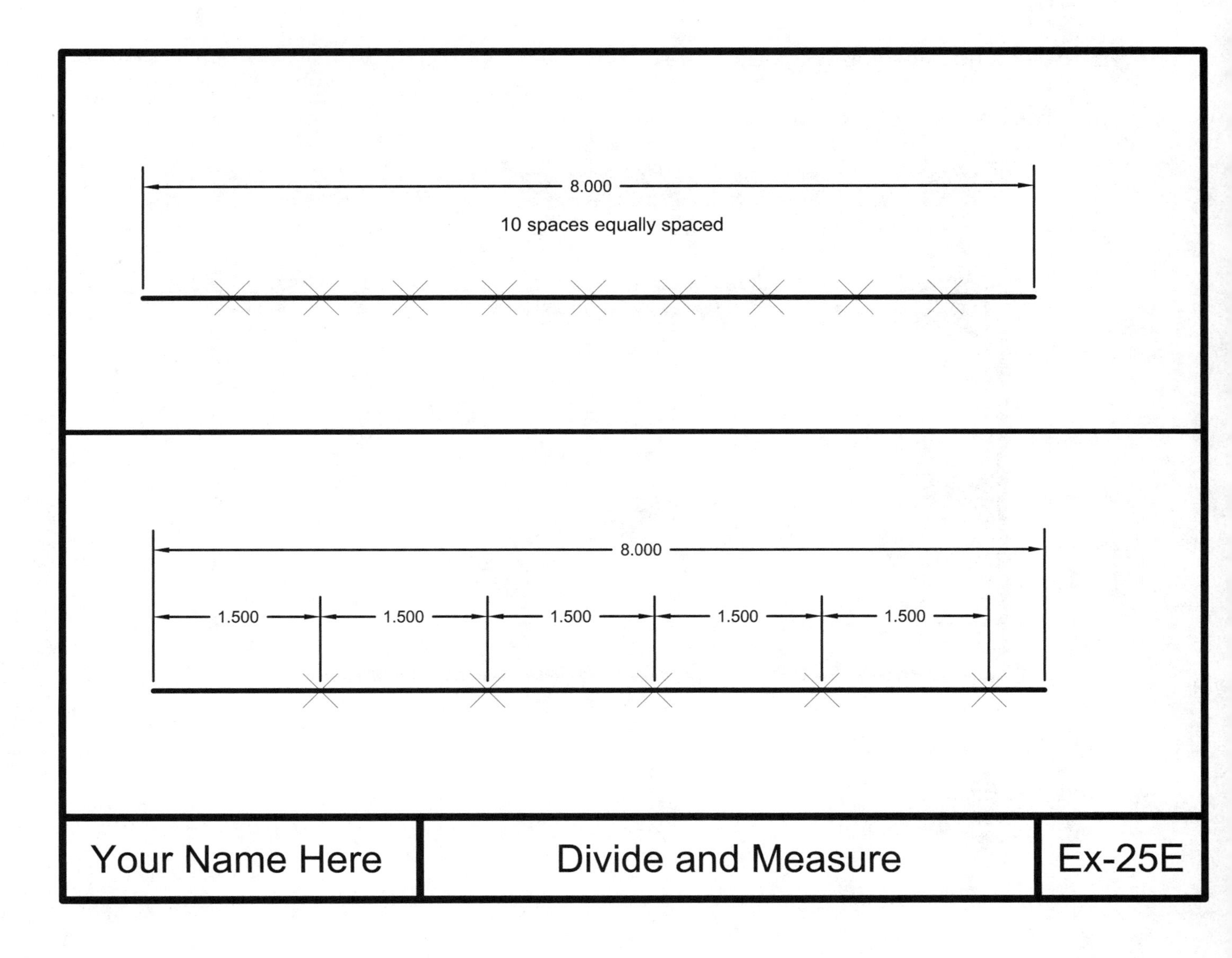

EXERCISE 25F

INSTRUCTIONS:

1. Start a **New** file using **Border A-2015.dwt**
2. Draw the "pretend" house roof plan with fence line shown below.

3. Refer to the next page for instructions and helpful hints.

4. Dimension using Dimension Style "Class Style" and Layer Dimension
5. Edit the Title and Ex-XX by double clicking on the text. Do not erase and replace.
6. Save as **EX25** and Plot using Page Setup **Class Model A**

(Note: You will learn to actually draw a full size house in Lesson 27.)

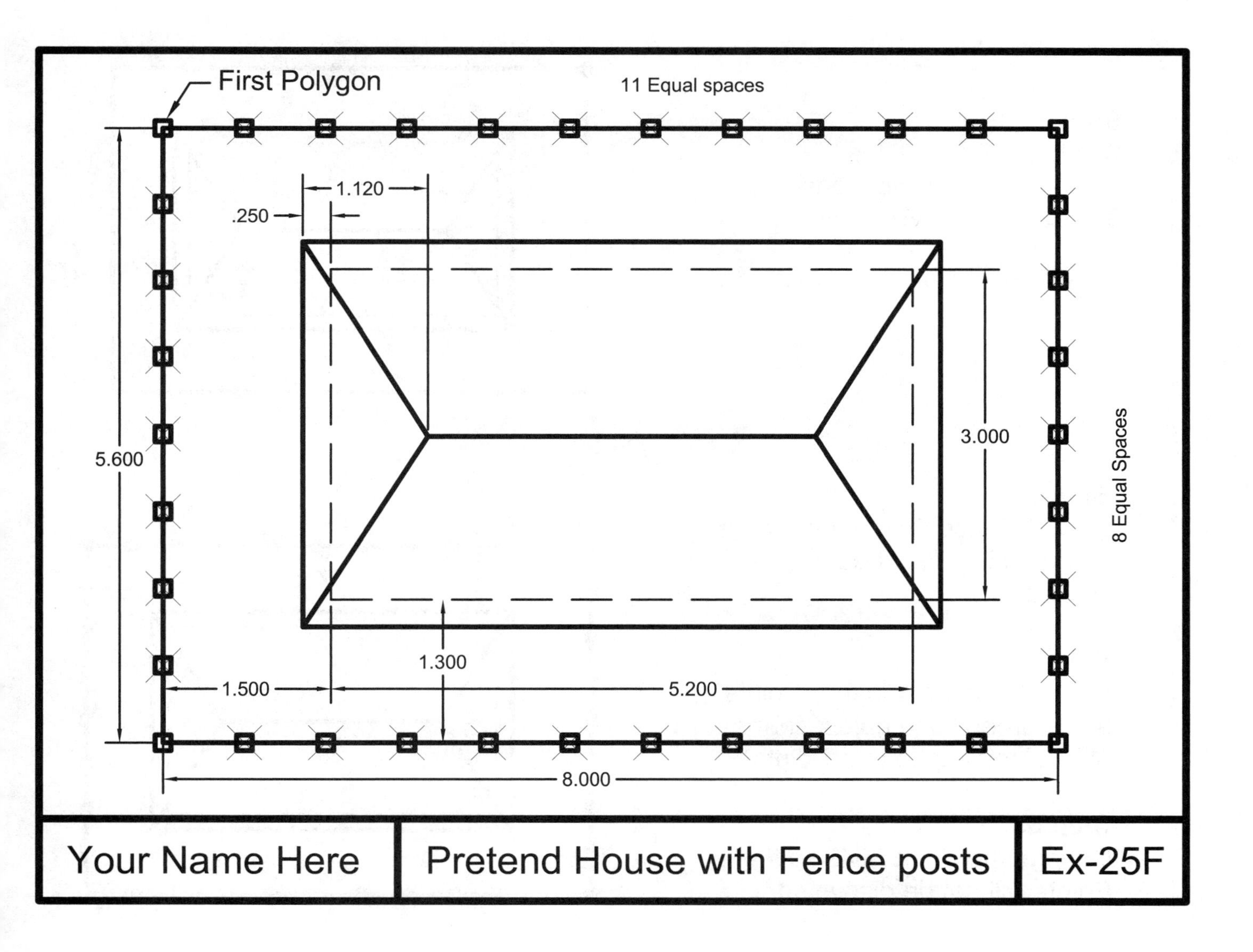

25F Helpful Hints

Step 1.
Draw the "pretend" house roof plan with fence line.
Use Layers Roof, Walls and Hidden Lines.

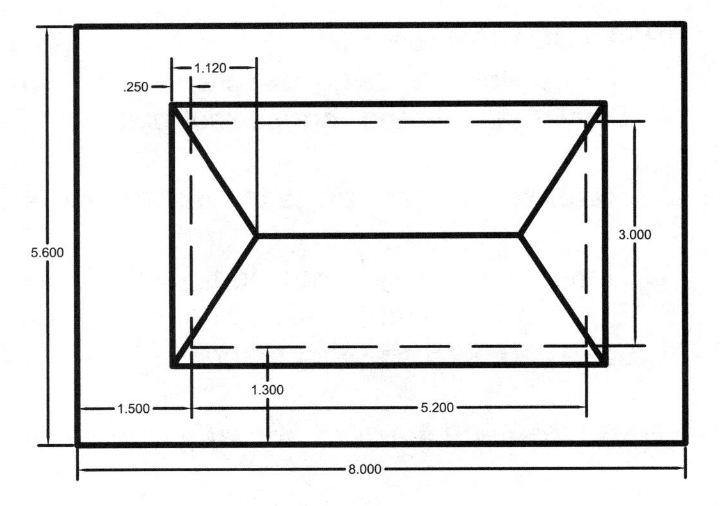

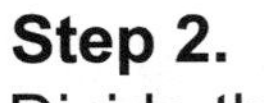

Step 2.
Divide the Fence lines.
11 equal spaces Horizontally
8 equal spaces Vertically

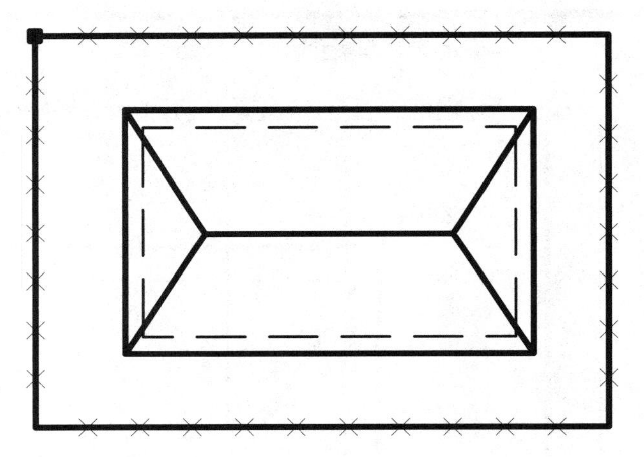

Step 3.
Draw 1 Fence Post at upper left corner using inscribed, 4 sided, .10 radius.

Step 4
Copy Fence Post to the "Points" and Corner. Use object snaps "Node" to snap to the "Points".

Step 5.
Change the Point Style so the Points will not be displayed.

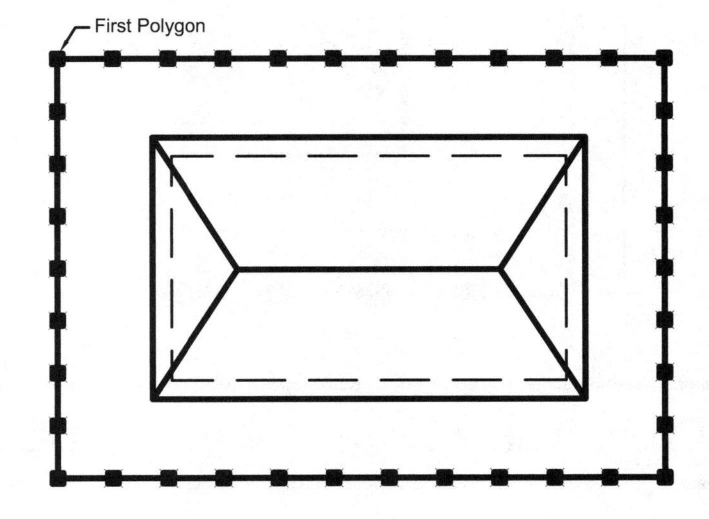

LEARNING OBJECTIVES

After completing this lesson, you will be able to:

1. Understand the difference between Model and Layout tabs
2. Create Viewports
3. Create a Page Setup
4. Create a Plot Page Setup
5. Create a Decimal Setup Master template
6. Create a new Border
7. Experiment with multiple viewports

LESSON 26

SERIOUS BUSINESS

In the previous lessons you have been having fun learning most of the basic commands that AutoCAD offers and you have been using a template that included preset layers and drawing settings etc. But now it is time to get down to the "serious business" of setting up your own drawing from "scratch".

Starting from scratch means you will need to set or create the following:

Items 1 through 5 you have learned in previous lessons

1. Drawing Units (Lesson 4)
2. Snap and Grid (Lesson 2)
3. Create Text styles (Lesson 25)
4. Create Dimension Styles (Lesson 16)
5. Create new layers and load linetypes.

Items 6 through 9 will be learned in this lesson

6. Create a "Layout" for plotting.
7. Create a "Floating Viewport" in the Layout.
8. Create a "Page Setup" to save plot settings.
9. Plot the drawing from Paper space.

After reading pages 26-3 through 26-22 start Exercise 26A and work your way through to 26D. When you have completed Exercise 26E you will have created a master template named, "**My Decimal Setup**". This master template will have everything set, created and prepared, ready to use each time you want to create a drawing using decimal units and to be plotted on an 8-1/2 X 11 inch sheet.

This means, for future drawings you merely select **File / New** and "**My Decimal Setup.dwt**" and start drawing. No time consuming setups. It is all ready to go.

In Lesson 27 you will create a master template for "feet and inches".

So take it one page at a time and really concentrate on understanding the process.

Note:
I am using sheet size 8-1/2 X 11 so you may print the exercises on your printer. After you understand the page setup concept you will be able to assign a larger sheet size to conform to any large format printer that you may use in the future.

MODEL and LAYOUT OPTIONS

Very important:
Before I discuss Model and Layout I need you to confirm ***Model and Layout tabs*** *are displayed.*

This will just take a minute.

1. Type **options <enter>**

   or

   Right click and select **Options**.

2. Select the **Display** tab.

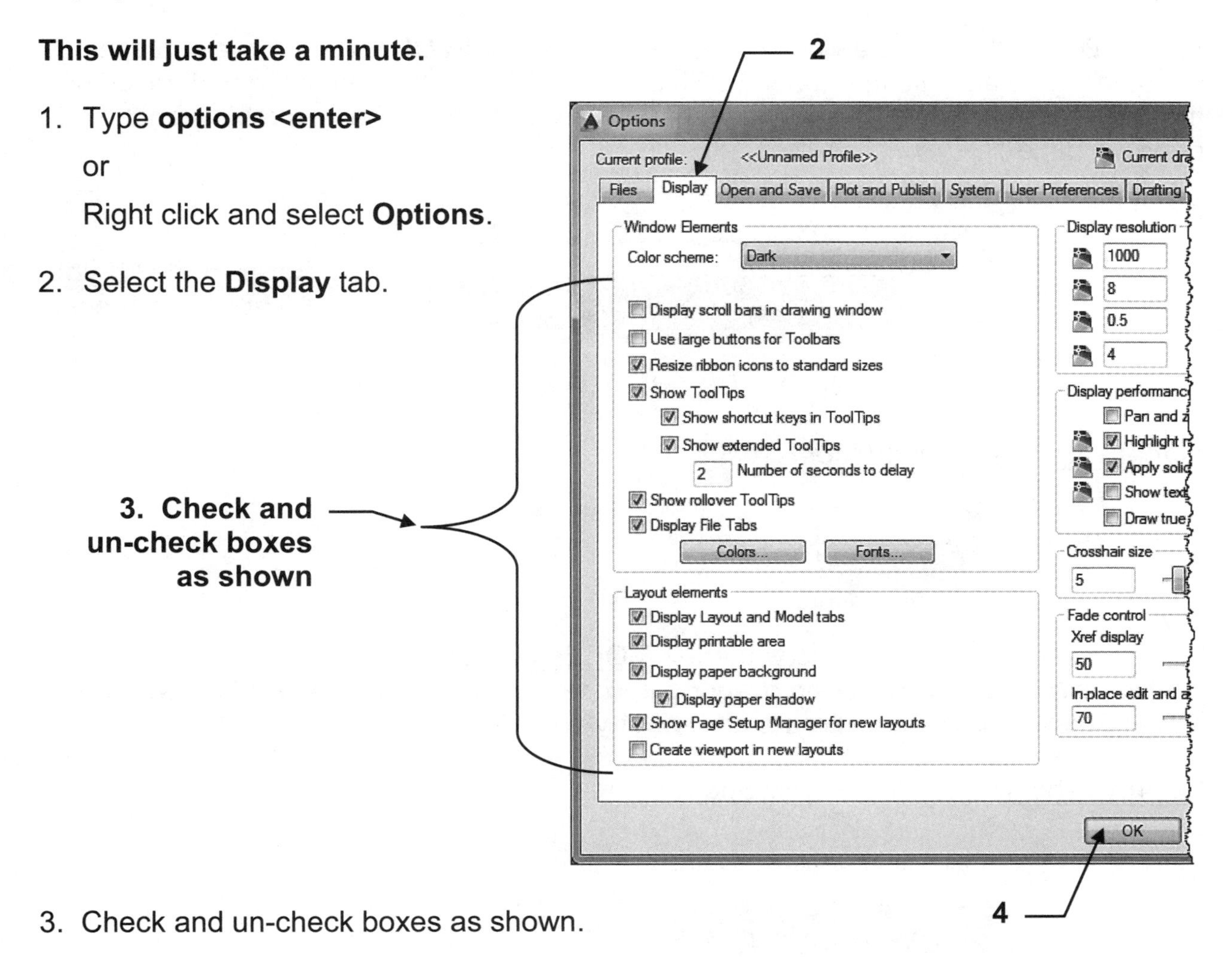

3. Check and un-check boxes as shown.

4. Select the **OK** button

5. The lower left corner of the drawing area should display the 3 tabs Model, Layout1 and Layout2 and a few tools should be displayed in the lower right corner above the command line.

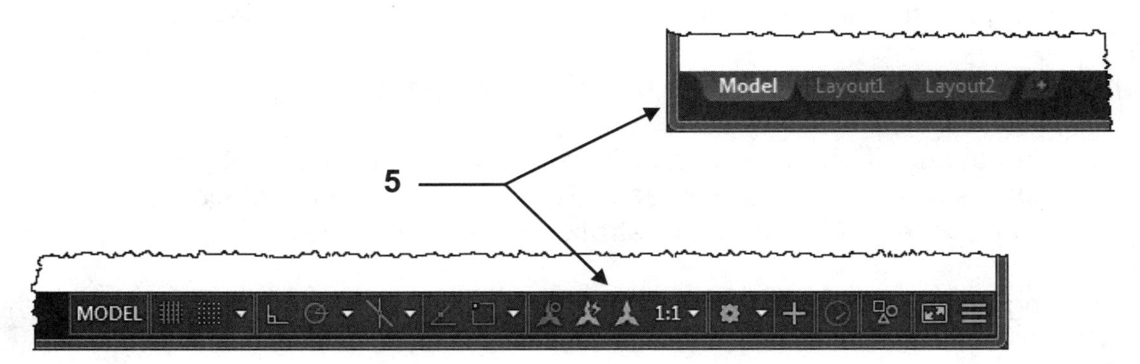

MODEL and LAYOUT tabs

Read this information carefully. It is very important that you understand this concept. More information on the following pages.

AutoCAD provides two drawing spaces, **MODEL** and **LAYOUT**. You move into one or the other by selecting either the MODEL or LAYOUT tabs, located at the bottom left of the drawing area. (If you do not have these displayed follow the instructions on the previous page.)

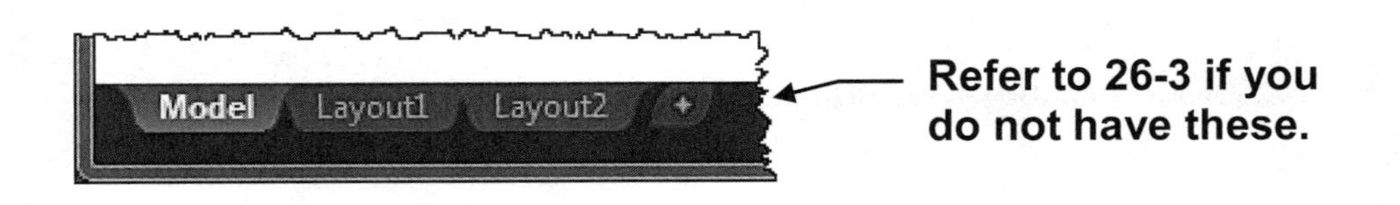

Model Tab (Also called ***Model Space***)

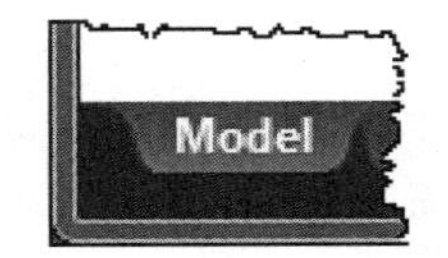

When you select the Model tab you enter MODEL SPACE.
(This is where you have been drawing and plotting from for the last 25 lessons)
Model Space is where you **create** and **modify** your drawings.

Layout Tabs (Also called ***Paper Space***)

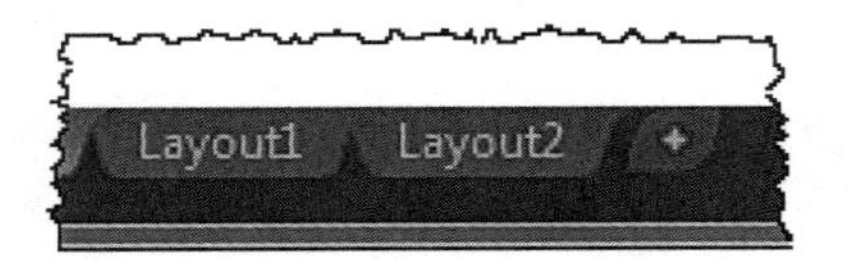

When you select a Layout tab you enter PAPER SPACE.
The primary function of Paper Space is to **prepare the drawing for plotting**.

When you select the Layout tab for the first time, the "Page Setup Manager" dialog box will appear. The Page Setup Manager allows you to specify the printing device and paper size to use.
(More information on this in "How to create a Page Setup" page 26-13)

When you select a Layout tab, Model Space will seem to have disappeared, and a blank sheet of paper is displayed on the screen. This sheet of paper is basically in front of the Model Space. (Refer to the illustration on the next page)

To see the drawing in Model Space, while still in Paper Space, you must cut a hole in this sheet. This hole is called a **"Viewport"**. *(Refer to "Viewports" page 26-6.)*

MODEL and LAYOUT tabs....continued

Try to think of this as a picture frame (paper space) in front of a photograph (model space).

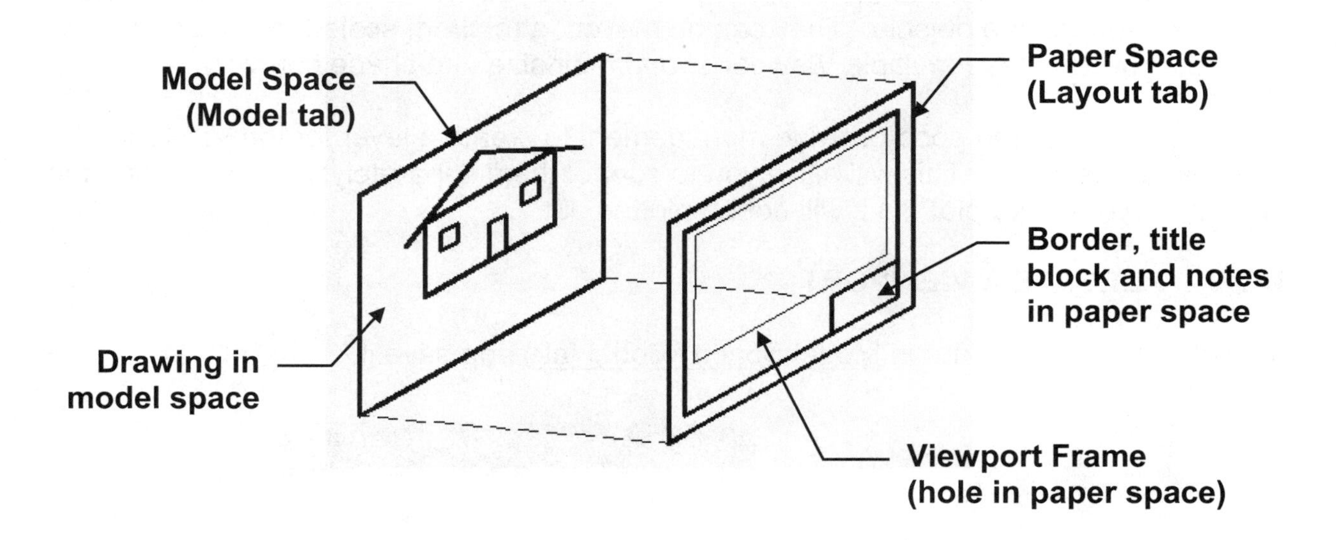

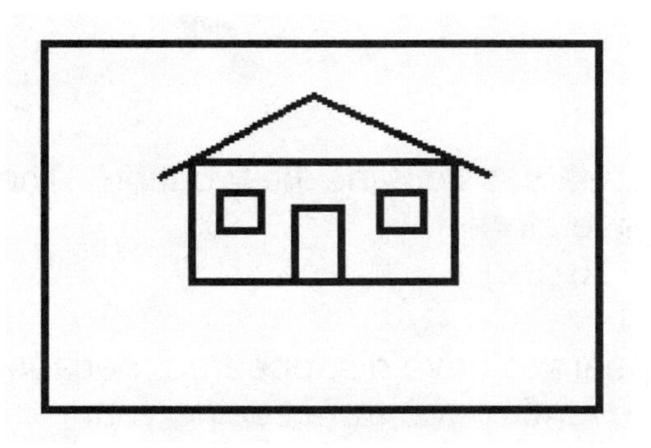

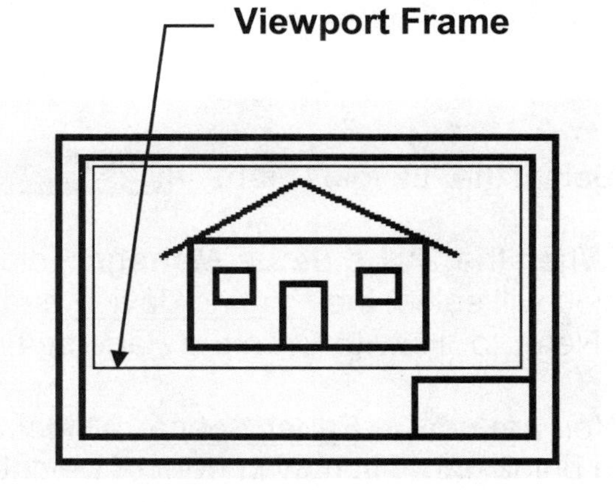

This is what you see when you select the Model tab.

You see only model space.

This is what you see when you select the Layout1 tab with a viewport.

You see through the Viewport to Model Space.

Now hold this thought.......more explanation coming on the next few pages.

VIEWPORTS

Note: This is just the concept to get you thinking. The actual step-by-step instructions will follow in the exercises.

Viewports are only used in Paper Space (Layout tab).
Viewports are holes cut into the sheet of paper displayed on the screen in Paper Space.
Viewports frames are objects. They can be moved, stretched, scaled, copied and erased. You can have multiple Viewports, and their size and shape can vary.

Note: It is considered good drawing management to create a layer for the Viewport "frames" to reside on. This will allow you to control them separately; such as setting the viewport layer to "No plot" so it will not be plotted out.

HOW TO CREATE A VIEWPORT

1. First, create a drawing in Model Space (Model tab) and save it.

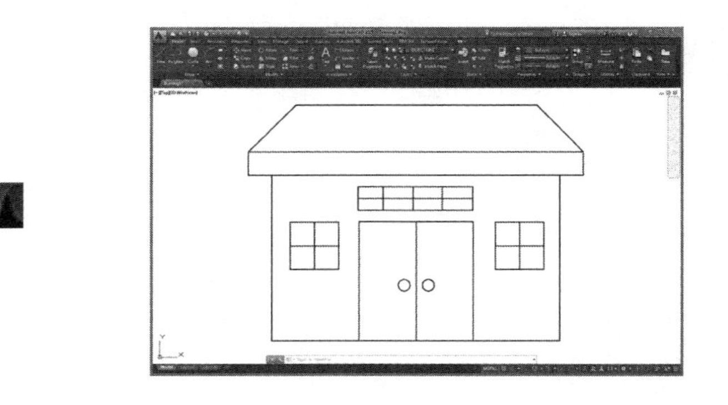

2. Select the "Layout1" tab.

   When the "**Page Setup Manager**" dialog box appears, select the **New** button. Then you will select the Printing device and paper size to plot on.
   (Refer to "How to Create a Page setup" on page 26-13.)

3. You are now in Paper Space. Model Space appears to have disappeared, because a blank paper is now in front of Model Space, preventing you from seeing your drawing. You designated the size of this sheet in the "page setup" mentioned in #2 above. (The Border, title block and notes will be drawn on this paper.)

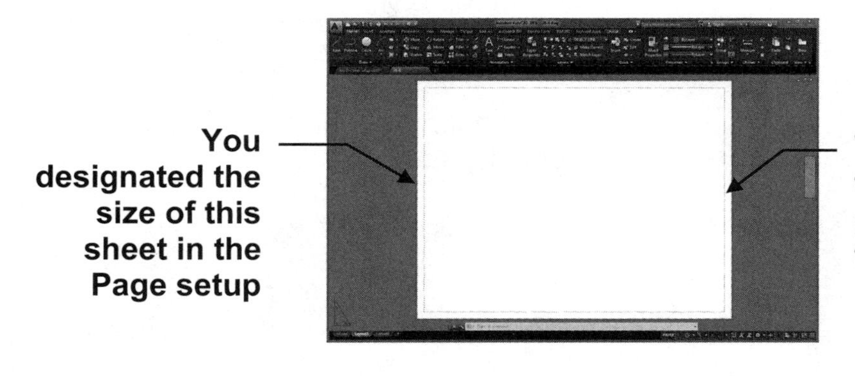

The dashed line represents the printing limits for the printer that you selected in the Page setup.

Continued on the next page...

VIEWPORTS....continued

4. Draw a border, title block and notes in Paper Space (Layout)

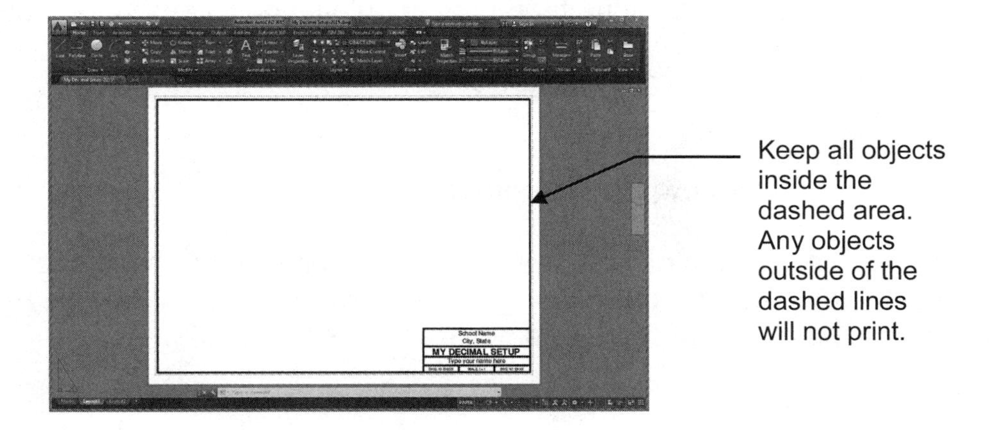

Now you will want to see the drawing that is in Model Space.

5. Select layer "**Viewport**"   (You want the viewport frame to be on layer viewport)

6. Select the Viewport command using one of the following:

   Ribbon = Layout tab / Layout Viewports panel /
   or
   Keyboard = MV <enter>

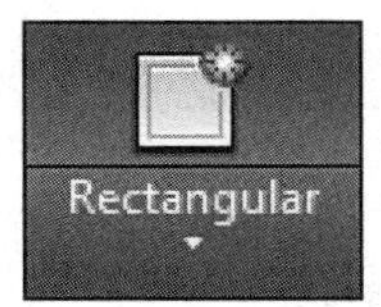

7. Draw a rectangular shaped Viewport "frame" by placing the location for the "first corner" and then the "opposite corner" using the cursor. (Similar to drawing a Rectangle, but **do not** use the Rectangle command. You must remain in the MV command)

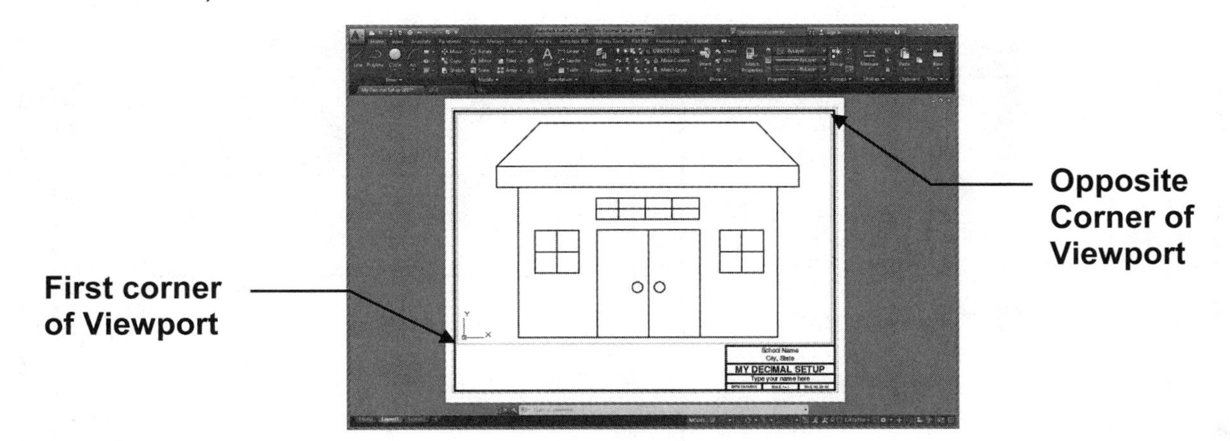

You should now be able to look through the Paper Space sheet to Model Space and see your drawing because you just cut a rectangular shaped hole in the sheet.

Note: Now you may go back to Model Space or return to Paper Space, simply by selecting the tabs, model or layout.
(Make sure your grids are ON in Model Space and OFF in Paper Space. Otherwise you will have double grids)

WHY LAYOUTS ARE USEFUL

I know you are probably wondering why you should bother with Layouts.
A Layout (Paper Space) is a great method to manipulate your drawing for plotting.

Notice the drawing below with <u>multiple viewports</u>.

Each viewport is a hole in the paper.
You can see through each viewport (hole) to model space.

Using Zoom and Pan you can manipulate the display of model space in each viewport.
To manipulate the display you must be <u>inside</u> the viewport.
Refer to the next page for instructions on how to achieve this.

Note: the dashed line indicates the maximum printing area for the printer and paper selected. Any object outside of this area will not print.

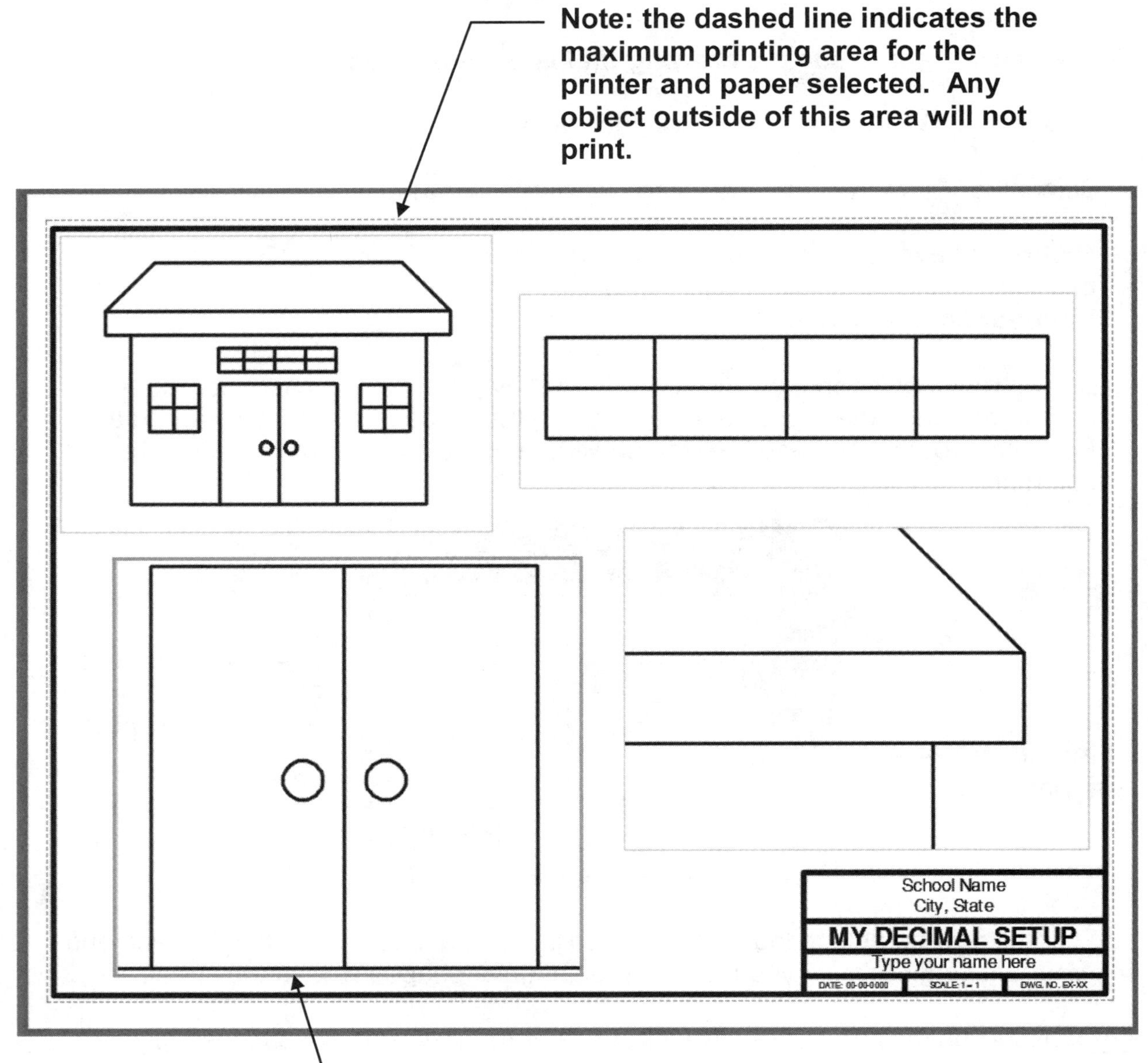

Active viewport indicated by heavier viewport frame.

HOW TO REACH INTO A VIEWPORT

Here are the rules:

1. You have to be in Paper Space (layout tab) and at least one viewport must have been created.

2. You have to be <u>inside a Viewport</u> to manipulate the scale or position of the drawing that you see in that Viewport.

How to reach into a viewport to manipulate the display.

<u>First, select a layout tab and cut a viewport</u>

At the bottom right of the screen on the status bar there is a button that either says Model or Paper. This button displays which space you are in currently.

When the button is PAPER you are working on the Paper sheet that is in front of Model Space. (Refer to the illustration on page 26 - 5)
You may cut a viewport, draw border, title block and place notes.

<u>If you want to reach into a viewport to manipulate the display</u>, double click inside of the viewport frame. Only one viewport can be activated at one time. The active viewport is indicated by a heavier viewport frame. (Refer to the illustration on the previous page. The viewport displaying the doors is active).
Also, the Paper button changed to Model.

While you are inside a viewport you may manipulate the scale and position of the drawing displayed. To return to the Paper surface click on the word Model and it will change to Paper

You may now work on the paper surface.

Note: Do not confuse the <u>Model / Layout tabs</u> with the <u>MODEL / PAPER button</u> .

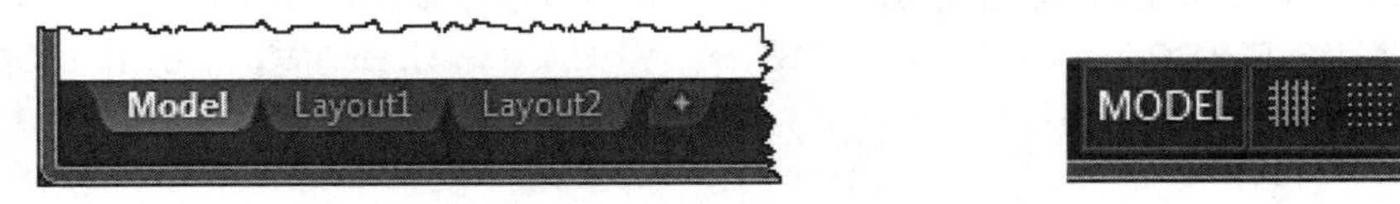

Here is the difference.

The **Model / Layout tabs** shuffle you from the actual drawing area (model space) to the Layout area (paper space). It is sort of like if you had 2 stacked pieces of paper and when you select the Model tab the drawing would come to the front and you could not see the layout. When you select the Layout tab a blank sheet would come to the front and you would not see model space……unless you have a viewport cut.

The **MODEL / PAPER button** allows you to work in model space or paper space without leaving the layout tab. No flipping of sheets. You are either on the paper surface or in the viewport reaching through to model space.

Don't worry it will get easier. This is the concept….but it will get more clear when you have completed the exercises in this lesson.

PAN

After you zoom in and out or adjust the scale of a viewport the drawing within the viewport frame may not be positioned as you would like it. This is where **PAN** comes in handy. **PAN** will allow you to move the drawing around, within the viewport, without affecting the size or scale.

Note: *Do not use the MOVE command. You do not want to actually move the original drawing. You only want to slide the viewport image, of the original drawing, around within the viewport.*

How to use the PAN command.

1. Select a layout tab (paper space)

2. Unlock the viewport if it is locked. (Refer to page 26-12)

3. Click inside a viewport.

4. Select the **PAN** command using one of the following:

   Ribbon = View tab / Navigate panel / Pan (Refer to pages 1-11 & 4-8)
   or
   Keyboard = P <enter>
   or
   Navigation Bar = **(turned off on page 1-23)**

   (Consider adding the PAN tool to the Quick Access tool bar. See page 1-10)

5. Place the cursor inside the viewport and hold the left mouse button down while moving the cursor. (Click and drag) When the drawing is in the desired location release the mouse button.

6. Press the **Esc** key or press **<enter>** to end the **PAN** command.

7. Lock the viewport.
   (Refer to page 26-12)

Refer to the Example on the next page.

PAN....continued

Before PAN

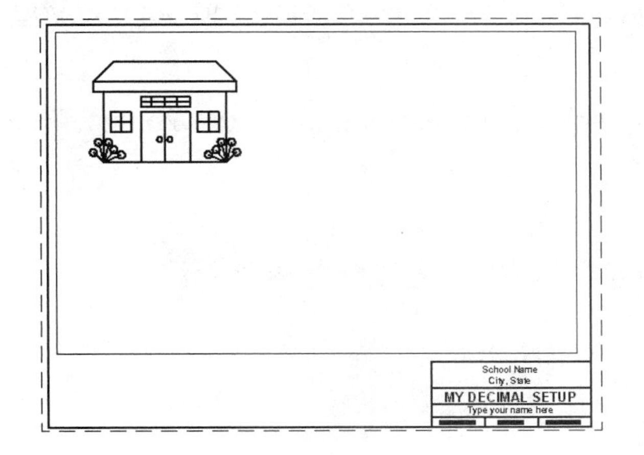

Double Click in the viewport to activate it.

Use PAN (Click, Drag, Release)

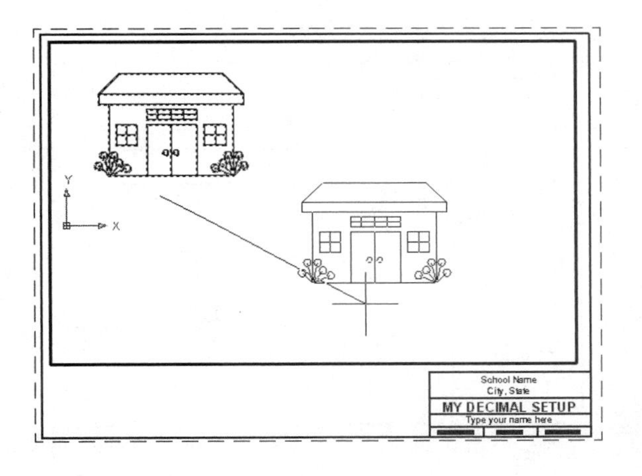

After PAN

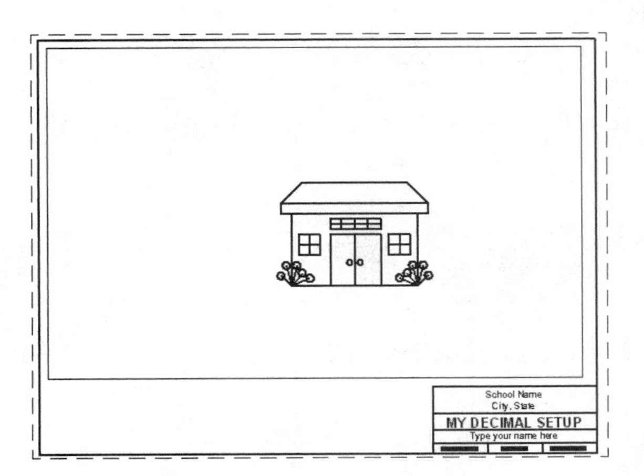

HOW TO LOCK A VIEWPORT

After you have manipulated the drawing within each viewport, to suit your display needs, you will want to **LOCK** the viewport so the display can't be changed accidentally. Then you may zoom in and out and you will not disturb the display.

Note:
Accurately adjusting the scale of a viewport will be discussed in detail in Lesson 27 .

1. Make sure you are in **Paper Space**.

2. Click once on a **Viewport Frame.**

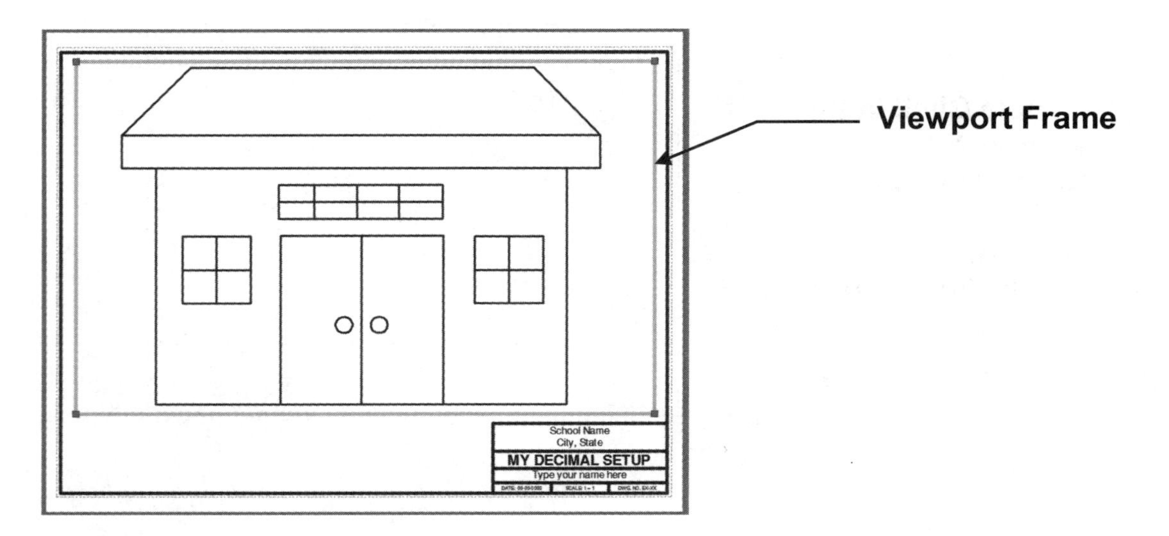

3. Click on the **<u>Open Lock tool</u>** located in the lower right corner of drawing area.

   The icon will change to a **<u>Closed Lock tool</u>**.

Viewport Unlocked

Viewport Locked

Now, any time you want to know if a Viewport is locked or unlocked just glance down to the <u>Lock tool</u> shown above.

HOW TO CREATE A PAGE SETUP

When you select a layout tab for the first time the Page Setup Manager will appear. The Page Setup Manager allows you to select the **printer/plotter** and **paper size**. These specifications are called the "**Page Setup**". This page setup will be saved to that layout tab so it will be available when ever you use that layout tab.

Note: The following is for concept only. The actual exercise starts with 26A.

1. **Open** the drawing you wish to plot.
   (The drawing must be displayed on the screen.)

2. Select a **Layout tab**. (Refer to page 26-3 if you do not have a layout tab)

Note: If the "Page Setup Manager" dialog box shown below does not appear automatically, right click on the Layout tab and select Page Setup Manager.

3. Select the **New...** button.

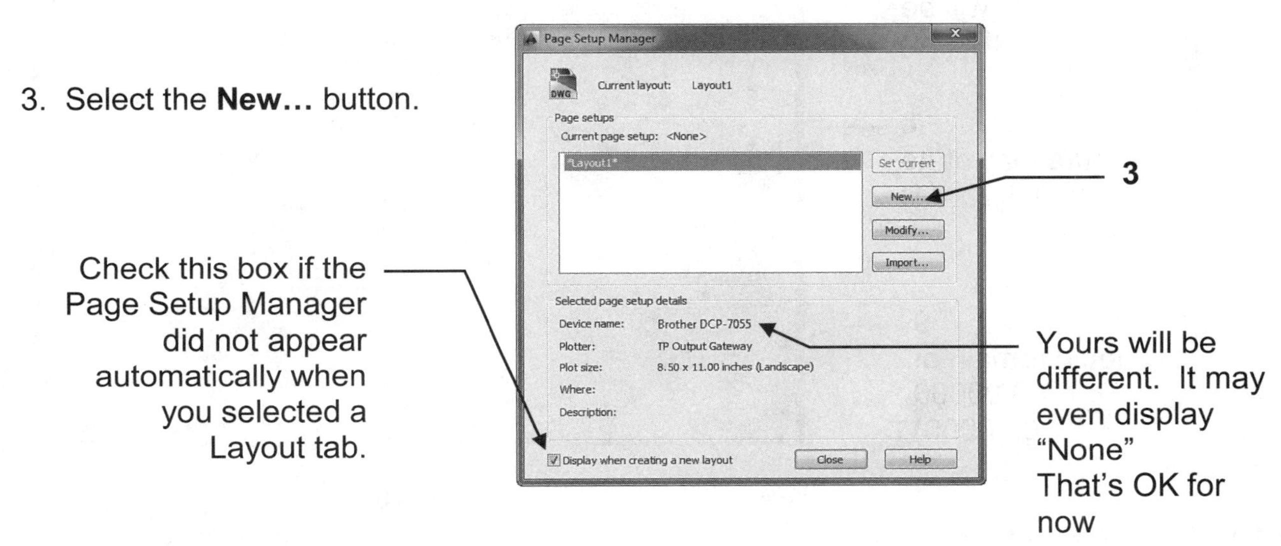

4. Select **<Default output device>** in the Start with: list.

5. Enter the New page setup name: **Setup A**

6. Select **OK** button.

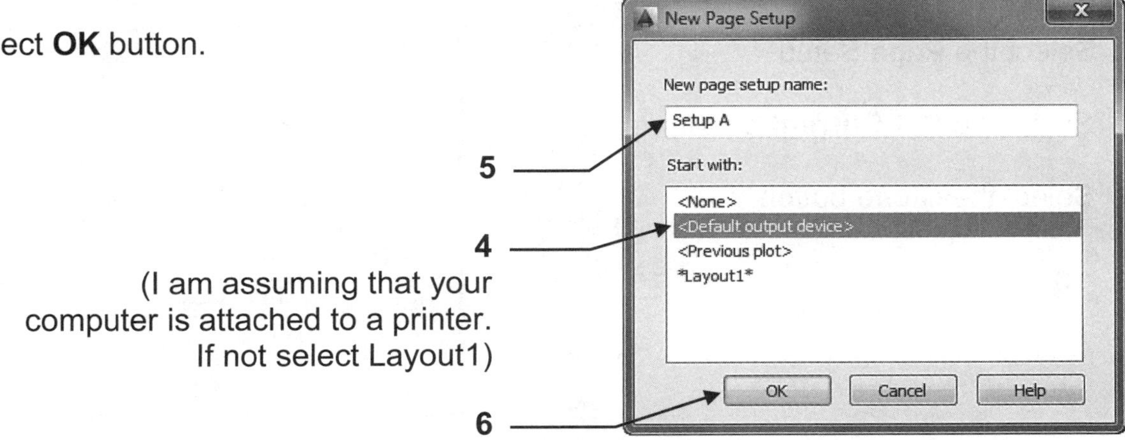

Continued on the next page...

HOW TO CREATE A PAGE SETUP....continued

This is where you will select the **printer / plotter**, **paper size** and the **plot offset**.

7. Select the **Printer / Plotter**
   Note: Your current system printer should already be displayed here. If you prefer another select the down arrow and select from the list. If the preferred printer is not in the list you must configure the printer. Refer to Appendix-A for instructions.)

8. Select the **Paper Size**

9. Select **Plot Offset**

Notice the name you entered is now displayed as the page setup name.

7 (Yours will be different)

8 (Yours may state 8-1/2 x 11)

9 (Should remain at 0.00000 0.00000)

10

10. Select **OK** button.

11. Select the Page Setup.

12. Select the **Set Current** button.

13. Select the **Close** button.

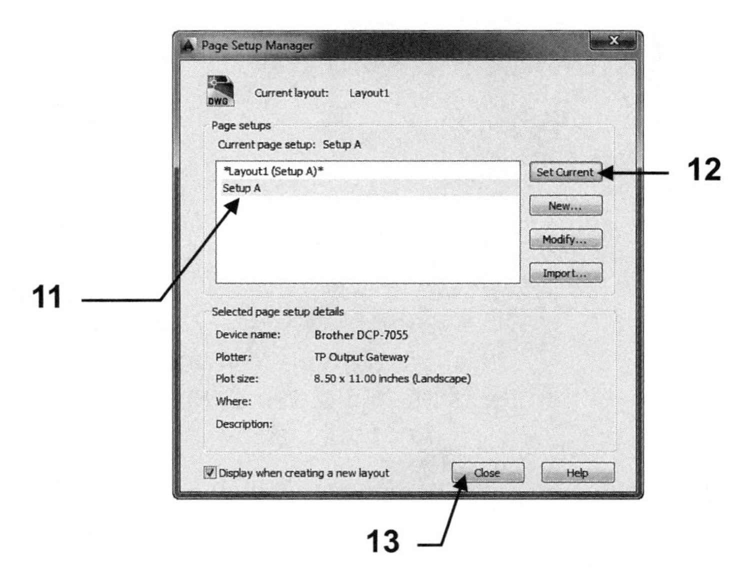

HOW TO CREATE A PAGE SETUP....continued

You should now have a sheet of paper displayed on the screen.
This sheet is the size you specified in the "Page Setup".
This sheet is in front of Model Space.

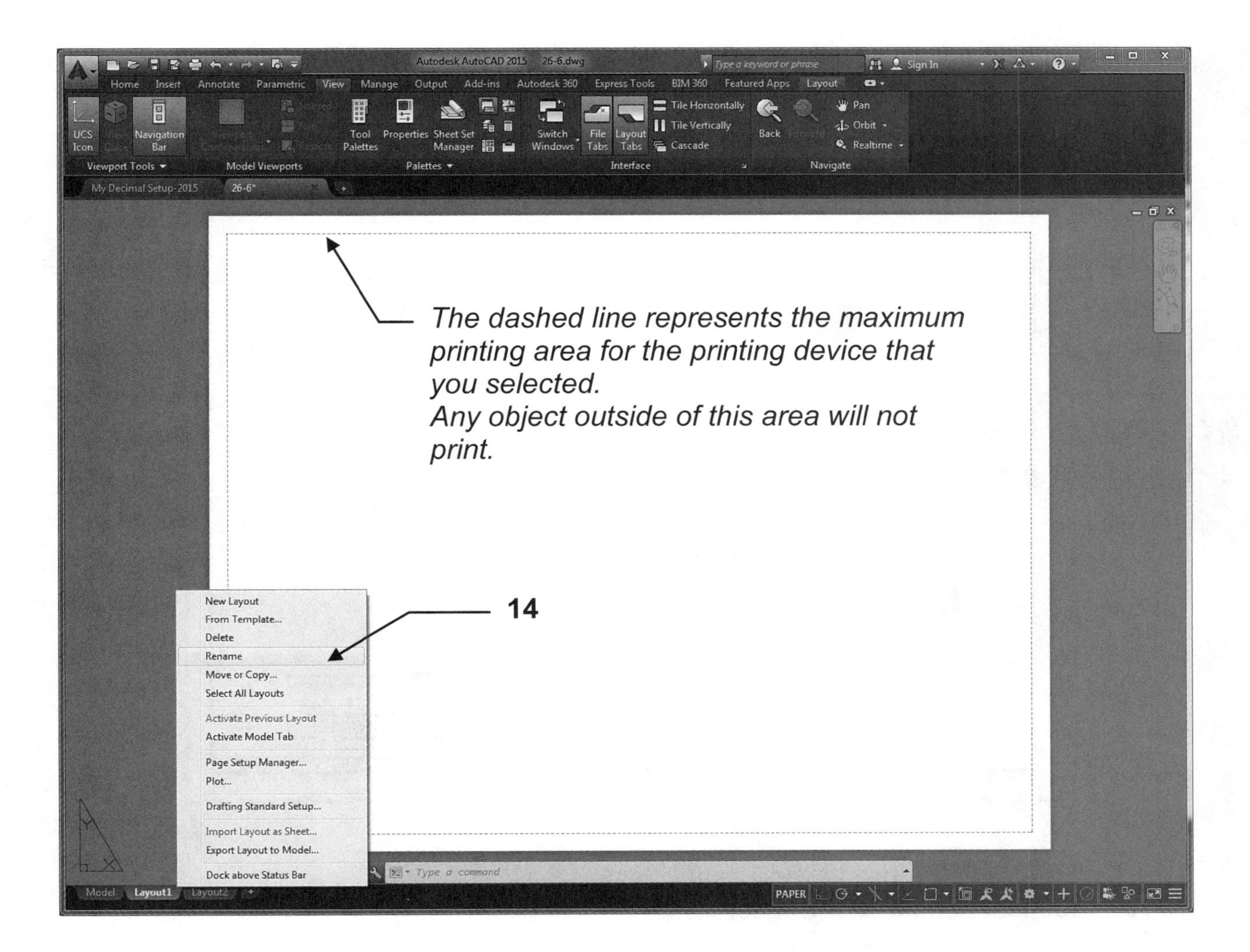

Rename the Layout tab

14. Right click on the active Layout tab and select **Rename** from the list.

15. Enter the new Layout name **A Size** <enter>

This was just the concept of Page Setup. You will actually create one in Exercise 26B.

USING THE LAYOUT

Now that you have the correct paper size on the screen, you need to do a little bit more to make it useful.

The next step is to:

1. Add a Border, Title Block and notes in Paper Space.

2. Cut a viewport to see through to Model Space. (Refer to page 26-6)

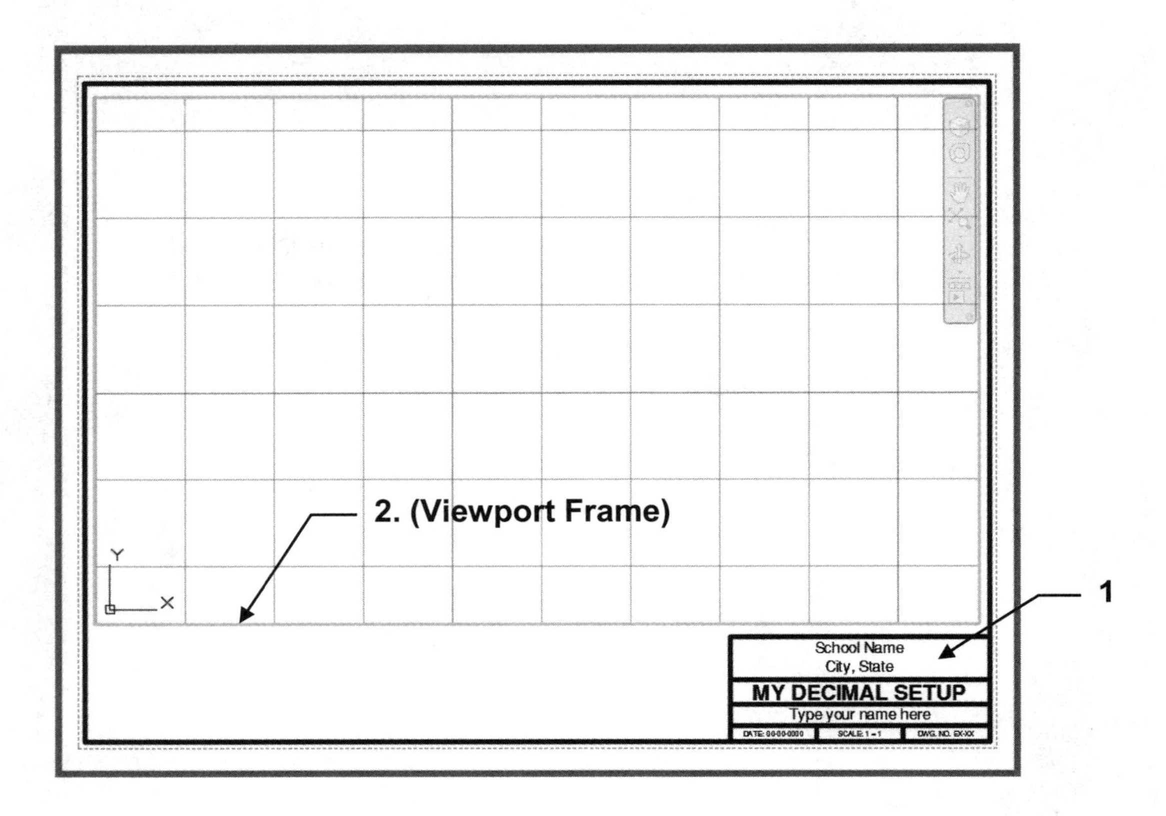

3. Adjust the scale of the viewport as follows.
   A. You must be in Paper Space.
   B. Click on the Viewport Frame.
   C. Unlock Viewport if it is locked.
   D. Select the <u>Viewport Scale</u> down arrow ▾.
   E. Select Scale (List shown on next page)

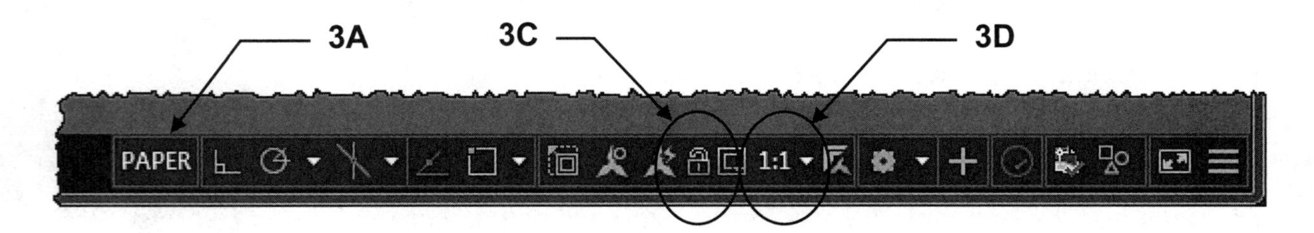

Continued on the next page...

USING THE LAYOUT....continued

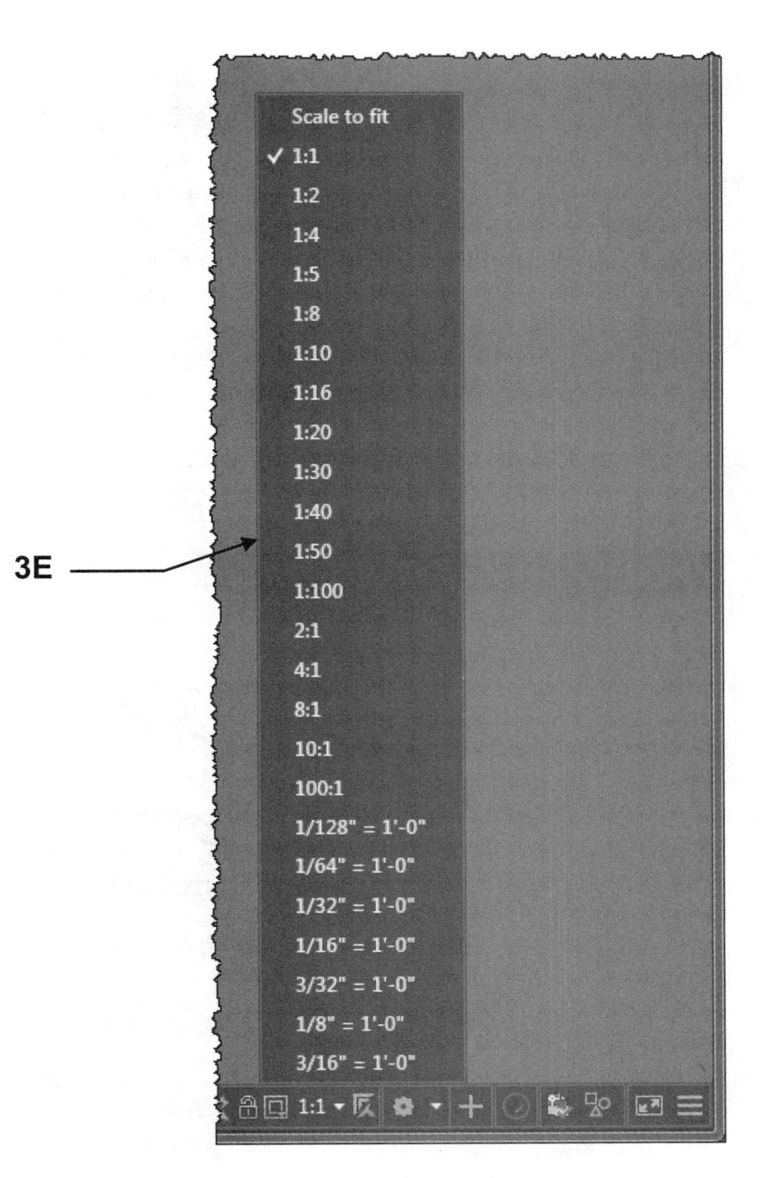

4. Lock the Viewport.

Now you may zoom as much as you desire and it will not affect the adjusted scale.

Note: Adjusting the scale of the viewport will be discussed more in lesson 27.

HOW TO PLOT FROM THE LAYOUT

Note: You must create a Page Setup before Plotting from a Layout tab.
If you have not created a Page Setup refer to pg 26-13 before proceeding.

The previous page setup instructions were to select the printer and paper size. Now you need to specify how you want to plot the drawing. You will find the PLOT dialog box almost identical to the Page Setup dialog box.

1. Open the drawing you wish to plot.
2. Select the layout tab you wish to plot.
3. Select the Plot command using one of the following:

   Quick Access tool bar =
   or
   Ribbon = Output tab / Plot panel /
   or
   Application Menu = Print / Plot
   or
   Keyboard = PLOT <enter>

The Plot dialog box shown below should appear.
Select the "More Options" button in the lower right corner if your dialog box does not appear the same as shown below

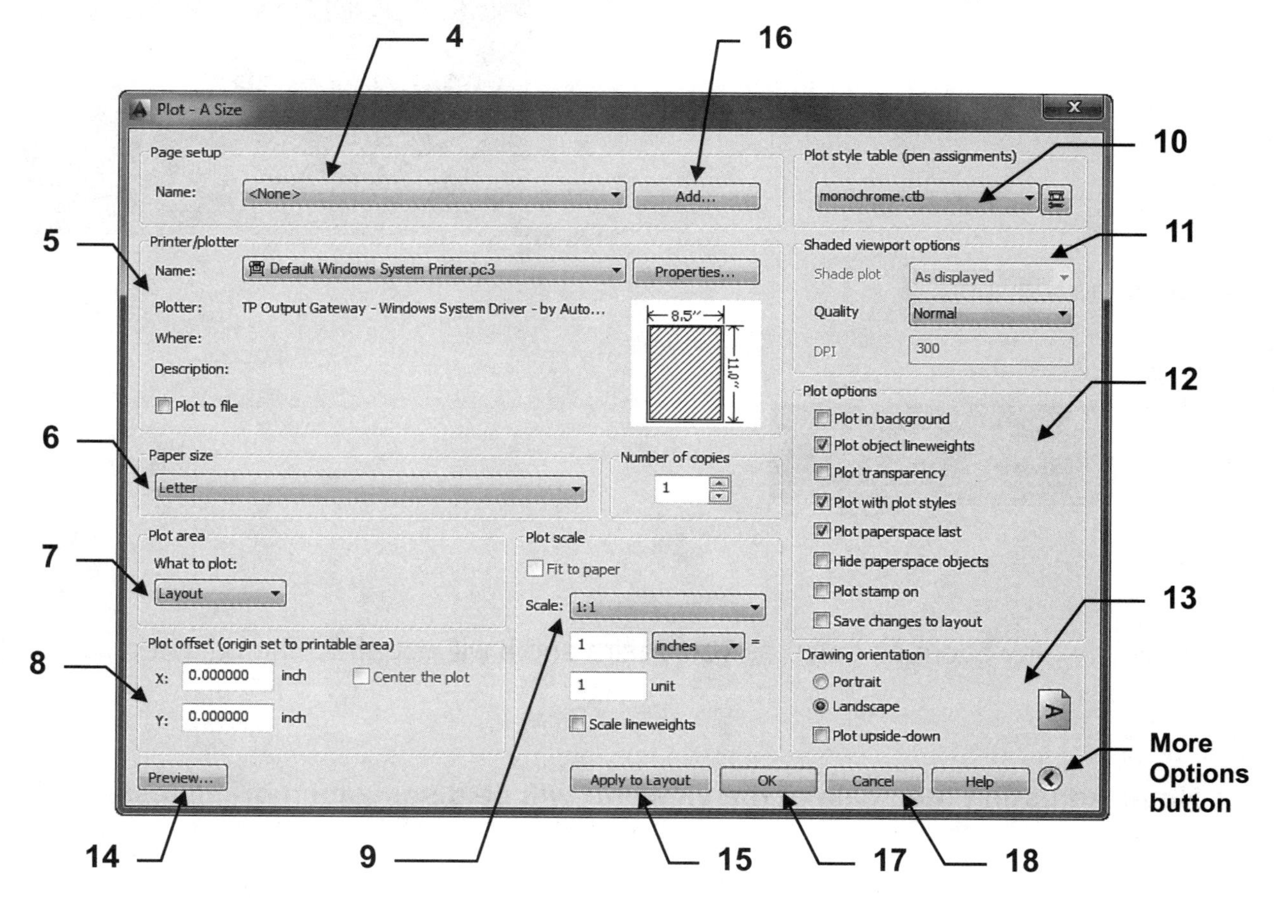

HOW TO PLOT FROM THE LAYOUT....continued

4. **Page Setup name**:
   After you have selected the desired settings you will save the new page setup and it will appear here. If you have previously created a page setup you may select it from the drop down list and all of the settings will change to reflect the previously saved page setup settings.

5. **Printer / Plotter:**
   Select the Printer that you wish to use. All previously configured devices will be listed here. (If your printer / plotter is not listed, refer to "Add a Printer / Plotter" Appendix A.)

6. **Paper Size:**
   Select the paper size. The paper sizes shown in the drop down list are the available sizes for the printer that you selected. If the size you require is not listed, the printer you selected may not be able to handle that size. For example, a letter size printer can not handle a 24 X 18 size sheet. You must select a large format printer.

7. **Plot Area:**
   Select the area to plot. Layout is the default.

   Limits plots the area inside the drawing limits.
   (This option is only available when plotting from model space)

   Layout plots the paper size
   (Select this option when plotting from a Layout)

   Extents plots all objects in the drawing file even if out of view.
   (This option only available if you have a viewport)

   Display plots the drawing exactly as displayed on the screen.

   Window plots objects inside a window. To specify the window, choose **Window** and specify the first and opposite (diagonal) corner of the area you choose to plot. (Similar to the Zoom / Window command)

8. **Plot offset:**
   The plot can be moved away from the lower left plot limit corner by changing the X and / or Y offset.
   If you have select **Plot area** "Display" or "Extents", select "**Center the plot**."

9. **Scale:** Select a **scale** from the drop down list or enter a custom scale.

   *Note: This scale is the Paper Space scale. The Model space scale will be adjusted within the viewport. If you are plotting from a "**LAYOUT**" tab, normally you will use **plot scale 1:1**. (I know this seems a little confusing right now. Scaling will be discussed more in Lesson 27)*

HOW TO PLOT FROM THE LAYOUT....continued

10. **Plot Style Table:** Select the Plot Style Table from the list. The Plot Styles determine if the plot is in color, Black ink or screened. You may also create your own.
    If you want to print in Black Ink only select Monochrome.ctb
    If you want to print in Color select Acad.ctb

11. **Shaded viewport options**
    This area is used for printing shaded objects when working in the 3D environment.

12. **Plot options**
    Plot background = specifies that the plot is processed in the background.

    Plot Object Lineweights = plots objects with assigned lineweights.

    Plot transparency = Plots any transparencies

    Plot with Plot Styles = plots using the selected Plot Style Table.

    Plot paperspace last = plots model space objects before plotting paperspace objects. Not available when plotting from model space.

    Hide Paperspace Objects = used for 3D only. Plots with hidden lines removed.

    Plot Stamp on = Allows you to print information around the perimeter of the border such as; drawing name, layout name, date/time, login name, device name, paper size and plot scale.

    Save Changes to Layout = Select this box if you want to save all of these settings to the current Layout tab.

13. **Drawing Orientation**.
    Portrait = the short edge of the paper represents the top of the page.

    Landscape = the long edge of the paper represents the top of the page

    Plot Upside-down = Plots the drawing upside down.

14. Select **Preview** button.
    Preview displays the drawing as it will plot on the sheet of paper.

    (Note: If you cannot see through to Model space, you have not cut your viewport yet)

    If the drawing appears as you would like it, press the **Esc** key and continue.

    If the drawing does not look correct, press the **Esc** key and re-check your settings, then preview again.

Note: If you have any of the layers set to "no plot" they will not appear in the preview display. The Preview Display only displays what will be printed.

HOW TO PLOT FROM THE LAYOUT....continued

15. **Apply to Layout** button
    This applies all of the settings to the layout tab. Whenever you select this layout tab the settings will already be set.

16. **Save the Page Setup**
    At this point you have the option of saving these settings as another page setup for future use on other layout tabs. If you wish to save this setup, select the **ADD** button, type a name and select **OK.**

17. If your computer **is** connected to the plotter / printer selected, select the **OK** button to plot, then proceed to **19**.

18. If your computer is **not** connected to the plotter / printer selected, select the **Cancel** button to close the Plot dialog box and proceed to **19**.

    *Note: Selecting Cancel will cancel your selected setting if you did not save the page setup as specified in **16** above.*

19. **Save the drawing**
    This will guarantee that the Page Setup you just created will be saved to this file for future use.

Note: This is the concept only.
The step by step instructions are shown in Exercise 26E

ANNOTATIVE PROPERTY

In Lesson 16 you learned to create a dimension style.
In Lesson 25 you learned to create a text style.
In this Lesson you will create a new dimension style and text style but this time you will include the **Annotative property** in both.

The Annotative property automates the process of scaling text, dimensions, hatch, tolerances, leaders and symbols. The height of these objects will automatically be determined by the annotation scale setting.

This will be discussed more in Lesson 27.
For now I just want you to know how to select it when creating your new styles.

TEXT STYLE

This symbol Indicates Annotative

Select the Annotative option here

DIMENSION STYLE

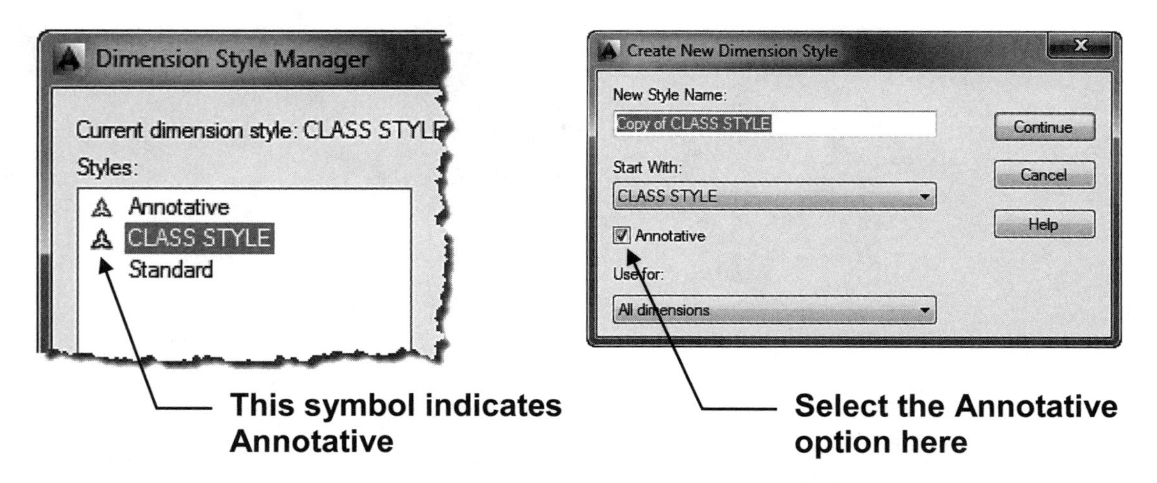

EXERCISE 26A

CREATE A MASTER DECIMAL SETUP TEMPLATE

The following instructions will guide you through creating a "Master" decimal setup template. The "2015-Workbook Helper and Border A" are examples of a Master setup template. Once you have created this "Master" template, you just open it and draw. No more repetitive inputting of settings.

NEW SETTINGS

A. Begin your drawing with a different template as follows:
   1. Select the **NEW** command..
   2. Select template file **acad.dwt** from the list of templates.
      (Note: Do not select "acad3D.dwt" by mistake)

B. Set the drawing **Units** as follows:
   1. Type **UNITS <enter>**
   2. Change the **Type: Decimal** and **Precision: 0.000** as shown below.
   3. Select **OK** button

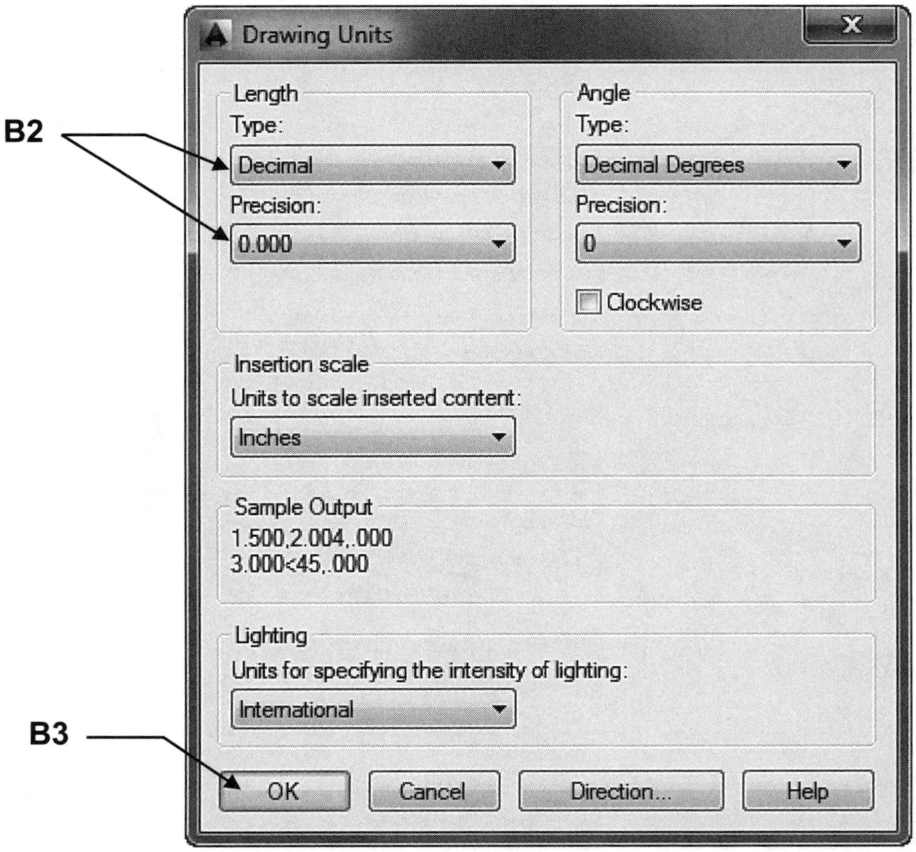

EXERCISE 26A....continued

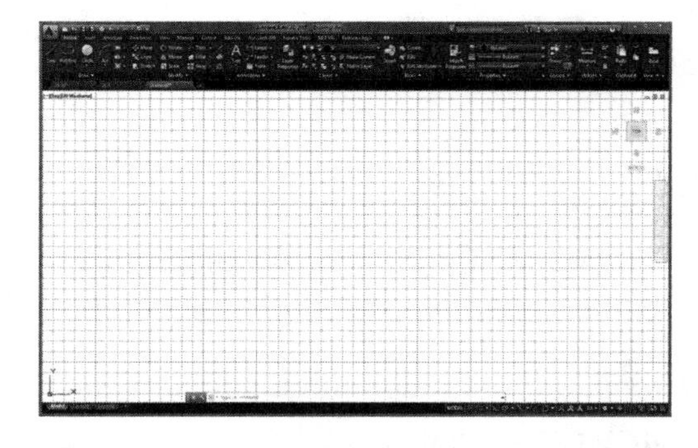

C. Set the **Drawing Limits**
(Size of the drawing area) as follows:
1. Type **Limits <enter>**
2. Enter Lower left corner = 0.000, 0.000
3. Enter Upper right corner = 11 , 8.5
4. Use **" ZOOM / ALL"** to view the new limits
5. Set your **Grids** to **ON** to display the limits

D. Set the Grids and Snap as follows:
1. Type **DS <enter>**
2. Select the **Snap and Grid** tab
3. Change the settings as shown below.
4. Select the **OK** button.

D2

Drafting Settings

Snap and Grid | Polar Tracking | Object Snap | 3D Object Snap | Dynamic Input | Quic

Snap On (F9)
Snap spacing
Snap X spacing: 0.250
Snap Y spacing: 0.250
Equal X and Y spacing
Polar spacing
Polar distance: 0.000
Snap type
Grid snap
Rectangular snap
Isometric snap
PolarSnap

Grid On (F7)
Grid style
Display dotted grid in:
2D model space
Block editor
Sheet/layout
Grid spacing
Grid X spacing: 1.000
Grid Y spacing: 1.000
Major line every: 4
Grid behavior
Adaptive grid
Allow subdivision below grid spacing
Display grid beyond Limits
Follow Dynamic UCS

Options... | OK | Cancel | Help

Important: Uncheck these

D4

EXERCISE 26A....continued

E. Change **Lineweight** settings as follows:
   1. Left click on Lineweight button down arrow located on the status line 
   2. Select **Lineweight Settings** (Refer to page 3-12 for instructions)
   3. Change to inches and adjust the Display scale as shown below.
   4. Select the **OK** button.

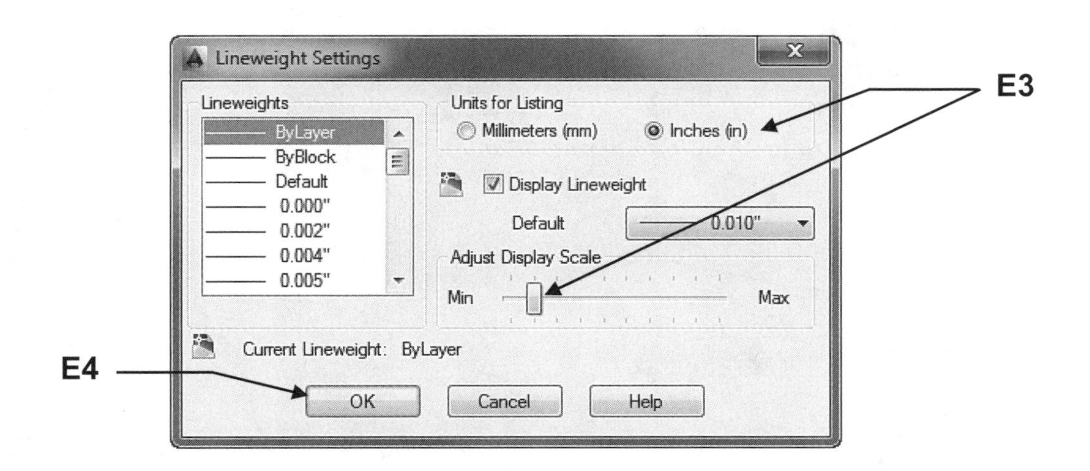

F. Create **new Layers** as follows:
   1. Load the linetypes listed below. (Refer to page 3-16 for instructions)
      Center2, Hidden, Phantom2
   2. Assign names, colors, linetypes, lineweights and plotability as shown below:
      (Refer to 3-15 for instructions)

Notice these

Current layer: OBJECT LINE

Search for layer

Filters: All, All Used Layers

Name	Color	Linetype	Lineweight	Trans...	Plot Style
0	white	Continuous	Default	0	Color_7
BORDERLINE	white	Continuous	0.039"	0	Color_7
CENTERLINE	cyan	CENTER2	Default	0	Color_4
DIMENSION	blue	Continuous	Default	0	Color_5
GRADIENT	white	Continuous	Default	0	Color_7
HATCH	green	Continuous	Default	0	Color_3
HIDDEN LINE	magenta	HIDDEN	Default	0	Color_6
OBJECT LINE	red	Continuous	0.028"	0	Color_1
PHANTOM LINE	white	PHANTOM2	Default	0	Color_7
SECTION LINE	white	PHANTOM2	0.031"	0	Color_7
SYMBOL	white	Continuous	0.020"	0	Color_7
TEXT	blue	Continuous	Default	0	Color_5
THREADS	green	Continuous	Default	0	Color_3
VIEWPORT	green	Continuous	Default	0	Color_3
WINDOW	green	Continuous	Default	0	Color_3
WIPE OUT	white	Continuous	0.035"	0	Color_7

Invert filter

All: 16 layers displayed of 16 total layers

LAYER PROPERTIES MANAGER

EXERCISE 26A....continued

G. Create a new **Text Style** as follows:
   1. Select **Text Style** (Refer to page 25-2)
   2. Create the text style **Text-Classic** using the settings shown below.
   3. When complete, select **Set Current** and **Close.**

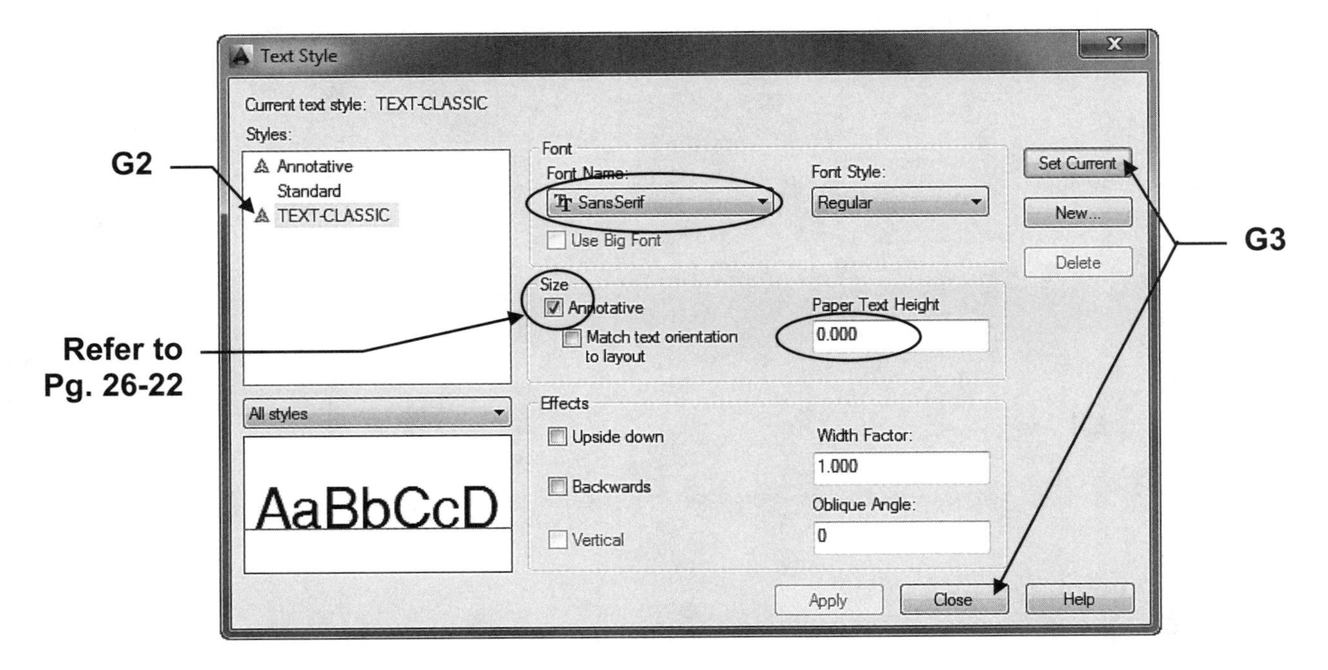

H. Create a **Dimension Style** as follows:

   1. **IMPORTANT: Follow the directions on page 16-8 through 16-14**

   2. All settings will be the same as page 16-8 through 16-14 except the following:
      Name = Dim-Decimal
      Annotative = this new dimension style will be Annotative (Refer to page 26-22)
      Text style: = select the newly created TEXT-CLASSIC

EXERCISE 26A....continued

I. Create a Dimension Sub-Style as follows:
   1. Follow the directions on page 18-9 and 18-10
   2. All settings will be the same except the following:
      <u>Annotative</u> = this new sub-style will be Annotative (Refer to page 26-22)

Your Dimension Style Manager should now appear as shown below:

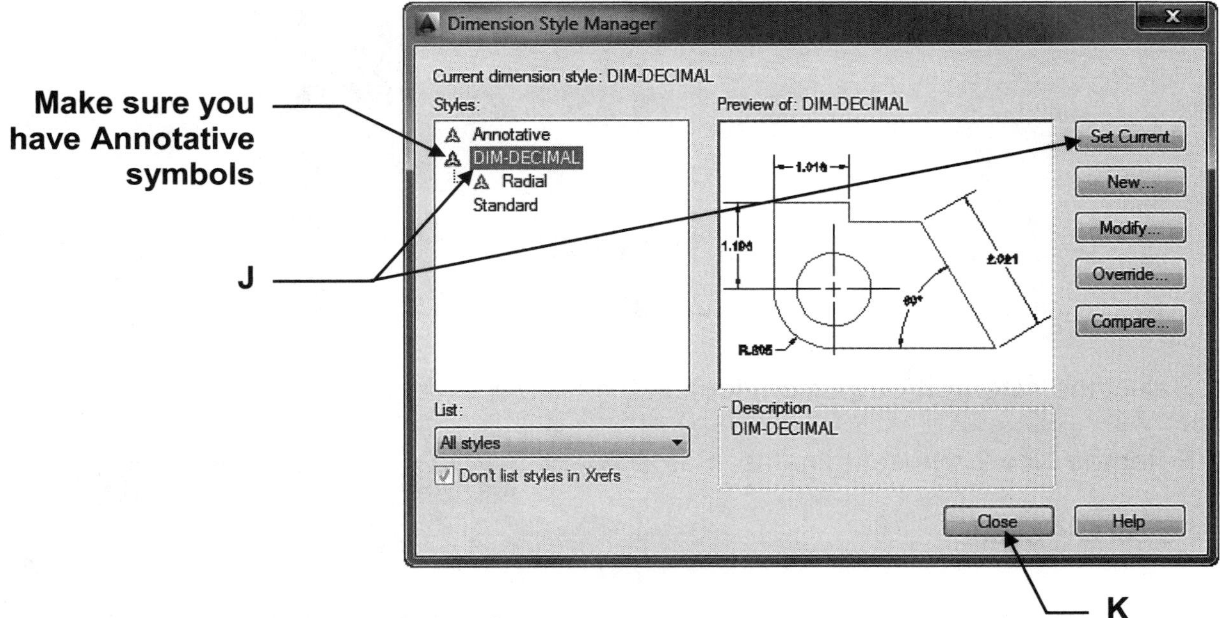

J. Select **DIM-DECIMAL** and **Set Current**,

K. Select **Close** button.

<u>THIS NEXT STEP IS VERY IMPORTANT</u>

L. **Save** all the settings you just created as follows:
   1. Save as / AutoCAD Drawing Template: **My Decimal Setup**

*Now continue on to **Exercise 26B**....you are not done yet.*

EXERCISE 26B

PAGE SETUP

Now you will select the printer and the paper size to use for printing.
You will use the Layout1 tab (Paper space).

A. Open **My Decimal Setup** (if not already open)

B. Select **Layout1** tab.

Refer to page 26-3 if you do not have these tabs.

Note: If the "Page Setup Manager" dialog box shown below does not appear automatically, right click on the Layout tab and select <u>Page Setup Manager</u>.

C. Select the **New...** button.

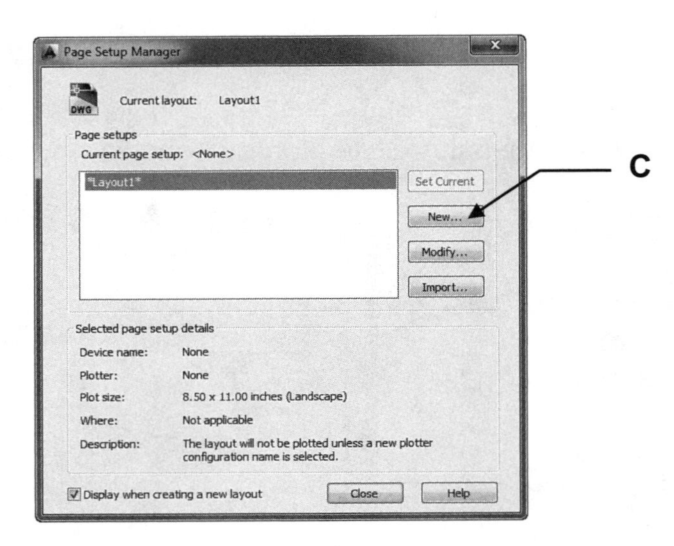

D. Select the **<Default output device>** in the Start with: list.

E. Enter the New page setup name: **Setup A**

F. Select **OK** button.

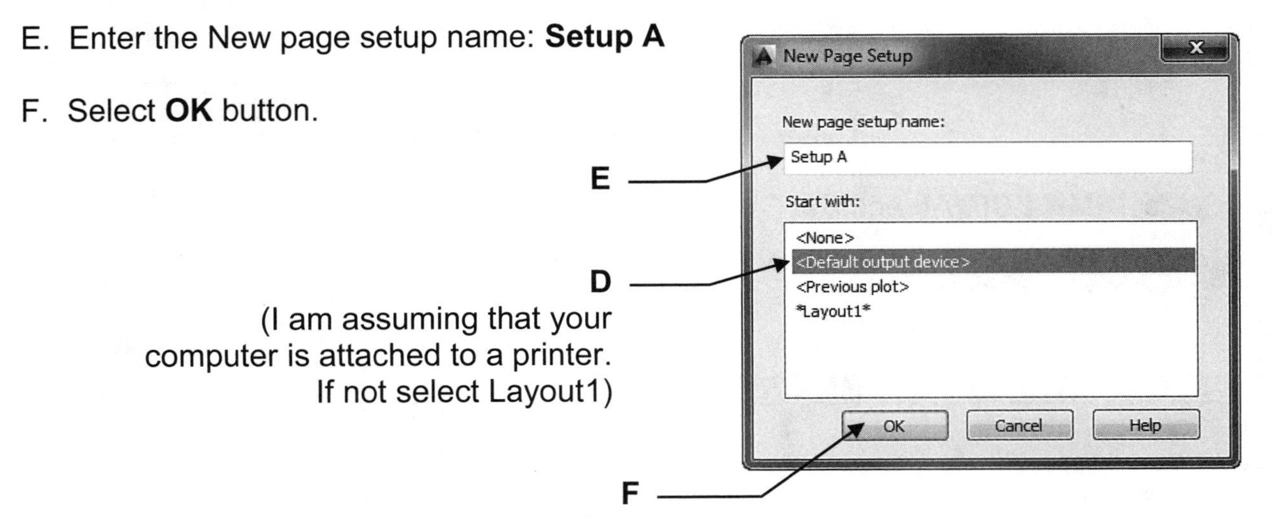

(I am assuming that your computer is attached to a printer. If not select Layout1)

Continued on the next page...

EXERCISE 26B....continued

This is where you will select the **print device**, **paper size** and the **plot offset**.

G. Select the **Printer / Plotter**
Note: Your current system printer should already be displayed here. If you prefer another select the down arrow and select from the list. If the preferred printer is not in the list you must configure the printer. Refer to Appendix-A for instructions.)

H. Select the **Paper Size**

I. Select **Plot Offset**

Notice the name you entered is now displayed as the page setup name.

G
(Yours may be different)

H
(Yours may be 8-1/2 x 11)

I
(Should stay at 0.00000)

J

J. Select **OK** button.

K. Select **Setup A**.

L. Select the **Set Current** button.

M. Select the **Close** button.

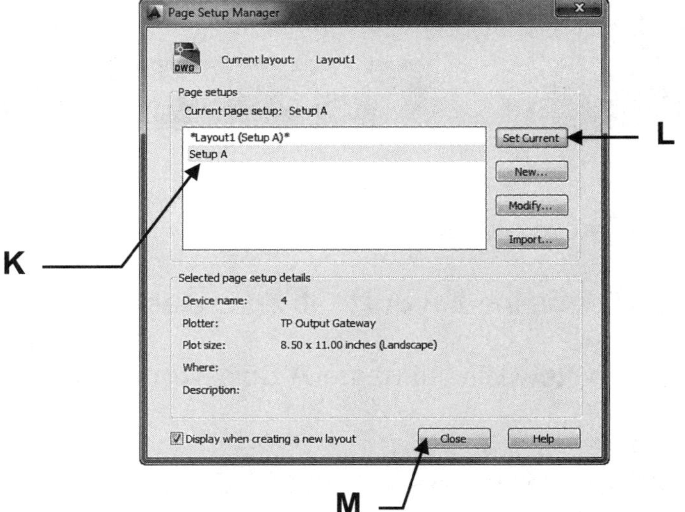

Continued on the next page...

EXERCISE 26B....continued

You should now have a sheet of paper displayed on the screen.
This sheet is the size you specified in the "Page Setup".
This sheet is in front of Model Space.

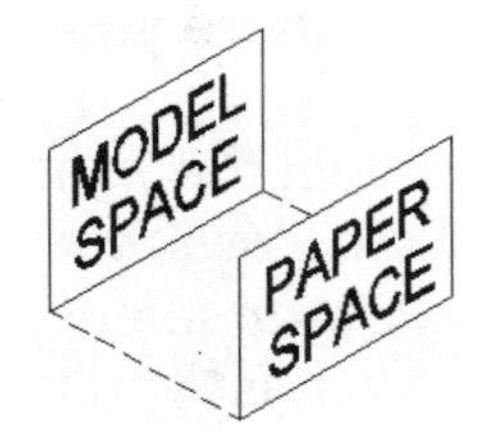

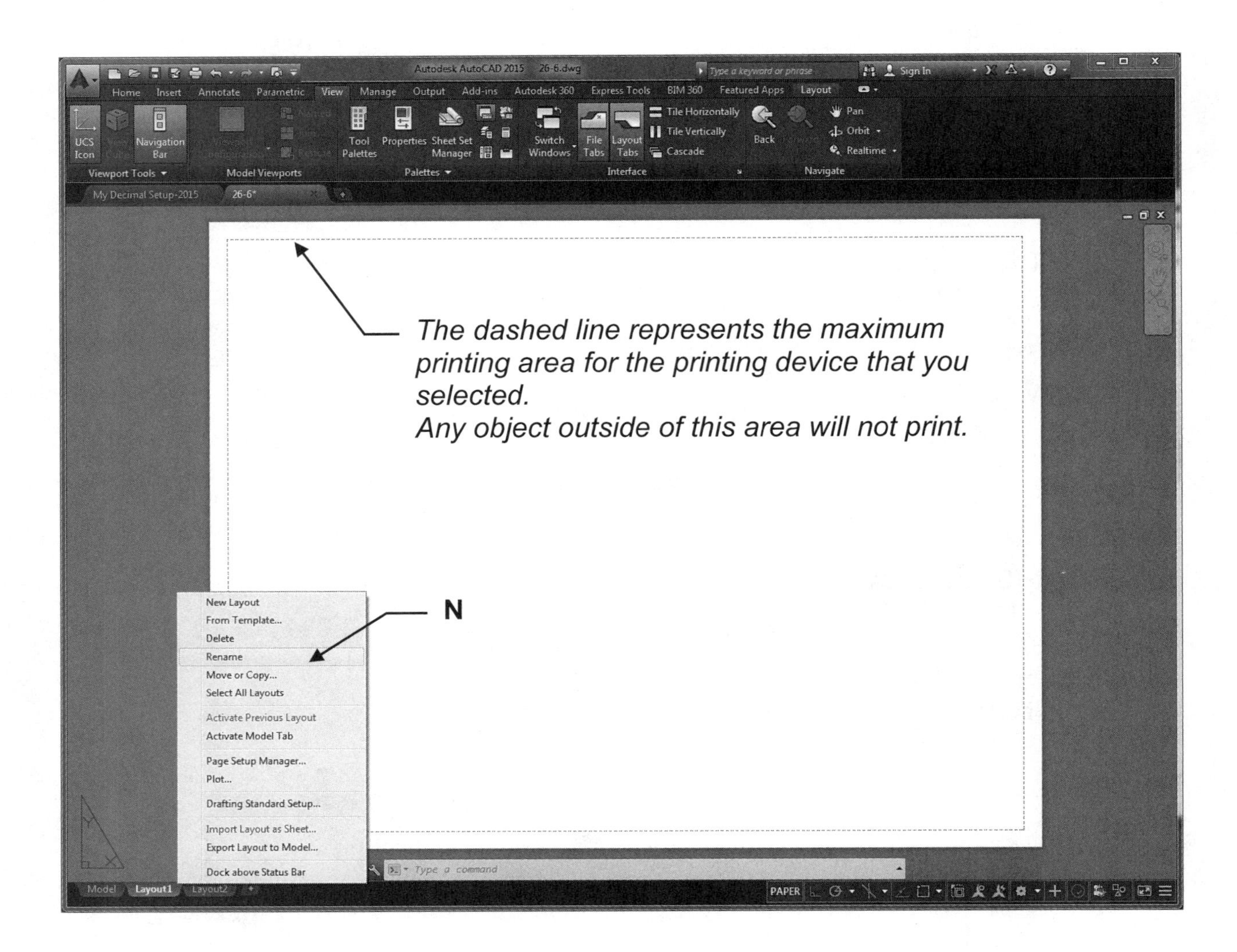

Rename the Layout tab

N. Right click on the **Layout1** tab and select **Rename** from the list.

O. Enter the New Layout name **A Size <enter>**

P. ***Very important:*** Save as / AutoCAD Drawing Template/ **My Decimal Setup** again.

Continue on to **Exercise 26C**.....you are not done yet.

EXERCISE 26C

CREATE A BORDER AND TITLE BLOCK

A. Draw the Border rectangle as large as you can within the dashed lines approximately as shown below using Layer BorderLine.

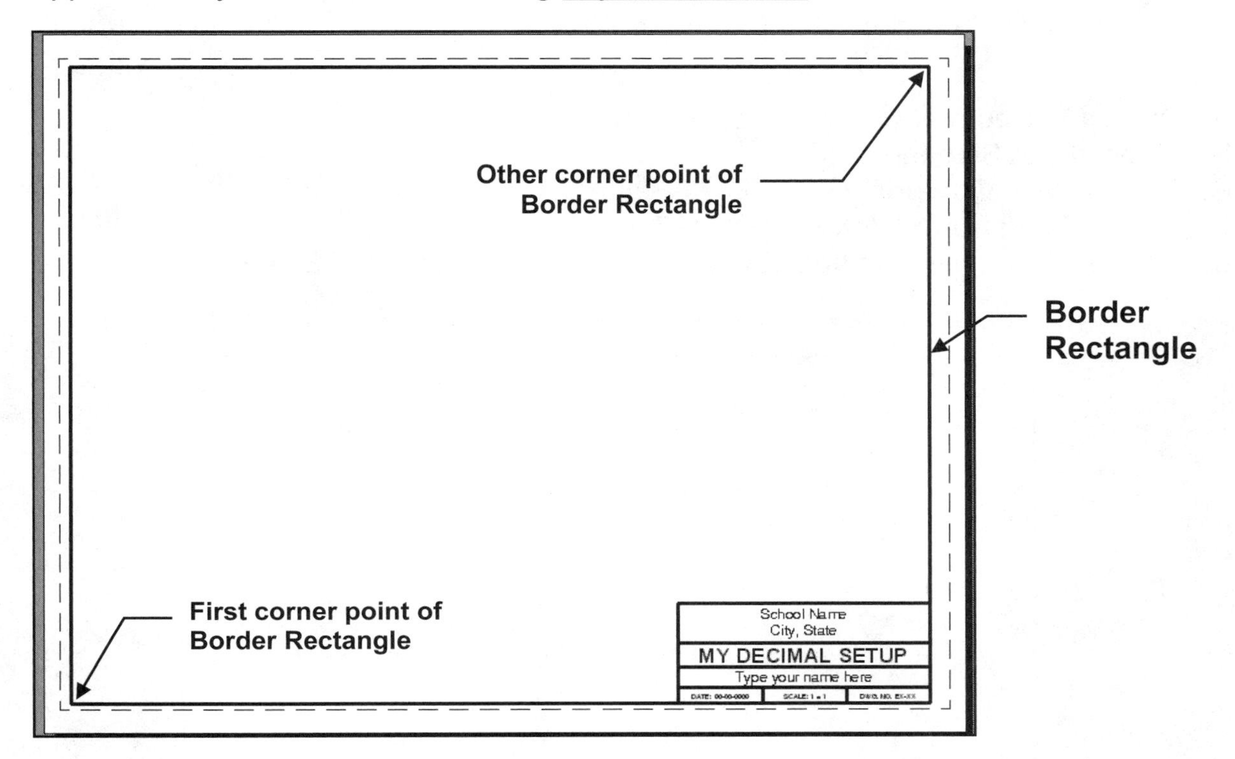

B. Draw the Title Block as shown using:
1. Layers BorderLine and Text.
2. Multiline Text ; Justify Middle Center in each rectangular area.
3. Text Style= Text-Classic Text height varies.

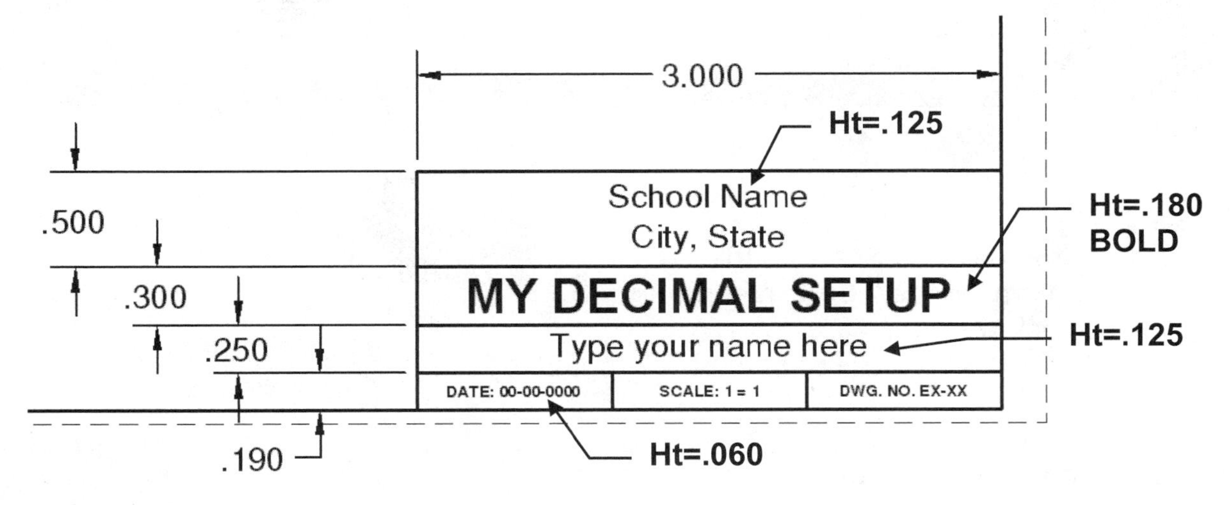

C. ***Very important:*** Save as / AutoCAD Drawing Template/ **My Decimal Setup** again.

Continue on to **Exercise 26D**.....you are not done yet.

EXERCISE 26D

CREATE A VIEWPORT

The following instructions will guide you through creating a VIEWPORT in the Border Layout sheet. Creating a viewport has the same effect as cutting a hole in the sheet of paper. You will be able to see through the viewport frame (hole) to Modelspace.

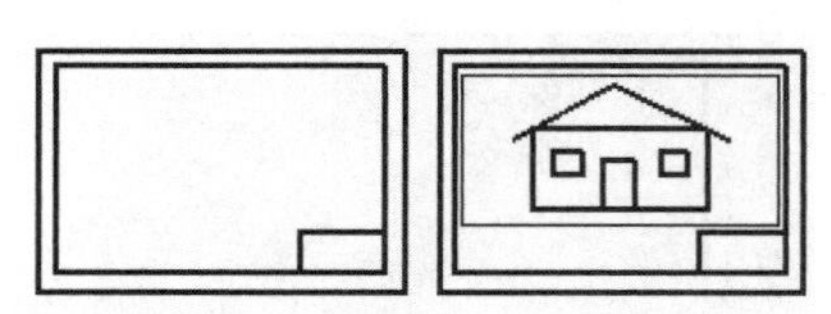

Before **After**

A. Open **My Decimal Setup** (If not already open)
B. Select the **A Size** tab.
C. Select layer **Viewport**
D. Type: **MV <enter>** (Refer to pg. 26-7)
E. Draw a Single viewport approximately as shown. (Turn off OSNAP)

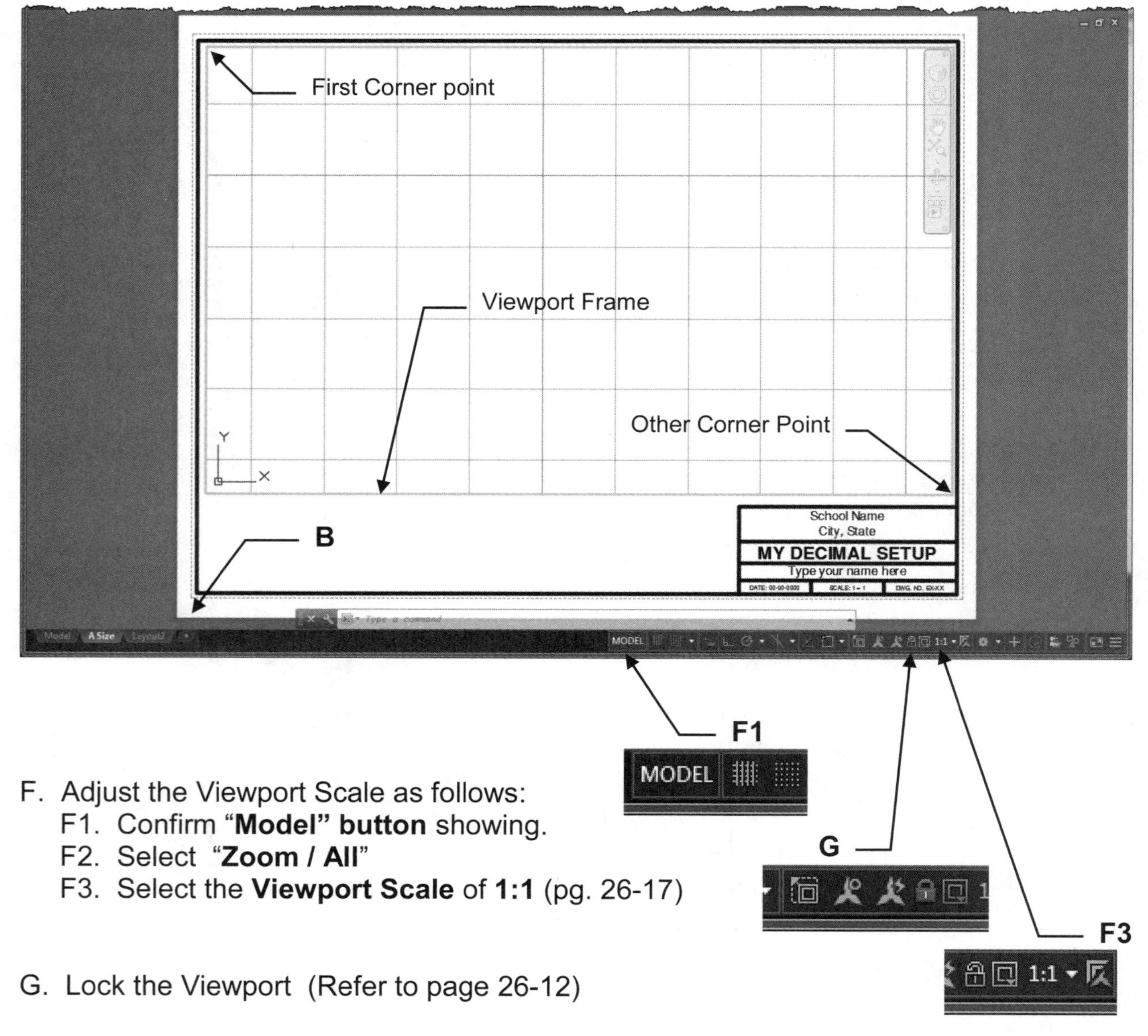

F. Adjust the Viewport Scale as follows:
   F1. Confirm "**Model" button** showing.
   F2. Select "**Zoom / All**"
   F3. Select the **Viewport Scale** of **1:1** (pg. 26-17)

G. Lock the Viewport (Refer to page 26-12)

H. ***Very important:*** Save as / AutoCAD Drawing Template/ **My Decimal Setup** again

Continue on to **Exercise 26E**....you are not done yet.

EXERCISE 26E

PLOTTING FROM THE LAYOUT

The following instructions will guide you through the final steps for setting up the master template for plotting. These settings will stay with **My Decimal Setup** and you will be able to use it over and over again.

A. Open **My Decimal Setup** (If not already open)

B. Select the **A Size** layout tab.

C. Select the **Plot** command

Note: Make sure that your settings match the settings shown below.

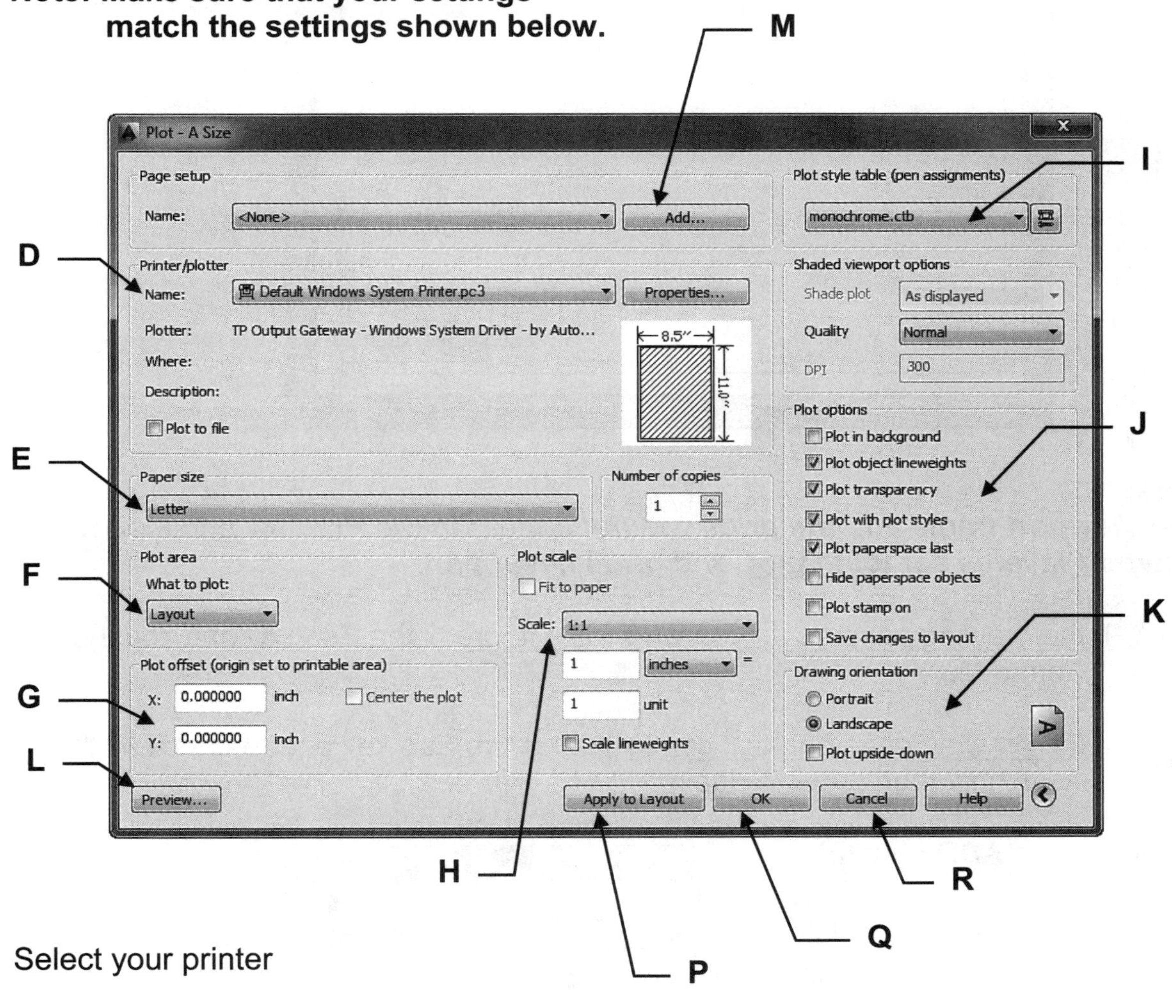

D. Select your printer

E. Select the Paper Size

F. Select the Plot Area

G. Plot offset should be 0.00000 for X and Y

EXERCISE 26E....continued

H. Select scale 1 : 1

I. Select the Plot Style Table **Monochrome.ctb**

J. Select the Plot options shown

K. Select Drawing Orientation: **Landscape**

L. Select **Preview** button.

Note:
The Viewport frame and the grids will not appear in the Preview because the viewport layer is set to no plot and grids never plot.

If the drawing appears as you would like it, press the **Esc** key and continue on to **M**.

If the drawing does not look correct, press the **Esc** key and re-check all the settings, then preview again.

M. Select the **ADD** button.

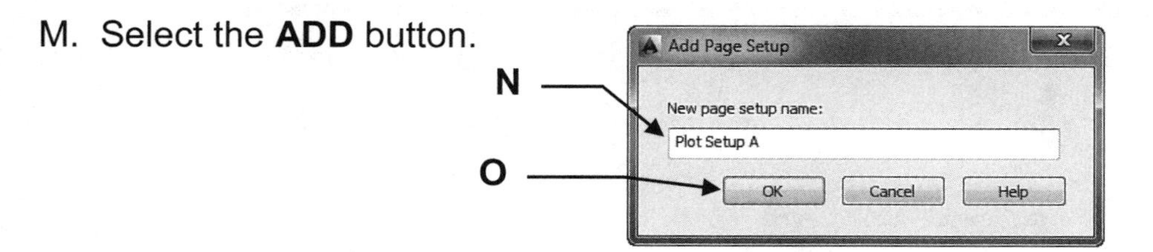

N. Type the New page setup name: **Plot Setup A**

O. Select the **OK** button.

EXERCISE 26E....continued

P. Select the "**Apply to Layout**" button.
The settings are now saved to the layout tab for future use.

Q. If your computer **is** connected to the plotter / printer selected, select the **OK** button to plot, then proceed to **S**.

R. If your computer is **not** connected to the plotter / printer selected, select the Cancel button to close the Plot dialog box and proceed to **S**. *Note: Selecting Cancel will cancel your selected setting if you did not **ADD** the page setup as specified in **M**.*

You are almost done

S. Select Layer **Object Line**.
(You don't want Layer Viewport to be the current layer)

T. Now Save all of this work as a **Template** one last time.
1. Select **Application Menu / Save As / AutoCAD Drawing Template**
2. Type: **My Decimal Setup**
3. Select **Save** button.

*Wow, I know that seemed like a lot of work but you have now completed the **Plotting Page Setup** for the **My Decimal Setup.dwt**.*

Now you are ready to use this master template to create and plot many drawings in the future. In fact, you have one on the very next page.

EXERCISE 26F

This exercise is in 2 parts. First you will draw the drawing in the model tab. Then you will create a layout to display parts of the drawing in 3 viewports. *Hopefully this will help you understand the differences between model space and paper space.*

Part 1

A. Select **New** and select **My Decimal Setup** template. (If it is not already open)

B. Select the Model tab.

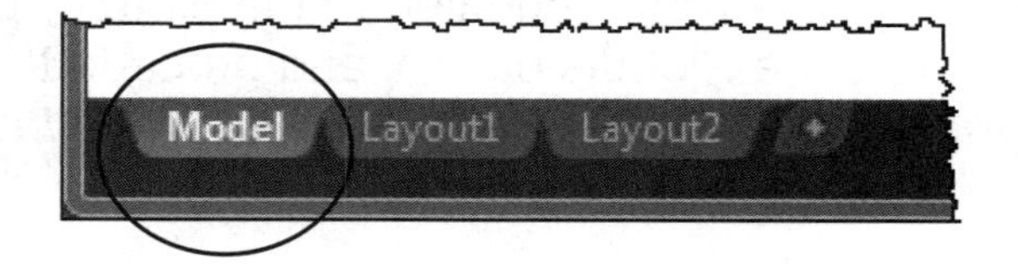

C. Draw the drawing shown below.

Use Layer Object Line.....do not use Layer Viewport

Do not dimension.

D. Save as: **EX-26F1**.

(Make sure that you save it as a drawing (.dwg) and not as a template.

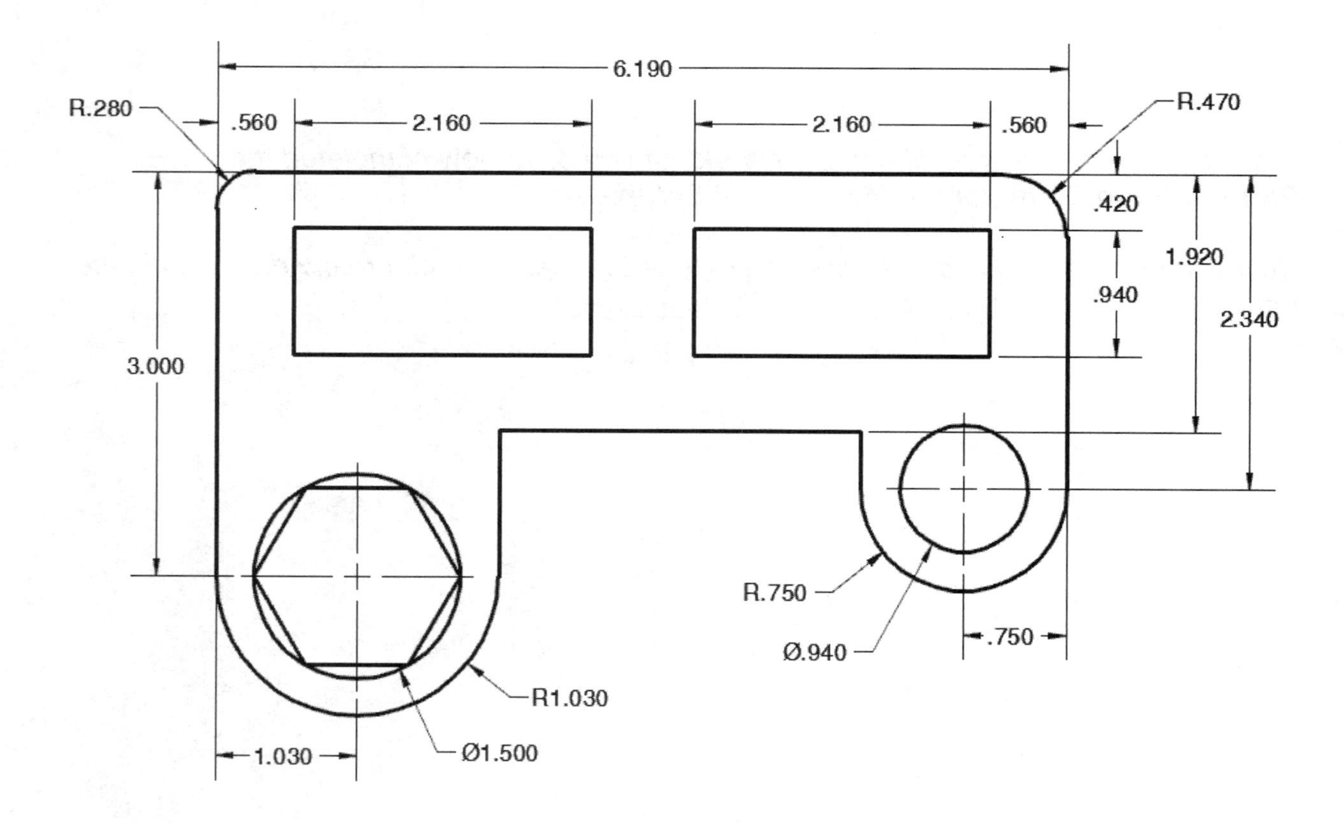

Continue to Part 2 on the next page.

EXERCISE 26F....continued

Part 2

A. Select the **A Size** tab.

B. Confirm the Model / Paper button says Paper.

C. Erase the one existing viewport frame. (Click on the frame and select Erase)

D. Select the Viewport Layer and create 3 new viewports approximately as shown below.

E. Using Zoom and Pan, inside each viewport, try to display each viewport content as shown below. (Note: Scale is not critical at this time but it will be in Lesson 27.)

F. Lock each viewport.

G. This time I would like you to print the Viewport Frames, so you need to make the Viewport layer **plottable**.
(Remove the **no plot sign**  on the printer symbol in the Layer Properties Manager)

H. Change the Title block: **Title**: VIEWPORTS **Scale**: NONE **Dwg no**: EX-26F

I. Save as: **EX-26F**

J. Plot using Plot Page Setup name: **Plot Setup A**. (Remember to view the Preview)

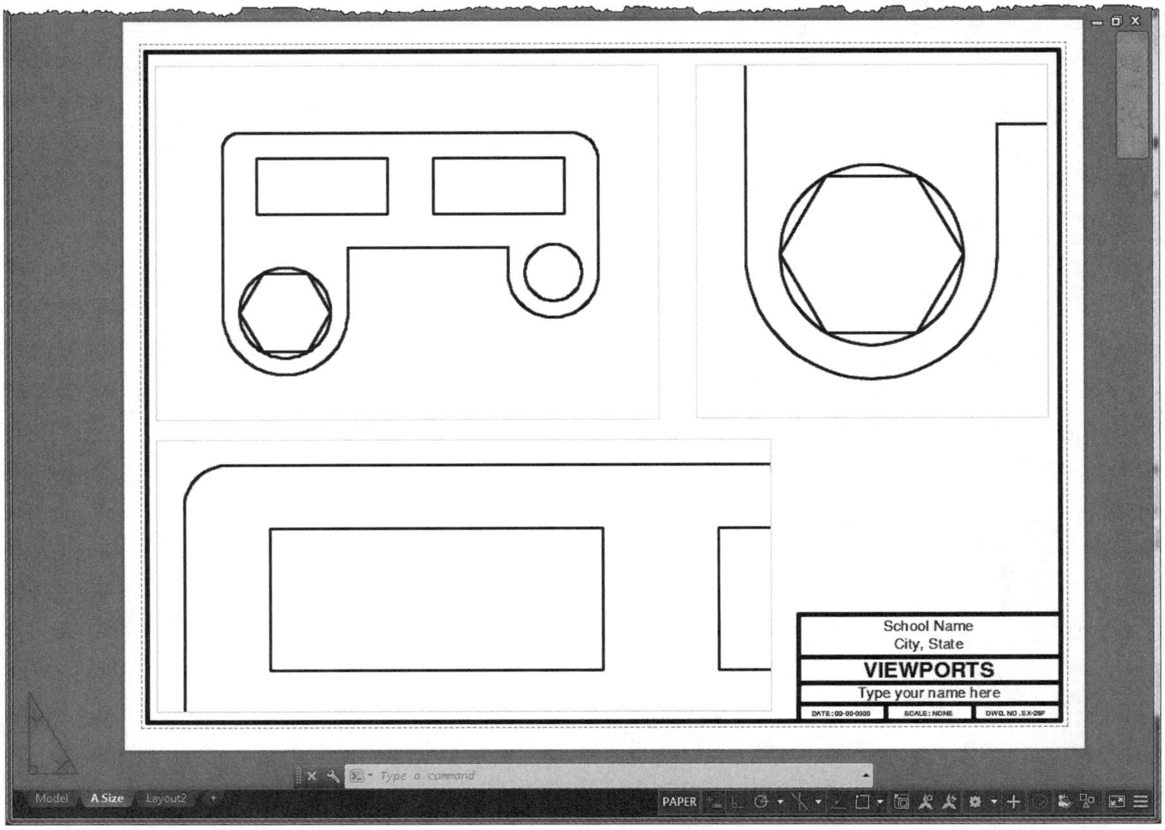

Notes:

LEARNING OBJECTIVES

After completing this lesson, you will be able to:

1. Understand Scaled Drawings
2. Adjust the scale within a Viewport
3. Understand Annotative Objects
4. Understand Paper space vs. Model Space dimensioning

LESSON 27

CREATING SCALED DRAWINGS

In the lessons previous to Lesson 26 you worked only in Model space. Then in Lesson 26 you learned that AutoCAD actually has another environment called Paper Space, or Layout. In this lesson we need to learn more about why we need 2 environments and how they make it easier to display and plot your drawings.

A very important rule in CAD you must understand is:

> ***"All objects are drawn full size"***

In other words, if you want to draw a line 20 feet long, you actually draw it 20 feet long. If the line is 1/8" long, you actually draw it 1/8" long.

Drawing and Plotting objects that are very large or very small.

In the previous lessons you created medium sized drawings. Not too big, not too small. But what if you wanted to draw a house? Could you print it to scale on an 8-1/2 X 11 piece of paper? How about a small paper clip. Could you make it big enough to dimension? Let's start with the house.

How to print an entire house on an 8-1/2 X 11 sheet of paper.

Remember the photo and picture frame example I suggested in lesson 26? (Refer to 26-5) This time try to picture yourself standing at the front door of your house with an empty picture frame in your hands. Look at your house through the picture frame. Of course the house is way too big to fit in the frame. Or is it because you are standing too close to the house?

Now walk across the street and look through the picture frame in your hands again. Does the house appear smaller? Can you see all of it in the frame? If you could walk far enough away from the house it would eventually appear small enough to fit in the picture frame in your hands. But....the house did not actually change size, did it? It only appears smaller because you and the picture frame are farther away from it.

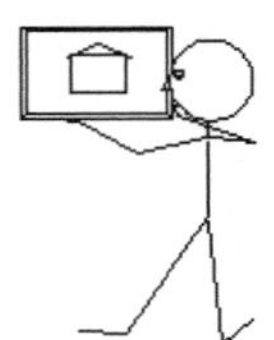

Adjusting the Viewport scale.

When using AutoCAD, walking across the street with the frame in your hands is called **Adjusting the Viewport scale.** You are increasing the distance between model space (your drawing) and Paper space (Layout) and that makes the drawing appear smaller. For example: A viewport scale of 1/4" = 1' would make model space appear 48 times smaller. But, when you dimension the house, the dimension values will be the actual measurement of the house. In other words, a 30 ft. line will have a dimension of 30'-0".

When plotting something smaller, like a paperclip, you have to move the picture frame closer to model space to make it appear larger. For example: 8 = 1 .

ADJUSTING THE VIEWPORT SCALE

The following will take you through the process of adjusting the scale within a viewport.

1. **Open** a drawing.
2. Select a **Layout** tab. (paper space)
3. Create a new **Page Setup**
4. Cut a new Viewport or unlock an existing Viewport.
5. Adjust the scale
   A. You must be in Paper Space. (See below)
   B. Select the Viewport Frame.
   C. Unlock Viewport, if locked.
   D. Select the Viewport Scale down arrow.
   E. Select the scale from the list of scales.

6. Lock the Viewport

Note: If you would like to add a scale that is not on the list:
1. Type **Scalelistedit**
2. Select **Add** button.

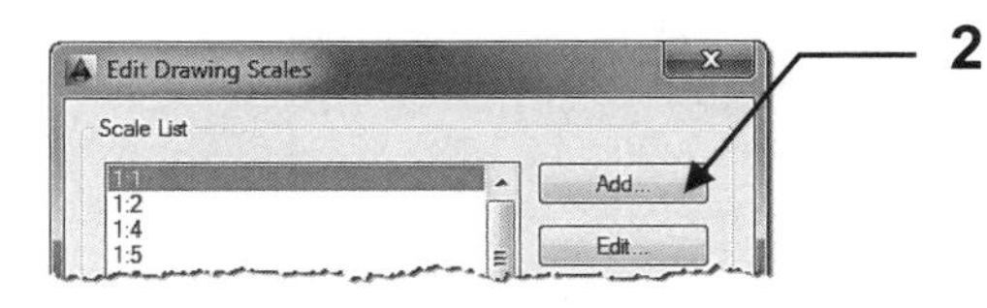

3. Enter Scale name to display in scale list.
4. Enter Paper and drawing units.
5. Select the **OK** button.

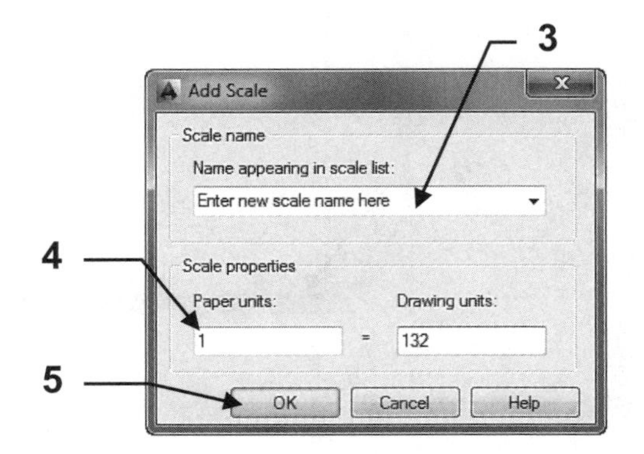

ANNOTATIVE OBJECTS

In lesson 26 you created a new text style and dimension style. You also added the "**Annotative" property** to both simply by placing a check mark in the Annotative box.

The Text style and Dimension Style shown above are now **Annotative**.
Notice the Annotative symbol beside the Style Name. (No symbol by "Standard".)

"**Annotative objects**" are scaled automatically to match the scale of the viewport.
For example, if you want the text, inside the viewport, to print .200 in height and the viewport scale is 1:2 AutoCAD will automatically scale the text height to .400. The text height needs to be scaled by a factor of 2 to compensate for the model space contents appearing smaller.

The easiest way to understand how **Annotative property** works is to do it.
So try this example. (<u>When complete you will save this file for use with Lesson 28</u>.)

1. Start a New file using **My Decimal Setup.dwt**
2. Select the **Model** tab.
3. Draw a rectangle 3.00 Long X 1.50 Wide ***Use layer Object line***.

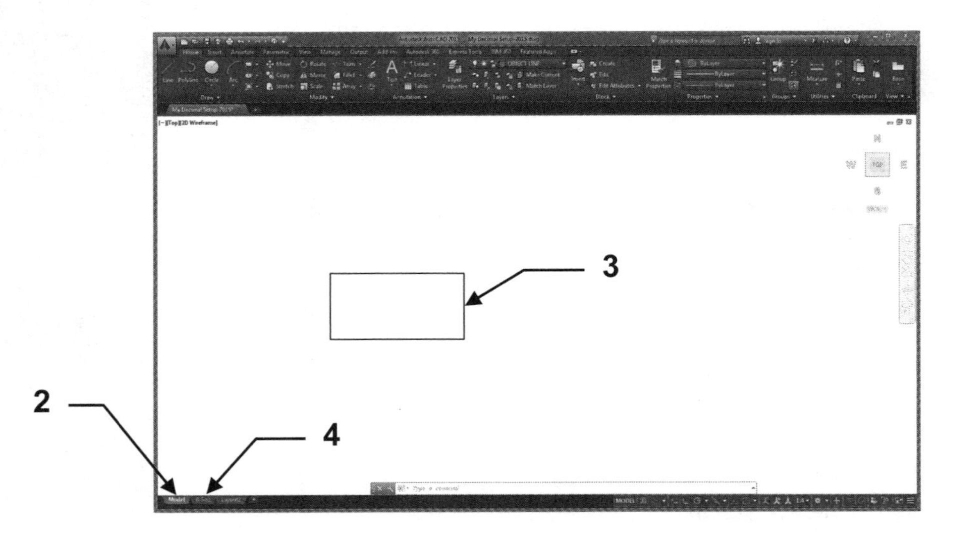

4. Select the **A Size** tab.

Continued on the next page...

ANNOTATIVE OBJECTS....continued

5. Erase the existing single viewport frame.

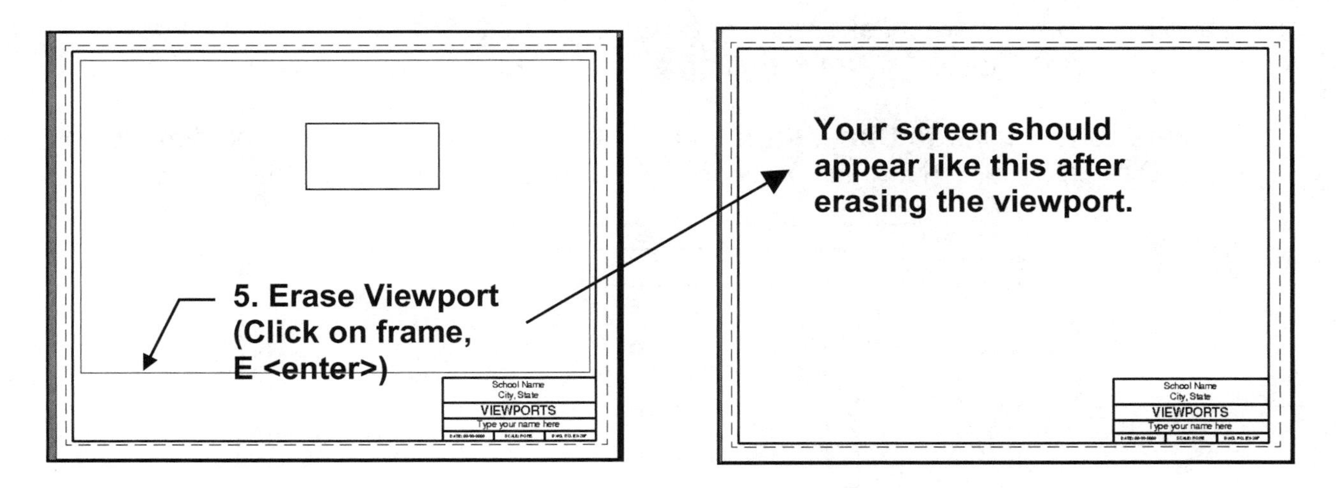

6. Change current layer to Viewport and cut 2 new viewports as shown below.(See 26-7)

7. Activate the left viewport (double click inside the viewport frame) and do the following:
   A. Adjust the scale of the viewport to **1 : 1** (Refer to page 27-3)
   B: PAN to place rectangle in the center of the viewport. (Refer to page 26-10)
   C. Lock the viewport (Refer to page 27-3)

8. Activate the right viewport (click inside the viewport frame) and do the following:
   A. Adjust the scale of the viewport to **1 : 4** (Refer to page 27-3)
   B: PAN to place rectangle in the center of the viewport. (Refer to page 26-10)
   C. Lock the viewport (Refer to page 27-3)

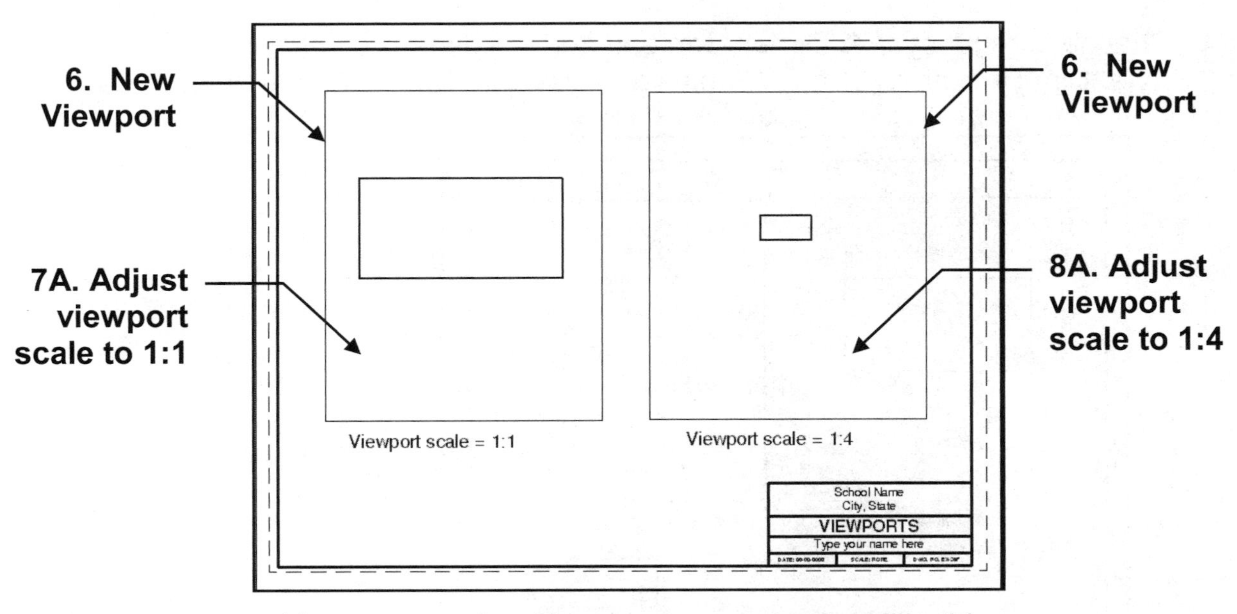

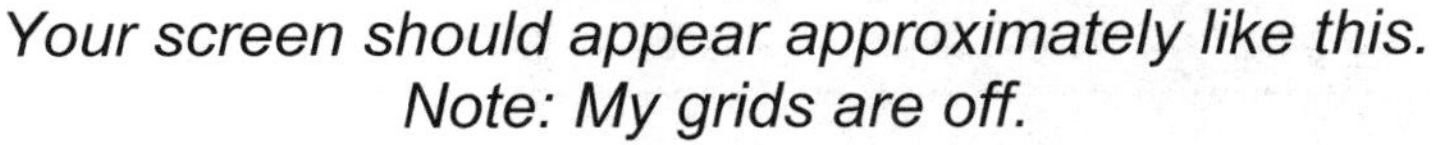

Your screen should appear approximately like this.
Note: My grids are off.

Continued on the next page...

ANNOTATIVE OBJECTS....continued

9. Activate the left hand viewport. (Double click inside left hand viewport)

10. Change current layer to **Text** and add .200 height text **Scale 1:1** as shown using text style "Text-Classic". (Note: Text style "Text-Classic" is annotative)

11. Change current layer to **Dimension** and add the dimension shown using dimension style "Dim-Decimal". (Note: Dimension style "Dim-Decimal" is annotative)

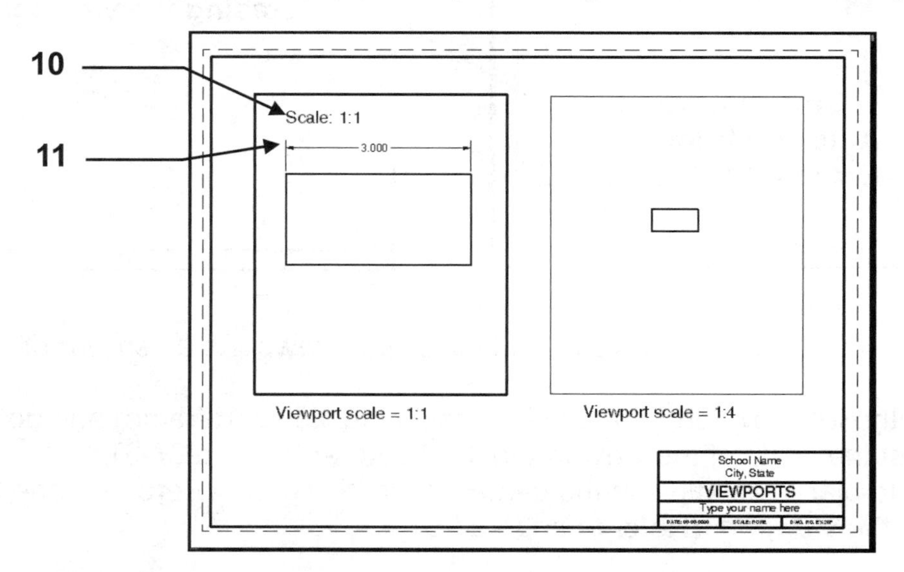

12. Activate the right hand viewport. (Click inside right hand viewport)

13. Change current layer to **Text** and add .200 height text **Scale 1:4** as shown using text style "Text-Classic". (Note: Text style "Text-Classic" is annotative)

14. Change current layer to **Dimension** and add the dimension shown using dimension style "Dim-Decimal". (Note: Dimension style "Dim-Decimal" is annotative)

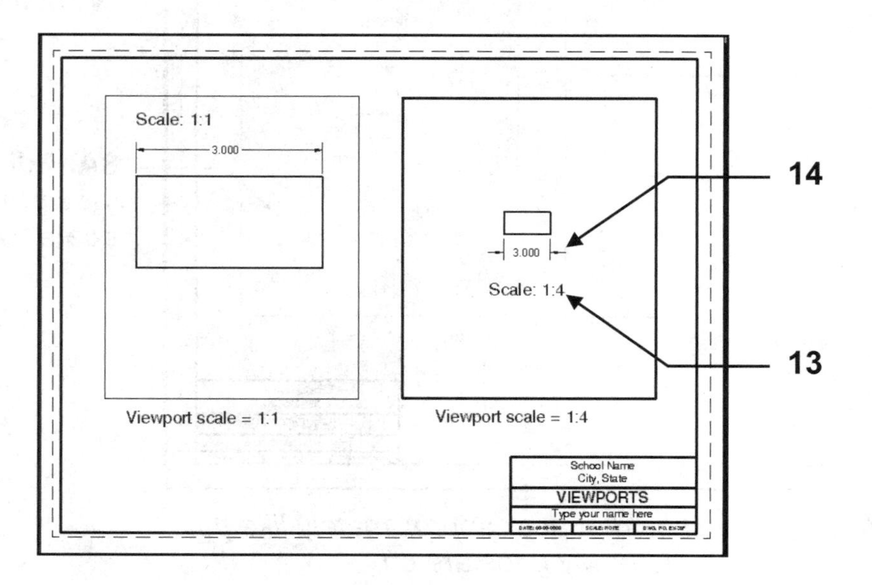

Continued on the next page...

Notice that the text and dimensions appear the <u>same size in both viewports</u>.

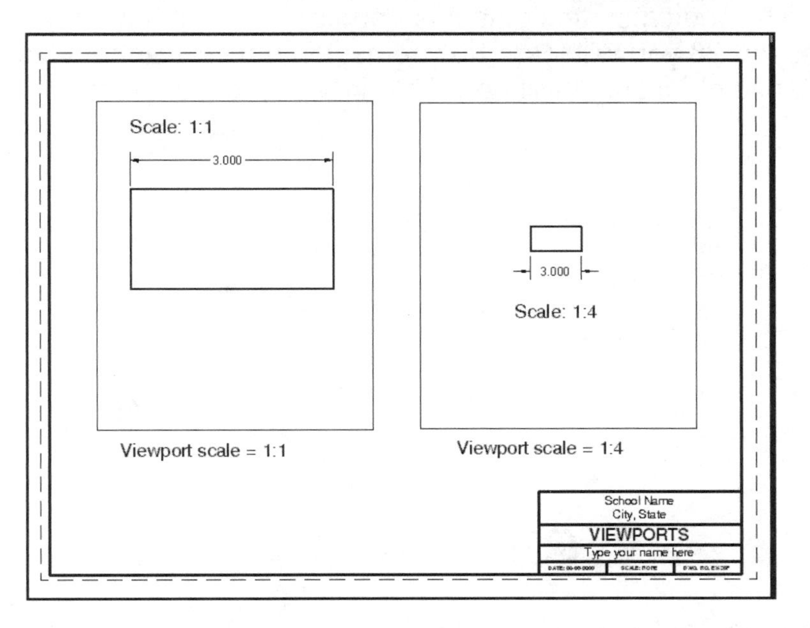

15. Now select the **Model** tab.

Notice there are 2 sets of text and dimensions.

One set has the annotative scale of 1:1 and will be visible only in a 1:1 viewport.
One set has the annotative scale of 1:4 and will be visible only in a 1:4 viewport.
But you see both sets when you select the model tab.

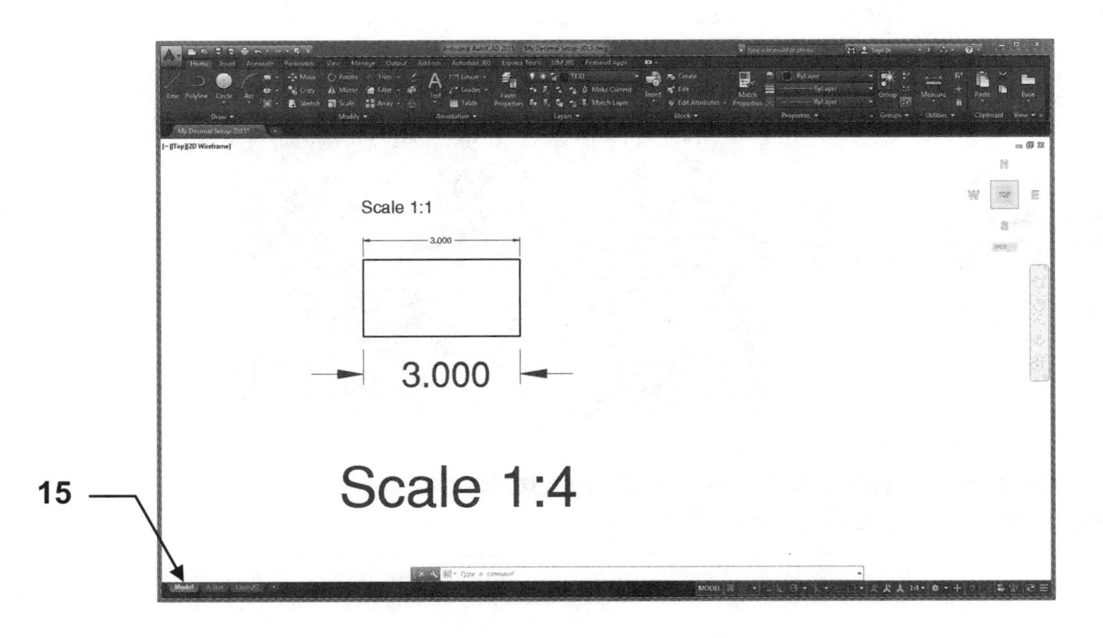

(In Lesson 28 you will learn how to assign multiple annotative scales to one set of text or dimensions so you will not have duplicate sets in the model tab.)

Continued on the next page...

16. Place your cursor on any of the text or dimensions. An "Annotative symbol" will appear. This indicates this object is annotative and it has only one annotative scale. (In lesson 28 you will learn how to assign multiple annotative scales to a single annotative object so it will be visible in multiple viewports.)

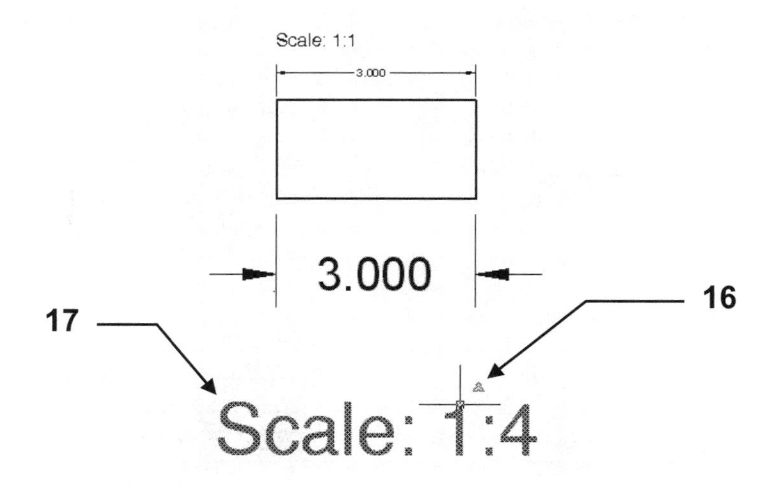

17. Click once on the **Scale: 1:4** text. The Quick properties box should appear. (**Quick Properties** button on the status bar must ON.)

Notice the text height is listed twice.

Paper text height = .200 This is the height that you selected when placing the text. When the drawing is printed the text will print .200.

Model text height = .800 This is the desired height of the text (.200) factored by the viewport scale (1:4). The viewport scale is a factor of 4. (4 X .200 = .800)

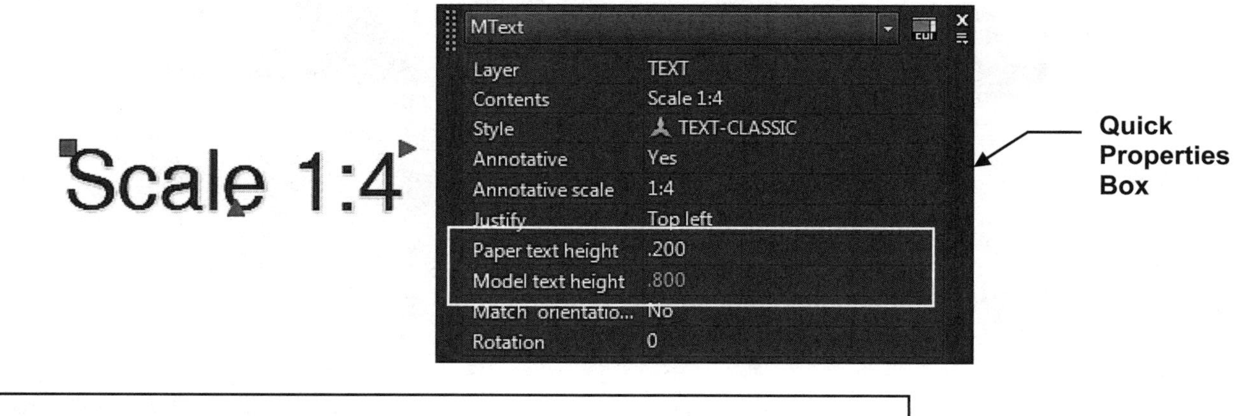

18. **IMPORTANT:** Save this drawing as: **Annotative objects (You will need it for Lesson 28)**

Summary:
If an object is annotative AutoCAD automatically adjusts the scale of the object to the viewport scale. The most commonly used Annotative Objects are: Dimensions, Text, Hatch and Multileaders. Refer to the Help menu for more. (In lesson 28 you will learn how to assign multiple annotative scales to a single annotative object.)

PAPER SPACE DIMENSIONING

Some people prefer to dimension in paper space. Paper space dimensions are Trans-spatial. Trans-spatial means that you may place the dimension in paper space while the object you are dimensioning is in model space. Even though the dimension is in paper space it is actually attached to the object in model space.

For example,
1. You draw a house in model space.
2. Now select the layout tab.
3. Cut a viewport so you can see the drawing of the house.
4. Go to model space, adjust the scale of the viewport (model space) and lock it.
5. Now go to paper space and dimension the house.

SHOULD YOU DIMENSION IN PAPER SPACE OR MODEL SPACE?

This is your choice. Personally, I dimension in either space depending on the situation. Here are some things to consider.

Paper Space dimensioning

Pro's
1. You never have to worry about the viewport scale.
2. All dimensions will have the same appearance in all viewports.
3. If you Xref the drawing the dimensions do not come with it.
   (Xref is not discussed in this workbook)

Con's
1. If you move the objects in model space, or adjust the viewport scale, sometimes the dimensions do not move with them. (If this happens type Dimregen <enter>)
3. Dimensions will not appear in the Model tab. They appear only in the layout tab in which they were placed. If you go to another layout tab they will not appear.

Annotative dimensioning in Model Space

Pro's
1. Dimensions always appear and move with the objects.
2. Dimensions will appear in all layout tabs within viewports.

Con's
1. If you Xref the drawing the dimensions come also.

EXERCISE 27A

CREATE A MASTER FEET-INCHES SETUP TEMPLATE

The following instructions will guide you through creating a “Master” Feet-inches setup template. The “2015-Workbook Helper” is an example of a Master setup template. Once you have created this “Master” template, you just open it and draw. No more repetitive inputting of settings.

NEW SETTINGS

A. Begin your drawing with a different template as follows:
   1. Select the **NEW** command.
   2. Select template file **acad.dwt** from the list of templates. (**Not** acad3D.dwt)

B. Set the drawing **Units** as follows:
   1. Type **UNITS <enter>**
   2. Change the **Type** and **Precision** as shown
   3. Select **OK** button

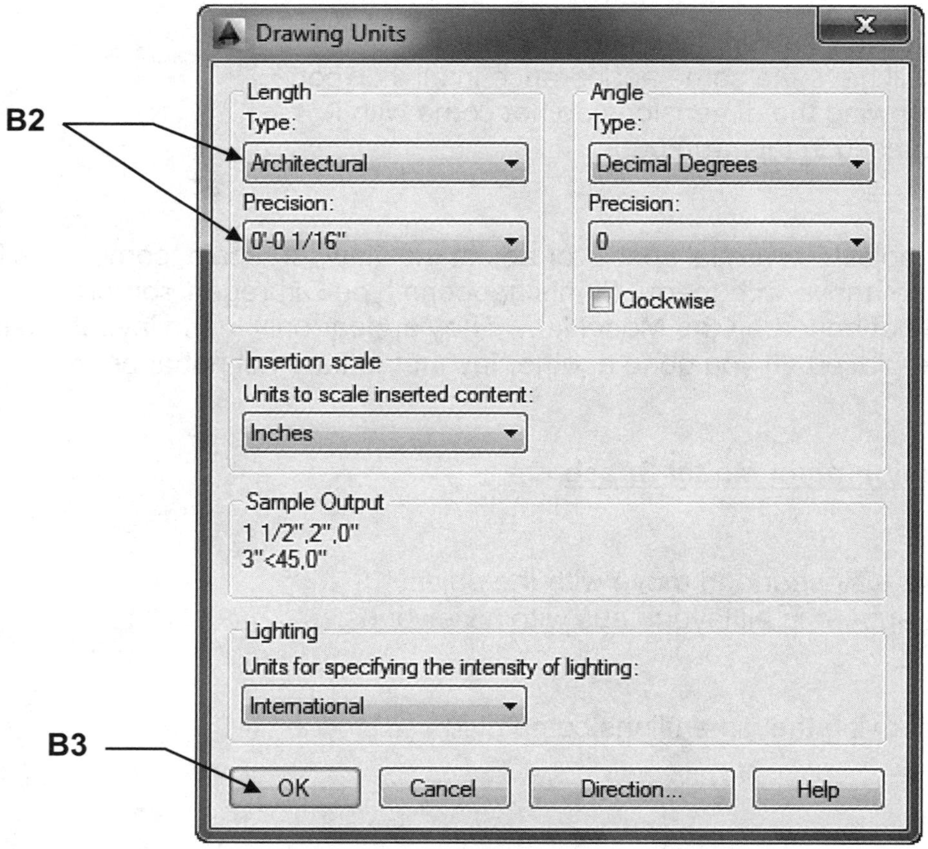

Continued on the next page...

EXERCISE 27A....continued

C. Set the Grids and Snap as follows:
   1. Type **DS<enter>**
   2. Select the **Snap and Grid** tab
   3. Change the settings as shown below.
   4. Select the **OK** button.

C2

Drafting Settings

Snap and Grid | Polar Tracking | Object Snap | 3D Object Snap | Dynamic Input | Quic

Snap On (F9)

Snap spacing
Snap X spacing: 3"
Snap Y spacing: 3"
Equal X and Y spacing

Polar spacing
Polar distance: 0"

Snap type
Grid snap
Rectangular snap
Isometric snap
PolarSnap

Grid On (F7)

Grid style
Display dotted grid in:
2D model space
Block editor
Sheet/layout

Grid spacing
Grid X spacing: 1'
Grid Y spacing: 1'
Major line every: 4

Grid behavior
Adaptive grid
Allow subdivision below grid spacing
Display grid beyond Limits
Follow Dynamic UCS

Options... | OK | Cancel | Help

Important:
1 <u>foot</u> not inches

Important:
Uncheck
these

C4

D. Set the **Drawing Limits** (Size of the drawing area) as follows:
   1. Type **Limits <enter>**
   2. Enter Lower left corner = 0'-0", 0'-0"
   3. Enter Upper right corner = **44' , 34'** (Notice this is feet, not inches)
   4. Use **" ZOOM / ALL"** to view the new limits
   5. Set your **Grids** to **ON** to display the limits (drawing area).

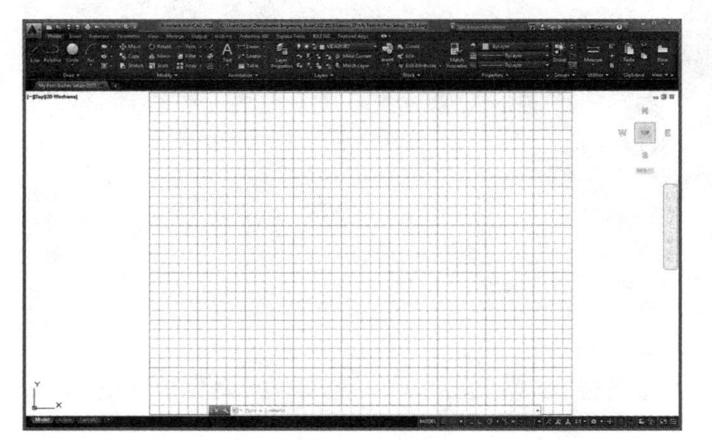

If your screen does not appear as shown here, check these:

- **Did you use, Zoom / All (D4)?**
- **Is your grid spacing feet or inches?**
- **Is the Grid ON?**

Continued on the next page...

EXERCISE 27A....continued

E. Change **Lineweight** settings as follows:
   1. Left click on Lineweight button down arrow ▼ on status line
   2. Select **Lineweight Settings**
   3. Change to inches and adjust the Display scale as shown below.
   4. Select the **OK** button.

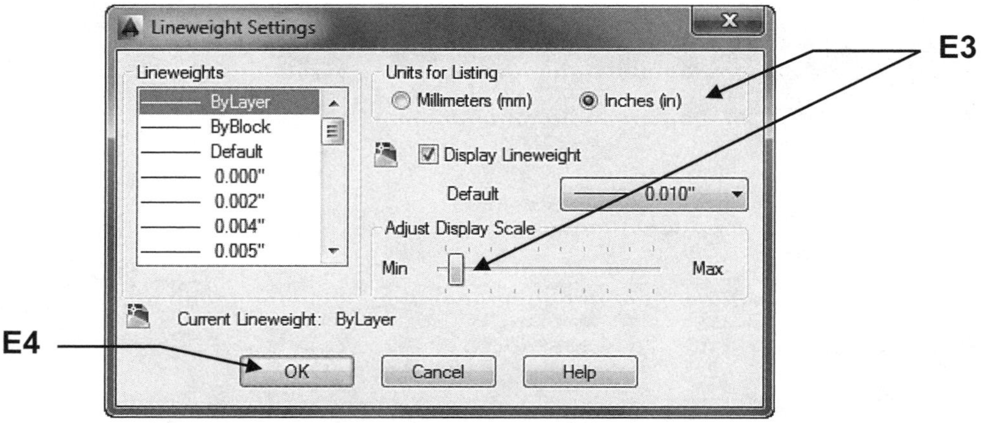

E3
E4
Lineweight Settings
Lineweights
ByLayer
ByBlock
Default
0.000"
0.002"
0.004"
0.005"
Units for Listing
Millimeters (mm)
Inches (in)
Display Lineweight
Default
0.010"
Adjust Display Scale
Min
Max
Current Lineweight: ByLayer
OK
Cancel
Help

F. Create **new Layers** as follows:
   1. Load the linetype **Dashed**. (Refer to page 3-16 for instructions)

   2. Assign names, colors, linetypes, lineweights and plotability as shown below: (Refer to 3-15 for instructions)

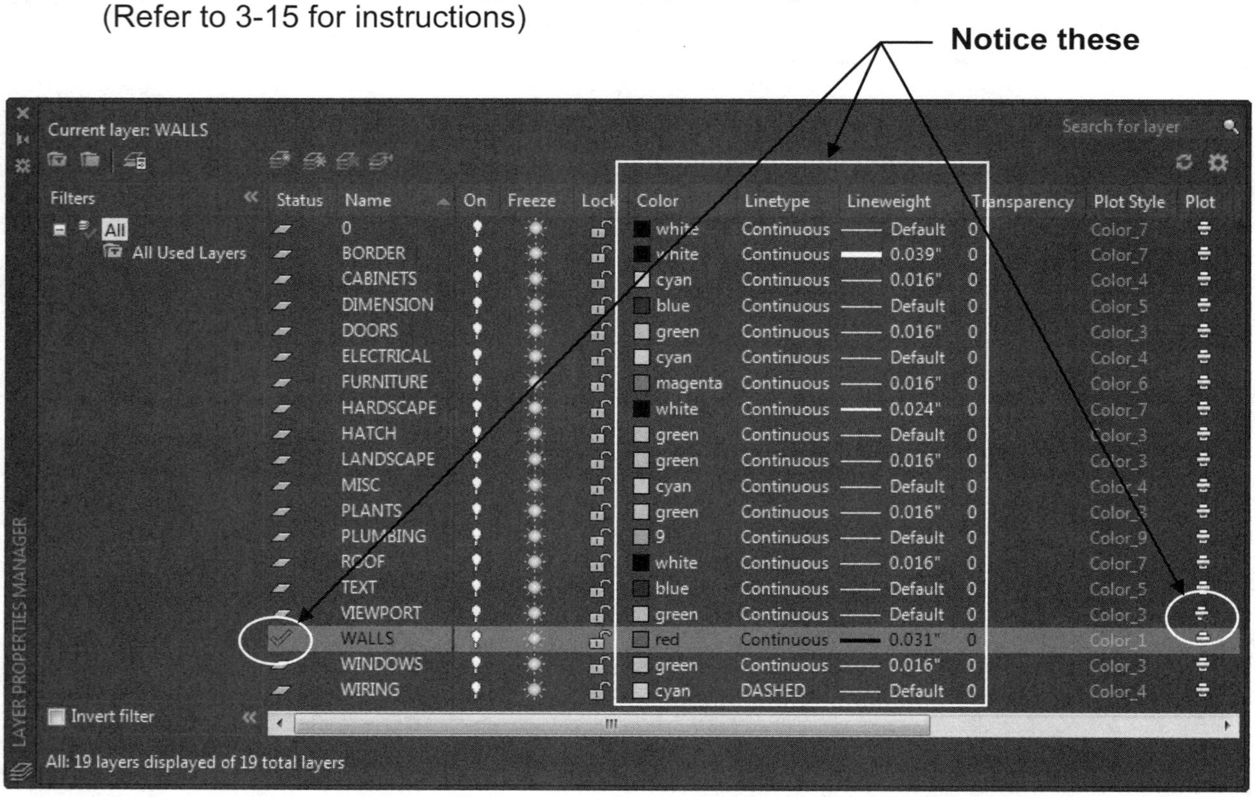

Notice these
Current layer: WALLS
Search for layer
Filters
All
All Used Layers
Status Name On Freeze Lock Color Linetype Lineweight Transparency Plot Style Plot
0 white Continuous Default 0 Color_7
BORDER white Continuous 0.039" 0 Color_7
CABINETS cyan Continuous 0.016" 0 Color_4
DIMENSION blue Continuous Default 0 Color_5
DOORS green Continuous 0.016" 0 Color_3
ELECTRICAL cyan Continuous Default 0 Color_4
FURNITURE magenta Continuous 0.016" 0 Color_6
HARDSCAPE white Continuous 0.024" 0 Color_7
HATCH green Continuous Default 0 Color_3
LANDSCAPE green Continuous 0.016" 0 Color_3
MISC cyan Continuous Default 0 Color_4
PLANTS green Continuous 0.016" 0 Color_3
PLUMBING 9 Continuous Default 0 Color_9
ROOF white Continuous 0.016" 0 Color_7
TEXT blue Continuous Default 0 Color_5
VIEWPORT green Continuous Default 0 Color_3
WALLS red Continuous 0.031" 0 Color_1
WINDOWS green Continuous 0.016" 0 Color_3
WIRING cyan DASHED Default 0 Color_4
Invert filter
All: 19 layers displayed of 19 total layers
LAYER PROPERTIES MANAGER

Continued on the next page...

EXERCISE 27A....continued

G. Create 2 new **Text Styles** as follows:
   1. Select **Text Style** (Refer to page 25-2)
   2. Create the text style **Text-Classic** and **Text-Arch** using the settings shown below.

<u>THIS NEXT STEP IS VERY IMPORTANT</u>

H. **Save** all the settings you just created as follows:
   1. Save as: **My Feet-Inches Setup**

I. Now continue on to **Exercise 27B**...you are not done yet.

EXERCISE 27B

PAGE SETUP

Now you will select the printer and the paper size to use for printing. You will use the Layout1 tab (Paper space).

A. Open **My Feet-Inches Setup** (If not already open)

B. Select **Layout1** tab.

Refer to page 26-3 if you do not have these tabs.

Note: If the "Page Setup Manager" dialog box shown below does not appear automatically, right click on the Layout tab and select Page Setup Manager.

C. Select the **New...** button.

D. Select the **<Default output device>** in the Start with: list.

E. Enter the New page setup name: **Setup B**

F. Select **OK** button.

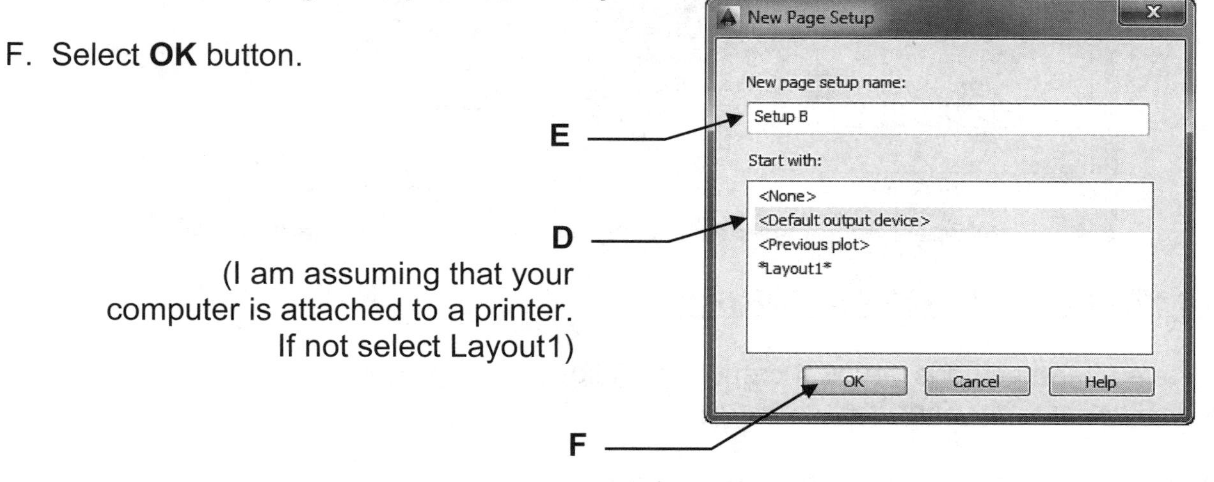

(I am assuming that your computer is attached to a printer. If not select Layout1)

Continued on the next page...

EXERCISE 27B....continued

This is where you will select the **print device**, **paper size** and the **plot offset**.

G. Select the **Printer / Plotter**
__Note__: Your current system printer should already be displayed here. If you prefer another, select the down arrow and select from the list. If the preferred printer is not in the list you must configure the printer. Refer to Appendix-A for instructions.)

H. Select the **Paper Size**

I. Select **Plot Offset**

Notice the name you entered is now displayed as the page setup name.

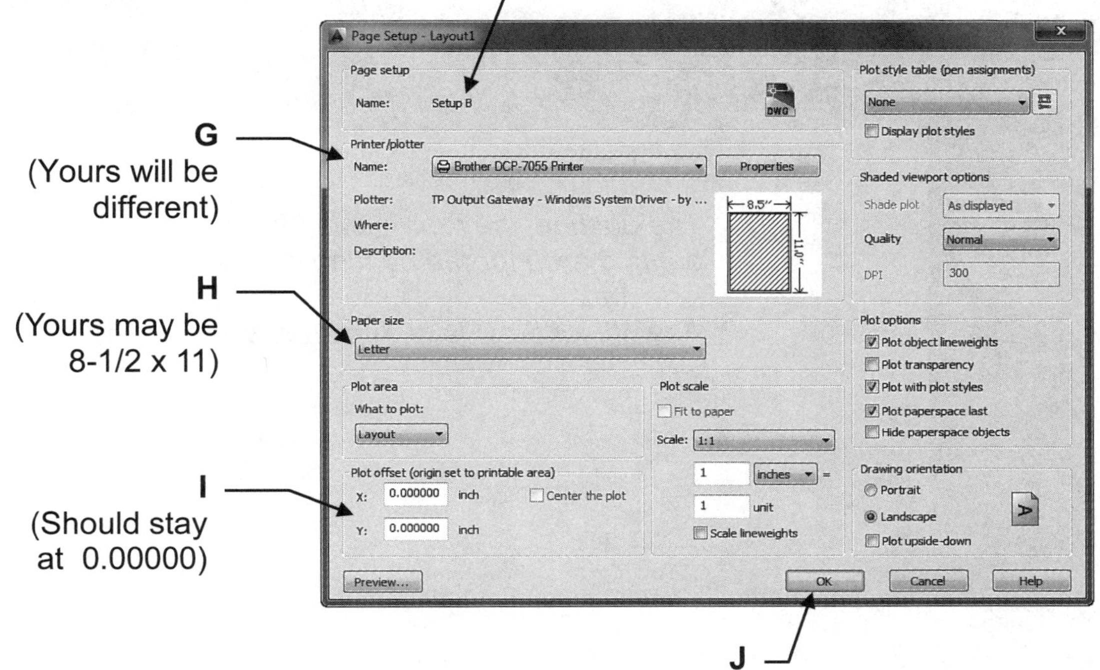

J. Select **OK** button.

K. Select **Setup B**.

L. Select the **Set Current** button.

M. Select the **Close** button.

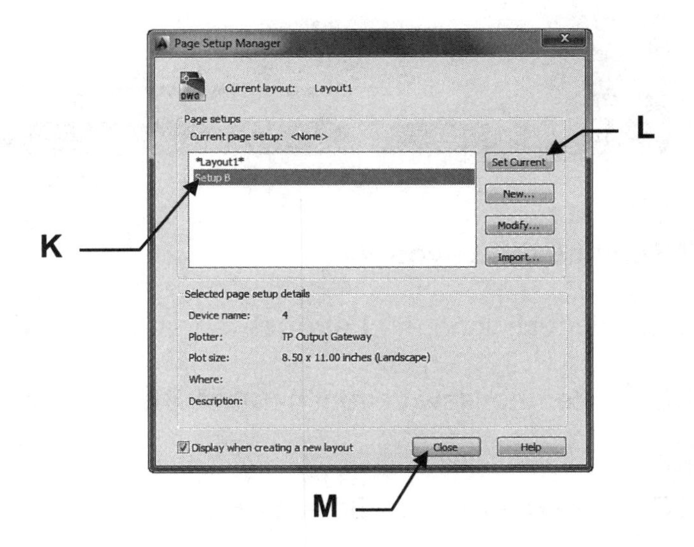

Continued on the next page...

EXERCISE 27B....continued

You should now have a sheet of paper displayed on the screen.
This sheet is the size you specified in the "Page Setup".
This sheet is in front of Model Space.

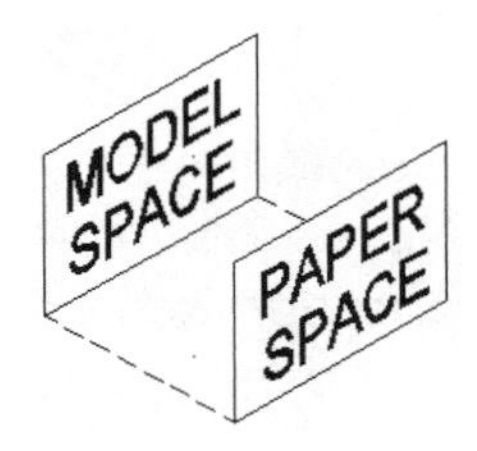

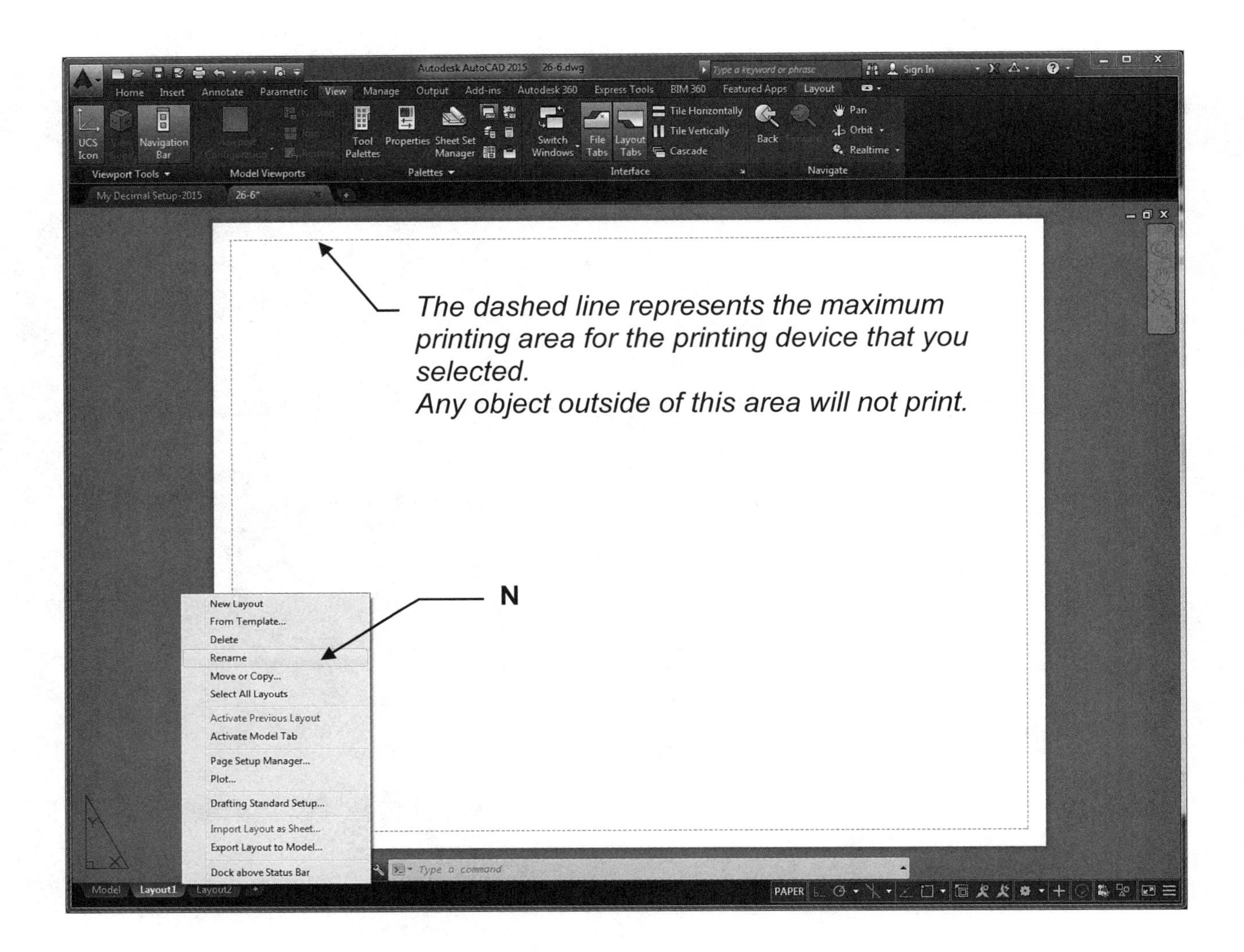

Rename the Layout tab

N. Right click on the **Layout1** tab and select **Rename** from the list.

O. Enter the New Layout name **A Size**

P. ***Very important:*** Save as **My Feet-Inches Setup** again.

Q. Continue on to **Exercise 27C**.....you are not done yet.

EXERCISE 27C

CREATE A BORDER AND TITLE BLOCK

A. Draw the Border rectangle as large as you can within the dashed lines approximately as shown below using Layer Border.

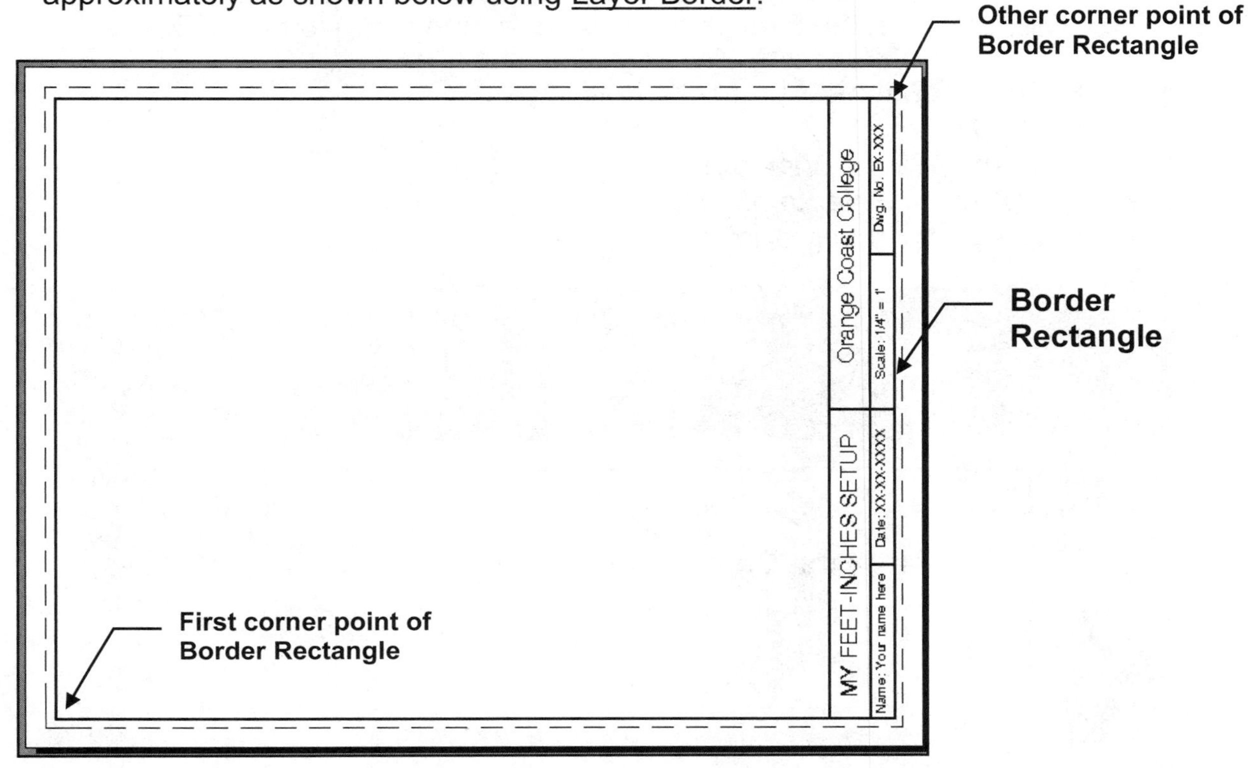

B. Draw the Title Block as shown using:
1. Layers Border and Text.
2. **Single Line Text** ; Justify **Middle** in each rectangular area. (Hint: Draw diagonal lines to find the middle of each area, as shown)
3. Text Style= Text-Classic   Text height varies.

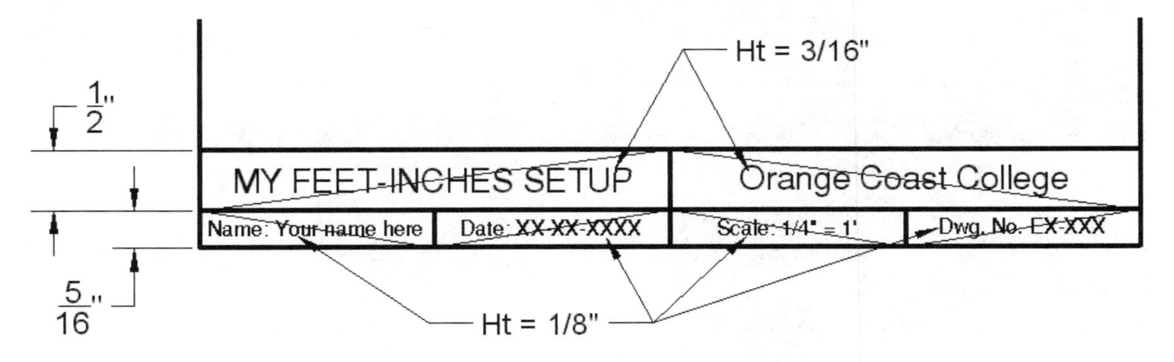

C. ***Very important:*** Save as **My Feet-Inches Setup** again.

D. Continue on to **Exercise 27D**.....you are not done yet.

EXERCISE 27D

CREATE A VIEWPORT

The following instructions will guide you through creating a VIEWPORT in the Border Layout sheet. Creating a viewport has the same effect as cutting a hole in the sheet of paper. You will be able to see through the viewport frame (hole) to Model Space.

A. Open **My Feet-Inches Setup** (If not already open)
B. Select the **A Size** tab.
C. Change Current layer to: **Viewport**
D. Type: **MV <enter>** (Refer to page 26-7)
E. Draw a Single viewport approximately as shown. (Turn off OSNAP temporarily)

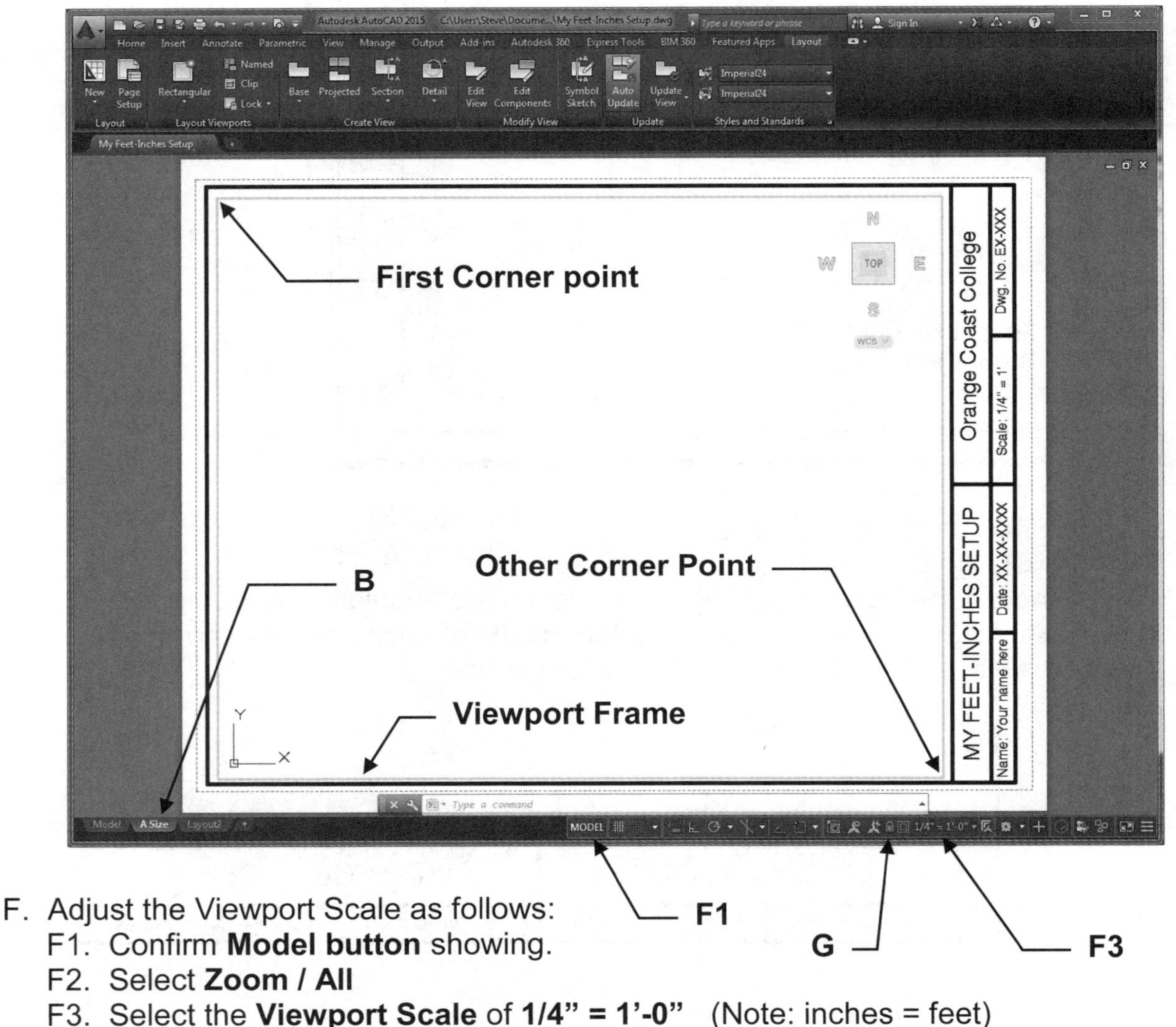

F. Adjust the Viewport Scale as follows:
   F1. Confirm **Model button** showing.
   F2. Select **Zoom / All**
   F3. Select the **Viewport Scale** of **1/4" = 1'-0"** (Note: inches = feet)

G. **Lock** the Viewport (Refer to page 27-3)

H. ***Very important....*** Save as **My Feet-Inches Setup** again.

I. Continue on to **Exercise 27E**....you are not done yet.

EXERCISE 27E

PLOTTING FROM THE LAYOUT

The following instructions will guide you through the final steps for setting up the master template for plotting. These settings will stay with **My Feet-Inches Setup** and you will be able to use it over and over again.

In this exercise you will take a short cut by **importing** the **Plot Setup A** from **My Decimal Setup**. (Note: If the Printer, Paper size and Plot Scale is the same you can use the same Page Setup.)

Note: If you prefer not to use Import, you may go to 26-33 and follow the instructions for creating the Plot- page setup.

A. Open **My Feet-Inches Setup** (If not already open)

B. Select the **A Size** layout tab.

You should be seeing your Border and Title Block now.

C. Select **Plot** command.

D. Select **Import...** from the Page Setup drop down menu

E. Find **My Decimal Setup.dwt** as follows:
   E1. Select Files of type: **Template [.dwt]**
   E2. Select **My Decimal Setup.dwt**
   E3. Select **Open** button.

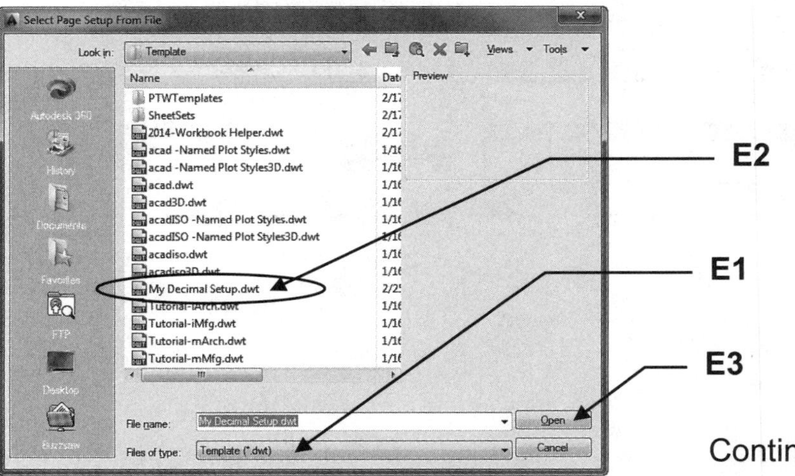

Continued on the next page...

EXERCISE 27E....continued

F. Select **Plot Setup A** from the Page Setups list. (Note: If you do not have a “Plot Setup A” in My Decimal Setup refer to page 26-33)

G. Select the **OK** button.

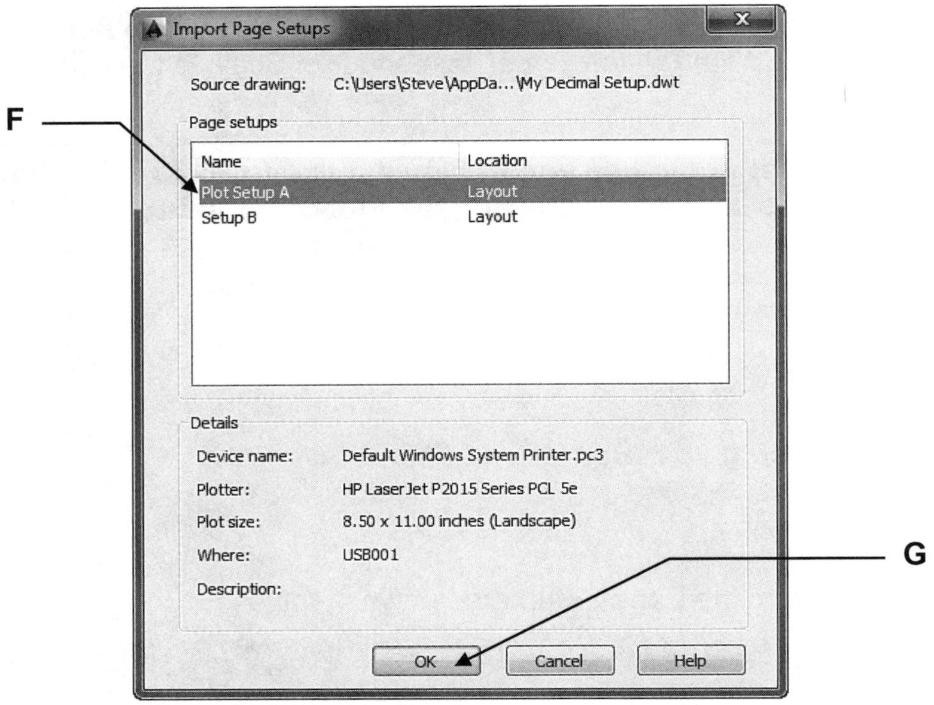

H. Select **Plot Setup A** from the Page Setup drop down list. (Note: It was not there before. You just imported it in from My Decimal Setup)

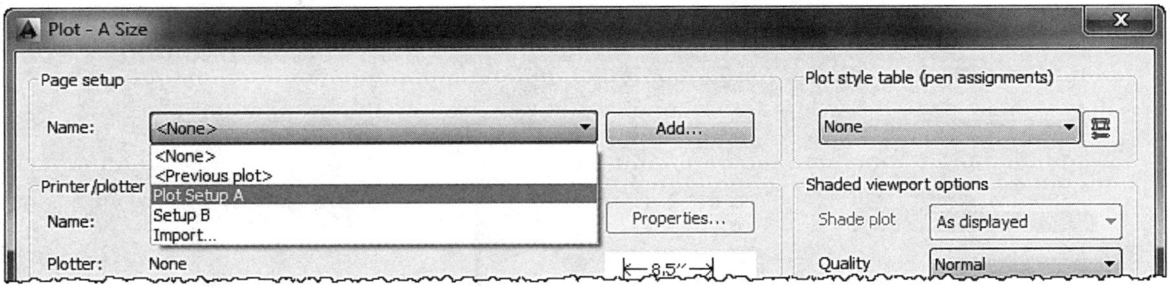

Continued on the next page...

EXERCISE 27E....continued

I. Check all the settings.

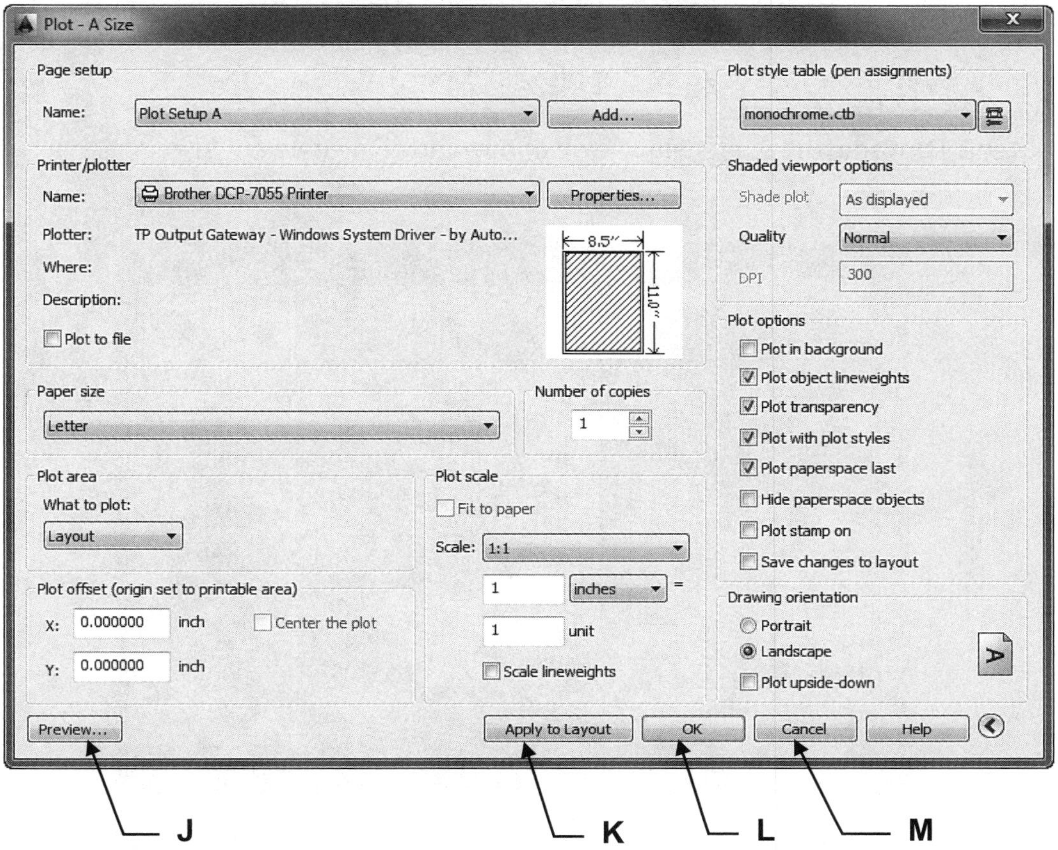

J. Select **Preview** button.

If the drawing appears as you would like it, press the **Esc** key and continue

If the drawing does not look correct, press the **Esc** key and check all your settings, then preview again.

Note:
The Viewport frame and the grids will not appear in the Preview because because the viewport is on a no plot layer and grids never plot.

K. Select the **Apply to Layout** button.
(The settings are now saved to the layout tab for future use.)

L. If your computer **is** connected to the plotter / printer selected, select the **OK** button to plot, then proceed to **N**.

M. If your computer is **not** connected to the plotter / printer selected, select the Cancel button to close the Plot dialog box and proceed to **N**.

N. ***Very important....*** Save as **My Feet-Inches Setup** again.

O. Continue on to **Exercise 27F**....you are not done yet.

EXERCISE 27F

CREATE A NEW DIMENSION STYLE

A. Open **My Feet-Inches Setup** (If not already open)

B. Select the **Dimension Style Manager** command (Refer to 16-8)

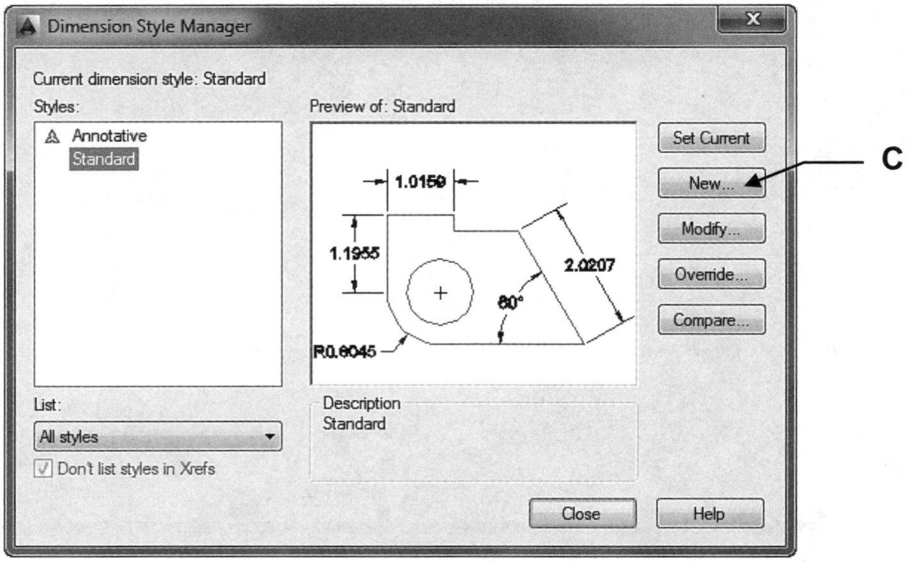

C. Select the **NEW** button.

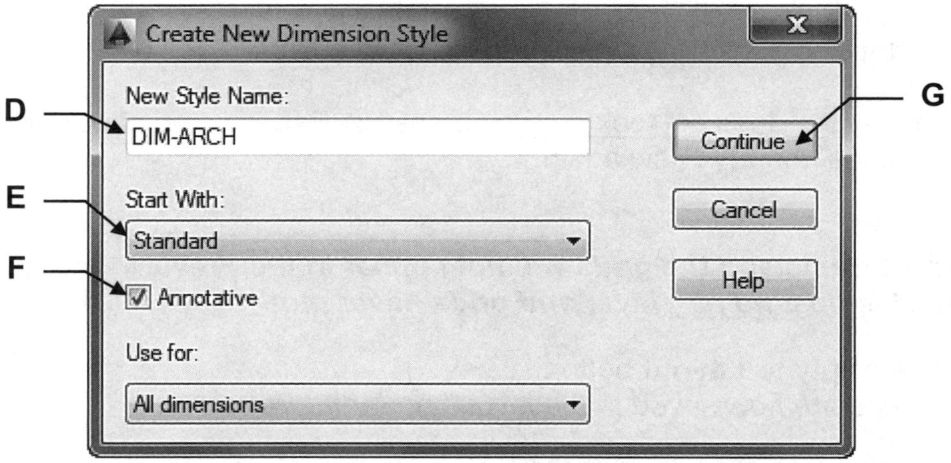

D. Enter **DIM-ARCH** in the “New Style Name” box.

E. Select **STANDARD** in the **“Start With:”** box.

F. Select **Annotative** box.

G. Select the **CONTINUE** button.

EXERCISE 27F ....continued

H. Select the **Primary Units** tab and change your settings to match the settings shown below.

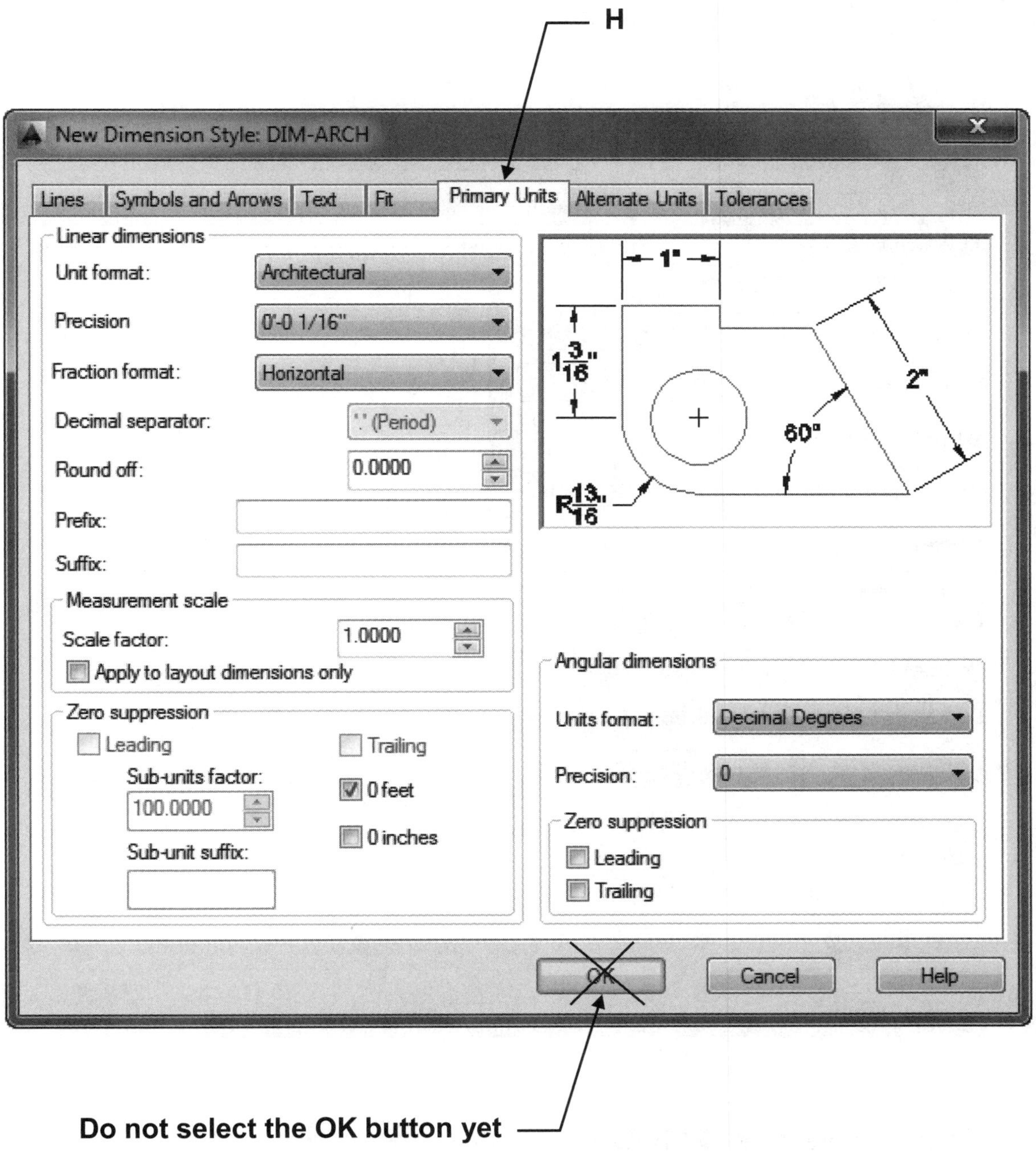

H

New Dimension Style: DIM-ARCH

Lines | Symbols and Arrows | Text | Fit | Primary Units | Alternate Units | Tolerances

Linear dimensions

Unit format: Architectural

Precision: 0'-0 1/16"

Fraction format: Horizontal

Decimal separator: '.' (Period)

Round off: 0.0000

Prefix:

Suffix:

Measurement scale

Scale factor: 1.0000

Apply to layout dimensions only

Zero suppression

Leading

Sub-units factor: 100.0000

Sub-unit suffix:

Trailing

0 feet

0 inches

Angular dimensions

Units format: Decimal Degrees

Precision: 0

Zero suppression

Leading

Trailing

OK | Cancel | Help

Do not select the OK button yet

EXERCISE 27F....continued

I. Select the **Lines** tab and change your settings to match the settings shown below.

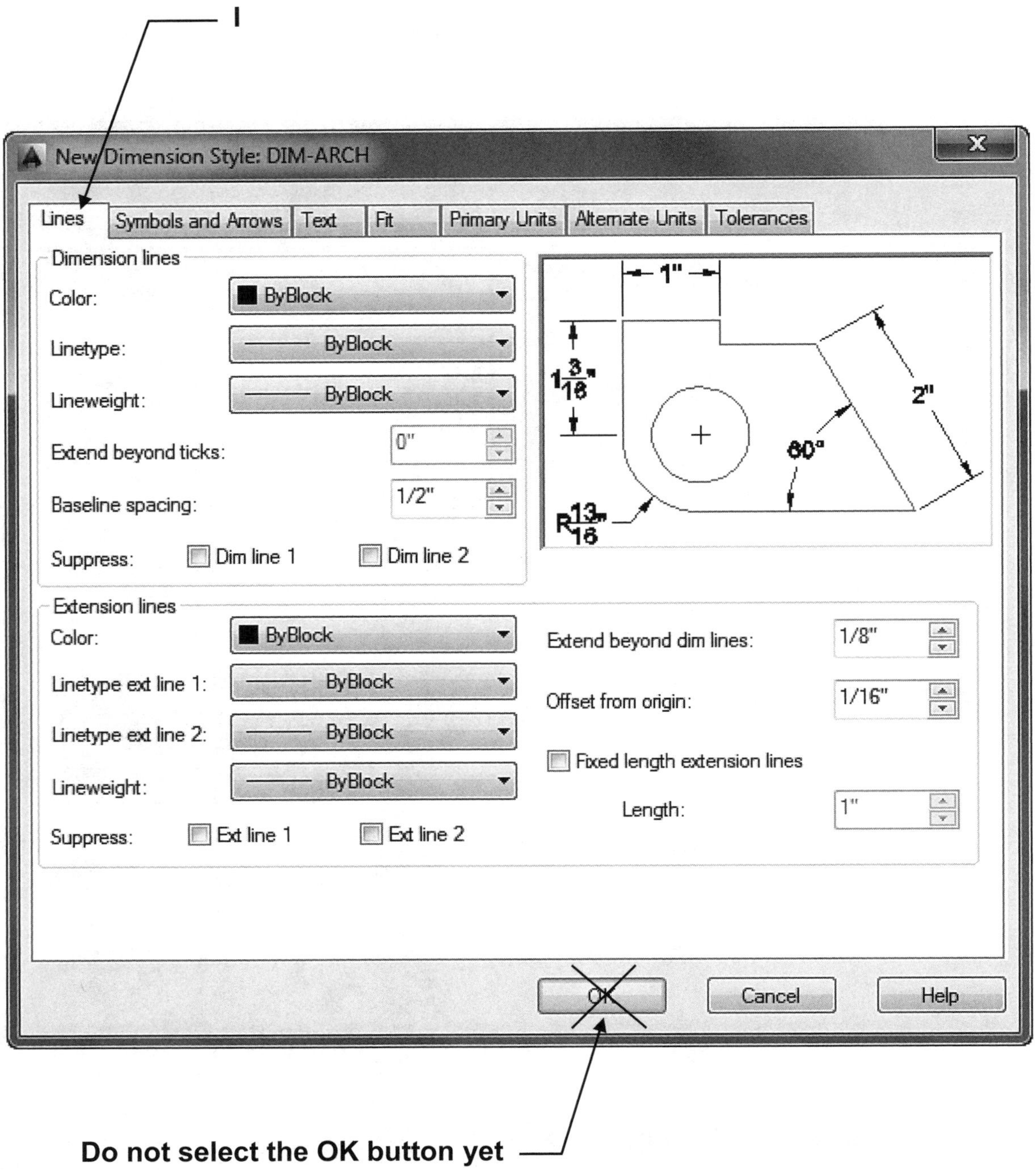

I

New Dimension Style: DIM-ARCH

Lines | Symbols and Arrows | Text | Fit | Primary Units | Alternate Units | Tolerances

Dimension lines

Color: ByBlock

Linetype: ByBlock

Lineweight: ByBlock

Extend beyond ticks: 0"

Baseline spacing: 1/2"

Suppress: Dim line 1 Dim line 2

Extension lines

Color: ByBlock

Linetype ext line 1: ByBlock

Linetype ext line 2: ByBlock

Lineweight: ByBlock

Suppress: Ext line 1 Ext line 2

Extend beyond dim lines: 1/8"

Offset from origin: 1/16"

Fixed length extension lines

Length: 1"

OK Cancel Help

Do not select the OK button yet

EXERCISE 27F....continued

J. Select the **Symbols and Arrows** tab and change your settings to match the settings shown below.

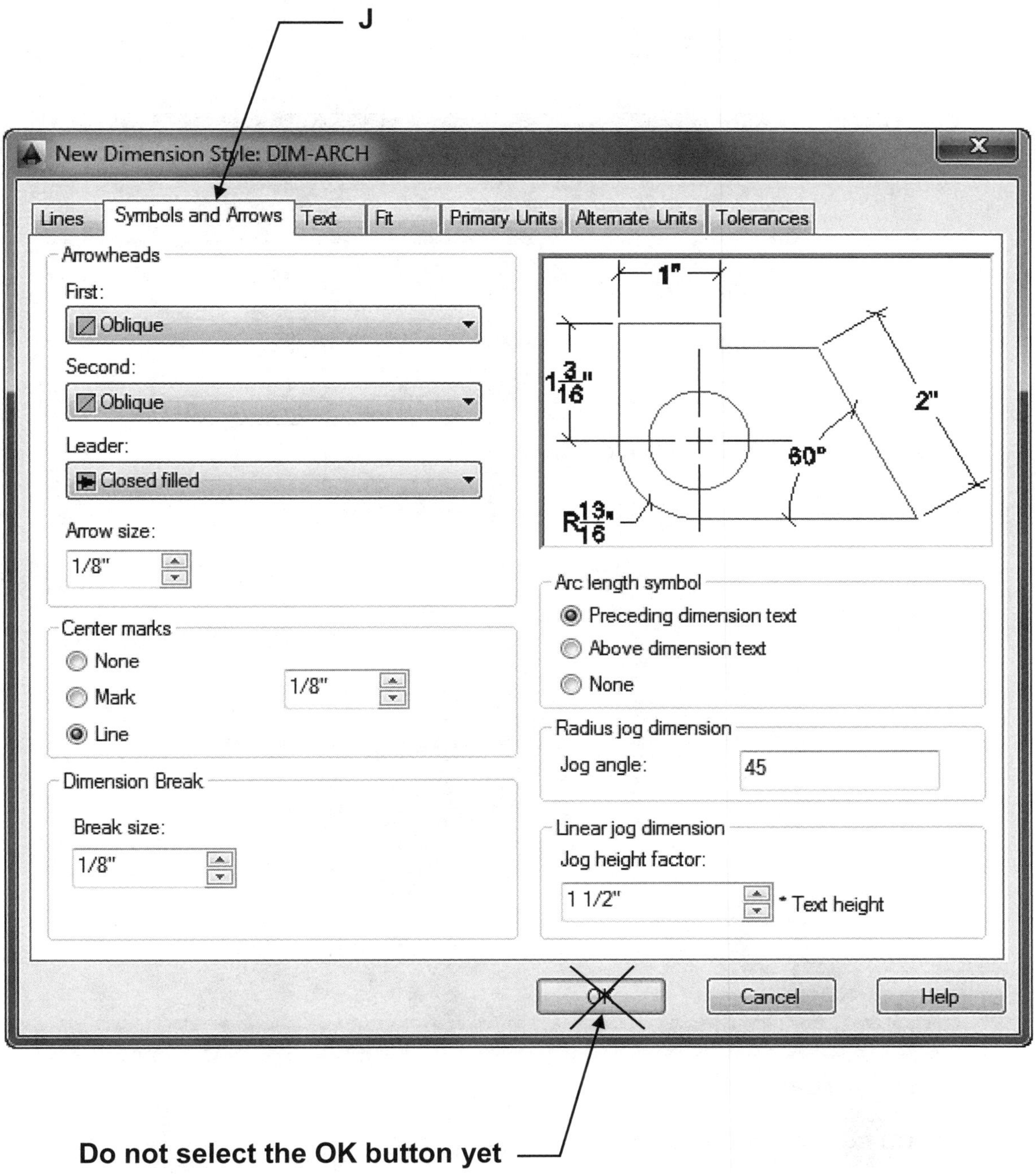

J

New Dimension Style: DIM-ARCH

Lines | Symbols and Arrows | Text | Fit | Primary Units | Alternate Units | Tolerances

Arrowheads
First: Oblique
Second: Oblique
Leader: Closed filled
Arrow size: 1/8"

Center marks
None
Mark
Line
1/8"

Dimension Break
Break size: 1/8"

1"
$1\frac{3}{16}$"
2"
60°
R$\frac{13}{16}$

Arc length symbol
Preceding dimension text
Above dimension text
None

Radius jog dimension
Jog angle: 45

Linear jog dimension
Jog height factor:
1 1/2" * Text height

OK | Cancel | Help

Do not select the OK button yet

EXERCISE 27F....continued

K. Select the **Text** tab and change your settings to match the settings shown below.

K

New Dimension Style: DIM-ARCH

Lines | Symbols and Arrows | Text | Fit | Primary Units | Alternate Units | Tolerances

Text appearance

Text style: TEXT-ARCH

Text color: ByBlock

Fill color: None

Text height: 1/8"

Fraction height scale: 1.0000

Draw frame around text

Text placement

Vertical: Above

Horizontal: Centered

View Direction: Left-to-Right

Offset from dim line: 1/16"

Text alignment

Horizontal

Aligned with dimension line

ISO standard

OK | Cancel | Help

Do not select the OK button yet

EXERCISE 27F....continued

L. Select the **Fit** tab and change your settings to match the settings shown below.

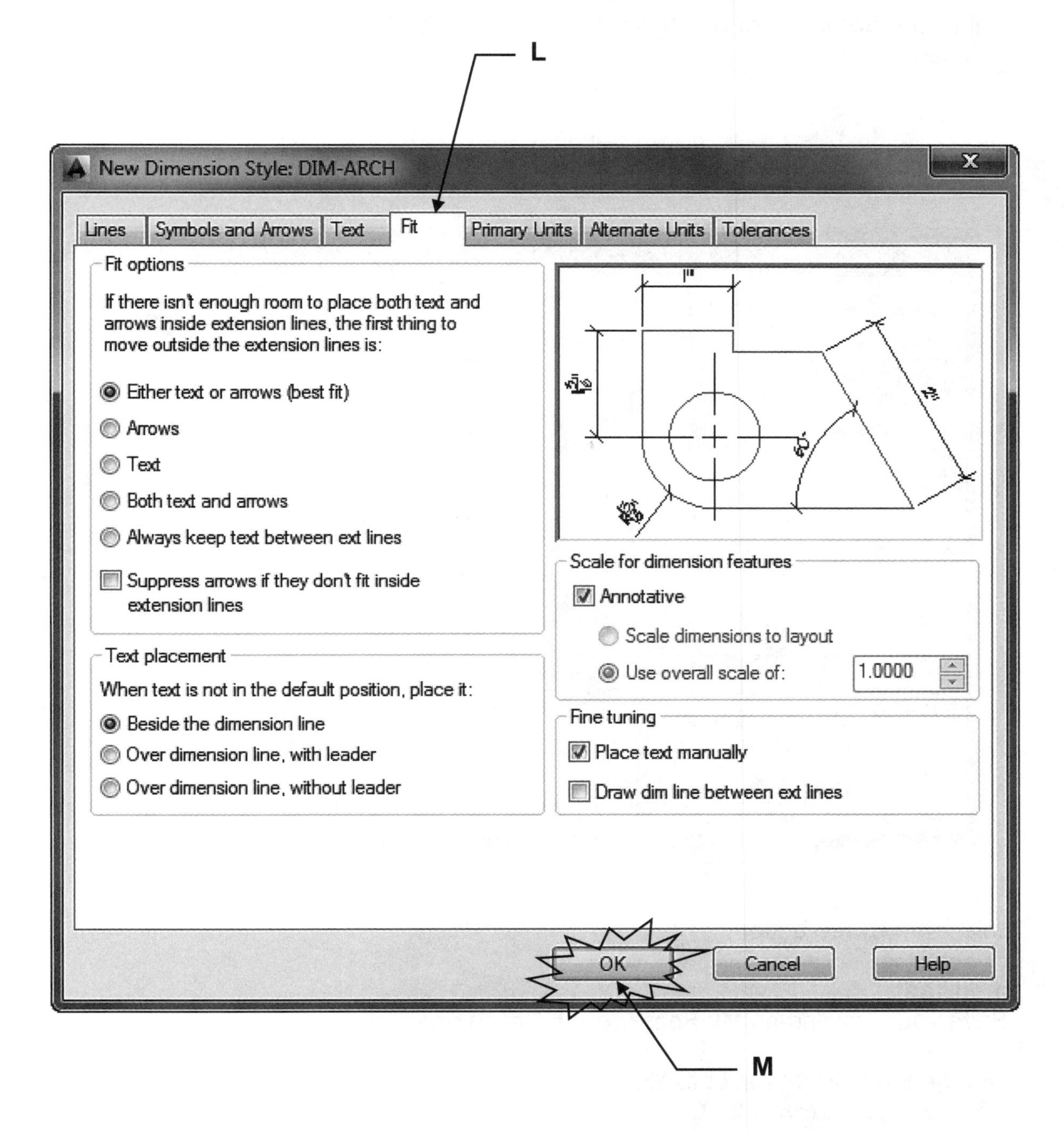

M. Now select the **OK** button.

EXERCISE 27F....continued

Your new **DIM-ARCH** dimension style should now be in the list.

N. Select the **Set Current** button to make your new style **DIM-ARCH** the style that will be used.

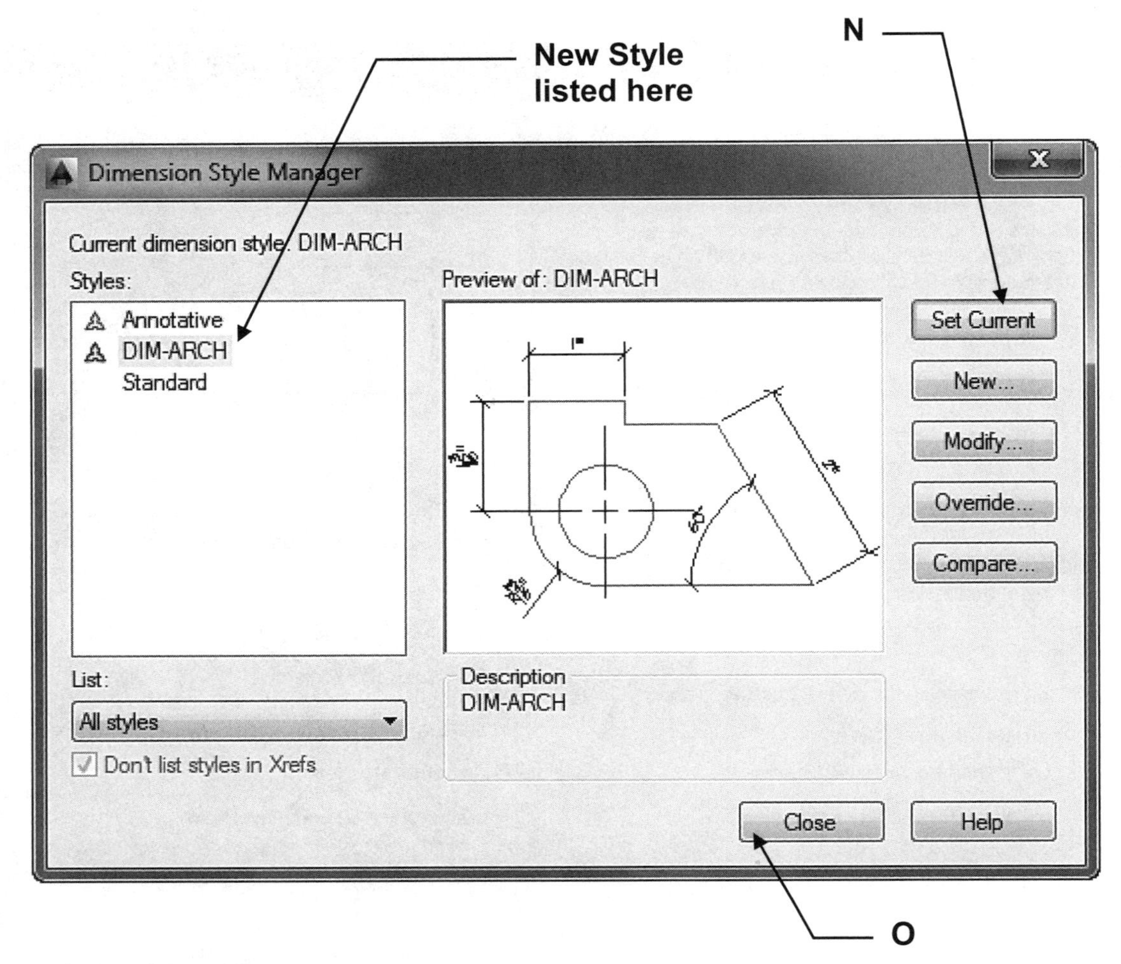

O. Select the **Close** button.

P. **Important:**
Save your drawing as **My Feet-Inches Setup** again.

Q. Change the Current Layer to **Walls**. (So layer Walls will always be the current layer when you start a New file)

R. You are almost done. Now Save all of this work as a **Template. (Refer to 2-3)**
   a. Select **Application Menu / Save As / AutoCAD Drawing Template**
   b. Type: **My Feet-Inches Setup**
   c. Select **Save** button.

Again, I know that seemed like a lot of work but you are now ready to use this master setup template to create and plot many drawings in the future.

EXERCISE 27G

INSTRUCTIONS:

1. Select **New** and select **My Feet-Inches Setup.dwt** (template)
2. Draw the classroom shown below.
3. Use Layers Walls, Furniture, Doors and Windows.
4. Dimension as shown inside the viewport.
5. Use Dimension Style "Dim-Arch" and Layer: Dimension
6. Edit the Title and Ex-XX by double clicking on the text. Do not erase and replace.
7. Save as **EX27G**
8. Plot using Page Setup **Plot Setup A** (Refer to page 27-21)

Drawing hints:

1. Consider using **Array** (Lesson 13) and **Offset** (Lesson 12).
2. Remember you have to Unlock the Viewport to use Pan inside the Viewport.
3. If you change the scale of the Viewport, change the viewport scale back to 1/4"=1' and lock the viewport.
4. If your dimensions are gigantic, you are not inside the viewport.

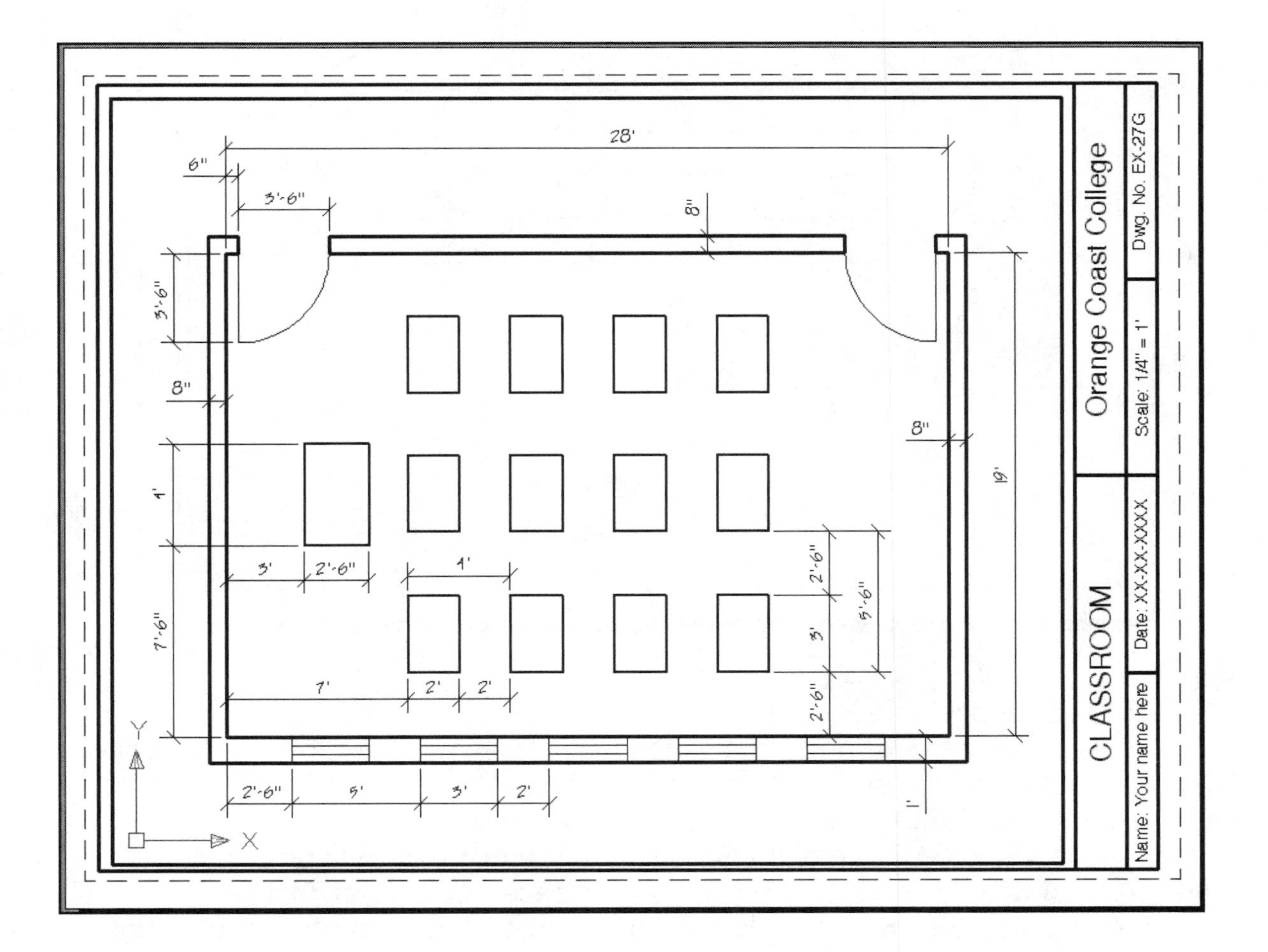

EXERCISE 27H

INSTRUCTIONS:

Step 1.

1. Select **New** and select **My Decimal Setup.dwt** (template)
2. Unlock the Viewport.
3. Change the Viewport Scale to 8 : 1 (Important: not 1:8)
4. Lock the Viewport
5. Turn **Snap** & **Grids** off or change your them relative to the new viewport scale.
6. Draw the Paperclip on layer Object line.
7. Dimension as shown using Dimension Style: Dim-Decimal
   (Dimensions should be inside the viewport, in Model space.
   Confirm "Model" button MODEL is displayed.)

Drawing hints: The entire Paper clip can easily be drawn using Offset (Lesson 12) and Fillet (Lesson 7) or Circles and trim.

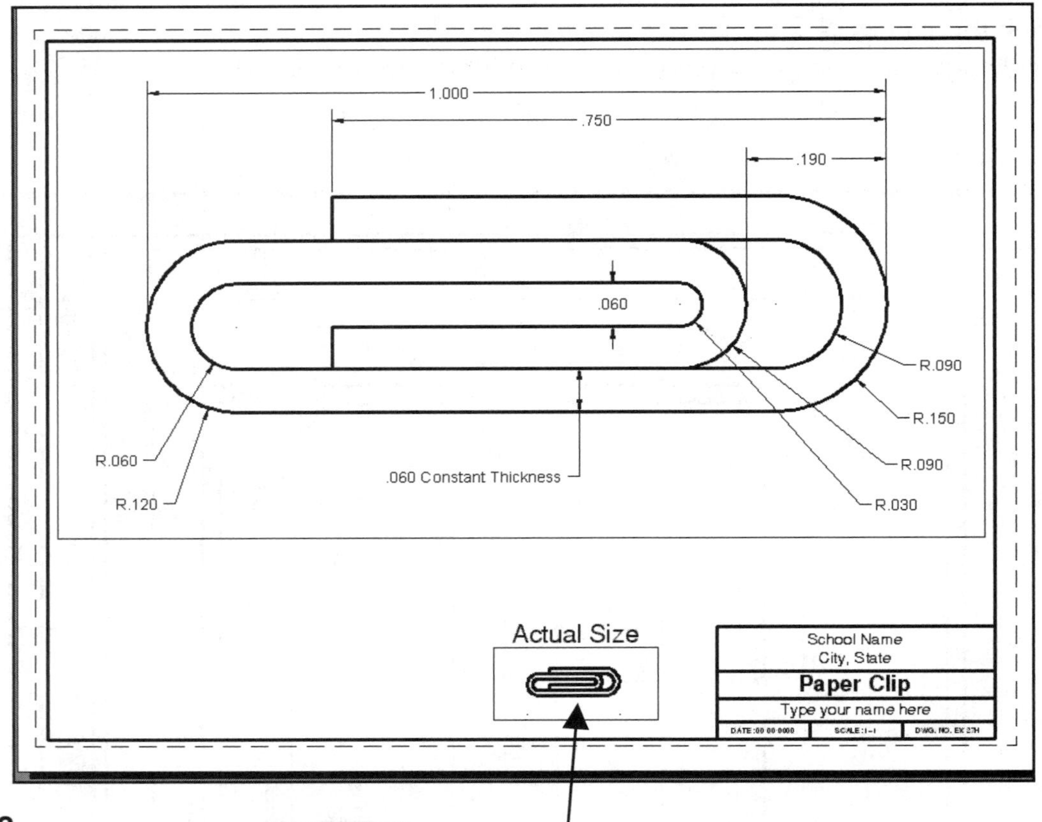

Step 2

1. Select Paper space. PAPER
2. Select the Viewport layer and cut a second small viewport next to the Title Block
3. Double click inside the small viewport to make it active.
   Use **Zoom / All** (You should now see the paper clip)
4. Change the Viewport Scale of the **small** Viewport to **1 = 1** and Lock.
5. Add "**Actual Size**" text .200 height in Paper Space. Use: layer Text.
6. Edit the Title and Ex-XX by double clicking on the text. Do not erase and replace.
7. Save as **EX27H**
8. Plot using Page Setup **Plot Setup A** (Refer to page 26-34)

LEARNING OBJECTIVES

After completing this lesson, you will be able to:

1. Assign multiple annotative scales to a single annotative object
2. Assigning annotative scales to Hatch sets

LESSON 28

ASSIGNING MULTIPLE ANNOTATIVE SCALES

In the previous lesson you learned how annotative text and dimensions are automatically scaled to the viewport scale. But in order to have an annotative object appear in both viewports you placed 2 sets of text and dimensions. In this lesson you will learn how to easily assign multiple annotative scales to a single text string or dimension so you need not duplicate them each time you create a new viewport. You will just assign an additional annotative scale to the annotative object.

Again, the easiest way to understand this process is to do it. The following is a step by step example.

1. Open **Annotative Objects.dwg** from the previous lesson and select the **A size** layout tab. *If you did not complete the example from the previous lesson go back and do it now. (Refer to page 27-4)*

2. Make the right hand viewport active. (Double click inside the viewport)

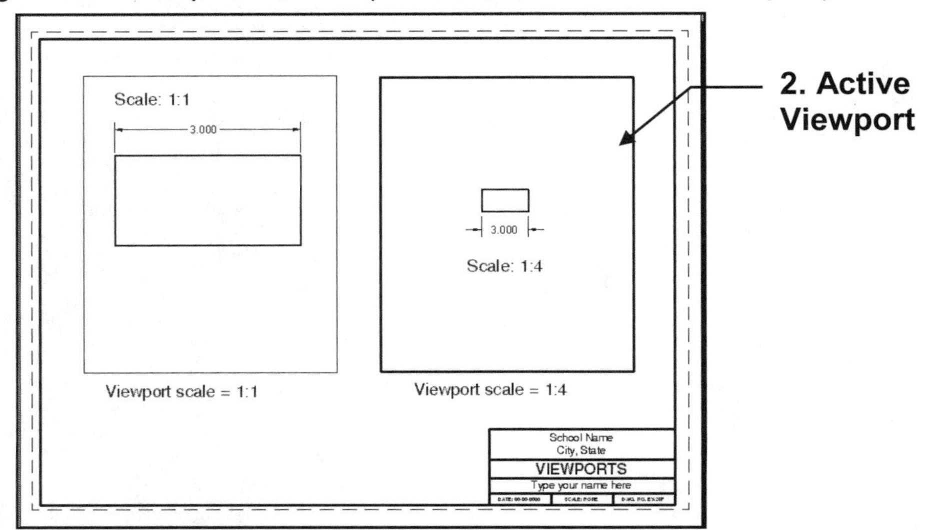

3. Erase the text and the dimension in the right hand viewport only.

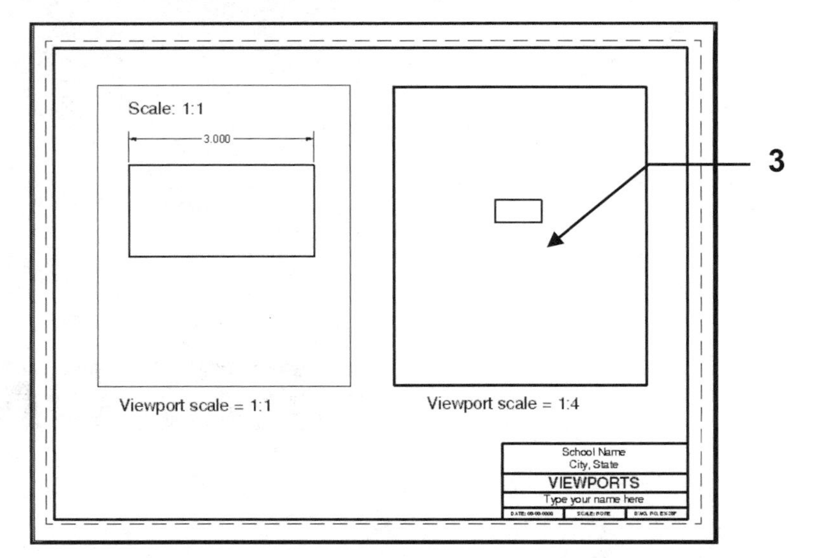

Continued on the next page...

ASSIGNING MULTIPLE ANNOTATIVE SCALES....continued

4. Display all annotative objects in all viewports as follows:
   A. Select the **Annotation Visibility** button located in the lower right corner of the drawing status bar.

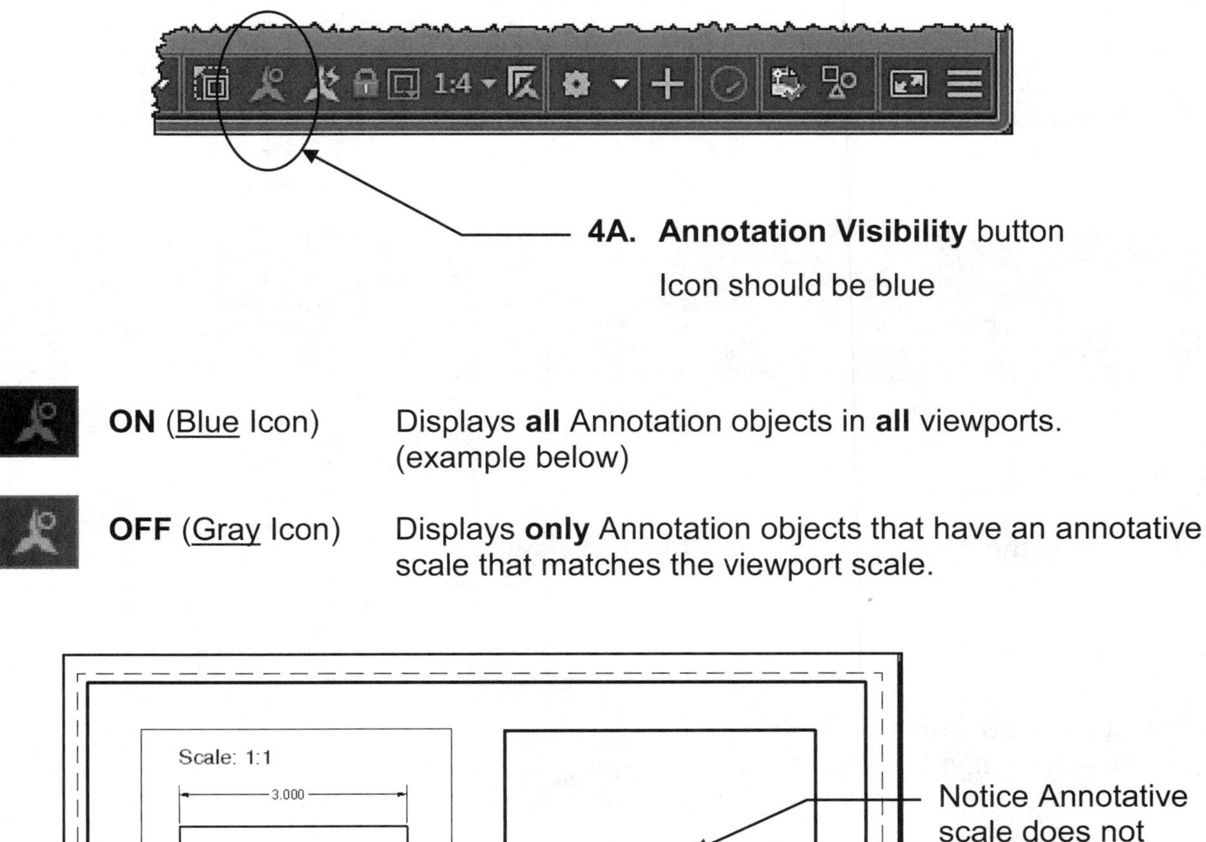

ON (<u>Blue</u> Icon) Displays **all** Annotation objects in **all** viewports. (example below)

OFF (<u>Gray</u> Icon) Displays **only** Annotation objects that have an annotative scale that matches the viewport scale.

The dimensions and text are now displayed in both viewports. But the annotative scale of the dimensions and text in the right hand viewport do not match the scale of the viewport. (Notice they are smaller) **The scale of annotative objects must match the scale of the viewport.** *<u>Follow the steps on the next page to assign multiple annotative scales to an annotative object.</u>*

Continued on the next page...

ASSIGNING MULTIPLE ANNOTATIVE SCALES....continued

5. Place your cursor near the **dimension** in the right hand viewport. Notice the <u>single</u> **Annotation symbol.** This single symbol indicates the annotative dimension has only one annotative scale assigned to it.

6. Select the **Annotate** tab.

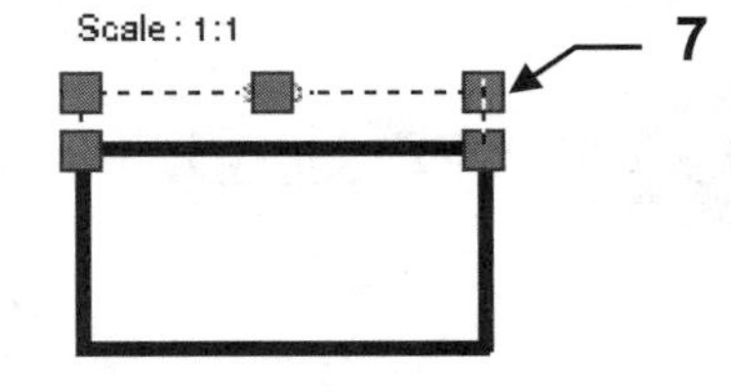

7. Select <u>only</u> the <u>dimension</u> in the <u>right-hand viewport</u>.

8. Select the **Add Current Scale** tool on the <u>Annotation Scaling</u> panel.

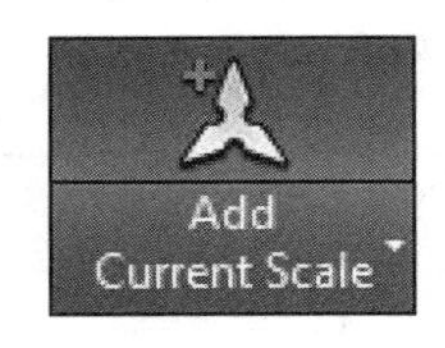

Increased in size

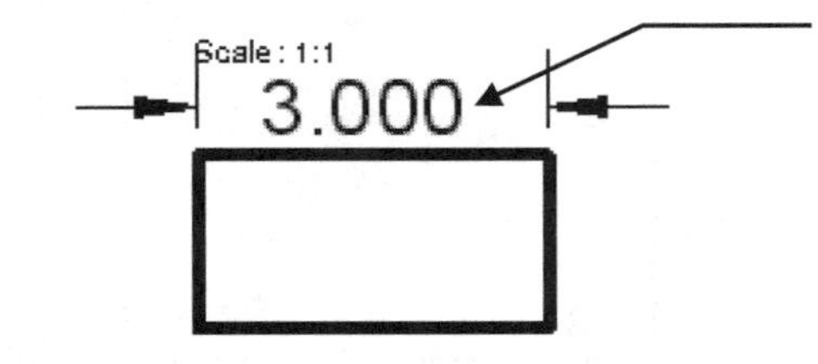

The annotative dimension should have increased in size as shown above.

9. <u>Turn OFF</u> the **Annotation Visibility.** (Click on button. The Icon should turn Gray)

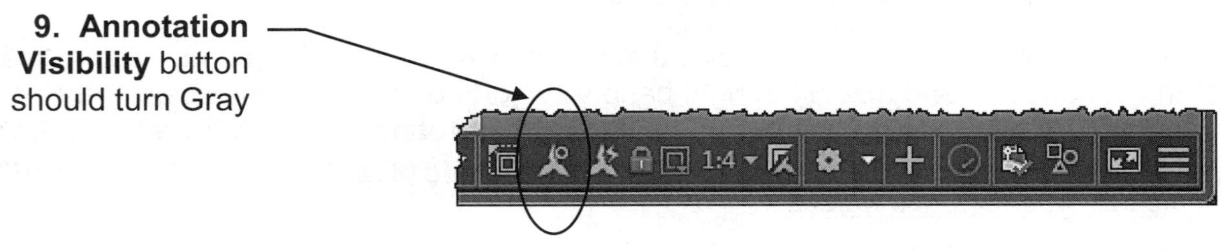

Continued on the next page...

ASSIGNING MULTIPLE ANNOTATIVE SCALES....continued

Notice the text in the right hand viewport is no longer visible. When the ***Annotative Visibility*** *is OFF only the annotative objects that match the viewport scale will remain visible. The dimension is the only annotative object that has the 1:4 annotative scale assigned to it.*

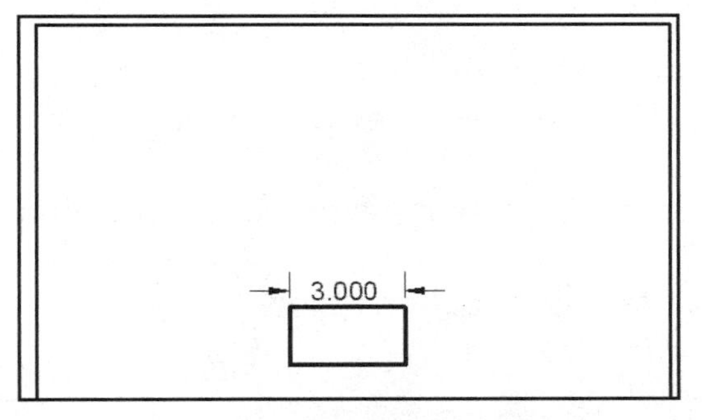

Annotative Visibility OFF

10. Place your cursor near the dimension. Notice 2 annotation symbols appear now. This indicates 2 annotation scales have been assigned to the annotative object.

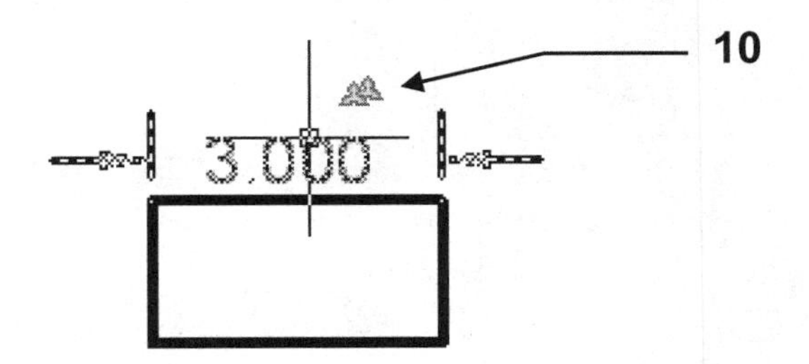

11. Click on the dimension to display the grips and drag the dimension away from the rectangle approximately as shown below. (The dimension was too close to the rectangle) <u>Notice the dimension in the left hand viewport did not move</u>. They can be moved individually.

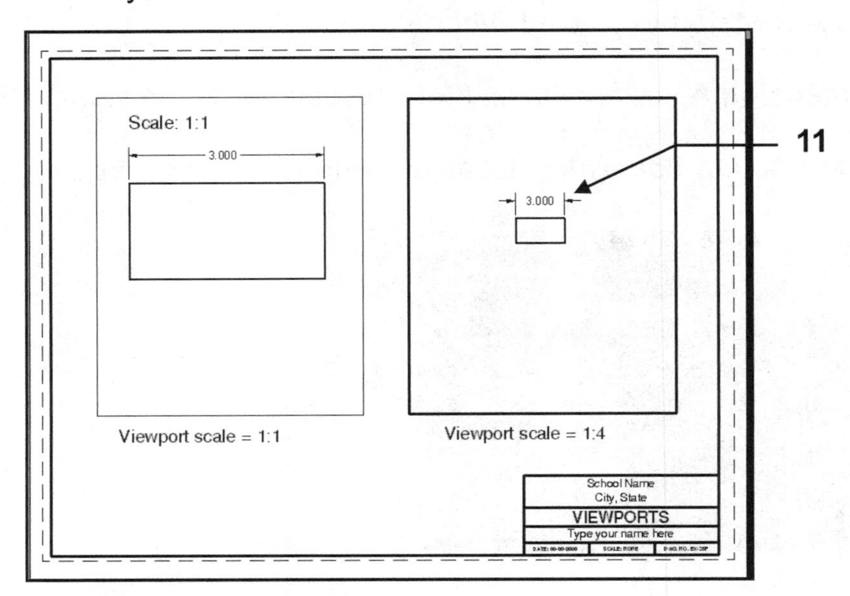

HOW TO REMOVE AN ANNOTATIVE SCALE

If you have an annotative object such as a dimension, that you would like to remove from a viewport, you must remove the annotative scale that matches the viewport scale. **Do not delete** the dimension because it will also be deleted from all of the other viewports. This sounds complicated but is very easy to accomplish.

Problem: I would like to remove **dimension A** from the right hand viewport but I do not want **dimension B** in the lower viewport to disappear.

Solution: I must remove the 1:4 annotation scale from **dimension A**.

(Refer to Step by step instructions below.)

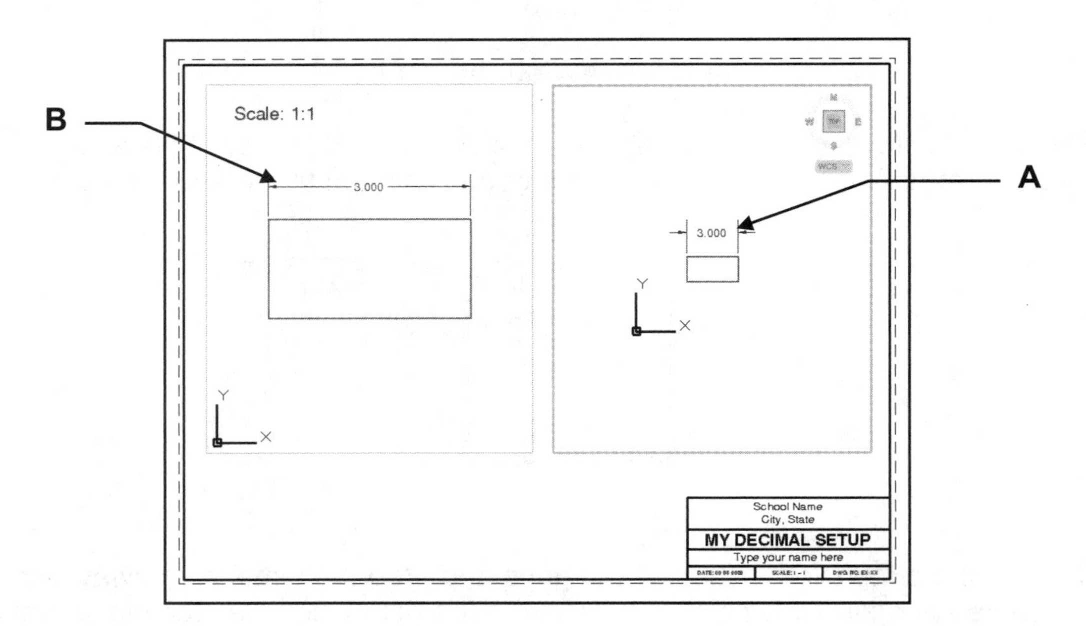

Step 1.

1. Select the **Annotate** tab on the Ribbon.

2. Select **dimension A** shown above. (You must be inside the viewport)

3. Select **Add / Delete Scales** tool located on the **Annotation Scaling** panel.

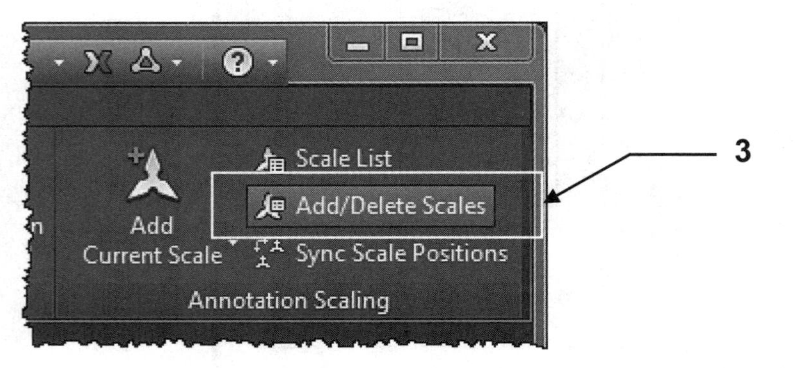

Continued on the next page...

HOW TO REMOVE AN ANNOTATIVE SCALE....continued

4. Select the annotation scale to remove. (1:4)

5. Select the **Delete** button.
   (Remember, you are deleting the annotative scale from the dimension. You are not deleting the dimension. The dimension still exists but it will not have an annotative scale of 1:4 assigned to it. As a result it will not be visible within any viewport that has been scaled to 1:4)

6. Select the **OK** button.

4

Annotation Object Scale

Object Scale List

1:1
1:4

Add...

Delete

5

1 paper unit = 4 drawing units

List all scales for selected objects

List scales common to all selected objects only

OK Cancel Help

6

Note: Dimension A has been removed and **Dimension B** remains.

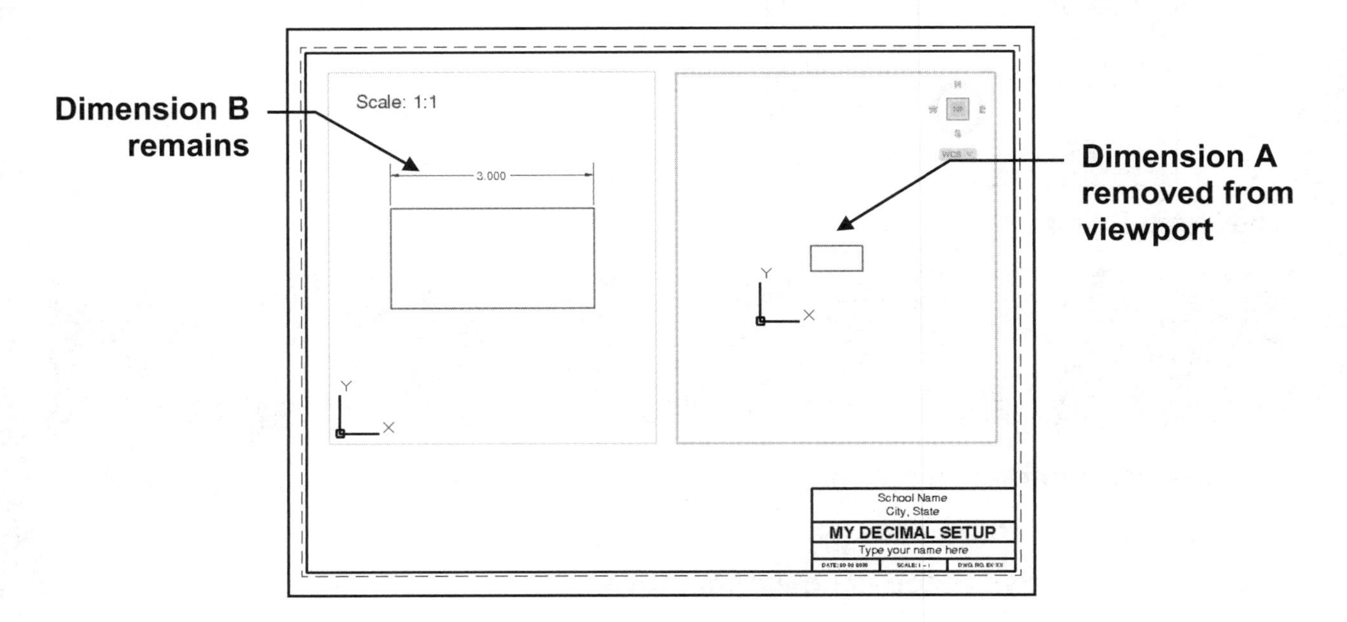

ANNOTATIVE HATCH

Hatch may be Annotative also. You may select the Annotative setting as you create the Hatch set or you may add the annotative setting to an existing Hatch set.

How to select the Annotative setting as you create the Hatch set.

1. Select the Hatch command (Refer to Lesson 15) and select the desired settings including **Annotative.** The Hatch set created will be annotative. (Blue is ON)

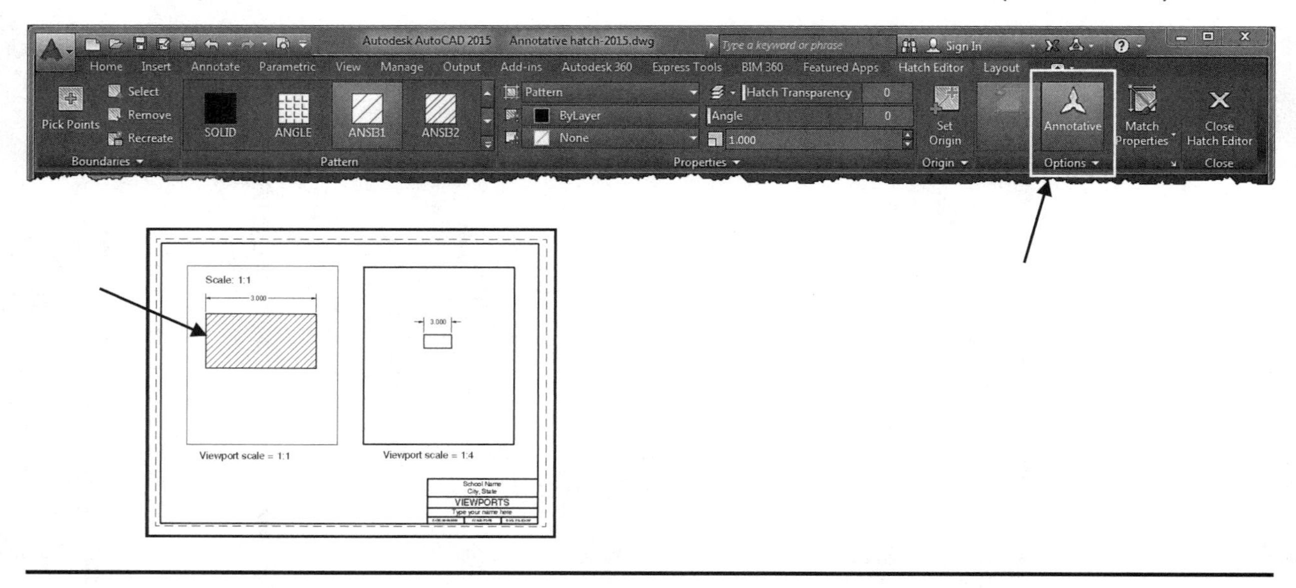

How to change an existing Non-Annotative Hatch set to Annotative.

1. Click on the existing Hatch set.

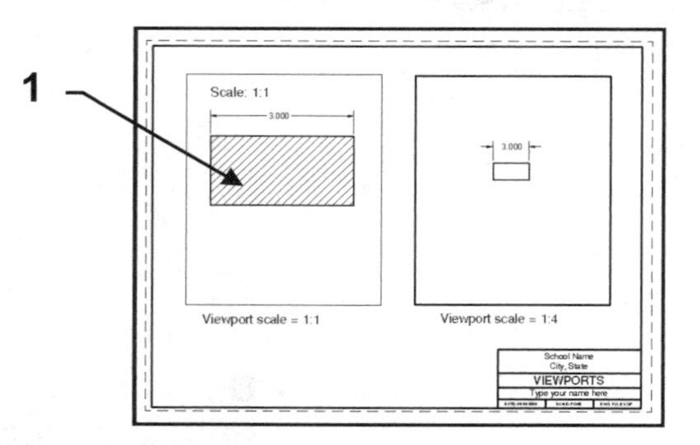

The Hatch Editor with appear.

2. Select the **Annotative** option.
3. Select **Close Hatch Editor**.

The Non-Annotative Hatch is now Annotative

ANNOTATIVE HATCH....continued

How to assign multiple Annotative scales to a Hatch set.

1. Draw the hatch in one of the viewports using **Annotative** hatch. (Refer to previous page)

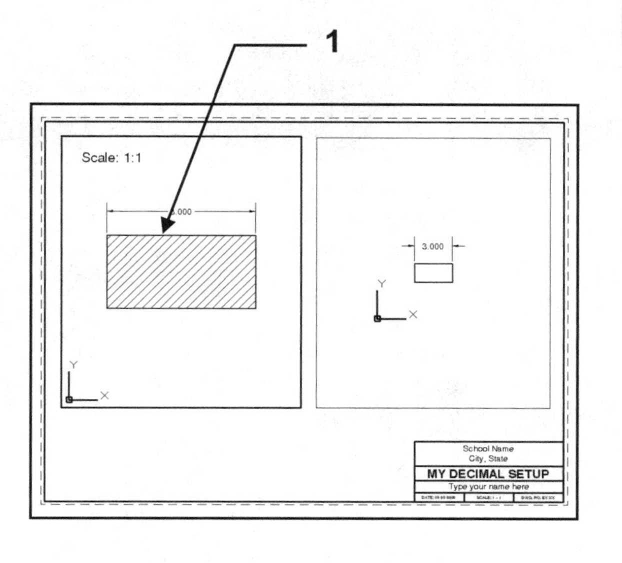

2. Turn ON **Annotation Visibility** **(Blue)** (Refer to page 28-3)

3. Select the Hatch Set to change.

Scale: 1:1

3.000

3

Viewport scale = 1:1

Viewport scale = 1:4

School Name

City, State

VIEWPORTS

Type your name here

Continued on the next page...

ANNOTATIVE HATCH....continued

4. Select the **Annotate tab**

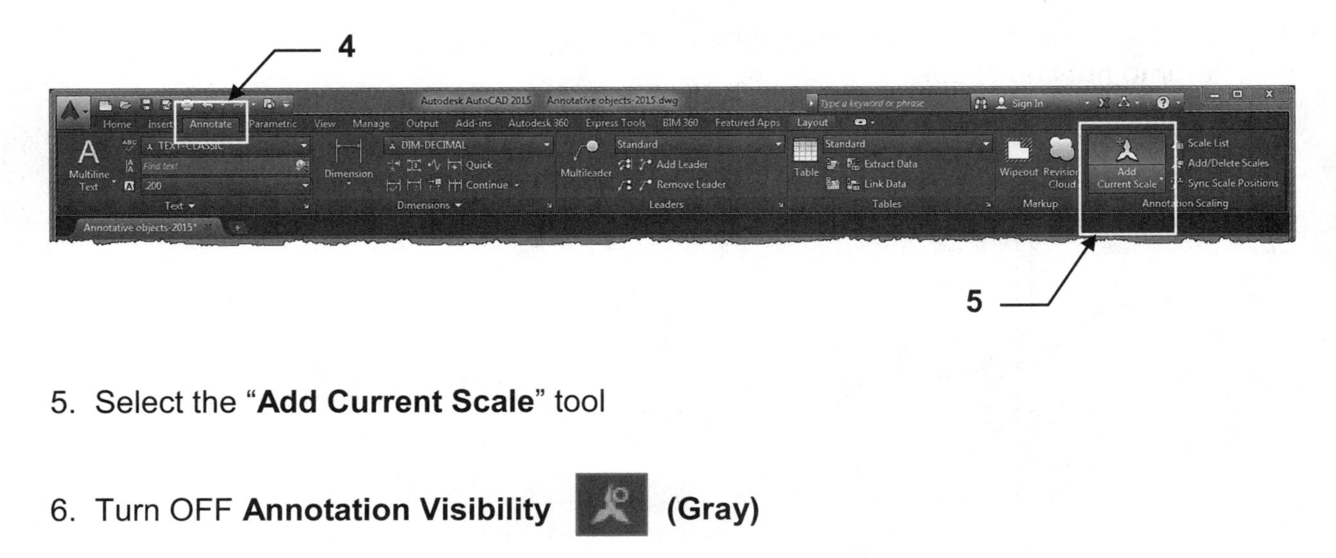

5. Select the "**Add Current Scale**" tool

6. Turn OFF **Annotation Visibility** **(Gray)**

Now the appearance of the hatch sets should be identical in both viewports.

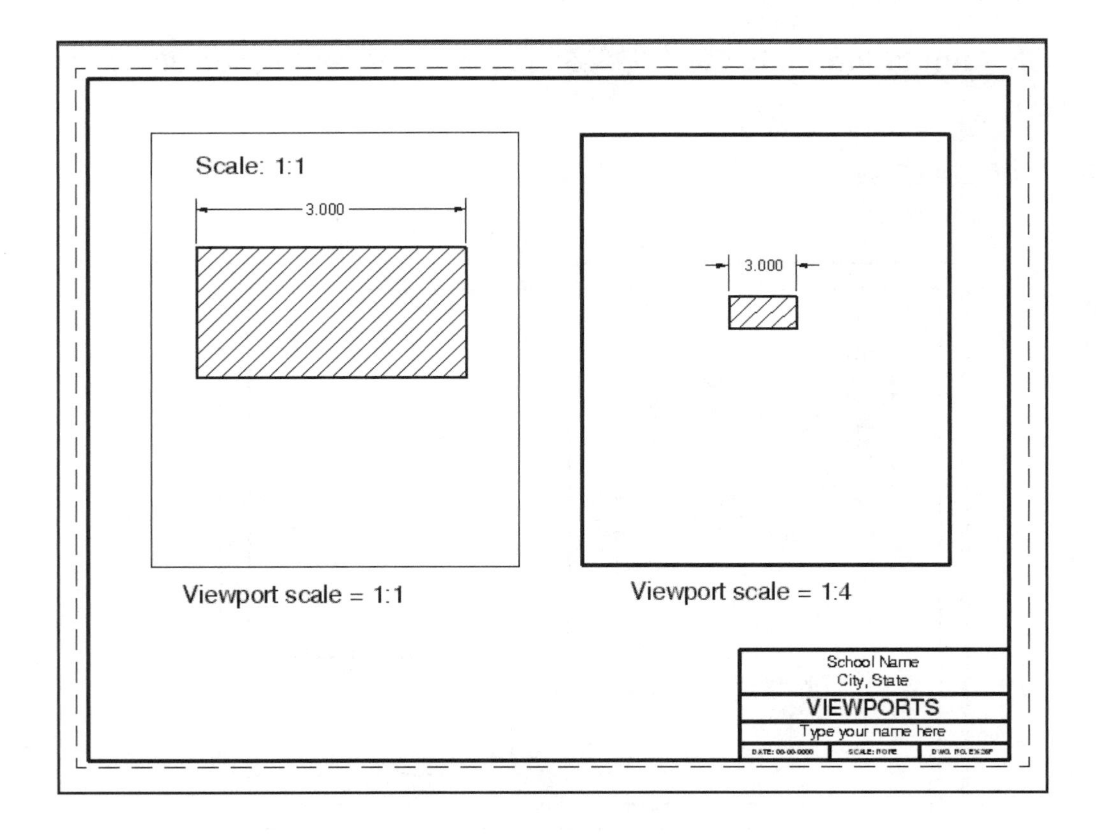

EXERCISE 28A

The following exercise will give you practice adding Annotative Object Scales to existing dimensions in order to display them within viewports with different scales. Follow the instructions below as you manipulate the drawing to appear as shown on page 28-14.

STEP 1

1. Open **EX-27G**
2. Select the **A size** tab.
3. **Erase** the 1 existing Viewport (click on it and select delete)
   (The classroom will disappear from paperspace because the hole is gone.)
4. Select the Viewport Layer and **Create** 4 new viewports approximately as shown below.

Notice each viewport is a hole in paperspace and you are looking through to modelspace. Also notice the dimension do not appear.

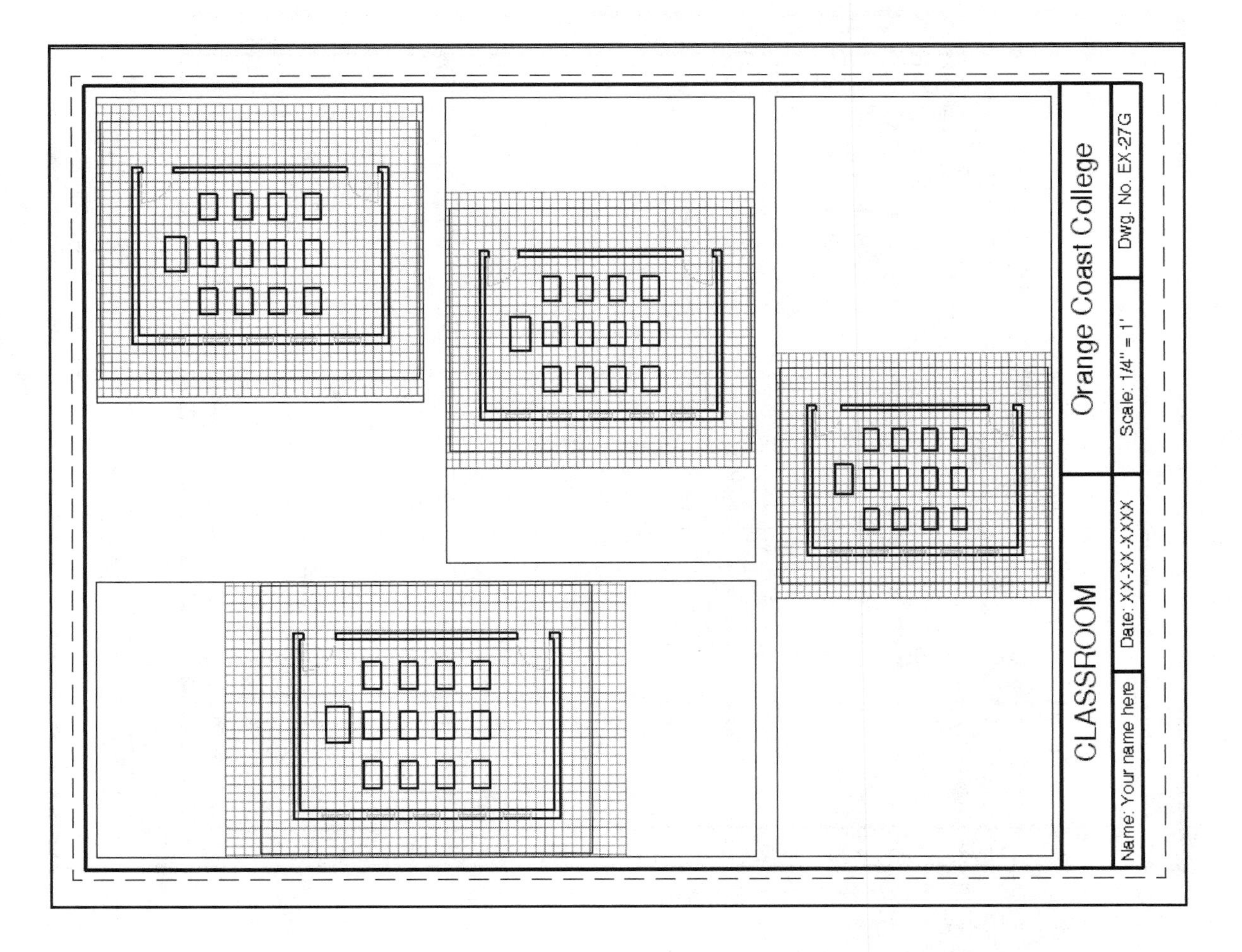

Continued on the next page...

EXERCISE 28A....CONTINUED

STEP 2

5. **Adjust the scale** of each Viewport as indicated below.

6. Use **Pan** to move the image within the viewport without changing the scale.

7. Add the **VP Scale** labels.
   Place them in Paper space. Text style: Text-Classic   1/8" height.

8. **Lock** each viewports.

Notice the annotative dimensions appear only in the 1/4" = 1' viewport.
The dimensions currently have only 1/4"=1' annotation scale.

Note: I turned the grids OFF for clarity.

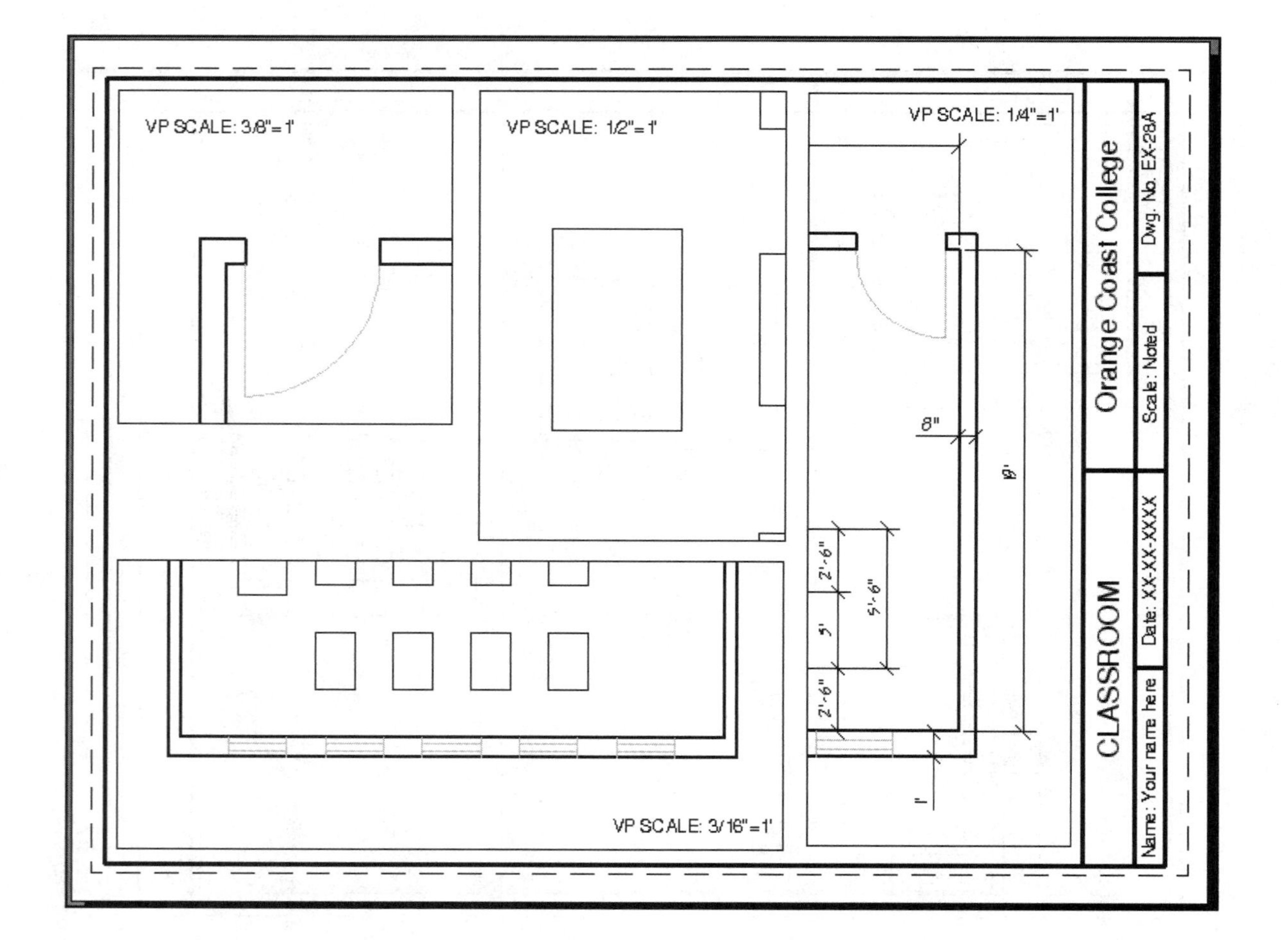

Continued on the next page...

EXERCISE 28A....CONTINUED

Step 3

1. Turn on **Annotation Visibility** 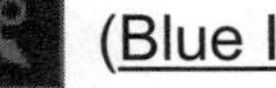 (Blue Icon Refer to 28-3)

You should see dimensions in all viewports now.
And they are all different sizes.

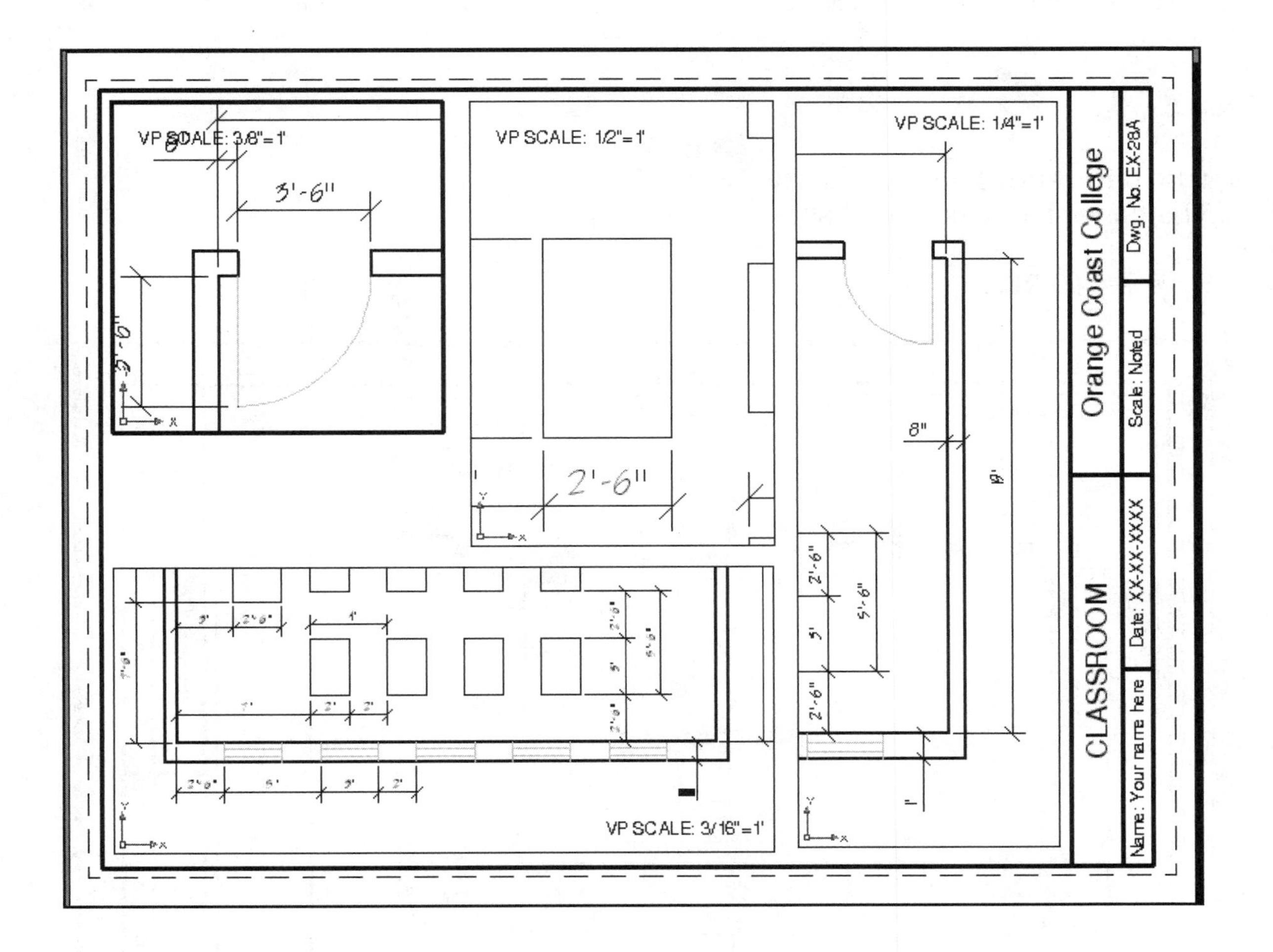

Continued on the next page...

EXERCISE 28A....CONTINUED

Step 4

1. Add **Annotative Object Scales** to the dimensions shown in the viewports below:
    A. Select a Viewport (double click inside viewport)

    B. Select the dimensions that you want to stay visible in that viewport.

    C. Select the **Annotate** panel. (Refer to 28-4)

    D. Select **Add Current Scale** button
    This will automatically <u>add</u> the <u>current viewport scale</u> to the <u>selected objects</u>.

    E. Go on to the next Viewport and repeat A thru D above until they are all done.

Step 5

1. Turn OFF **Annotative Visibility** (Gray Icon)
   The unwanted dimensions should have disappeared.

2. Save as: **EX28A**

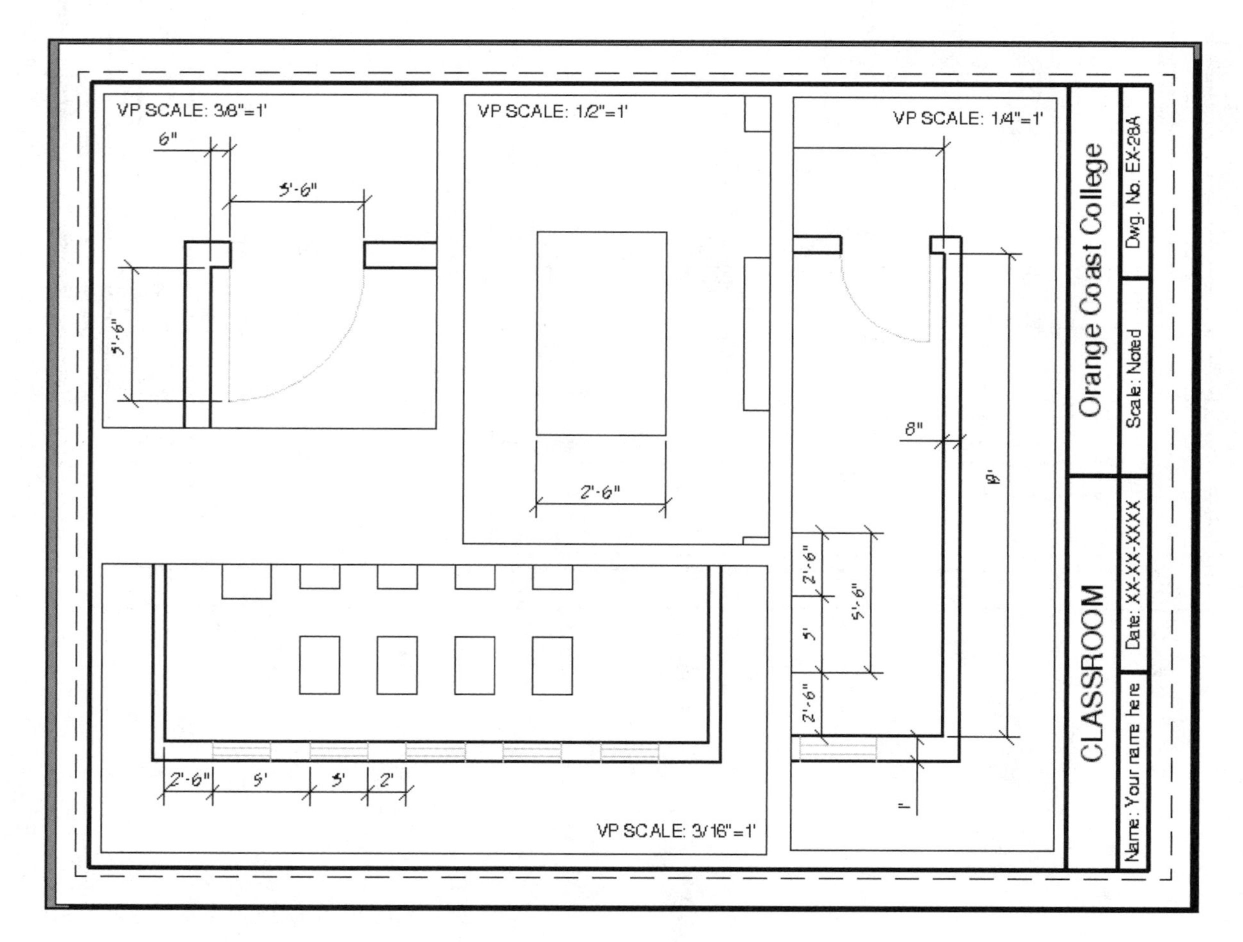

EXERCISE 28B

The following exercise will give you practice with Hatch within scaled viewports using **Annotative Object Scale** option. Follow the instructions below while you manipulate the drawing to appear as shown.

STEP 1

1. Start a NEW file using: **My Decimal Setup.dwt**
2. Select **A size** layout tab
3. **Erase** the 1 existing Viewport.
4. Select the Viewport Layer and **Create 3 new viewports** approximately as shown.
5. **Adjust the scale** of each Viewport as indicated.
6. **Lock** each viewport.
7. Activate the 1 = 1 viewport. (Double click inside the viewport)
8. **Draw** a 2" X 2" Rectangle inside the 1 = 1 Viewport as shown using Layer Object. (The rectangle will appear in the other 2 Viewports because all viewports display model space.)

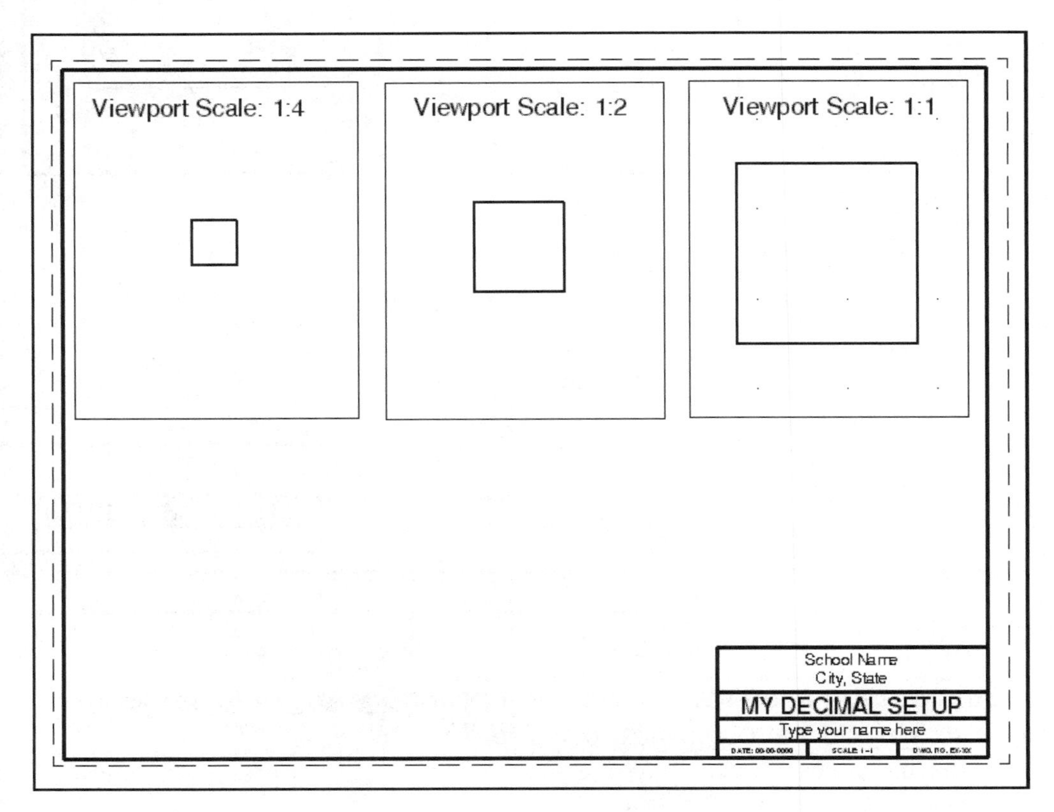

Continued on the next page...

EXERCISE 28B....CONTINUED

STEP 2

1. Select the layer **Hatch**

2. **Hatch** the 1 = 1 Rectangle using:
   A. Pattern **Ansi 31**
   B. Angle **0**
   C. Scale **1.000**
   D. Select the **<u>Annotative</u>** option.

3. Place Hatch inside the rectangle as shown and press <enter>

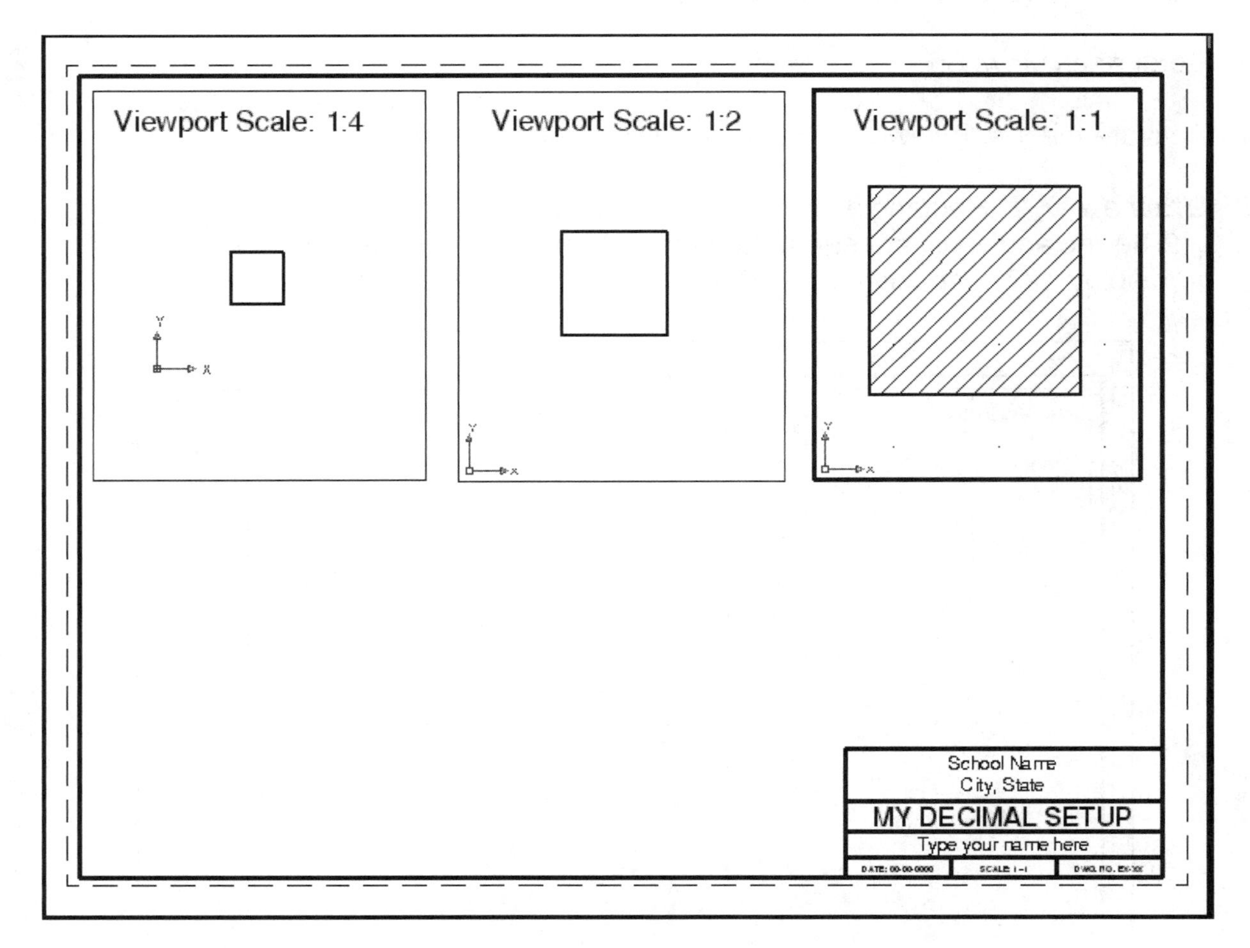

Note:
If a Hatch Set appears in all Rectangles you probably forgot to select the Annotative box <u>or</u> you have Annotation visibility ON.

Continued on the next page...

EXERCISE 28B....CONTINUED

STEP 3

1. Turn ON **Annotation Visibility**  (Blue Icon Refer to 28-3)

The Hatch set appears in all of the viewports.
But the sizes are all different.

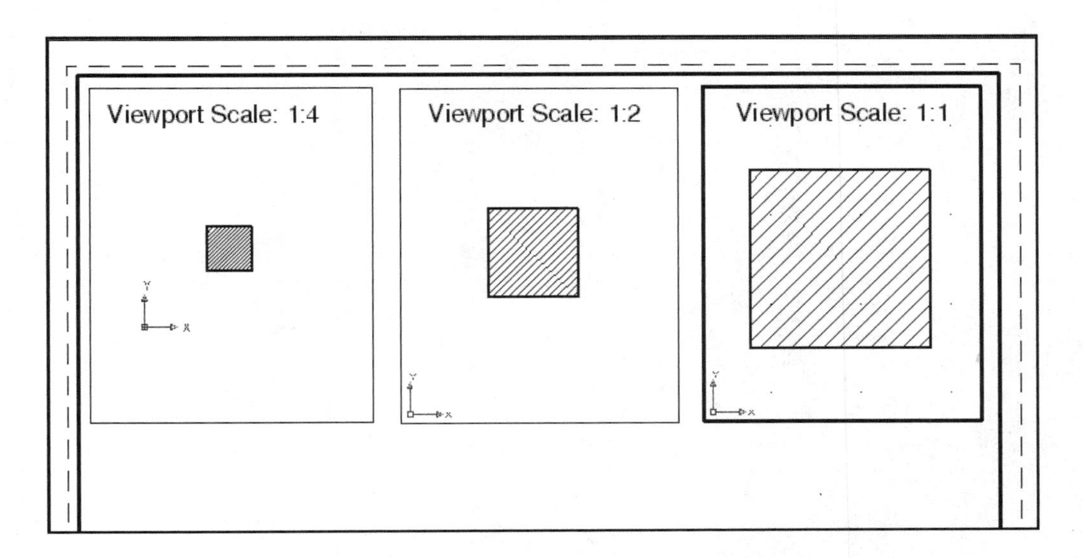

2. Assign the correct Annotative scale to the hatch sets in viewport 1:4 and 1:2.

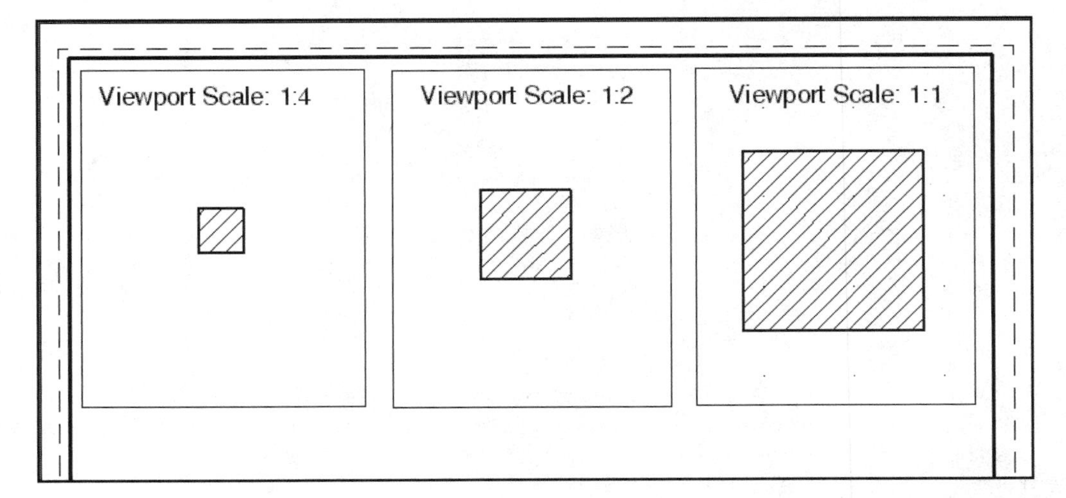

3. Turn OFF **Annotation Visibility**  (Gray Icon Refer to 28-3)

4. Save as: **EX28B**

Notes:

LEARNING OBJECTIVES

After completing this lesson, you will be able to:

1. Understand Blocks
2. Create a Block
3. Insert a Block
4. Re-define and Purge unused Blocks
5. Create Mulitleaders with blocks attached
6. Use the Collect multileader tool.

LESSON 29

BLOCKS

A **BLOCK** is a group of objects that have been converted into ONE object. A Symbol, such as a transistor, bathroom fixture, window, screw or tree, is a typical application for the block command. First a BLOCK must be created. Then it can be INSERTED into the drawing. An inserted Block uses less file space than a set of objects copied.

CREATING A BLOCK

1. First draw the objects that will be converted into a Block.

   For this example a circle and 2 lines are drawn.

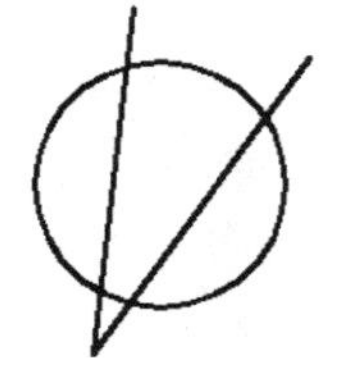

2. Select the **CREATE BLOCK** command using one of the following:

   Ribbon = Insert tab / Block Definition panel /
   or
   Keyboard = B <enter>

3. Enter the New Block name in the **Name** box.

3
7
Do not select
4
6

Block Definition
Name:
Type new block name here
Base point
Specify On-screen
Pick point
X: 0.0000
Y: 0.0000
Z: 0.0000
Objects
Specify On-screen
Select objects
Retain
Convert to block
Delete
No objects selected
Behavior
Annotative
Match block orientation to layout
Scale uniformly
Allow exploding
Settings
Block unit:
Inches
Hyperlink...
Description
Open in block editor
OK
Cancel
Help

Continued on the next page...

BLOCKS....continued

4. Select the **Pick Point** button. (Or you may type the X, Y and Z coordinates.)
   The Block Definition box will disappear and you will return temporarily to the drawing.

5. Select the location where you would like the insertion point for the Block.
   Later when you insert this block, the block will appear on the screen attached to the cursor at this insertion point. Usually this point is the CENTER, MIDPOINT or ENDPOINT of an object.

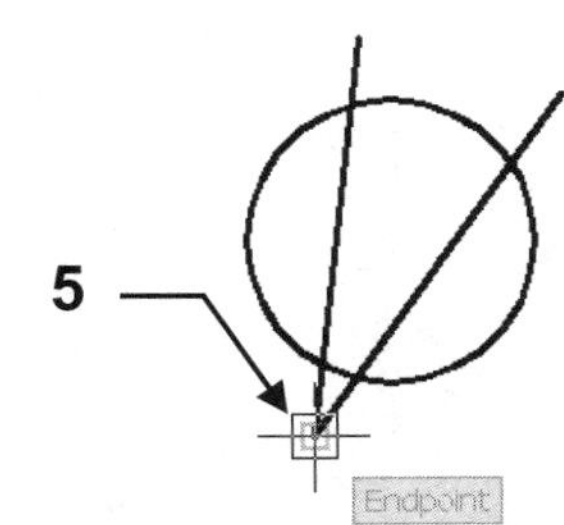

Notice the coordinates for the base point are now displayed. (Don't worry about this. Use Pick Point and you will be fine)

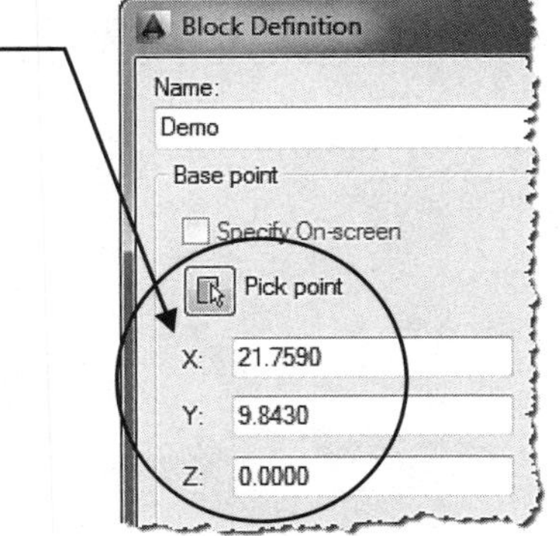

6. Select one of the options described below.

   It is important that you select one and understand the options below.

   Retain
   If this option is selected, the original objects will stay visible on the screen after the block has been created.

   Convert to block
   If this option is selected, the original objects will disappear after the block has been created, but will immediately reappear as a block. It happens so fast you won't even notice the original objects disappeared.

   Delete
   If this option is selected, the original objects will disappear from the screen after the block has been created. (This is the one I use most of the time)

7. Select the **Select Objects** button.

   The Block Definition box will disappear and you will return temporarily to the drawing.

8. Select the objects you want in the block, then press <enter>.

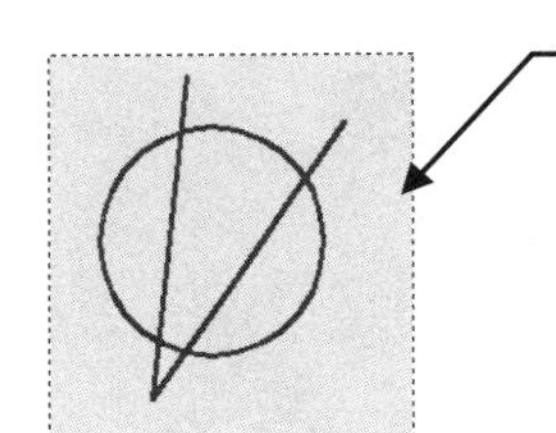

Selection Window

Continued on the next page...

BLOCKS....continued

The Block Definition box will reappear and the objects you selected should be displayed in the Preview Icon area.

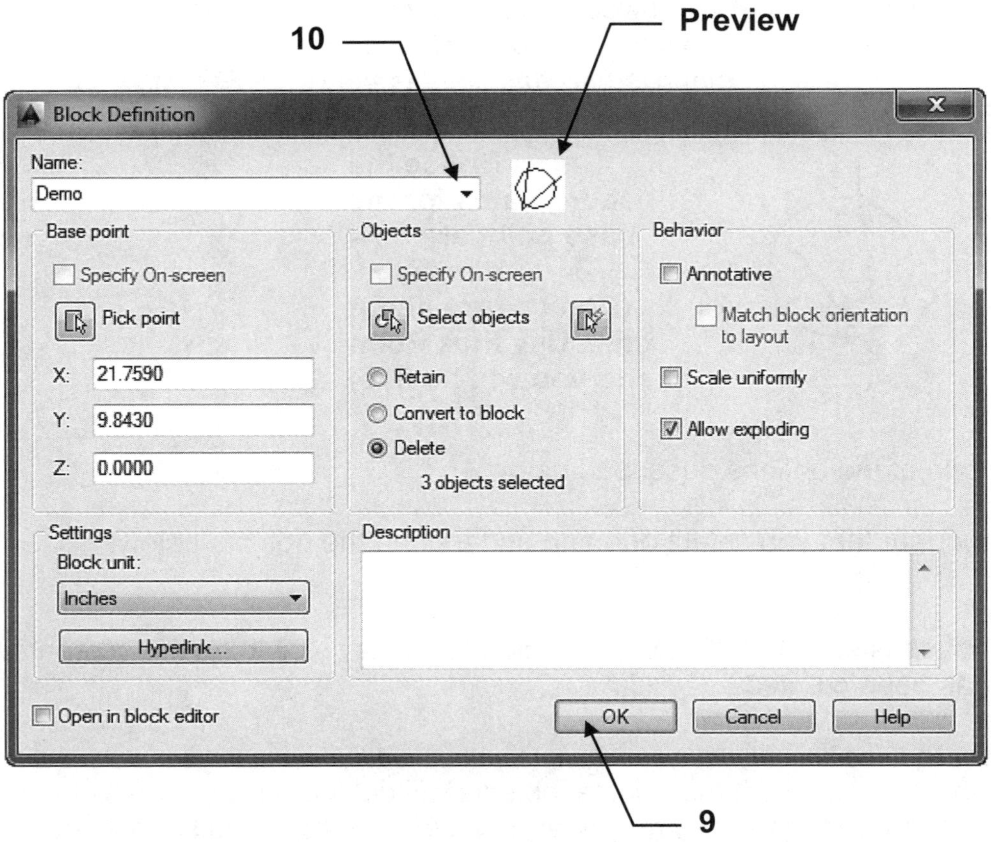

9. Select the **OK** button.
   The new block is now stored in the drawing's block definition table.

10. To verify the creation of this Block, select **Create Block** again, and select the Name (▾). A list of all the blocks, in this drawing, will appear.
    (Refer to page 29-6 for inserting instructions)

Continued on the next page...

BLOCKS....continued

ADDITIONAL DEFINITIONS OF OPTIONS

Block Units
You may define the units of measurement for the block. This option is used with the "Design Center" to drag and drop with Autoscaling. The Design Center is an advanced option and is not discussed in this book.

Hyperlink
Opens the **insert Hyperlink dialog box** which you can use to associate a hyperlink with the block.

Description
You may enter a text description of the block.

Scale Uniformly
Specifies whether or not the block is prevented from being scaled non-uniformly during insertion.

Allow Exploding
Specifies whether or not the block can be exploded after insertion.

HOW LAYERS AFFECT BLOCKS

If a block is created on Layer 0:

1. When the block is inserted, it will take on the properties of the current layer.
2. The inserted block will reside on the layer that was current at the time of insertion.
3. If you Freeze or turn Off the layer the block was inserted onto, the block will disappear.
4. If the Block is **Exploded**, the objects included in the block will revert to their original properties of layer 0.

If a block is created on Specific layers:

1. When the block is inserted, it will retain its own properties. It **will not** take on the properties of the current layer.
2. The inserted block **will reside** on the current layer at the time of insertion.
3. If you **freeze** the layer that was current at the time of insertion the block will disappear.
4. If you turn **off** the layer that was current at the time of insertion the block will not disappear.
5. If you **freeze** or turn **off** the blocks original layers the block will disappear.
6. If the Block is **Exploded**, the objects included in the block will go back to their original layer.

INSERTING BLOCKS

A **BLOCK** can be inserted at any location within the drawing. When inserting a Block you can **SCALE** or **ROTATE** it.

1. Select the **INSERT** command using one of the following:

   Ribbon = Insert tab / Block panel /
   or
   Keyboard = Insert <enter>

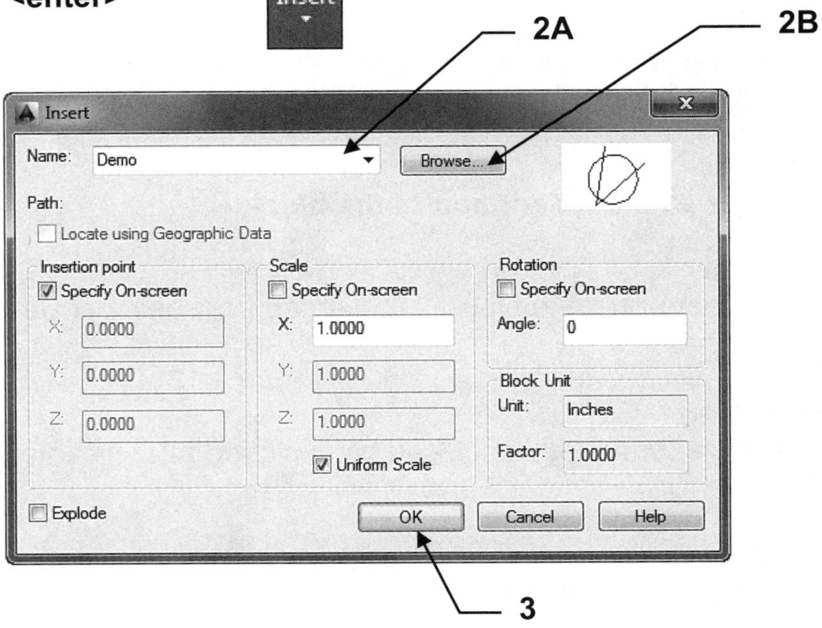

2A

2B

3

2. Select the **BLOCK** name.
   a. If the block is already in the drawing that is open on the screen, you may select the block from the drop down list shown above
   b. If you want to insert an entire drawing, select the Browse button to locate the drawing file.

3. Select the **OK** button.

   This returns you to the drawing and the selected block should be attached to the cursor.

4. Select the insertion location for the block by moving the cursor and pressing the left mouse button or typing coordinates.

   Command: _insert
   Specify insertion point or **[Basepoint/Scale/Rotate]:**

NOTE: If you want to change the **basepoint, scale** or **rotate** the block before you actually place the block, press the right hand mouse button and you may select an option from the menu or select an option from the command line menu shown above.

You may also "**preset**" the insertion point, scale or rotation. This is discussed on the next page.

Continued on the next page...

INSERTING BLOCKS....continued

PRESETTING THE INSERTION POINT, SCALE or ROTATION

You may preset the **Insertion point, Scale or Rotation** in the INSERT box instead of at the command line.

1. Remove the check mark from any of the **"Specify On-screen"** boxes.
2. Fill in the appropriate information describe below:

Insertion point
Type the X and Y coordinates from the Origin. The Z is for 3D only.
The example below indicates the block's insertion location will be 5 inches in the X direction and 3 inches in the Y direction, from the Origin.

Scale
You may scale the block proportionately by typing the scale factor in the X box and then check the Uniform Scale box. If you selected "Scale uniformly" box when creating the block this option is unnecessary and not available.
If the block is to be scaled non-proportionately, type the different scale factors in both X and Y boxes.
The example below indicates that the block will be scale proportionate at a factor of 2.

Rotation
Type the desired rotation angle relative to its current rotation angle.
The example below indicates the block will be rotated 45 degrees from its originally created angle orientation.

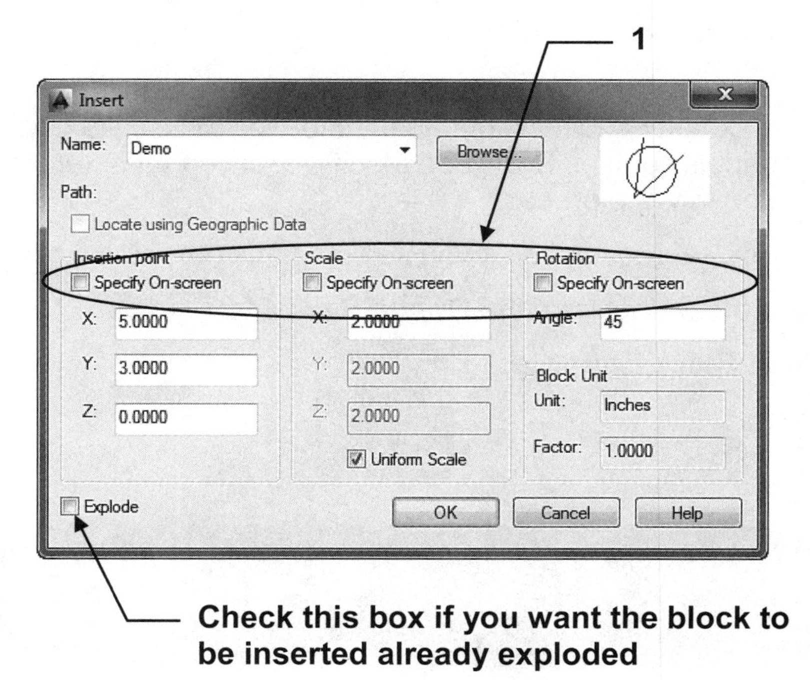

Check this box if you want the block to be inserted already exploded

RE-DEFINING A BLOCK

How to change the design of a block previously inserted.

1. Select **Block Editor** using one of the following:

Ribbon = Insert tab / Block Definition panel /
or
Keyboard = bedit

Block Editor

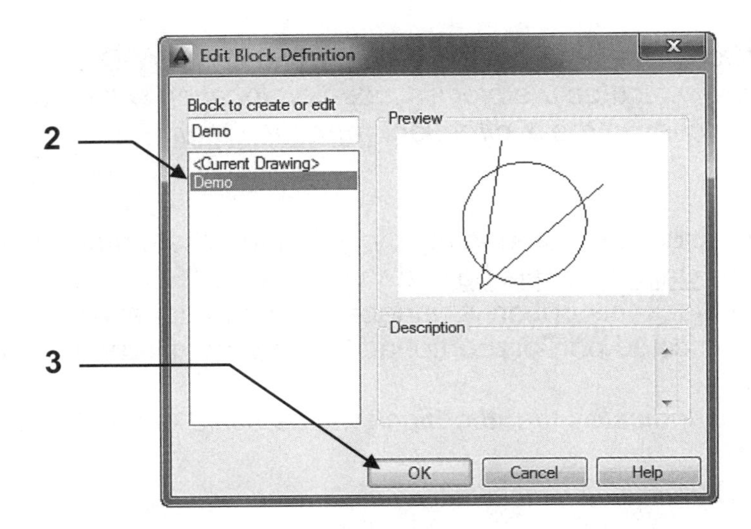

*The "**Edit Block Definition**" dialog box will appear.*

2. Select the name of the Block that you wish to change.
3. Select the **OK** button.
4. The Block that you selected should appear large on the screen.
   You may now make any additions or changes to the block. You can change tabs and use other panels such as Draw and Modify. But you must return to the **Block Editor** tab to complete the process.
5. Return to the **Block Editor** button if you selected any other tab while editing.
6. Select the **Save block** tool from the **Open/Save** panel.
7. Select the **Close Block Editor** tool.

Close Block Editor

Save Block

6

7

You will be returned to the drawing and **all previously inserted** blocks with the **same name** will be updated with the changes that you made.

PURGE UNWANTED and UNUSED BLOCKS

You can remove a block reference from your drawing by erasing it; however, the block definition remains in the drawing's block definition table. To remove any **unused** blocks, dimension styles, text styles, layers and linetypes you may use the **Purge** command.

How to delete unwanted and unused blocks from the current drawing.

1. Select the **Purge** command using one of the following:

   Ribbon = None
   or
   Application Menu = Drawing Utilities / Purge
   or
   Keyboard = Purge <enter>

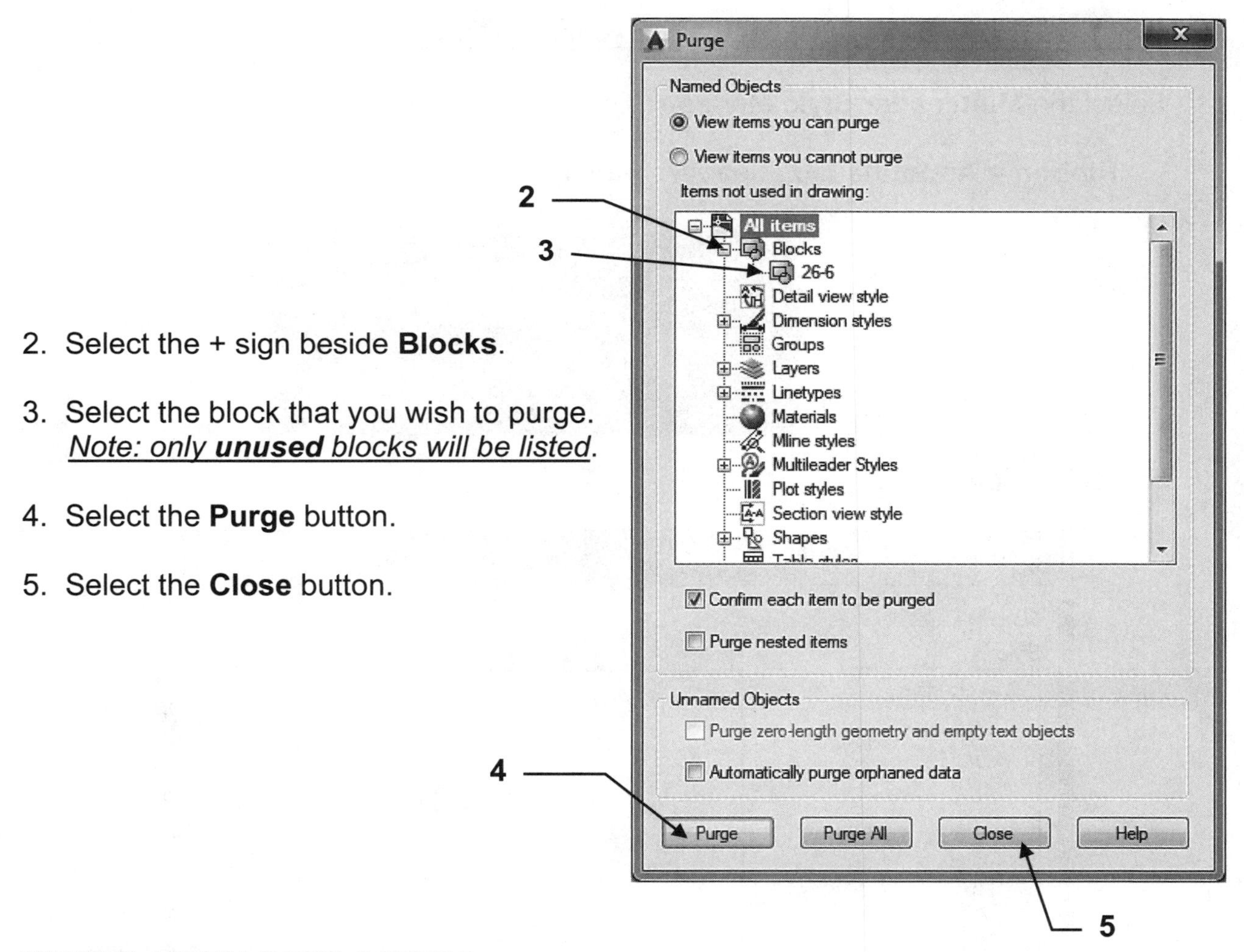

2. Select the + sign beside **Blocks**.

3. Select the block that you wish to purge.
   <u>*Note: only **unused** blocks will be listed*</u>.

4. Select the **Purge** button.

5. Select the **Close** button.

WHERE ARE BLOCKS SAVED?

When you create a Block it is saved **<u>within the drawing you created it in</u>**.

(If you open another drawing you will not find that block.)

MULTILEADER and BLOCKS

In Lesson 19 you learned about Multileaders and how easy and helpful they are to use. (*I saved this command for this lesson because it works best with Blocks attached to the Leader.)* In this lesson you will learn another user option within the Multileader Style Manager that allows you to attach a **pre-designed Block** to the landing end of the Leader. To accomplish this, you must first create the style and then you may use it.

Here are a few examples of multileaders with pre-designed blocks attached:

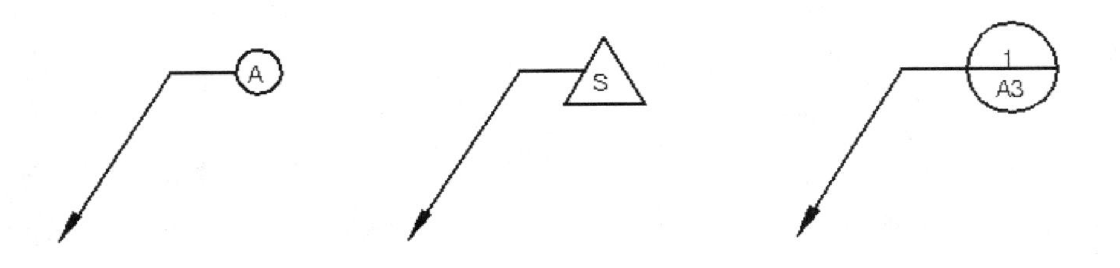

STEP 1. CREATE A NEW MULTILEADER STYLE

1. Select the **Multileader Style Manager** tool using

    Ribbon = Annotate tab / Leaders panel / ↘

Standard
Multileader
Add Leader
Remove Leader
Leaders
1

2. Select the **New** button.

Multileader Style Manager
Current multileader style: Standard
Styles:
Annotative
Standard
Preview of: Standard
Default Text
Set Current
New...
Modify...
Delete
List:
All Styles
Close
Help
2

MULTILEADER and BLOCKS....continued

3. Enter **New Style Name**
4. Select a style to **Start with**:
5. Select **Annotative** box
6. Select **Continue** button.

Create New Multileader Style
New style name: BALLOON
Start with: Standard
Annotative
Continue
Cancel
Help
3
4
5
6

7. Select the **Content** tab.
8. **Multileader Type**: Select **Block** from drop down menu.
9. **Source Block**: Select **Circle** from the drop down menu.
   Note: you have many choices here. These are AutoCAD pre-designed blocks <u>with Attributes.</u>

10. **Attachment**: Select **Center Extents**.
    Note: this selection works best with Circle but you will be given <u>different choices</u> depending on which <u>Source block you select.</u>
11. **Color**: Select **Bylayer**
    Bylayer works best. It means, it acquires the color of the current layer setting.
12. **Scale:** Select 1.000
13. Select **OK** button.

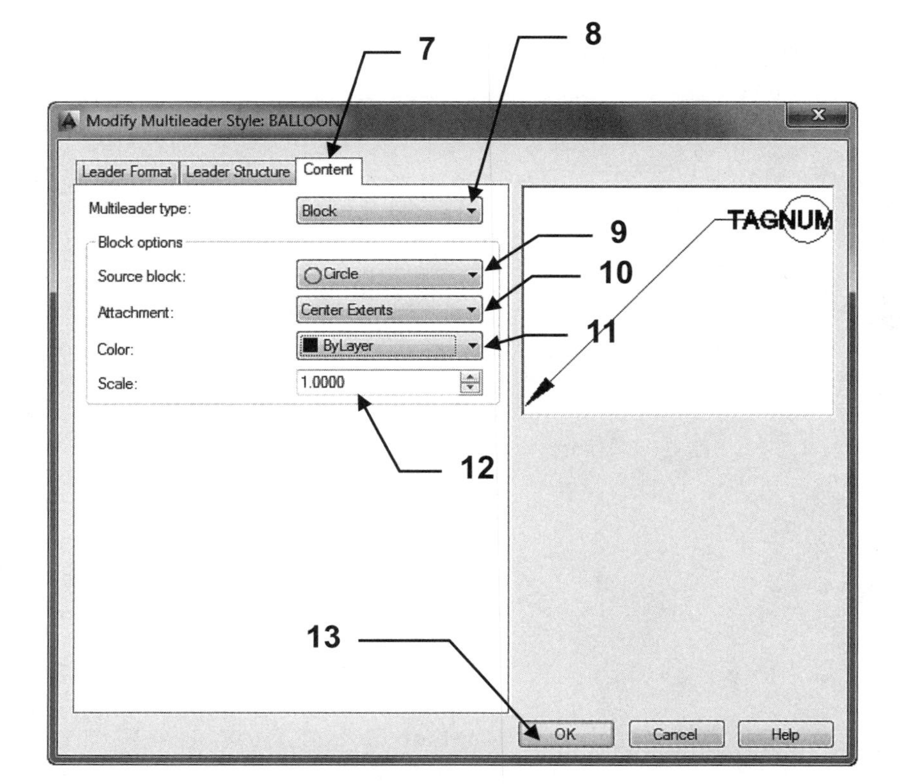

Your new multileader style should be displayed in the Styles list.

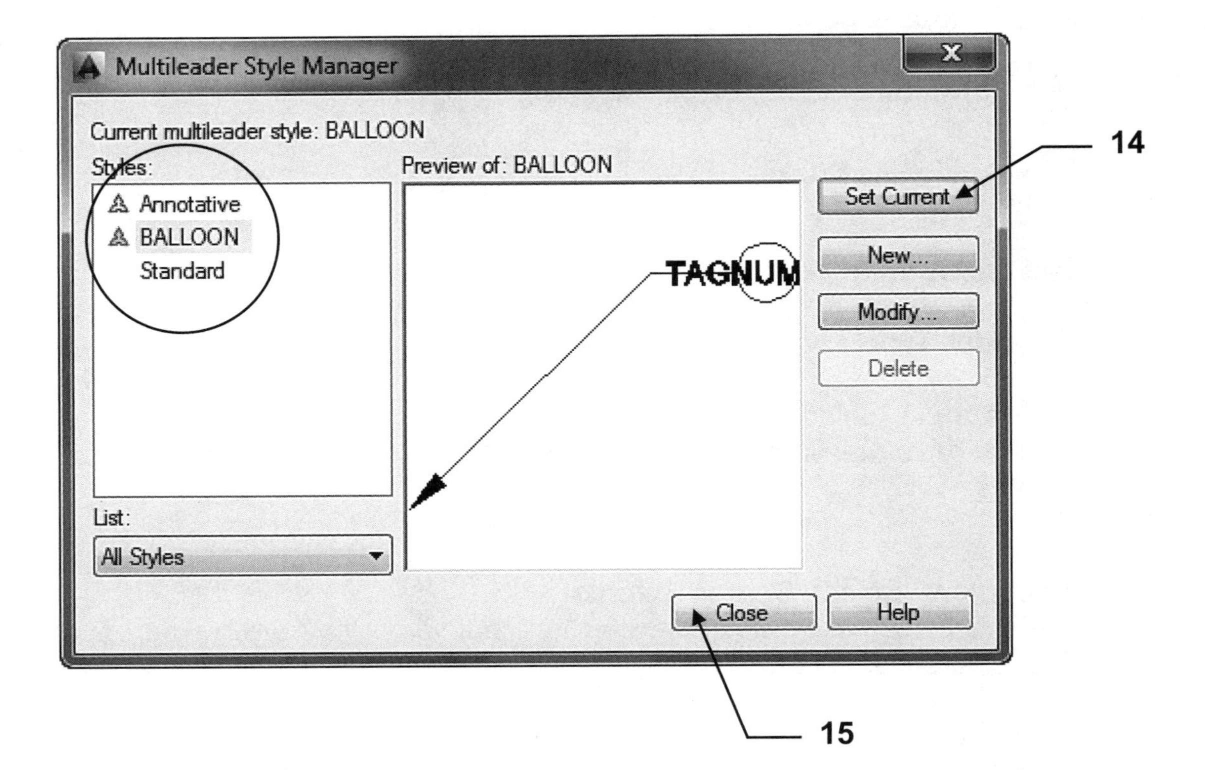

14. Select **Set Current**

15. Select **Close** button.

MULTILEADER and BLOCKS....continued

STEP 2. USING THE MULITLEADER WITH A BLOCK STYLE

1. Select the **Annotate tab / Leaders panel**

2. Select the **Style** from the Multileader drop down list

3
BALLOON
Multileader
Add Leader
Remove Leader
Leaders
2

3. Select the **Multileader** tool.

4. The **Select Annotation Scale** box may appear. Select **OK** for now.

Select Annotation Scale
You are creating an annotative object. Set the annotation scale to the scale at which the annotation is intended to display.
1:1
Don't show me this again
OK
4

5. Specify leader arrowhead location or [leader Landing first/Content first/Options] <Options>: ***place the desired location of the arrowhead***

6. Specify leader landing location: ***place the desired location of the landing***

The next step is where the pre-assigned "Attributes" activate.
Refer to the Advanced Workbook for Attributes.

7. Enter attribute values
   Enter tag number <TAGNUMBER>: ***type number or letter <enter>***

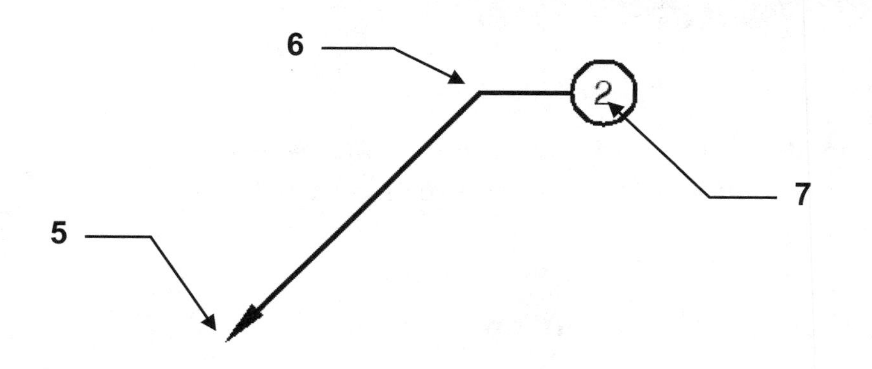

COLLECT MULTILEADER

In lesson 19 you learned how to ADD, REMOVE and ALIGN multileaders. Now you will learn how to use the **COLLECT** multileader tool.

If you have multiple Leaders pointing to the same location or object you may wish to COLLECT them into one Leader.

Example:

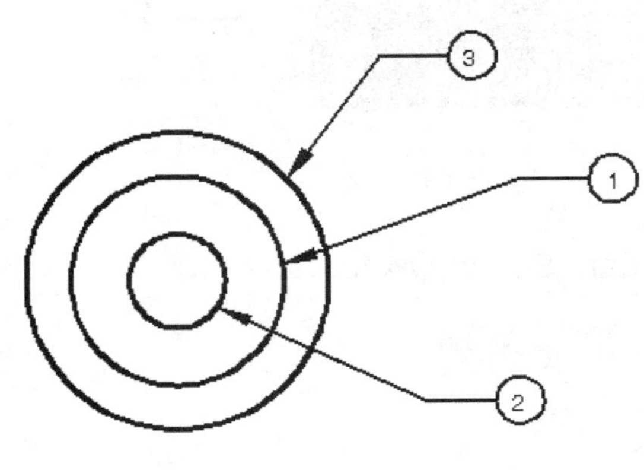

1 2 3

BEFORE **AFTER COLLECT**

HOW TO USE COLLECT MULTILEADER

1. Select the Collect Mutileader tool

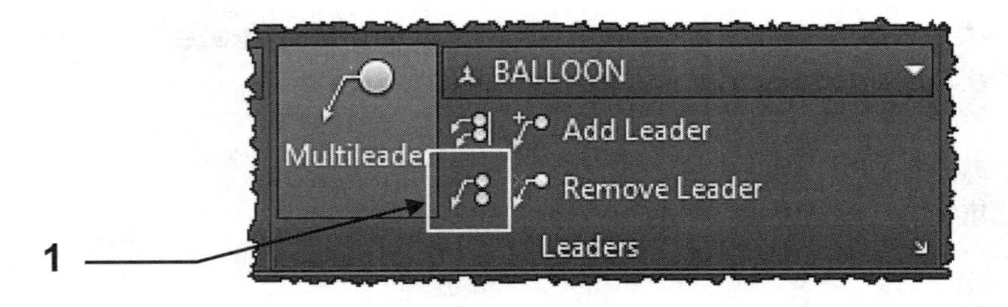

2. Select the Multileaders that you wish to combine then press <enter>.
   Note: Select them one at a time <u>in the order you wish them to display</u>.

   Such as (1, 2, 3, A, B, C, etc)

3. Place the combined Leader location. (**Orthomode** should be **OFF**)

EXERCISE 29A

STEP 1

1. Start a **NEW** file using **My Decimal Setup.dwt.**
2. Select the **Model** tab
3. Draw the objects shown below approximately as shown. Use Layer Symbol.
4. Do not dimension
5. Create a block of each.
   A. **Important:** Don't forget to select a "**Base Point**" somewhere on each object.
   B. Select "**Delete**" so they disappear as they are made.

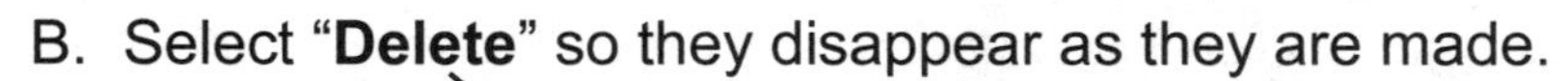

   C. Do not include the name when selecting the objects for each block.
6. Save as **EX-29A**

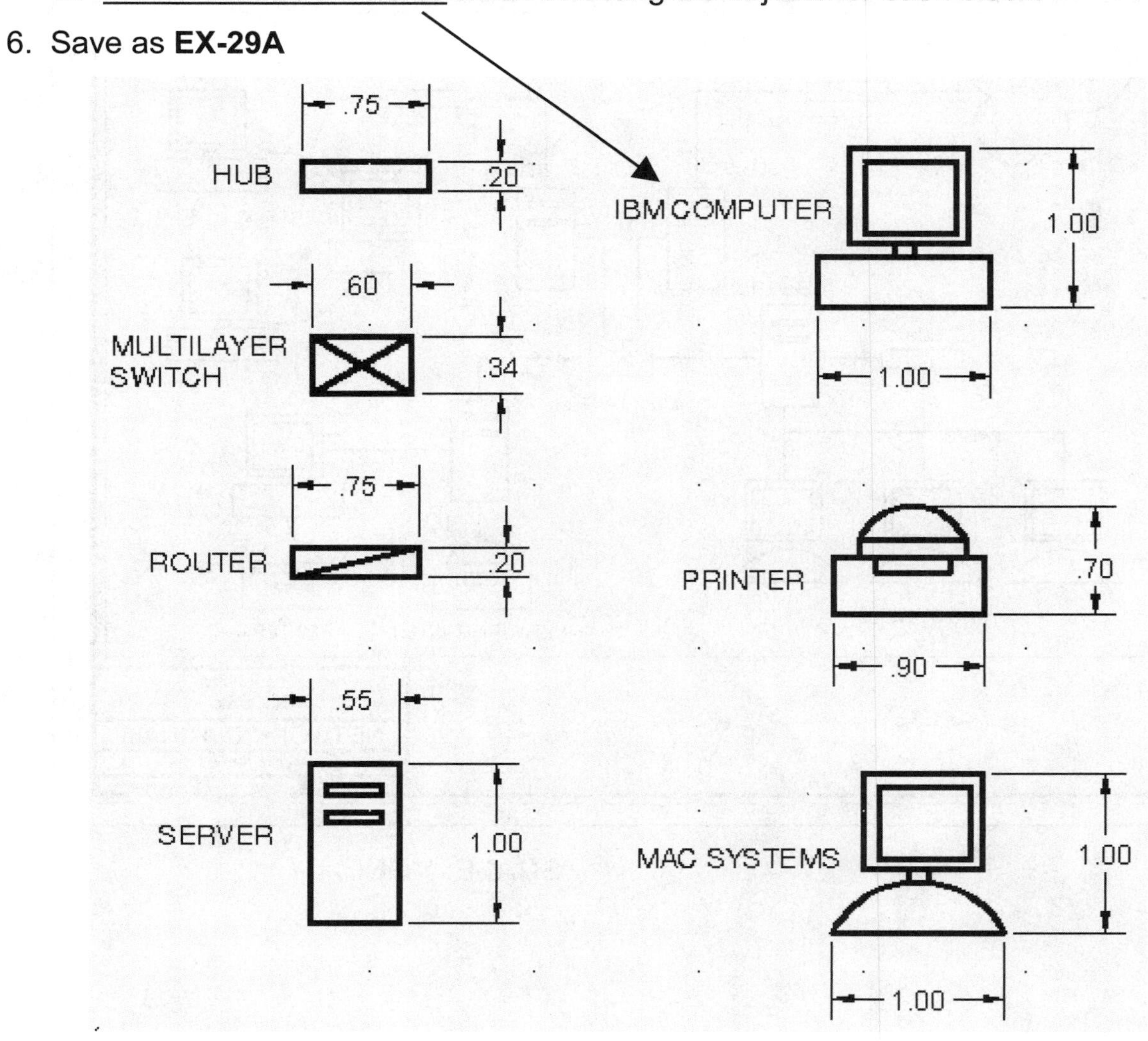

EXERCISE 29A....continued

STEP 2

1. Open **EX-29A** (If not already open)
2. Select the **A Size** tab and confirm **Model** is displayed.
3. Draw the Network Diagram approximately as shown within Model space.
4. Insert the Blocks you previously created in Step 1.
5. Add the Labels use: Text Style: Text-Classic Height: .125
6. Edit the Title Block
7. **Save** the drawing as: **EX-29A** again
8. **Plot** using page setup **Plot Setup A**

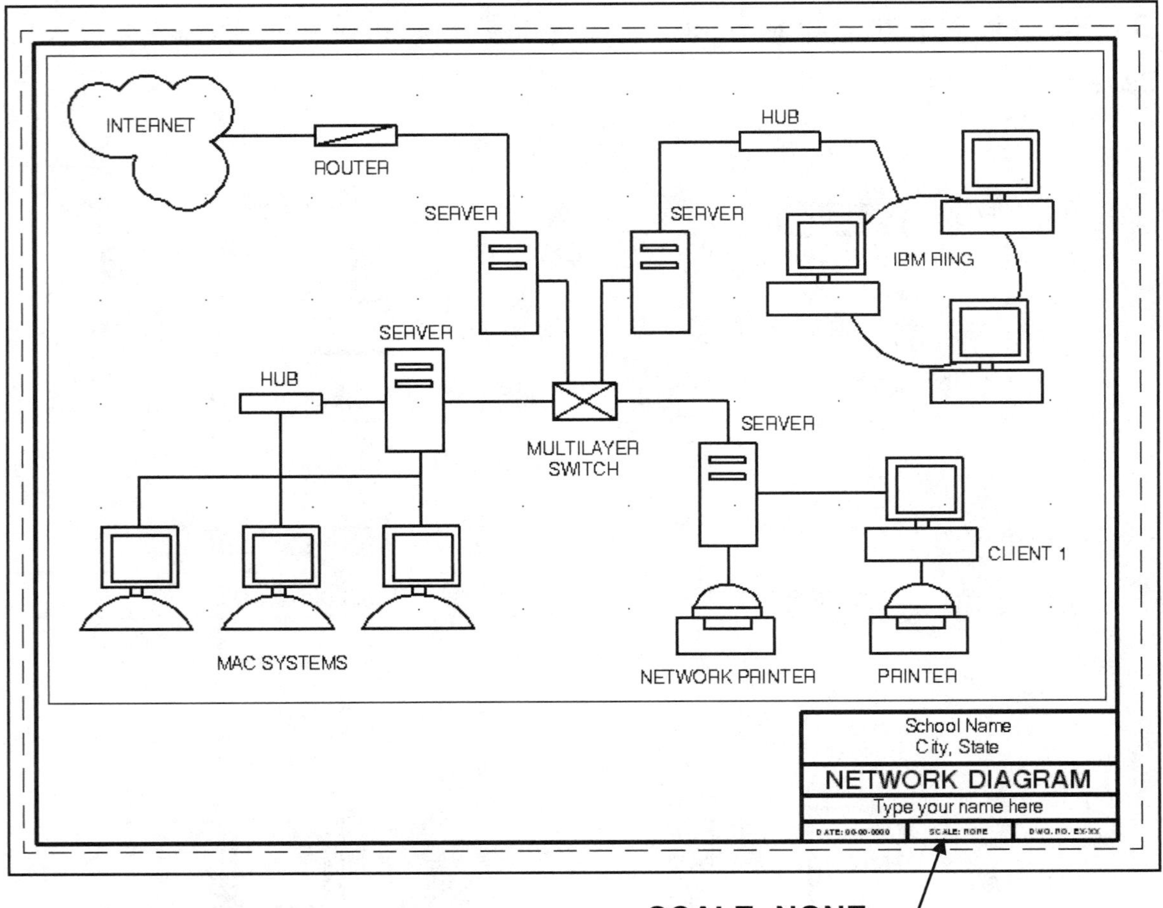

EXERCISE 29B

STEP 1

1. Start a **NEW** file using **My Feet-Inches Setup.dwt.**
2. Select the **Model** tab
3. Draw the objects shown below approximately as shown. **Use Layers specified**.
4. Do not dimension
5. Create a block for each.
   A. **Important:** Don't forget to select a "**Base Point**" on each object.
   B. Select "**Delete**" so they disappear as they are made.
   C. Use the Numbers for name. Do not include the number when selecting the objects for the block.
6. Save as **EX-29B**

Note: Your blocks will appear much smaller and thinner than the blocks shown below. These have been enlarged for clarity.

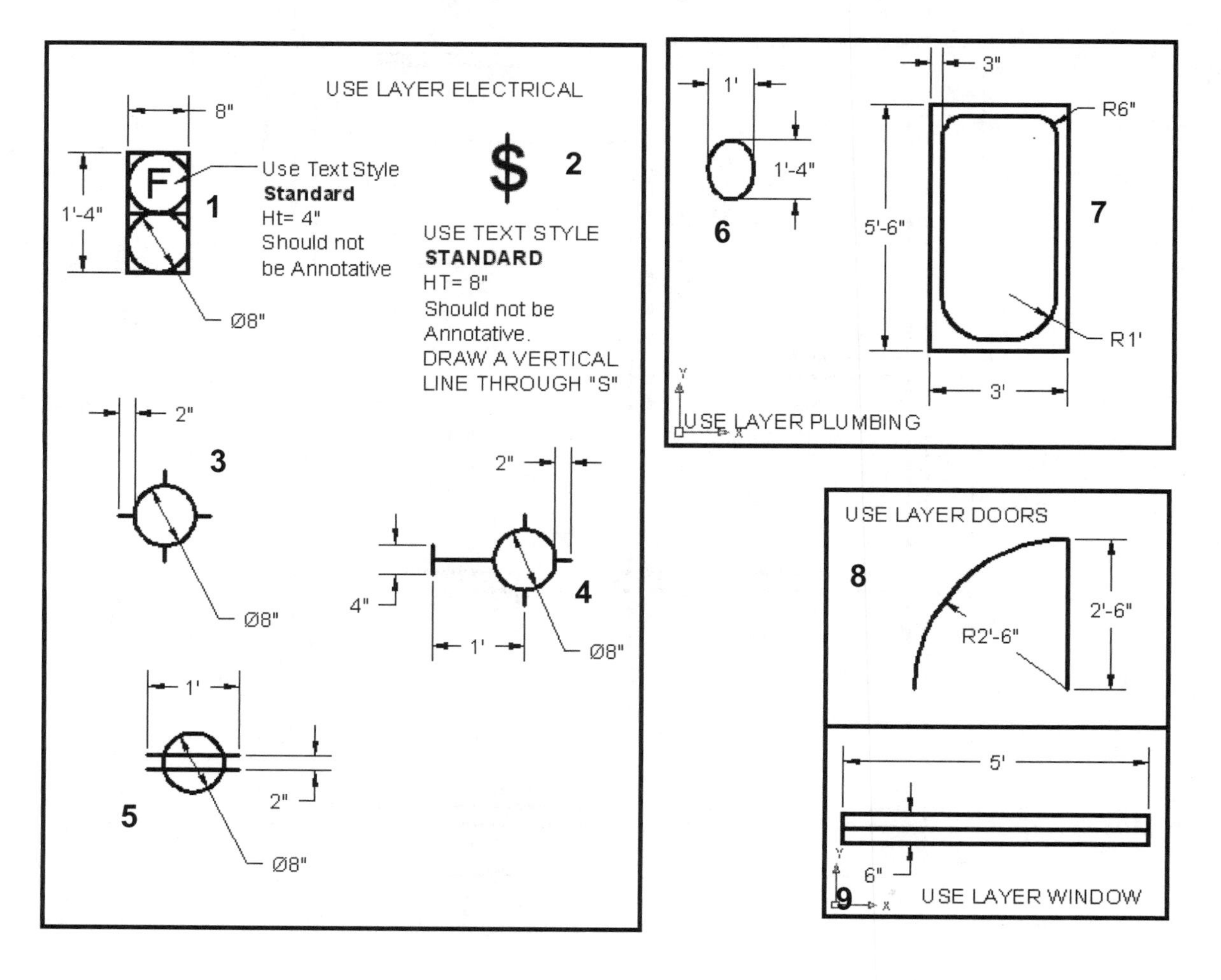

EXERCISE 29B....continued

STEP 2

1. Open **EX-29B** (If not already open)
2. Select the **A Size** tab
3. Draw the Floor plan approximately as shown below inside the viewport.
   (Note: the viewport scale should already be adjusted to 1/4" = 1')
   A. The walls are 6" wide.
   B. The space behind the door is 4" and counter depth is 24".
4. Insert the (8) DOOR Blocks on Layer Door
5. Insert the (9) WINDOW Blocks on Layer Window.
6. Dimension as shown using Dimension Style **Dim-Arch** (Your dimension text will appear different. I used text style Text-Classic because it is easier for you to read)
7. **Save** the drawing as: **EX-29B** again

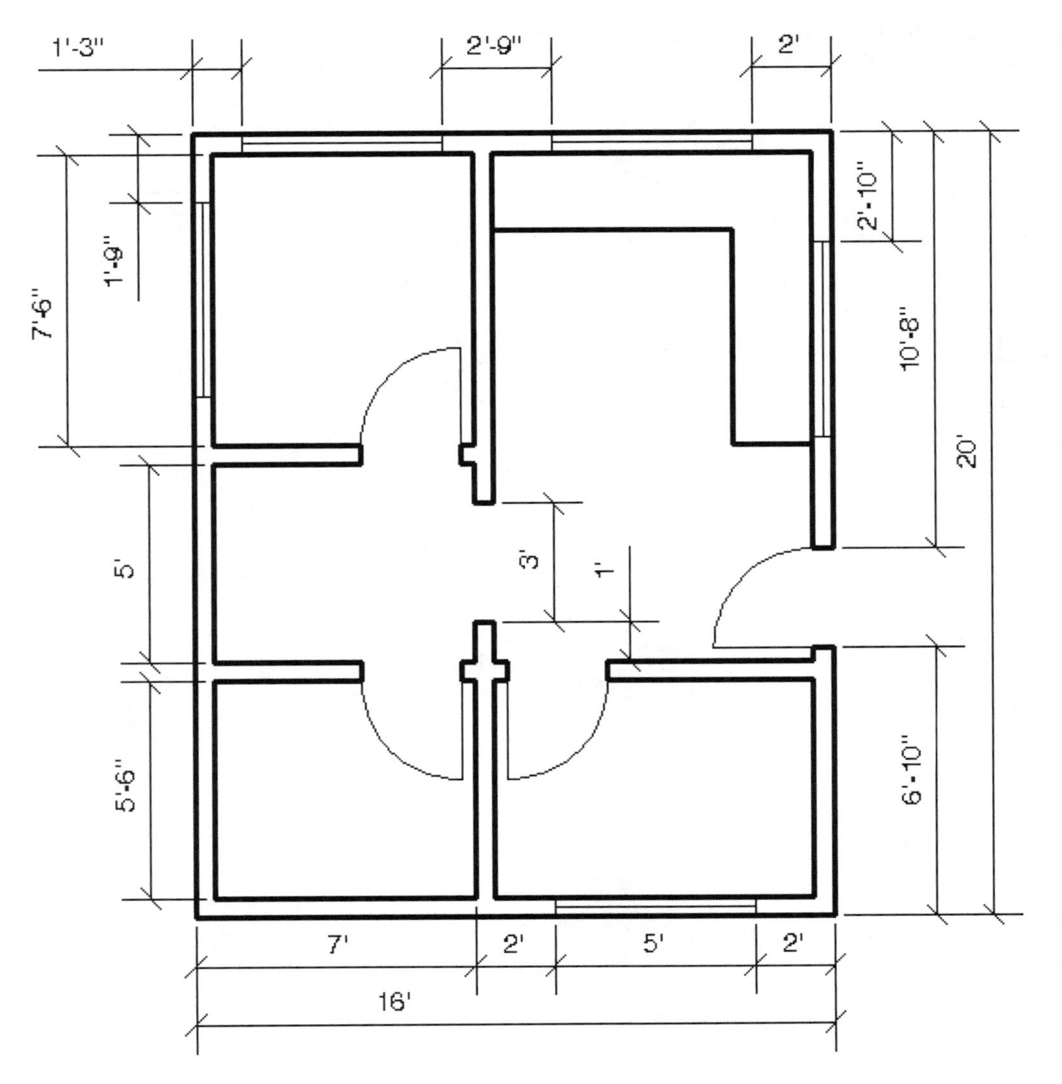

EXERCISE 29B....continued

STEP 3

1. Open **EX-29B** (If not already open)
2. Add the Electrical Blocks (1, 2, 3, 4 and 5) on Layer Electrical.
3. Add the Plumbing Blocks (6 and 7) on Layer Plumbing.
4. **Save** the drawing as: **EX-29B** again

Note: The dimension layer has been temporarily turned off so you can see the electrical and plumbing blocks easier.

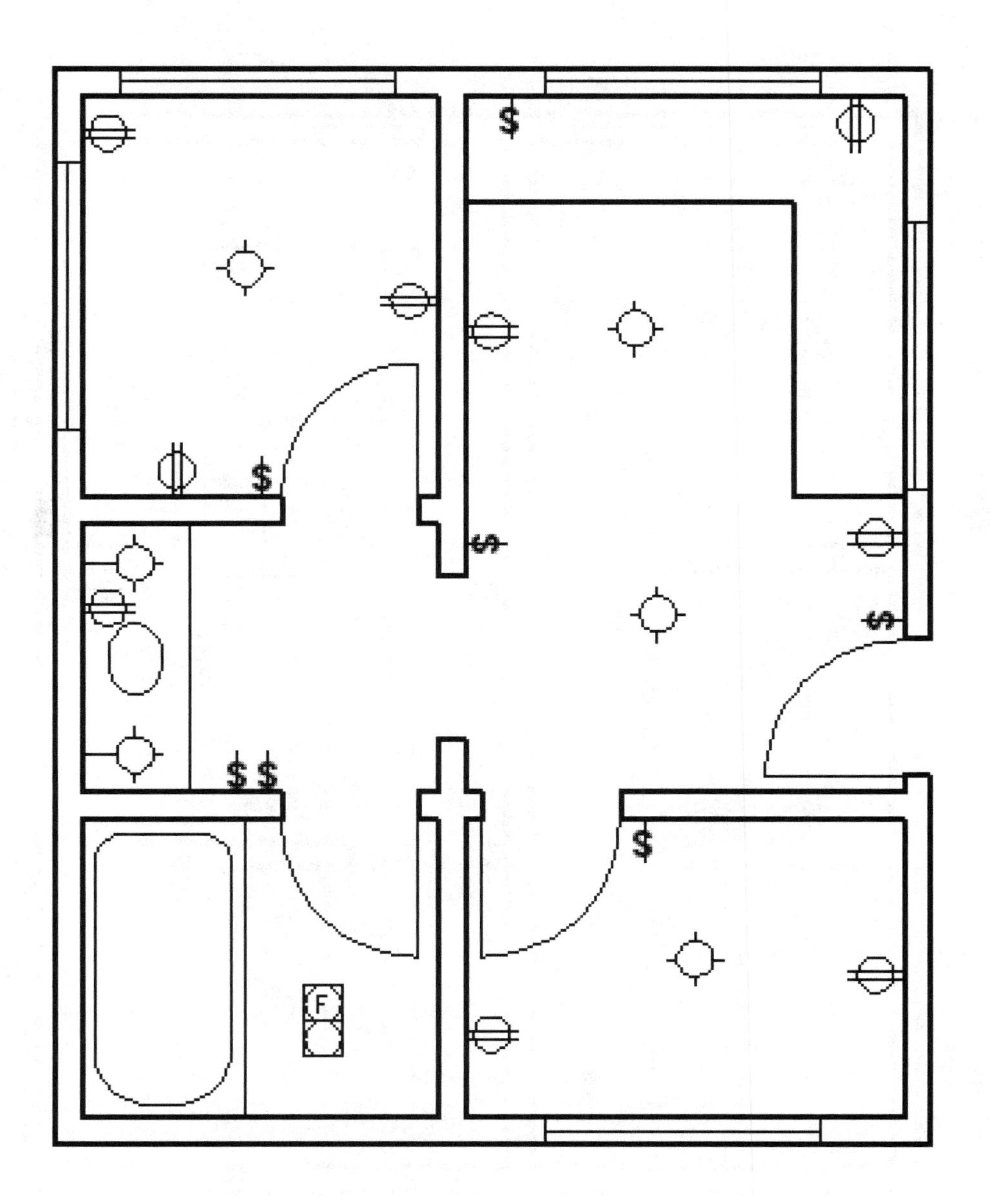

EXERCISE 29B....continued

STEP 4

1. Open **EX-29B** (If not already open)
2. Add the Wiring.
   A. Use Layer Wiring
   B. Use Arc, (Start, End, Direction)
3. Notice that your dashed lines do not appear like the example below.
   A. Type **LTS** (this is Linetype Scale setting)
   B. Type **.25** (This scales all linetypes)
4. **Save** the drawing as: **EX-29B** again

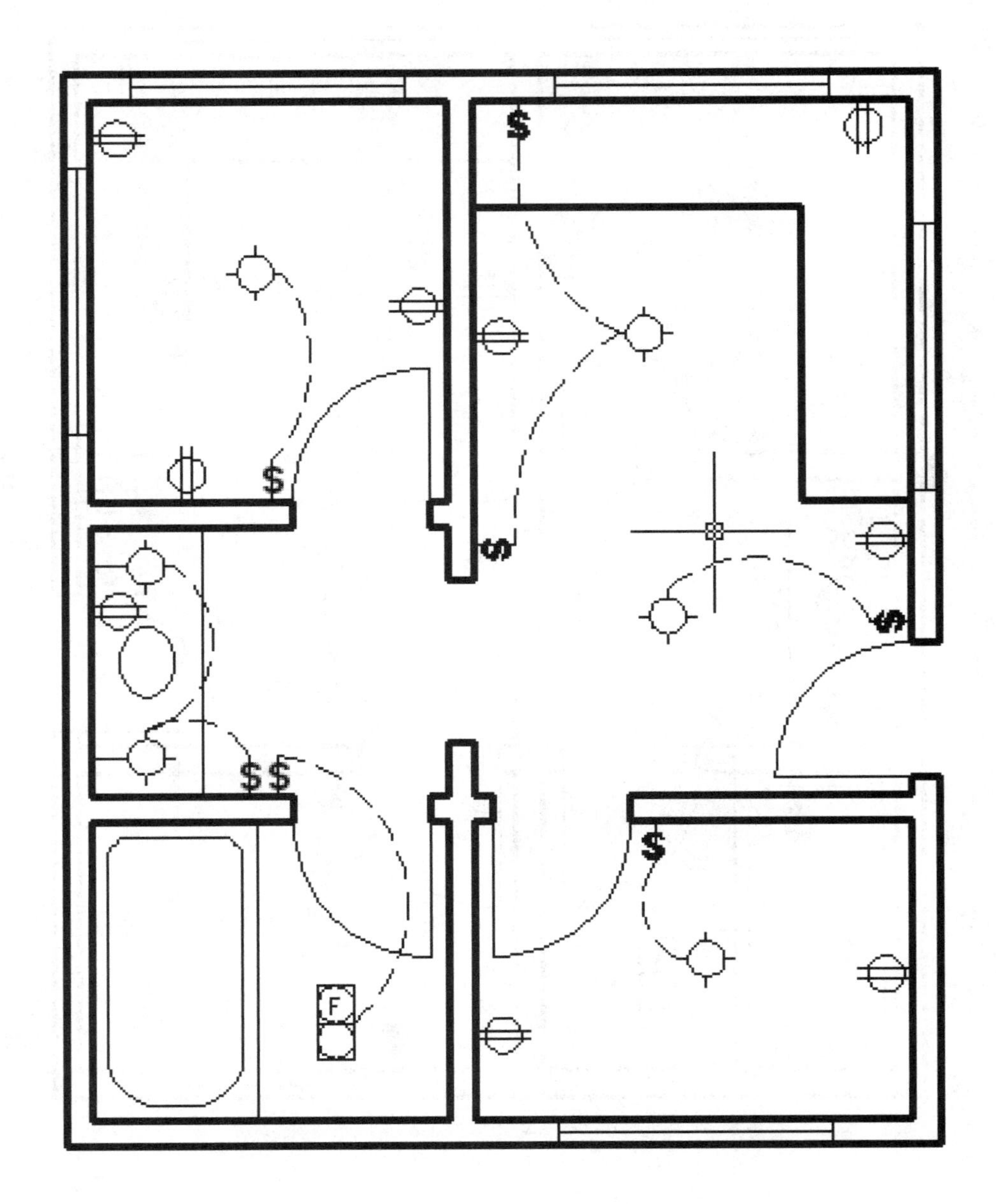

EXERCISE 29B....continued

STEP 5

1. Open **EX-29B** (If not already open)
2. Erase the existing Viewport Frame. (The Floor plan will disappear)
3. Select Viewport as Current layer and create **3 NEW** Viewports approx. as shown.
4. Adjust the scale of each Viewport.
5. Use **PAN** to find the area to view as shown below.
6. **Lock** each Viewport.
7. **Add** the dimensions in the 2 Viewports on the right
   by adding the current object scale. (Refer to page 28-3 and 28-4)
8. **Add** the Viewport Scale labels (1/8" height) in Paper space (Not Model space)
9. Make Viewport Frames plottable. (page 3-10)
10. **Save** as **EX-29B** **Edit** Title Block **Plot** using Plot Page Setup **Plot Setup A**

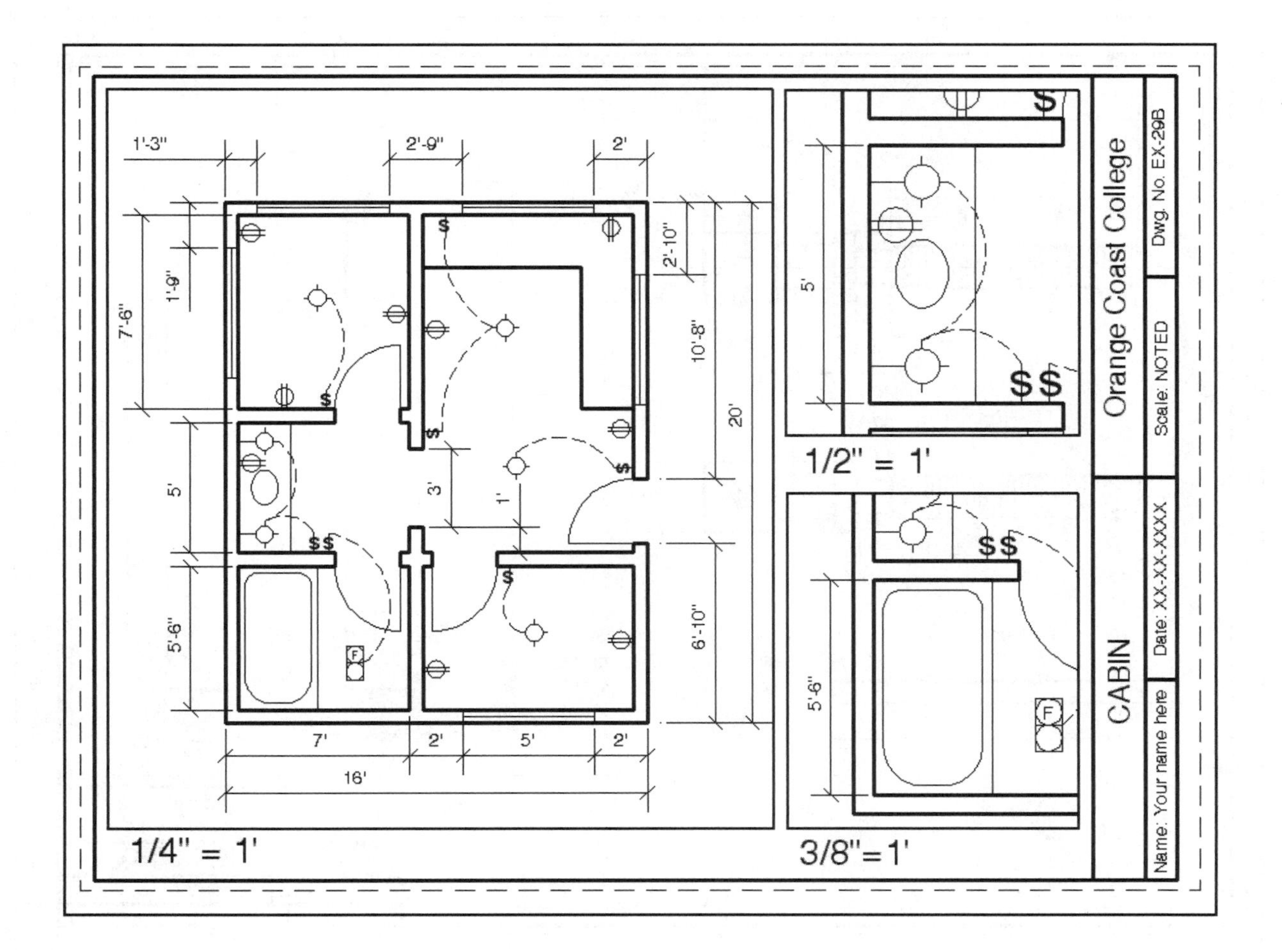

EXERCISE 29C

INSTRUCTIONS:
STEP 1

1. Start a **NEW** file using **My Decimal Setup.dwt.**
2. The Viewport should already be scaled to 1:1 and locked.
3. **Draw** the objects shown below.
4. Draw the **Break** line using Polyline with the Spline option (page 24-3)
5. Draw the Hatch using Pattern Ansi 31 and Layer Hatch.
6. Draw the Threads:
   A. Use Layer Threads
   B. Thread lines are .0625 distance apart.
7. Dimension as shown using Layer Dimension and Dimension Style = Dim-Decimal.
8. Save as **EX-29C**

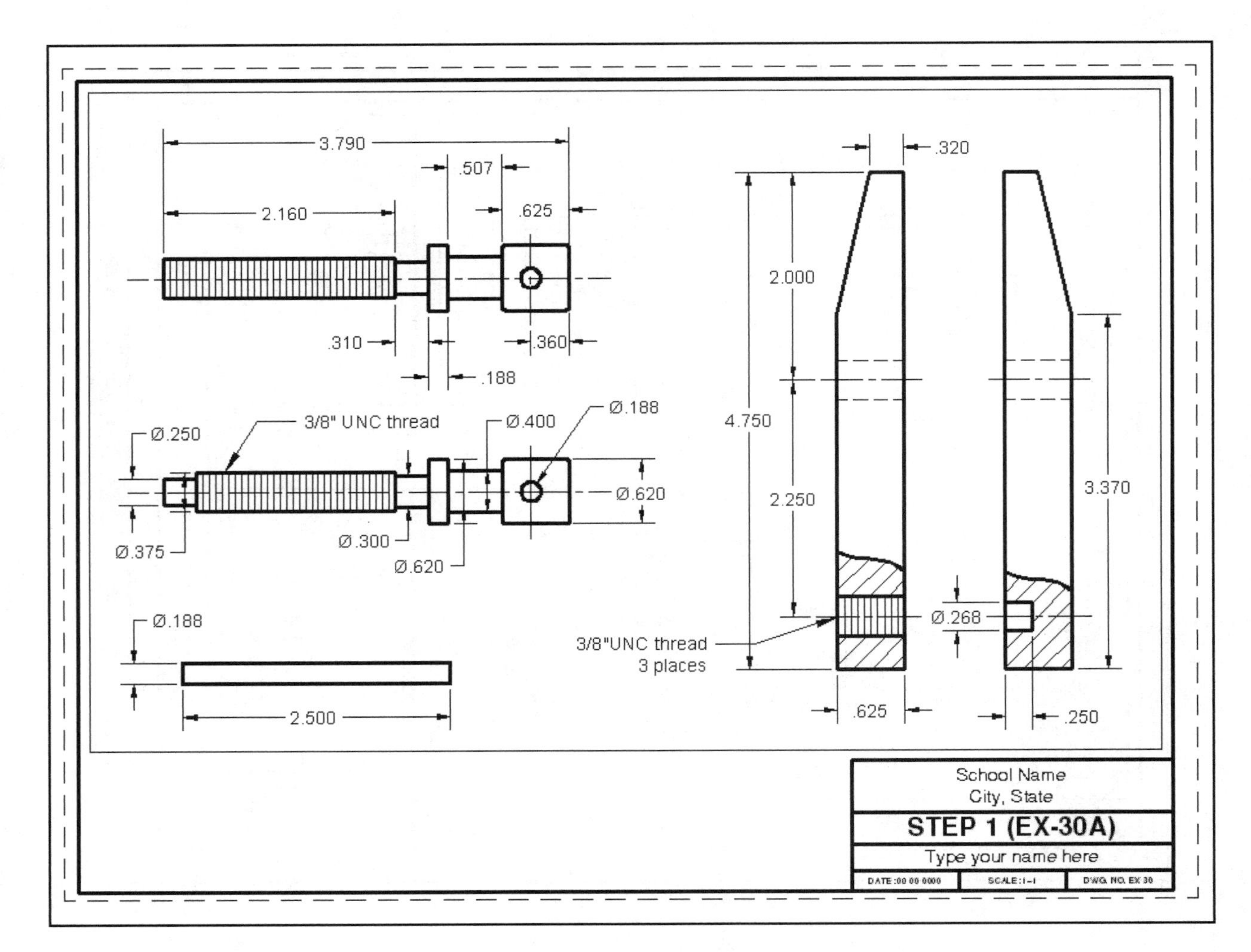

EXERCISE 29C....continued

STEP 2

1. **Freeze** the Dimension Layer.
2. **Assemble** the parts as shown below.
3. Use Move, Mirror, Rotate, Trim, Erase and any other command that you need.
4. The distance between the Jaws should be 1”
5. Save as **EX-29C**

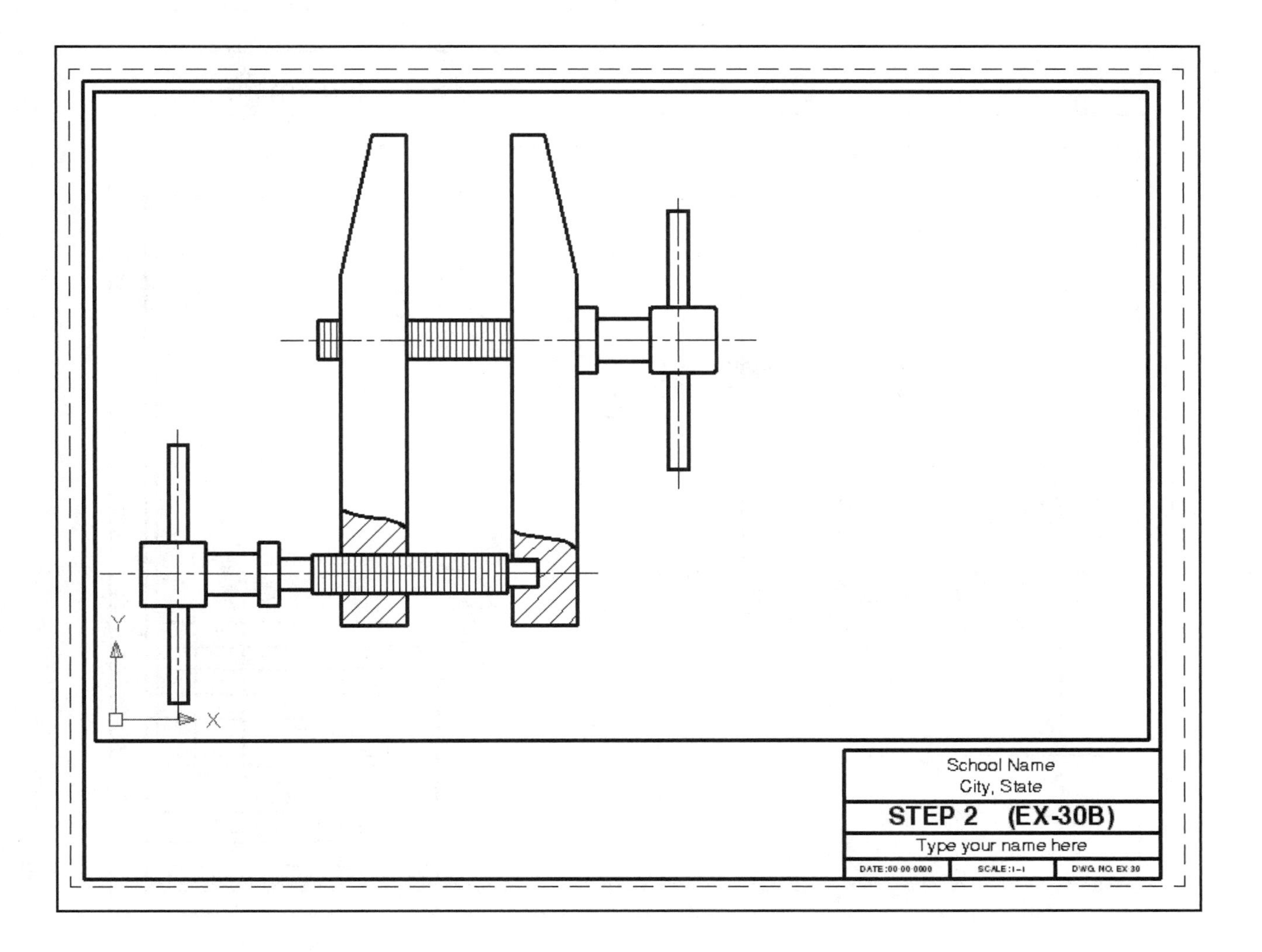

EXERCISE 29C....continued

STEP 3

1. Draw the List of parts **in Paper space** above the title block as shown.
2. Use Multiline Text to enter text. **Justify** = Middle Center **Text ht** = .125 and .062
3. Add Ballooned Leaders as shown **in Model Space**. Use Collect and Align.
4. Save as **EX-29C** and **plot** using plot page setup **Plot Setup A**

.500 .500 .500 .250

5	PIN	2	CRS
4	PILOT SCREW	1	CRS
3	SCREW	1	CRS
2	JAW-RIGHT	1	CRS
1	JAW-LEFT	1	CRS
ITEM	PART NAME	QTY	MATERIAL

School Name

List of Parts (enlarged)

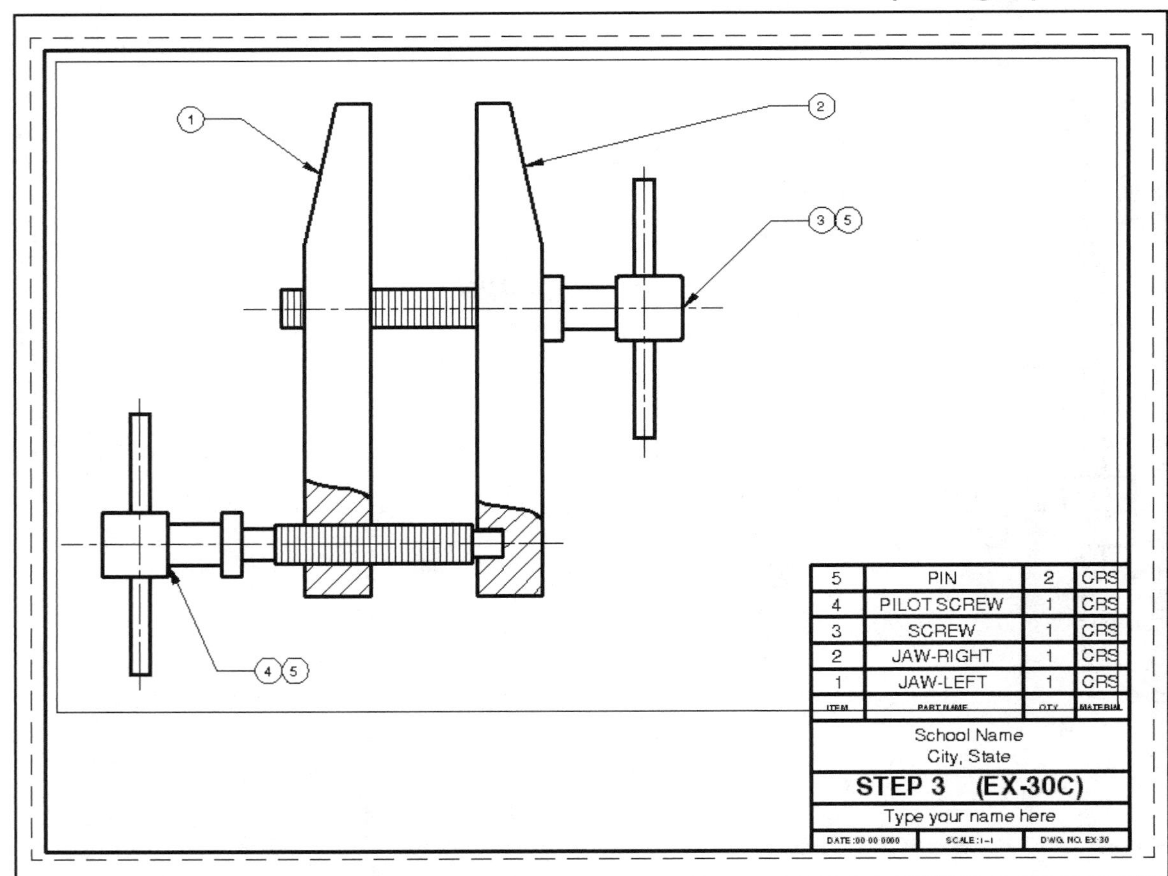

LEARNING OBJECTIVES

After completing this lesson, you will be able to:

1. Use 6 useful “Express Tools”.
2. Upload and Download documents to Autodesk ® 360.
3. Use the Design Feed to collaborate with colleagues.

LESSON 30

TEXT - Arc Aligned

ARCTEXT allows you to place Text along the arc's curve. You may control the appearance of the text easily using the ArcAlignedText Workshop Dialog box shown below.

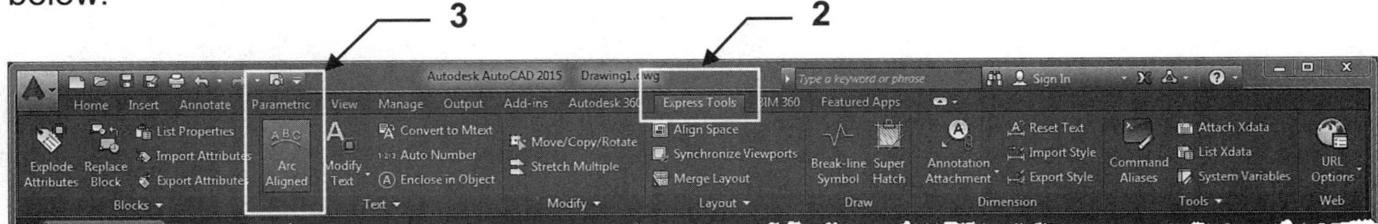

HOW TO CREATE ARCTEXT

1. Draw an Arc
2. Select **Express Tool** tab.
3. Select **Arc Aligned** tool.
4. Select the Arc.

Example of Arc Text

The **ArcAlignedTextWorkshop-Create** dialog box will appear.

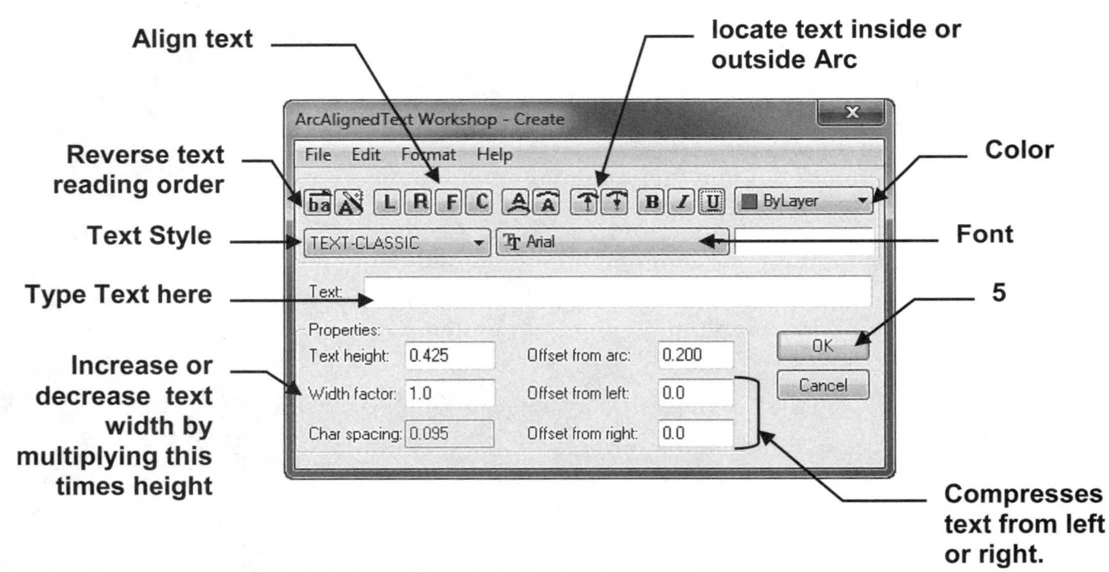

5. Select the options you desire and then select the **OK** button.

HOW TO EDIT THE ARCTEXT

1. Select the **ArcText** command.
2. Select the existing text.
   The **ArcAlignedTextWorkshop - Modify** dialog box will appear
3. Make changes and select the **OK** button.

TEXT - Modify Text

MODIFY TEXT has 5 options, **Explode, Change Case, Rotate, Fit** and **Justify,** to increase your manipulation of existing text.

1. Select the **Express Tools** tab.
2. Select the **Modify Text ▼** down arrow.
3. Select one of the options listed below.
4. Select the text to be modified.
5. Select **<enter>** to stop.

EXPLODE
Text explodes into Lines and Arcs

CHANGE CASE
Changes text case to one of the following:

Sentence Case: The first letter of the sentence is upper case. All others lower case.
Lowercase: All letters lower case.
Uppercase: All letters upper case
Title: The first letter of each word uppercase.
Toggle Case: Changes upper to lower and vice versa

ROTATE
Allows you to rotate the existing text "most readable" or a specific angle.

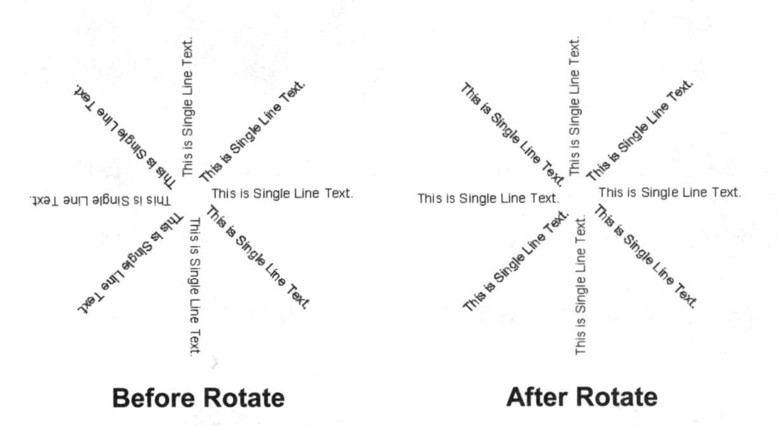

Before Rotate **After Rotate**

FIT
Allows you to compress or stretch text between two points.

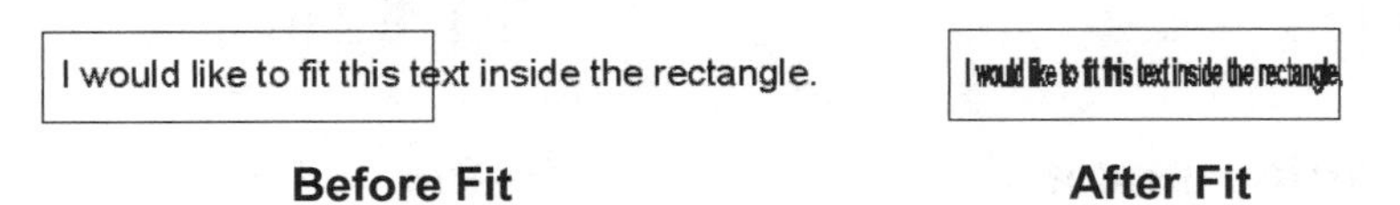

Before Fit **After Fit**

JUSTIFY
Allows you to change the justification but….you can do this with Properties also.

TEXT - Convert to MText and Auto Number

CONVERT TO MTEXT allows you to convert text created with Single Line text command to Multiline text. This is very helpful if you would like to modify the single line text because multiline text editor includes many additional editing options.

1. Select the **Express Tools** tab.

2. Select the **Convert to Mtext** tool.

3. Select the text to be converted.

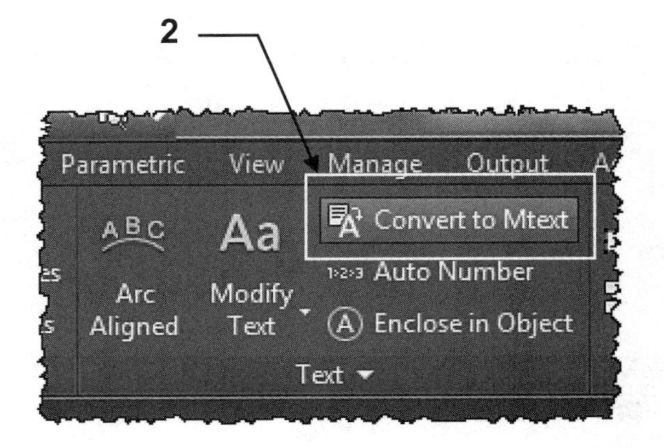

AUTOMATIC TEXT NUMBERING allows you to add sequential numbers to existing text.

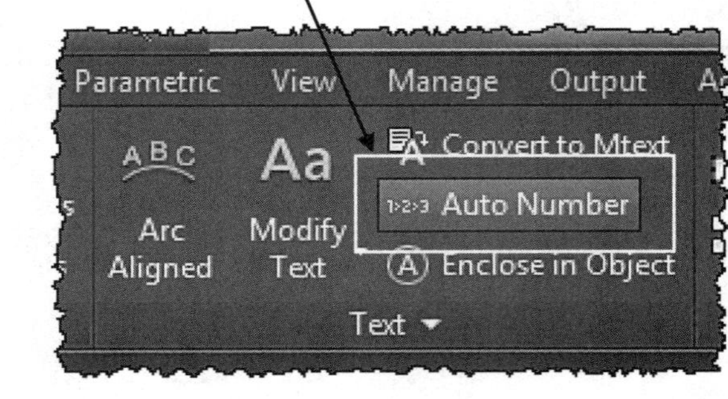

1. Select the **Express Tools** tab.

2. Select the **Auto Number** tool.

3. Select the text.

4. Sort selected objects by [X/Y/Select-order] <Select-order>: **<enter>**

5. Specify starting number and increment (Start,increment) <1,1>: **<enter>**

6. Placement of numbers in text [Overwrite/Prefix/Suffix/Find&replace..] < Prefix>:**<enter>**

This is line 1. This is line 2. This is line 3	1 This is line 1. 2 This is line 2. 3 This is line 3
Before Auto Number	**After Auto Number**

TEXT - Enclose in Object

ENCLOSE IN OBJECT allows you to place a Circle, Slot or Rectangle around text.

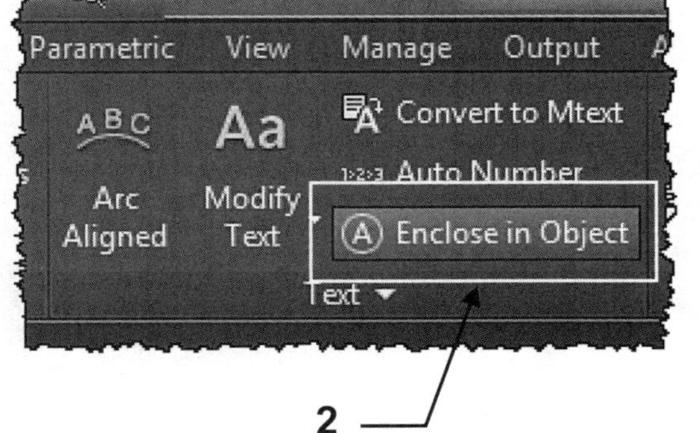

1. Select the **Express Tools** tab.

2. Select the **Enclose in Object** tool.

3. Select the Text.

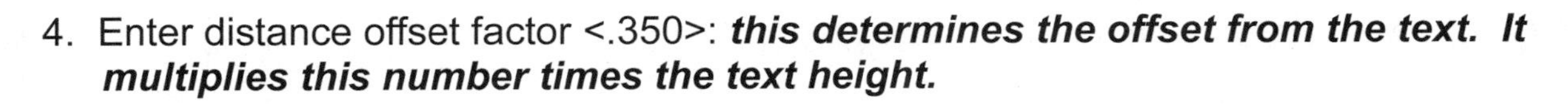

4. Enter distance offset factor <.350>: ***this determines the offset from the text. It multiplies this number times the text height.***

5. Enclose text with [Circles/Slots/Rectangles] <Circles>: ***select the enclosure***

6. Create circles of constant or variable size [Constant/Variable] <Variable>: ***if you select Constant, AutoCAD will select the largest text height and multiple it times the offset distance to determine the size for all. If you select Variable, each offset distance will be calculated based on the individual text height.***

EXAMPLES:

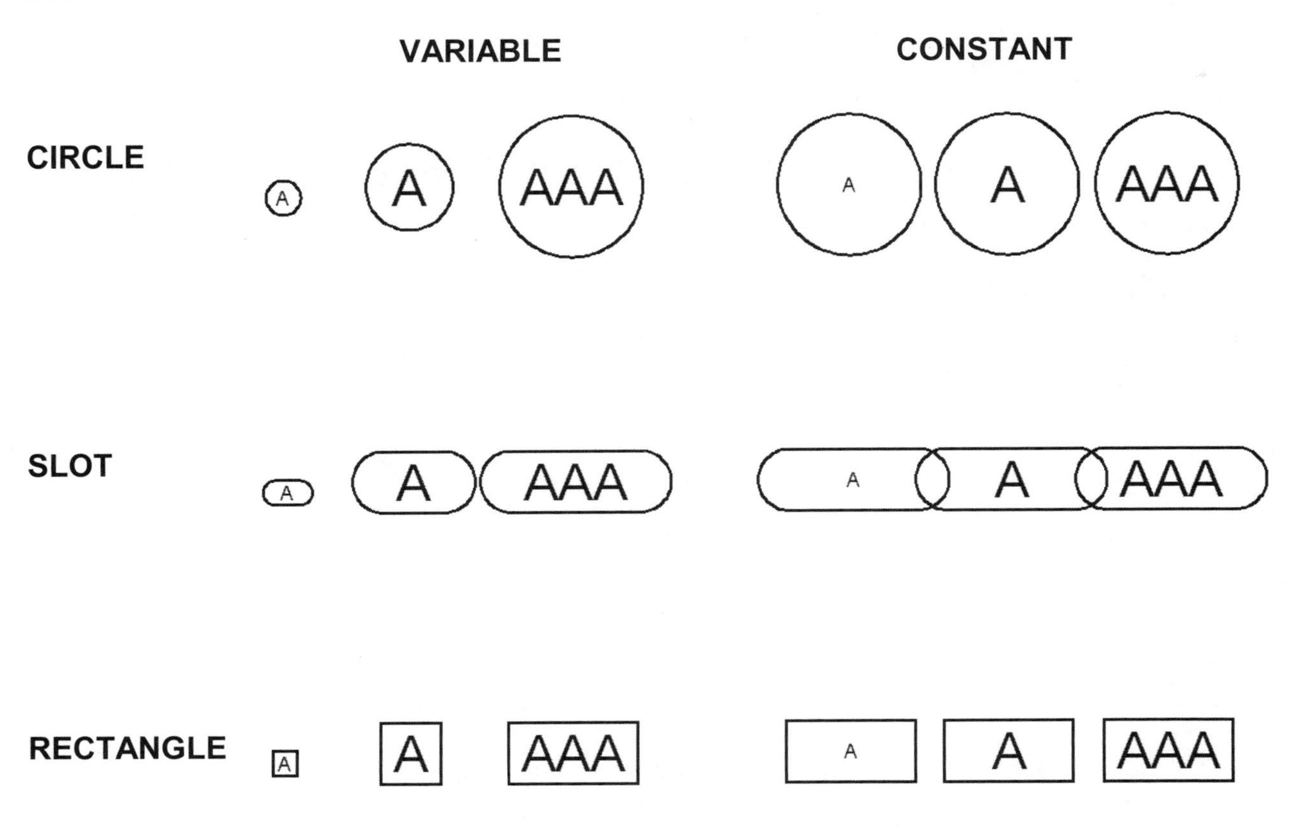

DRAW - Break-Line Symbol

BREAKLINE
Creates a polyline and inserts the break-line symbol.

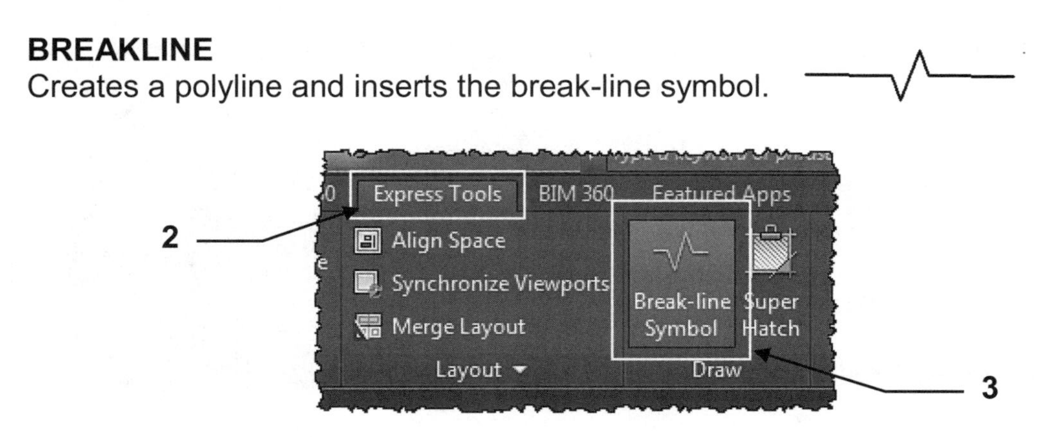

How to create a Break-Line Symbol

1. Draw 3 Lines as shown below.

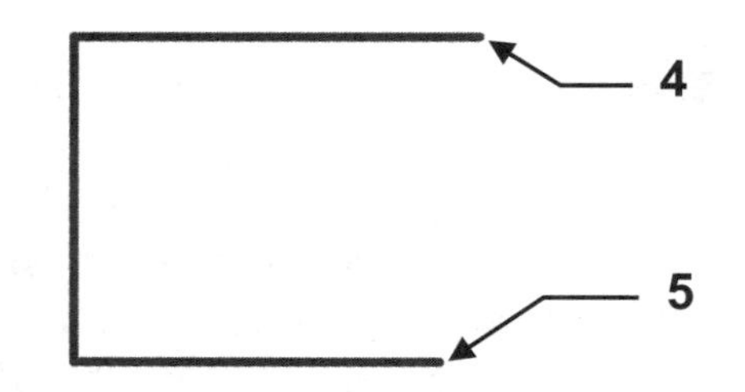

2. Select the **Express Tools** tab.
3. Select the **Break-Line Symbol** tool.

   Command: breakline
   Block= BRKLINE.DWG, Size= .500, Extension= .180

4. Specify first point for breakline or [Block/Size/Extension]: ***snap to 1st endpoint***
5. Specify second point for breakline: ***snap to 2nd endpoint***
6. Specify location for break symbol <Midpoint>: ***<enter>***

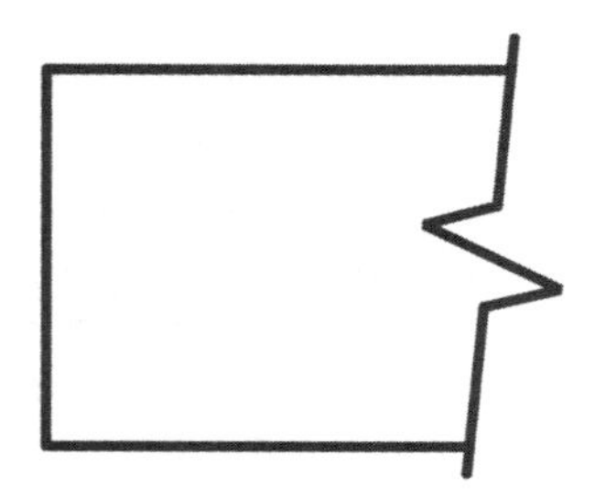

__Options:__

Block: You may select another block to be used instead of the Break-line symbol.

Size: You may designate the size of the Break-line symbol.

Extension: You may designate the amount the Break-line overlaps the existing line.

TOOLS - Command Aliases

COMMAND ALIASES

AutoCAD has many preset aliases for common commands. For example: **C** for **Circle**. These aliases are stored in the **acad.pgp** file. Using the **Command Aliases** tool you may **Add, Remove** or **Edit** an Alias.

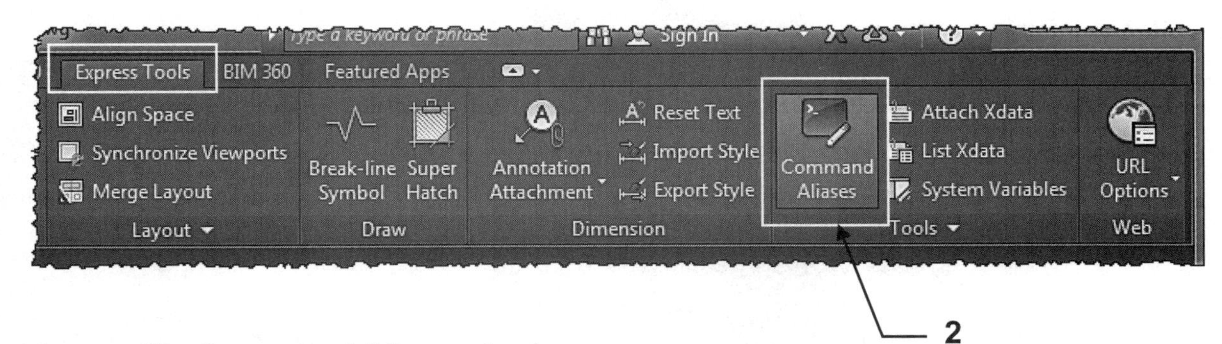

How to use the Command Aliases tool.

1. Select the **Express Tools** tab.

2. Select the **Command Aliases** tool.

*The **acad.pgp - AutoCAD Alias Editor** appears.*

3. Select the **Command Aliases** tab.

4. Select the **Add, Remove** or **Edit** button.
   Add: You will specify an **Alias** and select the corresponding command. Then **OK**
   Remove: Click on the Alias and select the **Remove** button.
   Edit: Enter the new Alias to replace the existing Alias. Then **OK**

5. Select **OK** button.

CREATING A GROUP

The Group tools allow you to group a set of objects and give them a name.
Select **Home tab / Groups panel**

You may:
1. Create a group
2. Ungroup
3. Edit: Add or remove objects from the group
4. On or Off: Temporarily turn the Group On or Off. (Blue is ON)

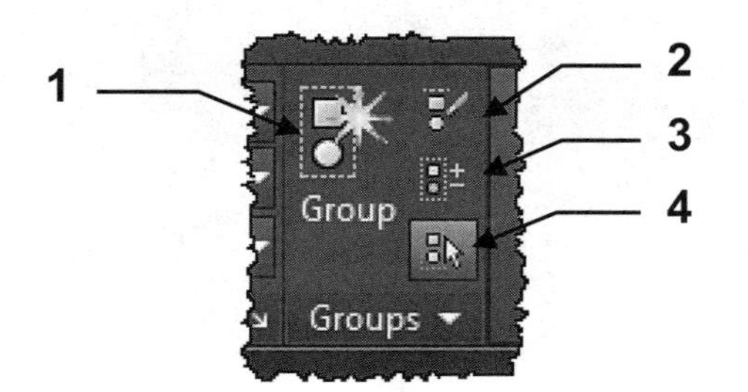

How to create a Group
1. Select the Create Group tool (1)
2.. Select the objects to Group.
3. You may give the group a name now or use the Group Manager. (See next page)

How to Ungroup
1. Select the Ungroup tool (2)
2. Select the Group to Ungroup

How to Add or Remove an object from the group.
1. Select the Edit tool (3)
2. Select the Group to edit
3. Select Add or Remove
4. Select the object(s) to Add or Remove. <enter>

How to turn a Group On or Off.
1. Select the On or Off tool (4)
2. Select the Group to turn off or on.

Note:
You may temporarily turn off a group, make a move an object within the group and then turn the group On. The moved object will remain in group but now in a different position.

Purge a Group
A Group, no longer needed, can be **purged**. (Refer to 29-9)

Continued on the next page...

CREATING A GROUP....continued

Managing the Group

1. Select the Groups down arrow ▼

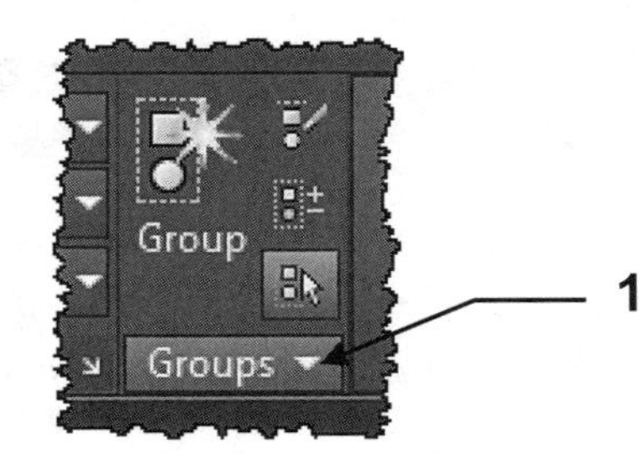

2. Select **Group Manager** or **Group Bounding Box**

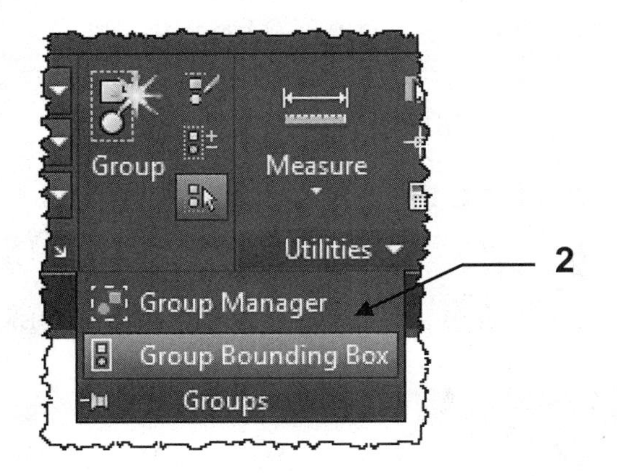

Group Manager
The Group Manager allows you to Identify, Create and Change a group.

Object Grouping
Group Name Selectable
Group Identification
Group Name:
Description:
Find Name < Highlight < Include Unnamed
Create Group
New < Selectable Unnamed
Change Group
Remove < Add < Rename Re-Order...
Description Explode Selectable
OK Cancel Help

Group Bounding Box
When you select a previously created group, AutoCAD identifies it using Grips or a Bounding Box. The default is Grips. If you want a Bounding Box select this option.

Bounding Box ON

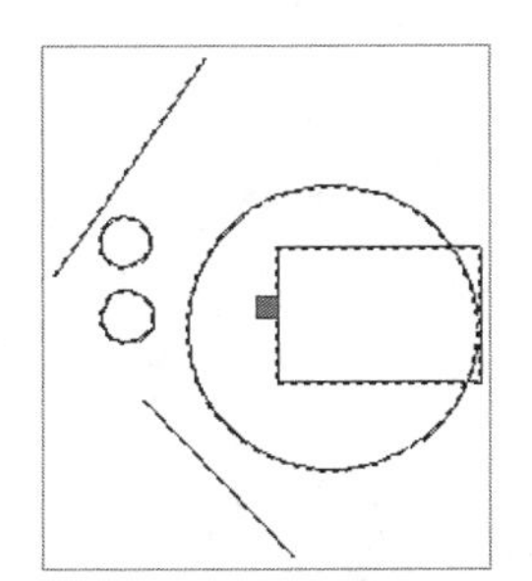

Bounding Box OFF

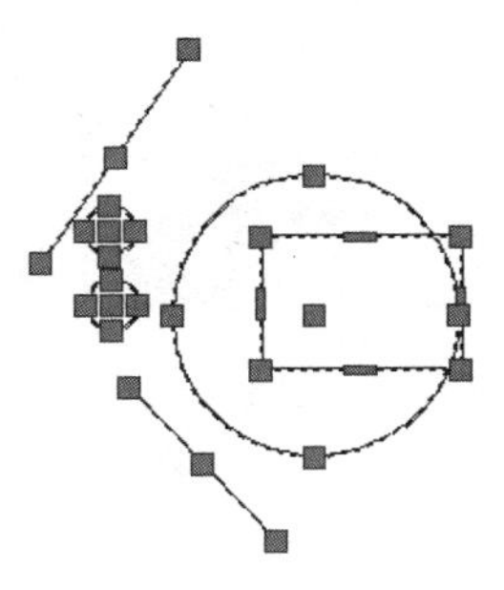

AUTODESK 360 CONNECTIVITY

Autodesk® 360 is a set of secure online servers that you can use to store, retrieve, organize, and share drawings and other documents.

Over the next few pages the Autodesk 360 "documents" will be discussed.
Use this service to store your design documents in the cloud, so you can access them anytime, anywhere and easily share them with colleagues, clients and other users. Viewing capabilities enable users to open and review 2D and 3D DWF files through a web browser, without the design software used to create the files. 5 GB of storage space is available for free. Autodesk subscription customers receive 25 GB of storage space for each seat of software on Subscription for the duration of their Subscription contract term.

In AutoCAD 2015 you can connect directly to the Autodesk 360 for online file sharing, customized file syncing and more. You can sign into the Autodesk 360 from the InfoCenter toolbar using your Autodesk single Sign-In account. If you do not yet have an account, you can create one.

After signing in, your user name is displayed and additional tools are displayed in the drop-down menu including the option to sync your settings with Autodesk 360, specify online options, access Autodesk 360 documents, sign out, and manage account settings. You may access the Autodesk 360 in various ways. Here are a few examples.

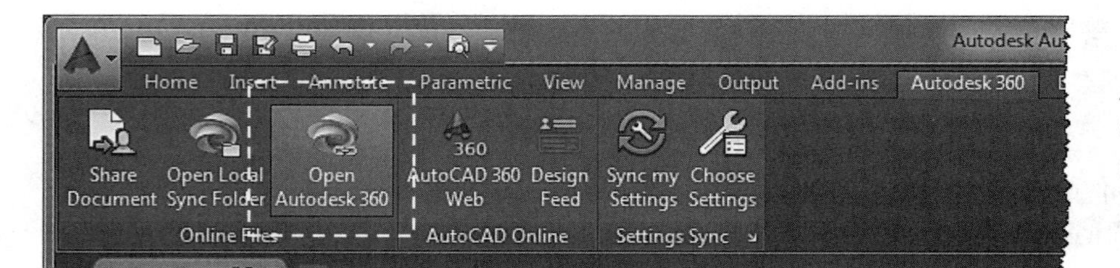

Autodesk 360 tab/
Online Files panel/
Open Autodesk 360

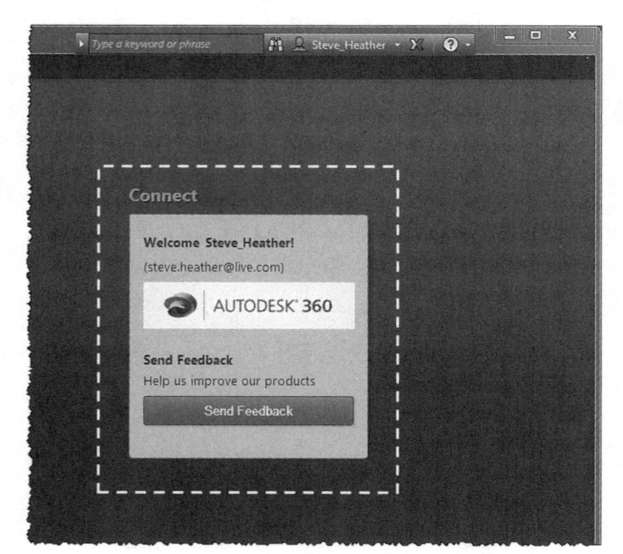

New Tab Page/ Create Page/ Connect

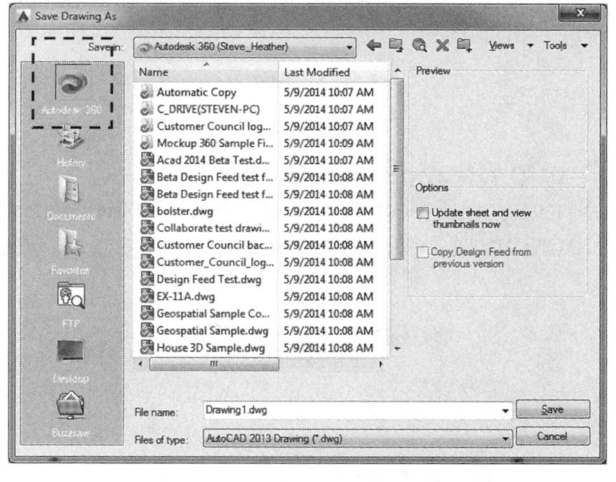

Save As or Open dialog boxes

Continued on the next page...

How to save a file to Autodesk 360

1. Select the **Application Menu**
2. Select **Save As ▶**
3. Select **Drawing to the Cloud**

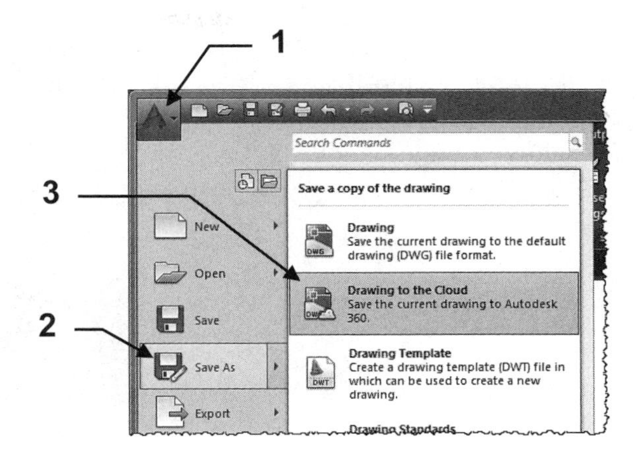

The **Account Sign In** box should appear.

4. If you have not previously created an account select **Need an Autodesk ID?**

If you have already created an account skip to 8.

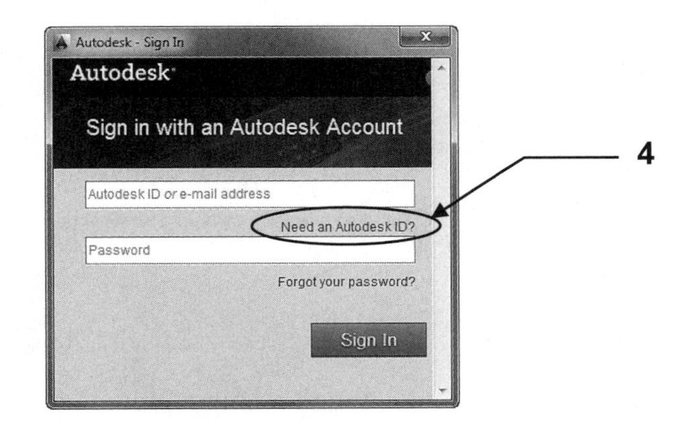

The **Create Account** box should appear.

5. Fill in the boxes
6. Select **Create Account** and then sign in.

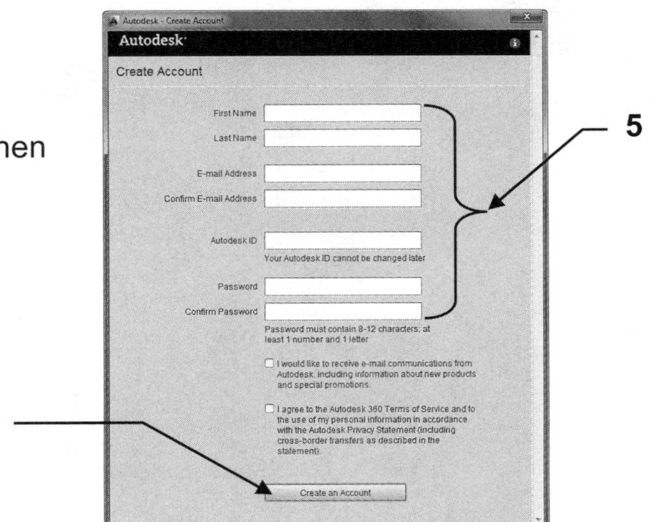

Continued on the next page...

How to save a file to Autodesk 360....continued

The first time you access the Autodesk 360, you have the opportunity to specify default Autodesk 360 settings. You may modify these selections later using the **Autodesk 360 ribbon tab.**

7. Make your selections then select **OK.**

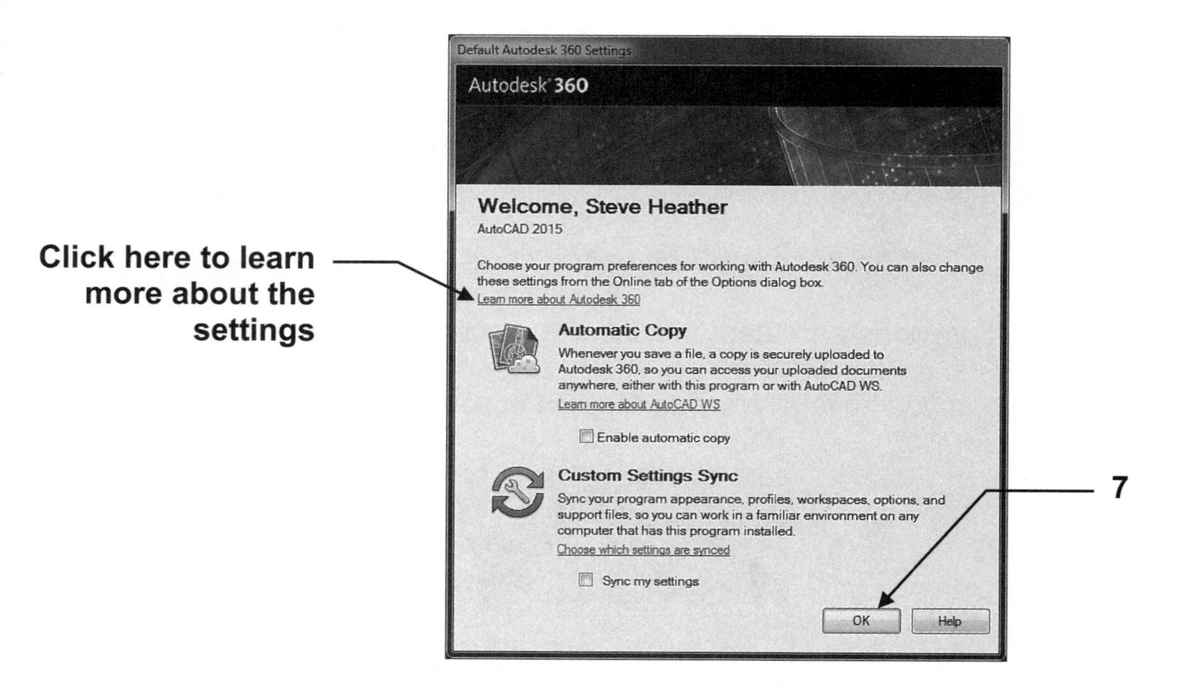

8. Enter the file name

9. Select **Save** button.

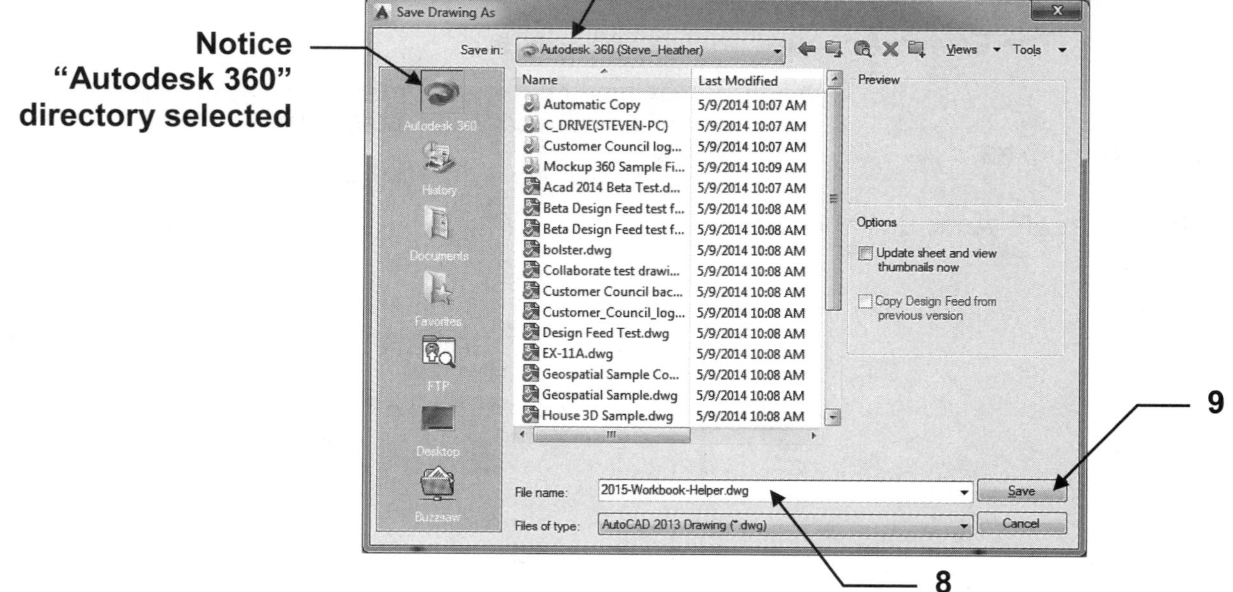

How to save a file to Autodesk 360 automatically

You may set AutoCAD to automatically save the file to the Autodesk 360 every time you save a file.

1. Select the **Autodesk 360** tab.

2. Select the arrow ↘ on the **Settings Sync** Panel.

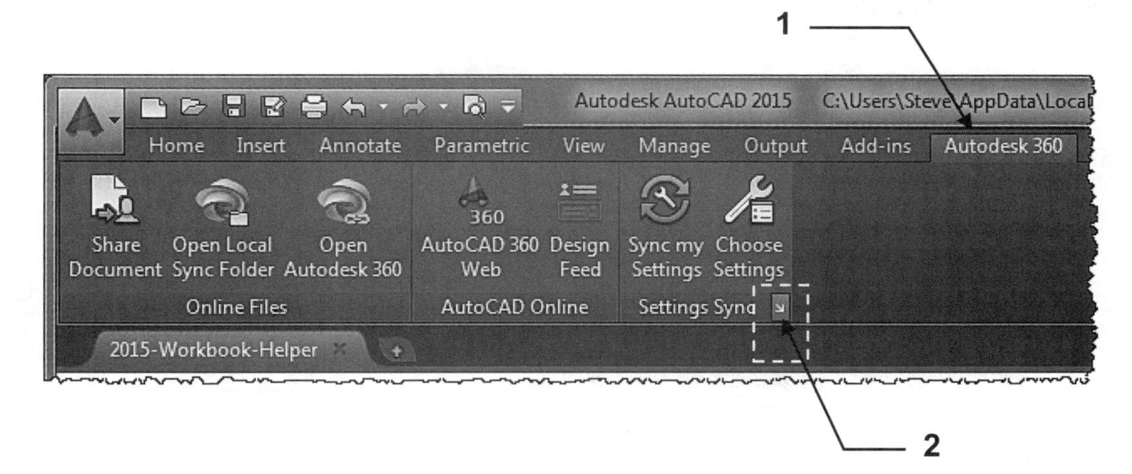

3. In the **Online** tab, check the box: **Enable automatic sync**.

Note: You must be signed in to access the settings in the **Online** tab.

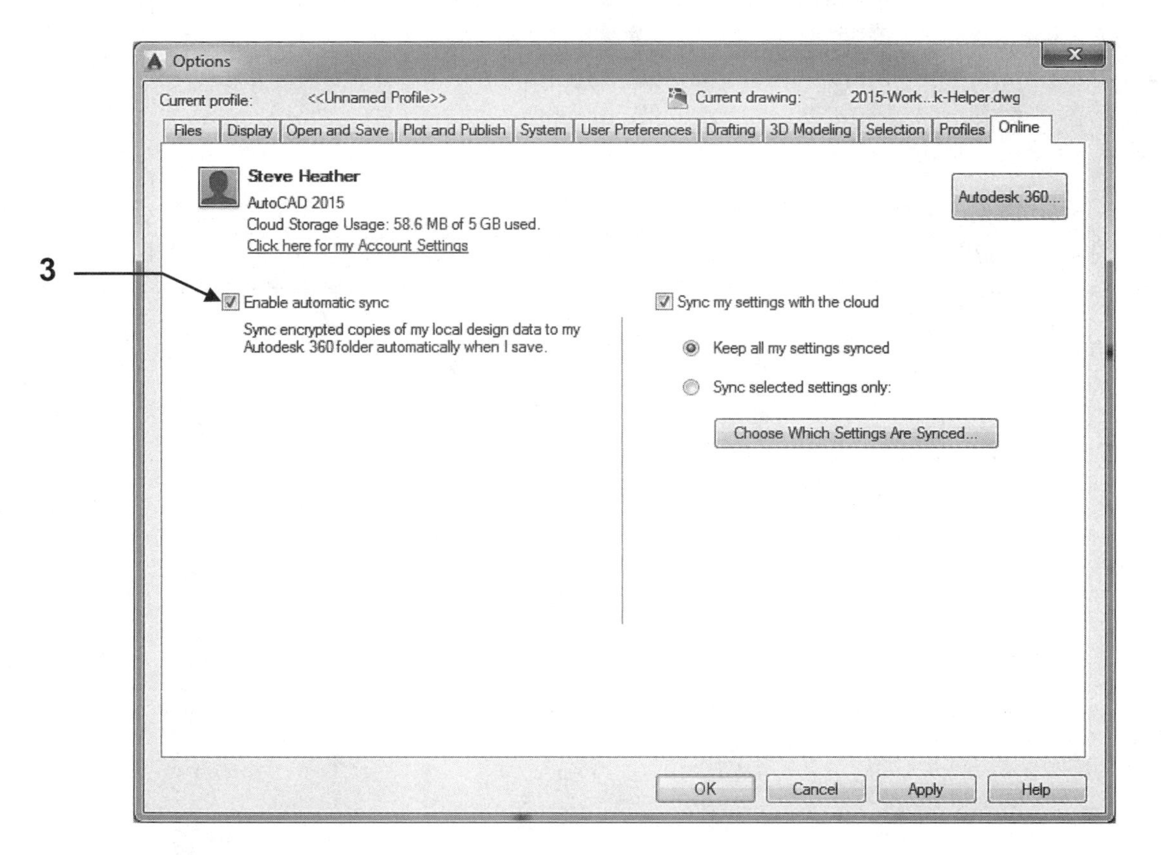

How to Open a file from Autodesk 360

1. Select **Open**
2. Select the **Autodesk 360** directory
3. Select the **file** to open.
4. Select the **Open** button.

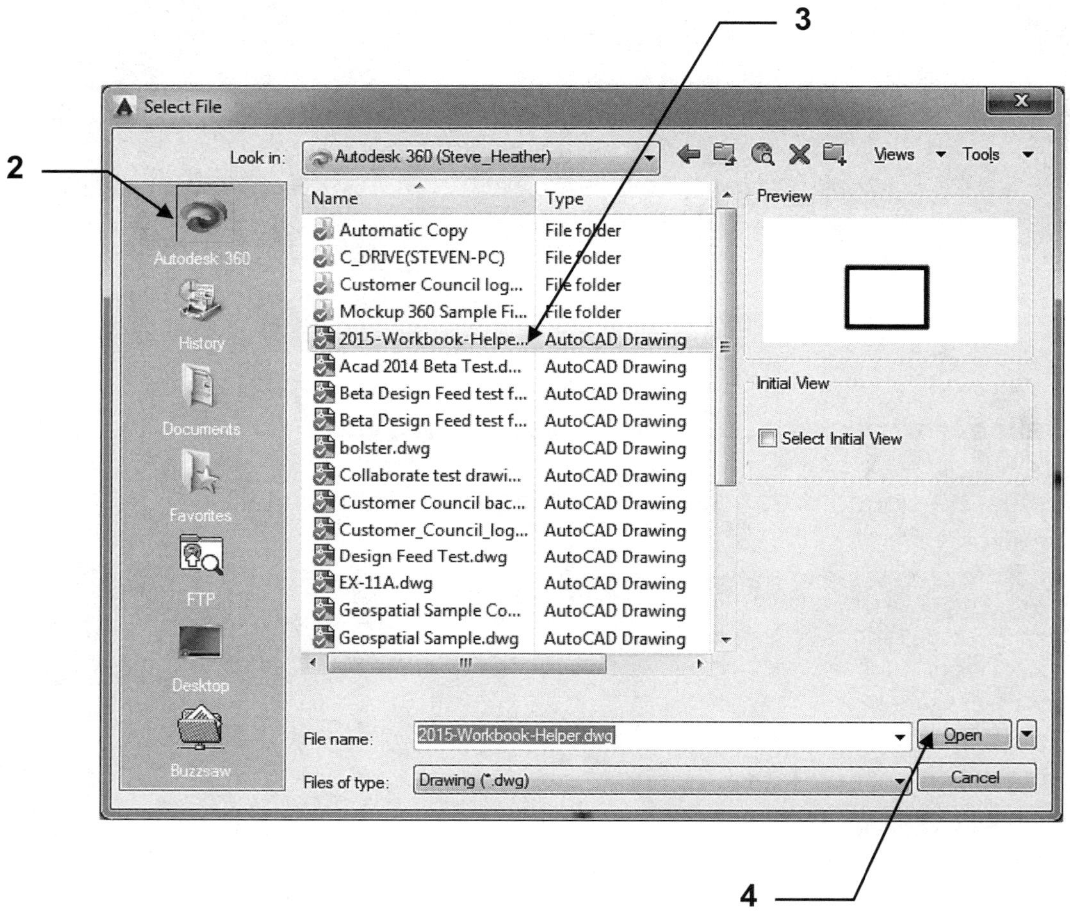

2
3
4

Select File
Look in: Autodesk 360 (Steve_Heather)
Views Tools
Autodesk 360
History
Documents
Favorites
FTP
Desktop
Buzzsaw

Name	Type
Automatic Copy	File folder
C_DRIVE(STEVEN-PC)	File folder
Customer Council log...	File folder
Mockup 360 Sample Fi...	File folder
2015-Workbook-Helpe...	AutoCAD Drawing
Acad 2014 Beta Test.d...	AutoCAD Drawing
Beta Design Feed test f...	AutoCAD Drawing
Beta Design Feed test f...	AutoCAD Drawing
bolster.dwg	AutoCAD Drawing
Collaborate test drawi...	AutoCAD Drawing
Customer Council bac...	AutoCAD Drawing
Customer_Council_log...	AutoCAD Drawing
Design Feed Test.dwg	AutoCAD Drawing
EX-11A.dwg	AutoCAD Drawing
Geospatial Sample Co...	AutoCAD Drawing
Geospatial Sample.dwg	AutoCAD Drawing

Preview
Initial View
Select Initial View
File name: 2015-Workbook-Helper.dwg
Open
Files of type: Drawing (*.dwg)
Cancel

How to Upload a document to Autodesk 360

You may **upload** your files to the Autodesk 360 to share with others.

1. Select the **Autodesk 360 tab / Online Files panel / Open Autodesk 360**

The Welcome area allows you to browse some of the options and view a video. This area can be temporarily closed by selecting the Close button in the upper right corner.

To re-open the Welcome area select the Help ▼ and select **Getting Started.**

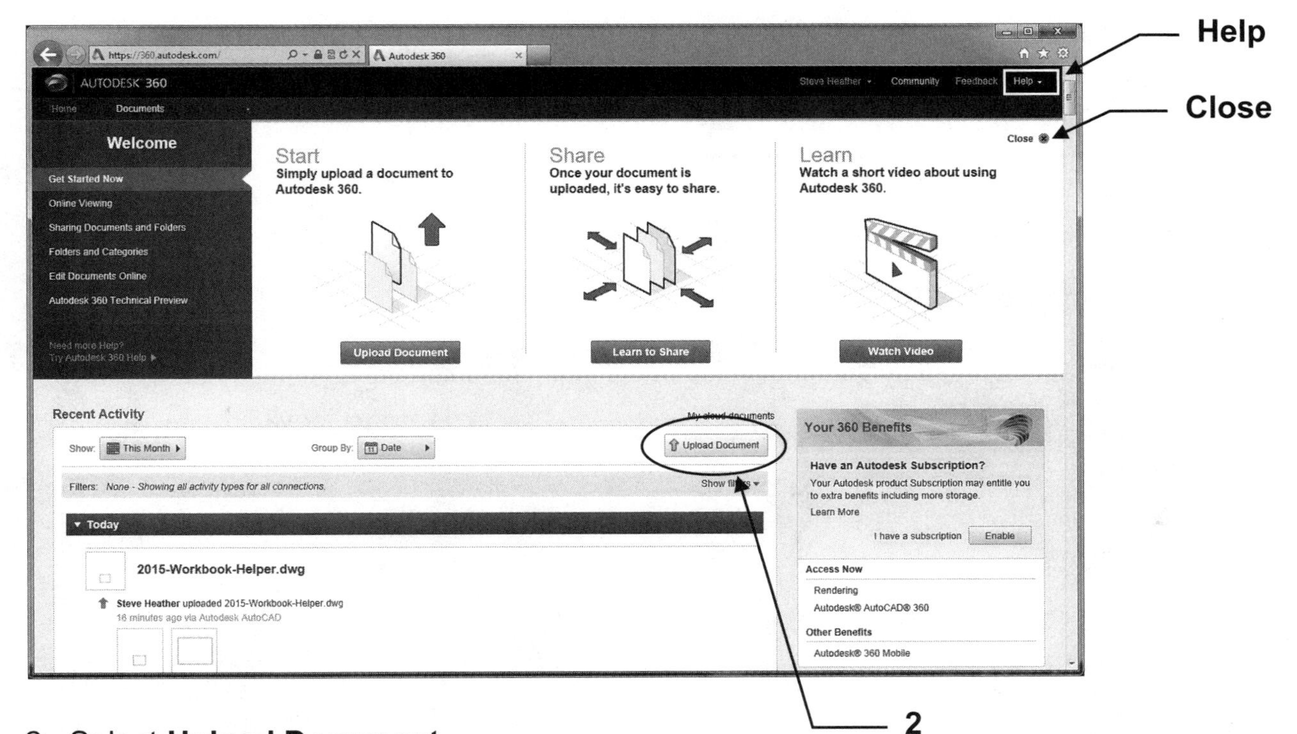

2. Select **Upload Document**

3. Select **Select Documents**

3

Upload Documents

Select documents to upload

Select Documents

Name	Size	Status

Cancel Upload Now

Continued on the next page...

How to Upload to Autodesk 360....continued

4. Locate the file to upload and select **open**

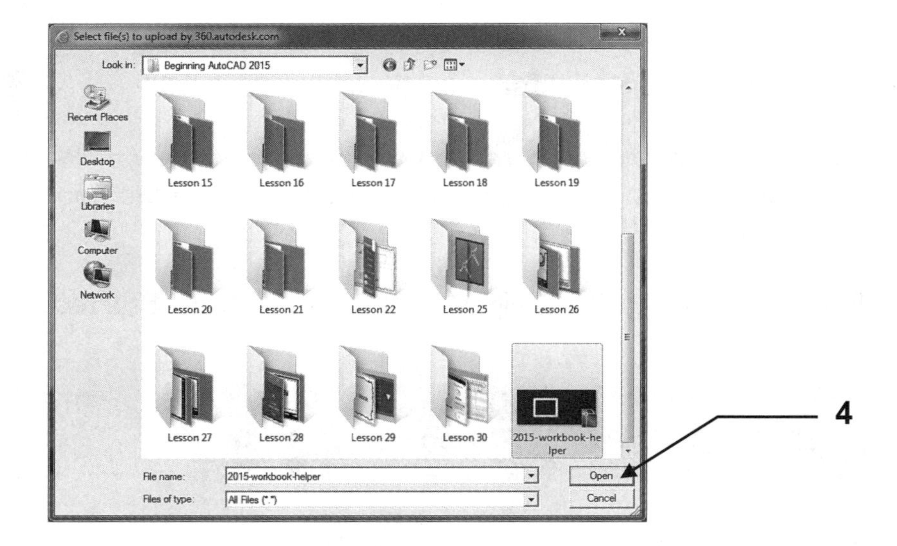

5. Select **Upload Now**

Upload Now

The uploaded document appears in the **My Cloud Documents** area

How to Delete an Uploaded document

How to Delete an uploaded file from My Cloud Documents

1. Select the document to be deleted

2. Select **Actions ▼**

3. Select **Delete** from the drop-down menu

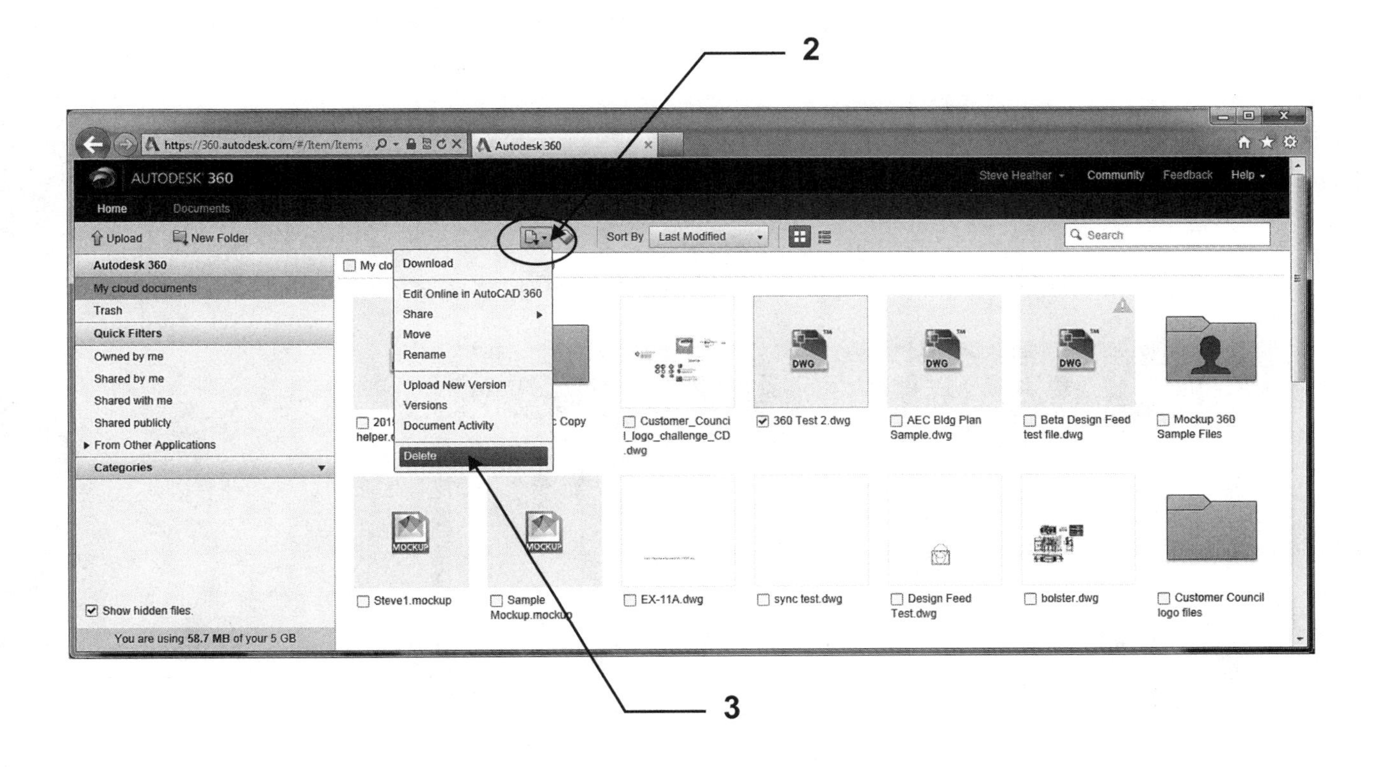

Note: You can also right-click on the document to access the same menu.

How to Download from Autodesk 360

1. Select the **Autodesk 360 tab / Online Files panel / Open Autodesk 360**
2. Select the document to download

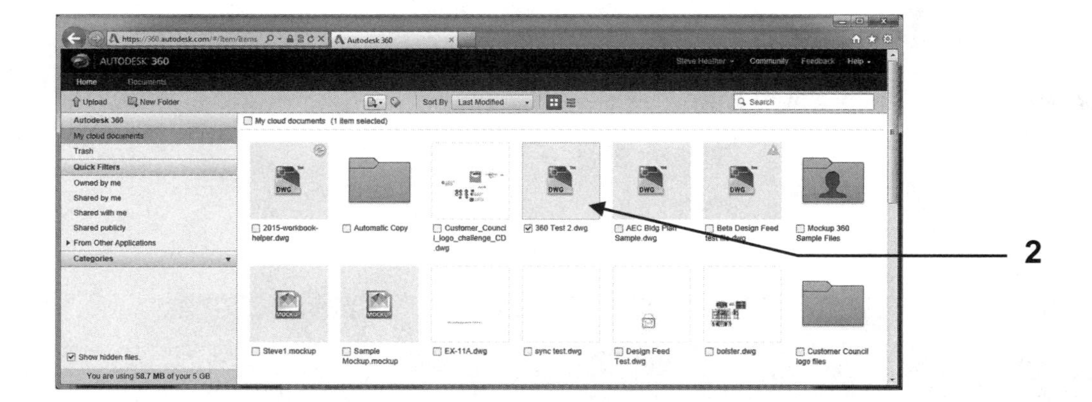

3. Select **Download** from the **Actions** drop-down menu.

4. Depending on your internet browser, select **Open**

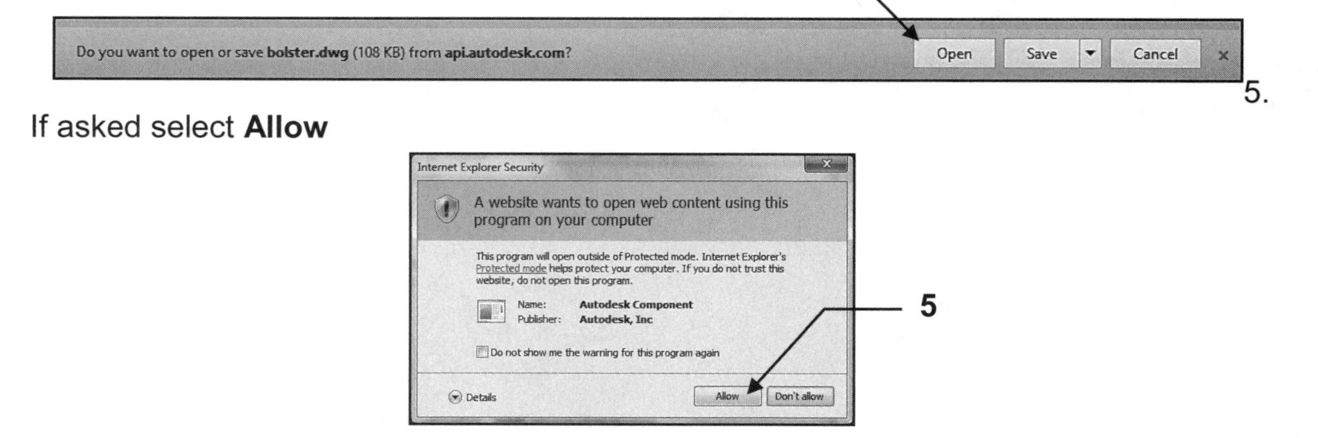

5. If asked select **Allow**

The document should have opened on to the AutoCAD screen. You may view, edit, print, save etc. You may place documents in the Autodesk 360 for others to download.

Autodesk 360 Design Feed Palette

The Design Feed palette allows you to enter text messages and attach images which can then be shared online with other users who are sharing your document. You can also use the Design Feed palette to post messages and images on a document that is shared with you by another user.

You can associate a message to an area in the drawing by using a location pin or by specifying a rectangular area.

To use the Design Feed palette you must first save the drawing to the Autodesk 360, you can then tag colleagues to be included in the discussion.

To open the Design Feed palette.

Ribbon = Autodesk 360 tab / AutoCAD Online panel / Design Feed
or
Keyboard = designfeedopen <enter>

Design Feed

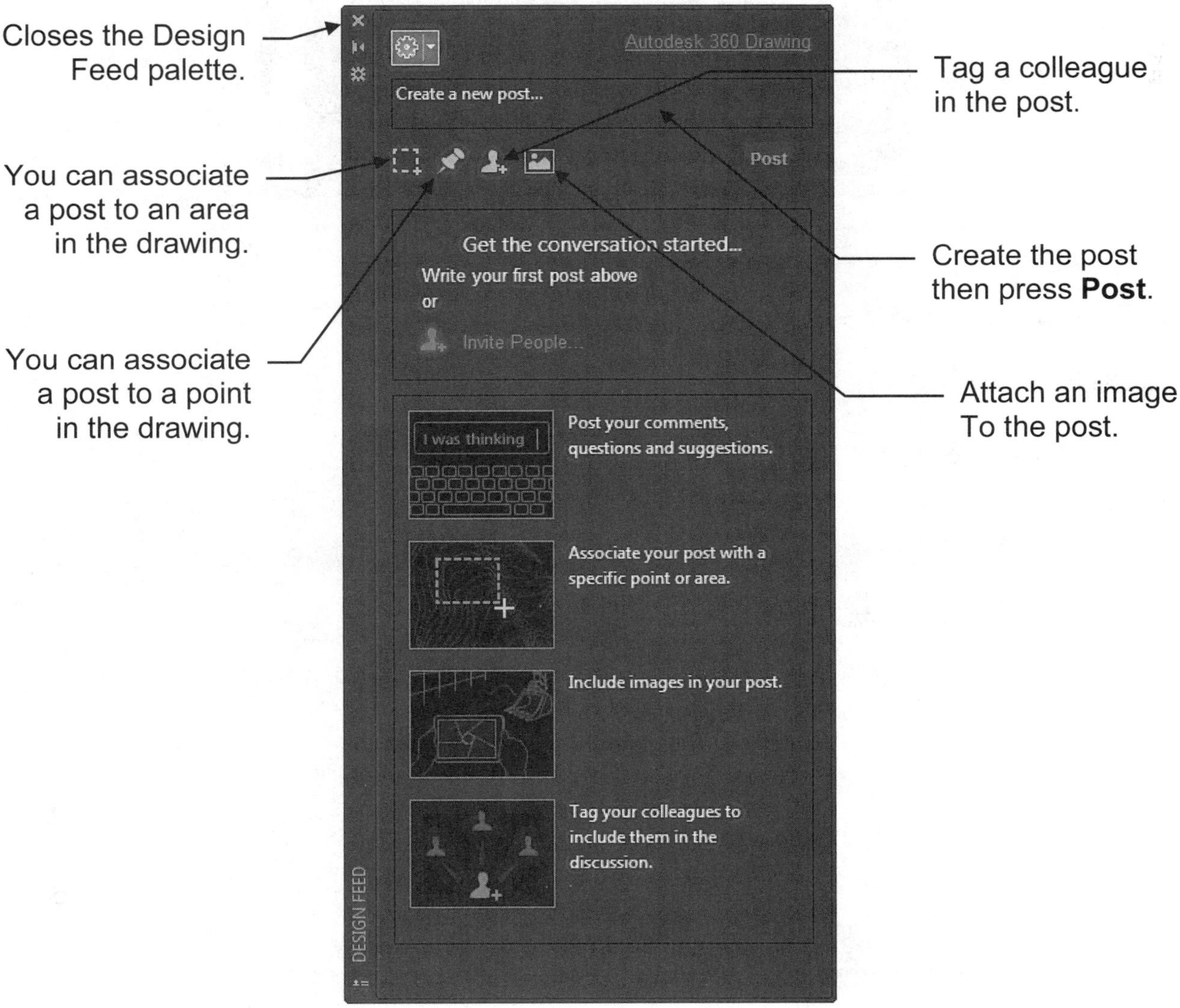

Closes the Design Feed palette.

Tag a colleague in the post.

You can associate a post to an area in the drawing.

Create the post then press **Post**.

You can associate a post to a point in the drawing.

Attach an image To the post.

How to tag a colleague in the Design Feed

You can tag as many people as you want to the Design Feed posts within your document, they will need to have access to AutoCAD or to the online AutoCAD 360. When you tag a colleague to a post they will be notified by e-mail.

To tag a colleague in the Design Feed.

1. In the Design Feed palette, select the **Tag in this post** command.
2. Select **+ Add People**.

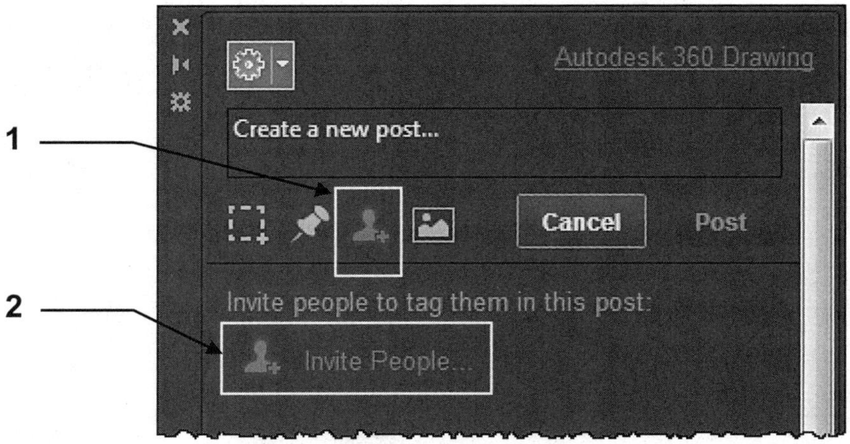

3. Enter the e-mail address of the colleague to be tagged.
4. Click on **Add**.
5. Select **Save & Invite**.

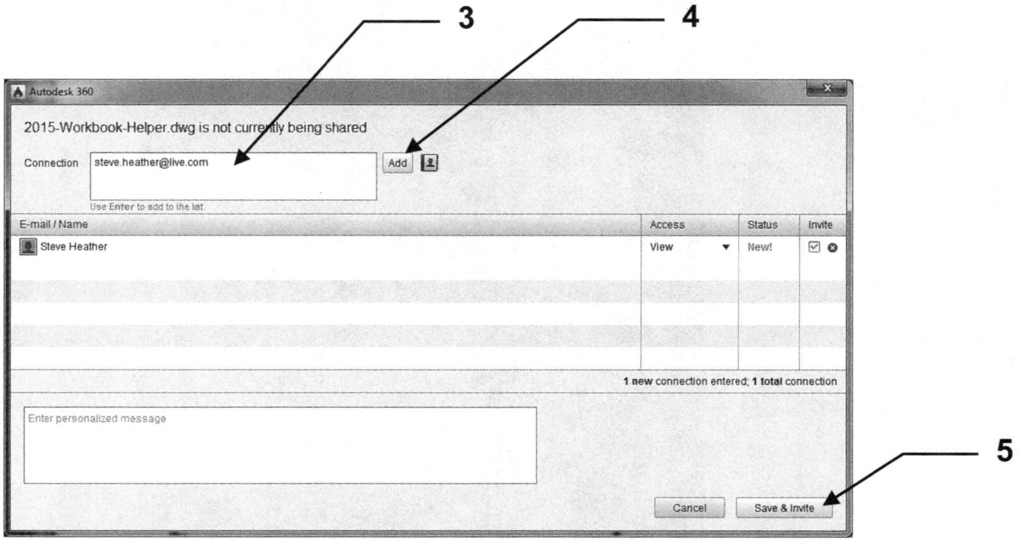

Continued on the next page...

How to tag a colleague in the Design Feed..continued

After you have selected **Save & Invite**, the **Invite Sent** confirmation dialog box appears informing you that an E-mail invitation has been sent to the colleague to be tagged, also known as Connections. Select **OK** to finish.

Autodesk 360

Invite Sent

E-mail invitation has been sent to **1** new connection.

OK

Select **OK** to close the dialog box.

A typical e-mail notification that is sent informing a colleague of being tagged in a Design Feed post is shown below.

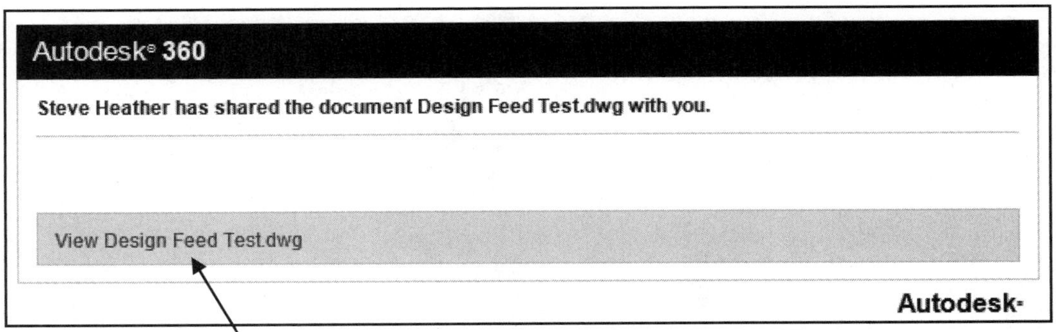

Autodesk® **360**

Steve Heather has shared the document Design Feed Test.dwg with you.

View Design Feed Test.dwg

Autodesk®

Clicking on the link will take the tagged colleague directly to the document at their Autodesk 360 account.

EXERCISE 30A

INSTRUCTIONS:

1. Start a **NEW** file using **My Decimal Setup.dwt.**
2. **Draw** the **Arc** shown below using "**Center, Start, End**"

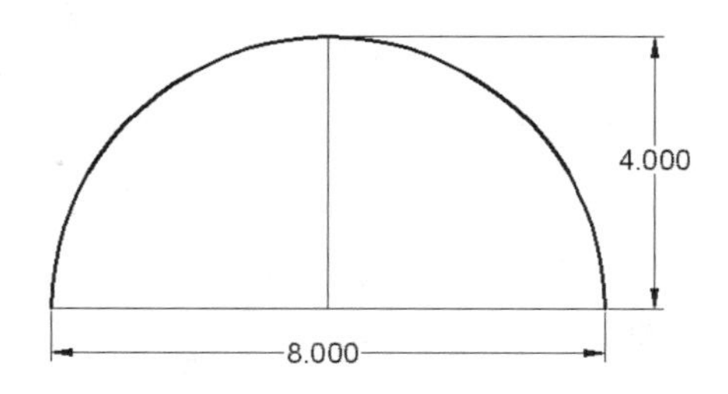

3. Create the **Arctext** shown below.

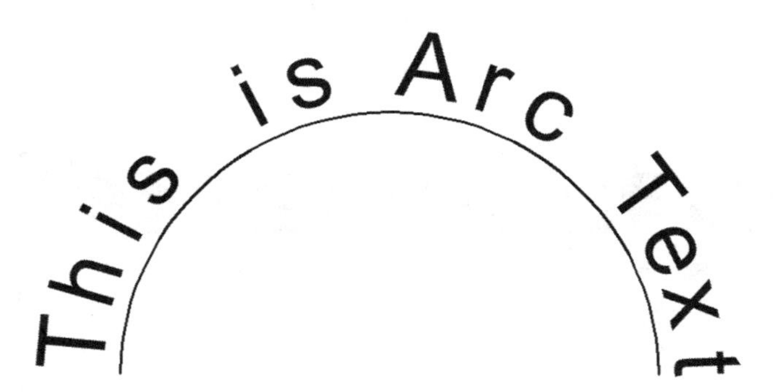

ArcAlignedText Workshop - Create
File Edit Format Help
White
Text-Classic
Arial
AaBbCcDd
Text: This is Arc Text
Properties:
Text height: 1
Offset from arc: 0.25
Width factor: 1.0
Offset from left: 0.0
Char spacing: 0.095
Offset from right: 0.0
OK
Cancel

4. **Save** the drawing as: **EX30A**

EXERCISE 30B

INSTRUCTIONS:

1. Open **EX30A** (if not already open).
2. Add the text on the inside of the Arc as shown below.

3. **Save** the drawing as: **EX30B**

EXERCISE 30C

INSTRUCTIONS:

1. Start a **New** file using **My Decimal Setup.dwt**
2. Draw the letters as shown below.
   a. Use text style **Text-Classic**
   b. Height shown

Height = .200 → A   A   AAA ← **Height = .400**

3. Copy the letters 3 times as shown below.
4. **Enclose the letters** as shown below.

VARIABLE **CONSTANT**

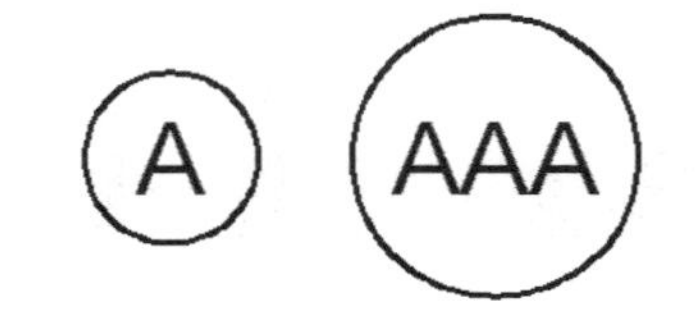

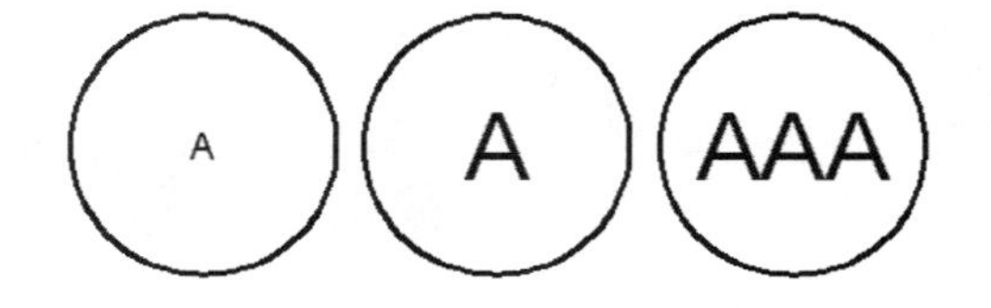

SLOT A A AAA A A AAA

RECTANGLE A

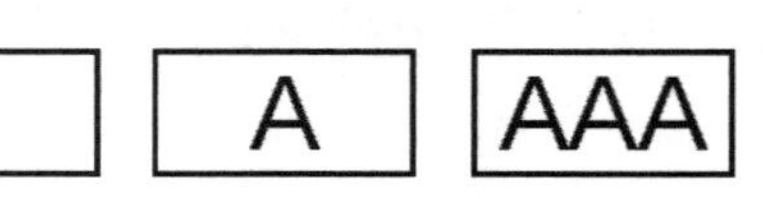

5. **Save** the drawing as: **EX30C**

EXERCISE 30D

INSTRUCTIONS:

1. Start a **New** file using **My Decimal Setup.dwt**
2. Draw the objects shown below approximately as shown.

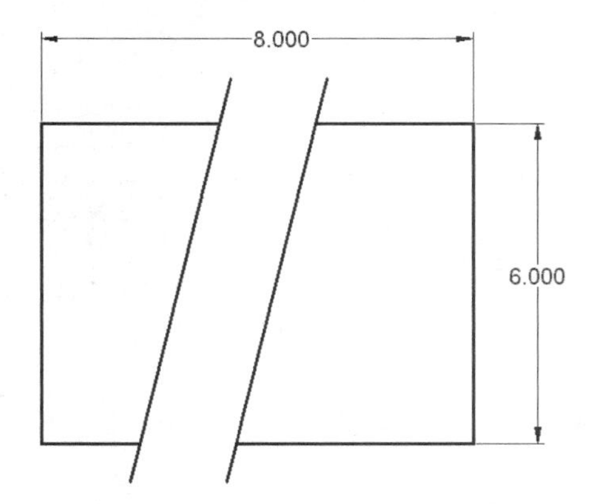

3. Add the **Break-Line** symbol to both lines as shown below

   Note: Use object snap **nearest** to place Break-LIne symbol accurately on the line.

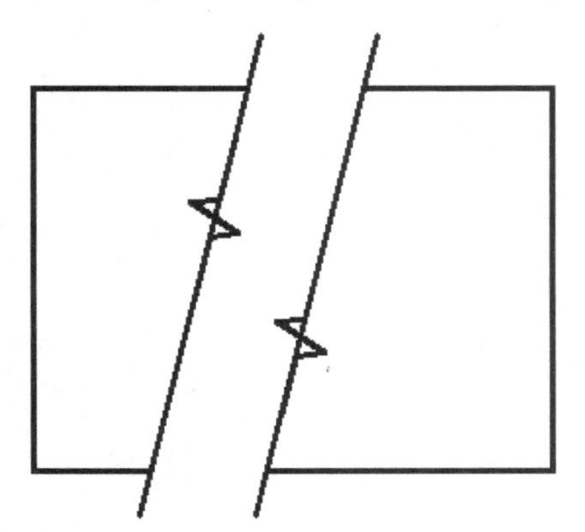

Size: .300
Extension: 0

4. Trim the Line under the Break-Line Symbol and add the Hatch as shown below.

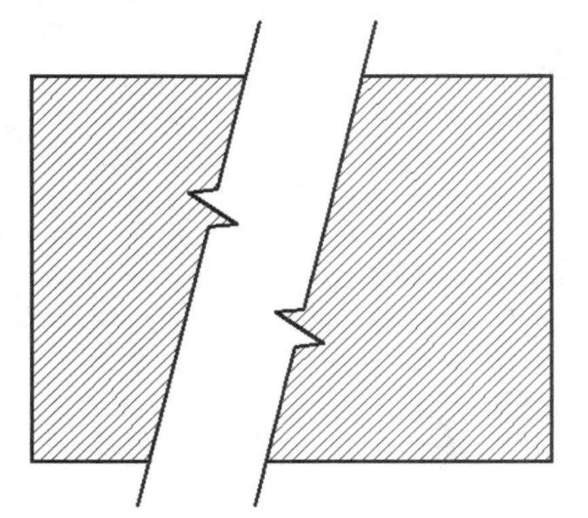

Hatch Type: Ansi 31

Trim before Hatch:
Trimming will take some thinking.

5. **Save** the drawing as: **EX30D**

Notes:

APPENDIX A
Add a Printer / Plotter

The following are step-by-step instructions on how to configure AutoCAD for your printer or plotter. These instructions assume you are a single system user. If you are networked or need more detailed information, please refer to your AutoCAD Help Index.

Note: You can configure AutoCAD for multiple printers. Configuring a printer makes it possible for AutoCAD to display the printing parameters for that printer.

A. Type: **Plottermanager <enter>**

B. Select **"Add-a-Plotter"** Wizard

Add-A-Plotter Wizard
Shortcut
1.20 KB

C. Select the **"Next"** button.

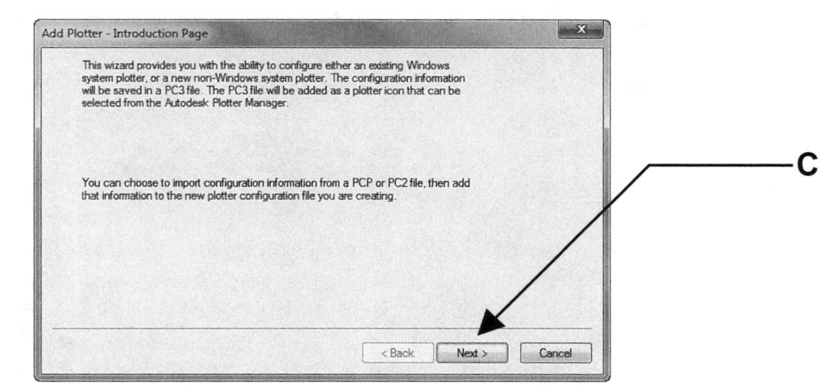

D. Select "**My Computer"** then **Next**.

E. Select the **Manufacturer** and the specific **Model** desired then N**ext**.

(If you have a disk with the specific driver information, put the disk in the disk drive and select “Have disk” button then follow instructions.)

F. Select the **“Next”** box.

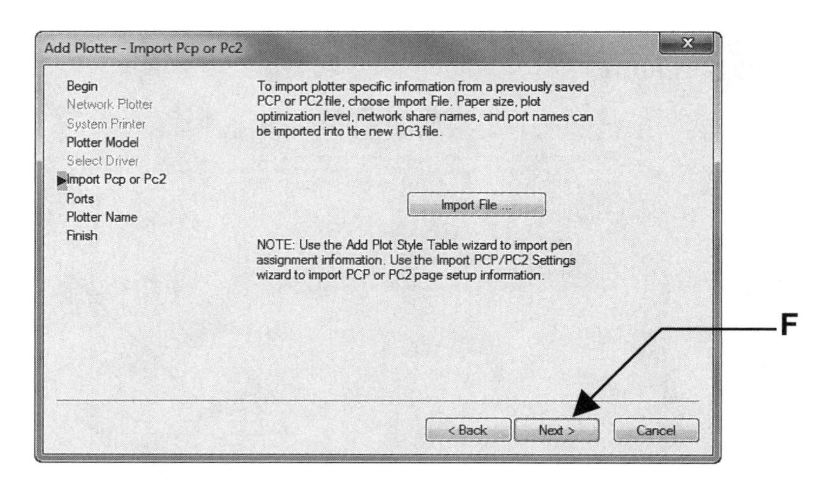

G1. Select **“Plot to a port”**.
G2. Then select **“Next”.**

Select LPT1 <u>only</u> if you are configuring your printer and it is attached to this computer.
<u>Not necessary to select any if selected printer is not attached to this computer.</u>

H. The Printer name that you previously selected should appear. Then select **“Next”**

I. Select the **“Edit Plotter Configuration…”** box.

J. Select:

1. Device and Document Settings tab.
2. Media: Source and Size
3. Size: (Select the appropriate size for your printer / plotter)
4. OK box.

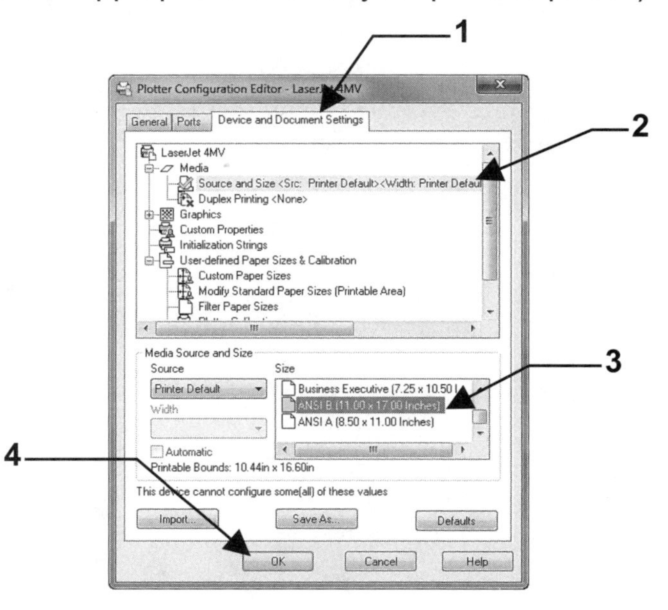

K. Select **"Finish".**

L. Type: **Plottermanager <enter>** again.

Is the printer / plotter there?

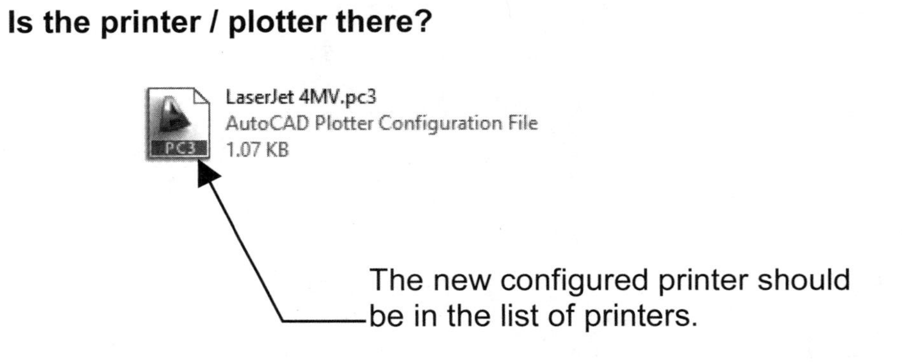

APPENDIX B
METRIC CONVERSION FACTORS

Multiply Length	By	To Obtain
centimeter	0.0328084	foot
centimeter	0.3937008	inch
foot	0.3048	meter (m)
foot	30.48	centimeter (cm)
foot	304.8	millimeter (mm)
inch	0.0254	meter (m)
inch	2.54	centimeter (cm)
inch	25.4	millimeter (mm)
kilometer	0.6213712	mile [U.S. statute]
meter	39.37008	inch
meter	0.5468066	fathom
meter	3.280840	foot
meter	0.1988388	rod
meter	1.093613	yard
meter	0.0006213712	mile [U. S. statute]
microinch	0.0254	micrometer [micron] (mm)
micrometer [micron]	39.37008	microinch
mile [U.S. statute]	1609.344	meter (m)
mile [U. S. statute]	1.609344	kilometer (km)
millimeter	0.003280840	foot
millimeter	0.03937008	inch

Multiply Length	By	To Obtain
rod	5.0292	meter (m)
yard	0.9144	meter (m)
Area		
acre	4046.856	meter2 (m2)
acre	0.4046856	hectare
centimeter2	0.1550003	inch2
centimeter2	0.001076391	foot2
foot2	0.09290304	meter2 (m2)
foot2	929.0304	centimeter2 (cm2)
foot2	92,903.04	millimeter2 (mm2)
hectare	2.471054	acre
inch2	645.16	millimeter2 (mm2)
inch2	6.4516	centimeter2 (cm2)
inch2	0.00064516	meter2 (m2)
meter2	1550.003	inch2
meter2	10.763910	foot2
meter2	1.195990	yard2
meter2	0.0002471054	acre
mile2	2.5900	kilometer2
millimeter2	0.00001076391	foot2
millimeter2	0.001550003	inch2
yard2	0.8361274	meter2 (m2)
fathom	1.8288	meter (m)

Symbols of SI units, multiples and sub-multiples are given in parentheses in the right-hand column.

APPENDIX C - DYNAMIC INPUT

To help you keep your focus in the “drawing area”, AutoCAD has provided a command interface called **Dynamic Input**. You may input information within the Dynamic Input instead of on the command line.

The **First point** displays the (Absolute: X, Y) distance from the Origin.
Enter the **X** dimension, <u>press the **Tab key**</u>, enter the **Y** dimension then **<enter>.**

The **Second and Next points** display (Relative: Distance and Angle) from the last point entered. Enter the **distance**, <u>press the **tab key**</u>, enter the **angle** then **<enter>**.

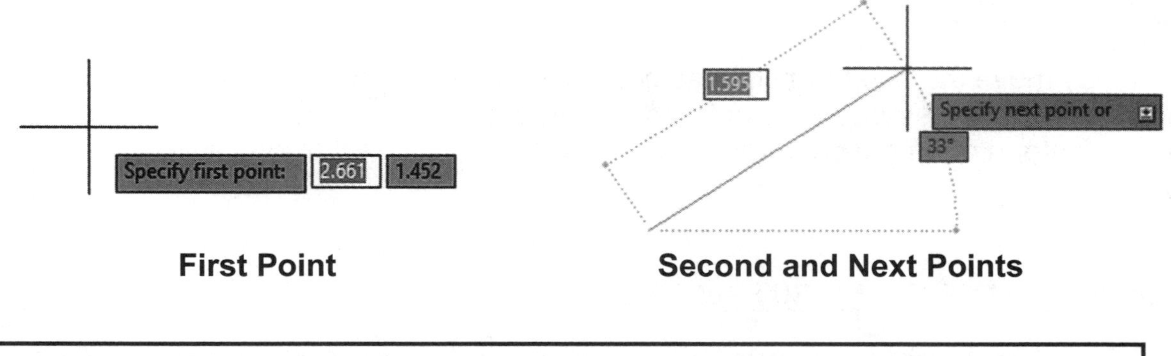

First Point **Second and Next Points**

<u>Note</u>: Some users find Dynamic Input useful some find it distracting. After completing this lesson you decide if you want to use it. <u>It is your choice</u>.

<u>How to turn Dynamic Input ON or OFF</u>

Select the **DYNAMIC INPUT** button on the status bar or use the **F12** key.

<u>Note</u>: Refer to page 1-12 for adding tools to the Status Bar.

<u>DYNAMIC INPUT has 3 components</u>

1. **Pointer Input**
2. **Dimensional Input**
3. **Dynamic Prompts**

You may control what is displayed by each component and turn each ON or OFF.

Refer to the following pages for more detailed descriptions.

POINTER INPUT

Pointer Input is only displayed for the **first** point.
When Pointer Input is enabled (ON) and a command has been selected, the location of the crosshairs is displayed as coordinates in a tooltip near the cursor.

Example:

1. Select the **LINE** command.

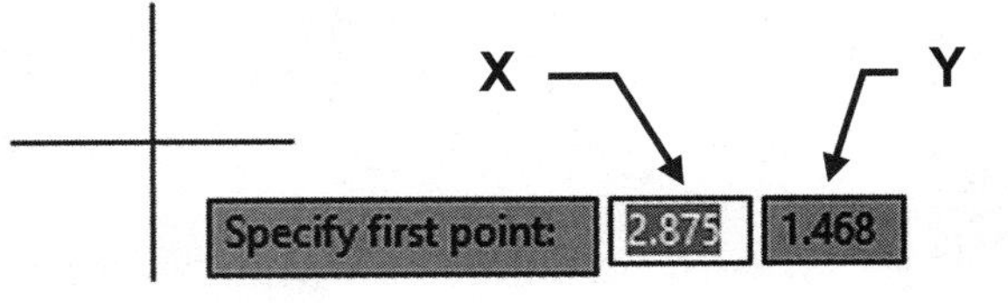

2. Move the cursor.

   The 2 boxes display the cursor location as X and Y coordinates **from the Origin**. (Absolute coordinates)

3. You may move the cursor until these coordinates display the desired location and press the left mouse button or enter the desired coordinate values in the tooltip boxes instead of on the command line.

How to change POINTER INPUT settings

1. Right-click the **Dynamic Input** button on the status bar and select **Dynamic Input Settings**.

2. In the Drafting Settings dialog box select the **Dynamic input** tab.

3. Under Enable Pointer Input select the **Settings** button.

4. In the Pointer Input Settings dialog box select:

   Format: (select one as the display format for second or next points)

   Polar or Cartesian format

   Relative or Absolute coordinate

   Visibility: (select one for tooltip display)

   - **As Soon As I Type Coordinate Data**. When pointer input is turned on, displays tooltips only when you start to enter coordinate data.

   - **When a Command Asks for a Point**. When pointer input is turned on, displays tooltips whenever a command prompts for a point.

   - **Always—Even When Not in a Command**. Always displays tooltips when pointer input is turned on.

5. Select **OK** to close each dialog box.

DIMENSIONAL INPUT

When Dimensional Input is enabled (ON) the tooltips display the distance and angle values for the **second** and **next points**.

The default display is Relative Polar coordinates.

Dimensional Input is available for: **Arc, Circle, Ellipse, Line** and **Polyline**.

Example:

1. Select the Line Command.
2. Enter the 1st point.
3. Move the cursor in the direction you desire and enter the **distance, Tab key,** and the **angle <enter>.**

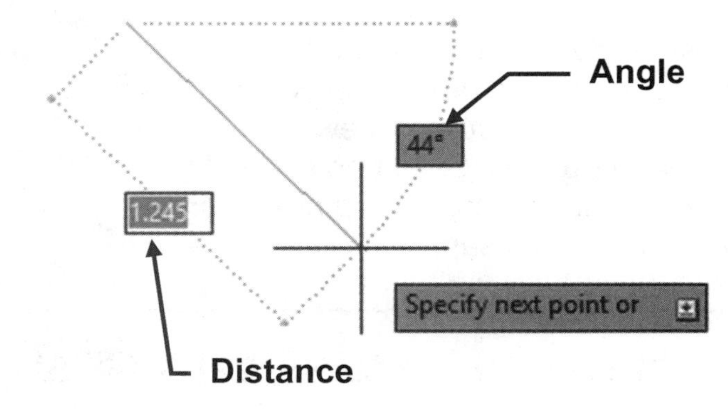

How to change DIMENSIONAL INPUT settings

1. Right-click the **Dynamic Input** button on the status bar and select **Dynamic Input Settings**.
2. In the Drafting Settings dialog box select the **Dynamic input** tab.
3. Under Dimension Input, select the **Settings** button.
4. In the Dimension Input Settings dialog box select:

   Visibility: (select one of the following options)

   - **Show Only 1 Dimension Input Field at a Time**. Displays only the distance dimensional input tooltip when you are using grip editing to stretch an object.
   - **Show 2 Dimension Input Fields at a Time**. Displays the distance and angle dimensional input tooltips when you are using grip editing to stretch an object.
   - **Show the Following Dimension Input Fields Simultaneously**. Displays the selected dimensional input tooltips when you are using grip editing to stretch an object. Select one or more of the check boxes.
5. Select **OK** to close each dialog box.

DYNAMIC PROMPTS

When Dynamic Prompts are enabled (ON) prompts are displayed.

You may enter a response in the tooltip box instead of on the command line.

Example:

1. Select the Circle command.

2. Press the right mouse button for command options or you may also press the down arrow to view command options. Select an option by clicking on it or press the up arrow for option menu to disappear.

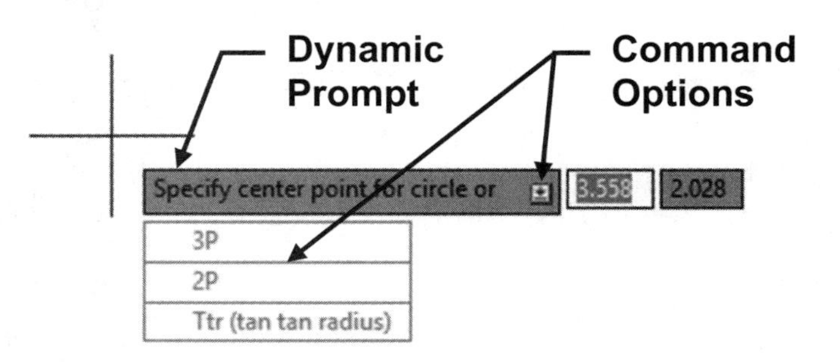

How to change the COLOR, SIZE, or TRANSPARENCY of tooltips

1. Right-click the **Dynamic Input** button on the status bar then select **Dynamic Input Settings**.

2. In the Drafting Settings dialog box select the **Dynamic input** tab.

3. At the bottom of the dialog box select **Drafting Tooltip Appearance** button.

4. Select "**Colors**" button. Under Context: select 2D Model Space. Under Interface Elements: select Drafting Tool Tip or Drafting Tool Tip Background. Select a color from the Color drop down list. Then select Apply and Close button.

5. Under **Size**, move the slider to the right to make tooltips larger or to the left to make them smaller. The default value, 0, is in the middle.

   (I like 2, a little larger than the default)

6. Under **Transparency**, move the slider. The higher the setting, the more transparent the tooltip.

   (I like 25, a little more transparent than the default)

7. Under **Apply To**, choose an option:

   - **Override OS Settings for All Drafting Tooltips**. Applies the settings to all tooltips, overriding the settings in the operating system.

   - **Use Settings Only for Dynamic Input Tooltips**. Applies the settings only to the drafting tooltips used in Dynamic Input.

8. Select **OK** to close each dialog box.

APPENDIX D – COMMAND LINE ENHANCEMENTS

The Command Line in AutoCAD has been extensively modified to further assist the user in searching for commands.

AutoCorrect

If you mistyped a command in versions prior to AutoCAD 2014, the system would respond with "Unknown command". AutoCAD will now AutoCorrect to the most relevant command.

In the example below, if you entered **CIRKLE**, the system will respond with **CIRCLE**, and any other commands that contain the word **CIRCLE**.

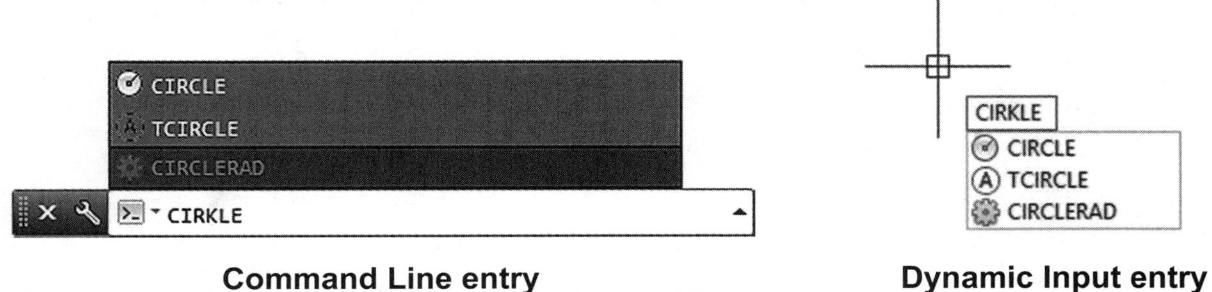

Command Line entry **Dynamic Input entry**

AutoCAD also has an **AutoCorrect List** which is stored in the system, if you mistype a command three times or more, that mistyped command will be stored in the **AutoCorrect List** along with the correct spelling of the command.

You can access the AutoCorrect List by selecting:

Ribbon = Manage tab / Customization panel / Edit Aliases ▼ / Edit AutoCorrect List

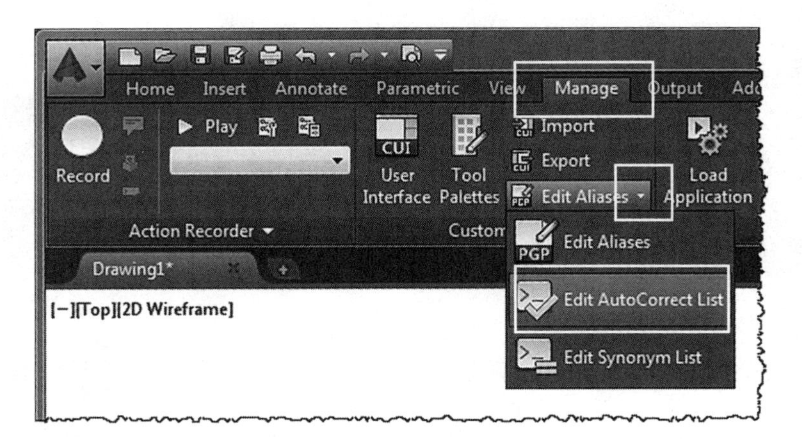

An example of the **AutoCorrect List** is shown below with two commands that have been mistyped, and with their correct spelling.

AutoCorrectUserDB.pgp - Notepad

File Edit Format View Help

```
CIRKLE, *CIRCLE
LYNE, *LINE
```

Mistyped Spellings

Correct Spellings

AutoComplete

The AutoComplete in AutoCAD has been further enhanced and now supports mid-string searches. In versions prior to AutoCAD 2014 the AutoComplete only displayed command suggestions beginning with the word you entered, it will now display command suggestions with the word you enter, anywhere within it.

In the example below, if you enter **SETTINGS**, AutoComplete will respond with various suggestions with the word **SETTINGS** anywhere within a command

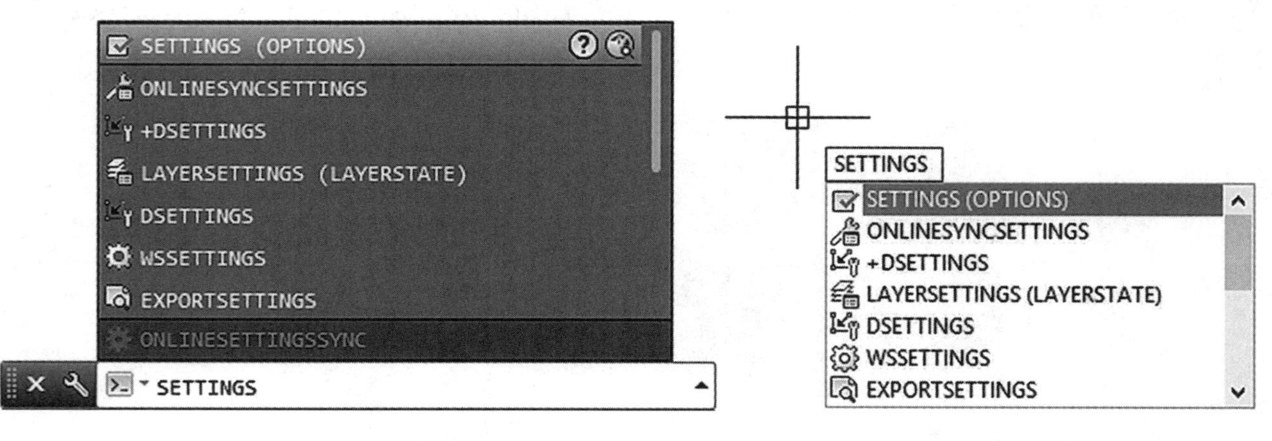

Command Line entry **Dynamic Input entry**

Adaptive Suggestions

When you first use AutoCAD 2015, commands in the suggestion list are displayed in the order of usage which is based on general customer data. As you use AutoCAD more and more, the commands will be displayed according to your usage, it adapts to your way of working, showing the commands you use most frequently in the suggestion list.

Synonym Suggestions

The command line in AutoCAD 2015 has a built in Synonym list. When you enter a word at the command line, AutoCAD returns a command if it can match it in the synonym list.

For example, if you enter **BREAKUP** at the command line, AutoCAD will return the command **EXPLODE**. Or if you want to type a paragraph of text and enter the word **PARAGRAPH** at the command line, AutoCAD will return the command **MTEXT**.

You can add your own synonym's to the list which can be accessed by selecting:

Ribbon = Manage tab / Customization panel / Edit Aliases ▼ / Edit Synonym List

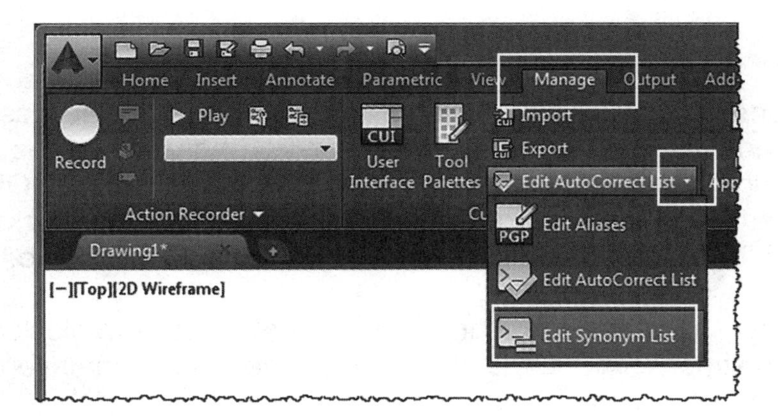

Internet Search

AutoCAD 2015 now allows you to search for more information on a command that is displayed in the suggestion list. If you move the mouse cursor over a command in the list, it will display a **Search on Internet** icon and a **Search in Help** icon.

You can click on either of these two options to get extended help on the command you entered. So for example, if you type in **ARRAY** at the command line and choose the **Search on Internet** option, your current internet browser will open and show internet suggestions for **AutoCAD ARRAY**.

Whichever command you choose to search for help on the internet, the word **AutoCAD** will always precede it. An example of the search and help icons is shown below.

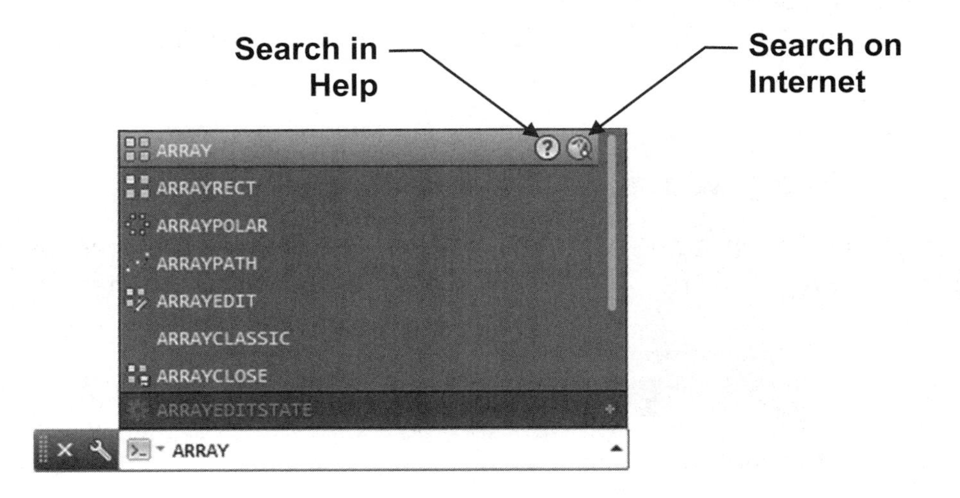

Content

The Command Line in AutoCAD 2015 also allows you to quickly access layers, blocks, hatch patterns/gradients, text styles, dimension styles and visual styles. For example, if you have a drawing open that has block definitions with the name **DOOR**, and you enter **DOOR** at the command line, the suggestion list will display all the blocks with that name in it, so you could then insert that block directly from the command line.

The example below shows the command line suggestion list with block definitions of **DOOR**, you would simply click on the block you needed, and then insert it into your drawing.

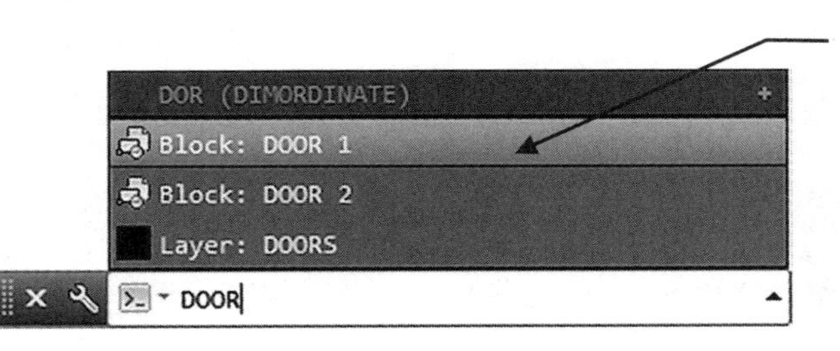

Categories

The command line suggestion list is made easier to navigate by organizing commands and system variables into categories. To see the results you can expand the category by clicking on the **+** sign, or you can press the **Tab** key to cycle through each category.

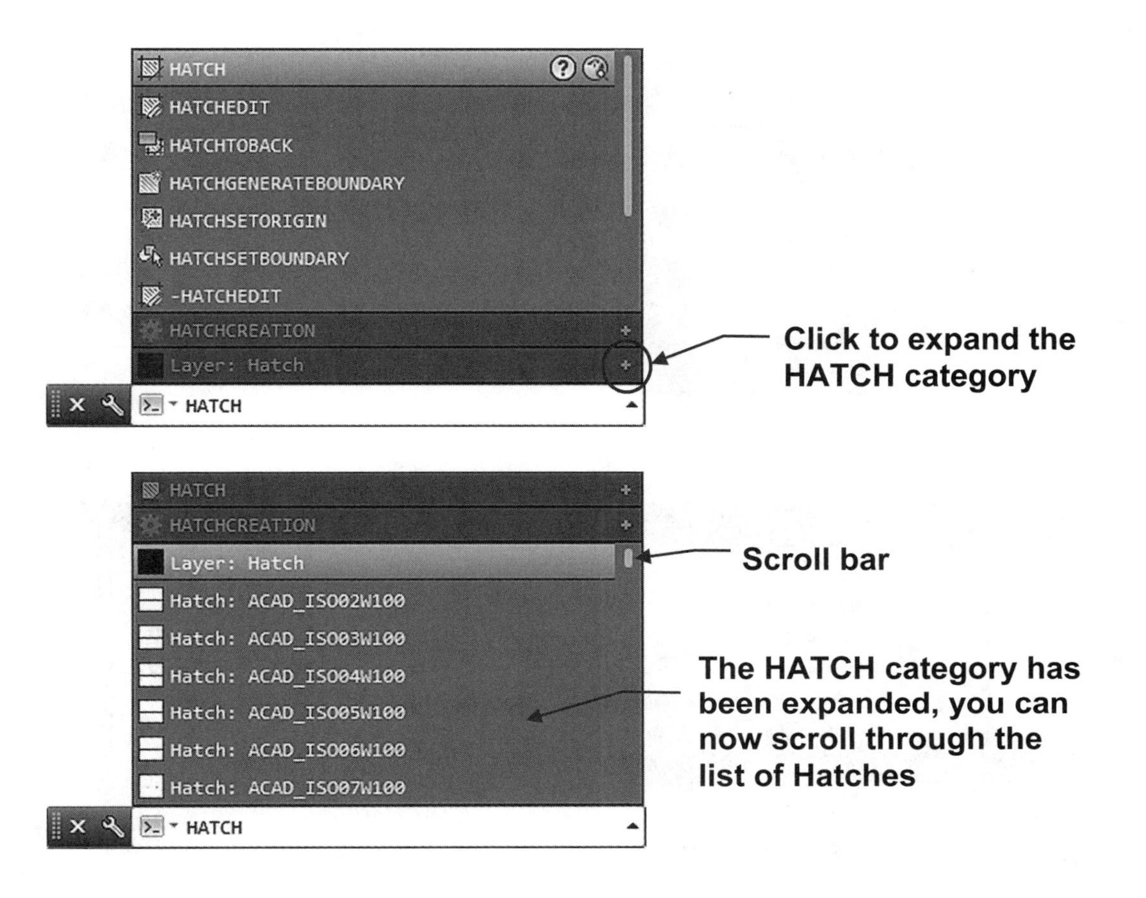

Input Settings

You can choose to turn on or off any of the new command line features by right-clicking on the command line and selecting **Input Settings**, you can choose between **AutoComplete**, **AutoCorrect**, **Search System Variables**, **Search Content**, and **Mid-string Search**.

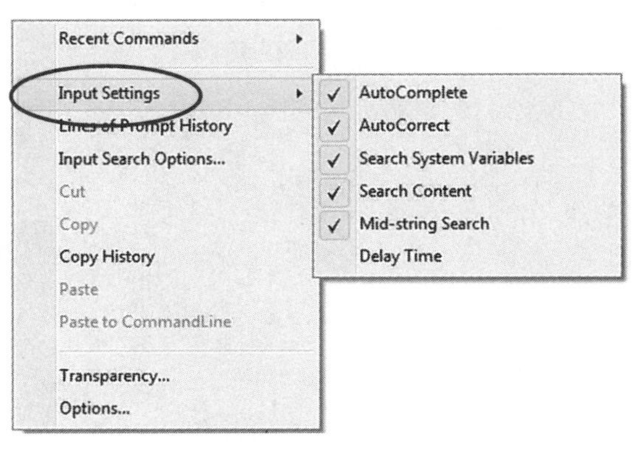

In addition to the **Input Settings**, you can further refine the settings by right-clicking on the command line and selecting **Input Search Options**. This will open the **Input Search Options** dialog box where you can change settings like the amount of times you can mistype a command before it gets entered into the **AutoCorrect List**.

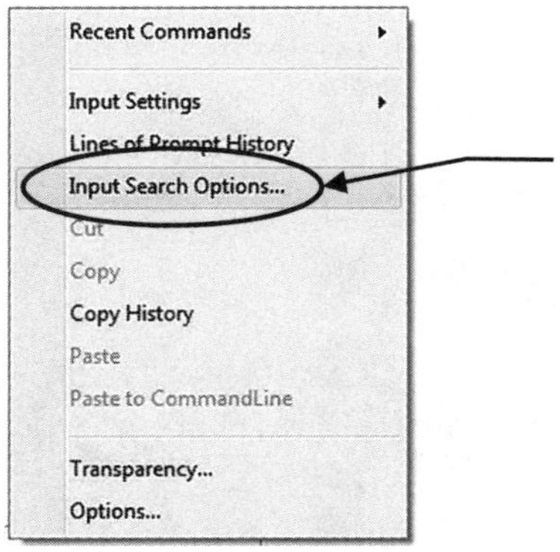

Select to open up the Input Search Options Dialog box

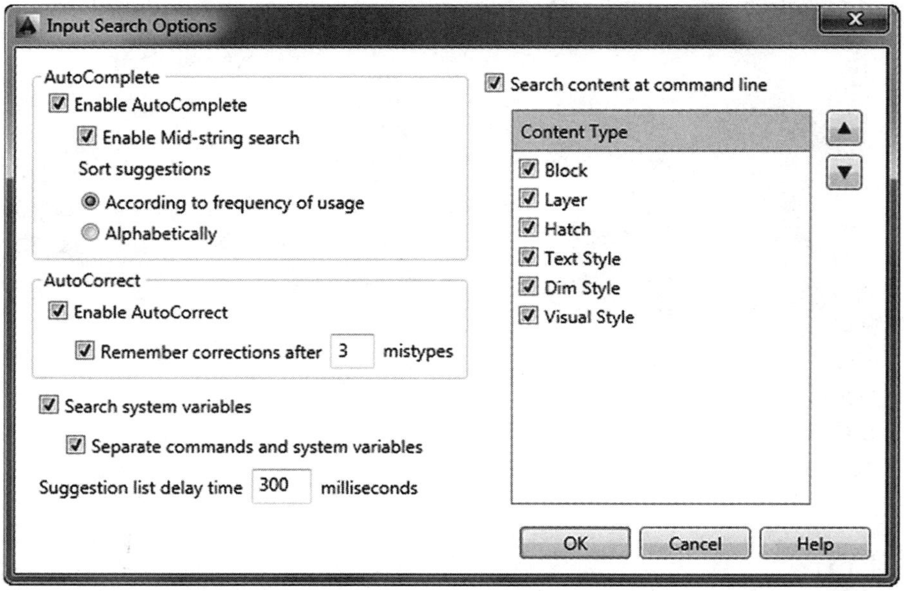

INDEX

A

B

C

D

E

F

G

H

I

J

K

L

M

N

O

P

Q

R

S

T

U

V

W

X

Y

Z

AutoCAD trial software

The enclosed disk contains a fully functioning 30-day trial version of AutoCAD software. This software has been provided to give you a preview of how easily AutoCAD installs and is used in conjunction with this Exercise Workbook. After 30 days, from the day you install the software, you will not be able to open the program. At that time you will have the opportunity to activate the software by purchasing the actual software license.

IMPORTANT: The 30-day trial version of AutoCAD can be loaded only once on a single computer. This applies to all 30-day trials regardless of where you obtained them, including from other Exercise Workbooks by Cheryl Shrock.

If you are attending a school, check with your instructor for educational discounts.
Or visit: www.autodesk.com for additional purchasing information.

Make efficiency a daily part of the job with AutoCAD® software. Meticulously refined with the drafter in mind, AutoCAD propels day-to-day drafting forward with features that increase speed and accuracy while saving time. Annotation scaling and layer properties per viewport minimize workarounds, while text and table enhancements and multiple leaders help deliver an unmatched level of aesthetic precision and professionalism.

AutoCAD system requirements

Refer to page Intro-8 and 9

How to install AutoCAD:

Note: AutoCAD can coexist with earlier versions of AutoCAD and AutoCAD LT.

1. Insert the AutoCAD Disk into your computer's drive.

2. Follow the on-screen menus.

3. When the installation is complete you will be prompted to:
   "Authorize AutoCAD" or "run AutoCAD without authorizing."

   Select "**run AutoCAD without Authorizing**"

Note: This trial software will stop working 30 days from the date of installation.